Second Edition

THE CELL

DON W. FAWCETT, M.D.

Hersey Professor of Anatomy
Harvard Medical School

1981
W. B. SAUNDERS COMPANY
Philadelphia London Toronto

W. B. Saunders Company: West Washington Square
Philadelphia, PA 19105

1 St. Anne's Road
Eastbourne, East Sussex BN21 3UN, England

1 Goldthorne Avenue
Toronto, Ontario M8Z 5T9, Canada

Library of Congress Cataloging in Publication Data

Fawcett, Don Wayne, 1917–

The cell.

Edition of 1966 published under title: An atlas of fine structure.

Includes bibliographical references.
1. Cytology—Atlases. 2. Ultrastructure (Biology)—Atlases. I. Title. [DNLM: 1. Cells—Ultrastructure—Atlases. 2. Cells—Physiology—Atlases. QH582 F278c]

QH582.F38 1981 591.8′7 80–50297

ISBN 0–7216–3584–9

Listed here is the latest translated edition of this book together with the language of the translation and the publisher.

German (*1st Edition*)—Urban and Schwarzenberg, Munich, Germany

The Cell ISBN 0-7216-3584-9

Last digit is the print number: 9 8 7 6 5 4 3 2 1

CONTRIBUTORS OF ELECTRON MICROGRAPHS

Dr. John Albright
Dr. David Albertini
Dr. Nancy Alexander
Dr. Winston Anderson
Dr. Jacques Auber
Dr. Baccio Baccetti
Dr. Michael Barrett
Dr. Dorothy Bainton
Dr. David Begg
Dr. Olaf Behnke
Dr. Michael Berns
Dr. Lester Binder
Dr. K. Blinzinger
Dr. Günter Blöbel
Dr. Robert Bolender
Dr. Aiden Breathnach
Dr. Susan Brown
Dr. Ruth Bulger
Dr. Breck Byers
Dr. Hektor Chemes
Dr. Kent Christensen
Dr. Eugene Copeland
Dr. Romano Dallai
Dr. Jacob Davidowitz
Dr. Walter Davis
Dr. Igor Dawid
Dr. Martin Dym
Dr. Edward Eddy
Dr. Peter Elias
Dr. A. C. Faberge
Dr. Dariush Fahimi
Dr. Wolf Fahrenbach
Dr. Marilyn Farquhar
Dr. Don Fawcett
Dr. Richard Folliot
Dr. Michael Forbes
Dr. Werner Franke
Dr. Daniel Friend
Dr. Keigi Fujiwara
Dr. Penelope Gaddum-Rosse
Dr. Joseph Gall
Dr. Lawrence Gerace
Dr. Ian Gibbon
Dr. Norton Gilula
Dr. Jean Gouranton
Dr. Kiyoshi Hama
Dr. Joseph Harb
Dr. Etienne de Harven
Dr. Elizabeth Hay
Dr. Paul Heidger
Dr. Arthur Hertig
Dr. Marian Hicks
Dr. Dixon Hingson
Dr. Anita Hoffer
Dr. Bessie Huang
Dr. Barbara Hull
Dr. Richard Hynes
Dr. Atsuchi Ichikawa
Dr. Susumu Ito
Dr. Roy Jones
Dr. Arvi Kahri
Dr. Vitauts Kalnins
Dr. Marvin Kalt
Dr. Taku Kanaseki
Dr. Shuichi Karasaki
Dr. Morris Karnovsky
Dr. Richard Kessel
Dr. Toichiro Kuwabara
Dr. Ulrich Laemmli
Dr. Nancy Lane
Dr. Elias Lazarides
Dr. Gordon Leedale
Dr. Arthur Like
Dr. Richard Linck
Dr. John Long
Dr. Linda Malick
Dr. William Massover
Dr. A. Gideon Matoltsy
Dr. Scott McNutt
Dr. Oscar Miller
Dr. Mark Mooseker
Dr. Enrico Mugnaini
Dr. Toichiro Nagano
Dr. Marian Neutra
Dr. Eldon Newcomb
Dr. Ada Olins
Dr. Gary Olson
Dr. Jan Orenstein
Dr. George Palade
Dr. Sanford Palay
Dr. James Paulson
Dr. Lee Peachey
Dr. David Phillips
Dr. Dorothy Pitelka
Dr. Thomas Pollard
Dr. Keith Porter

Dr. Jeffrey Pudney
Dr. Elio Raviola
Dr. Giuseppina Raviola
Dr. Janardan Reddy
Dr. Thomas Reese
Dr. Jean Revel
Dr. Hans Ris
Dr. Joel Rosenbaum
Dr. Evans Roth
Dr. Thomas Roth
Dr. Kogaku Saito
Dr. Peter Satir
Dr. Manfred Schliwa
Dr. Nicholas Severs
Dr. Emma Shelton
Dr. Nicholai Simionescu
Dr. David Smith
Dr. Andrew Somlyo
Dr. Sergei Sorokin
Dr. Robert Specian
Dr. Andrew Staehelin
Dr. Fumi Suzuki
Dr. Hewson Swift
Dr. George Szabo
Dr. John Tersakis
Dr. Guy de Thé
Dr. Lewis Tilney
Dr. Greta Tyson
Dr. Wayne Vogl
Dr. Fred Warner
Dr. Melvyn Weinstock
Dr. Richard Wood
Dr. Raymond Wuerker
Dr. Eichi Yamada

PREFACE

The history of morphological science is in large measure a chronicle of the discovery of new preparative techniques and the development of more powerful optical instruments. In the middle of the 19th century, improvements in the correction of lenses for the light microscope and the introduction of aniline dyes for selective staining of tissue components ushered in a period of rapid discovery that laid the foundations of modern histology and histopathology. The decade around the turn of this century was a golden period in the history of microscopic anatomy, with the leading laboratories using a great variety of fixatives and combinations of dyes to produce histological preparations of exceptional quality. The literature of that period abounds in classical descriptions of tissue structure illustrated by exquisite lithographs. In the decades that followed, the tempo of discovery with the light microscope slackened; interest in innovation in microtechnique declined, and specimen preparation narrowed to a monotonous routine of paraffin sections stained with hematoxylin and eosin.

In the middle of the 20th century, the introduction of the electron microscope suddenly provided access to a vast area of biological structure that had previously been beyond the reach of the compound microscope. Entirely new methods of specimen preparation were required to exploit the resolving power of this new instrument. Once again improvement of fixation, staining, and microtomy commanded the attention of the leading laboratories. Study of the substructure of cells was eagerly pursued with the same excitement and anticipation that attend the geographical exploration of a new continent. Every organ examined yielded a rich reward of new structural information. Unfamiliar cell organelles and inclusions and new macromolecular components of protoplasm were rapidly described and their function almost as quickly established. This bountiful harvest of new structural information brought about an unprecedented convergence of the interests of morphologists, physiologists, and biochemists; this convergence has culminated in the unified new field of science called cell biology.

The first edition of this book (1966) appeared in a period of generous support of science, when scores of laboratories were acquiring electron microscopes and hundreds of investigators were eagerly turning to this instrument to extend their research to the subcellular level. At that time, an extensive text in this rapidly advancing field would have been premature, but there did seem to be a need for an atlas of the ultrastructure of cells to establish acceptable technical standards of electron microscopy and to define and illustrate the cell organelles in a manner that would help novices in the field to interpret their own micrographs. There is reason to believe that the first edition of *The Cell: An Atlas of Fine Structure* fulfilled this limited objective.

In the 14 years since its publication, dramatic progress has been made in both the morphological and functional aspects of cell biology. The scanning electron microscope and the freeze-fracturing technique have been added to the armamentarium of the miscroscopist, and it seems timely to update the book to incorporate examples of the application of these newer methods, and to correct earlier interpretations that have not withstood the test of time. The text has been completely rewritten and considerably expanded. Drawings and diagrams have been added as text figures. A few of the original transmission electron micrographs to which I have a sentimental attachment have been retained, but the great majority of the micrographs in this edition are new. These changes have inevitably added considerably to the length of the book and therefore to its price, but I hope these will be offset to some extent by its greater informational content.

Twenty years ago, the electron microscope was a solo instrument played by a few virtuosos. Now it is but one among many valuable research tools, and it is most profitably used in combination with biochemical, biophysical, and immunocytochemical techniques. Its use has become routine and one begins to detect a decline in the number and quality of published micrographs as other analytical methods increasingly capture the interest of investigators. Although purely descriptive electron microscopic studies now yield diminishing returns, a detailed knowledge of the structural organization of cells continues to be an indispensable foundation for research on cell biology. In undertaking this second edition I have been motivated by a desire to assemble and make easily accessible to students and teachers some of the best of the many informative and aesthetically pleasing transmission and scanning electron micrographs that form the basis of our present understanding of cell structure.

The historical approach employed in the text may not be welcomed by all. In the competitive arena of biological research today investigators tend to be interested only in the current state of knowledge and care little about the steps by which we have arrived at our present position. But to those of us who for the past 25 years have been privileged to participate in one of the most exciting and fruitful periods in the long history of morphology, the young seem to be entering the theater in the middle of an absorbing motion picture without knowing what has gone before. Therefore, in the introduction to each organelle, I have tried to identify, in temporal sequence, a few of the major contributors to our present understanding of its structure and function. In venturing to do this I am cognizant of the hazards inherent in making judgments of priority and significance while many of the dramatis personae are still living. My apologies to any who may feel that their work has not received appropriate recognition.

It is my hope that for students and young investigators entering the field, this book will provide a useful introduction to the architecture of cells and for teachers of cell biology a guide to the literature and a convenient source of illustrative material. The sectional bibliographies include references to many reviews and research papers that are not cited in the text. It is believed that these will prove useful to those readers who wish to go into the subject more deeply.

The omission of magnifications for each of the micrographs will no doubt draw some criticism. Their inclusion was impractical since the original negatives often remained in the hands of the contributing microscopists and micrographs submitted were cropped or copies enlarged to achieve pleasing composition and to focus the reader's attention upon the particular organelle under discussion. Absence was considered preferable to inaccuracy in stated magnification. The majority of readers, I believe, will be interested in form rather than measurement and will not miss this datum.

Assembling these micrographs illustrating the remarkable order and functional design in the structure of cells has been a satisfying experience. I am indebted to more than a hundred cell biologists in this country and abroad who have generously responded to my requests for exceptional micrographs. It is a source of pride that nearly half of the contributors were students, fellows or colleagues in the Department of Anatomy at Harvard Medical School at some time in the past 20 years. I am grateful for their stimulation and for their generosity in sharing prints and negatives. It is a pleasure to express my appreciation for the forbearance of my wife who has had to communicate with me through the door of the darkroom for much of the year while I printed the several hundred micrographs; and for the patience of Helen Deacon who has typed and retyped the manuscript; for the skill of Peter Ley, who has made many copy negatives to gain contrast with minimal loss of detail; and for the artistry of Sylvia Collard Keene whose drawings embellish the text. Special thanks go to Elio and Giuseppina Raviola who read the manuscript and offered many constructive suggestions; and to Albert Meier and the editorial and production staff of the W. B. Saunders Company, the publishers.

And finally I express my gratitude to the Simon Guggenheim Foundation whose commendable policy of encouraging the creativity of the young was relaxed to support my efforts during the later stages of preparation of this work.

DON W. FAWCETT
Boston, Massachusetts

CONTENTS

CELL SURFACE

CELL MEMBRANE

The cell surface regulates the traffic of ions and macromolecules in and out of the cell. Although deceptively simple in microscopic appearance, it is remarkably complex in its molecular organization, possessing devices for attachment to other cells, specializations for cell-to-cell communication, antigenic macromolecules that are the basis for cell recognition and tissue specificity, ion pumps for regulating the internal milieu of the cell, receptors for hormones and other environmental signals, and mechanisms for generation of second messenger molecules that activate the cell's physiological responses. Understanding the biogenesis, structural organization, and functions of the cell surface is now one of the major challenges in cell biology.

The presence of a limiting *cell membrane* was inferred from indirect evidence nearly a century before it could be visualized microscopically. Nägeli in 1855 described the formation of a protective film where outflowing cytoplasm of an injured cell came into contact with water. He called this the *plasma membrane,* showed that it was semipermeable, and speculated that it was responsible for the osmotic phenomena exhibited by living cells. Forty years later, the permeability of living cells was exhaustively studied by Overton (1899), who noted a correlation between the lipid solubility of substances and their rate of entry into cells and suggested that the cell surface was probably lipid in nature. However, the reality of the cell membrane was not widely accepted by morphologists until Chambers (1926) developed microsurgical techniques that made it possible to deform, tear, or penetrate it with fine glass dissecting needles.

The modern concept of the biochemical nature of the cell membrane dates from Langmuir's (1917) demonstration that when fatty acids or phospholipids are dissolved in benzene and a few drops are placed on a large surface of water, the molecules orient with their hydrophilic ends inward to form a coherent layer at the air-water interface.

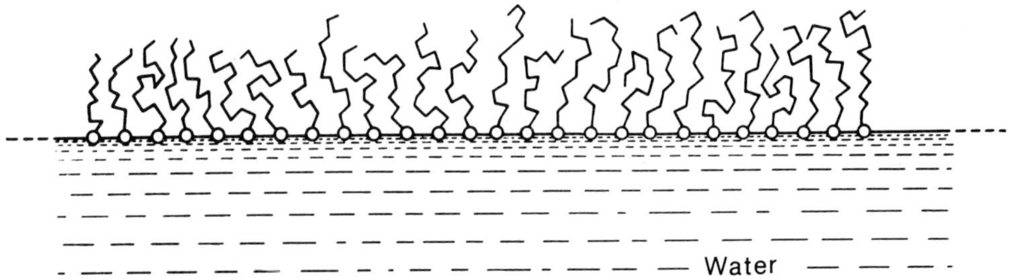

Drawing illustrating that when phospholipid is spread upon a large surface of water, the molecules orient with their hydrophilic ends inward to form a coherent layer at the air-water interface. (From Luzzati, Mussachia and Skoulios, Farad. Soc. Disc. No. 43, p. 43, 1958.)

Taking advantage of this property, Gorter and Grendel (1925) extracted the lipids from erythrocytes, measured the surface area covered when spread as a monolayer, and found it to be twice the calculated surface area of the original erythrocytes. Therefore it was suggested that the cells were enclosed by a lipid layer two molecules thick. This interpretation of the cell membrane as a bimolecular layer of mixed phospholipids has withstood the test of time. Later investigations by polarization optics, x-ray diffraction, measurement of their thickness, and assessment of their electrical properties have all been consistent with orientation of the hydrocarbons of the bilayer perpendicular to the plane of the membrane.

The observation that the surface tension of artificial membranes composed entirely of neutral lipids is high while that of natural membranes is relatively low led Davson and Danielli (1935) to postulate a layer of protein on both sides of the lipid bilayer. This model dominated physiological thought concerning the organization of the cell membrane for the next 20 years.

With the advent of the electron microscope, the cell surface membrane was visualized in thin sections as two dense lines separated by a less dense middle layer. This trilaminar appearance was also characteristic of membranes in the interior of the cell and was termed by Robertson (1957) the "unit membrane." The three layers were thought to be the visual expression of a structural organization common to all biological membranes. The two dense lines were attributed to deposition of osmium from the fixative on the hydrophilic ends of the phospholipid molecules, and in outer layers of protein while the unstained middle layer was believed to represent the saturated hydrocarbon chains of the lipid bilayer.

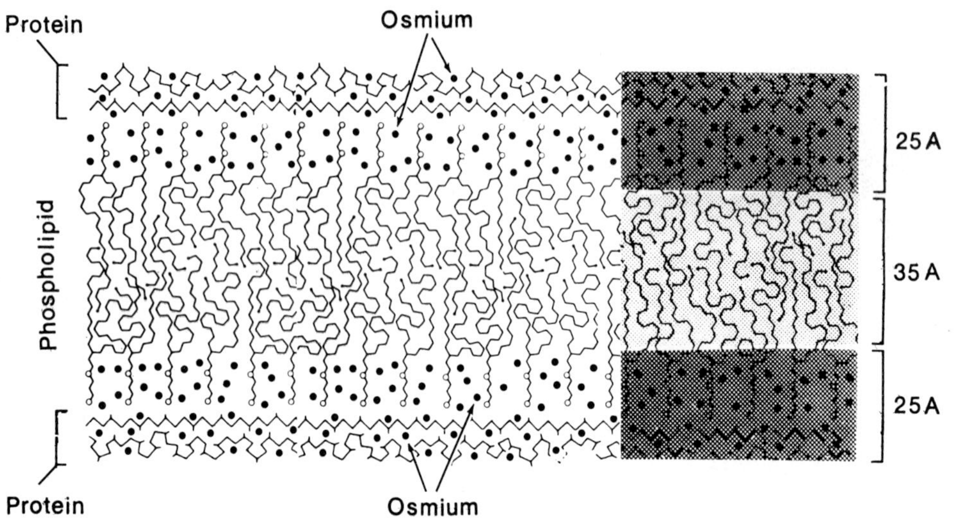

Diagram illustrating the initial interpretation of the trilaminar appearance of cell membranes in electron micrographs as a confirmation of the Davson-Danielli model. Deposition of osmium in layers of protein and in the hydrophilic ends of the phospholipids was thought to be responsible for the two dense lines. (Modified after W. Stoeckenius.)

Thus the trilaminar appearance of the unit membrane was widely accepted as electron microscopic confirmation of the general features of the Davson-Danielli models, even though the observation that myelin forms of pure phospholipid had a similar trilaminar appearance in section (Revel, Ito, and Fawcett, 1953) should have cast some doubt upon this interpretation.

Throughout the 1950s, there was great emphasis upon the universality of the unit membrane and the continuity of the membrane systems of the cell. Diagrams depicting continuity of the plasmalemma with the endoplasmic reticulum were widely reproduced, and a hypothetical schema suggesting the origin of mitochondria from invaginations of the plasmalemma found its way into textbooks. These concepts gained considerable acceptance even though convincing electron microscopic evidence was lacking. On the other hand, some investigators felt that this emphasis on the universality of the unit membrane was diverting attention from important biochemical differences that must exist in the membranes of organelles having distinctive functions. Improved methods for isolation of membrane fractions for biochemical analysis were revealing significant differences in lipid composition and enzymatic activities of membranes from various organelles. It became apparent that the biochemical heterogeneity and physiological specificity of membranes were not expressed in morphological differences detectable in electron micrographs of thin sections. In the late 1950s, it seemed unlikely that electron microscopy would contribute further to an understanding of the internal organization of membranes.

This pessimistic prediction proved to be quite wrong. The timely development of the method of freeze-fracturing by Steere (1957) and Moor and coworkers (1961) made the interior of the membrane accessible to morphological observation and ultimately led to an entirely new interpretation of membrane structure. The method is based upon the principle that by evaporation of platinum and carbon in a vacuum, a high fidelity replica can be made of the surface of frozen hydrated biological material. Small blocks of tissue fixed in glutaraldehyde are immersed in a solution of a cryoprotectant, such as glycerin, to avoid severe distortion from ice crystal formation during freezing. Freezing rate is maximized by freezing in fluids that have a high heat capacity and a low freezing point. In the commonest procedure, the tissue fragments are immersed in the fluid phase of partially solidified dichlorodifluoromethane (Freon) cooled at −150° C with liquid nitrogen. The rapidly frozen specimen is then fractured in a vacuum chamber by impact of the edge of a knife cooled at −196° C. A replica of the fracture surface is next prepared by evaporation of a heavy metal such as platinum from a source at an acute angle to the fracture surface. Metal is thus deposited on all surface elevations on the side toward the source. This results in an enhanced three-dimensional appearance in electron micrographs, comparable to the exaggeration of surface contours that results from oblique lighting or the shading used in graphic arts. To provide greater coherence and stability, carbon is then deposited uniformly on the entire surface from a separate electrode directly above the specimen. The vacuum is broken and the coated specimen is immersed in a solution of acid or sodium hypochlorite to dissolve the tissue. The replica remaining behind is gently washed and picked up on a specimen grid for examination with the transmission electron microscope. Regions of dense metal deposit on convex surfaces toward the source deflect or absorb electrons while the metal-free areas transmit electrons. After the resulting negatives are printed, a three-dimensional image of surface relief is seen in which the elevations are dark and their shadows are light. When this method was first developed, it was thought that frozen cells fractured along membranes revealing their true surface. It was later shown by Branton (1966) that the fracture preferentially follows a path of least resistance through the hydrophobic region of lipid bilayers, thus cleaving membranes in half and exposing extensive areas of their interior.

The Davson-Danielli model of the membrane envisioned no heterogeneity within the plane of the lipid bilayer and would have led to the prediction of two smooth, featureless fracture faces. Instead, freeze-fractured membranes presented two distinct appearances. The outwardly directed inner half-membrane, called the *P-face,* contained numerous randomly distributed globular particles 6 to 9 nm in diameter. The inwardly directed outer half-membrane, called the *E-face,* was relatively smooth, containing only about a fifth the number of particles found on the other fracture face.

The finding of globules of protein within the plane of the membrane provided compelling morphological evidence for the *fluid-mosaic model* of the cell membrane proposed by Singer (1971) on the basis of thermodynamic considerations and the known properties of proteins.

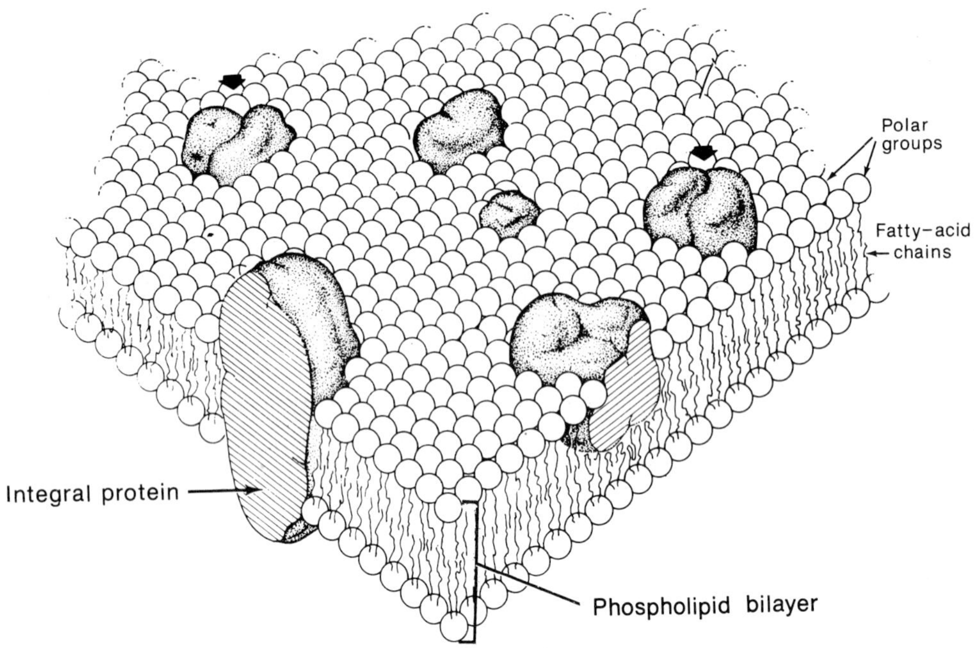

The current fluid-mosiac model of the cell membrane which envisions a two-dimensional solution of oriented lipids and globular proteins. The lipid bilayer is assumed to be fluid and the integral proteins are free to diffuse laterally within the plane of the membrane if not restrained by interaction with peripheral proteins in the underlying cytoplasm. (From S. J. Singer and G. L. Nicolson, Science 175, 720, 1972.)

According to this model, the membrane is a two-dimensional solution of oriented lipids and globular proteins. Like the phospholipid molecules of the bilayer, the protein molecules are assumed to be amphipathic. That is, they possess asymmetrical hydrophilic and hydrophobic regions. Thermodynamic considerations dictate that ionic amino acid and saccharide residues of integral glycoproteins would assume a position in the hydrophilic portion, and the non-ionic portions would preferentially localize in the hydrophobic interior of the membrane. Depending upon their secondary structure and the distribution of their hydrophilic and hydrophobic regions, the integral proteins would take up different positions in membrane. Some would have their oligosaccharide regions and polar amino acids exposed on the outer surface and their nonpolar region in the hydrophobic interior. Others with polar regions on either end and a nonpolar segment in the middle would extend through the entire thickness of the membrane. Freeze-fracture observations provided unambiguous evidence that protein particles are indeed embedded in membranes and not limited to layers on either side of the bilayer.

Using an electron microscopic technique for localizing specific antigenic proteins, Singer and Nicolson (1972) demonstrated that the integral proteins in unspecialized regions of membrane are essentially random in their distribution. Erythrocytes were treated with saturating amounts of antibody to Rh antigens and the membranes were then exposed to ferritin-labeled anti–human gamma globulin. The distribution of electron-opaque ferritin molecules visualized in the electron micrographs thus corresponded to the distribution of single antigen sites on the membrane. These were dispersed in a random two-dimensional array.

The fluid nature of the lipid bilayer and mobility of integral protein particles within the membrane were demonstrated in ingenious experiments by Frey and Edidin (1970). Human and mouse cells in coculture were induced to fuse, using Senai virus as the fusing agent. The distribution of human and mouse antigenic components was largely confined to their respective halves of the fused cell, but after 40 minutes they were intermixed and uniformly distributed over the entire surface, indicating that the proteins had diffused within a fluid membrane. Numerous additional demonstrations of lateral mobility of membrane proteins have since been reported.

In the 1960s, morphologists and biochemists seemed to be on diverging paths, with the structural uniformity implicit in the unit membrane concept at odds with the accumulating evidence of biochemical diversity among membranes. Freeze-fracturing and other technical advances have now brought about a gratifying convergence of the morphological and biochemical concepts of membrane structure. The number of particles per unit area in freeze-fracture replicas of membranes from different organelles correlates well with analyses of the protein content and degree of physiological activity of the same membranes. The fluid mosaic model is consistent with the great majority of morphological, biochemical, and immunological observations and is now generally accepted as the basis for further investigation of membrane-mediated phenomena.

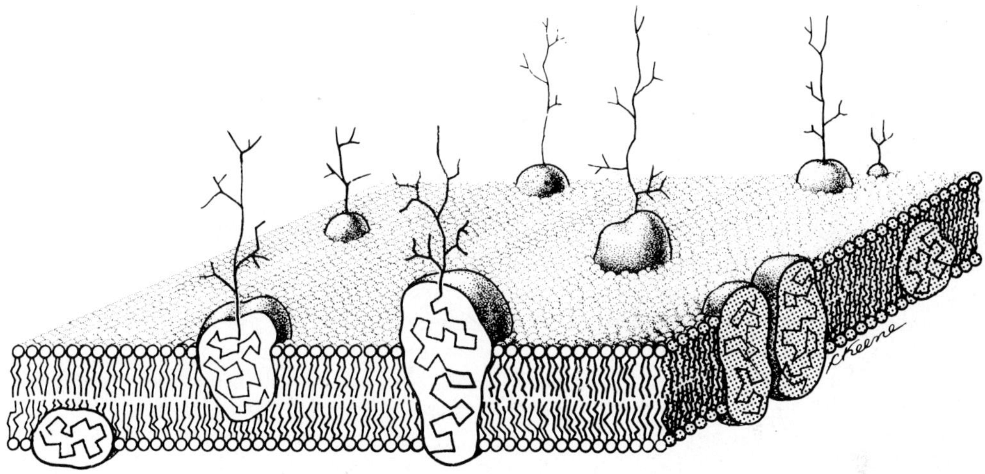

Schematic representation of amphipathic proteins taking up positions relative to the lipid bilayer that are determined by the distribution of their hydrophilic and hydrophobic residues. Saccharide residues of integral glycoproteins projecting from the outer surface contribute to the glycocalyx.

When it became possible to examine thin sections of cells with the electron microscope, the plasmalemma always appeared as a linear profile consisting of two dark lines, about 3.5 nm thick, separated by an intermediate light zone of similar width. This came to be called the *unit membrane* and was widely interpreted as a morphological confirmation of the Davson-Danielli model. The dense lines were thought to result from deposition of osmium in the protein and hydrophilic ends of the phospholipids, while the light line was believed to correspond to the hydrophobic region of the lipid bilayer. In light of the more recent demonstration by freeze-fracturing that the protein constituents do not form layers on either side of the lipid bilayer but occur in particulate form within the plane of the membrane, this interpretation has been modified. The protein of the membrane does not appear to contribute significantly to its density in electron micrographs. The assumption that osmium is deposited in the polar ends of the phospholipids but not in the hydrocarbon chains may still be valid.

Shown here are the surface membranes of two adjoining cells separated by a 15 nm intercellular cleft occupied by material of low electron density assumed to consist mainly of carbohydrate.

Figure 1. Boundary between two glial cells in the central nervous system of the annelid *Aphrodite.*

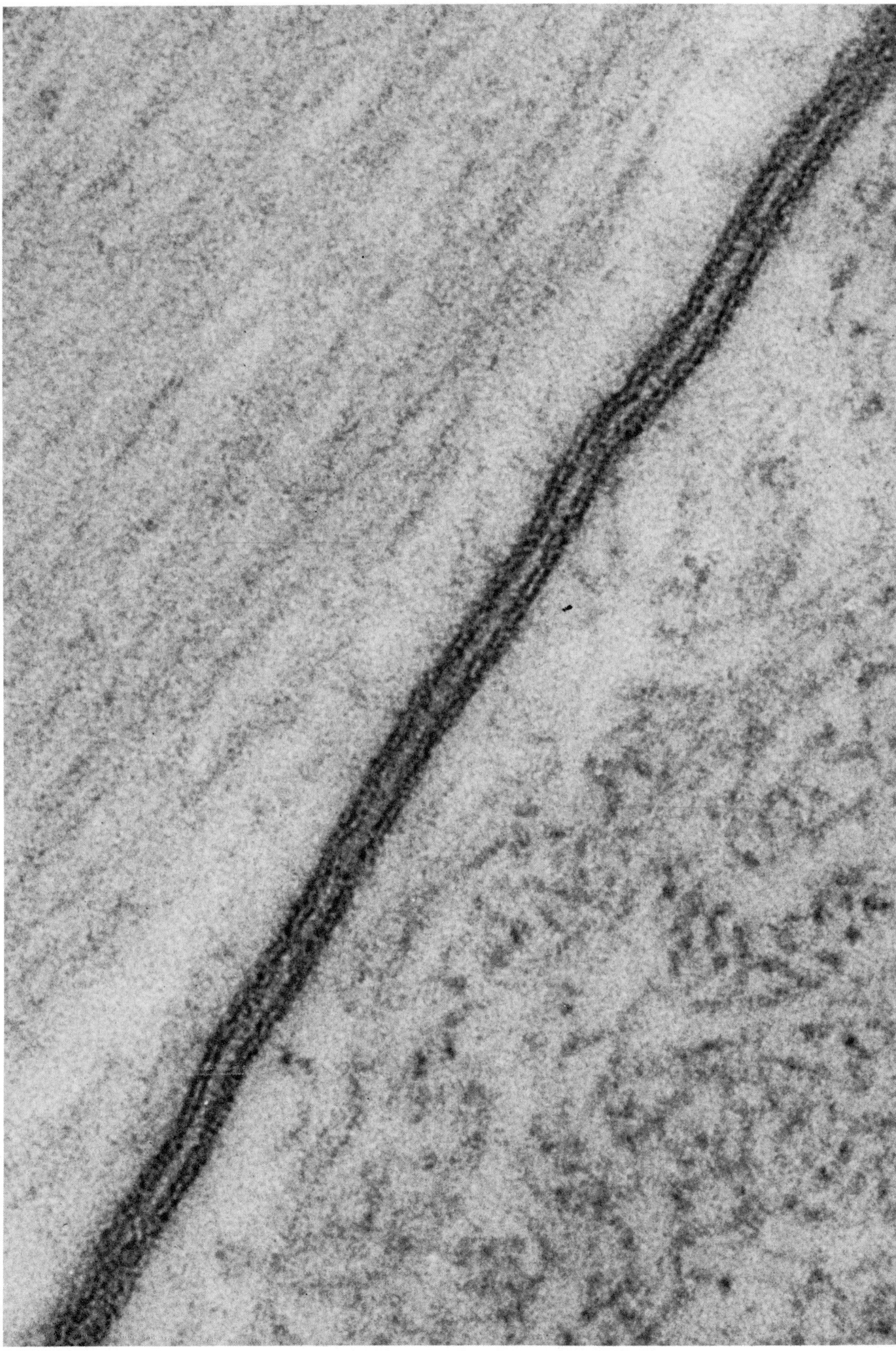

Figure 1

Where the plasmalemma has an uneven contour, its orientation with respect to the plane of the thin section changes and the trilaminar appearance of the membrane is not seen if the plane of section is slightly oblique. In the accompanying micrograph of the brush border of intestinal epithelium, the consistent orientation of the closely packed microvilli facilitated obtaining true transverse sections of their limiting membrane throughout the field. The cross section of each microvillus is bounded by two dense lines of similar thickness separated by a lighter intermediate zone. Not all cell membranes exhibit this degree of symmetry. In some the outer dense line is thinner than the inner.

Figure 2. Intestinal microvilli of a cat. (Micrograph courtesy of Susumu Ito.)

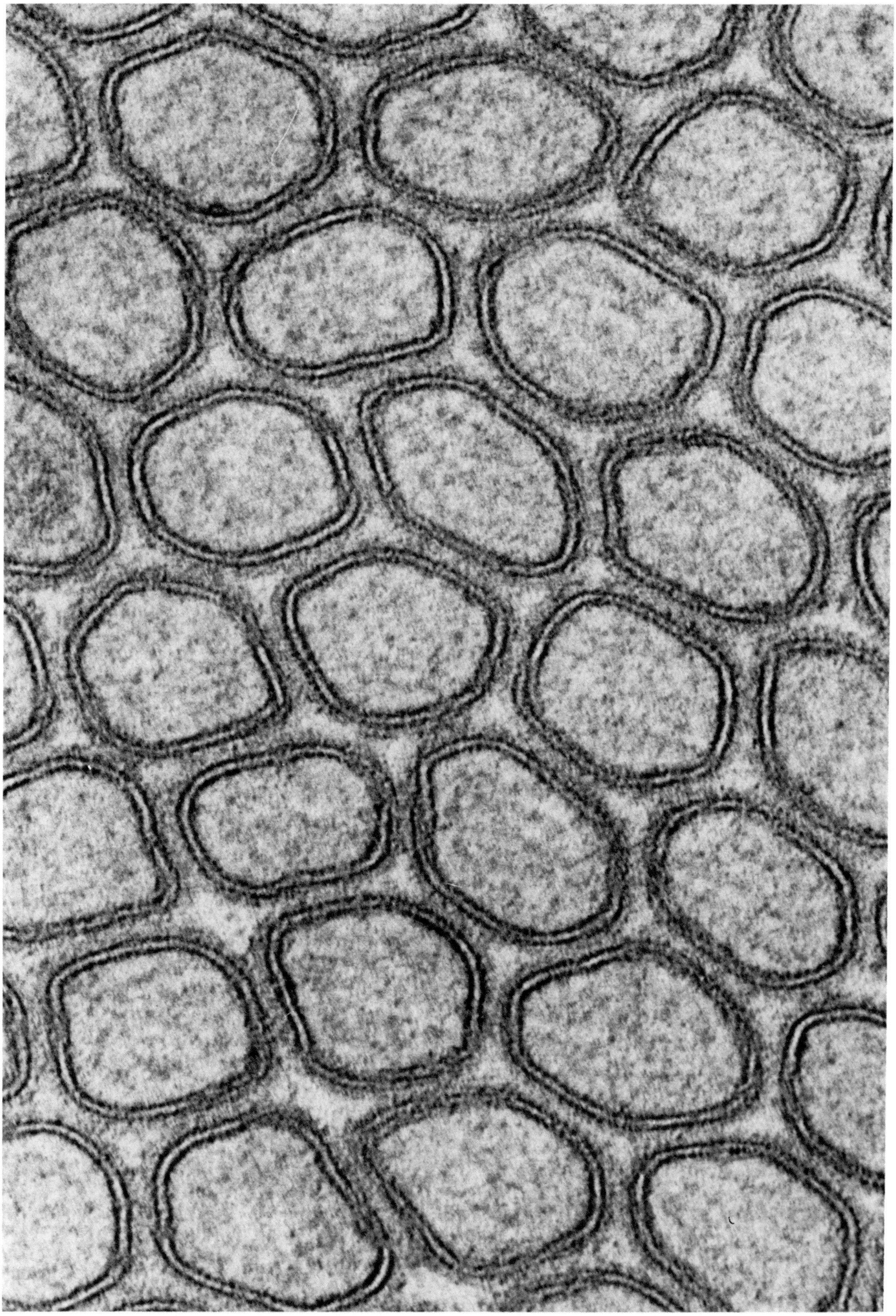

Figure 2

Replicas of frozen-fractured cell membranes present two distinctive appearances. The outwardly directed inner half-membrane, called the P-face, shows numerous 6 to 9 nm particles randomly distributed over the fracture face. The inwardly directed outer half-membrane, called the E-face, is relatively smooth, showing only occasional adherent particles. The globular particles vary in size and are believed to be the protein constituents of the membrane.

The upper half of the accompanying micrograph shows a typical E-fracture face. The plane of fracture then breaks across an intercellular space and cleaves the membrane of the adjacent cell, exposing its P-fracture face in the lower half of the figure.

Cell types vary in metabolic activity and their membranes vary in their protein content. Corresponding differences are observable in the number of intramembrane particles per unit area. For example, compare the concentration of P-face particles in this figure with that in the following illustration.

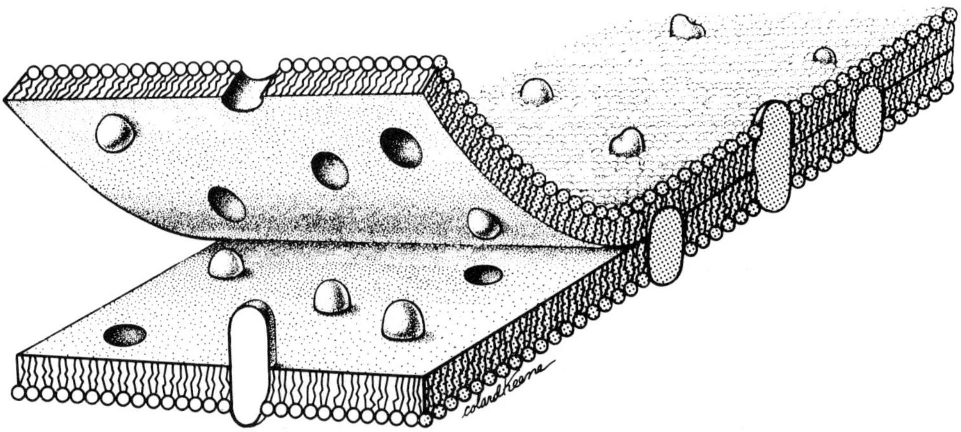

In the freeze-fracture method, the path of membrane cleavage is along the hydrophobic interior of the lipid bilayer resulting in two complementary fractures faces: (1) an outwardly directed inner half-membrane presenting the P-face from which the majority of the globular proteins project, and (2) an inwardly directed outer half-membrane presenting the E-face which is relatively smooth but shows occasional protein particles.

Figure 3. Freeze-fracture preparation of an unspecialized area of the membranes of adjacent Sertoli cells from guinea pig testis.

Figure 3

The replica on the facing page again shows an E-fracture face in the upper half of the figure and in the lower half the P-face of the membrane of another epithelial cell. The magnification is somewhat greater than in the previous figure; nevertheless it is evident that the number of intramembrane particles per unit area is much greater.

In the myelin sheath of peripheral nerves where the multiple layers of membrane are metabolically relatively inactive and mainly subserve an insulating function, intramembrane particles are very sparse.

Figure 4. Freeze-fracture preparation of ciliary epithelium of monkey eye. (Micrograph courtesy of Giuseppina Raviola.)

Figure 4

Regional Difference in Structure and Function

The fluidity of the lipid bilayer of biological membranes and the demonstrated capacity for lateral mobility of antigenic components of membranes might suggest that diffusion would distribute protein components uniformly over the cell surface. This is clearly not the case. The apparently random pattern of particles in unspecialized areas of cell membrane gives an impression of uniformity that may be misleading. There is morphological and cytochemical evidence for significant physiological differences between the membrane at the luminal surface of epithelia and the lateral and basal membranes. In addition to differences in thickness, degree of development of its glycocalyx, and its surface topography, the membrane at the free surface of epithelia differs from lateral and basal membranes in enzymatic properties. Sodium-potassium ATPase is demonstrable by cytochemical methods in the lateral and basal membranes of transporting epithelia but not in the apical surface (Ernst, 1975; Ernst and Mills, 1977). It has also been observed that the direction of budding of infecting viruses is a specific property of each virus. Different viruses may express opposite polarities in the same cell type. Some bud exclusively from the basal surface (Hess and Falcon, 1977), others from the apical surface, and still others from the lateral surface (Boulan and Sabatini, 1978). It seems undeniable that this specificity must depend not only upon properties of the respective viruses but also upon fundamental biochemical differences in the apical lateral and basal membranes of the same cell. Moreover, epithelial cells are able to express their functional polarization and membrane specificity even when removed from their normal environment and grown in tissue culture.

Regional specialization of the membrane for specific functions is not limited to epithelial cells. Indeed, one of the more dramatic examples is the mammalian spermatozoon. The plasma membrane overlying the anterior portion of the head is sensitive to environmental stimuli that induce the acrosome reaction, and it responds by fusion with membrane of the underlying acrosome. The posterior segment of the acrosome and the postacrosomal cell membrane are the unique sites of recognition and fusion with the oolemma during fertilization. The membrane enclosing the midpiece of the sperm flagellum is involved in access of substrate to the mitochondrial enzymes that generate the energy for locomotion. The membrane of the principal piece may play an ancillary role in propagation of bending waves along the sperm tail. In freeze-fracture preparations of sperm membranes, each of these regions exhibits a distinctive pattern of distribution of intramembrane particles (Friend and Fawcett, 1974; Fawcett, 1975).

Little is known about mechanisms of biogenesis and maintenance of regional differences in the cell membrane. Intramembrane protein particles of erythrocytes are believed to be associated with *spectrin,* a filamentous polypeptide polymer on the cytoplasmic surface of the membrane (Stick, 1974; Marchesi, Furthmayr and Tomita, 1976). It is speculated that the disposition of some transmembrane proteins may be determined by their attachment to spectrin, or actin filaments in the cortical zone of the underlying cytoplasm. This is currently an area of active investigation.

The mammalian spermatozoon exhibits an unusual degree of regional specialization of the cell membrane. The accompanying micrograph illustrates the P-face of the membrane in three regions of a guinea pig spermatozoon. The intramembrane particles in the midpiece are of uniform size and associated in linear arrays that are oriented circumferentially or oblique to the axis of the flagellum. In the principal piece, the particles are variable in size but generally larger than those of the midpiece and are randomly distributed. The physiological significance of the beadstrings of particles in the midpiece is not known, but this membrane is closely applied to the mitochondrial helix which provides the ATP necessary for sperm tail movements. It seems likely that these particle arrays are related in some manner to the permeability properties or specific enzymatic functions of this segment of the membrane.

Figure 5. Replica of the P-face of guinea pig sperm tail. (From Friend and Fawcett, J. Cell Biol. 63:641, 1974.)

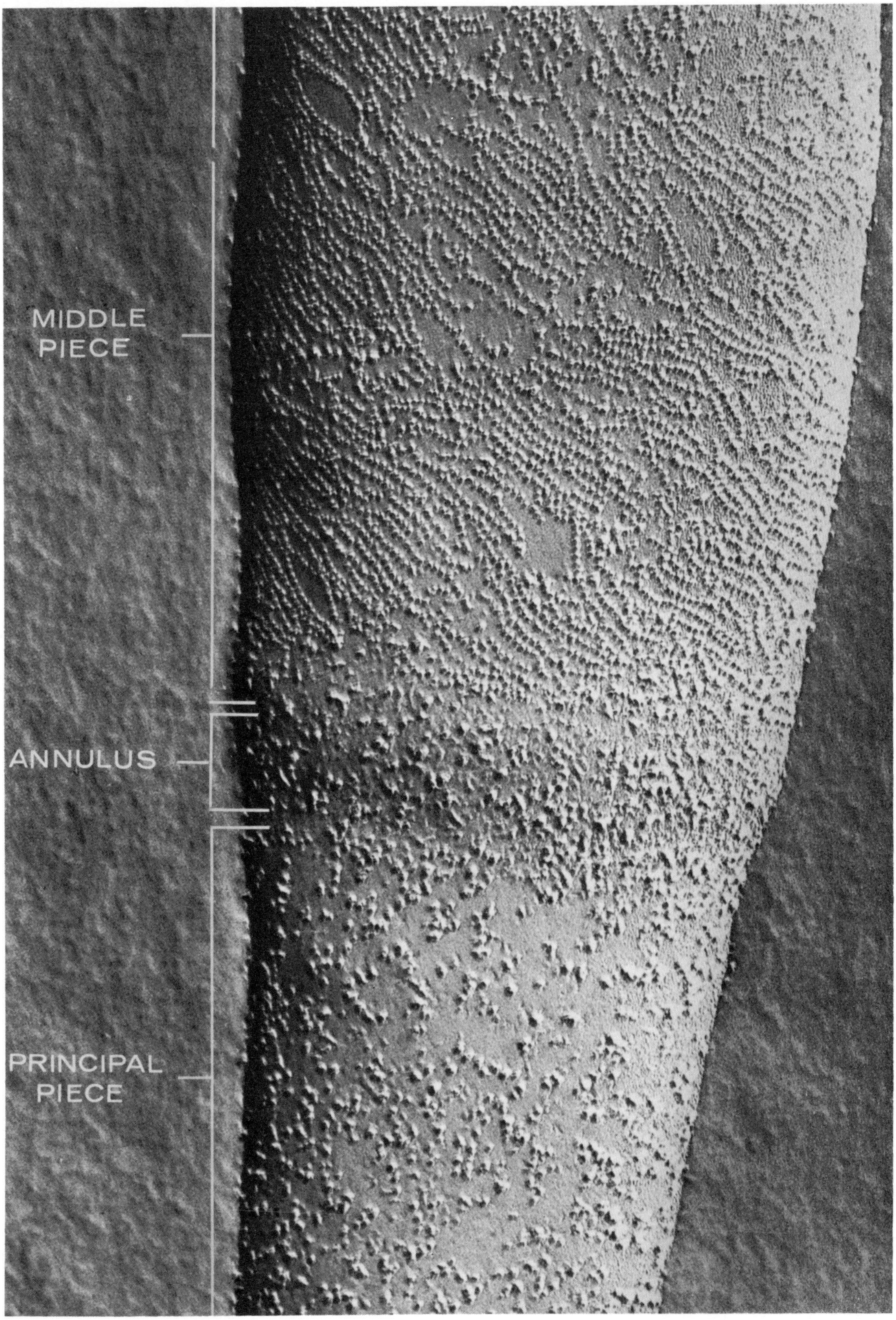

Figure 5

The basis for the alignment of particles in the midpiece membrane has not been established. Inasmuch as the particles do not always appear to be in close contact, their bonding to one another seems less likely than the possibility that they are linked to a filamentous component associated with the cytoplasmic surface of the membrane. Both the linear aggregation of the particles and the orientation of the resulting beaded strands appear to depend upon proximity of the membrane to the underlying mitochondria. In a fusiform thickening of the tail in the caudal portion of the normal midpiece (see inset), the membrane is separated from the mitochondria by an intervening layer of residual cytoplasm. As indicated in the inset, the area included in the accompanying micrograph shows a region of transition in membrane structure. In the lower half, circumferentially oriented beaded strands are concentrated where the membrane is closely applied to the gyres of the underlying mitochondrial sheath. In the upper part of the figure, the membrane diverges from the mitochondria and there individual particles are randomly dispersed or in very short rows. If the sperm are made to swell, elevating the membrane, the particle arrays are dissociated into individual particles throughout the length of the midpiece. This is one of the first examples of local variation in internal organization of a cell membrane that appears to be correlated with proximity to an underlying cell organelle.

Figure 6. Replica of the P-face of a small area of membrane on the midpiece of guinea pig sperm. (From Fawcett. *In* International Cell Biology 1976–1977, pp. 588–600, Rockefeller University Press, New York.

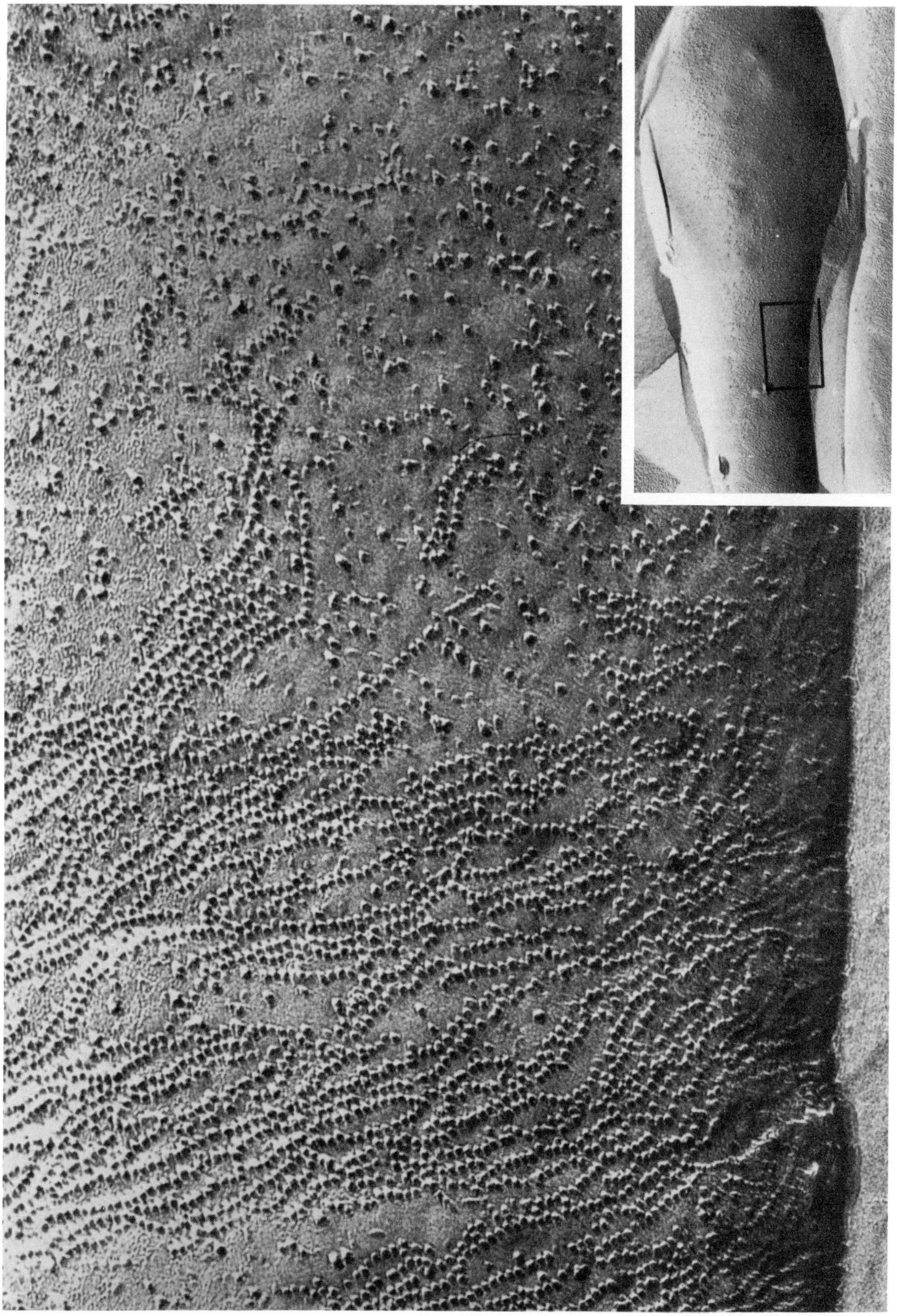

Figure 6

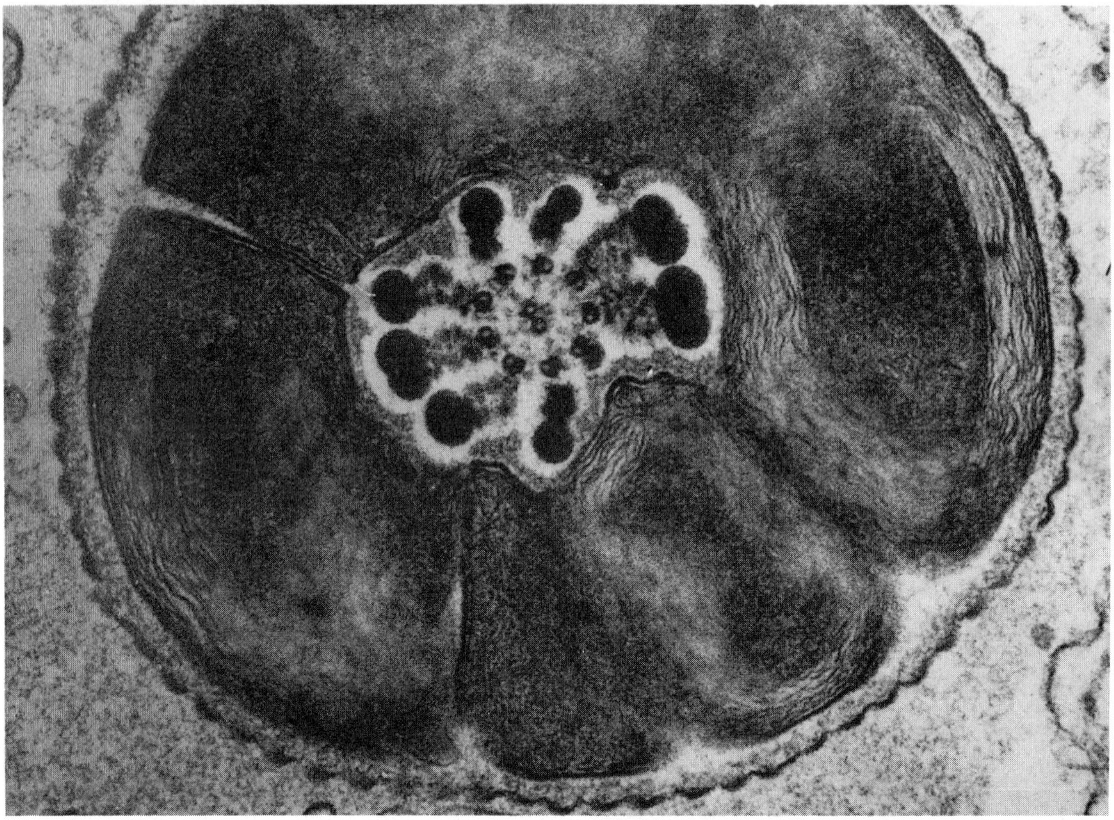

An electron micrograph of the midpiece of an opossum spermatozoon in cross section showing a filamentous layer of cytoplasm underlying regularly recurring thick and thin segments of membrane, which gives it a corrugated appearance.

In the midpiece of the opossum spermatozoon, there is a conspicuous layer of filamentous material associated with the inner aspect of plasma membrane. In addition there is a very regular pattern of variation in thickness and a corrugation of the membrane. A freeze-fracture replica of the P-face of this membrane, on the opposite page, shows intramembrane particles of uniform diameter closely aggregated in longitudinal bands three to five particles wide separated by particle-free aisles. Further study of the biochemistry of such membranes might shed further light upon the physiological significance of such particle arrays and the role of peripheral proteins in the assembly and stabilization of specific patterns of integral membrane proteins.

Figure 7. Freeze-fracture replica of the sperm midpiece of the opossum *Didelphis virginiana.* (Micrograph courtesy of Gary Olson.)

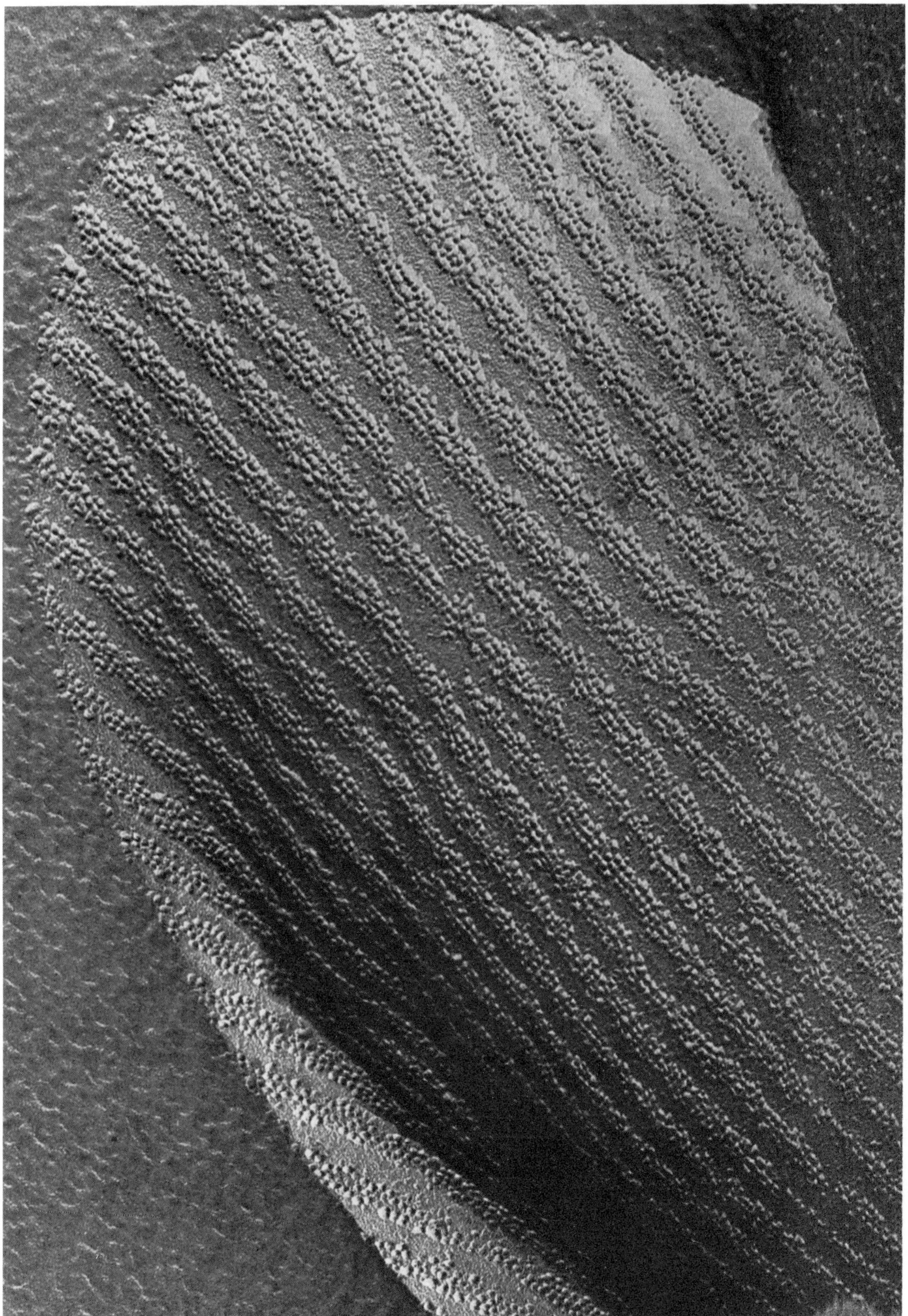

Figure 7

In ciliated cells, the P-face of the membrane around the base of the shaft of each cilium contains several circumferential rows of intramembrane particles called the "ciliary necklace." The membrane in this region is linked to the doublets of the axoneme by a radial pattern of densities. It is not known whether the particle rows are involved in anchoring the membrane to the axoneme or whether they represent sites of specific ion permeability (Gilula and Satir, 1972).

Linear arrays of intramembrane particles are also found on the microvilli of a small percentage of columnar epithelial cells in the colon and rectum of primates. The great majority of cells in this epithelium, however, have microvilli with the usual randomly dispersed individual particles on the P-face. The significance of these short rows of particles is not known, but it is speculated that they may represent assemblies of protein which are specific to a subpopulation of intestinal columnar cells that has a distinctive function not yet defined (Neutra, 1979).

Figure 8. Base of cilia on a cell in a ductus efferens of the rat. Freeze-fracture preparation. (Micrograph courtesy of Fumi Suzuki.)

Figure 9. Brush border of a surface columnar cell from human rectal mucosa. Freeze-fracture preparation. (Micrograph courtesy of Marian Neutra.)

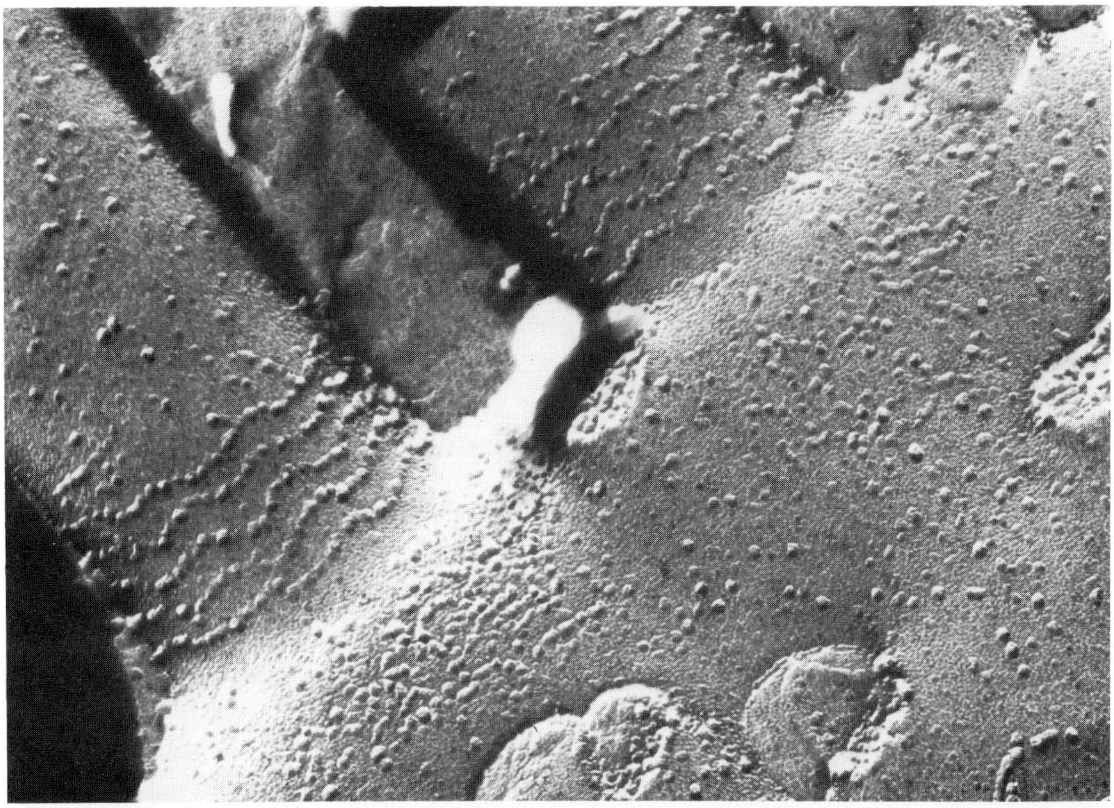

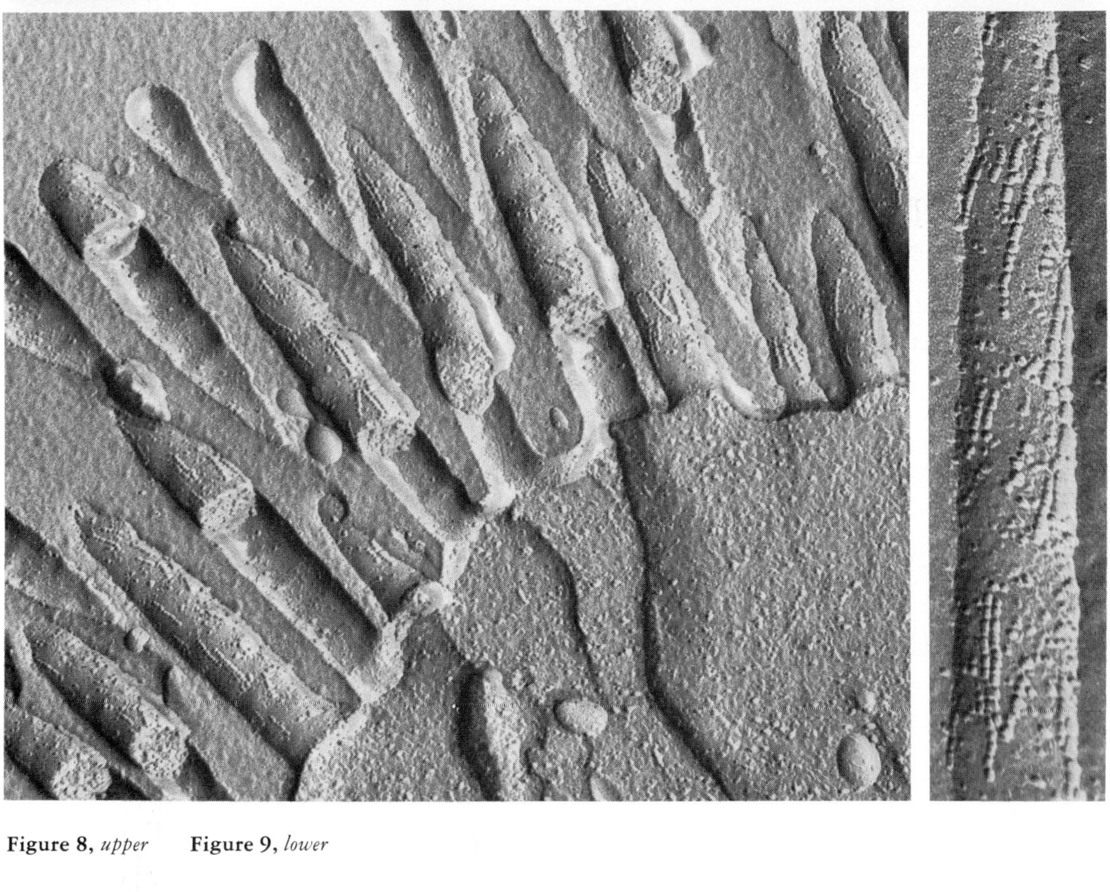

Figure 8, *upper* **Figure 9**, *lower*

A remarkable example of stable intramembrane specializations is found in neuronal synapses. Aggregates of intramembrane particles which remain associated with the P-face of the fractured plasmalemma are present in the presynaptic membrane. The postsynaptic membrane, on the other hand, has distinctive particle aggregates only in excitatory synapses; these particles remain associated with the P-face in peripheral cholinergic synapses and with the E-face in central synapses. The postsynaptic membrane of inhibitory synapses in the central nervous system is unspecialized. It has been suggested that these intramembrane particle aggregates contain ion channels or carry receptor molecules for the chemical transmitter.

The accompanying micrographs illustrate similar intramembrane specializations in the synapses of the primate retina. Above is a view of two invaginating synapses between a cone cell ending and processes belonging to horizontal and bipolar cells in the outer plexiform layer. The presynaptic aggregate of particles contained within the photoreceptor cell membrane is indicated by arrows. At the stars, the fracture process has exposed an internal specialization of the horizontal cell membrane whose significance is not known. In the micrograph below, the two postsynaptic processes of dyad synapse in the inner plexiform layer are characterized by an aggregate of particles which remain associated with the E-face (arrows).

Figure 10. Retina of *Macaca mulatta*. (Micrograph above courtesy of E. Raviola and N. B. Gilula, J. Cell Biol. 65:192–222, 1975. Micrograph below courtesy of E. Raviola and G. Raviola.)

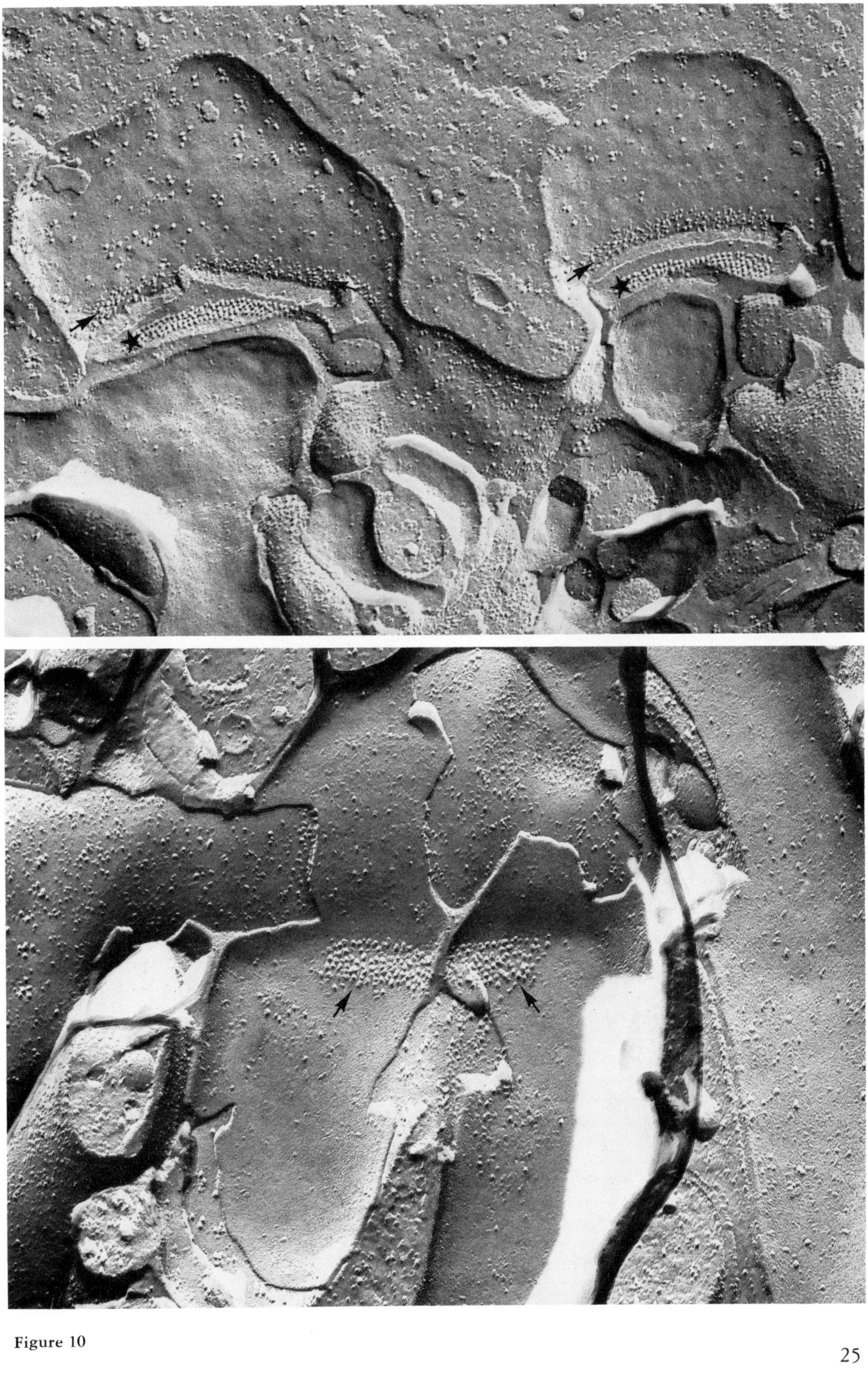

Figure 10

A unique example of a relatively stable differentiation within the plane of the cell membrane is found in the superficial cells of the transitional epithelium lining the mammalian urinary bladder. Rigid plaques within the membrane give the surface of the epithelium a peculiar angular contour when viewed in thin sections. In the accompanying micrograph, the brackets indicate the location of some of the plaques in the irregular surface of bladder epithelium of a rat. The plaque regions are 12 nm thick, whereas the interplaque areas are about 8 nm.

The lower figure is a scanning micrograph of bladder epithelium from a mouse. The polygonal plaques have receded during preparation, while the more flexible interplaque regions have been elevated into thin ridges, or folds, outlining the plaques. Although the irregularity of the surface may have been somewhat exaggerated during dehydration of the specimen, this scanning micrograph clearly demonstrates that the membrane is a mosaic of rigid plaques and more flexible interplaque regions.

Figure 11. Electron micrograph of a thin section of the superficial cells of uroepithelium from the rat. (Micrograph courtesy of Marian Hicks.)

Figure 12. Scanning electron micrograph of the luminal surface of mouse bladder. (Micrograph courtesy of Linda Malick.)

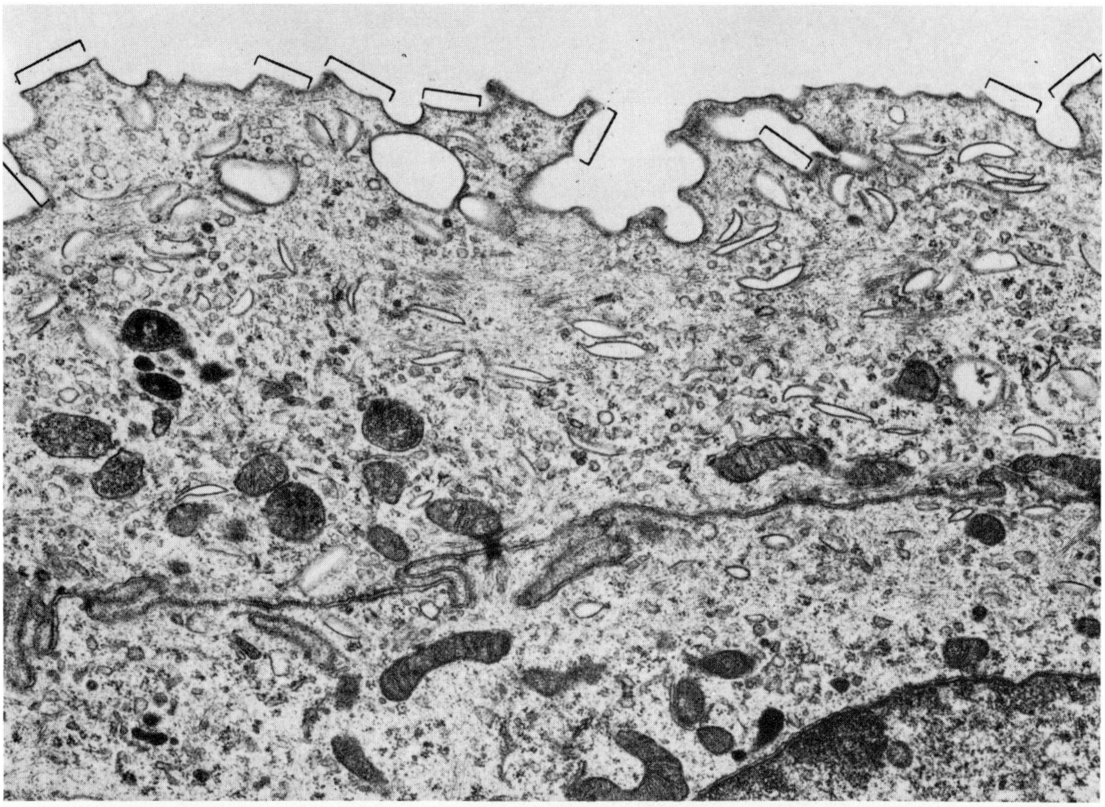

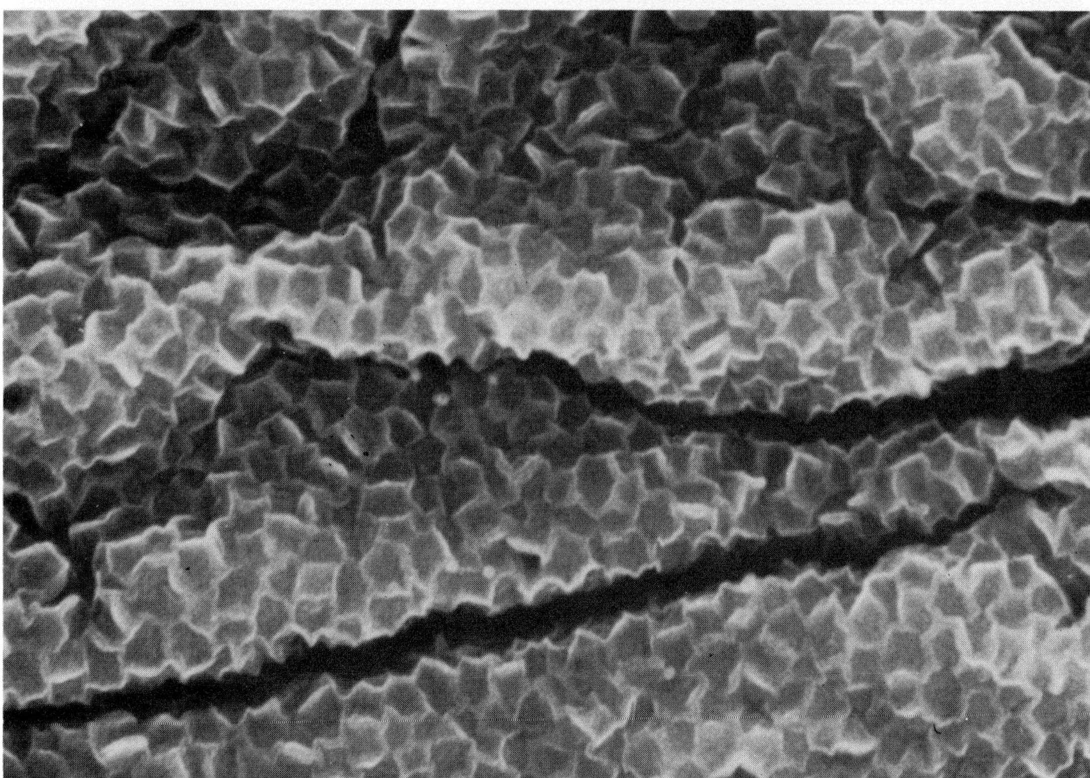

Figure 11, *upper* **Figure 12,** *lower*

Although the plaques in freeze-fracture replicas of bladder epithelium bear a superficial resemblance to gap junctions, they are, in fact, quite different. They are circular or polygonal and consist of an hexagonal array of subunits on the E-face (connexons of gap junctions are on the P-face). The subunits are about 12 nm in diameter and have a center-to-center spacing of 15.6 nm (Severs and Warren, 1978). On the P-face the plaques are relatively smooth but at high magnification show shallow pits complementary to the E-face particles. The interplaque regions of membrane have a few P-face particles.

Shown in the inset is a replica of a plaque in the lumenal membrane, not fractured but deep-etched after freezing. In favorable areas of such preparations, each subunit appears to be made up of a ring of six smaller particles around a central channel. There is, however, no evidence that the plaques are especially permeable. On the contrary, the special properties of the lumenal membrane may contribute to the barrier function of the bladder wall that prevents osmotic dilution of the urine by water from the bloodstream.

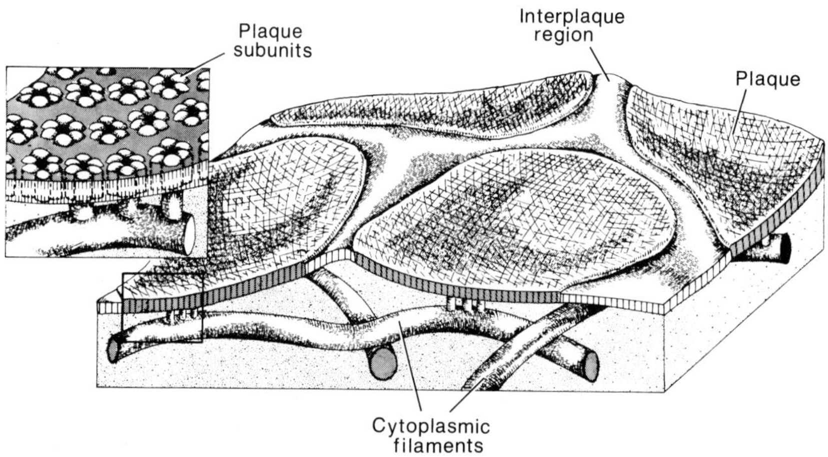

Drawing of the organization of the lumenal membrane of mammalian bladder epithelium. (From A. Staehelin et al., J. Cell Biol. 53:73–91, 1972.)

Figure 13. Freeze-fracture replica of the E-face of rat bladder lumenal membrane. Inset: replica of a frozen deep-etched lumenal membrane showing plaque subunits at high magnification. (Micrographs courtesy of Nicholas Severs and Marian Hicks.)

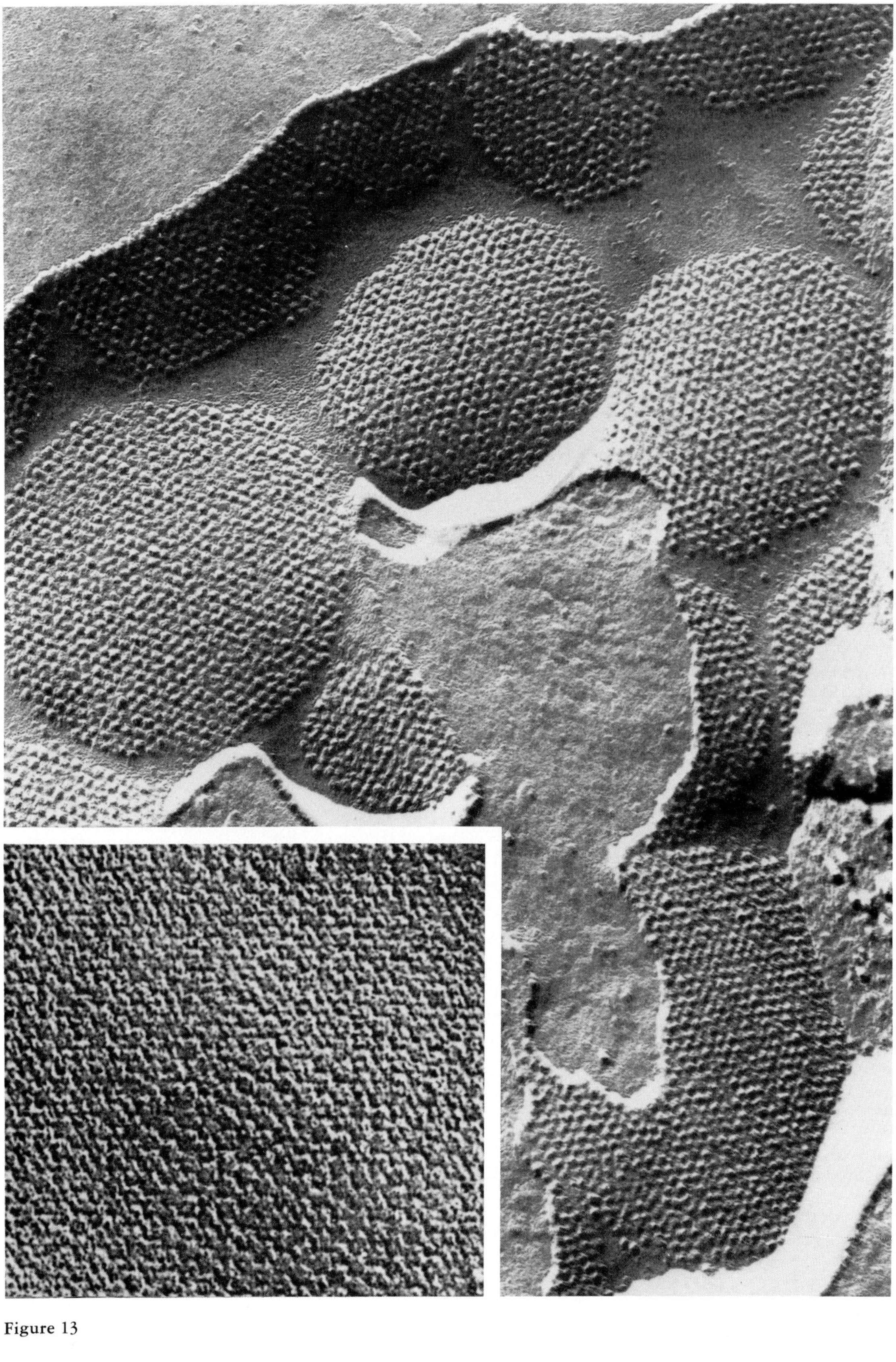

Figure 13

The myelin sheath of peripheral nerves is formed from the membrane of the associated Schwann cells wrapped in a tight spiral around the nerve axon. Early in development, the axon simply occupies a deep groove in the surface of the ensheathing cell. Where the rims of the groove meet to completely enclose the axon, the apposed segments of Schwann cell membrane form a structure called the *mesaxon*. Myelinization begins with a lengthening of the mesaxon and its wrapping in a loose scroll around the axon. As the process continues, the cytoplasm between successive turns is progressively extruded and the spiral is tightened to form compact myelin consisting of 20 to 40 layers of membrane. Much of our knowledge of the biochemistry and molecular organization of membranes has come from studies taking advantage of the concentration of highly ordered membrane in myelin.

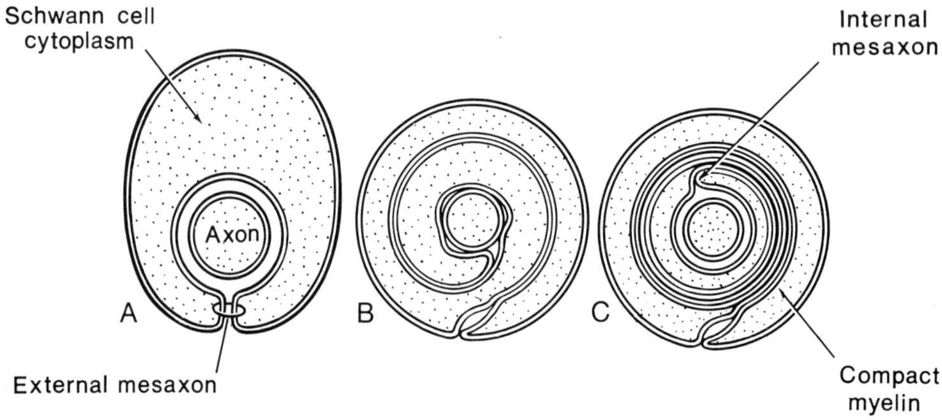

Diagram of successive stages in the formation of the myelin sheath from Schwann cell membrane. (Modified after J. D. Robertson, Prog. Biophys. 10:349, 1960.)

Figure 14. Myelin sheath in cochlear nerve of cat. (Micrograph courtesy of Enrico Mugnaini.)

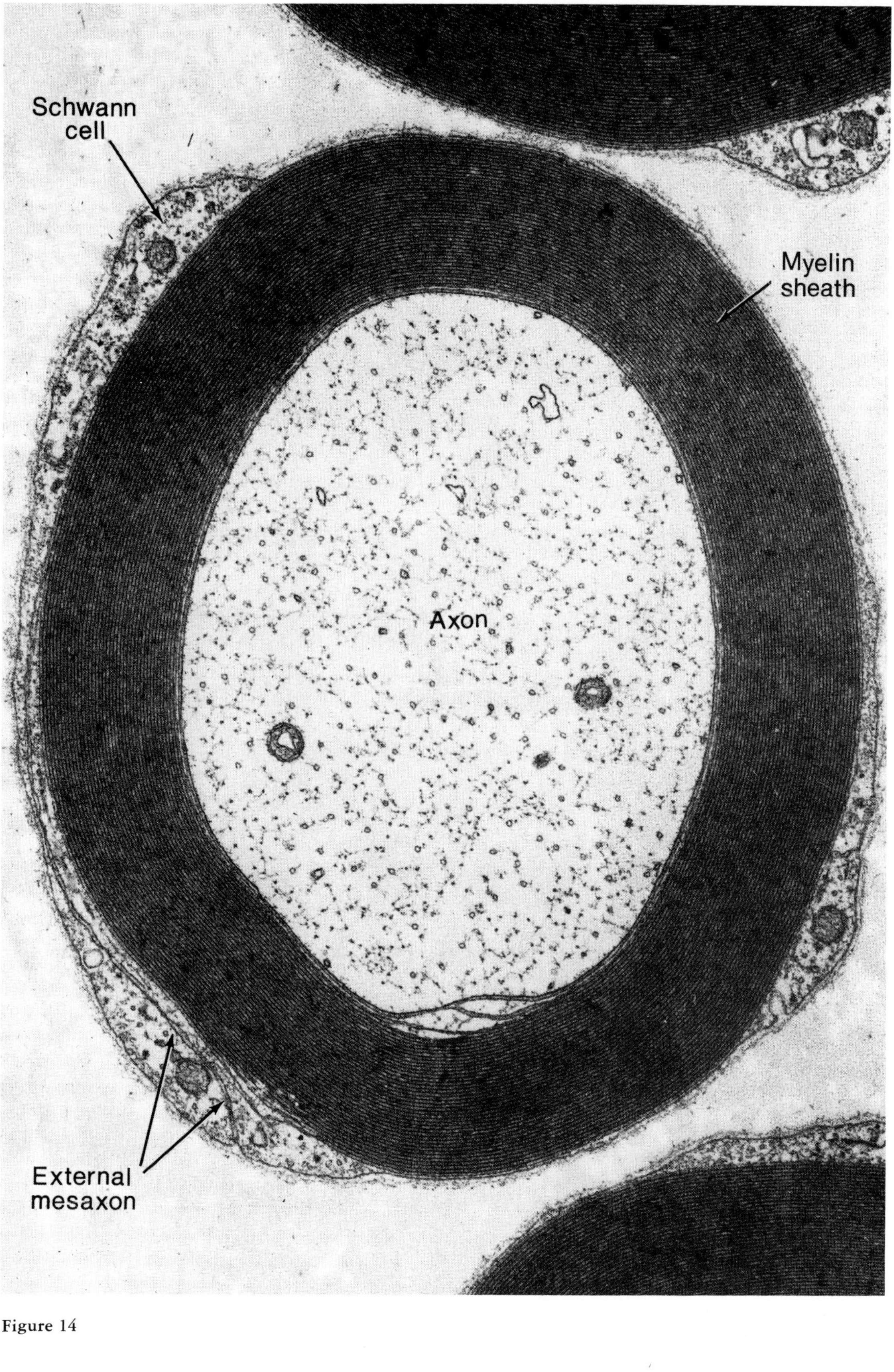

Figure 14

Where the internal surfaces of the apposed Schwann cell membranes are in contact, they form the *major dense lines* of myelin. The repeating period from one dense line to the next represents a pair of Schwann cell membranes. Where the outer surfaces of the two membranes are in contact, they form a faint *intraperiod line*. A myelin sheath in section therefore appears as a series of parallel dense lines alternating with less distinct intraperiod lines. At the magnification of the accompanying micrograph, only the major dense lines are visible. The membranes of the mesaxon are less closely apposed than those of the myelin.

Figure 15. Myelin sheath of nerve. (Micrograph courtesy of Enrico Mugnaini.)

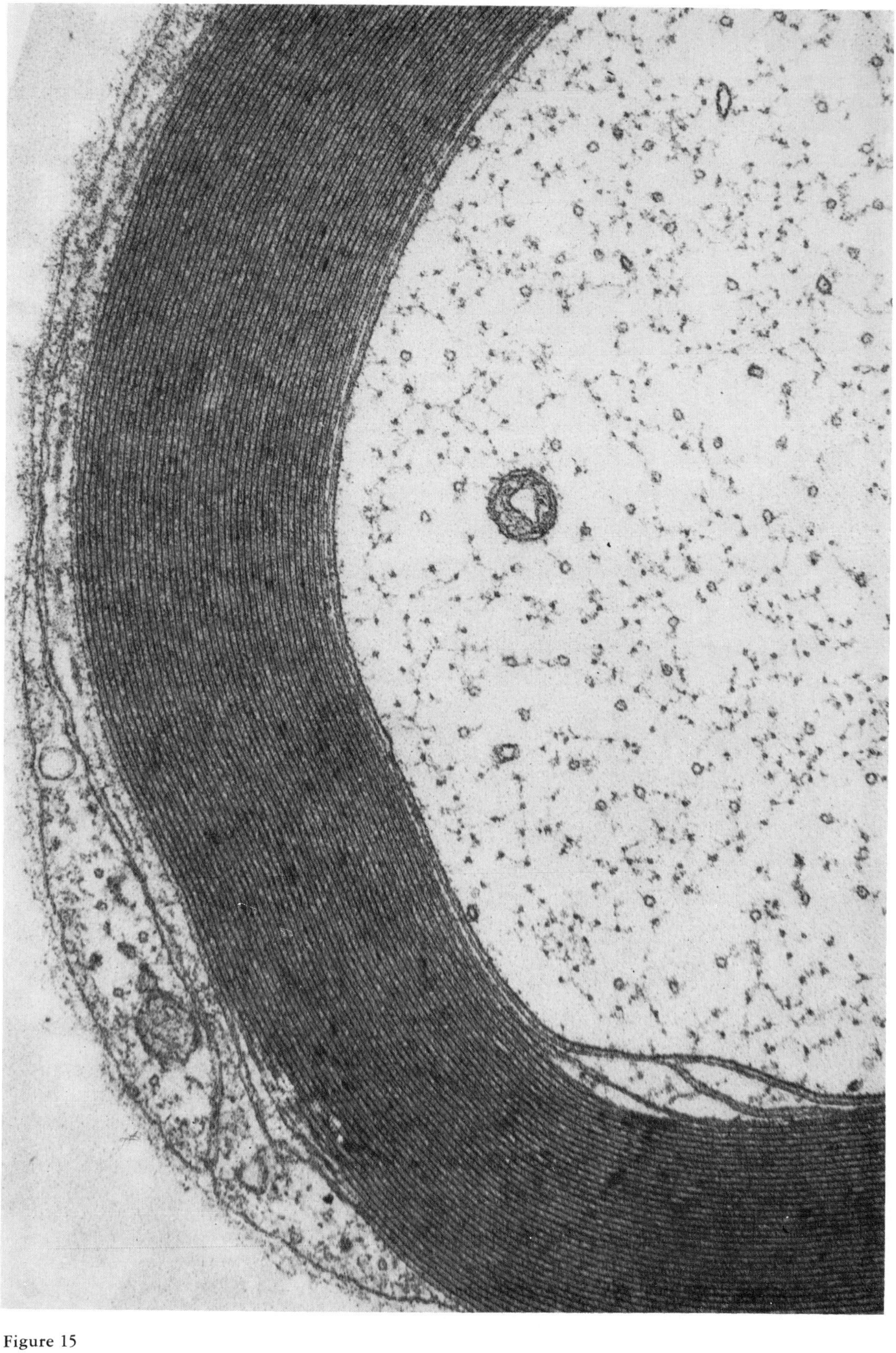

Figure 15

GLYCOCALYX OR SURFACE COAT

The membrane proper is not necessarily the outer limit of the cell surface. Transparent amorphous cell coats were detected by light microscopists employing microdissection techniques to study the ova of marine invertebrates and on various somatic cells (Chambers, 1940). A thin surface coat investing many cell types was detected with histochemical staining methods for localization of carbohydrates (Wislocki et al., 1951; Rambourg et al., 1966).

In an early electron microscopic study of gallbladder epithelium, Yamada (1955) described minute branching filaments or *antennulae* projecting from the membrane of the microvilli. In later investigations of intestinal epithelium, Ito (1965) observed branching filaments a few nanometers thick forming a more or less continuous fur-like layer over the tips of the microvilli. As more tissues were examined with the electron microscope, it became evident that a similar but thinner surface coat was of widespread occurrence on the free surface of cells. Taking cognizance of the high carbohydrate content of this component of the cell surface, Bennett (1963) proposed the descriptive term *glycocalyx* (from the Greek "sweet" and "husk"). As originally defined, the term embraced not only coats that are integral to the membrane but the cell walls of bacteria, fungi, and higher plants; mucous coats; intercellular matrices; and the basal lamina of epithelia. The diverse origin and chemical heterogeneity of the substances included in this broad definition diminishes its usefulness. In this book, devoted mainly to the cells of higher animals, the term glycocalyx will be limited to those layers intimately conforming to the contours of the membrane and continuous with its outer dense line.

The polysaccharide nature of the glycocalyx has been established by analysis of isolated cell membranes. Its high content of sialic acid contributes to the cell's strong negative charge. Removal of sialic acid from the surface with neuraminidase reduces the staining affinity of cells for cationic dyes and their anodic mobility when placed in a potential gradient (Wallach and Eylar, 1961; Wallach and Kamat, 1966). Studies on glycoproteins isolated from cell membranes provide strong evidence that their entangled polysaccharide chains projecting outward from the lipid bilayer form the glycocalyx. The glycocalyx may go undetected in routine electron micrographs but it can be intensely stained by electron opaque substances that bind to ionized carboxyl and sulfate groups of its acid carbohydrates. This can be accomplished by exposure of tissue blocks to colloidal thorium, or to Alcian blue during processing (Behnke and Zelander, 1970). Another common method involves the use of the inorganic dye ruthenium red, a hexavalent cation that binds firmly to acid polysaccharides (Luft, 1971).

The mat of delicate polysaccharide filaments that form the glycocalyx clearly plays a very important role in the interaction of cells with their environment. It constitutes a protective mechanical barrier or filter regulating which materials can come into contact with the membrane proper. Its polyanionic properties probably confer some degree of selectivity upon the adherence or binding of substances to cell surface.

In cells of the intestinal epithelium, there is evidence for the presence of hydrolytic enzymes in this layer that have an important role in the terminal steps of digestion of intraluminal nutrients.

The glycocalyx is exceptionally well developed on the brush border of intestinal absorptive cells. Each microvillus bears a tuft of fine, branching filaments. Those of neighboring microvilli intermingle to form a continuous mat over the villous tips of the brush border. Similar filaments are found on the sides of the microvilli but these are only a fraction the length of those at the tip. This surface layer acts as a barrier to penetration of large particles while allowing emulsified lipid, colloidal particles, and substances in solution to pass freely through its meshes and into the intervillous clefts. It is resistant to a variety of proteolytic and potent mucolytic agents. The prevailing radial orientation of the filaments and their continuity with the outer dense lamina of the underlying membrane indicates that they are an integral part of the cell surface and not merely an adherent layer of mucus. The morphological basis for this interpretation is more apparent in the higher magnification micrographs that follow.

Figure 16. Intestinal epithelium of cat. (Micrograph courtesy of Susumu Ito.)

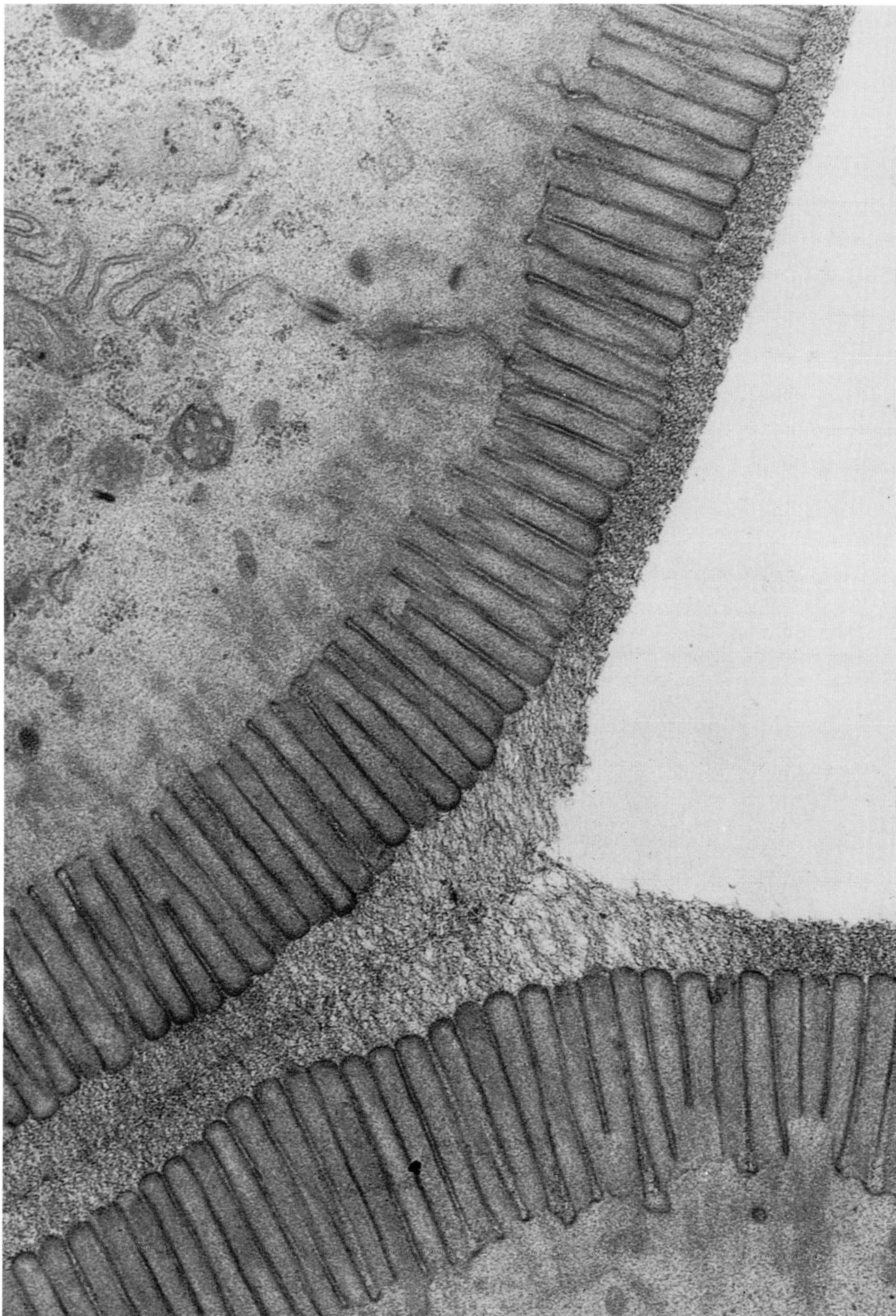

Figure 16

At high magnification, the filaments of the glycocalyx are 2.5 to 5 nm thick and may extend for 0.1 to 0.5 μm beyond the tips of the microvilli. They branch repeatedly and are generally somewhat thicker at sites of bifurcation. At their base, they appear to be continuous with the outer leaflet of the underlying membrane. This is especially evident at the arrows in the upper figure. The filaments radiating from adjacent microvilli intermingle to form a more or less continuous meshwork as shown in the lower figure. This appearance is due to superimposition of the images of the filaments rather than actual anastomosis.

Figures 17 and **18.** Microvilli on the intestinal absorptive cells of the bat, *Myotis lucifugus.* (Micrographs courtesy of Susumu Ito.) The upper figure reproduced from Fawcett, D. W., J. Histochem. Cytochem. 13:75–95, 1965.

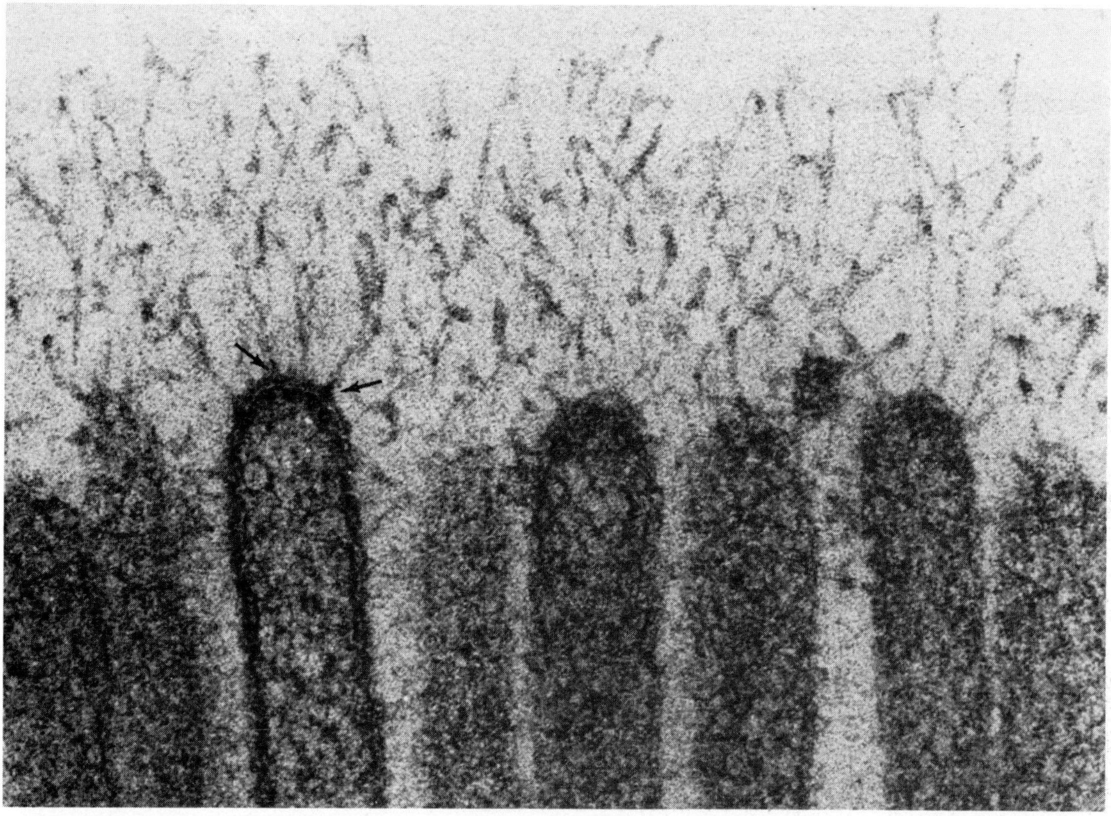

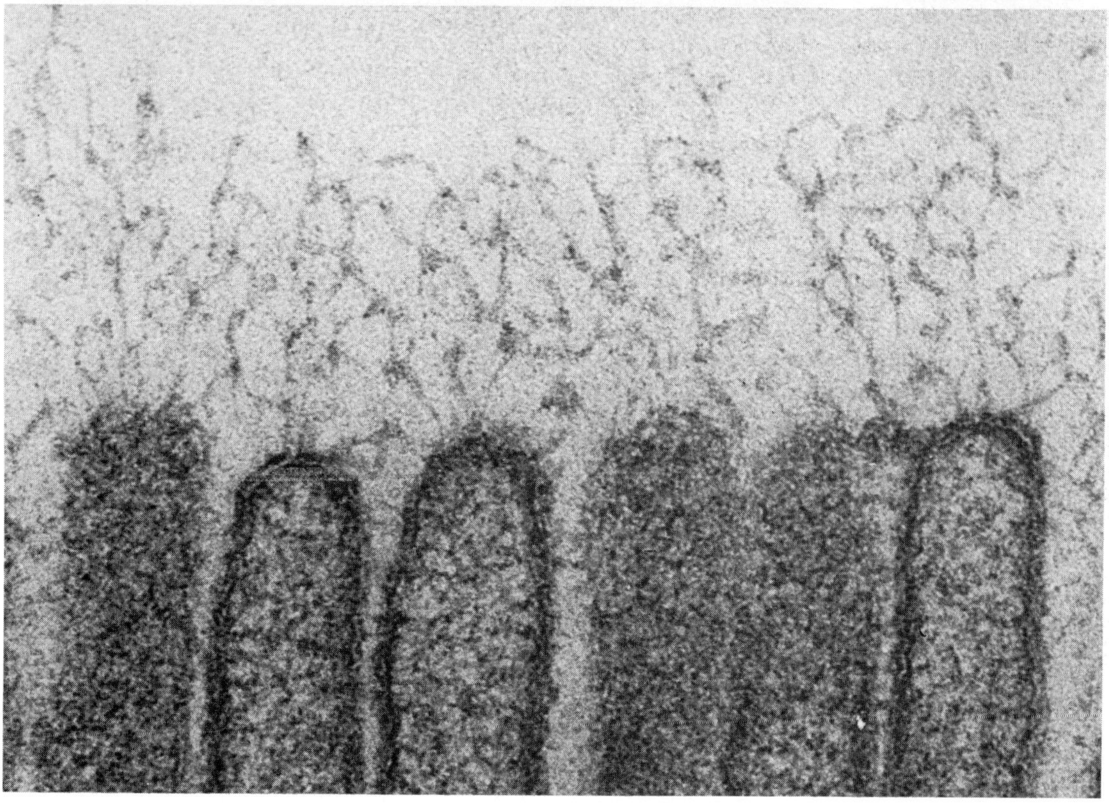

Figure 17, *upper* **Figure 18,** *lower*

Owing to the strong negative charge of its acidic polysaccharides, the glycocalyx binds colloidal thorium. In the micrograph on the facing page, of brush borders from tissue stained *en bloc* with thorium, the glycocalyx appears very dense because of electron scattering by the adsorbed heavy metal. With this enhancement of contrast, it can be seen that the glycocalyx is not confined to the tips of the microvilli but extends into the clefts between them.

Figure 19. Intestinal epithelium from cat. Stained *en bloc* with colloidal thorium. (Micrograph courtesy of Susumu Ito.)

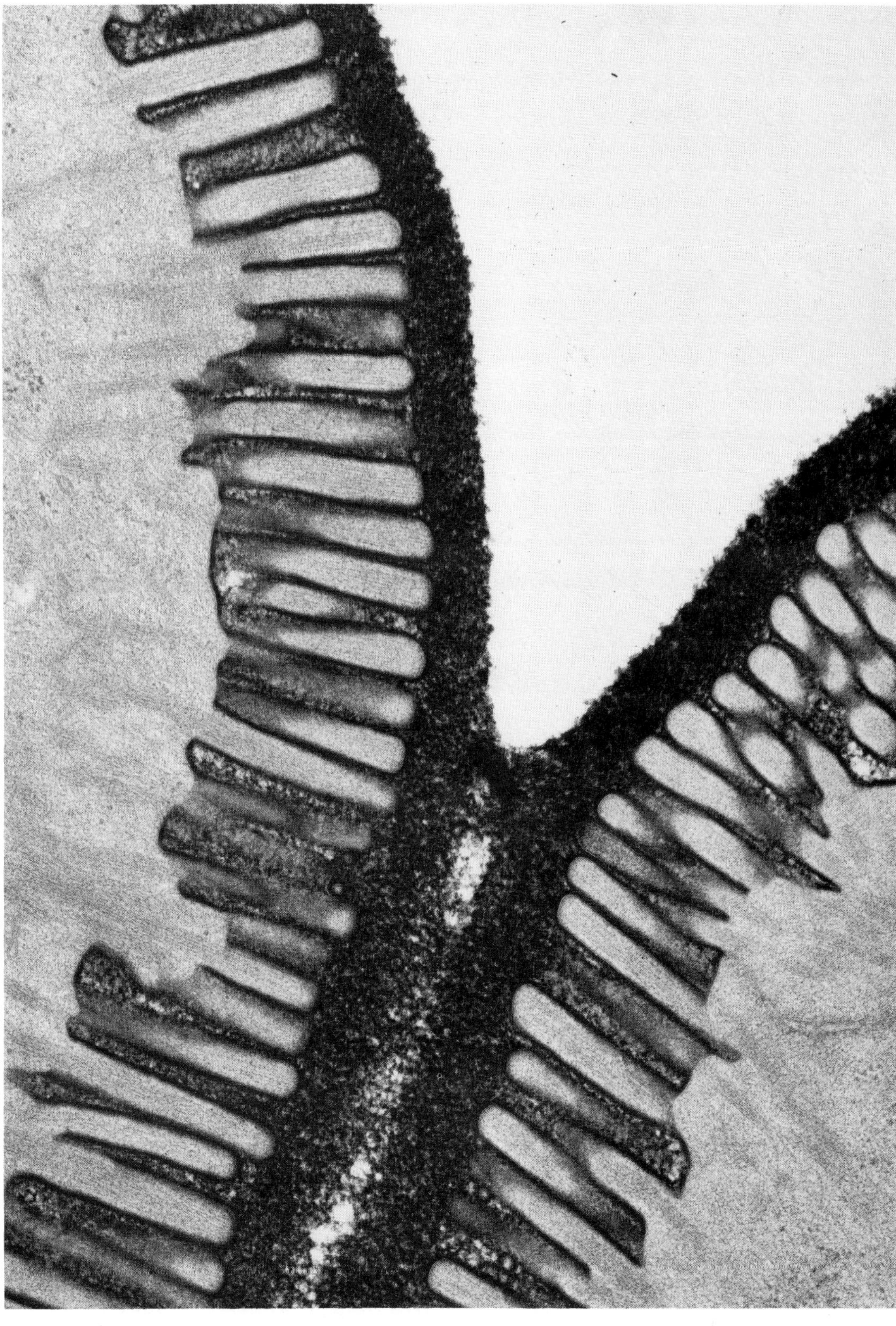

Figure 19

Another method for the electron microscopic visualization of the negative charges at the surface of cell membranes involves the use of a polycationic derivative of ferritin. The cationized ferritin molecules bind firmly to the glycocalyx of cells at physiological pH. The particles are electron scattering because of their high iron content and appear black in electron micrographs.

The upper micrograph on the facing page illustrates the binding of cationized ferritin to the glycocalyx of a cell in the intestinal mucosa. The lower figure shows a comparable preparation of a cell from the stomach.

Figures 20 and **21.** Electron micrographs of the lumenal surface of cells from the intestine (upper) and stomach (lower) exposed to cationized ferritin. (Micrographs courtesy of Susumu Ito.)

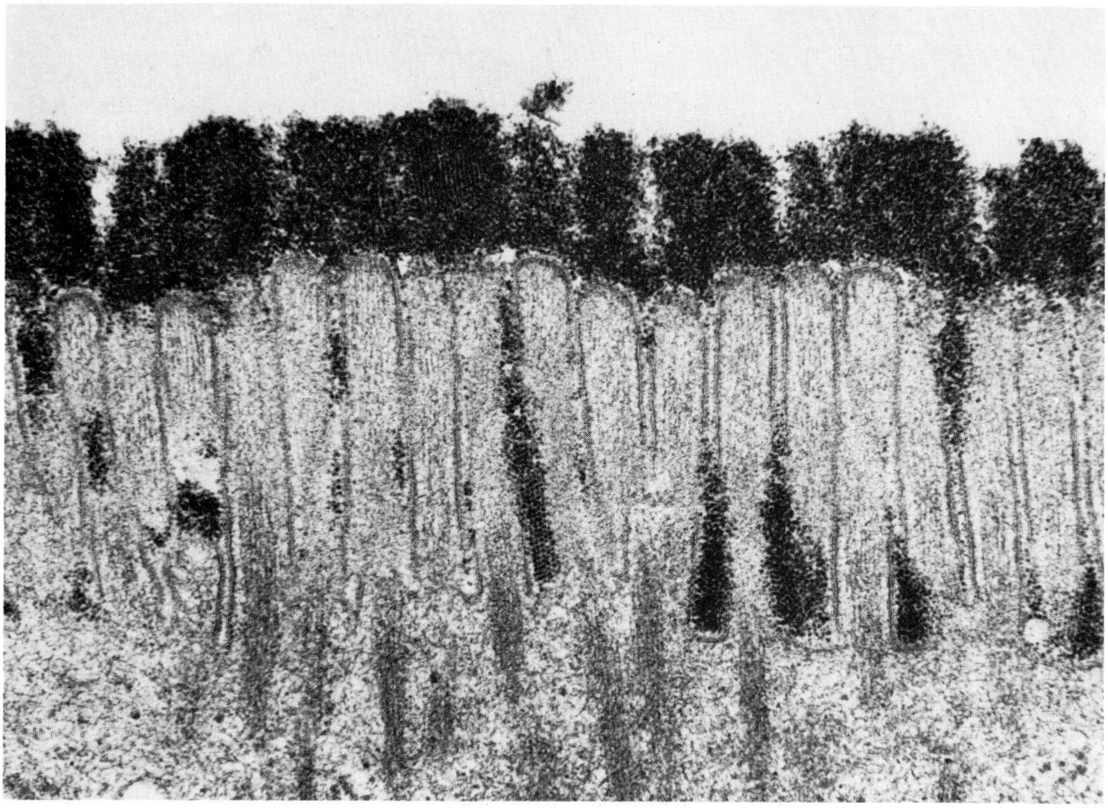

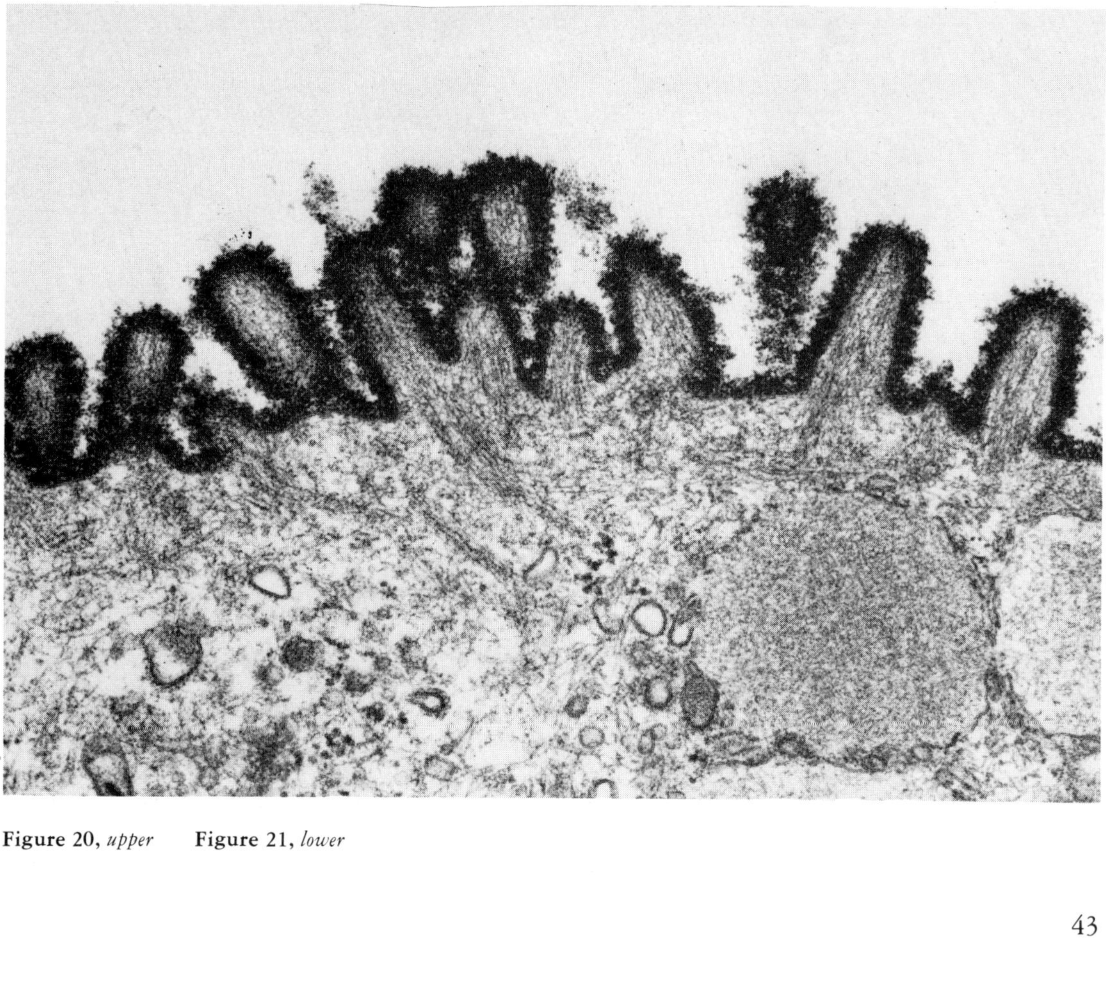

Figure 20, *upper* **Figure 21,** *lower*

BASAL LAMINA

The membrane at the base of epithelial cells does not appear to have a glycocalyx. Instead, all of the cells of the epithelium are associated with a continuous moderately dense extracellular layer ~ 100 nm thick, commonly called the *basal lamina,* or *basement lamina.* It is not contiguous with the cell membrane but separated from it by a clear zone of low density ~ 20 nm in thickness. Muscle fibers and certain other nonepithelial cells are completely enveloped by a layer of similar composition and ultrastructure.

The components of the basal lamina are synthesized and secreted by the epithelial cells (Hay and Dodson, 1973; Hay and Meier, 1974). Its substance is often described as amorphous, but at high magnification in favorable preparations it appears as a close meshwork of ~ 2 nm filaments. Biochemical analysis of isolated basal laminae indicates that these filaments consist of type IV collagens. This heterogeneous class of collagens is still poorly understood. Other types of collagen are secreted as procollagen molecules and polymerize to form fibers of varying size after being reduced in length by cleavage of a terminal peptide sequence. The type IV procollagen is believed to retain its telopeptide and therefore does not assemble to form fibers (Kefalides, 1975).

Varying amounts of glycosaminoglycans are associated with the feltwork of 2 nm collagen molecules comprising the basal lamina of embryonic epithelia. When stained with ruthenium red, these appear in micrographs as 10 to 20 nm particles regularly spaced ~ 55 nm apart. Whether these also occur in the basal laminae of mature animals is not known (Trelstad et al., 1974; Hay et al., 1978).

The basal lamina provides a structural substrate for the epithelial cells and serves as a barrier limiting access of macromolecules to the intercellular clefts. It also plays an important but poorly understood role in differentiation and repair of the epithelium.

In the accompanying micrograph, the basal lamina can be seen at the junction of the corneal epithelium (at the upper right) and Bowman's layer of the collagenous stroma (at the lower left). The dense lamina closely follows the irregular contour of the plasmalemma of the epithelial cells but is separated from it by a light zone of uniform width. At this magnification the basal lamina appears amorphous, but at very high magnification a filamentous substructure can be seen.

Figure 22. A small area of the base of the human corneal epithelium. (Micrograph courtesy of Toichiro Kuwabara.)

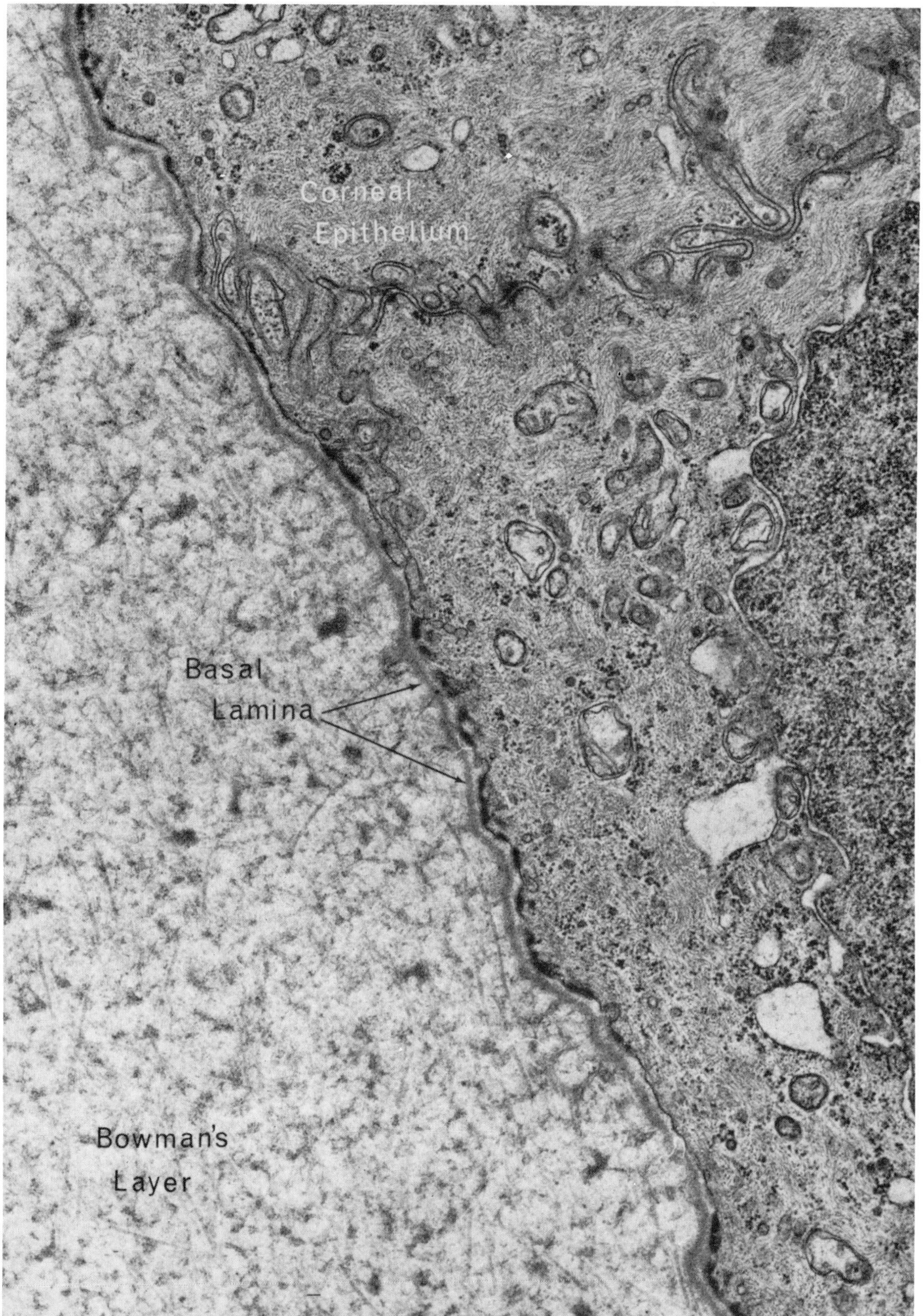

Figure 22

The basal lamina normally varies little in thickness, but in some organs, notably the renal glomerulus, seminiferous epithelium, and the endothelium of capillaries, it may become thickened in disease.

Descemet's membrane underlying the endothelium of the cornea appears to be an unusually thick and structurally specialized basal lamina. Its collagen is not all in the form of randomly oriented tropocollagen molecules, and in horizontal sections may exhibit an ordered array of evenly spaced nodes that are connected by radiating fibrils to form an hexagonal lattice. This pattern is especially prominent in older individuals. The accompanying micrograph illustrates the unusual thickness of Descemet's membrane but does not show the regular fibrillar pattern.

Figure 23. Endothelium and Descemet's membrane of human cornea. (Micrograph courtesy of Toichiro Kuwabara.)

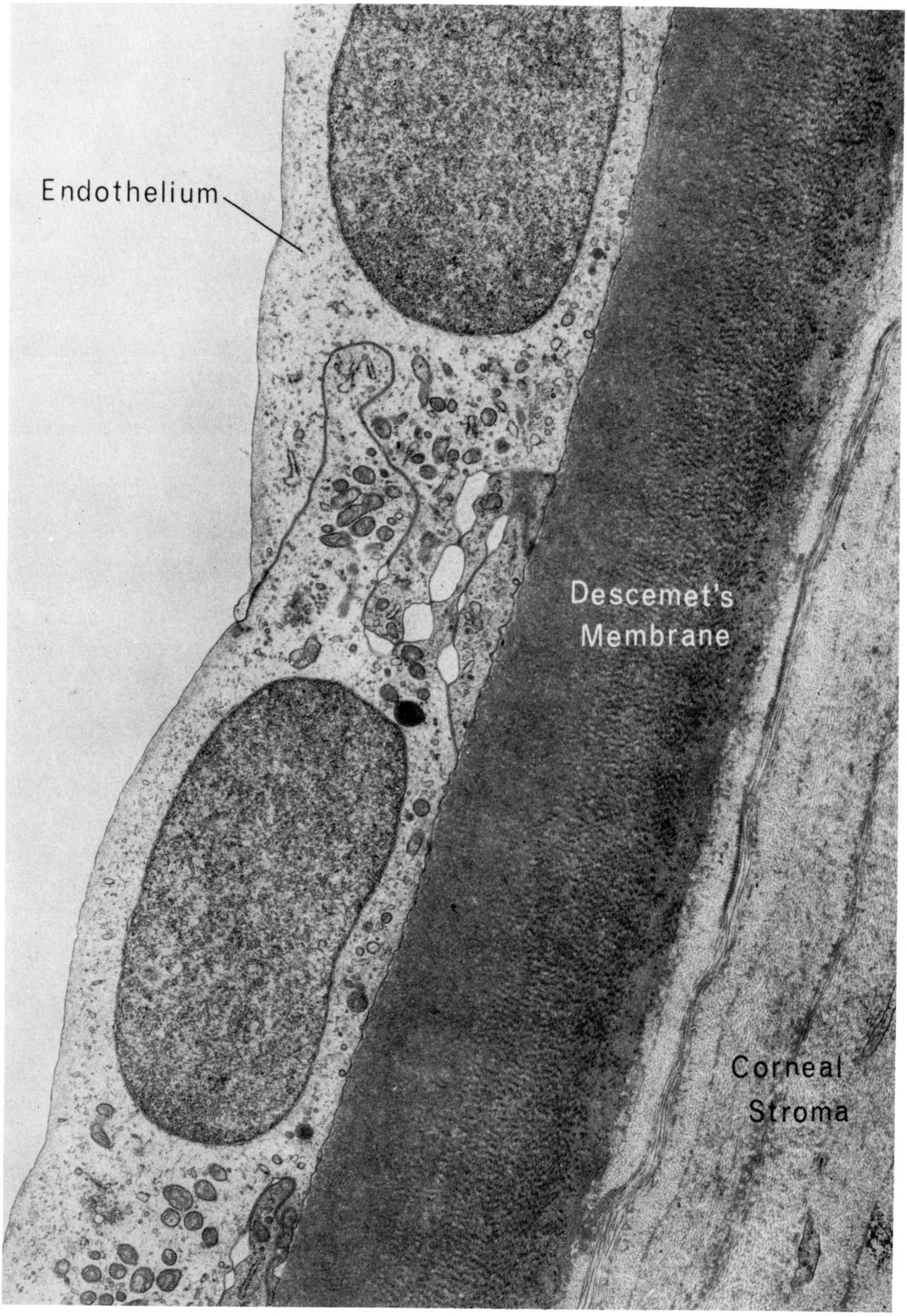

Figure 23

The basal lamina of epithelia in vertebrates appears homogeneous, but in invertebrates it occasionally shows periodic density differences in cross section. This is illustrated in the accompanying micrograph of midgut epithelium from a flea. When sectioned parallel to the base of the epithelium (inset at the lower left), it appears as a mosaic of elongated dense bodies separated by rows of smaller dense granules. The functional significance of this unusual pattern is not known.

Figure 24. Electron micrograph of the base of a cell in the gut epithelium of the flea. Inset is a horizontal section of the same. (Micrograph courtesy of Susumu Ito.)

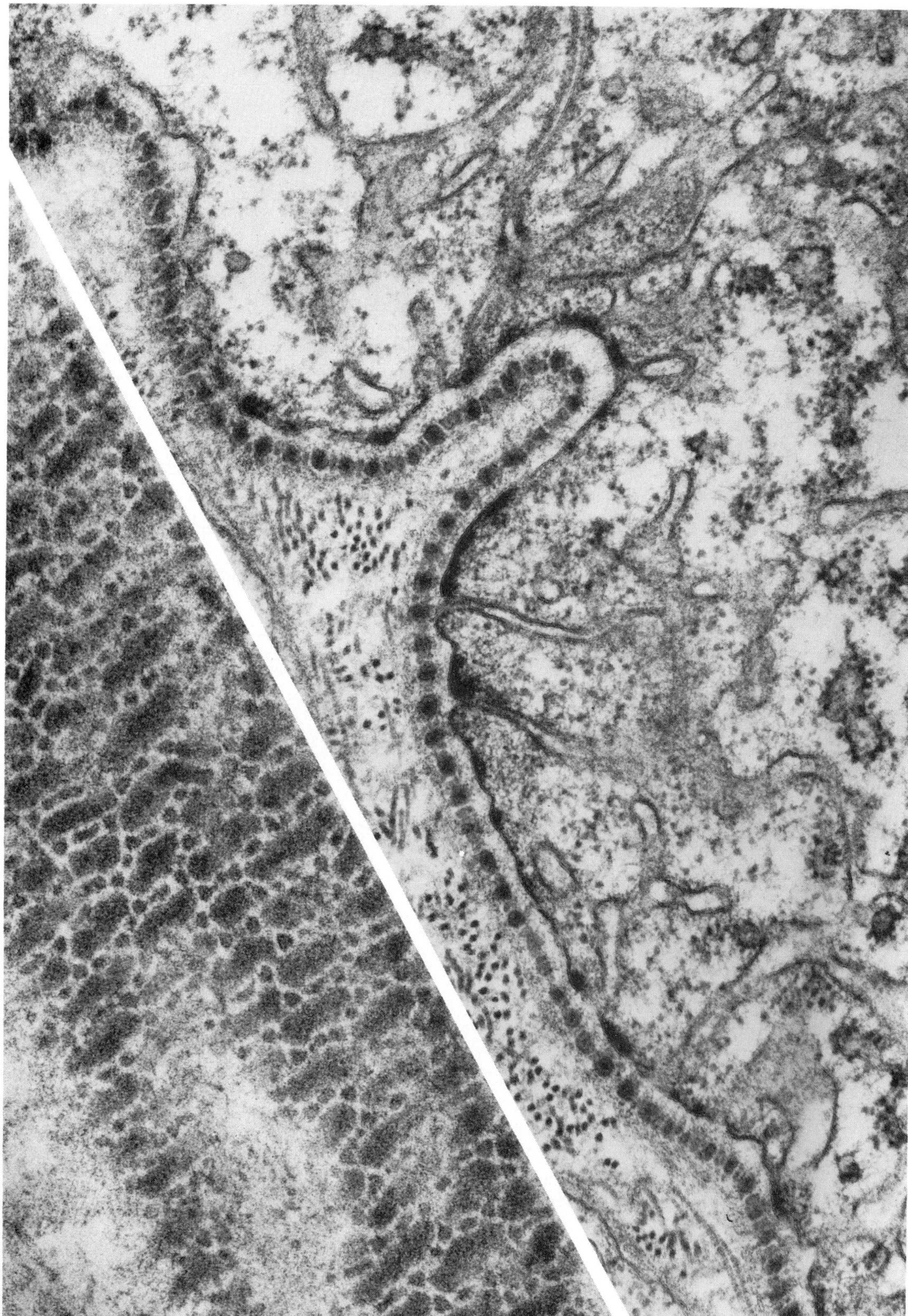

Figure 24

Lamina Externa

In classical histology, possession of a "basement membrane" was thought to be an exclusive property of epithelia, but electron microscopy has clearly established that an identical boundary layer completely invests certain sessile cell types of mesenchymal origin such as smooth and striated muscle fibers, pericytes of capillaries, and Schwann cells of peripheral nerves. The term "basal lamina" is often applied to this continuous mantle, but the adjective "basal" is clearly inappropriate for cells other than those of epithelia. Therefore the terms *lamina externa* and *boundary layer* are preferred.

The accompanying micrograph of a peripheral nerve in cross section shows groups of unmyelinated axons enclosed in deeply invaginated recesses in the surface of Schwann cells. Around each Schwann cell is a thin continuous *lamina externa*. It is especially apparent (at the arrows) where it crosses the sites of invagination of the Schwann cell membrane (mesaxons).

Figure 25. Nerve from mesentery of a rat.

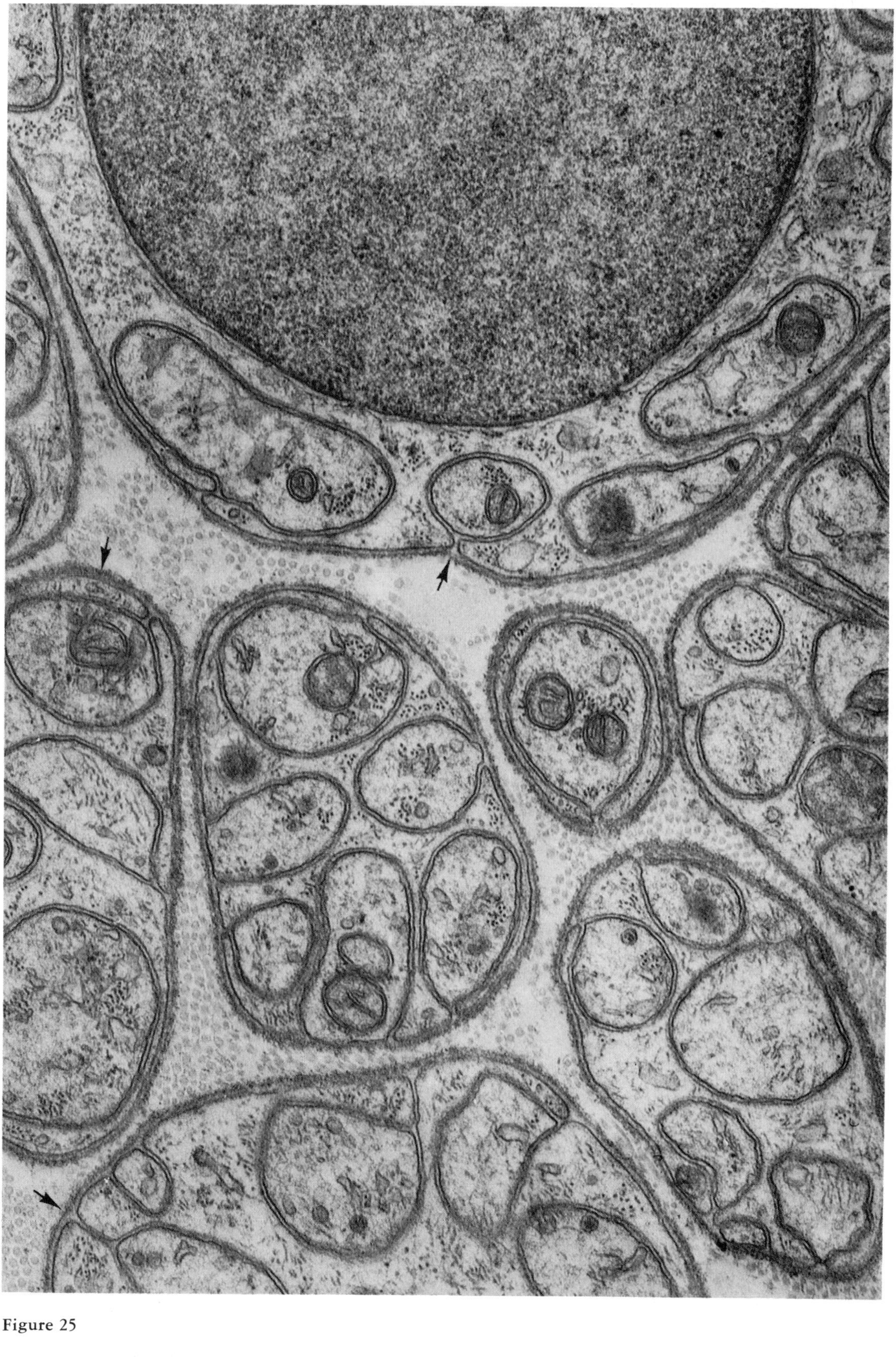

Figure 25

Illustrated here at higher magnification are myelinated and unmyelinated peripheral nerves, each invested by a continuous lamina externa. The outer limit of this layer is clearly visible in the discontinuity in density at the arrows.

Figure 26. Mixed nerve from cat pericardium.

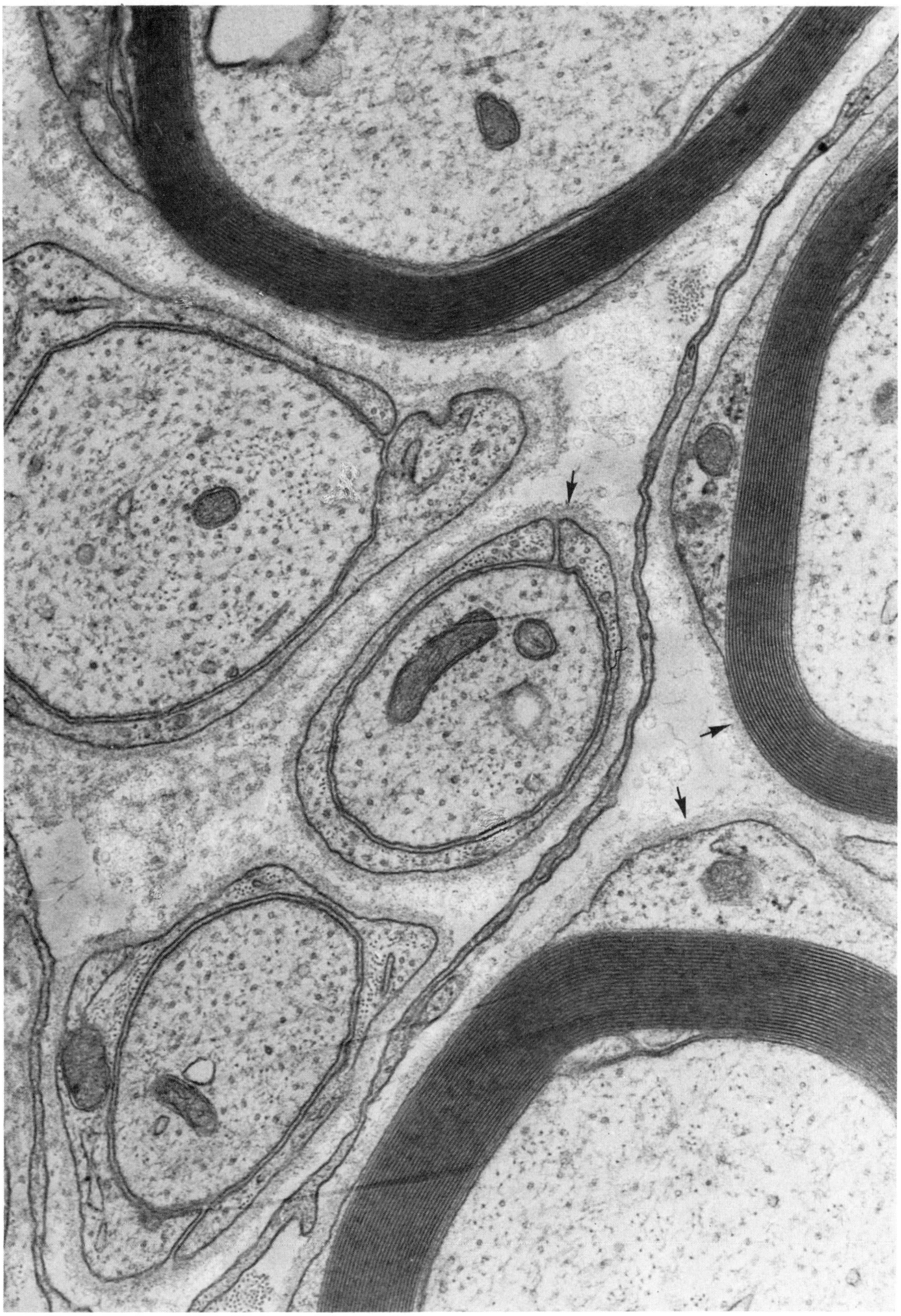

Figure 26

The cells of smooth muscle are separated by relatively wide intercellular spaces occupied by material with the staining properties of glycoprotein. In electron micrographs, small collagen fibrils occur singly in a light zone in the middle of the interspaces between adjacent cells or in small groups in the angular spaces between the diverging surfaces of three or more cells. The width of the space between cell membranes appears to be determined in large measure by the thickness of the lamina externa, which is greater on smooth muscle than on many other cell types.

Figures 27 and **28.** Micrographs of smooth muscle in the wall of a small artery.

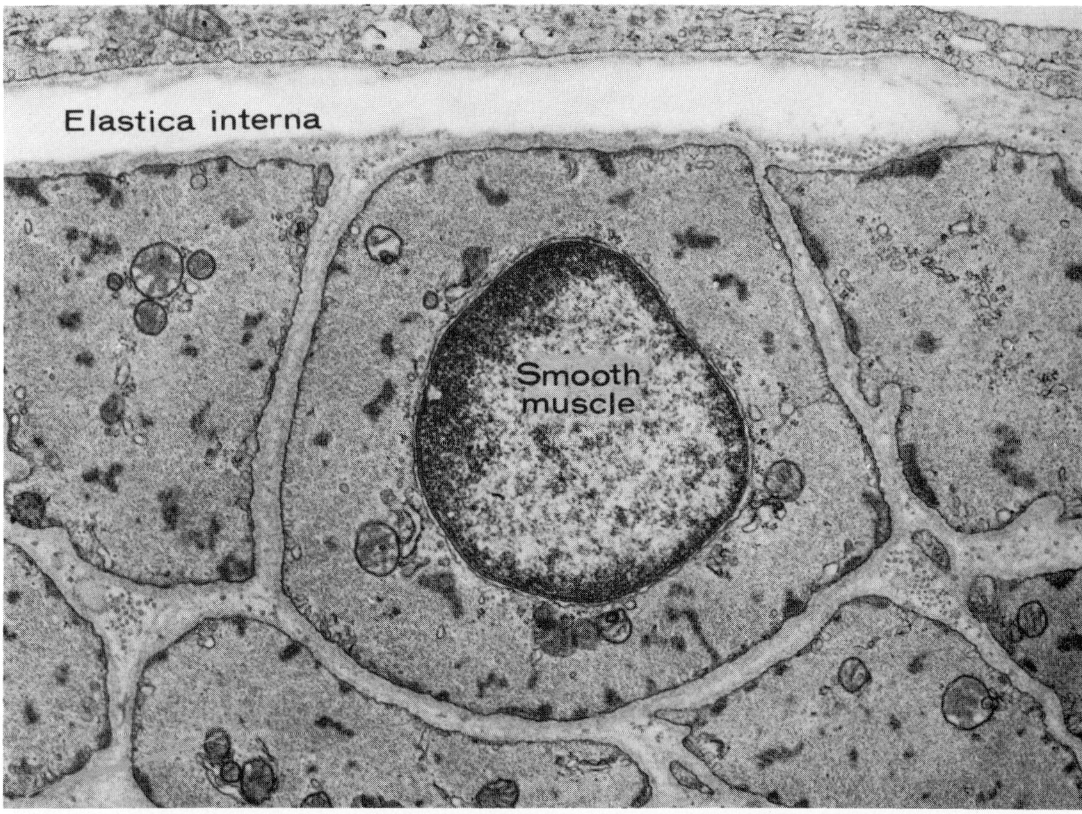

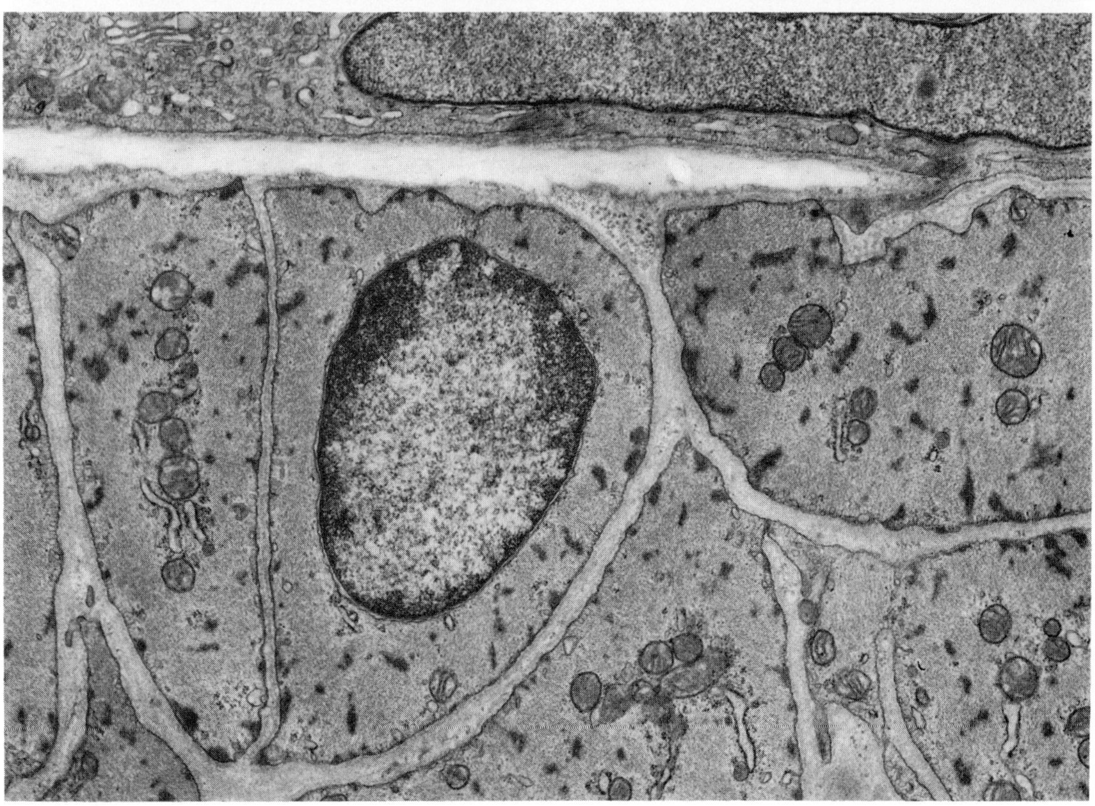

Figure 27, *upper* **Figure 28,** *lower*

The fine structure of the lamina externa is seldom resolved in electron micrographs of thin sections. A novel approach to the problem has been quick-freezing in liquid helium and preparation of replicas after deep etching. The accompanying micrograph shows a portion of the membrane of a muscle fiber and its associated lamina externa (here labeled basal lamina). It is composed of a meshwork of filaments of macromolecular dimensions. Some of the filaments traverse the clear zone and make contact with the cell membrane. Outside of the lamina externa are larger fibrils of collagen.

Figure 29. Quick-frozen deep-etched preparation of the sarcolemma of skeletal muscle and its associated lamina externa. (Micrograph courtesy of John Heuser.)

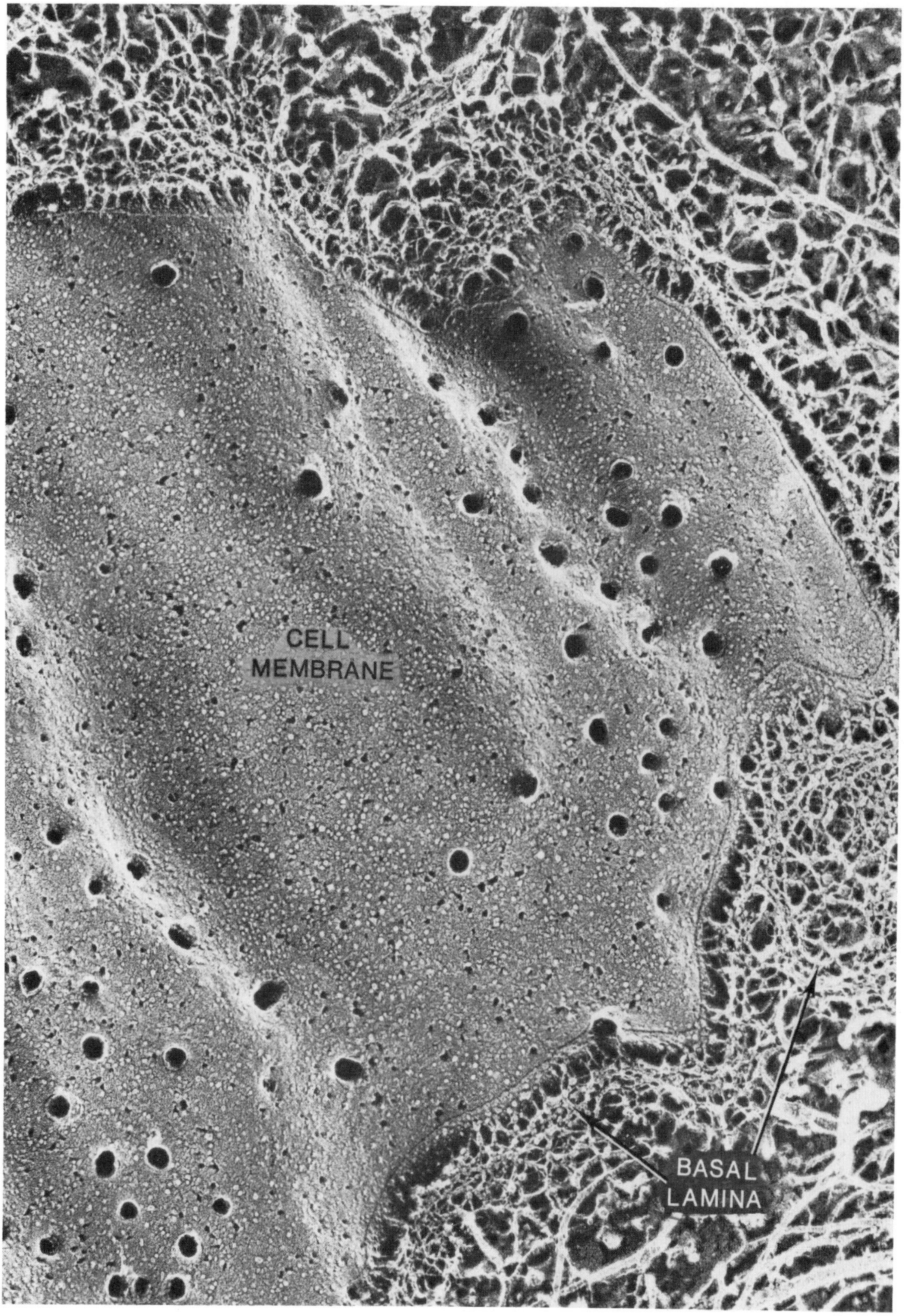

Figure 29

Shown here is another example of a quick-frozen deep-etched preparation of skeletal muscle and its associated lamina externa. Myofilaments are exposed at the upper right. The sarcolemma follows an undulant oblique course across the figure. Adjacent to it is the close meshed network of fine filaments constituting the basal lamina, or lamina externa. At the lower left are numerous collagen fibrils of the surrounding connective tissue. Such preparations are especially revealing when viewed as stereo pairs but are instructive even without the third dimension.

Figure 30. Quick-frozen deep-etched muscle fiber and adjacent extracellular components. (Micrograph courtesy of John Heuser.)

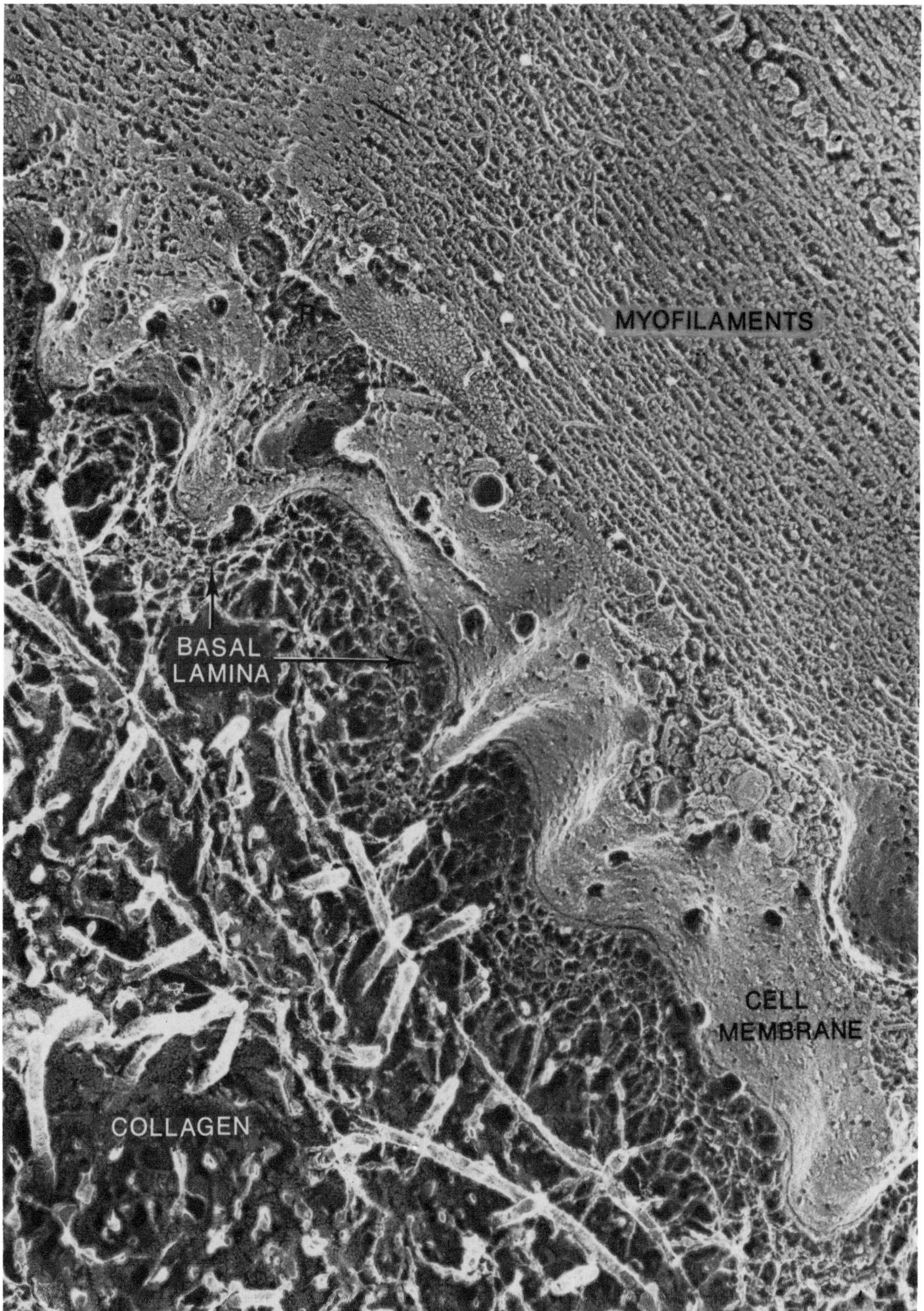

Figure 30

Intercellular Substance

The intercellular clefts of epithelia contain a small amount of material stainable with the periodic acid–Schiff reaction for carbohydrates. It is not clear whether this simply represents glycocalyx on the apposed cell surfaces or whether it is a glycosaminoglycan extracellular matrix.

Classical histologists postulated the existence of an "intercellular cement." Von Recklinghausen (1862) introduced a staining method that outlined the boundaries of epithelial cells with reduced silver. Silver was believed to be deposited from ammoniacal solutions of silver nitrate by reducing groups on carbohydrates in a cement substance occupying the intercellular clefts. At the turn of the century, it was shown that the stability of this intercellular material was dependent upon the presence of divalent cations and that cell dissociation and reaggregation could be caused by varying the calcium content of the medium (Herbst, 1900).

Interest in the intercellular substance of epithelia waned with the introduction of the electron microscope. The 15 to 20 nm space between cells was much narrower than expected from earlier light microscopic observations, and with the methods of specimen preparation then available, it was unstained and devoid of resolvable structure. The concept of an intercellular adhesive was nevertheless kept alive by Moscona's (1961) studies on reaggregation of embryonic chick cells after they had been dispersed by trypsin. He concluded that reaggregation was dependent upon synthesis and secretion of a substance (EOM) responsible for maintaining cell cohesion.

Interest in the mechanisms of cellular adhesion has recently been revived by the discovery of a class of glycoproteins called *fibronectins* that are secreted by cells in tissue culture and are believed to have a role in cell-to-cell and cell-to-substrate adhesion (Yamada and Weston, 1974; Hynes and Bye, 1974; Yamada, 1978; Yamada and Olden, 1978). The possibility that this class of glycoproteins constitutes the intercellular material of epithelia *in vivo* remains to be thoroughly explored.

REFERENCES

Membranes

Bangham, A. D. Lipid bilayers and biomembranes. Ann. Rev. Biochem. 41:753–776, 1972. (Review)

Branton, D. Fracture faces of frozen membranes. Proc. Nat. Acad. Sci. 55:1048–1056, 1966.

Branton, D. Membrane structure. Ann. Rev. Plant Physiol. 20:209–238, 1969.

Branton, D., S. Bullivant, N. B. Gilula, M. J. Karnovsky, H. Moor, K. Mühlethaler, D. H. Northcote, L. Packer, B. Satir, P. Satir, V. Spaeth, L. A. Staehelin, R. L. Steere and R. S. Weinstein. Freeze-etching nomenclature. Science 190:54–56, 1975.

Chambers, R. The nature of the living cell as revealed by microdissection. Harvey Lecture, 22:41, 1926.

Danielli, J. F. and H. Davson. A contribution to the theory of permeability of thin films. J. Cell Comp. Physiol. 5:495–508, 1935.

De Pierre, J. W. and M. L. Karnovsky. Plasma membranes of mammalian cells. A review of methods for their characterization and isolation. J. Cell Biol. 56:273–303, 1973.

Finian, J. B. The development of ideas on membrane structure. Subcell. Biochem. 1:363–373, 1972. (Review)

Gorter, E. and F. Grendel. On bimolecular layers of lipoids on the chromocytes of the blood. J. Exp. Med. 41: 439–443, 1925.

Guidotti, G. Membrane proteins. Ann. Rev. Biochem. 41:731–752, 1972. (Review)

Hendler, R. W. Biological membrane ultrastructure. Physiol. Rev. 51:66–97, 1971. (Review)

Korn, E. D. Structure of biological membranes. Science 153:1491–1498, 1966. (Review)

Langmuir, I. The constitution and fundamental properties of solids and liquids. II. Liquids. J. Am. Chem. Soc. 29:1848–1906, 1917.

Marchesi, V. T., H. Furthmayr and M. Tomita. The red cell membrane. Ann. Rev. Biochem. 45:667–698, 1976.

McNutt, N. S. Freeze-fracture techniques and applications to the structural analysis of the mammalian plasma membrane. *In* Dynamic Aspects of Cell Surface Organization (G. Poste and G. I. Nicolson, eds.), pp. 75–126. North Holland Publishing Company, Amsterdam, New York, 1977.

Moore, H. and K. Mühlethaler. Fine structure in frozen etched yeast cells. J. Cell Biol. 17:609–623, 1963.

Moor, H. Use of freeze-etching in the study of biological ultrastructure. Int. Rev. Exp. Pathol. 5:179–216, 1966.

Nägeli, K. and K. Cramer. Pflanzenphysiologische Untersuchungen. Schulters, Zurich, 1855.

Nicolson, G. L. and G. Poste. The cancer cell: Dynamic aspects and modifications in cell-surface organization. Part I. N. Engl. J. Med. 295:197–203, 1976; Part II. 295:253–258, 1976.

Overton, E. Uber die allgemeinen osmotischen Eigenschaften der Zelle, ihre vermutlichen Ursachen und ihre Bedeutung für die Physiologie. Vischr. Naturforsch. Ges. Zurich 44:88, 1899.

Robertson, J. D. New observations on the ultrastructure of the membranes of frog peripheral nerve fibers. J. Biophys. Biochem. Cytol. 3:1043–1047, 1957.

Robertson, J. D. The ultrastructure of cell membranes and their derivatives. *In* Biochem. Soc. Symp. No. 16, pp. 3–43. Cambridge University Press, London, 1959.

Steck, T. L. The organization of proteins in the human red blood cell membrane. J. Cell Biol. 62:1–19, 1974. (Review)

Steere, R. L. Electron microscope studies of structural detail in frozen biological specimens. J. Biophys. Biochem. Cytol. 3:45–60, 1957.

Singer, S. J. and G. L. Nicolson. The fluid mosaic model of the structure of cell membranes. Science 175:720–731, 1972.

Weissman, G. and R. Claiborne, editors. Cell Membranes. Biochemistry, Cell Biology and Pathology. HP Publishing Company, Inc., New York, 1975.

Regional Differences in the Plasma Membrane

Boulan, E. R. and D. D. Sabatini. Asymmetric budding of viruses in epithelial monolayers: A model system for study of epithelial polarity. Proc. Nat. Acad. Sci. 75:5071–5075, 1978.

Ernst, S. A. Transport ATPase cytochemistry: Ultrastructural localization of potassium dependent and potassium independent phosphatase activity in the rat kidney cortex. J. Cell Biol. 66:536–608, 1975.

Ernst, S. A. and J. W. Mills. Basolateral plasma membrane localization of ouabain-sensitive sodium transport sites in the secretory epithelium of the avian salt gland. J. Cell Biol. 75:74–94, 1977.

Fawcett, D. W. Unsolved problems in morphogenesis of the mammalian spermatozoon. *In* International Cell Biology 1976–1977, pp. 588–600, Rockefeller University Press, New York.

Friend, D. S. and D. W. Fawcett. Membrane differentiations in freeze-fractured mammalian spermatozoa. J. Cell Biol. 63:641–644, 1974.

Gilula, N. B. and P. Satir. The ciliary necklace. A ciliary membrane specialization. J. Cell Biol. *53*:494–509, 1972.

Hess, R. T. and L. A. Falcon. Observations on the interaction of bacterioviruses with the plasma membrane. J. Gen. Virol. 36:525–530, 1977.

Neutra, M. Linear arrays of intramembrane particles on microvilli in primate large intestine. Anat. Rec. 193:367–382, 1979.

Olson, G., M. Lifsics, D. W. Hamilton and D. W. Fawcett. Structural specializations in the flagellar plasma membrane of opossum spermatozoa. J. Ultrastr. Res. 59:207–221, 1977.

Glycocalyx or Surface Coat

Behnke, O. and T. Zelander. Preservation of intercellular substances by the cationic dye, Alcian blue, in preparation procedures for electron microscopy. J. Ultrastr. Res. 31:424–438, 1970.

Bennett, H. S. Morphological aspects of extracellular polysaccharides. J. Histochem. Cytochem. 11:14–23, 1963.

Chambers, R. The relation of the extraneous coats to the organization and permeability of cell membranes. Cold Spring Harbor Symposium on Quantitative Biology 8:144, 1940.

Ito, S. The surface coat of enteric microvilli. J. Cell Biol. 27:475–491, 1965.

Luft, J. H. Ruthenium red and violet. II. Fine structural localization in animal tissues. Anat. Rec. 171:369–416, 1971.

Rambourg, A., M. Neutra and C. P. Leblond. Presence of a cell coat rich in carbohydrate at the surface of cells in rat. Anat. Rec. 154:41, 1966.

Wallach, D. F. H. and E. H. Eylar. Sialiac acid in the cellular membranes of Ehrlich ascites tumor cells. Biochim. Biophys. Acta 52:594–596, 1961.

Wallach, D. F. and V. B. Kamat. The contribution of sialic acid to the surface charge of fragments of plasma membrane and endoplasmic reticulum. J. Cell Biol. 30:660, 1966.

Wislocki, G. B., D. W. Fawcett, and E. W. Dempsey. Staining of stratified squamous epithelium of mucous membranes and skin of man and monkey with the periodic acid–Schiff method. Anat. Rec. 110:359, 1951.

Yamada, E. The fine structure of the gallbladder epithelium of the mouse. J. Biophys. Biochem. Cytol. 1:445–458, 1955.

Basal Lamina

Hay, E. D. and J. W. Dodson. Secretion of collagen by corneal epithelium. I. Morphology of the collagenous products produced by isolated epithelia grown on frozen-killed lens. J. Cell Biol. 57:190–213, 1973.

Hay, E. D. and S. Meier. Glycosaminoglycan synthesis by embryonic inductors: Neural tube, notochord, and lens. J. Cell Biol. 62:889–898, 1974.

Kefalides, N. A. Structure and biosynthesis of basement membranes. Int. Rev. Connect. Tissue Res. 6:63–104, 1973.

Kefalides, N. A. Basement membranes: Structural and biosynthetic considerations. J. Invest. Dermatol. 65:85–92, 1975.

Trelstad, R. L., K. Hayashi, and B. P. Toole. Epithelial collagens and glycosaminoglycans in the embryonic cornea. Macromolecular order and morphogenesis in the basement membrane. J. Cell Biol. 62:815–830, 1974.

Intercellular Material

Herbst, C. Uber der Auseinander gehen von Furchungs — und Gewebzellen in Kalkfreiem Medium. Arch. f. Entwicklungemechanik 9:424, 1900.

Hynes, R. O. and J. M. Bye. Density and cell cycle dependence of cell surface proteins in hamster fibroblasts. Cell 3:113–120, 1974.

Moscona, A. A. Effect of temperature on adhesion to glass and histogenetic cohesion of dissociated cells. Nature 190:408–409, 1961.

Moscona, A. A. Studies on cell aggregation: Demonstration of materials with a selective cell binding activity. Proc. Nat. Acad. Sci. 49:742–747, 1963.

Wislocki, G. B., D. W. Fawcett and E. W. Dempsey. Staining of stratified squamous epithelium of mucous membranes and skin of man and monkey by the periodic acid–Schiff method. Anat. Rec. 110:359, 1951.

Yamada, K. M. and J. A. Weston. Isolation of a major cell surface glycoprotein from fibroblasts. Proc. Nat. Acad. Sci. 71:3492–3496, 1974.

Yamada, K. M. Immunological characterization of a major transformation sensitive fibroblast cell surface glycoprotein. Localization, redistribution and role in cell shape. J. Cell Biol. 78:520–541, 1978.

Yamada, K. M. and K. Olden. Fibronectins — adhesive glycoproteins of cell surface and blood. Nature 275:179–184, 1978.

SPECIALIZATIONS OF THE FREE SURFACE

In living organisms, most physiologically important events take place at cell surfaces or at interfaces between intracellular compartments. The rate of activity per unit area of surface probably cannot be increased above a certain limit. Therefore at all levels of organization, we see structural devices for increasing available surface area without appreciable increase in overall size.

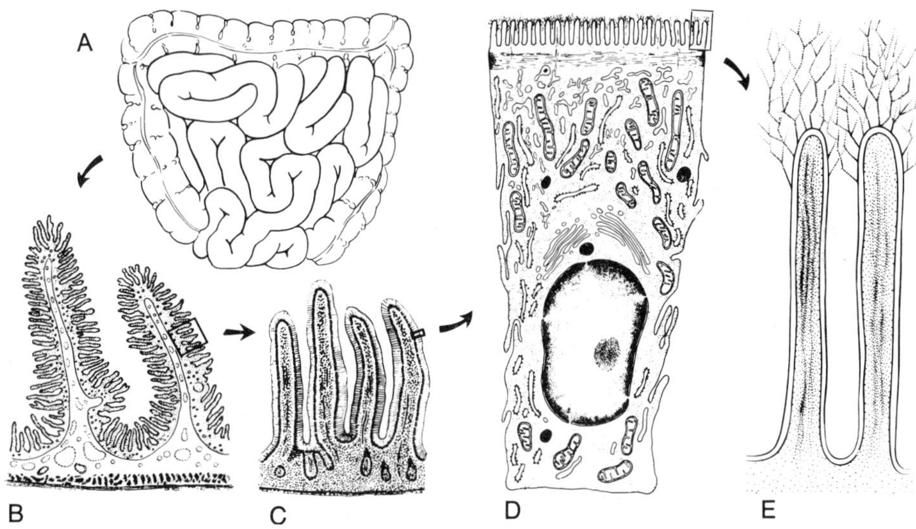

Drawing illustrating the hierarchy of levels of surface amplification in the intestinal tract. *A*, elongation and convolution of the tubular bowel; *B*, mucosa plication; *C*, formation of intestinal villi; *D*, microvilli on the individual cells; *E*, saccharide residues of membrane glycoproteins projecting from the microvilli. The same devices repeat at the subcellular level in elongation and convolution of the endoplasmic reticulum, and plication of the inner mitochondrial membranes.

In the alimentary system, for example, the absorptive surface is increased by elongation and convolution of the tubular gut. If the small intestine is opened, one sees that the mucosal surface is further increased by folding to form the *plicae circulares*. At higher magnification, the surface is found to be further amplified by the presence of 30 to 40 *intestinal villi* per square millimeter. Electron microscopic examination of any intestinal absorptive cell in the epithelium covering a villus reveals that its luminal surface is again augmented by the presence of several hundred *microvilli* and that polysaccharide residues of membrane glycoproteins extending from the tip of each microvillus form a tuft of delicate branching filaments of molecular dimensions. The architectural devices of convolution, plication, villus formation, and arborizing cell processes are found in other organs throughout the body. These represent Nature's stratagems for increasing the efficiency of the metabolic machinery with minimal mass. Impressive as are man's recent efforts at miniaturization through development of minute printed circuits on computer chips, they still fall short of Nature's achievements by several orders of magnitude.

Increase of surface area is also achieved by division of a given volume into very small cellular units. Thus in the case of the erythrocytes of blood, although they have a relatively simple biconcave form with an individual surface area of about 140 μm^2, their aggregate surface area in man is about 3820 square meters or 2000 times the surface of the body as a whole.

The infinite variety of shapes which we normally use only for identification of cell type are associated with significant quantitative differences in surface area that we seldom take into account. To illustrate the point, Garven presented a set of simple geometric simulacra of epithelial cells with their calculated surface areas while their volume remained constant at 1000 μm^3. In view of the nearly threefold increase in surface from a spherical form to a squamous cell of the same volume and the fivefold increase in a sphere with 50 broad processes, it is intriguing to contemplate the relatively enormous surface presented by neurons in the brain, such as the Purkinje cell (F, below), which occurs in a density of 300 per sq mm and has a remarkably complex arborization of dendrites extending 300 μm from a cell body, some 50 μm in diameter. The total length of dendrites is some 10,500 μm, over 95 per cent of which bears small lateral processes called *thorns,* whose additional contribution to surface area has not been estimated.

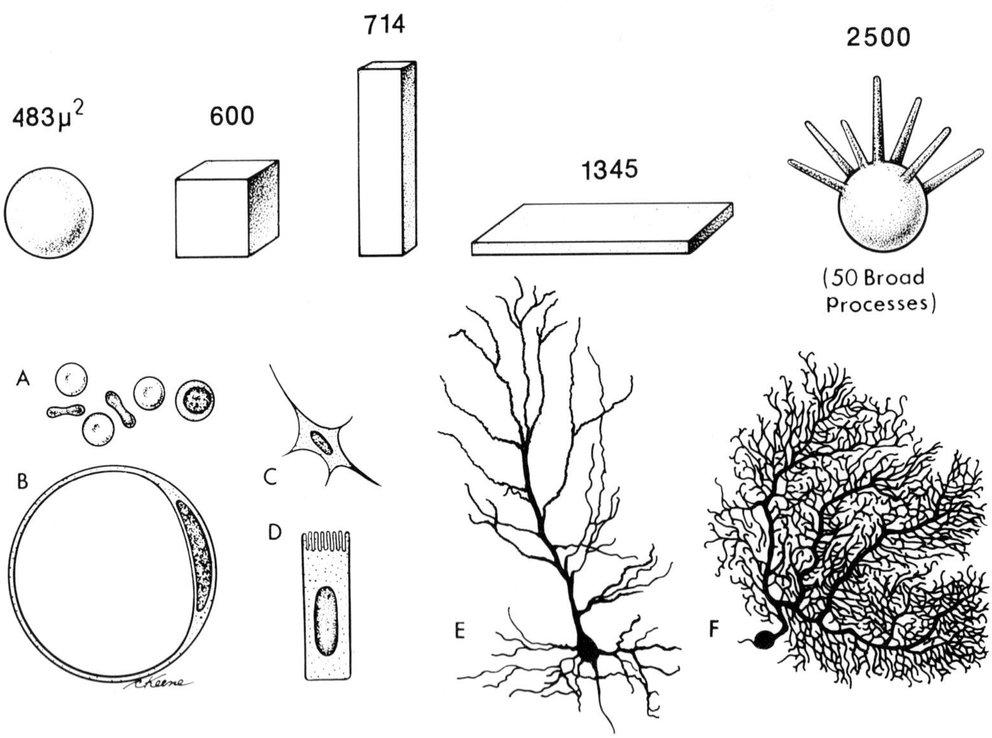

Above, surface areas of geometric figures of constant volume (1000 μm^3) approximating the shapes of cells (after Garven). Below, shape variations of some cell types. *A*, erythrocytes; *B*, adipose cell; *C*, fibroblast; *D*, absorptive epithelial cell; *E* and *F*, cells from brain with elaborate dendritic arborization.

One of the serious limitations of transmission electron microscopy (TEM) is the requirement that the beam of electrons penetrate the specimen. The specimen interposed in the path of the electrons must therefore be very thin (50 to 90 nm) — about a thousandth the thickness of an average cell. The exploitation of this instrument for research in cell biology required the development of methods for tissue embedding in plastic and the cutting of ultrathin sections. The images produced are essentially two dimensional and although they permit very high resolution (5 Å) and effective magnification up to half a million times, they provide little insight into the three-dimensional configuration of cells.

Examination of ultrathin sections of cells by transmission electron microscopy was not the method of choice for studying their surface topography. Reliance upon this method alone led to gross underestimates of the number of microvillous processes, owing to the small sample of the surface that is included in a thin section. The length of these processes was not always appreciated because, if curved, they often pass out of the plane of section and part of their length was not recorded in the micrograph. The shape of cell processes was often misjudged because it is not always possible to distinguish the sectional profile of a surface ridge or fold from a longitudinal section of a villus.

The scanning electron microscope (SEM) does not require penetration of the specimen, which therefore need not be sectioned. The instrument has great depth of field and produces three-dimensional images at magnifications of 15 to 20,000 times at a resolution of about 100 Å. The introduction of scanning electron microscopy, with its capacity to record images of surface relief with a remarkable three-dimensional quality, overcame one of the limitations of transmission microscopy and has greatly stimulated interest in the topography of cell surfaces and the changes it undergoes in different physiological states.

To exploit the potential of this instrument, it was necessary to develop methods of specimen preparation that bring the cells to the dry state before introduction into the vacuum and accomplish this with minimal shrinkage or distortion of surface contours. As the water film boundary passes through a tissue sample in an air-drying procedure, surface tension forces of many tons per square inch are brought to bear upon the specimen. To avoid the distortional forces of a boundary film, the so-called *critical point method* of drying is usually employed. This takes advantage of the fact that in heating any fluid confined in a closed system, there is a critical point at which the decreasing density of the fluid and the increasing density of the vapor become identical and the liquid surface vanishes. In applying this principle in specimen preparation, water in the tissue is replaced by immersion in an appropriate transition fluid such as liquid carbon dioxide or Freon 13, and it is heated in a sealed chamber to slightly above the critical point where the liquid phase disappears. The vapor is then released while holding the apparatus above the critical temperature, thus leaving a completely dry specimen which has not been subjected to surface tension forces. Even very slender processes of fixed cells are left projecting from the surface in much the same configuration that prevailed in the living state.

In the pages that follow, some of the surface specializations of cells will be illustrated. Some of these are relatively stable structures serving passively to amplify the surface area; others are transient differentiations actively involved in locomotion or in uptake of material from the environment.

SPECIALIZATIONS FOR SURFACE AMPLIFICATION

Cells in culture are particularly useful for scanning microscopy. In the accompanying micrograph, the free surfaces of HeLa cells are seen to be covered by large numbers of long slender microvilli which escape detection by light microscopy and appear to be short and relatively sparse in micrographs of thin sections. These are highly responsive to the nutritional state of the cells and vary in abundance in different phases of the cell cycle.

Figure 31. Scanning micrograph of HeLa cells in culture. (Micrograph courtesy of Keith Porter.)

Figure 31

Microvilli cover the free surface of many cell types, including the mammalian ovum. They are labile structures rapidly increasing or decreasing in number in response to environmental conditions or to changes in metabolic activity of the cell.

In mammalian fertilization, illustrated here, the postacrosomal region of the sperm head contacts the microvillous surface of the ovum and the two cell membranes coalesce. As the sperm head sinks into the ooplasm, microvilli rapidly re-form on the hybrid membrane overlying the caudal half of the sperm head, as shown in the lower figure. The remainder of the head and flagellum are ultimately engulfed and the sperm nucleus decondenses to form the male pronucleus.

Figure 32. Scanning micrograph of an early stage of *in vitro* fertilization of a hamster ovum.

Figure 33. A lateral view of the site of sperm penetration showing the oocyte membrane covered with microvilli over the caudal portion of the sperm head. (Micrographs courtesy of David Phillips.)

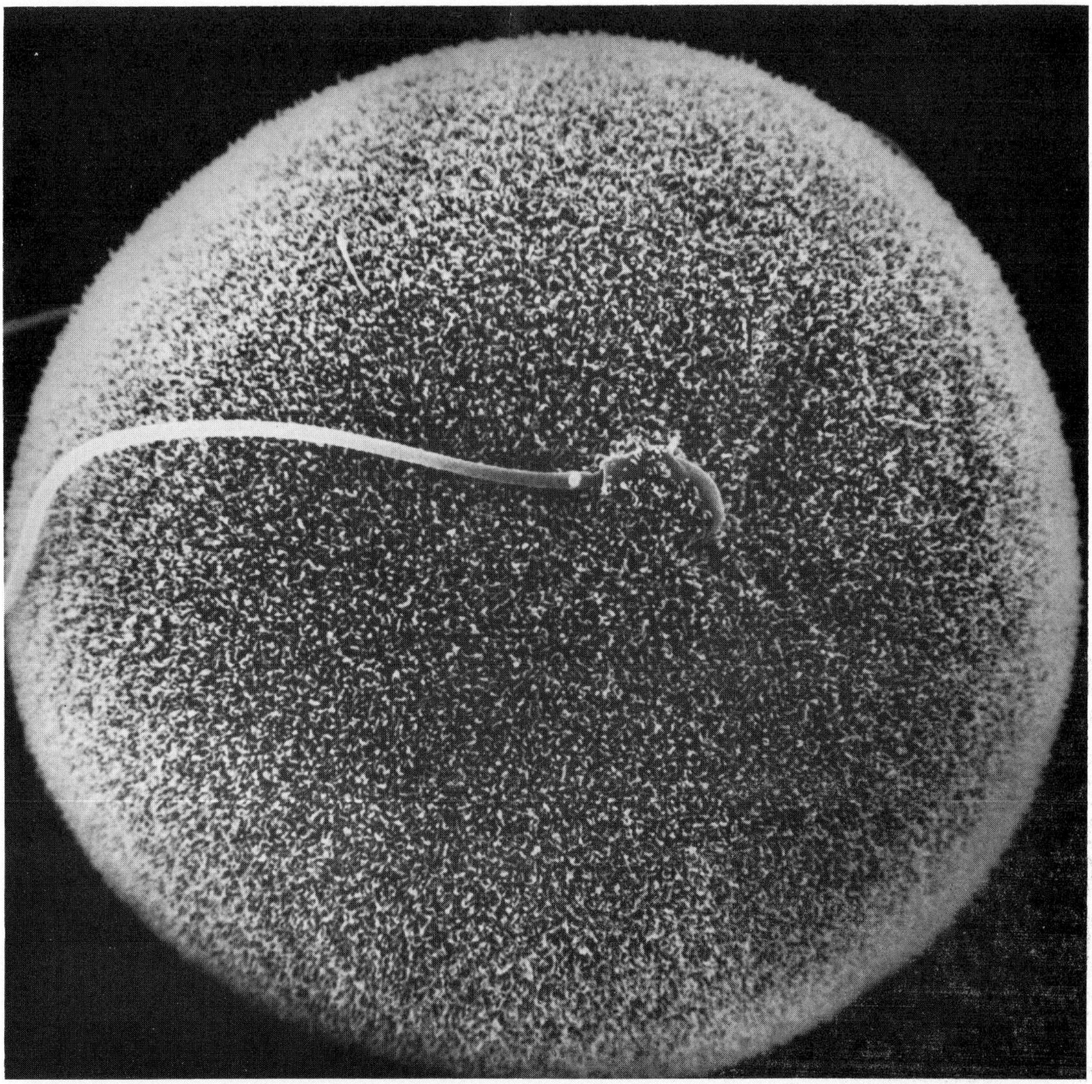

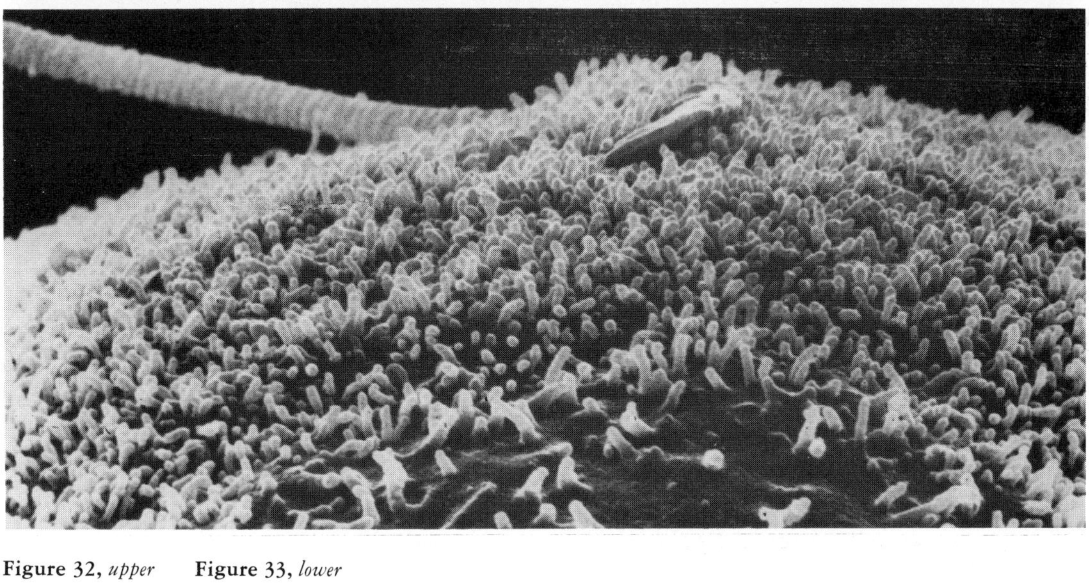

Figure 32, *upper* Figure 33, *lower*

The surface topography of free cells of the peritoneal cavity can be studied to advantage by allowing them to settle onto and attach to the moist surface of a millipore filter, which is then processed for scanning electron microscopy.

The characteristic pattern of surface processes is useful in identification of cell types. The accompanying micrograph shows a peritoneal macrophage and two lymphocytes. The surface of the macrophage is entirely covered by short folds and microvilli. The lymphocytes, on the other hand, show expanses of smooth surface between sparse microvilli. These features generally serve to distinguish these two cell types.

The cells evidently sense the presence of another cell within 0.5 μm and respond by extending a few longer microvilli in that direction (see at arrows). Altered environmental conditions may also result in rapid changes in the character of the cell surface as illustrated in the next figure.

Figure 34. Cells of peritoneal exudate of mouse. (Micrograph courtesy of Emma Shelton and Jan Orenstein.)

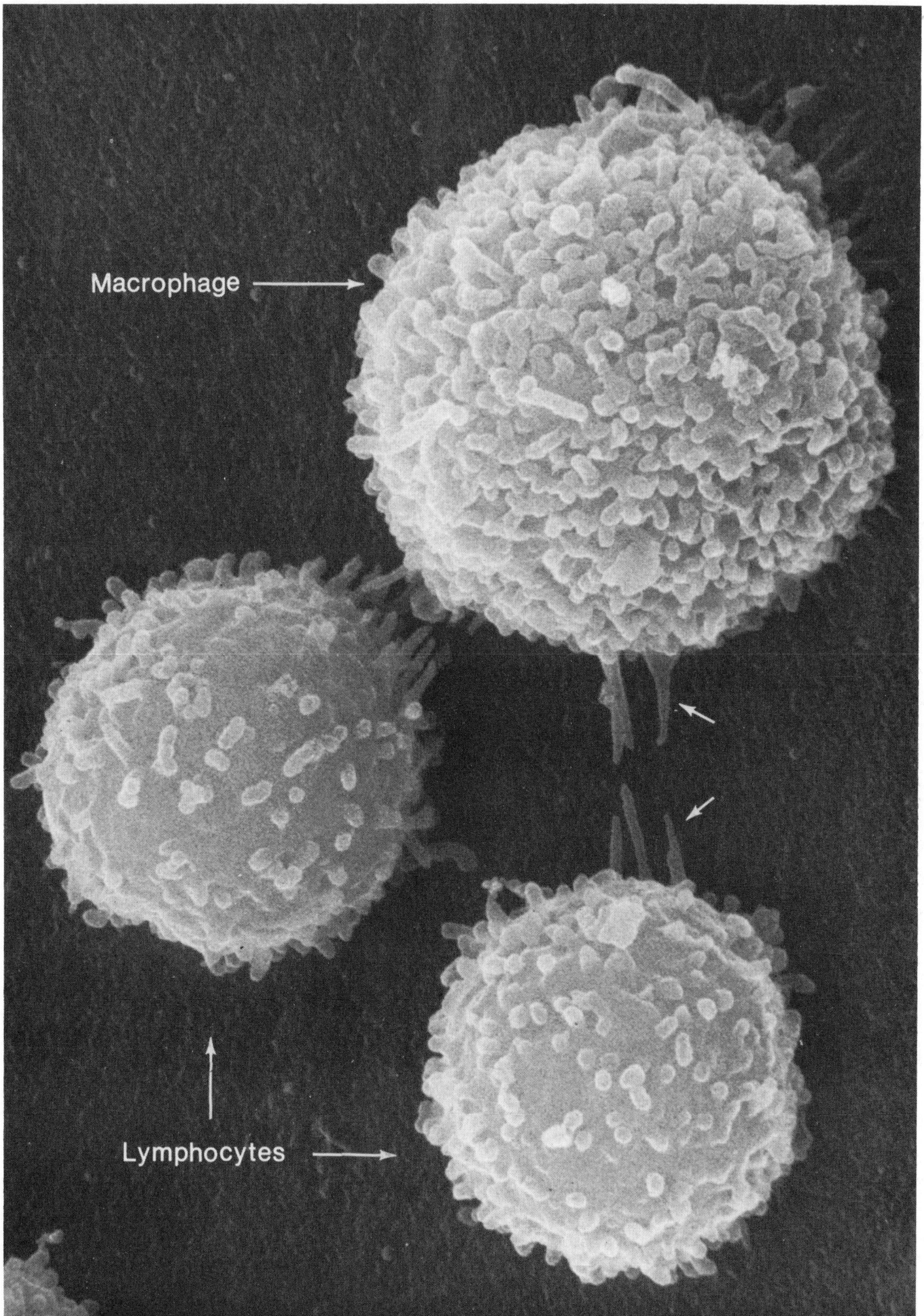

Figure 34

In a scanning micrograph of peritoneal cells exposed to sodium azide for 20 minutes at 22° C, the macrophage has responded to the lower temperature and presence of metabolic inhibitor by extending very long microvilli. The lymphocytes are unchanged, but at 37° C they too display many long microvilli. These changes are reversible and serve to illustrate the responsiveness of the cell surface to internal metabolic and external environmental conditions.

Figure 35. Cells of mouse peritoneal exudate exposed to 10^{-2} M sodium azide for 22 minutes at 22° C. (Micrograph courtesy of Emma Shelton.)

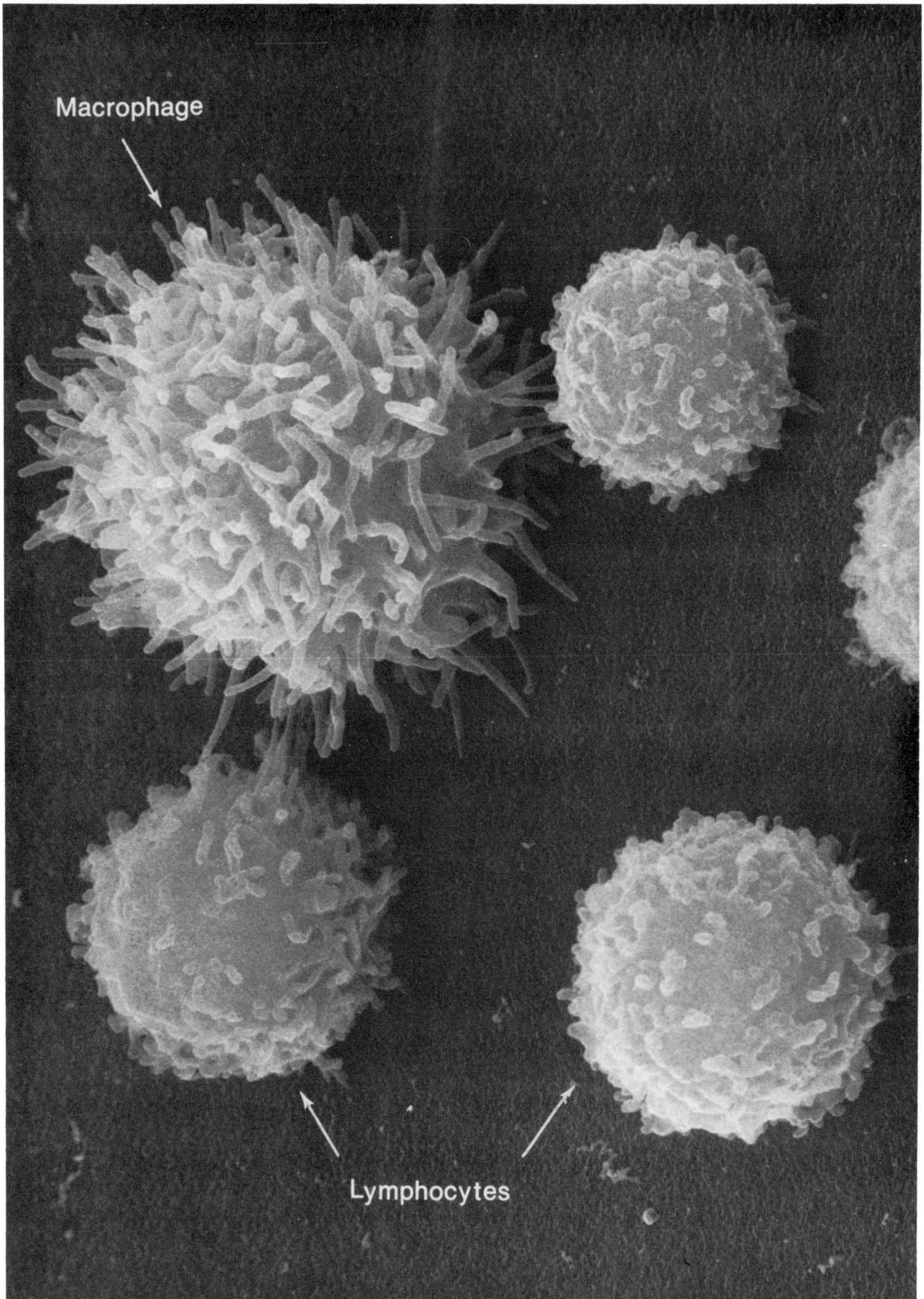

Figure 35

At higher magnification, one can appreciate the differences in length of the microvilli of the lymphocytes above and of the macrophage below. The capacity of the macrophage to form these long delicate processes in large numbers in a short time in the presence of a metabolic inhibitor raises some interesting questions about the mechanisms of formation of new membrane to accommodate rapid changes in surface area.

Figure 36. Peritoneal macrophage and neighboring lymphocytes exposed to sodium azide for 20 minutes at 22° C. (Micrograph courtesy of Emma Shelton.)

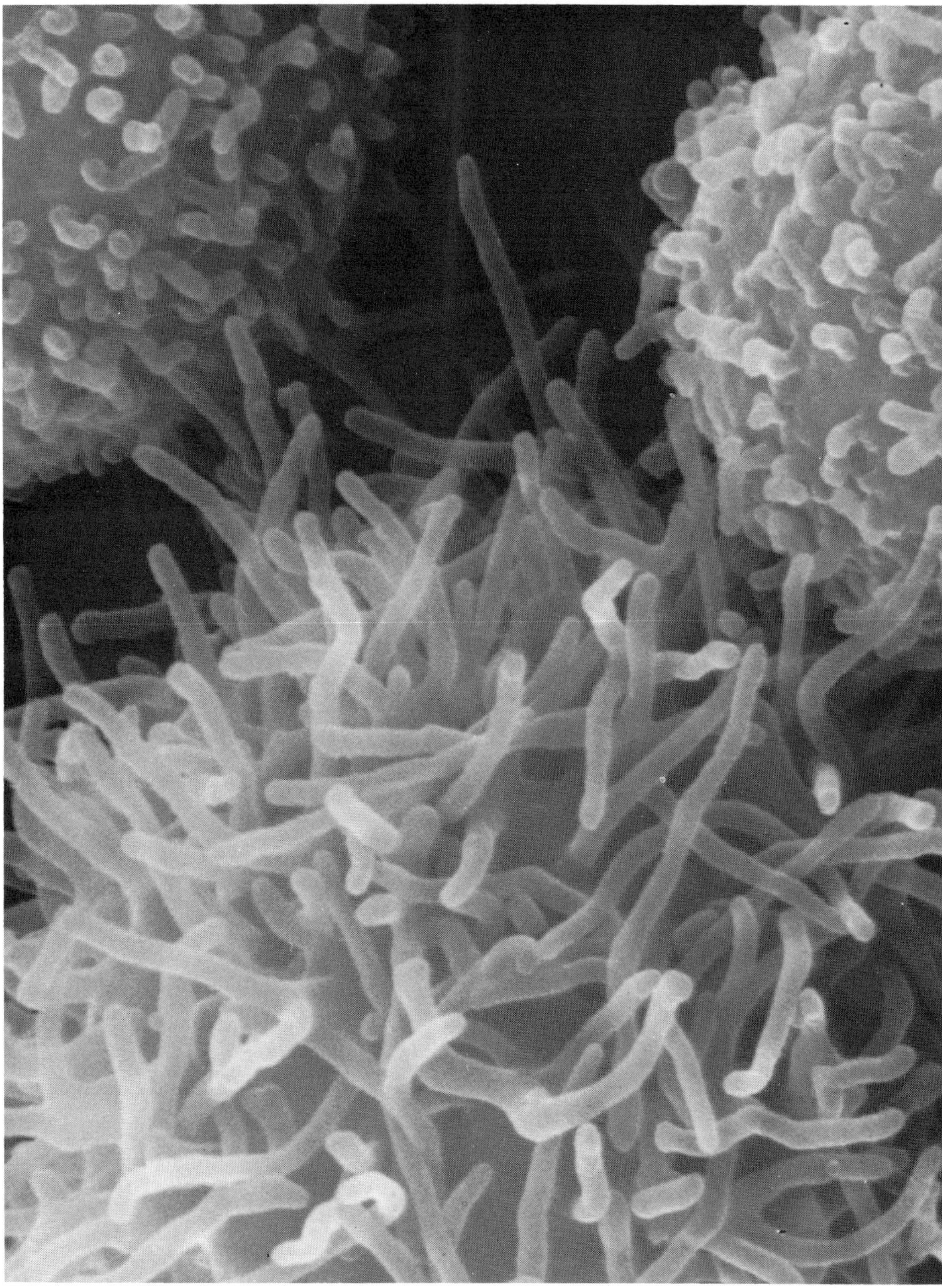

Figure 36

The surface of mast cells often shows undulating folds which would be mistaken for microvilli in thin sections. These may become confluent to form a complex pattern of anastomosing plications. Very short microvilli are also present in limited numbers.

Figure 37. Peritoneal mast cell from mouse. (Micrograph courtesy of Emma Shelton.)

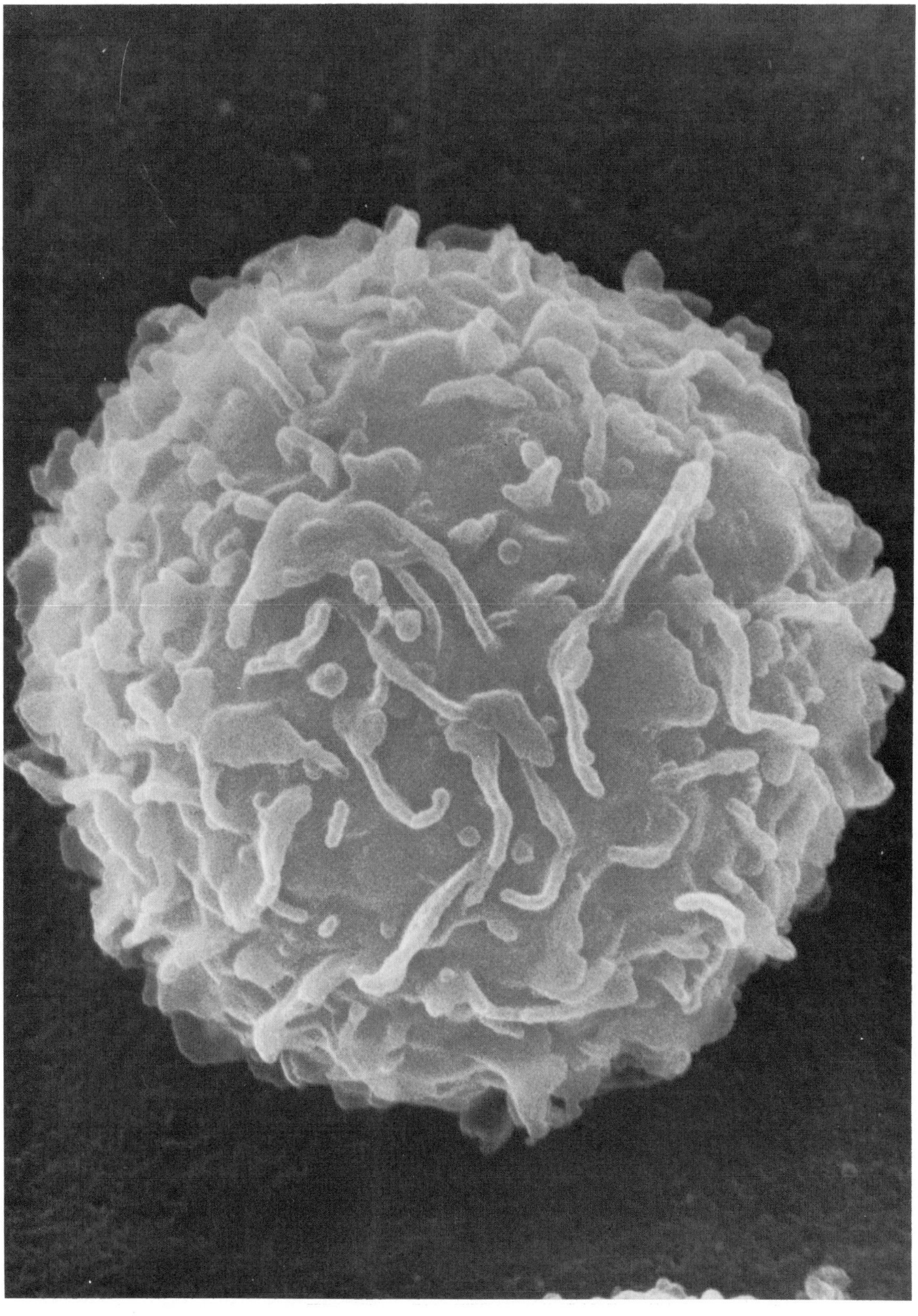

Figure 37

RELATIVELY STABLE SURFACE SPECIALIZATIONS

Cells specialized for absorption such as those of the intestinal epithelium and the renal proximal convoluted tubule have a *brush border* consisting of closely packed, parallel microvilli of uniform length. Those in the intestine are 0.1 μm wide and 1.0 μm in length, while those in the kidney vary from 1.5 to 3 μm in length in the various segments of the proximal tubule. There are approximately 60 per μm^2, and they increase the lumenal surface 15- to 40-fold compared to a flat apical surface.

Brush borders are enduring specializations, with each microvillus stabilized by a core of about 20 actin filaments which extend downward into the apical cytoplasm and mingle with the transversely oriented filaments of the terminal web. That these borders are not merely a device for increasing the area of the absorptive surface is indicated by their rich complement of enzymes. The membrane of the intestinal brush border contains sucrase, maltase, isomaltase, lactase, cellobiase, trihalase, aminopeptidases, lipase, cholesterol esterase and mono- and diglyceride acylases — enzymes that are involved in the terminal digestion of proteins, carbohydrates, and lipids.

The less highly ordered, sparse microvilli on many other unspecialized cell types appear to be more transient structures, generally lacking a conspicuous core of actin filaments. There is no evidence that their enzymatic activities differ from those common to the plasmalemma of cells in general.

Figure 38. Intestinal epithelium of rat in a period of lipid absorption. (Micrograph courtesy of Jean Paul Revel and Sanford Palay.)

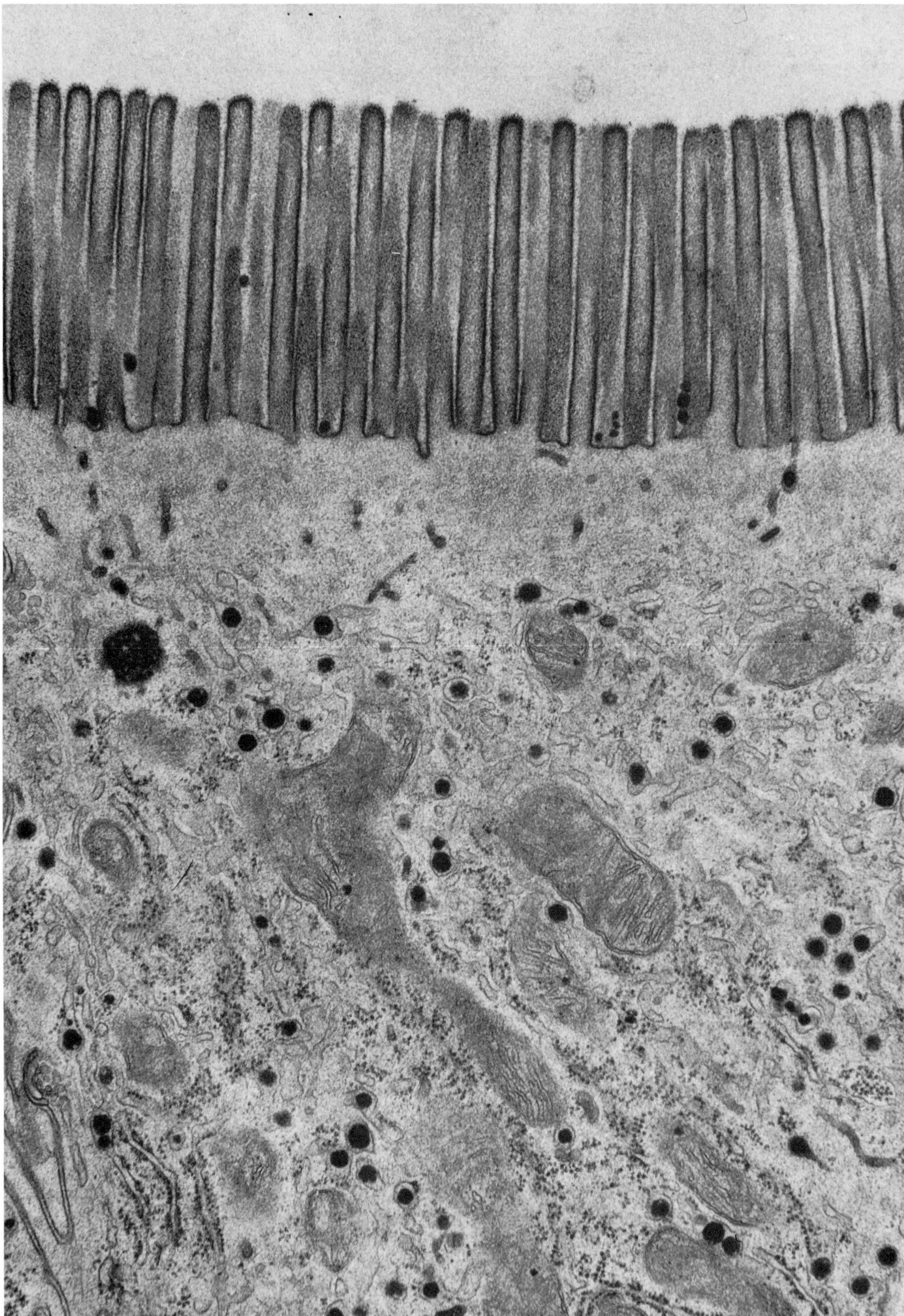

Figure 38

The lumenal surface of the cells lining the toad bladder was described from electron micrographs of thin sections as bearing short microvilli, but scanning microscopy reveals that the surface specializations are in fact plications or ridges that have a meandering labyrinthine pattern. The boundaries of the polygonal squamous cells are marked by juxtaposed continuous ridges on adjacent cells. The elaborate surface of these so-called granular cells is in contrast to that of the small mitochondria-rich cell in the center of the figure, which bears short globular microvilli.

Although relatively stable, this surface pattern does respond to changing physiological conditions. When stimulated with arginine vasopressin, the surfaces of the granular cells become convex and the pattern of ridges is transformed into one of discrete projections resembling microvilli. Physiological evidence suggests that this change in surface configuration is attended by a change in permeability of the membrane to both sodium and water. Under the same conditions, the mitochondria-rich cells do not undergo any significant change in surface configuration.

Figure 39. Scanning electron micrograph of a "mitochondria-rich" cell of the toad bladder surrounded by portions of four squamous "granular" cells. (Micrograph from Walter L. Davis, J. Cell Biol. 61:544–547, 1974.)

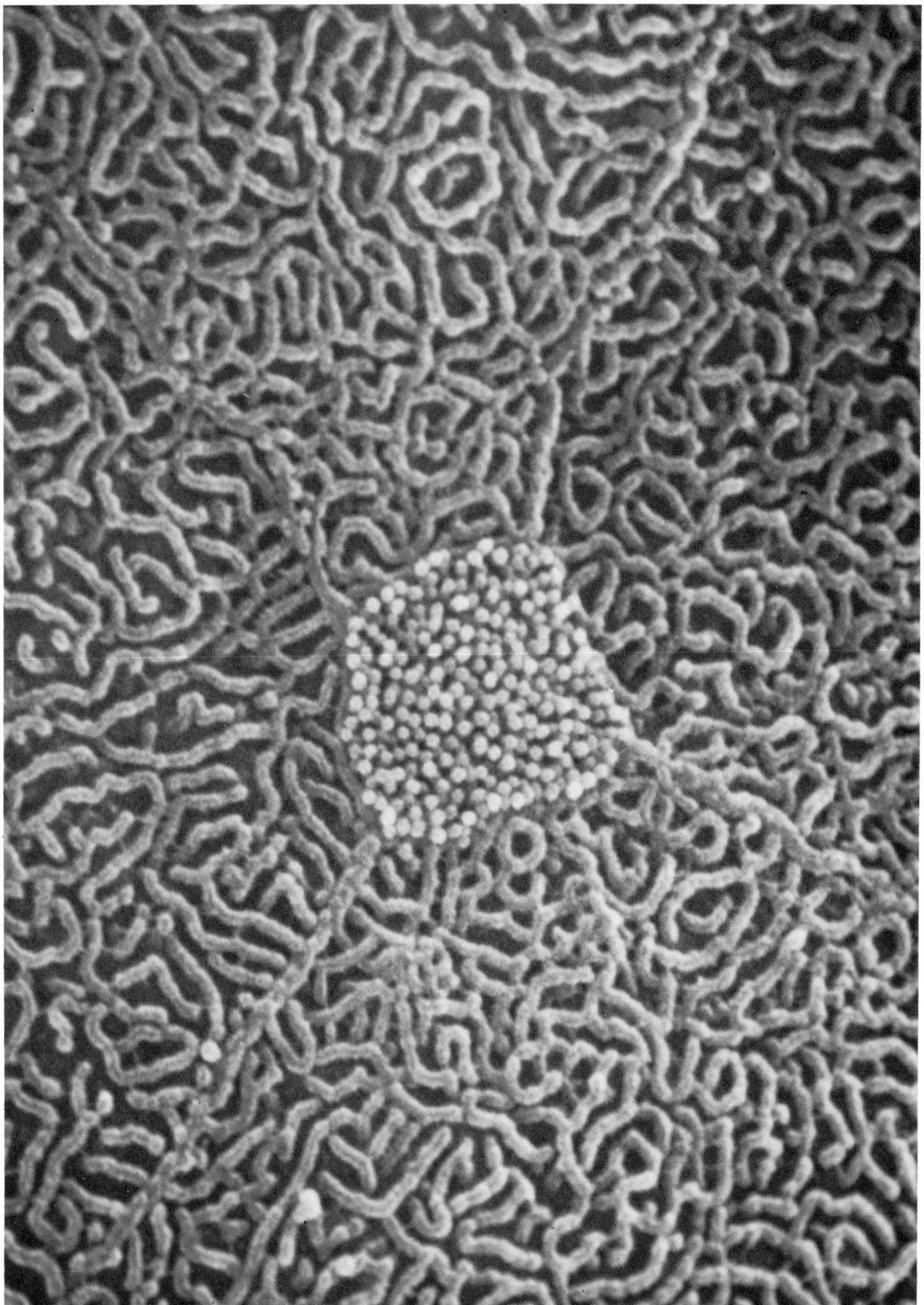

Figure 39

Fingerprint-like patterns of ridges are often found on the surface of cells where maintenance of a protective layer of mucus is advantageous, as on the squamous cells of the mammalian vagina, cervix, and oral cavity.

In the skin of fish, epidermal cells are directly exposed to an aqueous environment. The surface of the cells is specialized in various ways to accommodate to osmotic pressure differences between the cell interior and the aqueous environment and to physical and mechanical stresses to which they may be subjected.

The projections from these cell surfaces were interpreted as microvilli when examined in thin sections such as that in the micrograph below. But scanning microscopy reveals that they are low ridges, or microplicae. The pattern of ridges exhibits a wide range of interspecific variation. Especially striking is the labyrinth of surface ridges on the flatfish illustrated on the facing page. Whatever their pattern, they probably provide mechanical reinforcement against trauma, and, more significantly, they hold in place an abundant mucous layer which lowers frictional drag and serves as one of the most important protective devices associated with fish skin.

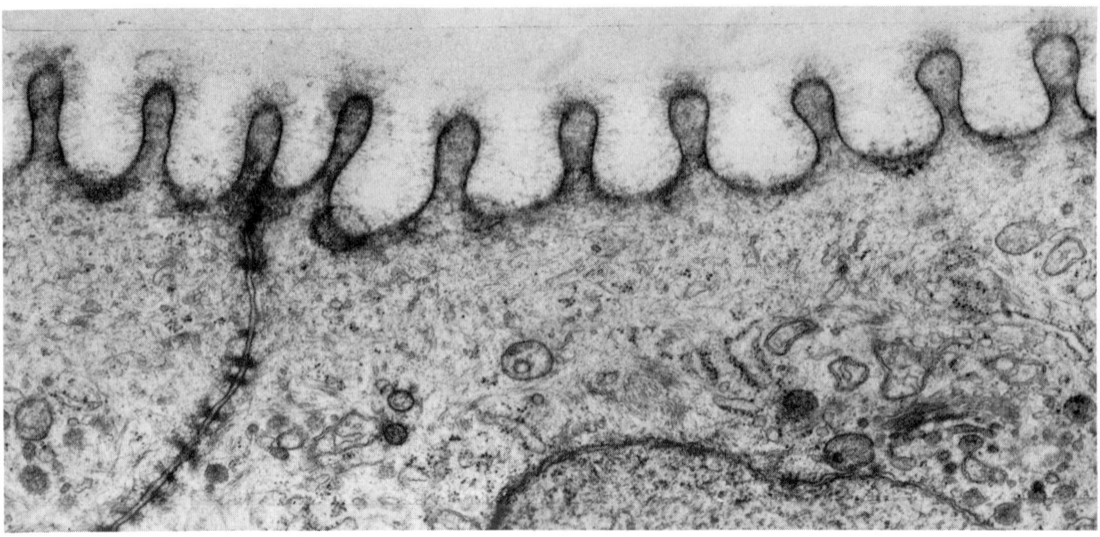

Figure 40

Figure 40 (above). Thin section of a surface epidermal cell from the neon tetra (*Hyphesobryon innesi*). (Micrograph courtesy of Nancy Alexander.)

Figure 41 (facing page). Scanning micrograph of the epidermal surface of the sole *Parophrys detulus*. (Micrograph courtesy of Wolf Fahrenbach, J. Invest. Dermatol. 65:39–44, 1975.)

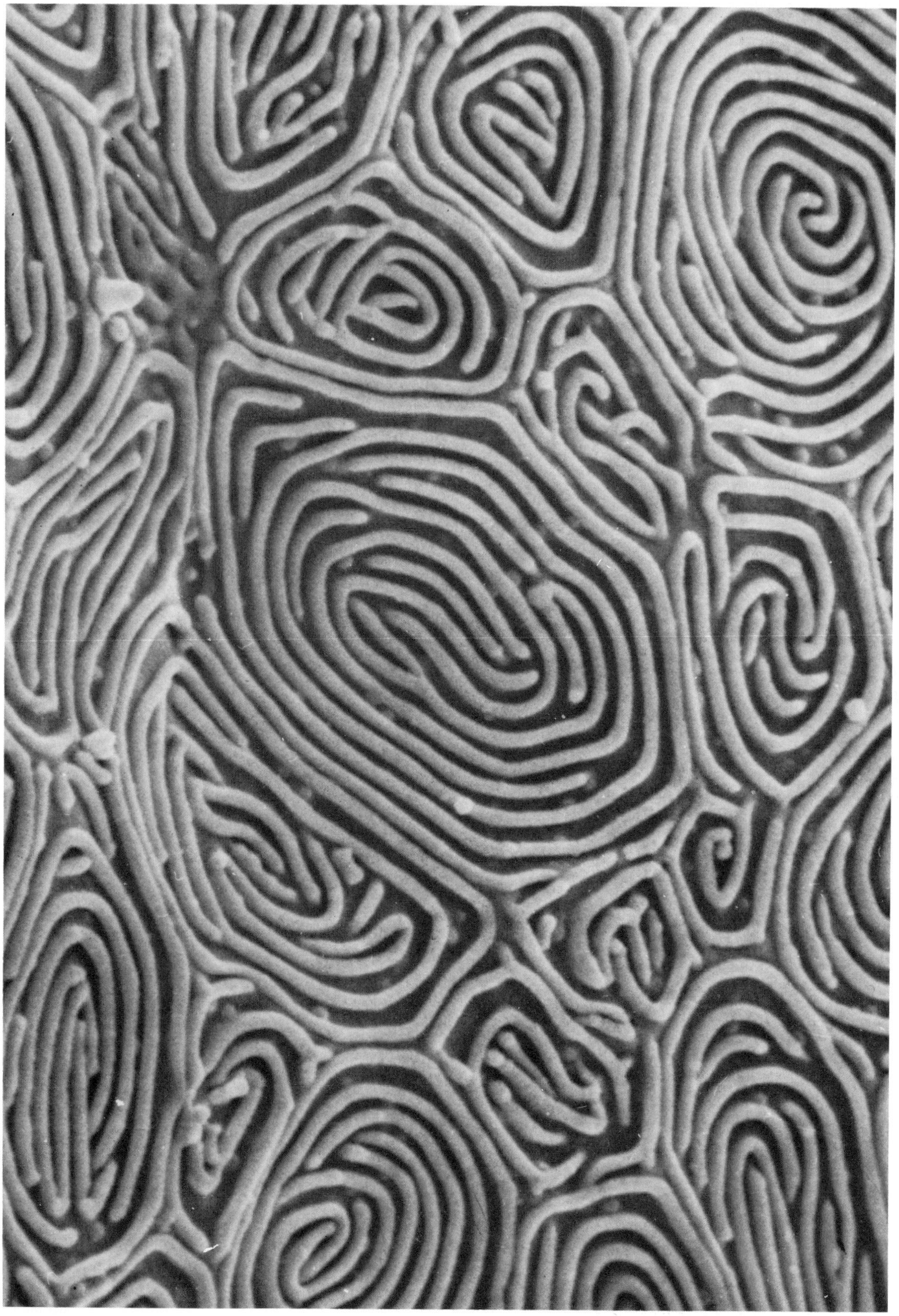

Figure 41

The epidermal surface of the Pacific coho salmon displays a labyrinthine pattern of ridges or microplicae with a more or less continuous circumferential ridge demarcating the boundaries of the polygonal cells.

Figure 42. Scanning electron micrograph of epidermis of the coho salmon. (Micrograph from Hawkes and Fahrenbach, Cell Tissue Res. 149:147–158, 1974.)

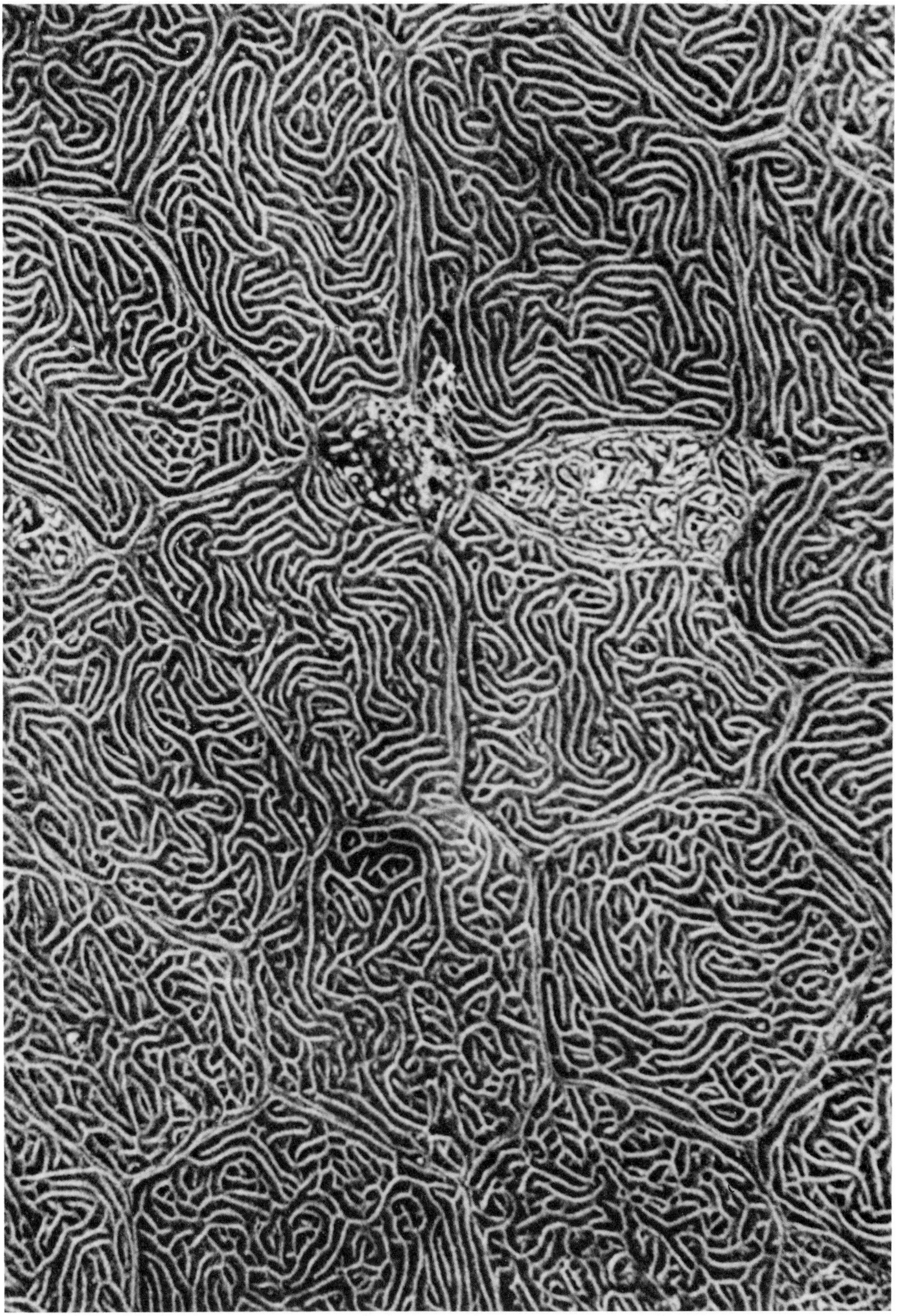

Figure 42

Epidermal cells of lamprey larvae exhibit a close meshed reticular pattern of low ridges bearing short microvilli. A row of microvilli also outlines the polygonal boundaries of each cell. Although the pattern is different from that of flatfish or salmon, the functional significance of this surface specialization is no doubt the same.

Figure 43. Scanning electron micrograph of the epidermal surface of a larval lamprey (*Entosphenus tridentatus*). (Micrograph courtesy of Wolf Fahrenbac, J. Invest. Dermatol. 65:39–44, 1975.)

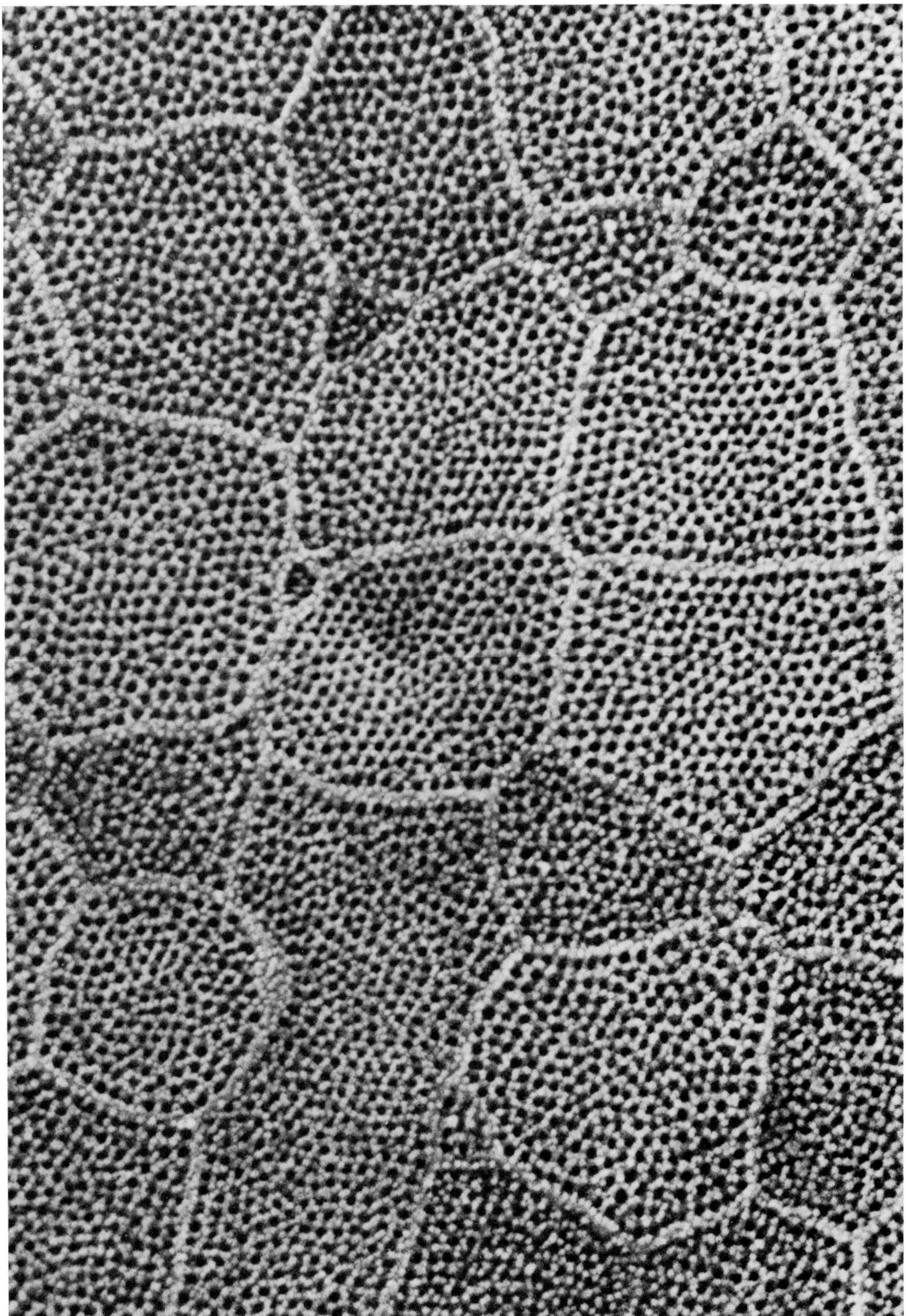

Figure 43

Amplification of surface area by microvilli and microplicae is not confined to the exposed lumenal surfaces. In many epithelia such processes also project into intercellular clefts on the lateral aspects of the cells, as shown in this micrograph of parotid gland. With routine preparative procedures for electron microscopy, these intercellular clefts are often narrowed and with villi compressed against the side of the cells. The preparation shown here was processed by quick freezing and freeze substitution, preserving the tissue in a more lifelike state with conspicuous intercellular clefts. These cells are firmly attached only at juxtalumenal junctional complexes and along the commissures of intercellular secretory canaliculi. Neither of these features is illustrated here.

Figure 44. Parotid gland of gerbil. (Micrograph courtesy of Atsushi Ichikawa.)

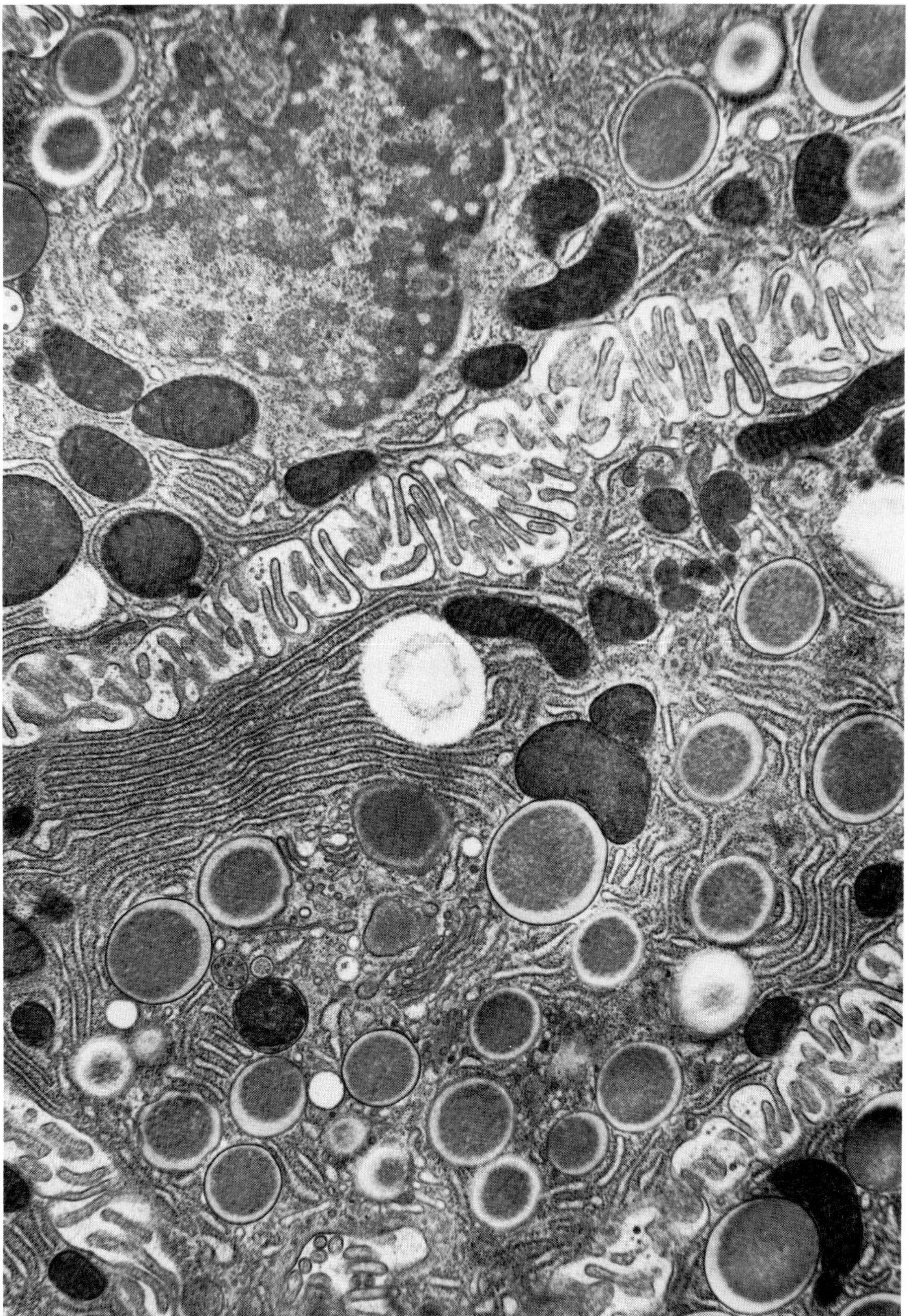

Figure 44

SPECIALIZATIONS INVOLVED IN ENDOCYTOSIS

Endocytosis is the general term applied to the uptake by cells of fluid in vacuoles formed by plasmalemmal invagination or of particles by encirclement with cell processes. If the material taken up is fluid, the process is called *pinocytosis* (drinking by cells); if it is particulate, the process is termed *phagocytosis* (eating by cells). All eukaryotic cells probably engage in pinocytosis and many are capable of phagocytosis. In cell types such as leucocytes and macrophages, which are involved in immunological reactions and defense of the organism against invasion by bacteria, the capacity for phagocytosis is highly developed.

Phagocytosis consists of ingestion of particles large enough to be visible with the light microscope. Under physiological conditions, these include bacteria, senescent erythrocytes, and other moribund cells and cell debris. Ingestion involves such close apposition of the plasma membrane to the surface of the particle that most, if not all, of the surrounding fluid is excluded. In general, normal cells and naturally occurring extracellular elements are immune to phagocytosis, but degenerating cells whose membranes have been modified or other tissue components whose surface proteins have been denatured are readily ingested. Nonpathogenic bacteria are rapidly taken up *in vitro,* but many pathogenic bacteria have capsules that inhibit their binding to the membranes of phagocytic leucocytes. However, the host invaded by a pathogen responds by producing blood-borne *antibodies* and *complement,* which coat the surface of the bacteria and stimulate their ingestion by phagocytes.

Of the several categories of immune globulins, only one, IgG, promotes phagocytosis by interacting with specific receptors (Fc receptors) on the leucocyte membrane. Serum complement is a complex series of a dozen or more enzymatic proteins that play an important role in the body's defense mechanisms. The third component, C3, is the principal phagocytosis-promoting member of the series. A derivative of this molecule, C3b, formed by proteolytic abscission of a small fragment from C3, is the ligand that binds to specific receptors in the leucocyte membrane. In addition to IgG and C3 receptors, leucocytes evidently have nonspecific receptors of unknown nature that enable them to ingest latex beads, zymosan, and other unnatural particles that are widely used in experimental studies.

The process of phagocytosis is considered to have two distinct phases, *attachment* by the interaction of a ligand on the particle with surface receptors in the membrane of the phagocyte, followed by *ingestion* of the particle. Attachment requires no expenditure of energy and can take place at low temperature. Ingestion requires energy and is temperature dependent. It was formerly thought that the cell simply extended pseudopodia that impounded a droplet of fluid containing the particle. It is now realized that the cell membrane is always directly applied to the particle and closely conforms to its surface topography as it is enveloped and ultimately interiorized. The current interpretation of these events assumes that an initial interaction of ligands on the particle with the receptors on the phagocyte starts a zipper-like process which spreads laterally from the point of initial contact and involves progressive apposition of increasing numbers of receptors and ligands until the particle is completely enclosed within a phagocytic vacuole of conforming shape.

To account for the energy-dependent flow of cytoplasm around the particle, it is postulated that the initial ligand-receptor interaction generates a signal which results in local aggregation of actin and other contractile proteins in the cell cortex leading to the formation of pseudopodia. Their extension increases the area of contact with the particle. This entails further receptor-ligand interaction and in turn induces more aggregation of contractile proteins. This self-propagating process continues until the plasma membranes of the pseudopodia meet and fuse beyond the particle, forming a closed phagocytic vacuole.

Pinocytosis describes the vesicular uptake of fluid containing low molecular weight solutes, soluble macromolecules, and colloidal particles too small to be visualized with the light microscope. Under physiological conditions, the substances imbibed include lipoproteins, ferritin, immune complexes, yolk proteins, hormones, and under experimental conditions peroxidase and electron-dense probes such as ferritin, colloidal gold, and thorium. Since pinocytosis takes different forms, it is useful to distinguish *macropinocytosis,* by which undulating folds capture droplets visible with the phase contrast microscope, from *micropinocytosis,* in which fluid is taken up in minute invaginations visible only with the electron microscope.

Two mechanisms of micropinocytosis are recognized: *fluid-phase pinocytosis* and *adsorptive pinocytosis.* The former is nonselective, the solutes being taken up in proportion to their concentration in the fluid medium, and the membrane involved is morphologically unspecialized. In adsorptive pinocytosis, on the other hand, the vesicular invaginations are coated on their cytoplasmic surface by delicate bristle-like projections and on the luminal side by a thin local glycocalyx. The nature and concentration of the material taken up depend upon the number and affinity of binding sites in the specialized membrane lining the vesicle. Large amounts of specific solutes can thus be taken up by this mechanism without imbibing a correspondingly large volume of solution.

The amount of membrane interiorized in endocytosis may be very large, especially in phagocytosis. Studies *in vitro* on the uptake of latex spheres by tissue culture cells indicate an hourly intake of some 30 per cent of the total surface area of the plasma membrane (Hubbard and Cohn, 1975). Macrophages *in vitro* may interiorize by pinocytosis 186 per cent of their surface membrane per hour (Steinman et al., 1972). Since the surface area of the cells remains constant, there must be an efficient mechanism for membrane recycling.

The importance of endocytosis in the economy of the normal organism can be appreciated from consideration of a few examples. It is estimated that in man 3×10^{11} senescent erythrocytes are removed from the blood by mononuclear phagocytes in the spleen each day. In the testis producing some 30 million spermatozoa daily, the Sertoli cells dispose of an equal number of residual spermatid cell bodies. In the eye the pigment epithelium is continually ingesting the senescent tips of the outer segment of the retinal rods, and failure of this process may lead to blindness (Young and Droz, 1968).

Pseudopodia in Phagocytosis and Locomotion

Free cells of the blood, peritoneal cavity, and connective tissues exhibit amoeboid motility. This mode of progression involves the formation and retraction of blunt processes called *pseudopodia*. In their formation, there is a local thickening of the ectoplasm which is rich in actin and other contractile proteins. As seen in the accompanying micrograph, the organelles and inclusions of the endoplasm are initially excluded, but as the pseudopod enlarges and the ectoplasm thins out, they flow into its core.

Figure 45. Polymorphonuclear leucocyte from guinea pig with three forming pseudopodia.

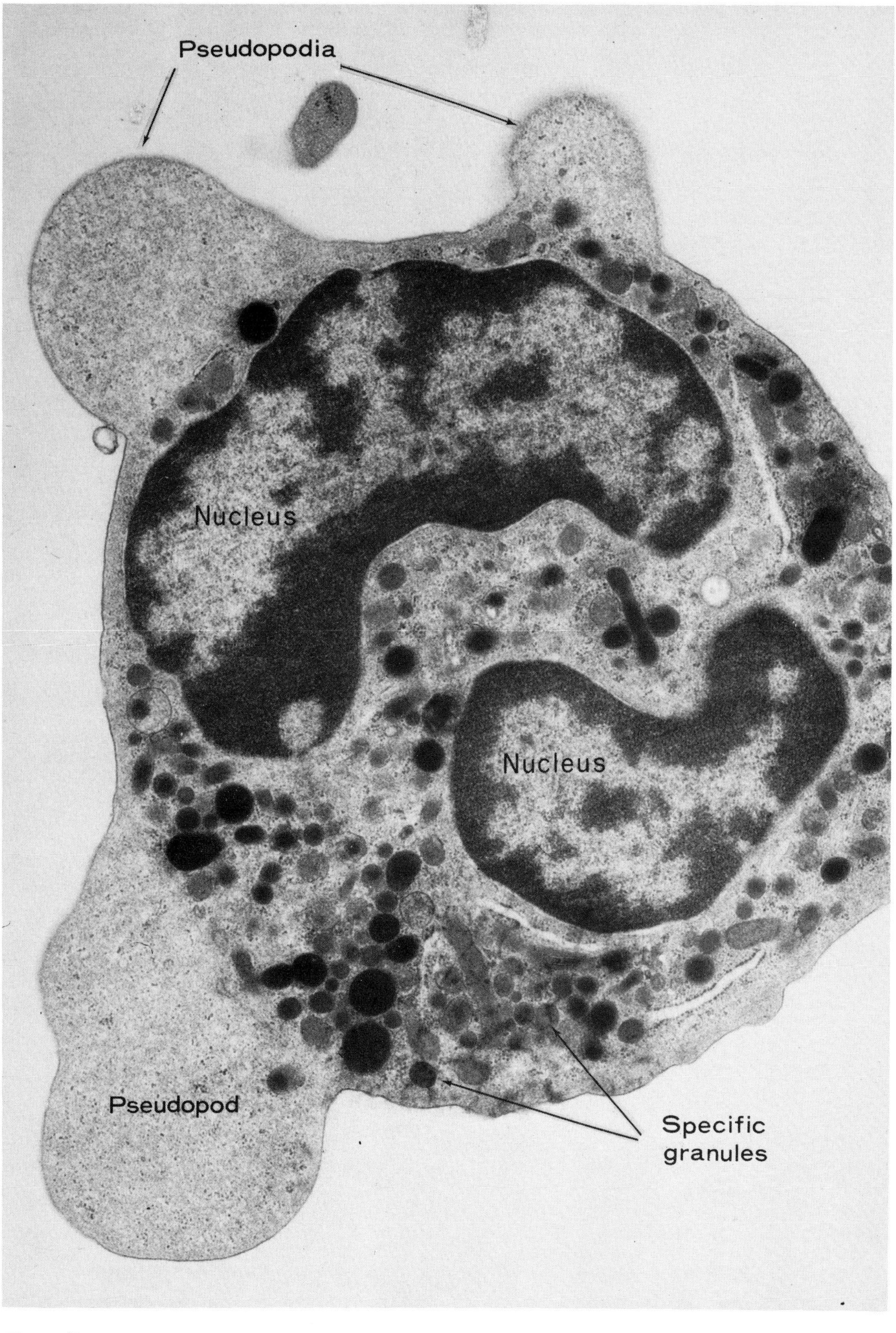
Pseudopodia
Nucleus
Nucleus
Pseudopod
Specific
granules

Figure 45

Lamellipodia (Undulating Membranes) in Pinocytosis

Living cells observed in tissue culture are observed to take into the cytoplasm droplets of the fluid medium. This activity, pinocytosis, was first described by Lewis (1931). It occurs at the thin peripheral portion of cells flattened against the solid substrate. In phase contrast images such as those presented here, thin undulating folds or ruffles projecting upward from the cell surface appear as dark sinuous lines. These are in continuous slow movement, and in the course of their undulations the margins of neighboring folds may meet and fuse, impounding a droplet of fluid which then moves into the cytoplasm as a clear vacuole.

In the upper photomicrograph, a number of undulating folds at the cell periphery are indicated by arrows. At the star a droplet of fluid has just been enclosed. Numerous vacuoles have accumulated in the cytoplasm near this area of active pinocytosis.

The lower figure is a phase contrast photomicrograph of an active region at the periphery of another cell showing a profusion of surface folds. Phase contrast microscopy of living cells is valuable for lapsed-time cinematographic recording of pinocytotic activity, but it does not reveal the vertical extent or three-dimensional configuration of the undulating membranes as dramatically as scanning electron micrographs of fixed cells, illustrated in the next two figures.

Figures 46 and **47.** Phase contrast photomicrographs of fields at the periphery of two cells in culture. (Courtesy of Charles Pomerat. From Fawcett, J. Histochem. Cytochem. 13:75–91, 1965.)

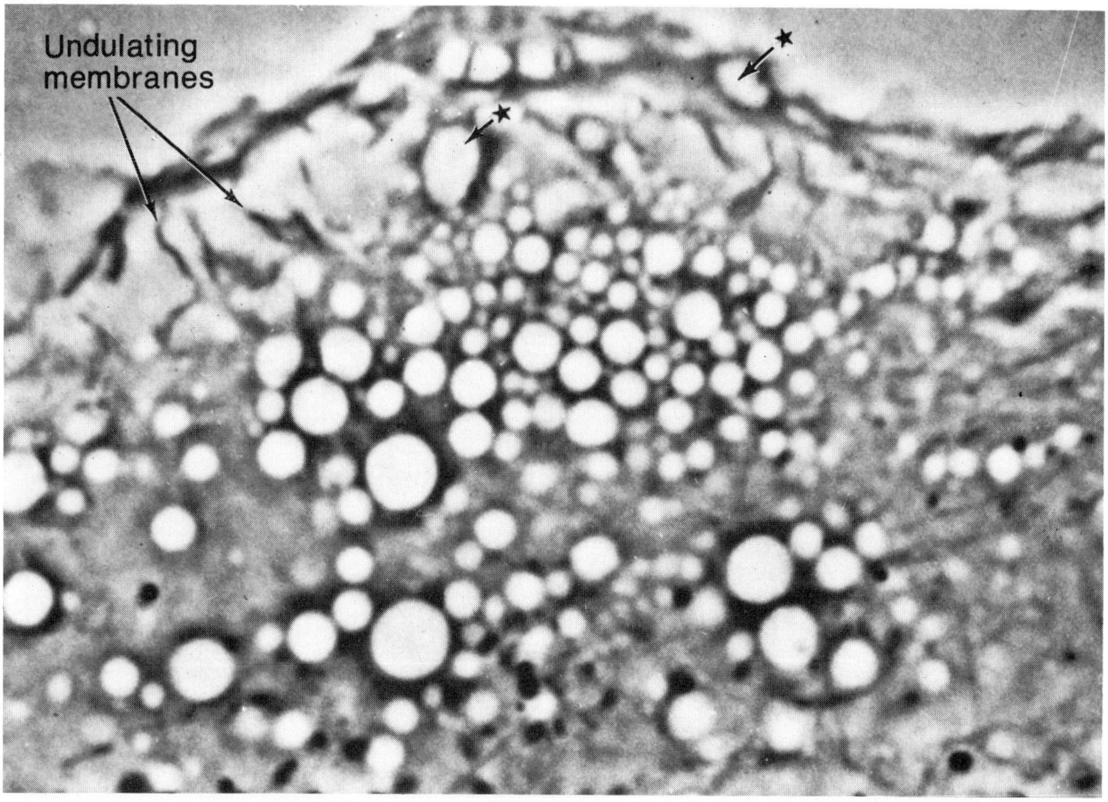

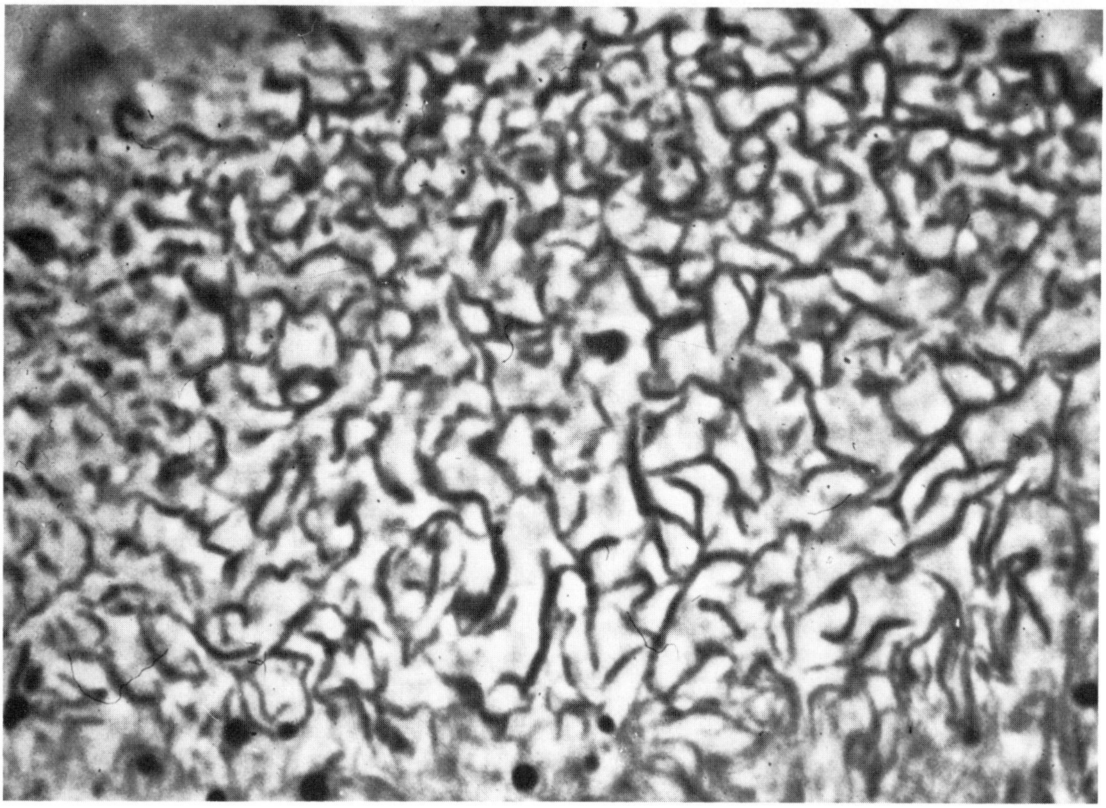

Figure 46, *upper* **Figure 47,** *lower*

With scanning microscopy, fluid-filled vesicles in the interior of cells cannot be seen, but the differentiations of the cell surface involved in the pinocytosis are clearly visible. At the ends of three of the processes of this cell, there are conspicuous undulating membranes or ruffles. The remainder of the cell surface is smooth or bears sparse microvilli of varying length.

Figure 48. Scanning micrograph of 3T3 cell in culture. (Micrograph courtesy of Susan Brown.)

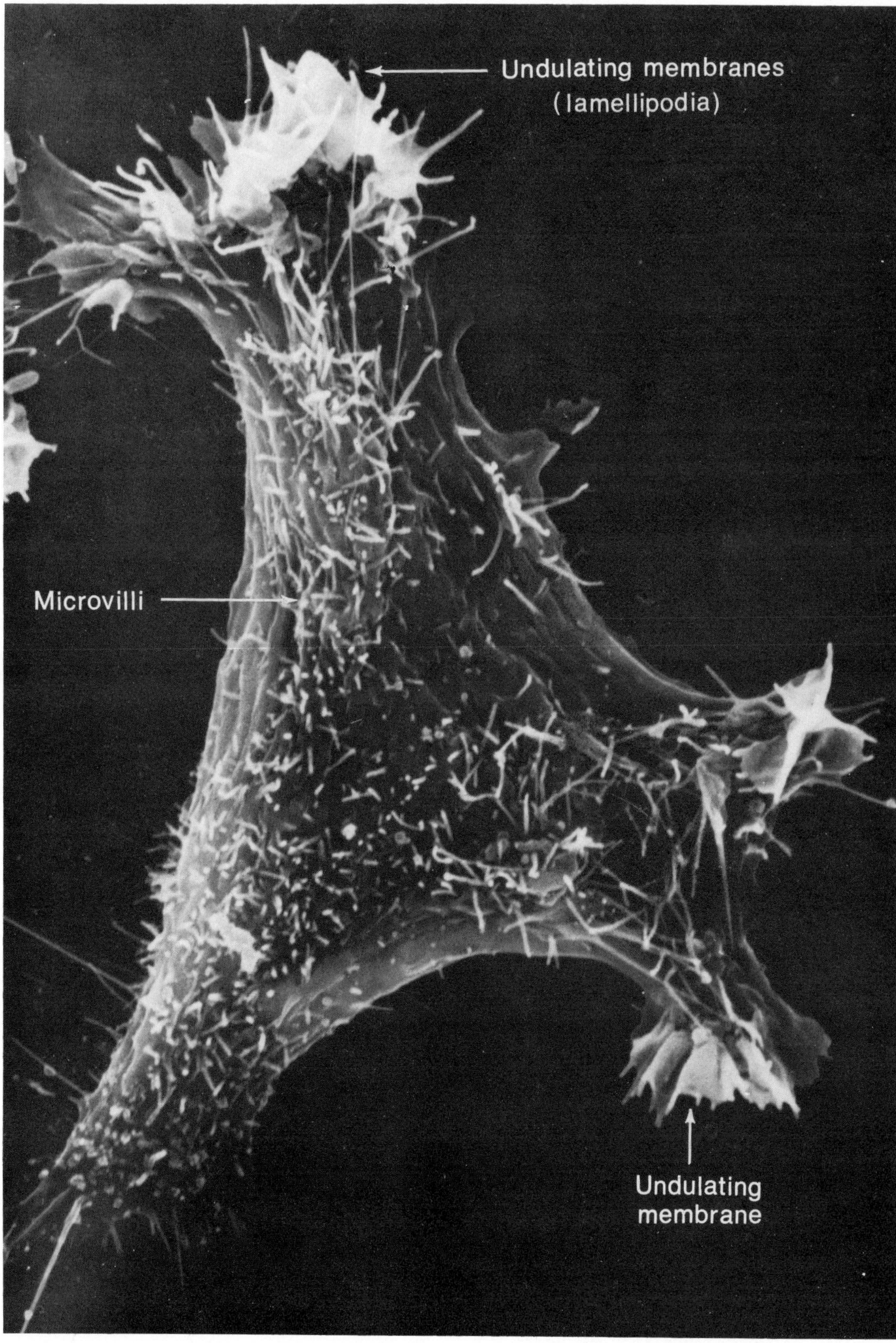
Undulating membranes
(lamellipodia)
Microvilli
Undulating
membrane

Figure 48

Scanning electron micrograph of daughter cells of a recent division, still joined by a slender strand (white arrow). Spreading upon the substrate, the thin periphery of each cell exhibits spectacular ruffles or undulating folds while the convex central area over the nucleus is studded with many microvilli. It is easy to envision the funnel configuration (at black arrows) or the petal-like arrangement of folds (at the star) closing around a quantum of fluid medium which would then be drawn into the cell as a membrane-bounded pinocytosis vesicle.

Figure 49. Scanning micrograph of cells. (Micrograph courtesy of Keith Porter. From Porter and Bonneville. The Fine Structure of Cells and Tissues, Lea and Febiger, Philadelphia, 1973.)

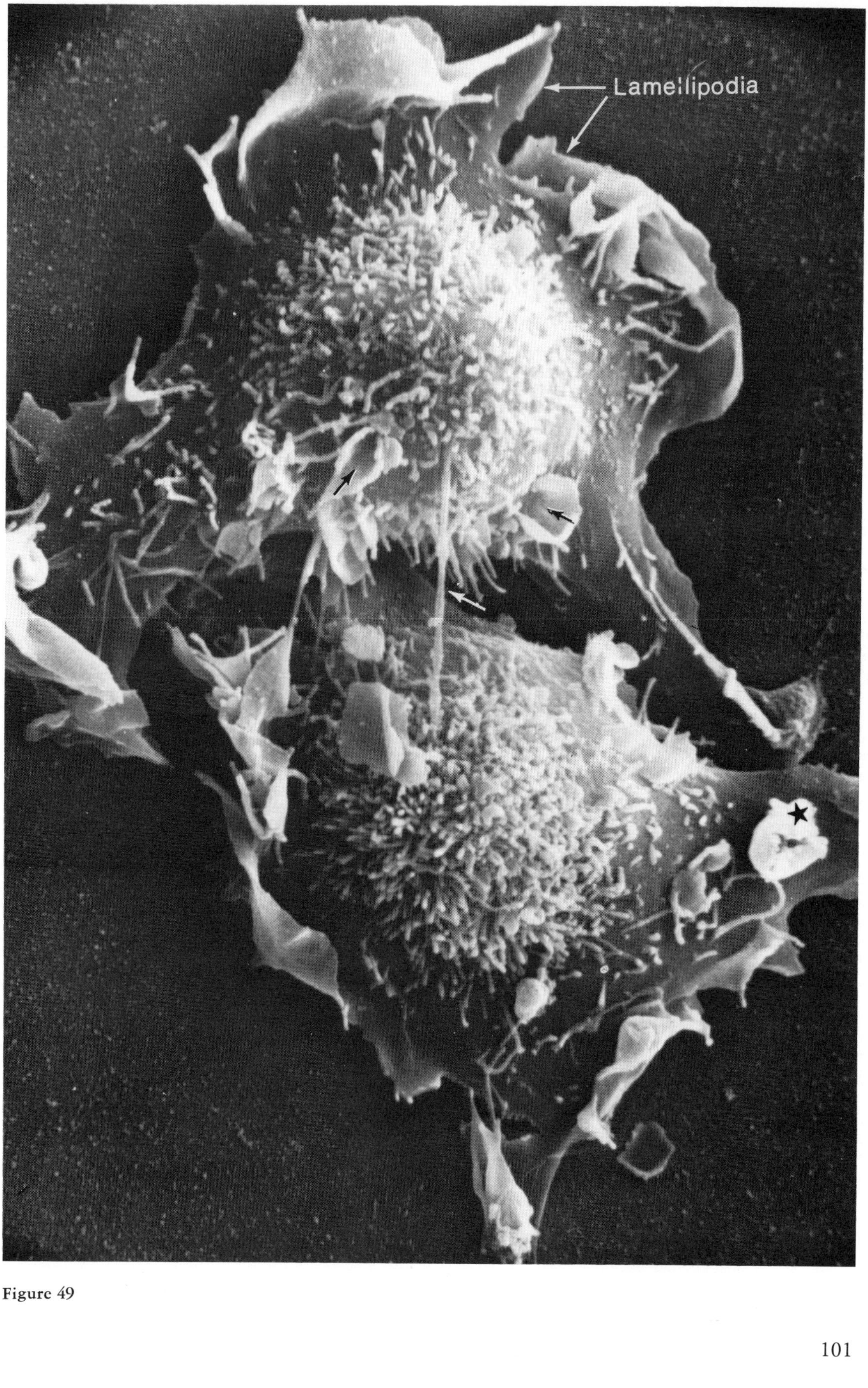

Figure 49

For many years following the initial description of pinocytosis, it was debated whether this kind of surface activity occurred in the intact organism or whether it was simply a reaction of cells to the abnormal environment in tissue culture. Although it may be more common *in vitro,* there is no longer any reason to doubt that it is a physiologically significant process *in vivo*. Electron microscopy provides many examples of such surface activity. The accompanying micrograph of a macrophage in the interstitial connective tissue of the testis shows an extraordinary array of undulating surface folds, some capturing droplets of fluid.

Figure 50. Macrophage in the interstitium of the testis in the opossum *Didelphis virginiana.*

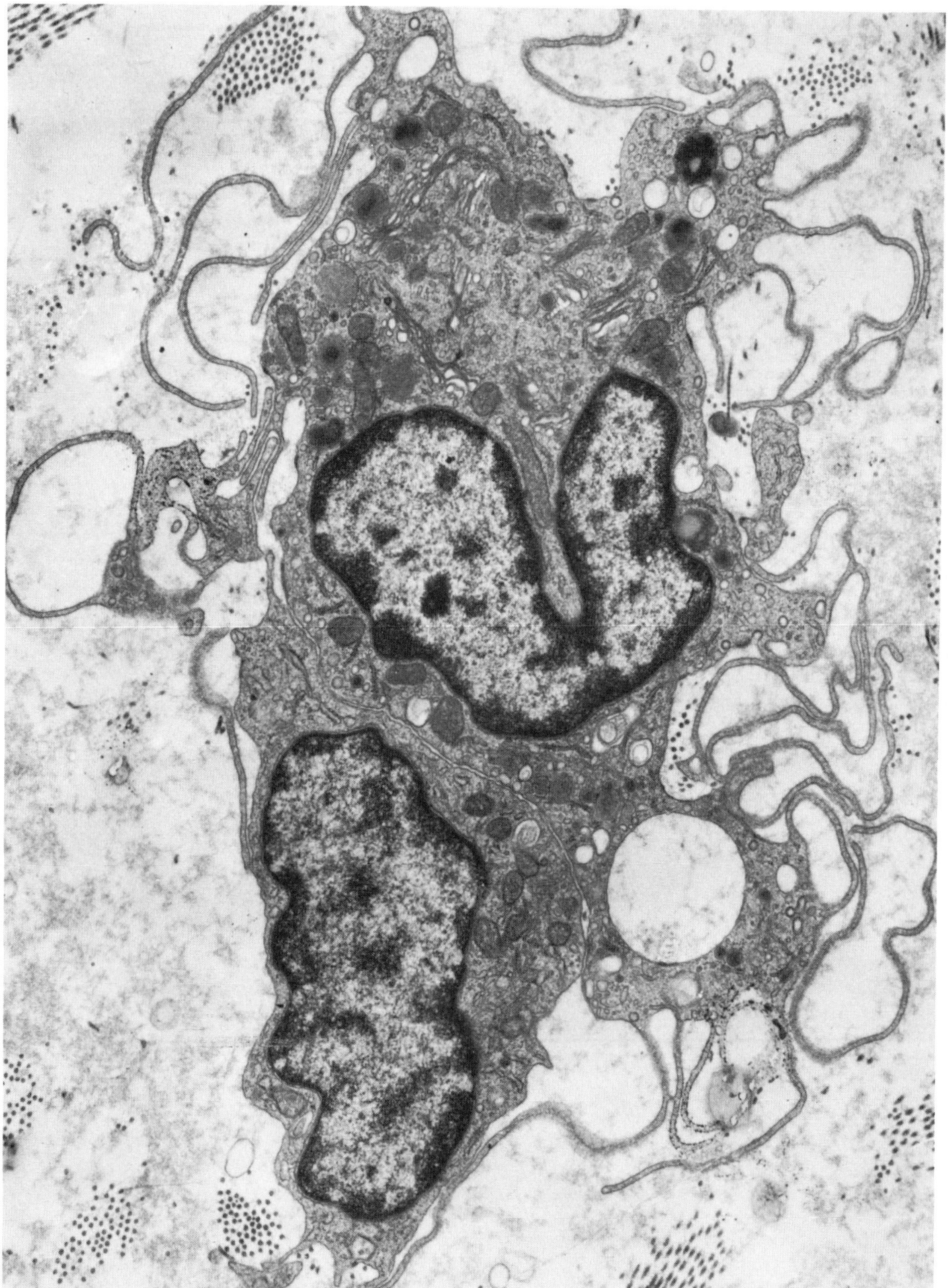

Figure 50

Pinocytosis is not confined to cells of mesenchymal origin but occurs also at the free surface of some epithelial cells. In endothelium of blood vessels, a thin fold often projects from the margins of the cells into the lumen. These may create local turbulence and mixing of the boundary layer, but it also seems likely that these marginal folds are capable of active undulant movements similar to those seen on cells in tissue culture. The accompanying plate assembles images suggesting that the recurring free margins may contact the cell surface and coalesce with it in such a way as to impound a droplet of plasma. *B* to *F* suggest a possible sequence of stages in the entrapment of a droplet of fluid by the marginal fold. The vacuole formed then appears to move toward the cell center and a new fold forms at the edge of the cell as indicated in *G*.

A process similar to this is observed in cells of the thyroid stimulated by thyrotropic hormone. Droplets of thyroglobulin taken into the cells from the lumen of the follicle are then processed by lysosomes to liberate thyroxin.

Figure 51. Capillary endothelial cell junctions. *A, B, C,* and *H* from cat myocardium. *D, E, F, G* from the choroid rete of *Amia calva*.

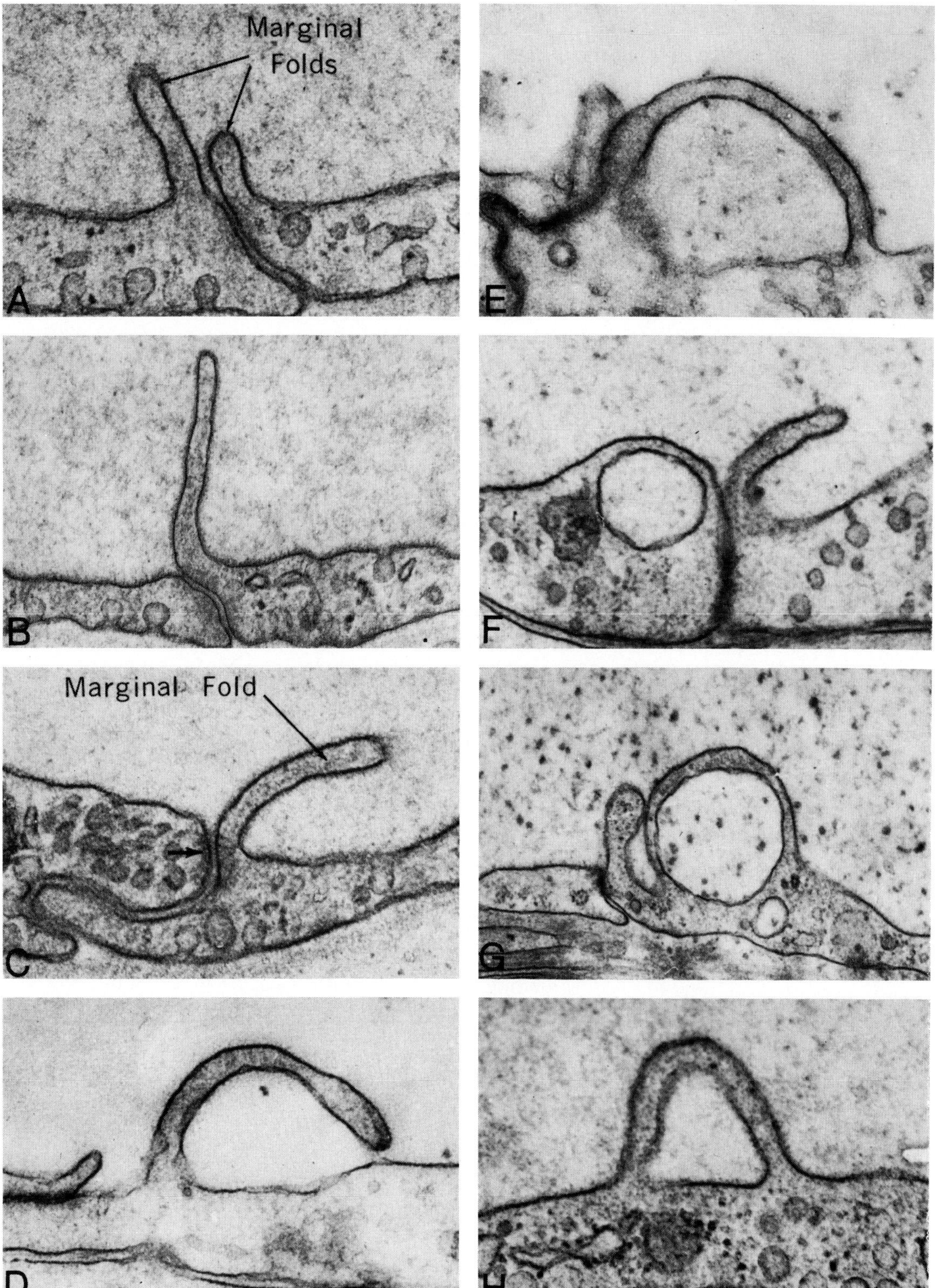

Figure 51

Fluid-Phase Micropinocytosis

Fluid-phase micropinocytosis occurs in nearly all cell types but at varying rates. Its morphological features are well shown in sections of capillaries. Numerous small flask-shaped pits or caveolae are seen along both the lumenal and basal surfaces of the endothelial cells. The membrane invagination involved in their formation is evidently a rapid event since intermediate stages are rarely observed.

It is not possible to determine from static images whether a given flask-shaped vesicle is taking up fluid or discharging it. The fact that these open-necked pits are seen at the cell surface in considerable number while rather few closed vesicles are found free in the subjacent cytoplasm is puzzling. Chemical fixation is a relatively slow process. *In vivo,* the collapse of vesicles, once they have fused with the plasmalemma, probably occurs very quickly. However, it is not unlikely that during fixation vesicle fusion may continue for a while after the surface membrane has been partially fixed, preventing collapse of the vesicles and incorporation of their membrane into the plasmalemma. If this is true, the number of open-necked vesicles seen in micrographs may be deceptively high, owing to their accumulation during the process of tissue preservation.

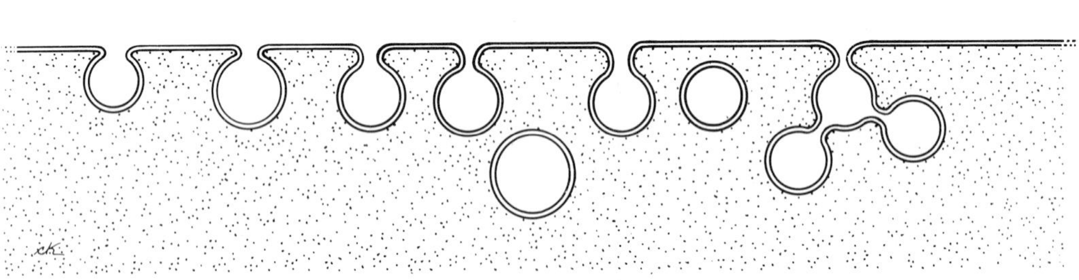

Fluid-phase micropinocytosis via smooth surfaced caveolae and interiorized vesicles. Intermediate stages in formation of the flask-shaped invaginations are rarely seen.

This form of endocytosis is nonselective, taking up solutes in the same concentration as that prevailing in the surrounding fluid. It is mainly a device for imbibition of fluid, but small colloidal particles may also be taken up with the fluid. Taking advantage of this, a number of electron dense probes such as ferritin, colloidal gold, and peroxidase have been widely used to study the origin and fate of pinocytosis vesicles. In the majority of cell types, the vesicles fuse with each other to form larger vacuoles and subsequently with multivesicular bodies or lysosomes where the vesicular contents are enzymatically degraded.

In capillary endothelium, the vesicles formed at the lumenal surface traverse the cell and discharge their content at the basal surface. Micropinocytosis vesicles thus shuttle small quanta of plasma from the capillary lumen to the surrounding tissue space (Bruns and Palade, 1968; Simionescu et al., 1972, 1975).

Figure 52. Micropinocytosis in capillary endothelium from mammalian cardiac muscle.

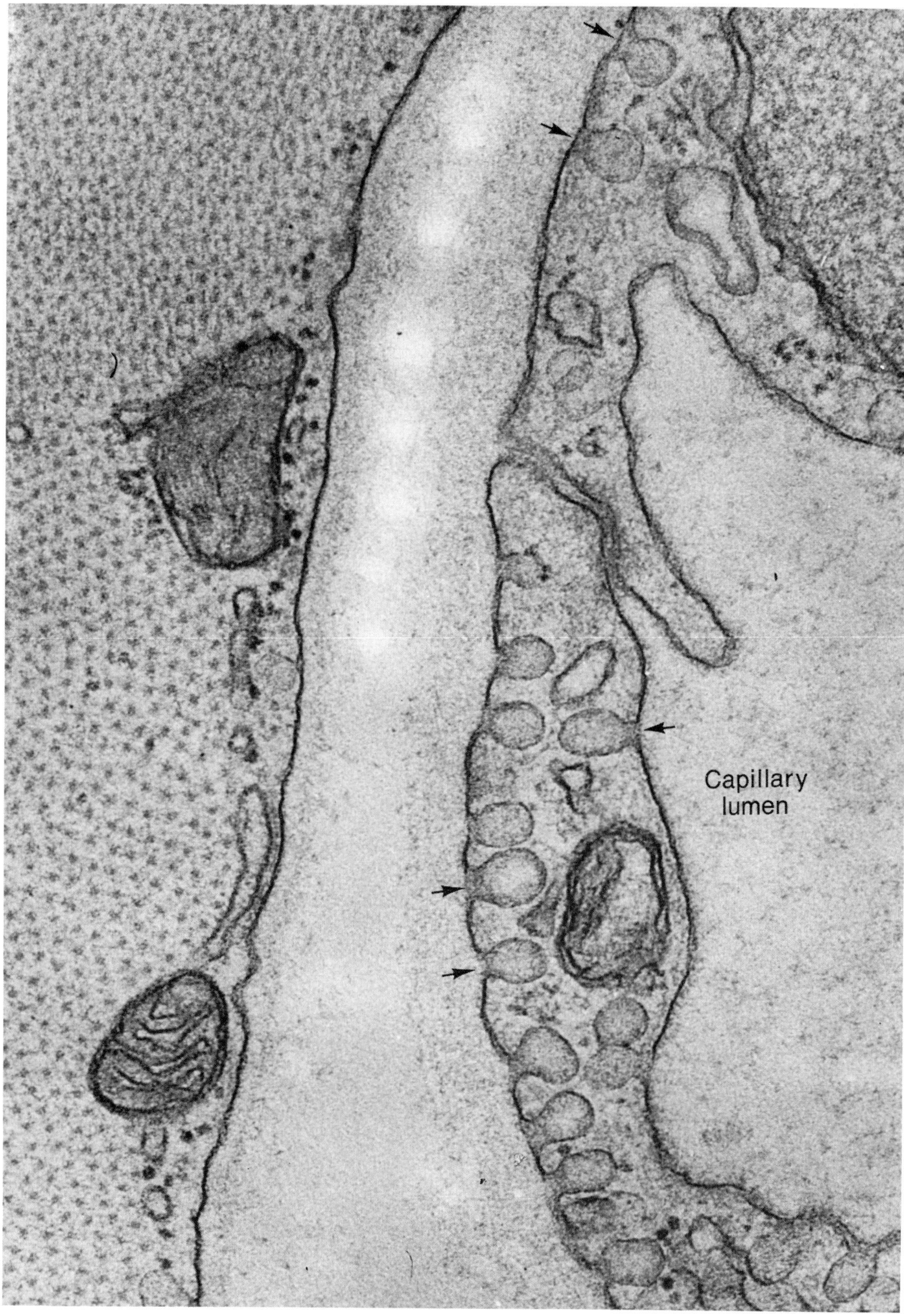

Figure 52

Adsorptive Micropinocytosis via Coated Vesicles

When morphologists in the 1960s obtained electron microscopic images apparently confirming Lewis's observations of 30 years earlier on the imbibition of fluid in bulk by pinocytosis, physiologists were slow to accept this as an important mechanism because it offered no satisfactory explanation of the high degree of selectivity that characterizes the uptake of some substances by cells. However, the demonstration of a carbohydrate-rich glycocalyx on cells soon led to the suggestion that such a layer might have properties suitable for selective binding of particular classes of molecules to the cell surface prior to its invagination to form pinocytosis vesicles (Bennett, 1963).

Observations on amoeba were cited in support of this concept. The entire plasmalemma of this organism is covered with a dense nap of fine filaments forming a continuous surface coat that has the staining properties of carbohydrates. It was shown by Pappas and Brandt (1960) that when amoebae are exposed to solutions containing thorium dioxide, ferritin, or other electron opaque markers, these colloidal particles were adsorbed onto the filamentous coating of the plasmalemma in high concentration and could be traced in their passage into the cell. Pinocytosis in amoeba was therefore considered to be a two-stage process involving a binding of material on the membrane coat followed by invagination of the surface to form endocytosis vesicles or channels.

As methods of fixation for electron microscopy improved, two categories of small pits, or caveolae, were recognized on the surface of all cells — the smooth-surfaced micropinocytosis vesicles that have already been described and a second type lined by a thin surface coat. In addition to a modest glycocalyx on their concave extracellular surface, there are short bristle-like projections on the convex cytoplasmic surface of the membrane. These invaginations were variously described as *coated vesicles, bristle-coated vesicles, acanthous vesicles* or *acanthosomes*. The term now most widely used is coated vesicle. In addition to the ultrastructural features that distinguish them from the smooth-surfaced micropinocytosis vesicles, there seems to be a significant difference in the time course of events in their formation, if one assumes that the relative frequency of images of intermediate stages is a reliable index of the rate of the process.

In the case of smooth micropinocytosis vesicles, early stages of their formation are rarely seen. In contrast, a series of images of intermediate stages in the invagination of coated vesicles is easily assembled. Small patches can be found along the smooth contour of the cell where the membrane appears thickened by local acquisition of a thin surface layer and inward-projecting bristles. Equally common are sites where such specializations form shallow depressions. Progressive deepening of these latter results in flask-shaped invaginations communicating with the extracellular space through a narrow neck. Constriction and fusion of membranes in the neck region restores a smooth contour to the plasmalemma and detaches a spherical coated vesicle, which is then free in the cytoplasm. Since there is no accumulation of flask-shaped coated invaginations along the cell membrane, it can be inferred that their formation and detachment proceed without pause.

In an electron microscopic study of the surface of mosquito oocytes after a blood meal, Roth and Porter (1962, 1964) presented persuasive evidence that these coated

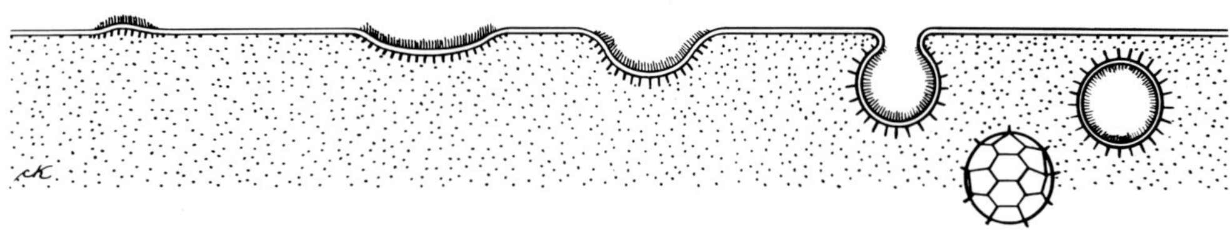

Adsorptive micropinocytosis in coated vesicles. Specialized areas of membrane with an external glycocalyx and internally projecting spines form shallow depressions which deepen to become coated pits, and finally detach as coated vesicles. In surface view the membranous vesicle is enclosed in a polygonal lattice of clathrin.

invaginations are specialized sites for uptake of protein. In many invertebrates and some vertebrates, yolk protein is synthesized outside of the ovary, secreted into the extracellular fluid compartment, and removed from the circulating blood by the developing egg. In the mosquito egg, development is initiated by feeding. Examination of the oocyte surface in the hours immediately following a blood meal showed a 15-fold increase in coated pits. A conspicuous dense layer on their concave extracellular surface was interpreted as protein which had accumulated by selective adsorption from the extraoocyte space. Concurrent observations on cells of the mammalian bone marrow suggested that a similar process is involved in selective uptake of ferritin by cells of the erythropoietic series (Fawcett, 1964). Small plaques were noted on the surface of erythroblasts where the membrane appeared thickened by a furry external layer. Particles of ferritin were found adhering to these plaques but not to other unspecialized areas of the cell surface. These invaginated to form a vesicle that detached from the surface, carrying its charge of ferritin into the cytoplasm of the erythroblast. Observation of the natural uptake of ferritin into erythropoietic cells by this process was taken as further evidence that local development of a surface coat may confer a measure of selectivity upon a process of bulk uptake of fluid that would otherwise be undiscriminating and physiologically inefficient (Fawcett, 1965).

Interest in the structure and composition of coated vesicles was stimulated by their successful isolation and the demonstration by negative staining that they consist of a membranous vesicle enclosed by an extremely regular polygonal network. In thin sections, the appearance of spines or bristles projecting from their convex surface is apparently due to superimposition of the hexagonal and pentagonal elements comprising this peripheral lattice (Kanaseki and Kadota, 1969). Biochemical studies of coated vesicles isolated from several sources show that a large proportion of their total protein consists of a 180,000 dalton subunit called *clathrin* (Pearse, 1976). Later investigations have identified associated minor proteins of 125,000 and 55,000 daltons (Woodward and Roth, 1978). The preparations isolated from brain consist of a mixture of intact coated vesicles and separate coat lattices without a central vesicle. The proportion of clathrin varies directly with the concentration of lattices, and it appears therefore that the polygonal coat is composed of clathrin and that the other proteins are involved in coat-vesicle interaction. The lattice can be solubilized by urea, high pH, or magnesium chloride. Upon removal of the dissociating agent by dialysis, the coat lattices reassemble and if vesicles are present will form around them, suggesting that there may be vesicle-to-coat association sites (Woodward and Roth, 1978).

Interest in the role of coated vesicles in absorptive endocytosis has recently been revived by biochemical and genetic studies on the uptake of low-density lipoproteins (LDL), the major cholesterol-carrying protein in human plasma. The surface of normal human fibroblasts possesses specific receptors for this lipoprotein. When LDL labeled with ferritin is presented to cultured human fibroblasts at 4° C, it binds to the cells with high affinity, and the ferritin label is found to be concentrated on coated segments of the cell surface and in shallow invaginations of the plasma membrane (Anderson, Goldstein and Brown, 1976). At 37° C, the ferritin is taken into the cells in coated vesicles.

Fibroblasts from patients with familial hypercholesterolemia lack functional LDL receptors and show no binding or receptor-mediated uptake. In electron micrographs, the cells contain the same number of coated segments and caveolae as normal fibroblasts but no ferritin-labeled LDL binds to these regions of the mutant cells (Anderson et al., 1976). These findings provide further support to the concept that coated regions of the plasma membrane are sites specialized for receptor-mediated binding and uptake of specific macromolecules (Anderson, Brown and Goldstein, 1977). The postulated steps in this process are (1) synthesis of receptors, (2) their insertion at random in the cell membrane, (3) aggregation of receptors in clathrin-coated pits, (4) internalization of the receptors in coated vesicles, possibly followed by (5) recycling of receptors to the plasma membrane (Goldstein et al., 1979).

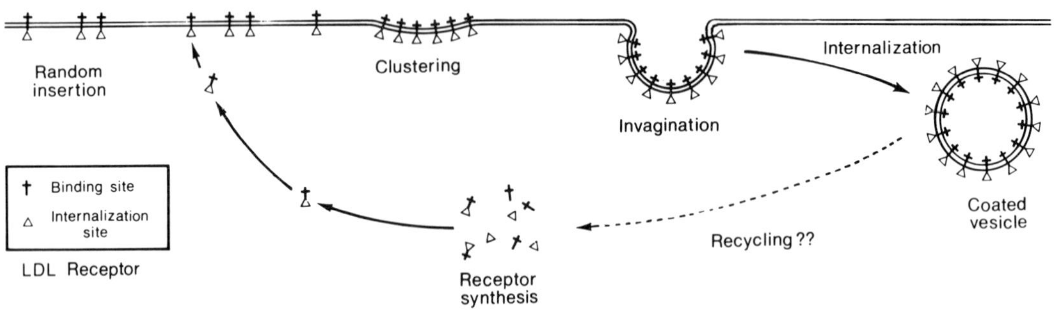

Schematic representation of the proposed mechanism by which LDL receptors become localized in coated pits and later in coated vesicles of human fibroblasts. (Redrawn from R. W. Anderson, J. L. Goldstein and M. S. Brown, Nature 270:695–697, 1977.)

Although the uptake of low-density lipoprotein has been the most thoroughly studied, there are other examples in which transport proteins are internalized by this mechanism. Endocytosis of the iron transport protein *transferrin* by erythroblasts has already been mentioned (Sullivan et al., 1976). The yolk proteins *phosvitin* and *lipovitellin* are similarly taken up and delivered to the yolk granules of oocytes (Roth et al., 1976). The protein hormones *insulin, nerve growth factor,* and *epidermal growth factor* are reportedly taken up in coated vesicles and delivered to lysosomes for degradation (Marsh et al., 1979; Bradshaw, 1978; Maxfield et al., 1978). *Maternal immunoglobulins* are ingested in coated vesicles in the fetal yolk sac and neonatal intestinal mucosa and discharged at the base of the epithelium into the circulation (Rodewald, 1973; Linden and Roth, 1978). Uptake of receptor-bound proteins in coated vesicles is common to all of these examples, but the ultimate disposition of the protein varies, some being degraded in lysosomes, others deposited in specific organelles, and still others transported across the cell and released into the bloodstream. Although it is not known how endocytotic vesicles are directed to specific delivery sites, receptor-mediated endocytosis appears to be an important mechanism for the cellular uptake of nutritional and regulatory proteins (Goldstein et al., 1979).

Membrane Recycling Via Coated Vesicles

In secretory cells the product is usually packaged in membrane-limited secretory granules that are formed in the Golgi complex. When the product is released by exocytosis, the membrane of the secretory granule is incorporated in the plasma membrane. During active secretion, large amounts of membrane are added to the cell surface. It is believed that unlimited expansion is prevented by invagination and interiorization of excess membrane in small vesicles that form at the lumenal surface and return membrane to the Golgi cisternae and condensing vacuoles, where it is reutilized in the formation of new secretory granules.

The most convincing experimental evidence for membrane recycling has come from studies on transmitter release at nerve terminals. At the myoneural junction, transmitter accumulates in numerous synaptic vesicles at the nerve terminals and is released by exocytosis in quanta corresponding to the content of the individual synaptic vesicles (Katz and Miledi, 1965). Morphometric analysis shows that when the nerves of isolated frog muscle are electrically stimulated the synaptic vesicles are depleted and there is a concurrent increase in the area of the plasma membrane, such that the total amount of membrane in the terminal remains constant. After prolonged stimulation, the extensive depletion of synaptic vesicle membrane is counterbalanced by the appearance of membrane-bounded cisternae in the interior of the terminals. After cessation of stimulation, the reappearance of synaptic vesicles is associated with a decrease in surface area of these cisternae and of the plasma membrane. These observations suggest that there is a redistribution and recycling of membrane, and this is strongly supported by experiments in which the electron opaque tracer horseradish peroxidase is introduced into the extracellular space. During nerve stimulation, numerous coated vesicles appear and take up the tracer in peripheral regions of the nerve terminals covered by Schwann cells, and later these can be seen coalescing with the cisternae that form during prolonged stimulation. The peroxidase accumulates in the cisternae and subsequently appears in the reconstituted synaptic vesicles. Thus the formation of coated vesicles appears to be a mechanism for retrieval of membrane added to the surface in synaptic vesicle exocytosis. It is assumed that the associated baskets depolymerize upon fusion of the coated vesicles with the cisternae and that the clathrin is recycled back to the surface for reassembly in the formation of new coated vesicles. Thus it is reasonable to conclude from observations on the myoneural junction that the membrane added to the surface in stimulus-induced exocytosis is retrieved by coated vesicles and recycled to new synaptic vesicles via the cisternae in the interior of the nerve terminal (Heuser and Reese, 1973).

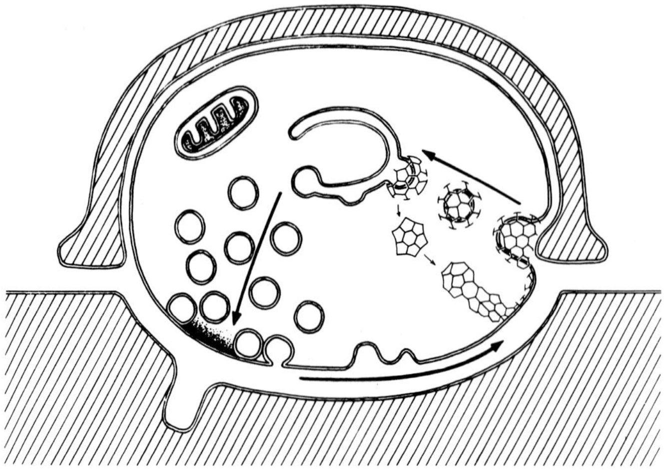

Diagram of the myoneural junction illustrating involvement of coated vesicles in membrane recycling. Membrane added to the presynaptic surface is believed to be retrieved by formation of coated vesicles and recycled via cisternae in the interior of the nerve ending to form new synaptic vesicles. (Slightly modified from J. Heuser and T. Reese, J. Cell Biol. 57:315–344, 1973.)

This interpretation has derived strong support from studies on turtle retina in conditions of varying synaptic activity in dark and light adaptation and at low temperature (Schaeffer and Raviola, 1978). Neural processes form invaginated synapses with the cone cells. The depth of their penetration into the photoreceptor and the complexity of outline appear to depend upon variations in surface area of the nerve ending that reflect the state of balance between addition of synaptic vesicle membrane to the surface and its retrieval. In the dark, the rate of exocytosis is increased; the nerve endings penetrate deeply into the photoreceptor and form a highly digitated interface. In the light, membrane retrieval is dominant and surface area of the interface and depth of invagination decrease. Exposure to cold blocks membrane retrieval, but upon rewarming, numerous vacuoles and cisternae appear in the presynaptic cytoplasm and large numbers of coated vesicles form at the surface. When exposed to peroxidase during rewarming in the light, the tracer is rapidly taken up in coated vesicles, accumulates in cisternae, and is ultimately found in the regenerated synaptic vesicles. Thus there seems little doubt that coated vesicles, cisternae, and vacuoles play a major role in recycling of the plasma membrane at nerve terminals.

Exocytosis may not always involve transient increase in the area of the surface membrane followed by retrieval in coated vesicles at a distance from the site of membrane addition. There is suggestive evidence that in the secretory cells of the anterior pituitary, membrane retrieval may take place at the site of exocytosis. The membrane of the secretory granules is smooth surfaced until it contacts and fuses with the plasmalemma. The empty caveola left by discharge of the granule then appears to acquire a clathrin coat. Its neck presumably constricts and the membrane of the secretory granule is retrieved directly without ever being completely incorporated in the plasmalemma (Fujita, 1979). Similar bristle coats covering casein-containing secretory vesicles were reported in earlier studies on mammary gland (Franke et al., 1976). The bristle coat on those vesicles was believed to correspond to a microfilamentous web lining the entire plasmalemma. The incorporation of coated caveolae into the surface was interpreted as a means of replacing the considerable amounts of membrane lost by those same cells in apocrine secretion of the fat globules of milk. In the light of more recent reports of bristle coats — presumably clathrin — on secretory caveolae of pituitary cells which do not undergo any significant loss of membrane in secretion and on the large pinocytosis vesicles of thyroid epithelium taking up thyroglobulin (Fujita, 1979), it seems likely that all of these examples represent stages in interiorization or retrieval of membrane for recycling instead of additions to the plasmalemma.

Micropinocytosis by coated vesicles is a widespread phenomenon, but it is especially well illustrated in the erythropoietic cell line. Indicated by arrows on the accompanying micrograph of an orthochromatic erythroblast are shallow depressions of the surface which stand out clearly because the cell membrane at these sites appears noticeably thicker than elsewhere. The nature of these local specializations of the cell membrane responsible for this appearance can be seen better at higher magnification in the micrographs that follow.

Figure 53. Late orthochromatic erythroblast (normoblast) from human bone marrow.

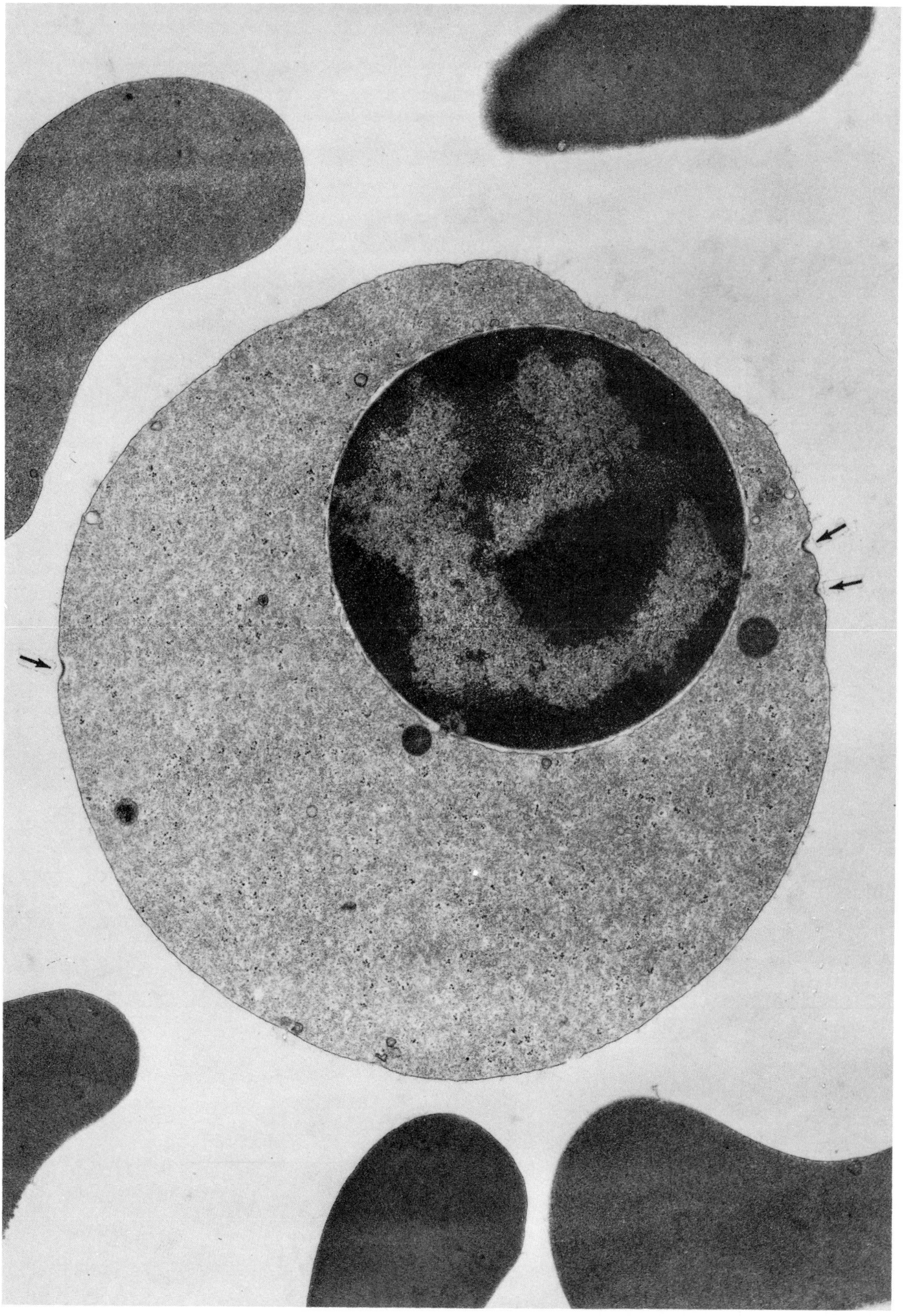

Figure 53

Whereas smooth-surfaced micropinocytosis vesicles are rarely seen in the process of invagination, intermediate stages in the formation of coated vesicles are common. Some of these are illustrated in the accompanying micrographs of erythroblasts. Small thickened areas of membrane acquire an inner coat and a fuzzy external layer to which particles of ferritin adhere (1). These first form shallow depressions (2) and then deeper pits (3). The neck of the pits narrows (4) and is ultimately severed by fusion of the membrane at its rim, releasing a spherical, coated vesicle into the cytoplasm and restoring the continuity and smooth contour of the plasmalemma. The ferritin adsorbed onto the lining of the vesicle is carried into the cell, where its iron is probably reutilized in the synthesis of hemoglobin. The binding of endogenous ferritin exclusively to specialized areas of the cell surface that are subsequently interiorized suggests that adsorptive micropinocytosis is a highly selective physiologically significant process in the erythropoietic cell line.

Figures 54, 55, and 56. Portions of the surface of polychromatophilic erythroblasts from guinea pig bone marrow. (From Fawcett, J. Histochem. Cytochem. 13:75–91, 1965.)

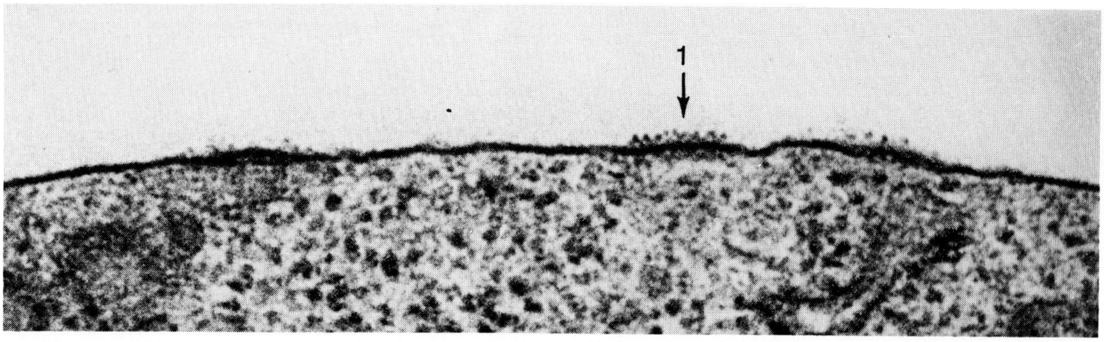

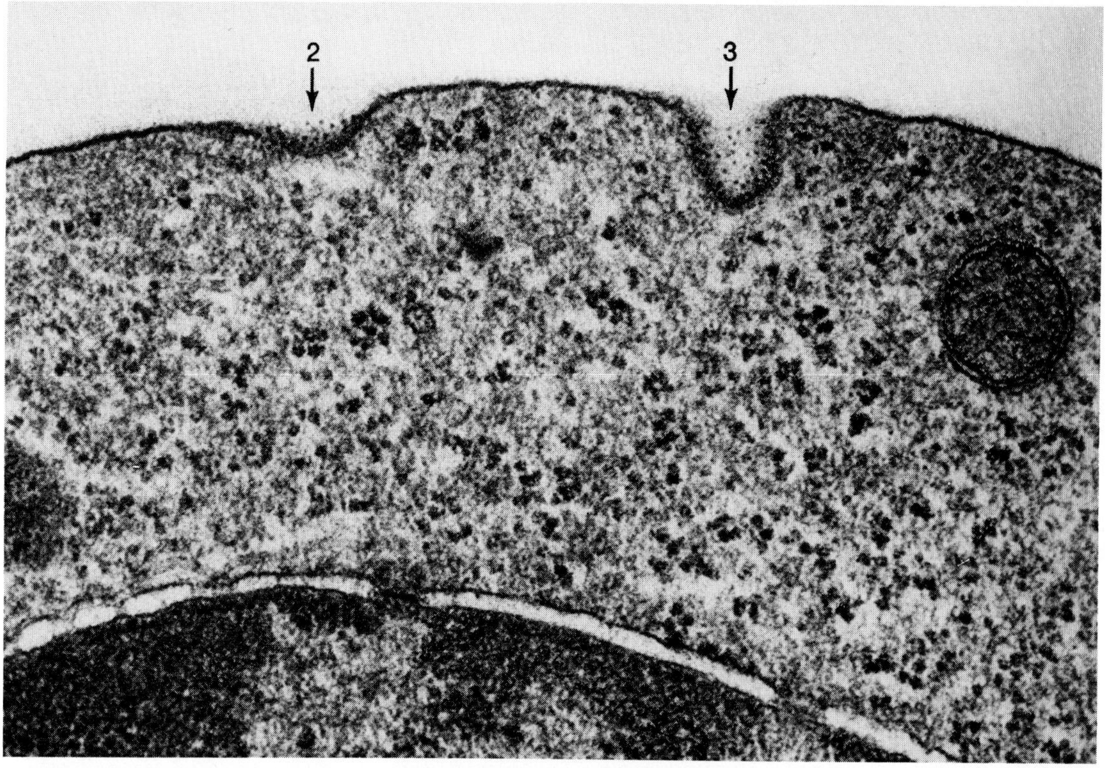

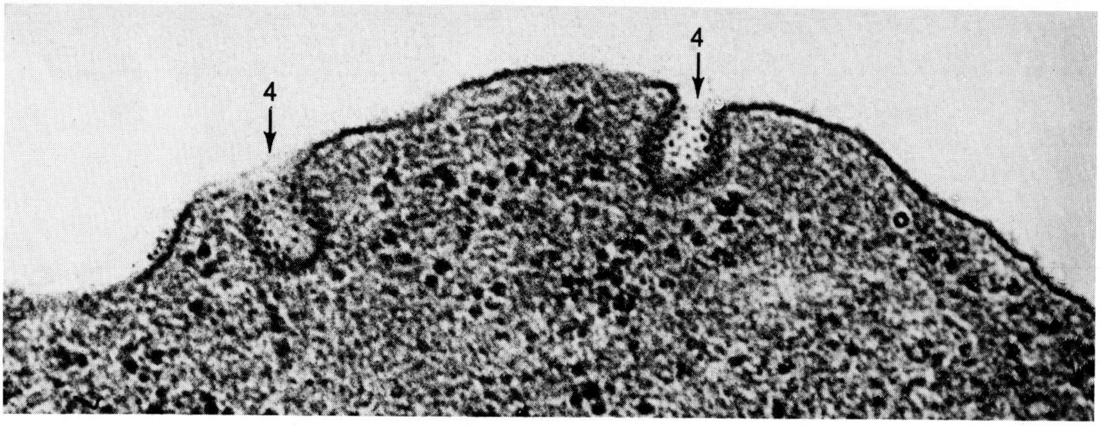

Figure 54, *upper* **Figure 55**, *center* **Figure 56**, *lower*

One of the more dramatic examples of selective uptake of material in coated vesicles is that found in the oocytes of insects during vitellogenesis. Yolk protein is synthesized in the midgut and transported in the hemolymph to the ovary, where it is adsorbed onto areas of the oocyte membrane having a prominent glycocalyx. These areas then invaginate to form coated vesicles. The process is similar to adsorptive micropinocytosis in the cells of vertebrates but is especially demonstrative because the yolk protein bound to the glycocalyx is osmiophilic and its density in electron micrographs makes it unusually conspicuous. The short spine-like projections on the cytoplasmic surface of the invaginations identifies them as typical coated vesicles (see at arrows).

An atypical feature of endocytosis in oocytes is the fact that the vesicles fuse not with lysosomes but with membrane-bounded structures that accumulate and store yolk proteins.

Figures 57 and 58. Areas at the periphery of oocytes from the cockroach *Periplaneta americana.* (Micrographs courtesy of Everett Anderson.)

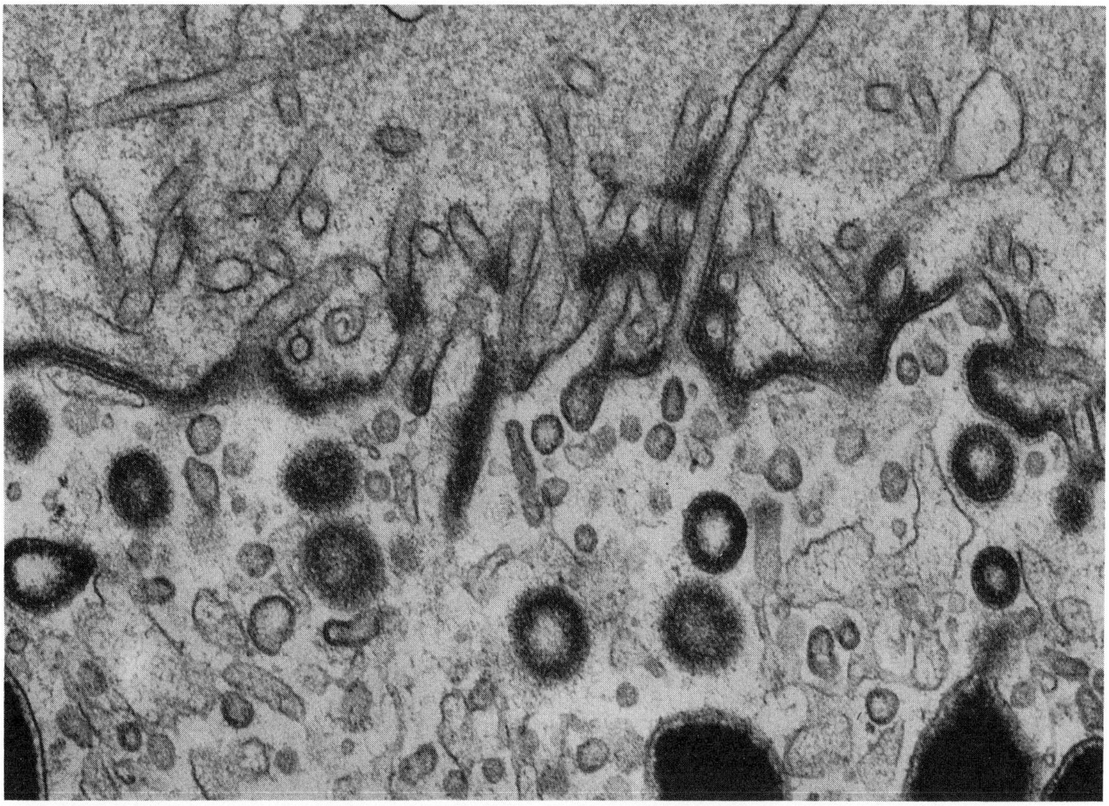

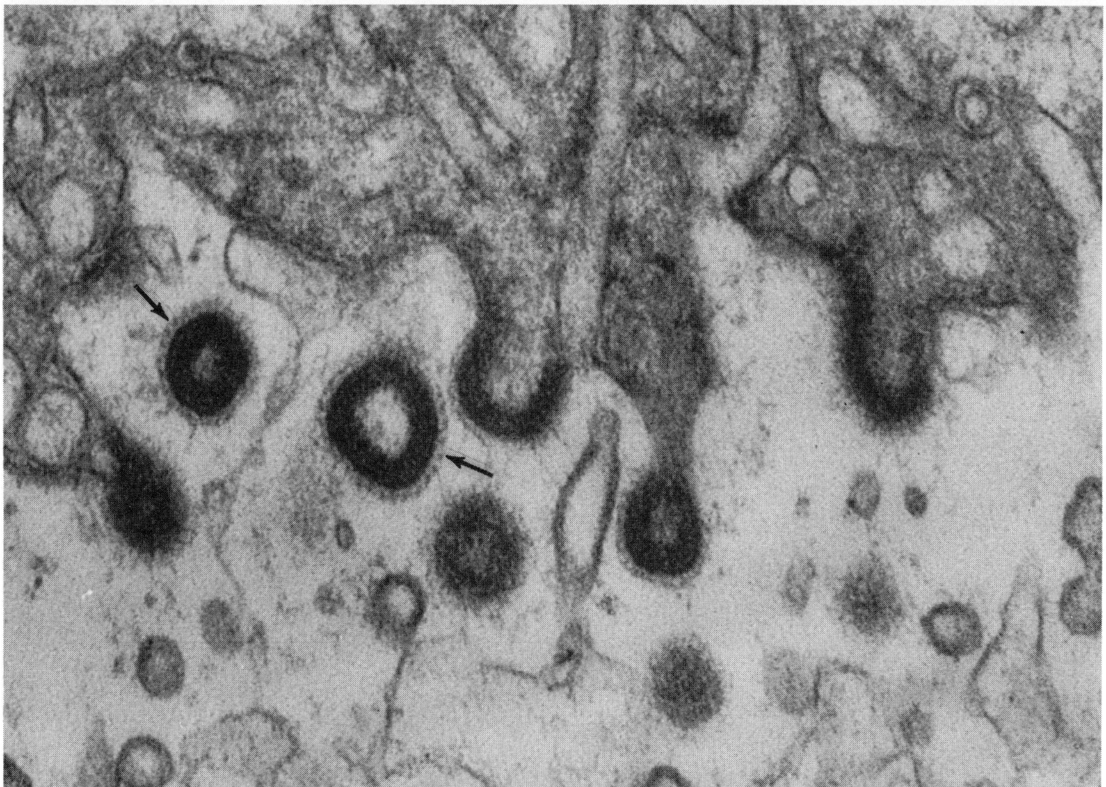

Figure 57, *upper* **Figure 58,** *lower*

The accompanying micrographs (Figs. 59*A*, *B*; 60*A*, *B*) are negatively stained preparations of coated vesicles, isolated from brain, illustrating the clathrin basket that encloses the membranous vesicle. The inset to 59*A* is a model depicting a lattice as composed of regular hexagons and pentagons comparable to those comprising the spherical surface of a soccer ball. Figure 60*C* depicts the clathrin basket of a coated vesicle negatively stained and *D* shows the same after image reinforcement by Markam rotation. A central hexagon is bounded by six polygons, of which three would be hexagons and three pentagons.

Figures 59 and **60.** Negatively stained coated vesicles from guinea pig brain (59*A* and *B*, × 300,000; 60*A* and *B*, × 400,000). (From Kanaseki and Kadota, J. Cell Biol. 42:202–220, 1969.) (Micrographs 60*C* and *D* courtesy of Thomas Roth.)

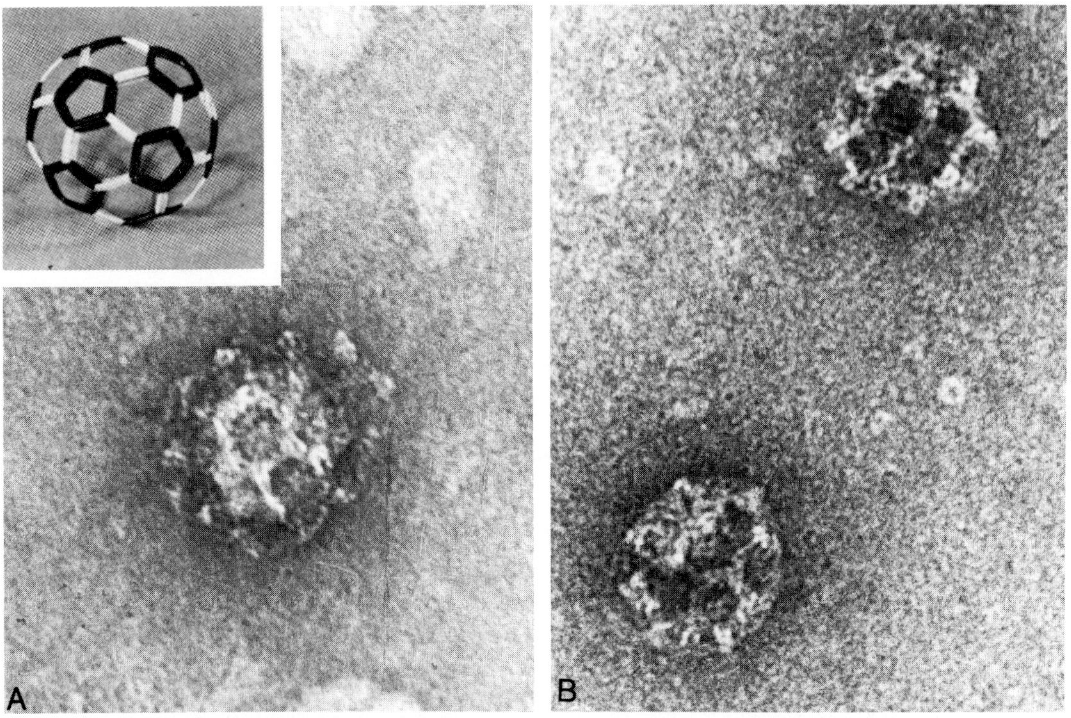

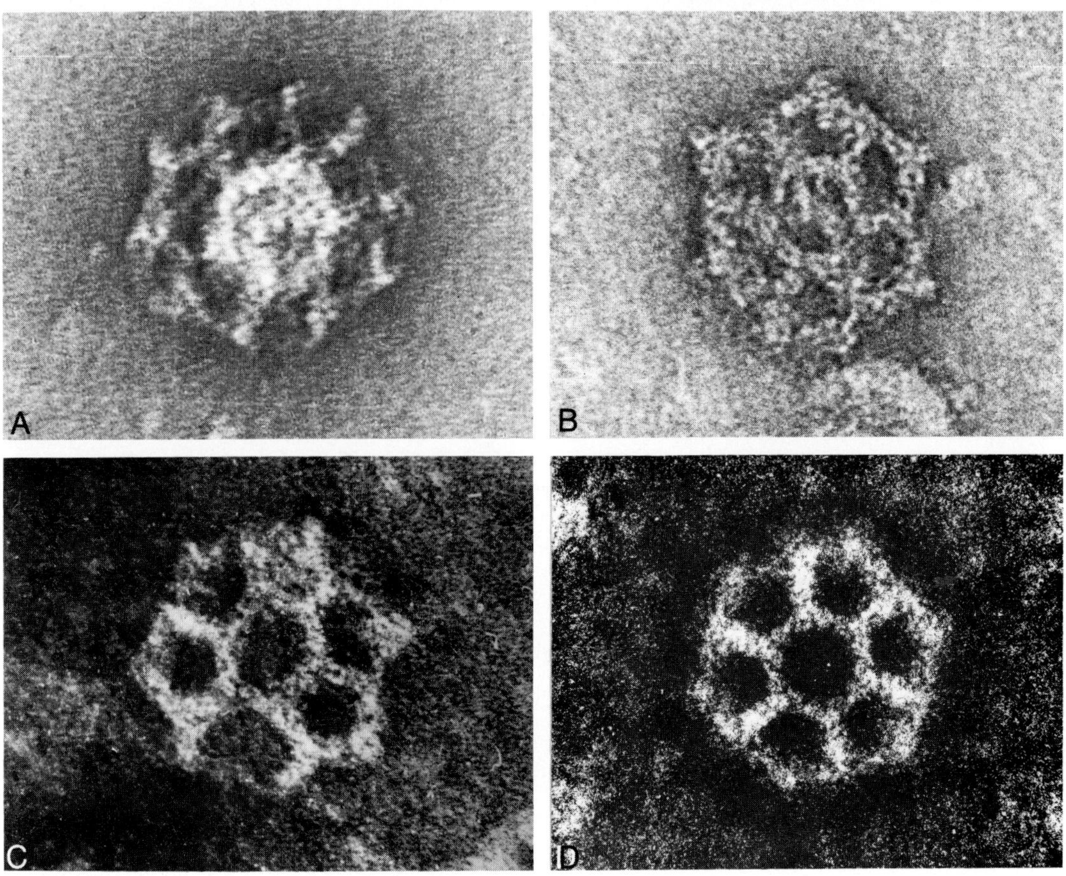

Figure 59, *upper, A, B* **Figure 60,** *lower, A-D*

REFERENCES

Specializations Involved in Endocytosis

Allison, A. C. and P. Davies. Mechanisms of endocytosis and exocytosis. Symp. Soc. Exp. Biol. 28:419–446, 1974. (Review)

Anderson, E. Oocyte differentiation and vitellogenesis in the roach, *Periplanata americana*. J. Cell Biol. 20:131–155, 1964.

Anderson, R. G. W., J. L. Goldstein and M. S. Brown. Localization of low density lipoprotein receptors on the plasma membrane of normal human fibroblasts and their absence in cells from a familial hypercholesterolemia homozygote. Proc. Nat. Acad. Sci. 73:2434–2438, 1976.

Anderson, R. G. W., M. S. Brown and J. L. Goldstein. Role of the coated endocytic vesicle in the uptake of receptor-bound low density lipoprotein in human fibroblasts. Cell 10:351–364, 1977.

Bennett, H. S. Morphological aspects of extracellular polysaccharides. J. Histochem. Cytochem. 11:2–13, 1963.

Bradshaw, R. A. Nerve growth factor. Ann. Rev. Biochem. 47:191–216, 1978.

Brandt, P. W. and G. D. Pappas. An electron microscopic study of pinocytosis in ameba. J. Biophys. Biochem. Cytol. 8:675–687, 1960.

Bruns, R. R. and G. E. Palade. Studies on blood capillaries. II. Transport of ferritin across the wall of muscle capillaries. J. Cell Biol. 37:277–299, 1968.

Butz, A. L., R. E. Fine and P. A. Tosselli. Evidence that coated vesicles isolated from brain are calcium-sequestering organelles resembling sarcoplasmic reticulum. J. Cell Biol. 75:135–147, 1977.

Fawcett, D. W. Local specialization of the plasmalemma in micropinocytosis vesicles of erythroblasts. Anat. Rec. 148:369, 1964.

Fawcett, D. W. Surface specializations of absorbing cells. J. Histochem. Cytochem. 13:75–90, 1965.

Franke, W. W., M. R. Luder, J. Kartenbeck, J. Zerbon and T. W. Keenan. Involvement of vesicle coat material in casein secretion and surface regeneration. J. Cell Biol. 69:174–195, 1976.

Friend, D. S. and M. G. Farquhar. Functions of coated vesicles during protein absorption in the rat vas deferens. J. Cell Biol. 36:357–376, 1967.

Goldstein, J. L., R. G. W. Anderson and M. S. Brown. Coated pits, coated vesicles, and receptor mediated endocytosis. A review. Nature 279:679–685, 1979.

Heuser, J. E. and T. S. Reese. Evidence for recycling of synaptic vesicle membrane during transmitter release at frog neuromuscular junctions. J. Cell Biol. 57:315–344, 1973.

Hirsch, J. G. Phagocytosis. Ann. Rev. Microbiol. 19:339–350, 1965. (Review)

Hubbard, A. L. and Z. A. Cohn. Externally disposed plasma membrane proteins. II. Metabolic fate of iodinated polypeptides of mouse L cells. J. Cell Biol. 64:461–479, 1975.

Kanaseki, T. and K. Kadota. The vesicle in a basket. A morphological study of coated vesicles isolated from the nerve endings of guinea pig brain. J. Cell Biol. 42:202–220, 1969.

Katz, B. and R. Miledi. Propagation of electric activity in motor nerve terminals. Proc. Roy. Soc. Lond. B. Biol. Sci. *161*:453, 1965.

Korn, E. D. Biochemistry of endocytosis. *In* MTP International Review of Science, Vol. 2 (C. F. Fox, ed.), pp. 1–26. Butterworths, London, 1975. (Review)

Lagunoff, D. and D. E. Curran. Role of bristle-coated membrane in uptake of ferritin by rat macrophages. Exp. Cell. Res. 75:337–346, 1972.

Lewis, W. H. Pinocytosis. Bull. J. Hopkins Hosp. 49:17, 1931.

Linden, C. D. and T. F. Roth. IgG receptors on foetal chick yolk sac. J. Cell Sci. 33:317–328, 1978.

Marsh, J. W., C. Hofmann and D. F. Steiner. Characterization of insulin interaction with a minimal deviation rat hepatoma. Fed. Proc. 38:301, 1979.

Pearse, B. M. F. Clathrin: A unique protein associated with intracellular transfer of membrane by coated vesicles. Proc. Nat. Acad. Sci. 73:1255–1259, 1976.

Roth, T. F. and K. R. Porter. Yolk protein uptake in the oocyte of the mosquito, *Aedes aegypti*. J. Cell Biol. 20:313–332, 1964.

Rodewald, J. Intestinal transport of antibodies in the newborn rat. J. Cell Biol. 58:189–211, 1973.

Roth, T. F., J. A. Cutting and S. B. Atlas. Protein transport: A selective membrane mechanism. J. Supramol. Struct. 4:527–548, 1976.

Schaeffer, S. F. and E. Raviola. Membrane recycling in the cone cell endings of turtle retina. J. Cell Biol. 79:802–825, 1978.

Silverstein, S. C., R. M. Steinman and Z. A. Cohn. Endocytosis. Ann. Rev. Biochem. 46:669–722, 1977.

Simionescu, N., M. Simionescu and G. E. Palade. Permeability of intestinal capillaries. J. Cell Biol. 52: 365–392, 1972.

Simionescu, N., M. Simionescu and G. E. Palade. Permeability of muscle capillaries to small heme peptides. Evidence for the existence of patent transendothelial channels. J. Cell Biol. 64:586–607, 1975.

Steinman, R. J., M. Silver and Z. A. Cohn. Pinocytosis in fibroblasts: Quantitative studies in vitro. J. Cell Biol. 63:949–969, 1974.

Steinman, R. J., S. E. Brodie and Z. A. Cohn. Membrane flow during pinocytosis. A stereologic analysis. J. Cell Biol. 68:665–687, 1976.

Sullivan, A. L., J. A. Grasso and L. R. Weintraub. Micropinocytosis of transferrin by developing red cells: an electron microscope study utilizing ferritin-conjugated transferrin and ferritin conjugated antibody to transferrin. Blood 47:133–143, 1976.

Woods, J. W., M. P. Woodward and T. F. Roth. Common features of coated vesicles from dissimilar tissues: composition and structure. J. Cell Sci. 80:87–99, 1978.

Woodward, M. P. and T. F. Roth. Coated vesicles: Characterization, selective dissociation and reassembly. Proc. Nat. Acad. Sci. 75:4394–4398, 1978.

JUNCTIONAL SPECIALIZATIONS

The need for cell-to-cell communication was recognized by Schwann (1839), who postulated protoplasmic connections between cells. Structures were soon observed in stratified squamous epithelia that were interpreted by some investigators as "intercellular bridges." But Bizzozero (1870), studying the stratum spinosum of epidermis, observed that processes projecting from the adjacent cells met end to end in small dense nodules. He concluded that there was no cytoplasmic continuity between the cells and that the dark nodules he observed in the so-called intercellular bridges were bipartite structures to which both cells contributed. This perceptive interpretation was accorded very limited acceptance owing in large measure to the prestige and influence of Ranvier, who held a contrary view. Impressed by the seemingly uninterrupted passage of tonofibrils throughout the epidermis, Ranvier (1879) rejected Bizzozero's interpretation and insisted upon cell-to-cell continuity. The *nodes of Bizzozero* were interpreted as elastic swellings along the course of the filaments as they passed from cell to cell surrounded by a thin continuous layer of protoplasm. For the next 40 years there was no serious challenge to Ranvier's interpretation. Kolossow (1893) developed a method of fixation that purported to demonstrate intercellular bridges in nearly all epithelia. In retrospect, it seems likely that this procedure resulted in shrinkage of cells so that they remained attached only at the nodes, creating a spurious appearance of intercellular bridges traversing the expanded intercellular spaces. However, the concept of intercellular bridges became firmly established in the histological literature (Carlier, 1895).

In the 1920s, Schaffer espoused Bizzozero's interpretation that the node in each "intercellular bridge" was a bipartite surface specialization of the adjacent cells serving an adhesive function. He proposed the term *desmosome* from the Greek words "desmos" (a ligament or bond) and "soma" (a body) — a connecting body. Despite Schaffer's careful studies, Ranvier's interpretation continued to dominate the teaching of histology until the first electron micrographs of amphibian epidermis by Porter (1954) resolved the controversy. Electron micrographs clearly demonstrated that there was no protoplasmic continuity through the "intercellular bridges." As postulated by Bizzozero, 80 years earlier, the dense nodes appeared to be local thickenings of the membranes at the ends of abutting cell processes.

As better methods of specimen preparation became available, analysis of the finer structure of desmosomes progressed rapidly. In all epithelia examined, they were found to consist of a pair of dense circular or elliptical plaques on the membranes of contiguous cells serving as devices for cell attachment and sites of binding of cytoplasmic tonofilaments to the cell surface. A densitometric traverse across a desmosome identified seven dark and light zones, the two attachment plaques, the two membranes, and an intermediate dense line bisecting the intercellular space. As resolution improved and the trilaminar unit membrane was identified, the number of resolvable layers increased to 11.

Other devices for maintaining the cohesion of epithelial cells had been described by light microscopists. von Recklinghausen (1862) developed a silver impregnation method that blackened the cell boundaries in squamous epithelia. The silver was believed to be deposited in an *intercellular cement* (*Kittsubstanz*) that bound the cells together. For the next hundred years, cohesion of epithelial cells was attributed primarily to the presence of a viscous intercellular substance, with desmosomes playing a secondary role. Bonnet (1895) and others using the iron hematoxylin stain observed dark dots on the boundary between cells near the free surface of the epithelium. In horizontal sections at this level, these structures appeared as dense bars or rods that intersected to form a polygonal network outlining the apices of the epithelial cells. These dense lines were called *terminal bars* (*Schlussleisten; bandes de fermeture*) and were interpreted as local thickenings of the intercellular cement which served to bind the cells together and to seal up the intercellular clefts, preventing material in the lumen from passing between cells (Zimmerman, 1911). Other investigators regarded them as band-like specializations of the cell surface comparable to desmosomes. In support of this view, they cited observation of tonofibrillae converging upon terminal bars in much the same manner as they did at desmosomes. As further evidence of their bipartite nature, it was noted that where cells were forced apart, half of the terminal bar went with each cell.

In early electron microscopic investigations of epithelia, a cytoplasmic density was observed adjacent to the membranes of neighboring cells in the region of the terminal bar. The membranes were separated by an empty-appearing intercellular cleft, only 15 to 20 nm wide. This provided little support for the concept of an intercellular cement and favored the view that the juxtalumenal terminal bars were attachment devices comparable to desmosomes but in the form of circumferential bands instead of localized plaques (Fawcett, 1958). The juxtalumenal specializations for attachment were studied in detail in several epithelia by Farquhar and Palade (1963), who defined a "junctional complex" consisting of three components. Closest to the lumen was a *zonula occludens* (tight junction) characterized by fusion of opposing cell membranes over a variable distance and resulting in obliteration of the intercellular space. Within this zone, the membranes converged one or more times and their outer leaflets fused, forming short pentalaminar segments. On the cell boundary below this junction, they described a *zonula adhaerens* (intermediate junction) where the membranes coursed parallel for 0.2 to 0.5 μm at a distance of ~ 20 nm, with a dense band of dense fibrillar material associated with the cytoplasmic surfaces of the membranes. The zonula occludens and zonula adhaerens were both described as circumferential, belt-like specializations of the cell surface. The third component of Farquhar and Palade's junctional complex was called the *macula adhaerens,* a descriptive term offered as a synonym for the classical "desmosome." These dense plaques were believed to be spaced at more or less regular intervals in a circumferential row parallel to the zonula adherens. Elsewhere on the adjoining cell surfaces, similar maculae were distributed at random.

The new terminology of Farquhar and Palade focused attention upon important local differences in surface relationships within the region of the terminal bar — differences of which contemporary investigators had been vaguely aware but had not clearly articulated. Most significant was their finding that the outermost component of the complex was not merely an attachment device but also achieved an obliteration of the intercellular space and hence constituted an effective permeability barrier that excluded material which might otherwise pass through the epithelium between cells. Although this important concept seemed novel at the time, it actually represented an ultrastructural validation of the classical interpretation of terminal bar function that was implicit in the early terms *bandes de fermeture* and *bandellettes obturantes.* What was distinctive in these early electron microscopic studies was the clear demonstration that the structure responsible for the intense staining of the terminal bar region was a local specialization of the membranes and superficial cytoplasm and not an intercellular component. The efficacy of the occluding junctions as a barrier to diffusion was soon demonstrated by use of electron opaque probes of the extracellular space (Miller, 1960; Graham and Karnovsky, 1965). Hemoglobin or horseradish peroxidase introduced either into the lumen or into the subepithelial connective tissue entered the intercellular clefts but was invariably barred from further penetration by the juxtalumenal occluding junctions.

Karrer (1960) had observed in stratified squamous epithelium of the cervix, some distance from its free surface, small areas of close membrane apposition and apparent fusion. In sections these appeared as three dense lines separated by two intervening layers of lower density and were therefore called *quintuple-layered cell interconnections*. Similar junctions were found on cells of cardiac muscle and since these were the only sites where the muscle cells came into contact without an intervening extracellular layer, Karrer and Cox (1960) speculated that the quintuple-layered areas might be of significance in conduction of excitation throughout the myocardium. This prophetic suggestion attracted little attention at the time, as similar contacts were reported in a number of nonexcitable tissues. They were found on the boundary between glial cells in the central nervous system (Peters, 1962), between endothelial cells of capillaries (Muir and Peters, 1962), and between cultured fibroblasts (Davis and James, 1962) and described as *pentalaminar or quintuple-layered membrane junctions*. Dewey and Barr (1962) reported similar sites of close membrane contact on the cell boundaries of smooth muscle and proposed for them the descriptive term *nexus*. These investigators also speculated that these contacts might permit electrolytes and metabolites to move from cell to cell and might therefore be the structural basis for spread of current through excitable tissues.

For several years following the 1963 publication of Farquhar and Palade's influential paper, there was a widespread tendency to equate all junctional specializations of cells to one or another of the three categories in their junctional complex (zonula occludens, zonula adherens, macula adherens). This led to considerable terminological and conceptual confusion. Specializations resembling the zonula adherens in fine structure were identified in the intercalated discs of cardiac muscle, but were plaques of varying size instead of circumferential bands. Since the term *macula adherens* had been preempted for the desmosome, *fascia adherens* was suggested for these, although etymologically inappropriate. The term *tight junction* came to be broadly applied to regions of apparent membrane fusion, whether they occurred in the form of circumferential belts around the cell apex or as isolated plaques elsewhere on the cell boundaries. Since isolated areas of membrane fusion obviously could not be effective in sealing off the intercellular spaces of the epithelium, some investigators sought to avoid the misleading implication of the term "tight junction" by use of "close junction" or by a return to "pentalaminar junction" or "nexus." Uncertainty prevailed as to the function of this type of junctional specialization where it occurred in configurations which could not serve as a barrier to diffusion through the intercellular spaces.

Electrical coupling of smooth muscle cells had been reported by physiologists some 15 years earlier (Bozler, 1948; Proiser et al., 1955). Furshpan and Potter had demonstrated in 1957 that injection of a depolarizing current into the presynaptic element of the giant motor neuron of the crayfish resulted in a rapid depolarization of the postsynaptic element, and they concluded that this could best be explained by electrical transmission across the synaptic membranes. Confirmatory observations of electrical transmission in the invertebrate nervous systems rapidly followed and the concept of electrical coupling of excitable tissues had gained wide acceptance before its morphological basis was established.

Robertson (1963), studying the nerve endings on the Mauthner cells of fish, with potassium permanganate fixation, observed "synaptic discs" where the unit membranes were apparently in contact, forming a dense lamina which showed periodic densities 8.5 nm apart. In oblique or tangential sections, the synaptic discs exhibited a regular hexagonal pattern of subunits. Robertson entertained the possibility that this was either an artifact of permanganate fixation or an example of a hitherto overlooked subunit structure of cell membranes in general. A few years later Revel and Karnovsky (1967), using lanthanum nitrate as an electron opaque probe of the extracellular space, demonstrated a 2 nm gap between the apposed membranes in the "tight junctions" of liver and cardiac muscle. In tangential sections, the lanthanum clearly outlined hexagonally packed subunits with a diameter of about 7 nm and a center-to-center spacing of 9 nm. The lanthanum tracer also penetrated the core of each subunit, producing an electron opaque dot in its center. They recognized that this pattern was similar to that described earlier by Robertson in electrical synapses and suggested that such junctions might prove to be characteristic of all electrically coupled cells. This work clearly established that the widely used term "tight junction" had previously been applied to two structurally and functionally distinct specializations: (1) the zonula occludens, in which the intercellular cleft is obliterated and impenetrable to lanthanum, and (2) a type of junction into which lanthanum penetrates and outlines a regular array of hexagonally packed subunits that traverse a 2 nm gap between the apposed unit membranes. The term *gap junction* came into common use, replacing the "pentalaminar junction" and "nexus" of earlier investigators.

The uncertainty that formerly prevailed as to the number of types of junctional specialization and their functional significance has now been largely eliminated. Four principal types are recognized in vertebrates. The *occluding junction* (*zonula occludens, tight junction*) prevents small molecules from passing through the intercellular spaces of epithelia and thus enables the organism to maintain an internal environment that is chemically distinct from its surroundings. The *adhering junctions* (*zonula adherens* and *desmosomes*) provide sites for attachment of fibrillar cytoskeletal and contractile elements onto the cell membrane and maintain cell cohesion. The *gap junction (communicating junction)* permits passage of ions and small molecules from cell to cell and thus functions in coordinating the activities of groups of cells.

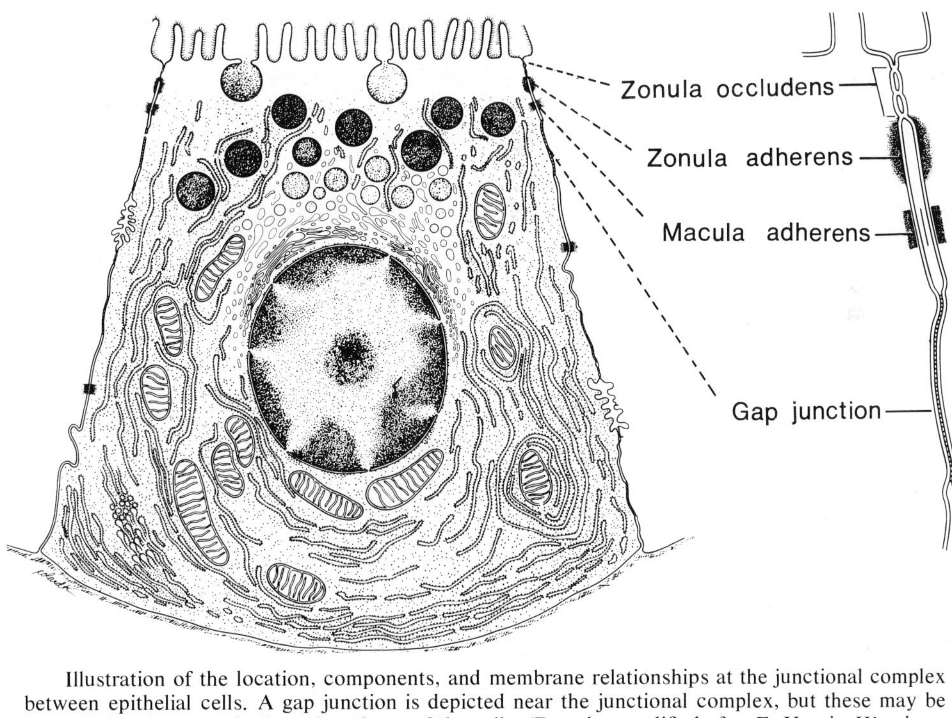

Illustration of the location, components, and membrane relationships at the junctional complex between epithelial cells. A gap junction is depicted near the junctional complex, but these may be located elsewhere on the lateral surfaces of the cells. (Drawing modified after E. Hay in *Histology*, R. O. Greep, ed., second edition, McGraw-Hill Book Company, 1965.)

TIGHT JUNCTION (ZONULA OCCLUDENS)

This type of junction, found on the boundary between epithelial cells near the luminal surface, forms a continuous belt-like region of intimate membrane contact encircling the apex of the cells. In thin sections, it appears as a series of punctate contacts where the dense outer leaflets of adjoining membranes converge and fuse to form a single line, or not infrequently the outer dense line of the unit membranes is interrupted at these sites of membrane fusion. The width of the fused membranes is slightly less than the combined width of the two unit membranes involved. Although the tight junction is commonly described as a belt-like zone of fusion, actually the membranes are held together only along a series of lines of attachment approximately parallel to the apical cell surface.

In freeze-fracture replicas of epithelial cells, the junction region appears as a band-like meshwork of branching and anastomosing thin ridges on the P-face and a corresponding pattern of grooves on the E-face (Kreutziger, 1968; Staehelin et al., 1969; Chalcroft and Bullivant, 1970). At high magnification, the ridges may appear as a row of particles, but after glutaraldehyde fixation, they are usually cross-linked to form continuous strands or fibrils. These linear structures seen on the P-face of freeze-fracture replicas correspond to the punctate sites of membrane fusion seen in thin sections. Between the lines of fusion, the membranes diverge and converge again, resulting in bow-shaped profiles in sections and shallow "hills" and "valleys" on the P-fracture face. The ridges are on top of the hills on the P-face and the grooves are in the valleys on the E-face (Chalcroft and Bullivant, 1970).

The physiological significance of tight junctions resides in the fact that they are barriers to paracellular diffusion across the epithelium. The most direct evidence for their barrier function is the demonstration that high molecular weight, electron opaque tracers penetrating the intercellular clefts either from the base or the lumen are stopped at the level of the focal membrane fusions seen in thin sections (Farquhar and Palade, 1963; Reese and Karnovsky, 1967; Goodenough and Revel, 1970). Changes in permeability experimentally induced by exposure to a high osmotic gradient from the luminal side results in distention of the compartments within the junction and disruption of the intramembrane strands or fibrils. Further indirect evidence that the fibrils are involved in the barrier function comes from the observation that there is a correlation between the number of parallel rows of intramembrane strands and the degree of impermeability of the epithelium (Claude and Goodenough, 1973; Humbert et al., 1975).

Interpretations differ as to the relationship of the ridges seen on the P-face to the two membranes involved in the junction. Some investigators propose that the network of fibrils in one membrane is in register with that in the other. The fibrils are considered to be more strongly bound to the cytoplasmic halves of their membranes than to their counterparts in the adjacent membrane. The fracture is thus assumed to go around the fibril in one membrane leaving it as a ridge on the P-face, but not to make an excursion out of the membrane and around the in-register fibril in the opposing membrane (Chalcroft and Bullivant, 1970).

Other investigators assume that each of the interconnected lines of attachment in tight junctions consists of two adhering rows of adhesion particles or continuous strands, one in each membrane. The particles which are cross-linked by glutaraldehyde to form fibrils or strands are believed to bridge the width of the adjoining membranes and to be strongly bonded together in the intercellular space, like a zipper. In this model, the fracture plane is assumed to make an excursion into the adjoining membrane and around both of the apposed fibrils, leaving a prominent ridge on the P-face (Staehelin, 1973, 1974).

Still others suggest that there is a single set of fibrils which is shared by the adjacent membranes instead of two in-register fibrils, one for each membrane. This model also assumes that the fracture makes an excursion around fibrils in adjoining membranes (Wade and Karnovsky, 1974).

Careful study of complementary replicas at the transitions from one fracture face to the other has led to proposal of an offset, two-fibril model of the tight junction. In this two fibrils or particle rows, one in each membrane, form an offset pair and the seal results from their side-to-side contact (Bullivant, 1976, 1978). A choice among these alternative interpretations is difficult, but the bulk of the evidence at present seems to favor one of the two-fibril models over the shared fibril model.

ADHERING JUNCTION (ZONULA ADHERENS)

The zonula adherens is primarily involved in maintenance of cell cohesion — a role that it shares with the desmosomes. It is a circumferential belt linking adjacent epithelial cells just below the zonula occludens. Because of its similarity in function, some authors refer to it as a belt desmosome. It is most prominent in cells with a well-developed brush border. Below the microvilli of the border, the apical cytoplasm is traversed at the level of the zonula adherens by a mat of interwoven filaments called the *terminal web*. At its periphery, the transverse filaments of the web mingle with the circumferential filaments that are responsible for the local density of the cytoplasm on the inner aspect of the zonula adherens. The filaments in the cores of the microvilli of the brush border extend downward into the apical cytoplasm and mingle with the transverse filaments of the terminal web. Among these 7 to 10 nm filaments, the majority are actin. Myosin can also be localized in this region immunocytochemically but it is not preserved as identifiable filaments in electron micrographs. It has been shown, however, that isolated brush borders and their subjacent cytoplasm containing the terminal web contract upon addition of calcium and ATP, presumably as a consequence of actin-myosin interaction (Mooseker and Tilney, 1972). Thus in such cells the terminal web is involved in movements of the brush border, and the zonula adherens constitutes the peripheral anchorage of the web as well as a band of attachment to the neighboring epithelial cells. The terminal web is present but less well developed in epithelia lacking a brush border.

The components of the epithelial junctional complex in apical-based sequence are the *zonula occludens* (tight junction), the *zonula adherens* (intermediate junction), and the *macula adherens* (desmosome).

The zonula occludens is a belt-like specialization in which the adjacent unit membranes converge and fuse, obliterating the intercellular space over variable distances. The membranes commonly fuse at multiple levels within the zonula to form local pentalaminar regions separated by short segments in which the membranes are closely apposed but not fused.

At the zonula adherens, the membranes course parallel over a distance of 0.2 to 0.5 μm and are separated by a 25 nm intercellular space. The cytoplasm immediately subjacent to the membranes is relatively dense and is the site of insertion of the transversely oriented filaments comprising the so-called *terminal web*.

The desmosome or macula adherens is the third component of the complex. It consists of parallel segments of the opposing membranes reinforced by dense plaques and separated by an interspace of about 25 nm, which is often bisected by a thin dense line. Bundles of *tonofilaments* converge upon and appear to terminate in the dense plaques associated with the cytoplasmic surface of the opposing membranes.

Desmosomes associated with the junctional complex are deployed in a circumferential row below and parallel to the zonula adherens. Since the desmosomes are plaques spaced at intervals instead of continuous bands, some planes of section will pass between them. In the resulting micrographs, the junctional complex may therefore appear to consist only of the two zonulae.

Figure 61. Junctional complex of the intestinal epithelium of rat. (From Farquhar and Palade, J. Cell Biol. 17:375–412, 1963.)

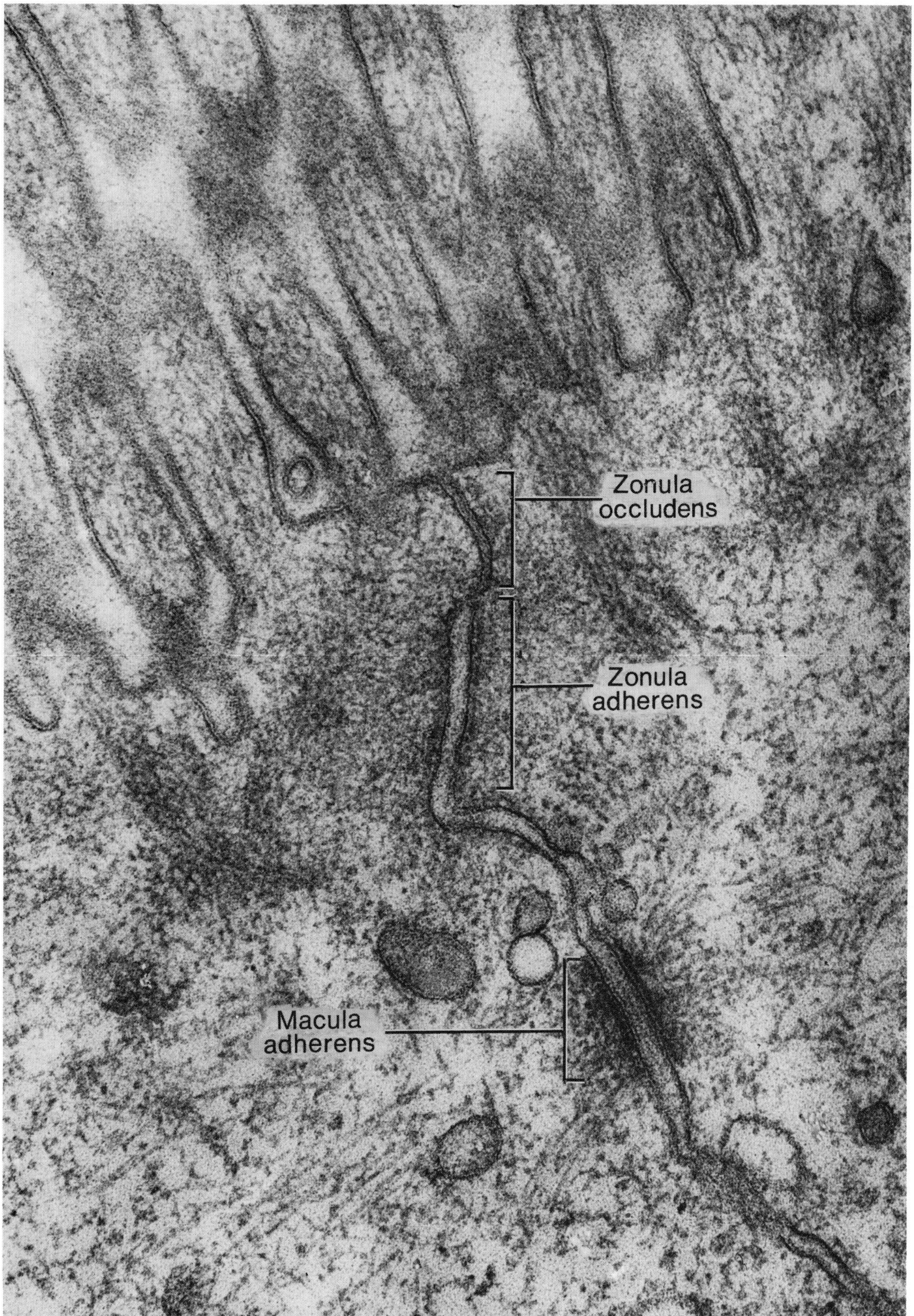

Figure 61

When the junctional region of epithelial cells is examined in freeze-fracture preparations, the zonula occludens appears as a meshwork of ridges on the P-face of the membrane or a corresponding pattern of grooves on the E-face. The ridges seen in replicas are variously described as rods or strands and are interpreted as linear arrays of integral protein particles within the membranes. Some of the linear elements forming these patterns in the opposing membranes are in register and are firmly bonded to one another. These adherent bridging elements correspond to the sites of membrane fusion seen in thin sections of the zonula occludens.

The physiological significance of this junctional specialization resides in the fact that it seals the intercellular cleft and constitutes a permeability barrier limiting diffusion through the epithelium via a paracellular route. It has not been established that the honeycomb patterns in the adjoining cells are identical. It is likely that only a portion of the elements in the opposing membranes are in register and are attached. In general, however, there appears to be a correlation between the number of rows of ridges seen in freeze-fracture replicas of the junction and the tightness of the permeability barrier as measured by transepithelial electrical resistance.

In the upper micrograph on the facing page, the intersection of short strands with long sinuous strands results in highly irregular shapes of the areas enclosed in the reticulum. In the lower figure, the zonula appears as a regular network of similarly shaped polygons. In some instances, of which this is an example, the ridges or rods adhere to the E-face, leaving grooves on the P-face.

Figure 62. Replica of an occluding junction from the large intestine of a tadpole.

Figure 63. Replica of an occluding junction from intestine of a postmetamorphic toad. (Both figures from B. Hull and A. Staehelin, J. Cell Biol. 68:688–704, 1976.)

Figure 62, *upper* **Figure 63,** *lower*

The geometrical patterns formed by the ridges and grooves in replicas of the zonula occludens vary considerably from epithelium to epithelium. In the upper figure, the ridges intersect to form a network of nearly equilateral polygons which exhibit no predominent orientation relative to the free surface of the epithelium. In the center micrograph, the grooves on the E-face have a gently curving course and are intersected at long intervals by connecting grooves, thus delimiting polygons that are elongated in a direction parallel or slightly oblique to the apical cell surface. In the lower figure, the P-face ridges are predominantly parallel to each other and to the lumenal surface and show few intersections.

The functional significance of these geometric variations is not entirely clear. They may be due in part to intraepithelial stresses exerted on the membrane during junction formation. It seems likely that those patterns with a predominance of intramembrane strands oriented transverse to the axis of the columnar cell constitute more effective permeability barriers than those with many elements oriented parallel to the cell axis.

Figure 64. Zonula occludens from small intestine of a *Xenopus laevis* tadpole. (From Hull and Staehelin, J. Cell. Biol. 68:688–704, 1976.)

Figure 65. E-face of zonula from epithelium of the large intestine of an adult toad *Xenopus laevis.* (From Hull and Staehelin, J. Cell Biol. 68:688–704, 1976.)

Figure 66. P-face of junction from epididymal epithelium of mouse, *Mus musculus.* (Courtesy of Fumi Suzuki and Toichiro Nagano.)

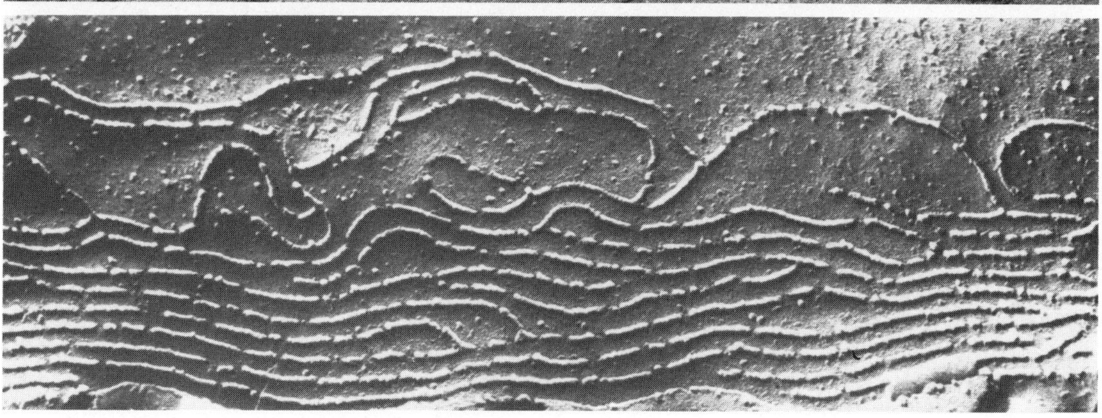

Figure 64, *upper* **Figure 65**, *center* **Figure 66**, *lower*

SERTOLI CELL JUNCTIONS

An exception to the general occurrence of occluding junctions near the free surface is found in the seminiferous epithelium of mammals. There, a variant of the zonula occludens is found on the contact surface of adjacent Sertoli cells in the lower third of the epithelium.

Unique features of the zonulae shown in the diagram below are (1) the unusual width of the junctional specialization which includes numerous sites of membrane fusion, (2) the occurrence of circumferential bundles of actin filaments adjacent to the membranes, and (3) the presence of a cistern of the endoplasmic reticulum in each cell coursing parallel to the junctional membrane deep to the filaments.

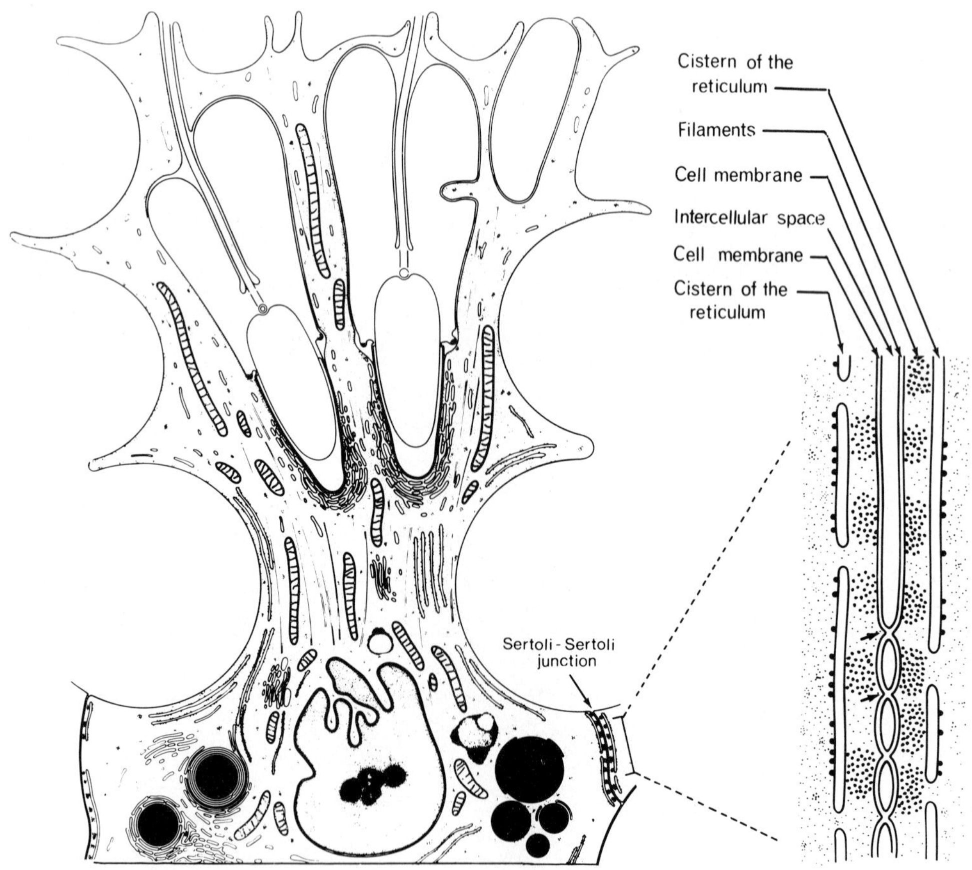

Diagram of a Sertoli cell from mammalian testis showing the occluding junction near the base of the epithelium. The components of the junction are illustrated at the right. Bundles of filaments and cisternae of the endoplasmic reticulum are consistently associated with the junctional region and are features peculiar to Sertoli cell junctions. (From Fawcett in *Handbook of Physiology,* Section 7, Endocrinology, Vol. V, pp. 21–55, American Physiological Society, Washington, D.C., 1975.)

The occluding Sertoli junctions form a permeability barrier that divides the epithelium into two compartments — a *basal* or *peripheral compartment*, containing the spermatogonia and preleptotene spermatocytes, and a *central* or *adlumenal* compartment that contains the meiotic and postmeiotic stages of germ cell development.

The developing mammalian germ cells occupy expanded intercellular spaces in a columnar epithelium of Sertoli cells. The diagram below shows how occluding junctions between adjacent Sertoli cells create two distinct compartments. All but a few late spermatids have been omitted to show more clearly the location of the junctions and the labyrinthine system of intercellular niches in which the developing germ cells reside. The existence of the intraepithelial permeability barrier enables the supporting cells that form the walls of the adlumenal compartment to maintain in it a special microenvironment favorable to the differentiation of the more advanced germ cells. The early stages of germ cell development in the peripheral compartment are outside of the barrier and exposed to the general extracellular fluid environment of the testis.

The blood-seminiferous tubule barrier also serves to impound specific germ cell antigens in the adlumenal compartment of the epithelium and in the lumen of the tubules, preventing their access to the bloodstream where they would induce an autoimmune aspermatogenesis.

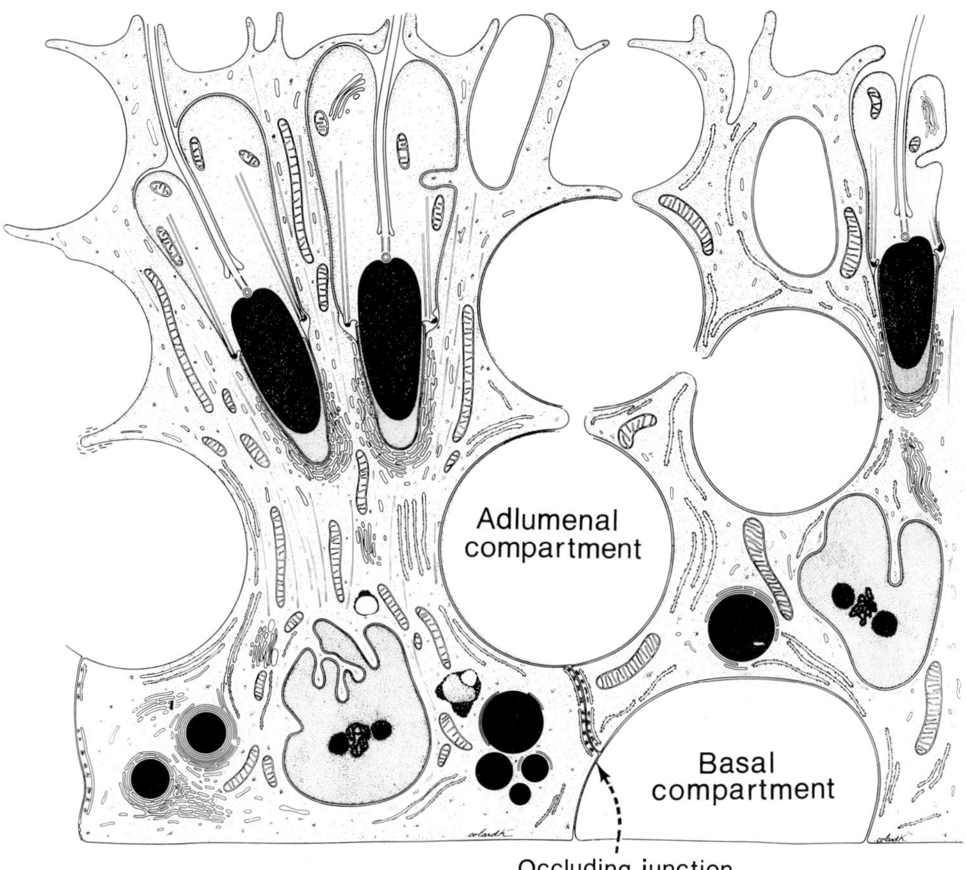

Occluding junctions between lateral processes of adjacent Sertoli cells form a permeability barrier that separates the intercellular spaces of the epithelium into a basal (peripheral) compartment and an adlumenal (central) compartment. Isolation of postmeiotic germ cells in the central compartment enables the Sertoli cells forming its walls to create a special microenvironment favorable for germ cell differentiation. (From Fawcett in *Handbook of Physiology*, Section 7, Endocrinology, Vol. V, pp. 21–55, American Physiological Society, Washington, D.C., 1975.)

The accompanying micrographs offer two examples of Sertoli cell occluding junctions, illustrating the narrowing of the intercellular space in the junctional region and parallel cisternae of the endoplasmic reticulum separated from the apposed cell membranes by a zone of cytoplasm rich in 6 to 7 nm filaments. Occasional ribosomes are associated with the cytoplasmic aspect of the cistern but are never found on the membrane facing the filaments. Sites of membrane fusion are not seen in these junctions, owing to the plane of the section, but will be shown in subsequent micrographs.

Figures 67 and **68.** Sertoli-Sertoli junctions from guinea pig testis.

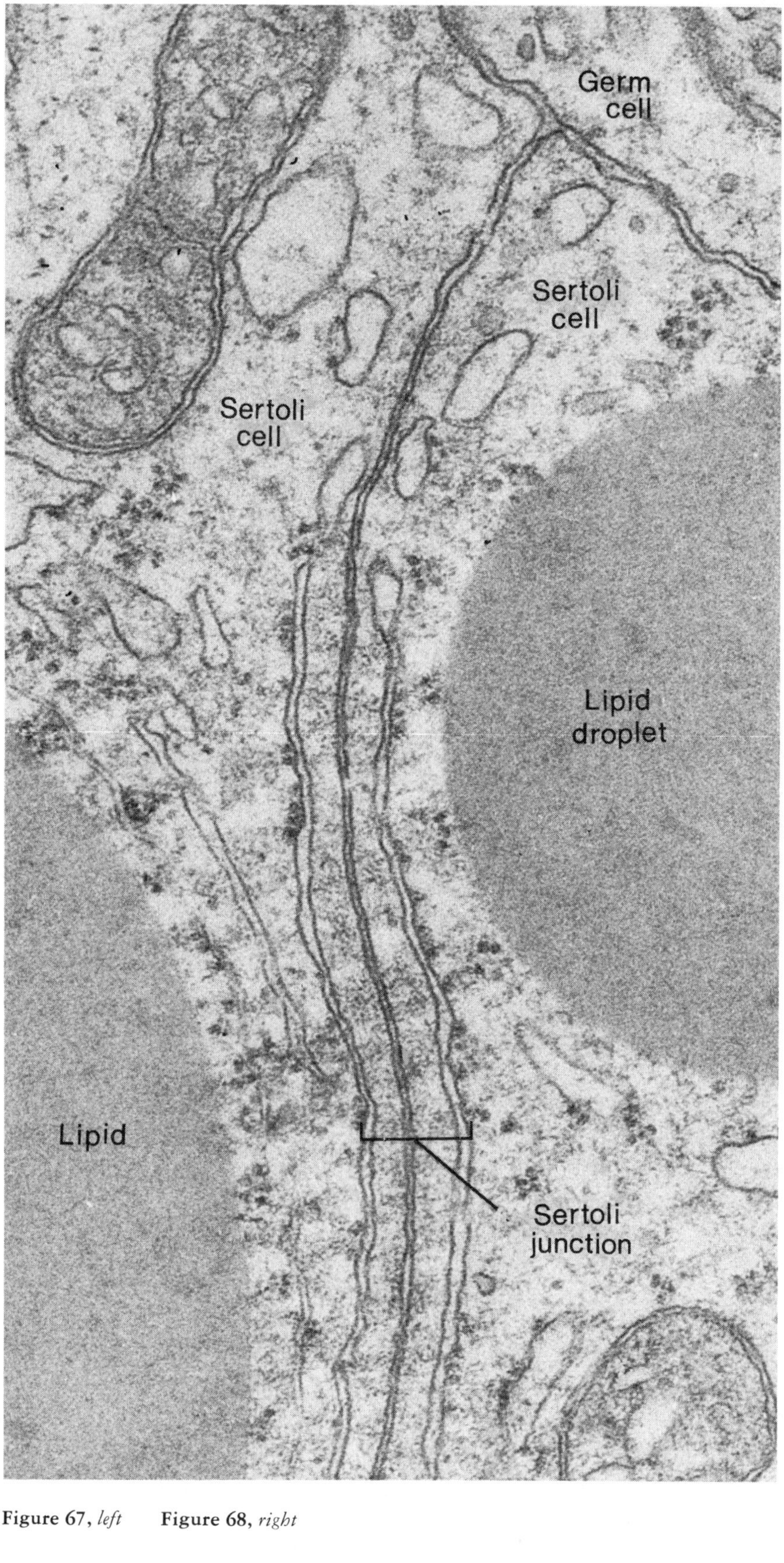

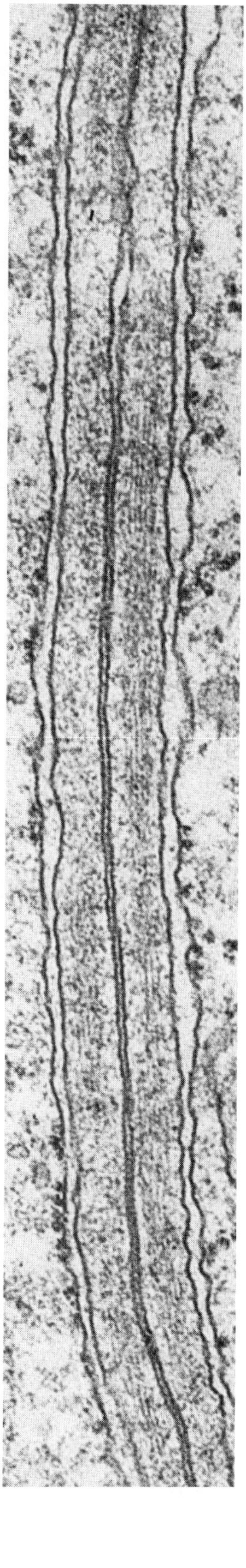

Figure 67, *left* **Figure 68**, *right*

On the facing page are four examples of Sertoli junctions in thin section showing multiple focal sites of membrane fusion (at arrows), associated bundles of filaments (in *A*, *B*, and *D*), and typical subsurface cisternae.

There has been some disagreement as to whether the rods or strands seen in freeze-fracture replicas of occluding junctions are a single set, bridging the gap and shared by the opposing membranes, or whether separate strands in the two membranes are in register and bonded to one another at the sites of membrane fusion. These alternative interpretations are depicted in the insets. Upon close examination, the micrograph (*C*) shows negative images of two distinct intramembrane elements (at asterisks), thus favoring alternative *A* in the inset.

Figure 69. *A* to *D*, occluding Sertoli junctions from ram and rat testis. (*B* and *C* from Gilula, Fawcett and Aoki, Dev. Biol. 50:142–168, 1976. Inset from Wade and Karnovsky, J. Cell Biol. 60:168–191, 1974. *D* from Dym and Fawcett, Biol. Reprod. 3:308–326, 1970.)

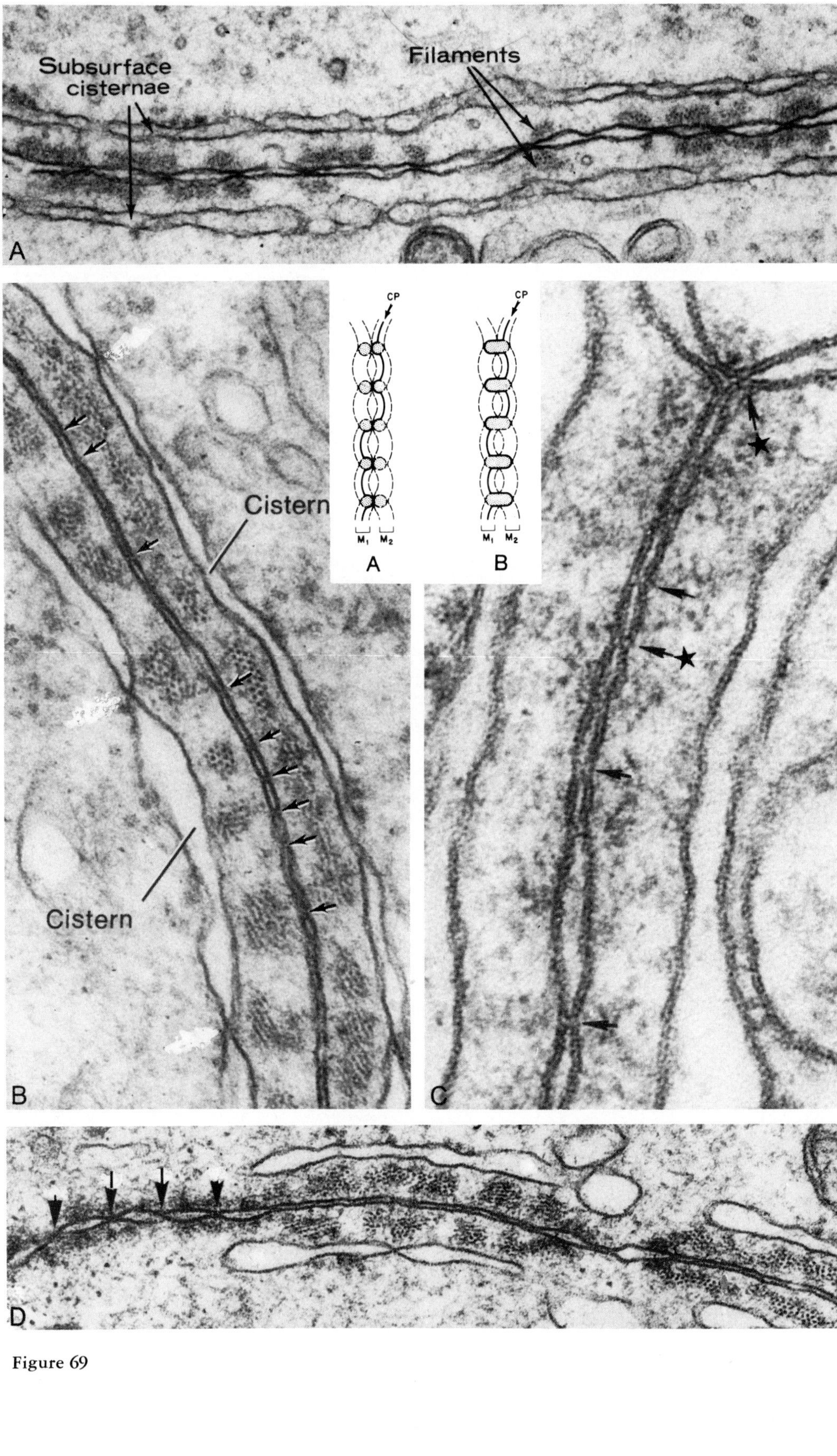

Subsurface cisternae
Filaments
A
CP
M1
M2
A
B
Cistern
Cistern
B
C
D

Figure 69

In freeze-fracture preparations, the Sertoli cell junctions differ from the juxtalumenal occluding junctions of other epithelia in several respects. The upper figure on the facing page shows a typical zonula occludens from intestinal epithelium with a network of intersecting strands on the P-face and a complementary pattern of grooves on the E-face. This can be compared with the lower figure, which includes a portion of a Sertoli junction from human testis consisting of a large number of parallel rows of intramembrane particles. The particles adhere preferentially to the E-face and the complementary grooves course along the tops of low ridges on the P-face. The most remarkable feature of these junctions is the very large number of particle rows. Since the tightness or leakiness of epithelial permeability barriers is roughly correlated with the width of the zonulae, the Sertoli junctions would be expected to constitute a very tight barrier. Experimental evidence seems to bear out this prediction.

Figure 70. Zonula occludens of rat intestinal epithelium. (Micrograph courtesy of Jean Paul Revel.)

Figure 71. Sertoli junction of human seminiferous epithelium. (From F. Suzuki and T. Nagano, Cell Tissue Res. 166:37, 1976.)

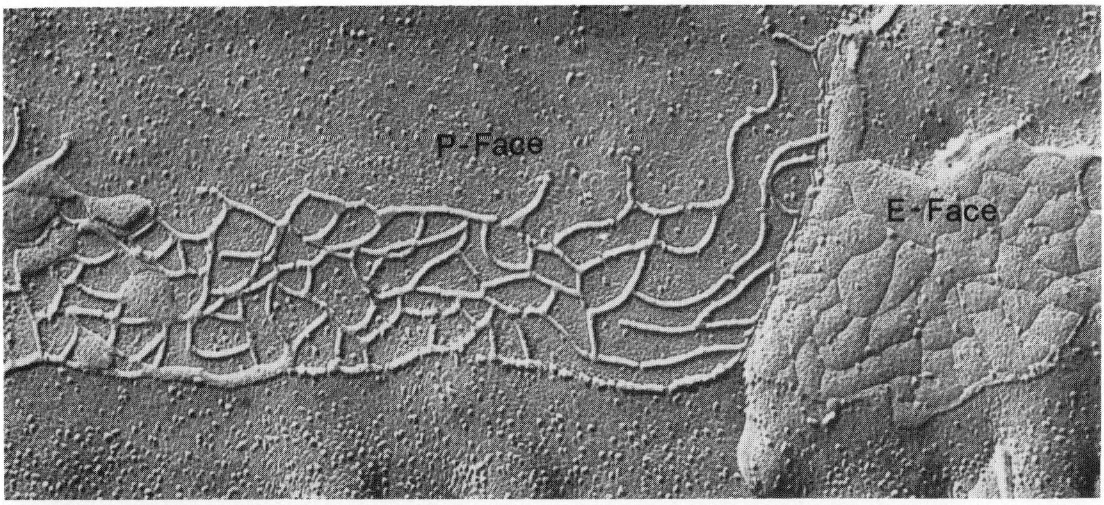

Figure 70, *upper* **Figure 71,** *lower*

A replica of a Sertoli cell junction from rat testis showing more than 50 particle rows on the E-face. Although the intramembrane particles in these junctions generally adhere to the E-face, some come away with the P-face. Discontinuities in the E-face particle rows in this replica are represented by short rows of particles on the P-face (lower right). The distribution of particles between the two fracture-faces and whether they appear in replicas as strands or rows of discrete particles are influenced by the concentration of glutaraldehyde and duration of fixation prior to freezing.

Figure 72. Sertoli cell junction from rat testis. (From Gilula, Fawcett and Aoki, Dev. Biol. 50:142–168, 1976.)

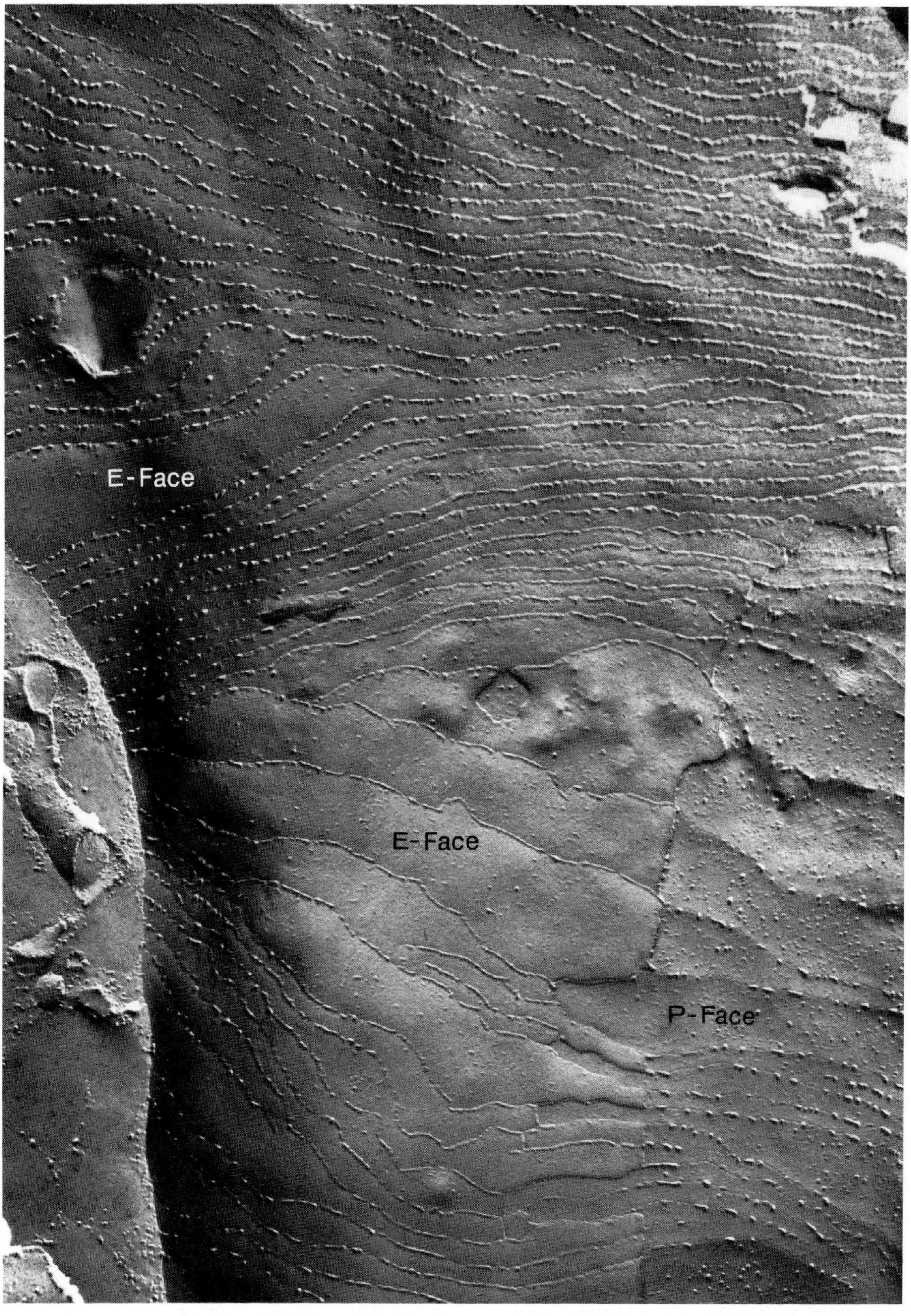

Figure 72

Although relatively rare, there are epithelia that lack zonulae occludentes and appear to present no effective barrier to penetration of the intercellular clefts. The ependymal epithelium of the brain shown in the accompanying micrographs was the first example studied (Brightman and Palay, 1963). Instead of the usual junctional complex consisting of a juxtalumenal zonula occludens followed by a zonula adherens and occasional desmosomes, there appeared to be no consistent order of the junctional specializations. As illustrated in the upper two figures, a zonula adherens was often found nearest to the free surface and one or more pentalaminar junctions were situated some distance below it. Occasionally a gap junction was found in the usual position of a zonula occludens, as illustrated in the lower figure. When electron opaque probes were used to test the patency of the intercellular cleft (Brightman and Palay, 1965), the tracer was found both above and below the pentalaminar segments of the junction. These were therefore considered to be plaques instead of continuous zonulae and were designated *maculae occludentes* in the belief that they were structurally similar to occluding junctions but of more limited extent. It is now apparent that the pentalaminar segments illustrated are *gap junctions* on the cell boundaries of this epithelium.

Figures 73, 74, and 75. Junctional specializations of cells in the ependymal epithelium of rat brain. (Micrographs courtesy of Enrico Mugnaini.)

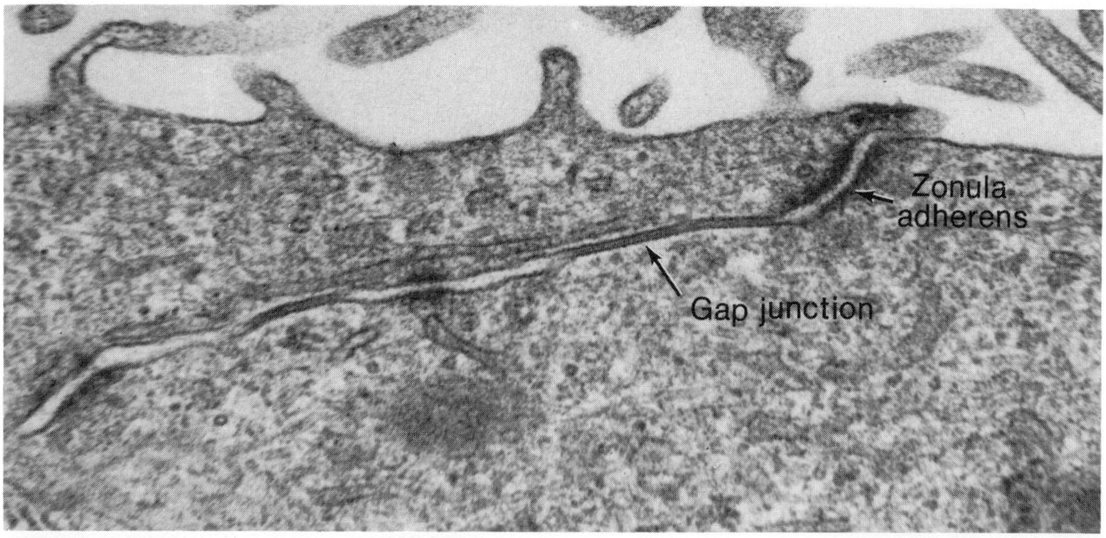

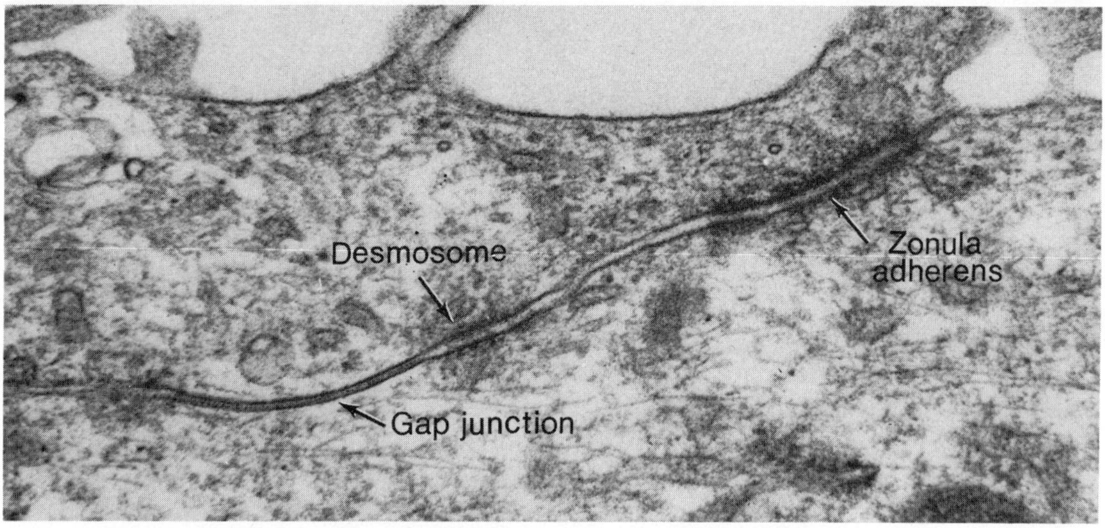

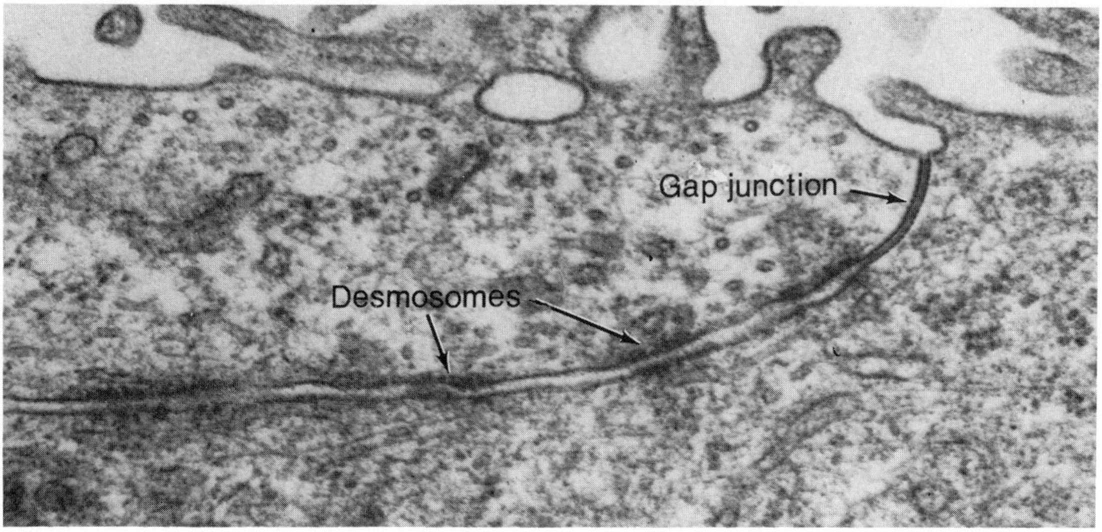

Figure 73, *upper* **Figure 74,** *center* **Figure 75,** *lower*

ZONULA CONTINUA AND SEPTATE JUNCTIONS OF INVERTEBRATES

Two types of junctional specializations are found in the epithelia of invertebrates that are not observed in mammalian tissues. These are the *zonula continua* and the *septate junction*. In thin sections of certain insect epithelia (left figure), the 17 to 18 nm space between precisely parallel membranes of adjacent cells is filled with a continuous layer of material of appreciable density, rich in glycoprotein. This junctional specialization has been called the *zonula continua* (Noirot and Noirot-Timothee, 1967; Dallai, 1970). Electron lucent transverse elements interrupting the continuity of the dense intercellular material can sometimes be resolved. None are evident in the accompanying micrograph, possibly because of slight obliquity of the section.

The septate junction (at the right) is a circumferential band around the apex of the cell wherein the 17 to 18 nm space between adjacent cells is traversed at regular intervals by thin septa which result in a characteristic ladder-like appearance of the junction in section.

Figure 76. Zonula continua from gut epithelium of a flea. (Micrograph courtesy of Susumu Ito.)

Figure 77. Septate junction from midgut epithelium of an insect. (Micrograph courtesy of Jean Gouranton.)

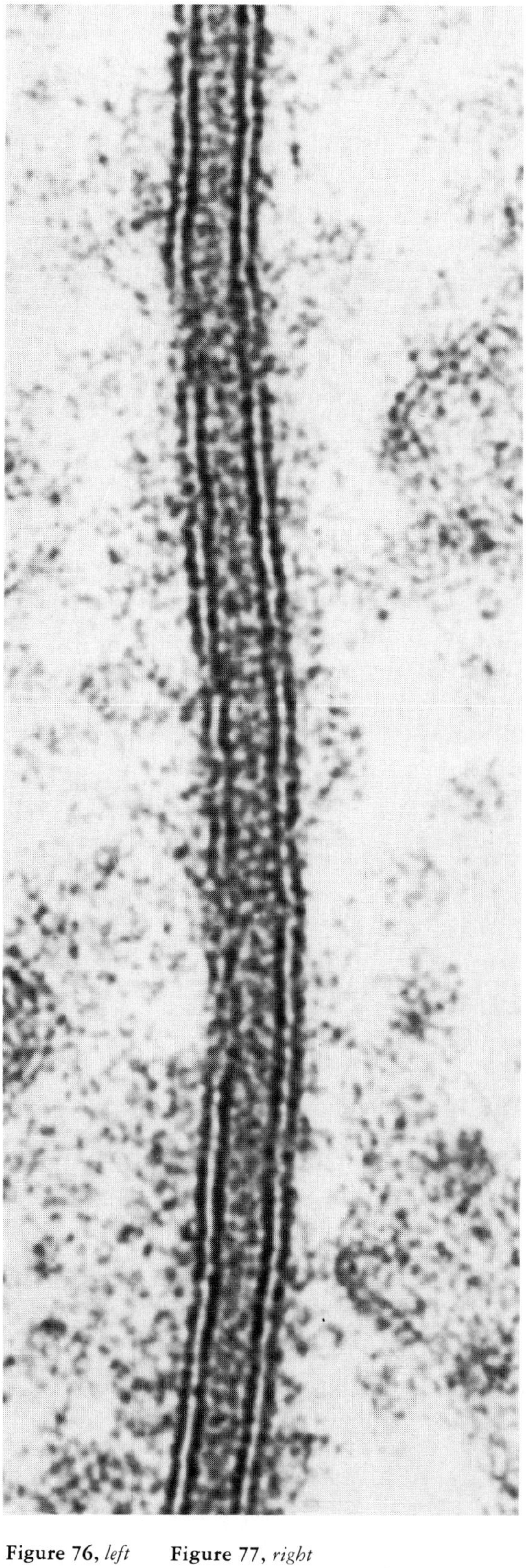

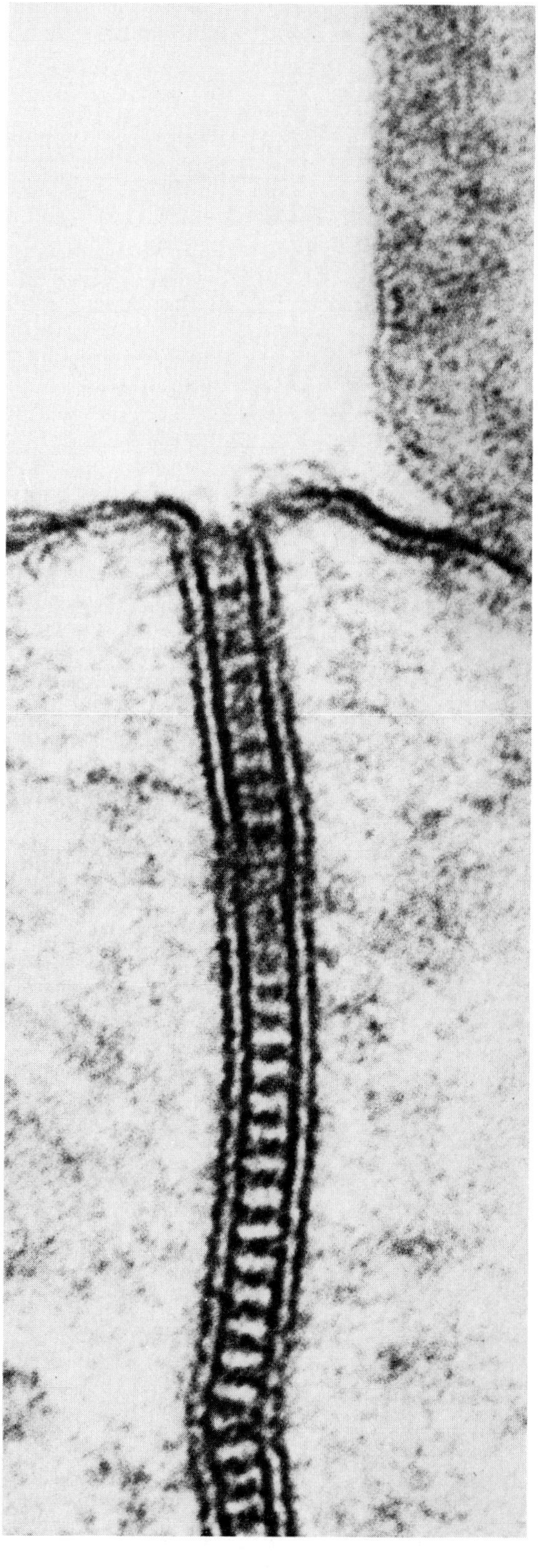

Figure 76, *left* **Figure 77,** *right*

In freeze-fracture preparations, the zonula continua shows periodic ridges on the E-face and complementary furrows on the P-face. The ridges may exhibit a particulate substructure. These junctions bear a superficial resemblance to the zonula occludens of vertebrate epithelia but are far more extensive and the ridges do not branch and anastomose.

The accompanying micrographs illustrate continuous junctions on epithelial cells of insect midgut. The ridges are associated with the E-face and the grooves are on the P-face. The linear arrays of particles frequently converge and run together as double rows, as seen in the lower figure.

Figures 78 and **79.** Freeze-fracture images of continuous junctions from midgut of *Rhodnius prolixus*. (Micrographs courtesy of Nancy Lane.)

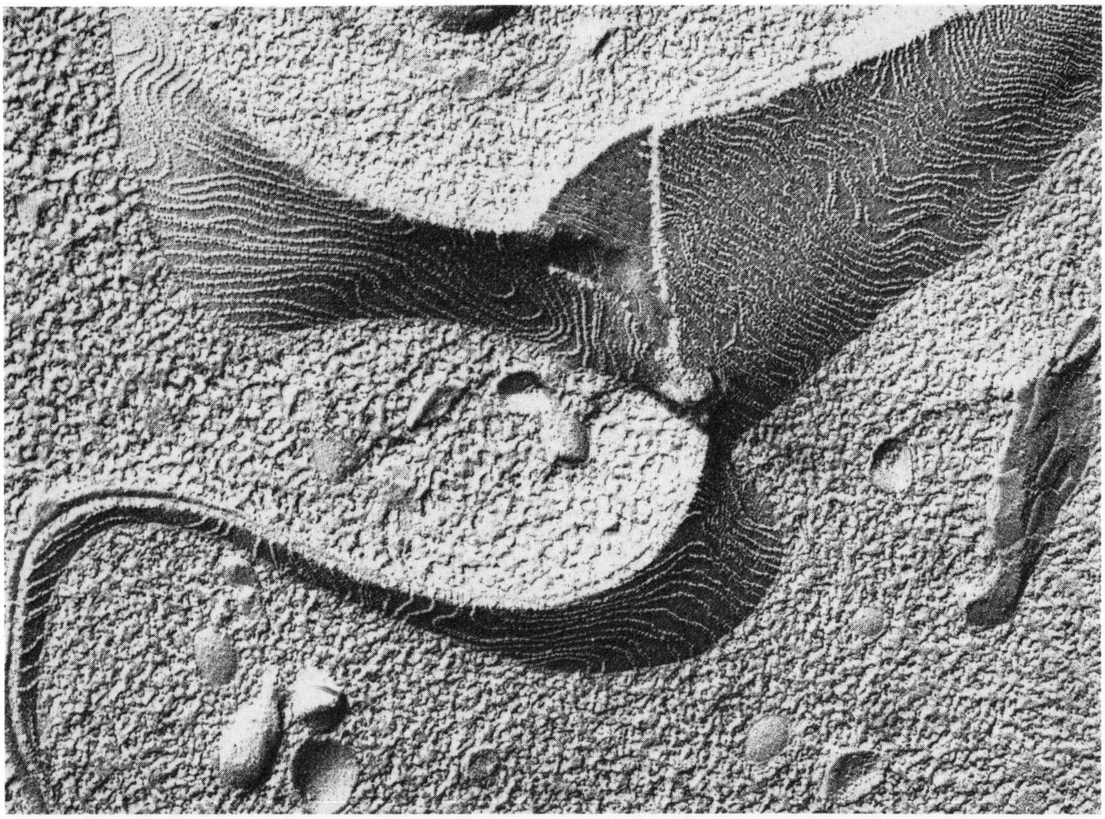

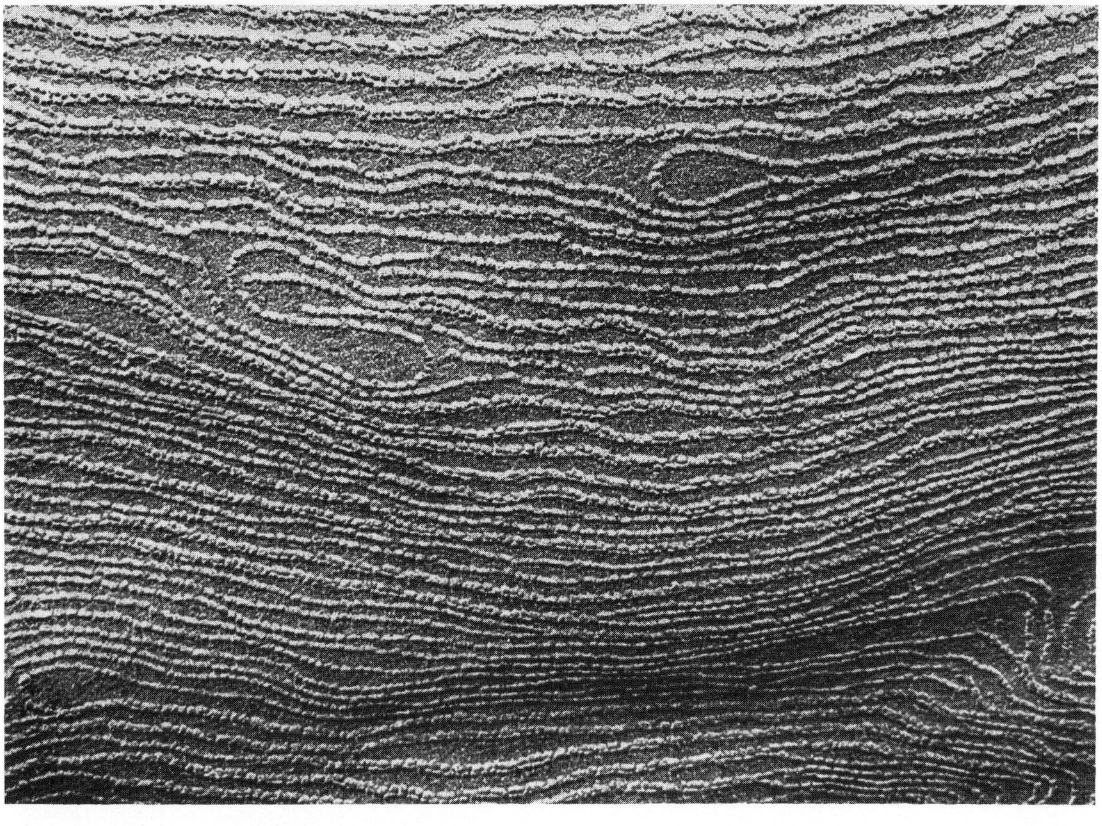

Figure 78, *upper* **Figure 79,** *lower*

Presented here is another example of a continuous junction on the lateral surface of epithelial cells of the gut. The ridges here are unusually meandering and appear to be on the P-face as judged by the smooth transition to the convex particle-studded surface of the microvillus (at the arrow).

Figure 80. Freeze-fracture preparation of gut epithelium from the flea. (Micrograph courtesy of Susumu Ito.)

Figure 80

In freeze-fracture preparations, septate junctions appear as rows of 6 to 10 nm particles on the P-face and complementary rows of shallow pits on the E-face. These rows may be relatively loose and undulating, as shown in the upper micrograph, or closely parallel, as in the lower micrograph. Typical zonulae occludentes are rarely observed in invertebrates. Septate junctions appear to be their counterparts. They constitute a deterrent to entry of large molecules into the intercellular clefts, but they are relatively ineffective permeability barriers since peroxidase and lanthanum can pass through them (Lane and Treherne, 1972). They are probably more concerned with maintenance of structural integrity of epithelia than with barrier function.

Figure 81. Septate junction from the rectum of the cockroach, *Periplaneta americana*.

Figure 82. Septate junction from testis of *Periplaneta americana*. (Both micrographs courtesy of Nancy Lane.)

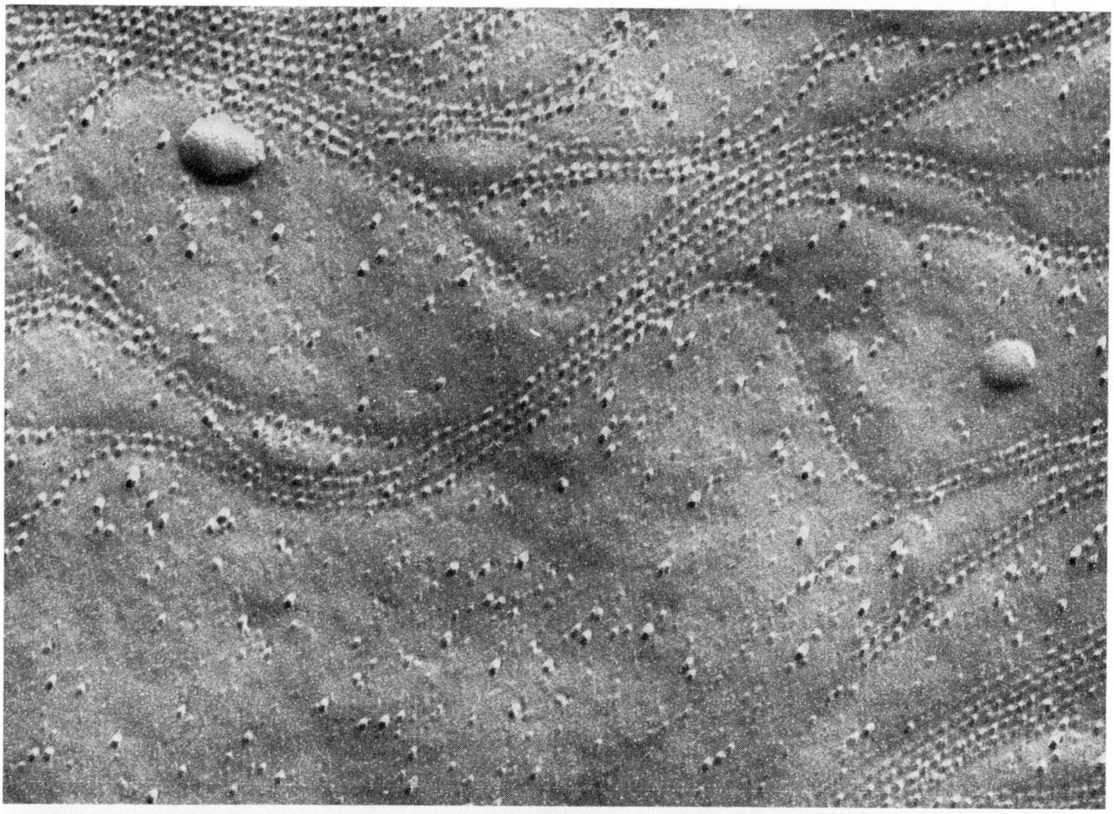

Figure 81, *upper* **Figure 82,** *lower*

DESMOSOMES

At desmosomes the adjacent cell membranes are strictly parallel and separated by a distance of about 30 nm. On the cytoplasmic surface of each membrane is a dense *attachment plaque* about 20 nm thick and up to 500 nm in diameter separated from the inner leaflet of the cell membrane by a 10 to 15 nm layer of lower density. Cytoplasmic 10 nm tonofilaments converge upon the attachment plaques. Examination of micrographs in stereo pairs suggests that the tonofilaments converging upon the desmosomes form loops near or within the attachment plaque and then return to the filament tracts forming the general cytoskeleton of the cell (Kelly, 1966). Thinner filaments are described as arising within the plaque and mingling with the tonofilaments to form a dense mat in the cytoplasm adjacent to the plaque. These filaments also traverse the membrane and project into the intercellular space (Staehelin and Hull, 1978). When outlined by lanthanum and examined with a tilting stage, they appear to be arranged in a quadratic pattern and to branch into side arms that meet and join a staggered array of similar linkers and side arms from the opposing cell (Rayns et al., 1969). Superimposition of the densities of the resulting lattice is responsible for the central layer or intermediate dense line seen in electron micrographs of thin sections (Staehelin and Hull, 1978). This line often has a narrow zigzag course. The current view of the organization of the material in the 30 nm gap at desmosomes is depicted in the illustration at right, but it must be admitted that there is considerable risk of error in interpreting the geometry of structures so near the limits of resolution of the methods. Evidence for the existence of transmembrane linkers rests heavily upon freeze-fracture replicas which do not show the usual intramembrane particles at sites of desmosomes, but show irregularly shaped and oriented structures which suggest minute filaments broken off at different levels within the membrane and variously deformed in the fracture process. The transmembrane linker filaments are believed to provide a direct mechanical coupling between the tonofilament networks of adjacent epithelial cells. A functionally continuous filamentous framework for the entire epithelium is thus created even though it is composed of distinct cellular units.

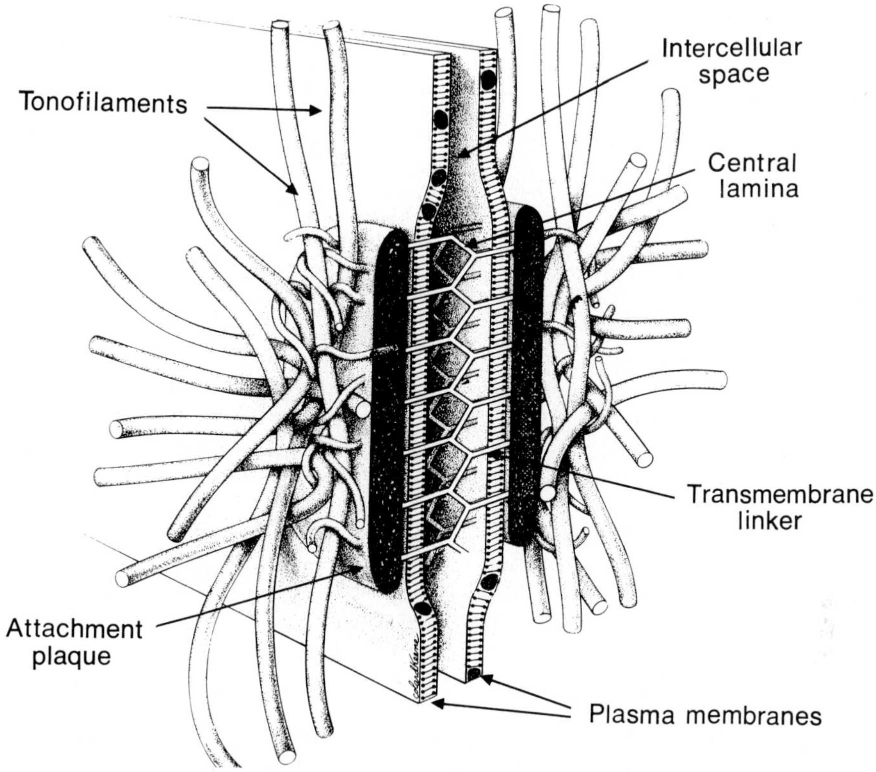

Diagrammatic representation of the structure of a desmosome (macula adherens). (Redrawn from L. A. Staehelin and B. E. Hill. Cell Junctions. In Scientific American, 238, No. 5, 1978. Copyright Scientific American, Inc. All rights reserved.)

Desmosomes are inconspicuous in freeze-fracture replicas. Typical intramembrane particles are lacking, but slightly elevated or depressed circular areas can be recognized in which there are small punctate or elongate elevations. These are interpreted as broken and plastically deformed fine filaments that traverse the membrane and bond the attachment plaques and tonofilaments to it.

Desmosomes are unusually abundant on stratified squamous epithelial cells, such as those of the skin, cervix, and oral cavity, which are normally subject to attrition and shearing forces. These cells are also abundantly provided with a cytoskeleton of bundles of 10 nm tonofilaments. The desmosomes maintain cell-to-cell adhesion and the attachment of tonofilaments to them serves to transmit stresses from the cytoskeleton of one cell to those of adjacent cells, thereby distributing more widely forces which might otherwise be locally disruptive. In other epithelia less subject to mechanical stress, the cytoskeleton is less well developed and fewer desmosomes are found.

When cells are experimentally dissociated by tryptic digestion of the intercellular material linking the junctional membranes, the half desmosomes and associated membrane invaginate, detach from the cell surface, and are taken into the cytoplasm in the wall of a vesicle which is then subject to lysosomal degradation (Overton, 1968; Berry and Friend, 1969). The observation of interiorized half-desmosomes in invasive tumors suggests that cells may be able to dissociate and dispose of desmosomes without experimental interference.

Desmosomes are morphologically similar from tissue to tissue, but differences in their sensitivity to dissociation by chelating agents and enzymatic digestion suggest that there are significant differences in the biochemical composition of the material in the junctional extracellular space (Rosenbluth, 1972).

The desmosome is a bipartite structure formed by cooperation of two cells. Half-desmosomes are rarely observed on the lateral surfaces of epithelial cells. Evidently, initiation of the formation of a half-desmosome in one cell immediately induces differentiation of a complementary specialization in the neighboring cell. The cohesion of the two halves depends upon relations established between minute linking filaments that meet in the intercellular cleft. Their chemical composition and the nature of their bonding are unknown, but the site of their interaction is marked by the zigzag course of the central lamina or intermediate dense line.

Half-desmosomes are found spaced at regular intervals along the membrane at the base of some stratified squamous epithelia, as illustrated in the lower figure on the facing page. Tonofilaments converge upon a typical dense attachment plaque. Delicate filaments extend from the attachment plaque into the basal lamina and no doubt serve to attach the cell to its substrate. It is likely that these filaments are similar to the linking filaments that traverse the intercellular cleft at desmosomes on the lateral surfaces of cells.

Figure 83. Adjoining portions of two cells in the stratum spinosum of hamster cheek pouch epithelium. Uranyl acetate staining. (Micrograph courtesy of J. T. Albright and M. A. Listgarten.)

Figure 84. A portion of the basal surface of a cell in the epidermis of a larval *Amblystoma punctatum.* (Micrographs courtesy of Elizabeth Hay.)

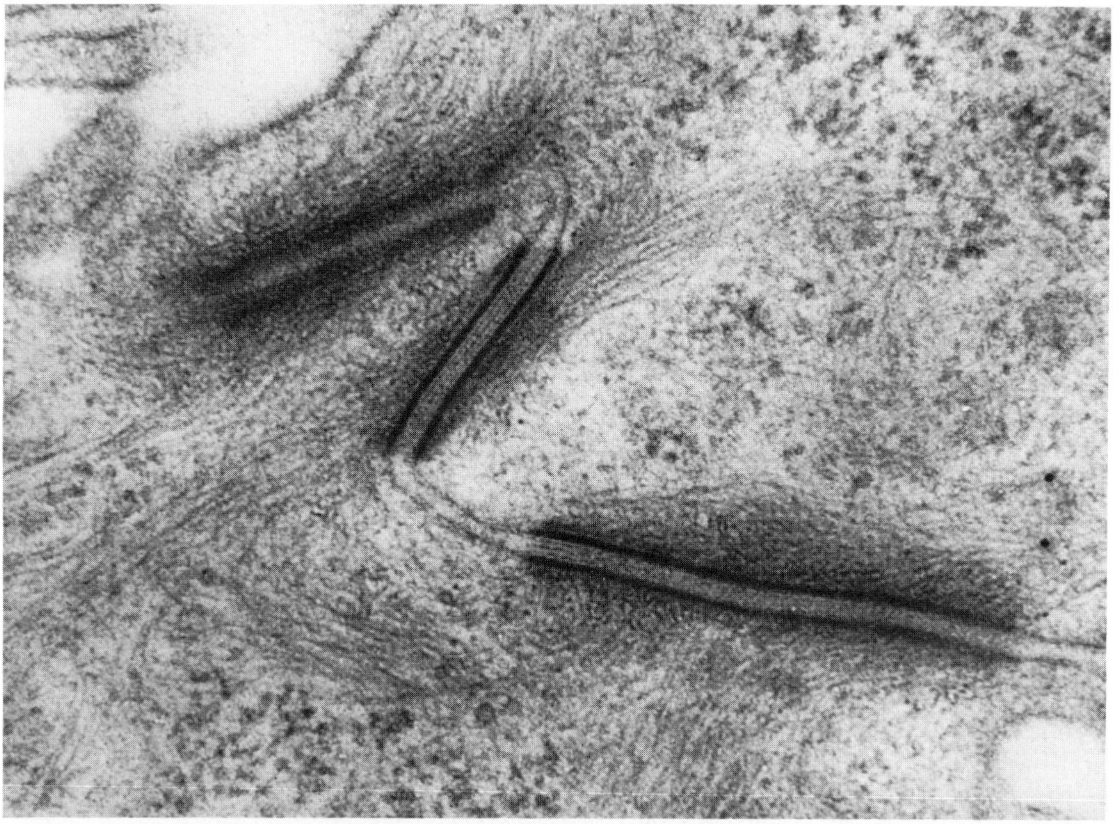

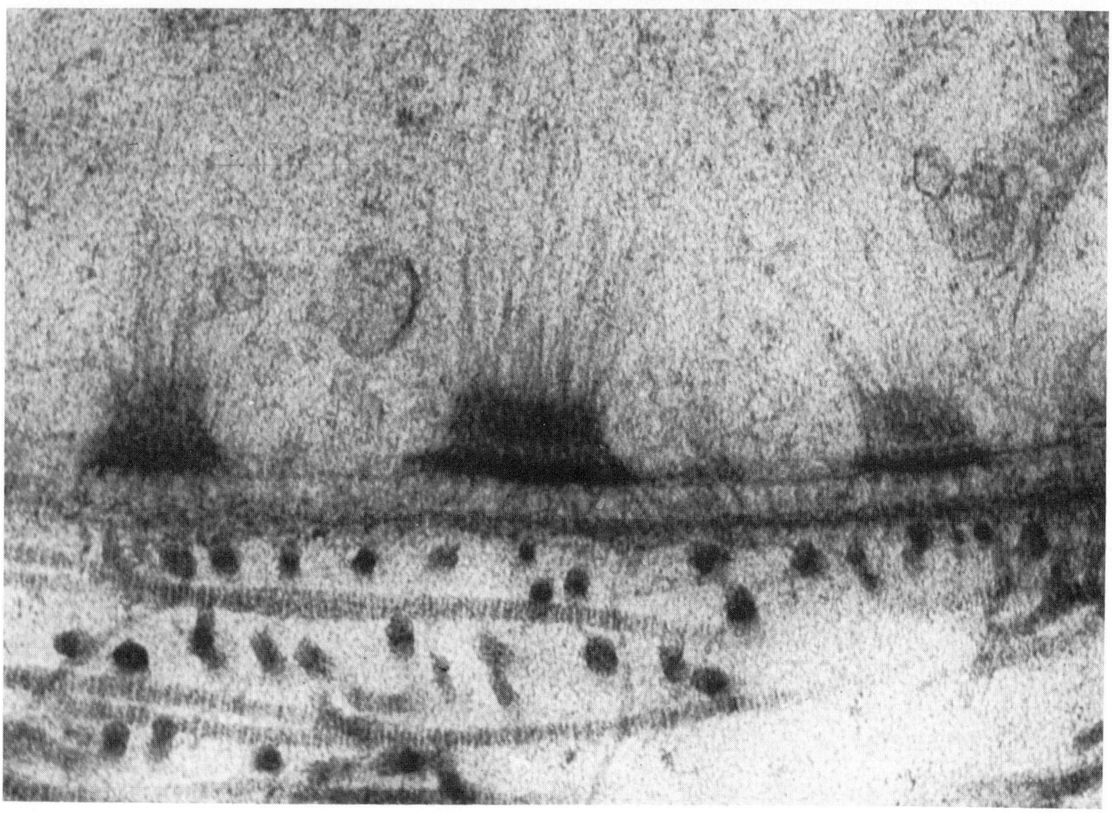

Figure 83, *upper* **Figure 84,** *lower*

Defining characteristics of the desmosome are the dense intracellular plaques and the central dense line in the intercellular space. Smaller, less highly organized densities in opposing cell membranes are encountered in various tissues. These probably also function as sites of adhesion but usually have few or no associated tonofilaments and are relatively weak attachments. Two examples marked by asterisks are shown. These structures are sometimes erroneously called desmosomes or more commonly "desmosome-like" specializations. To maintain the distinction between these and true desmosomes the term *punctum adherens* may be more appropriate.

Desmosomes are very numerous in stratified squamous epithelia that are subject to severe mechanical stress. The associated network of tonofilaments serves to limit the stretching of cells and distributes potentially disruptive shearing forces throughout the epithelium. An extreme example of abundant desmosomes is illustrated in a section through the interdigitating cell processes in the stratum spinosum of the epidermis on the bovine muzzle — an epithelium which is constantly subjected to flexion and attrition during grazing.

Figure 85. Capillary endothelial cell junction in the rete mirabile of the gas bladder of the toadfish *Opsanus tau*.

Figure 86. Epidermis from the muzzle of the cow *Bos taurus*. (Micrograph courtesy of Gida Matoltsy.)

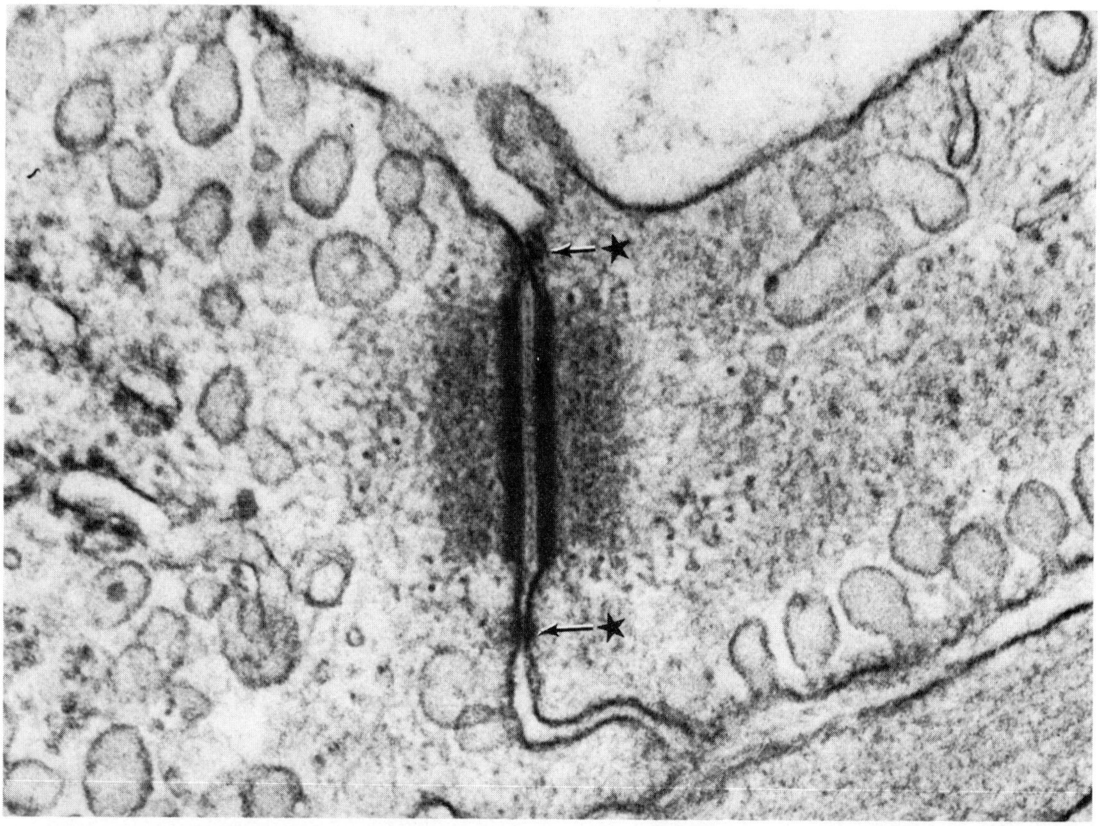

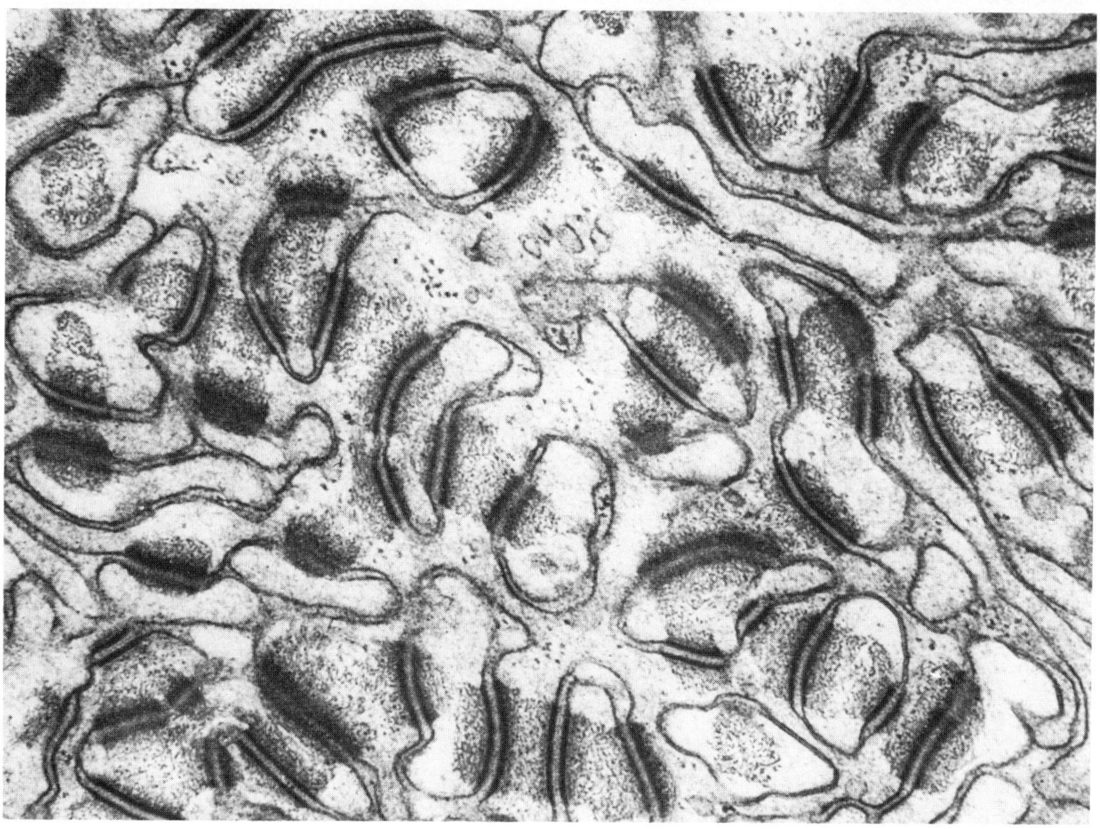

Figure 85, *upper* **Figure 86**, *lower*

The esophagus is lined by a stratified squamous epithelium. In the basal cell layer, which is subjected to relatively mild stresses, tonofilaments (at arrows) are sparse and the desmosomes relatively few. One of these shown at high magnification in the inset shows clearly the dense plaques, the lucent middle layer of the cell membranes, and the central dense stratum of the desmosome.

Figures 87 and **88.** Rat esophageal epithelium, basal layer. (Micrograph courtesy of Scott McNutt.)

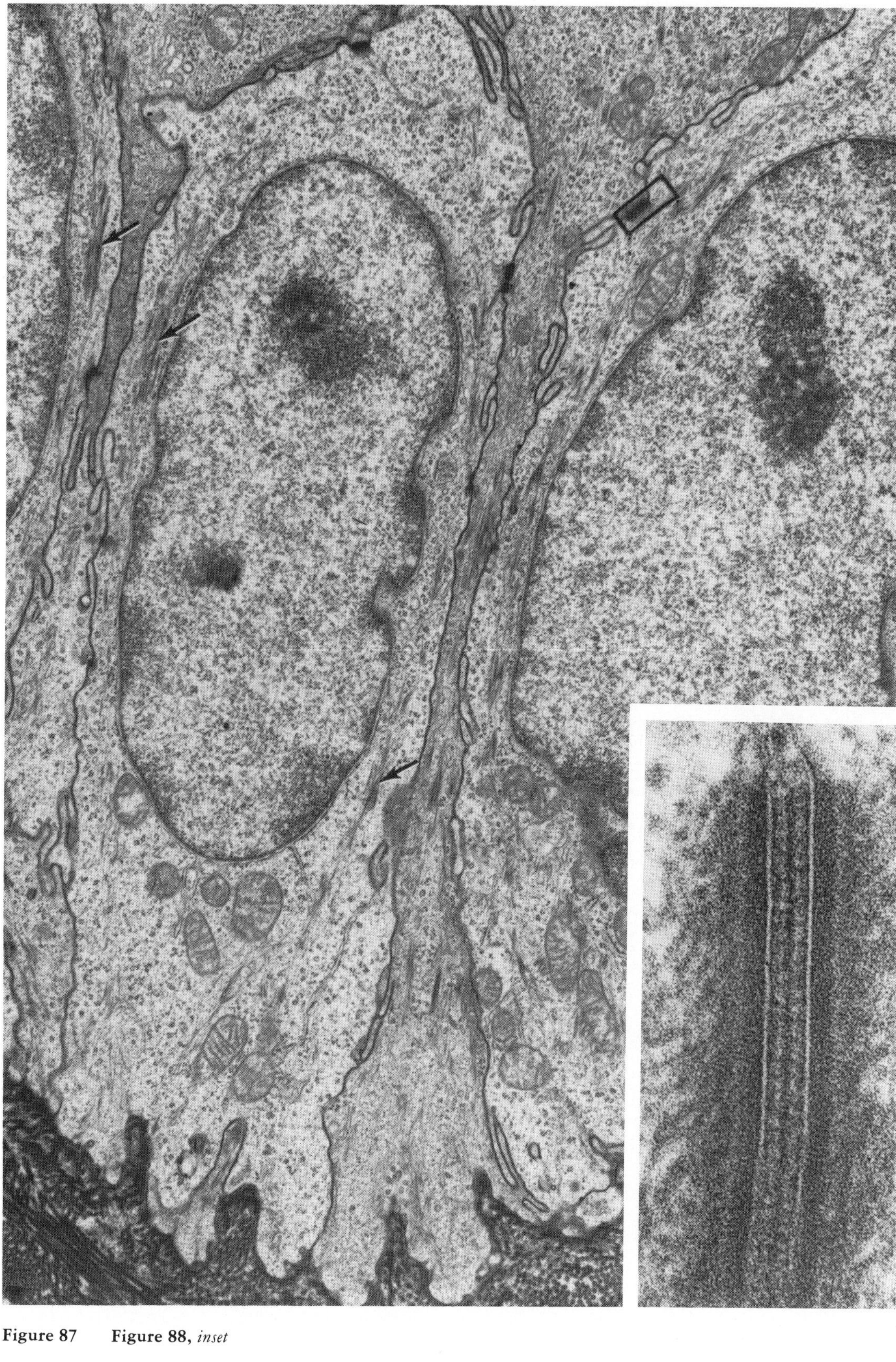

Figure 87 **Figure 88,** *inset*

The structure of desmosomes is remarkably similar in a wide range of animal species. Such differences as do exist probably involve the content of the interspace between the two halves of the desmosome. The material occupying this space is stainable with ruthenium red and digestible with trypsin and therefore is probably glycoprotein in nature. Variations from species to species and from tissue to tissue in the susceptibility of desmosomes to chelation and enzymatic digestion suggest that there may be significant biochemical differences that are not always reflected in their ultrastructure.

In some invertebrates, however, the dimensions and staining pattern of the extracellular portion of the desmosomes are distinctly different from those commonly seen in vertebrates. The interspace between half-desmosomes may be twice the usual width. The extracellular zones adjacent to the membranes are of a density comparable to the intracellular plaques and a zone of lower density is found in the center of the intercellular space where the intermediate dense line is usually found. It may be that the basic structure is the same as in vertebrate desmosomes but that the transmembrane linker filaments are obscured by a dense matrix.

The cytoplasm of the glial cells in the nerve cord of annelids is very rich in 10 nm filaments, and desmosomes are abundant. It is possible that the unusual degree of development of these structures is related to the fact that the central nervous system in these animals is subjected to more deformation and mechanical stress during locomotion than is the case in vertebrates, in which the cord is protected by the axial skeleton.

Figure 89. Glial cells in the nerve cord of the annelid worm, *Aphrodite aculeata.*

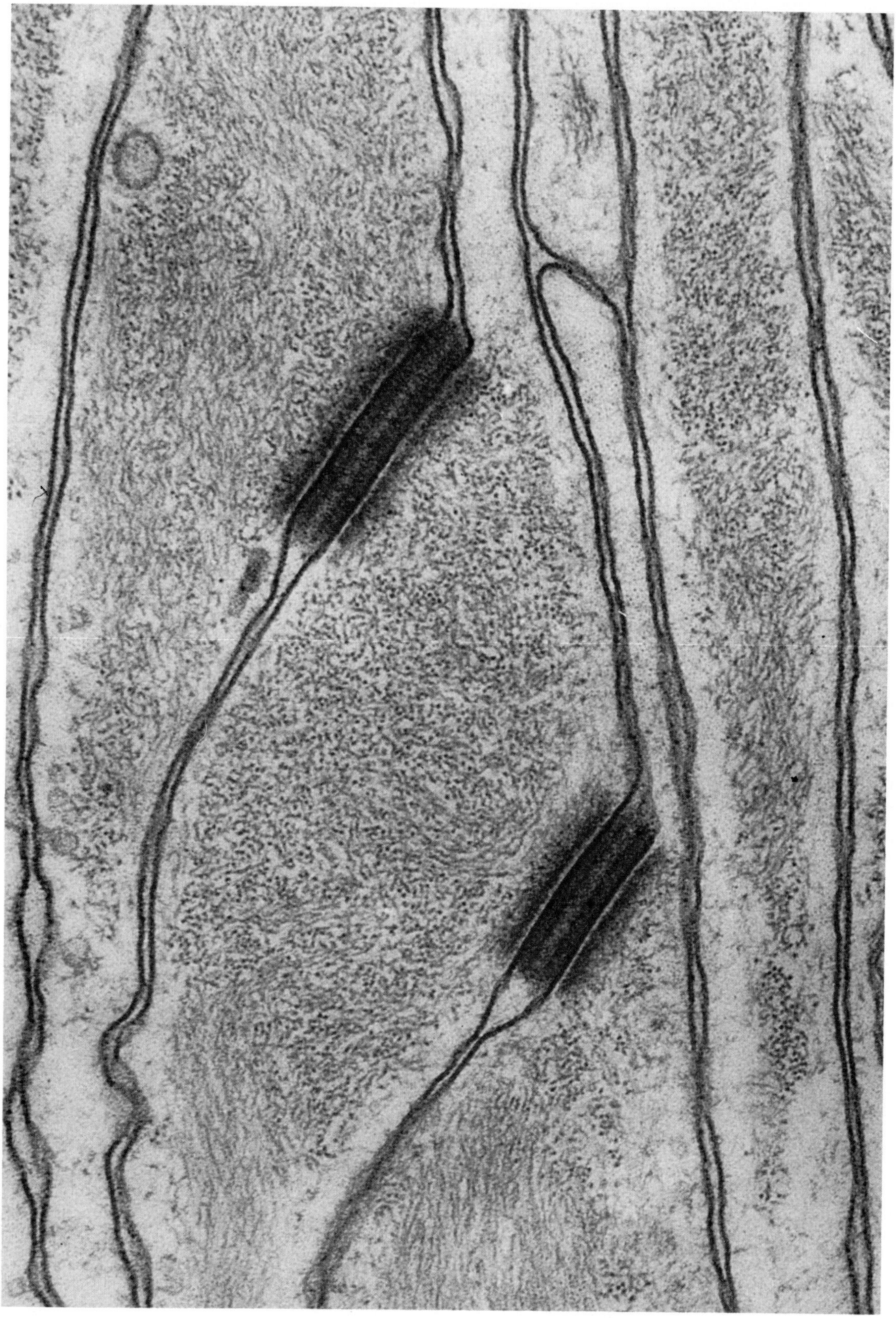

Figure 89

The freeze-fracturing method has provided no new information on the structure of desmosomes, but it confirms previous interpretations based on thin sections. On the facing page, a thin section of a typical desmosome is compared with a similar field as seen in a cross fracture. Associated with the clustered elevations interpreted as broken tonofilaments are smaller elements which may represent the postulated linking filaments that are believed to be interwoven with loops of tonofilaments.

In the replica shown in the lower figure, the membrane of an epidermal cell has been cleaved. Four desmosomes are seen on the exposed E-face. Unlike the familiar globular intramembrane particles of gap junctions, the irregularly shaped elevations seen in fractures of the membrane at desmosomes are interpreted as fragments of fine filaments that have broken off in longer or shorter lengths and have assumed varying orientations. Such images provide the most persuasive evidence for the existence of transmembrane filaments linking the two half-desmosomes.

Figure 90. Section of a cell boundary in the ciliary epithelium of the eye. (Micrograph courtesy of Giuseppina Raviola.)

Figure 91. Replica of a cross fracture across the boundary between two ciliary epithelial cells. (Micrograph courtesy of Giuseppina Raviola.)

Figure 92. Replica of a portion of the plasma membrane of an epidermal cell from newborn mouse. (Micrograph courtesy of Peter Elias and Daniel Friend.)

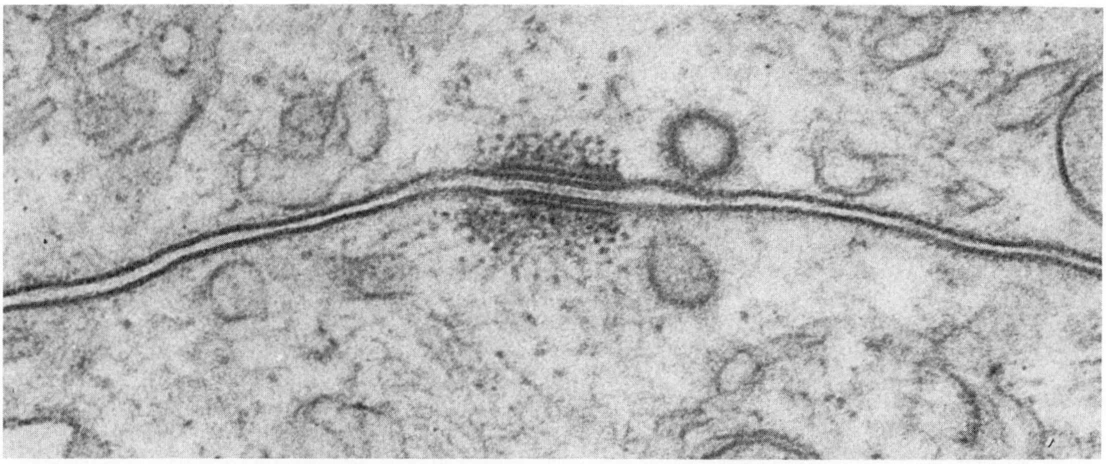

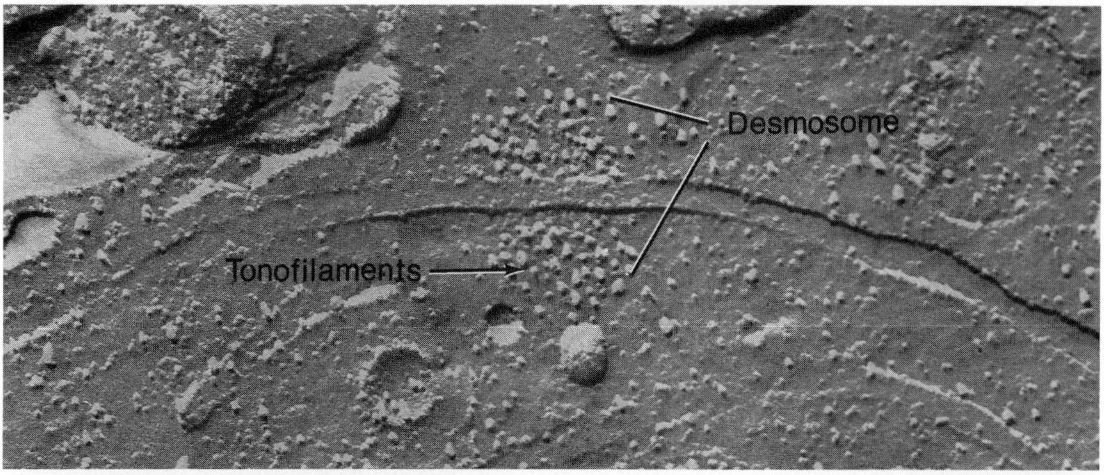

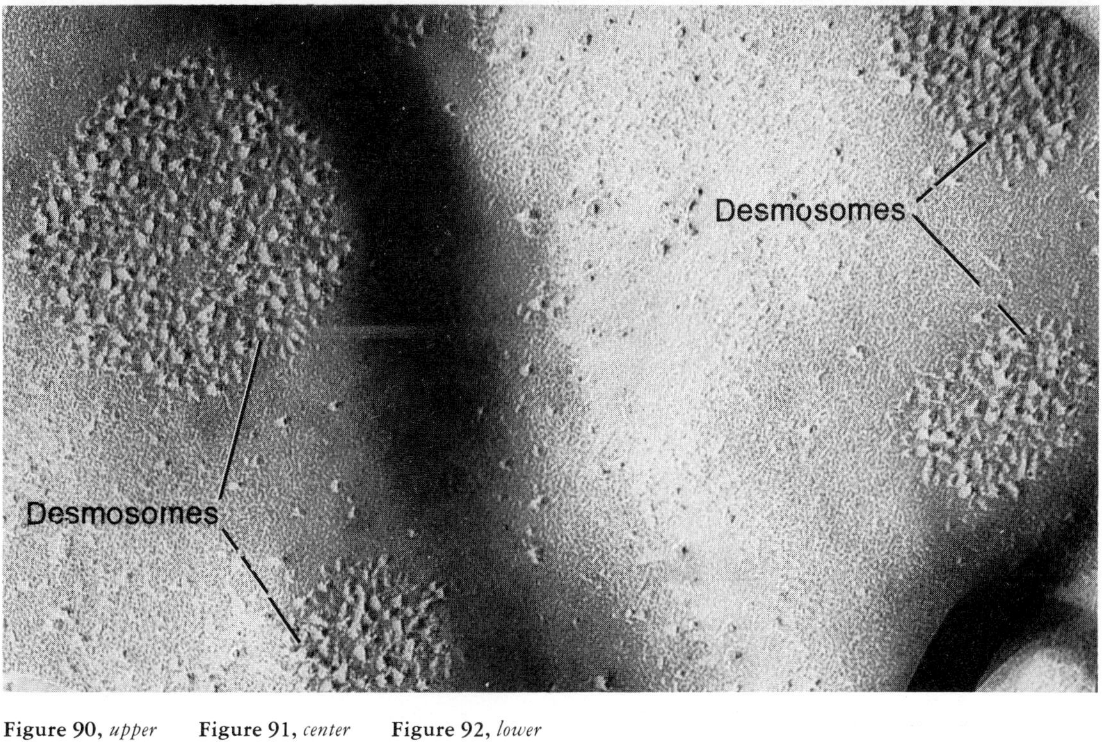

Figure 90, *upper* **Figure 91**, *center* **Figure 92**, *lower*

GAP JUNCTIONS (NEXUSES)

The *gap junction* is a differentiated area of the plasma membranes of adjacent cells, specialized to facilitate diffusion of ions and small molecules from cell to cell through low resistance pathways. Synonymous descriptive terms are *nexus* and *communicating junction*. The intercellular cleft is narrowed at these sites to only 2 nm. If the tissue is exposed to lanthanum or some other electron opaque probe of the extracellular space, the narrow gap between the opposing junctional specializations is filled, and it is seen to be traversed by bridging elements about 7 nm in diameter and with a center-to-center spacing of about 10 nm. In electron micrographs of sections that coincide with the plane of the interspace, the junctional area presents a regular polygonal pattern delineated by the dense lanthanum outlining a hexagonal array of structures that cross the intercellular gap. In favorable preparations the lanthanum may penetrate into the interior of these bridging elements, revealing the presence of a minute pore that appears as a central dot when seen on end or as a slender dense line in sections perpendicular to the plane of the membranes.

In freeze-fracture preparations of gap junctions, an aggregation of closely packed 8 nm globular particles is found on the P-face of the cleaved membrane and a complementary pattern of shallow pits on the E-face. In replicas examined at high magnification, a minute central depression can be detected on each gap-junctional particle. This corresponds to the central pore that is demonstrated by lanthanum penetration into each of the elements bridging the 2 nm intercellular gap. Each particle of the gap junction is believed to extend through the lipid bilayer of the membrane and to project into the intercellular gap, where it is joined to a corresponding particle in the opposing membrane. The alignment and end-to-end bonding of these junctional particles form units called *connexons* (Goodenough, 1974). A 1.5 to 2 nm hydrophilic channel passes from cell to cell through each connexon (Chalcroft and Bullivant, 1970; Goodenough and Gilula, 1972). Gap junctions can be isolated from mammalian liver, myocardium, and lens epithelium by using selective detergent solubilization techniques (Goodenough, 1974; Gilula, 1974), and it has been possible to obtain x-ray diffraction patterns from isolated gap junctions stacked in oriented pellets by high speed centrifugation (Caspar et al., 1977; Makowski et al., 1977). The results of these analyses suggest a model in which the connexons are composed of two particles end to end, each consisting of six protein subunits arranged around a central aqueous channel. This interpretation is consistent with experimental evidence that ions and small molecules pass from cell to cell to maintain their electrical and metabolic coupling (Lowenstein, 1976; Lawrence et al., 1978).

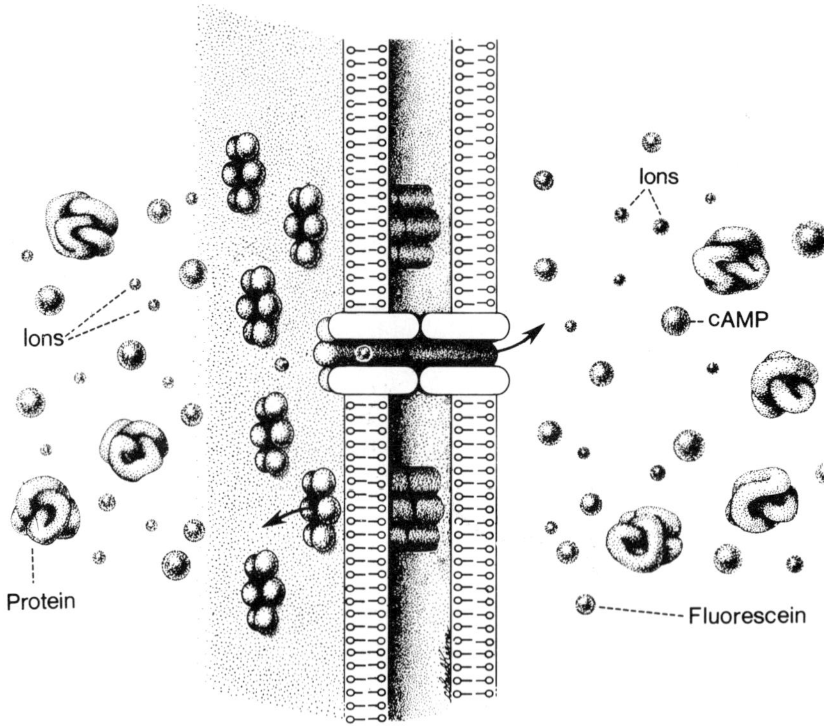

Schematic depiction of the relationship of the connexons and their subunits in a gap junction. The hydrophilic pore permits passage of ions and small molecules such as cyclic AMP or fluorescein but excludes macromolecules. (Redrawn after B. Tagawa from Lowenstein Cellular Communication by Permeable Membrane Junctions. Hospital Practice, 9, No. 11 and from Cell Membranes: Biochemistry Cell Biology and Pathology, G. Weissmann and R. Claiborne, eds., H. P. Publishing Company, Inc., New York, 1975.)

Upon chemical analysis, isolated gap junctions consist of lipid and protein. Protein profiles visualized by SDS-polyacrylamide gel electrophoresis identify a characteristic major polypeptide of about 27,000 molecular weight, called *connexin*. This is believed to be the principal constituent of the six subunits comprising the globular particles seen in freeze-fracture replicas of the junctional membrane (Goodenough, 1974; Gilula, 1974). The junctions are known to be sensitive to the proteolytic enzymes used in the isolation procedure, and protein components of greater than 27,000 molecular weight have been reported (Duguid and Revel, 1977; Ehrhart and Chauveau, 1977). Some uncertainty remains as to whether the 27,000 molecular weight polypeptide represents the native polypeptide or is a fragment of a larger subunit.

The permeability properties of gap junctions have been thoroughly studied. Microinjected fluorescent dyes have been shown to pass freely between conjoined cells. Passage of amino acids, sugars, cyclic AMP, and other nucleotides has also been demonstrated, but proteins, nucleic acids, and other macromolecules do not traverse gap junctions. Molecules of 1000 to 1200 daltons have been defined as the upper limit of size that can pass through the junctional channels, suggesting a pore size of 1.0 to 1.4 nm (Simpson et al., 1977). Nucleotides have been estimated to move between cells at a rate of 10^6 molecules per second per cell pair — a rate that would permit rapid equilibration of small molecules throughout sizeable populations of cells and enable them to share metabolites (Pitts and Simms, 1977). It has been shown that cyclic AMP released in response to hormone can pass through low resistance pathways to neighboring cells (Lawrence et al., 1978). Thus dissemination of second messenger via gap junctions may play an important role in coordination and amplification of the response of groups of cells to hormonal stimulation.

Gap junctions of most cell types have the capacity to change from a low resistance to a high resistance state, effectively isolating the cells from communication with their neighbors. Oxygen deprivation and aldehyde fixation result in a change from the low to the high resistance state. Therefore the majority of ultrastructural studies of gap junctions describe their appearance in the high resistance state. To preserve their normal configuration, it is necessary to maintain oxygenation of the tissue and rapidly freeze with liquid helium (Raviola et al., 1978).

Electrophysiological investigations indicate that a rise in intracellular calcium or a lowering of intracellular pH is associated with uncoupling of the cell (Rose and Lowenstein, 1975; Turin and Warner, 1977). A variety of other chemical treatments result in a switch of the gap junctions from the low to the high resistance state, but all of these experimental manipulations are highly unnatural. Nevertheless, it is believed that the intercellular channels may open and close in physiological circumstances *in vivo*. The conditions under which the change to high resistance occurs normally are still poorly understood, but there are a few examples in which embryonic cells appear to become uncoupled at specific stages of development (Furshpan and Potter, 1968; Blackshaw and Warner, 1976). Uncoupling of oocytes from granulosa cells in response to hormonal stimulation has also been reported (Gilula et al., 1978).

Presented on the facing page is the appearance of typical communicating or gap junctions as seen in thin sections and in replicas of freeze-fractured cell membranes.

In thin section, the trilaminar unit membranes are separated by a very narrow interspace (2 nm). There is a suggestion of periodic densities in the interspace corresponding to the connexons of the junction, but these are evident only in very thin sections heavily stained.

In freeze-fracture preparations of gap junctions the outwardly directed half-membrane (P-face) shows an aggregation of 8 nm globular intramembrane particles. The closeness and pattern of their packing depends, to some extent, upon the rapidity of fixation and freezing. The junctional particles are very uniform in size, whereas other intramembrane particles randomly distributed on the P-face are of more variable dimensions. On the inwardly directed outer half-membrane (E-face), not shown here, the junctional area exhibits a pattern of shallow pits complementary to the pattern of particles seen on the P-face.

In replicas of the P-face examined at high magnification a small light spot can sometimes be seen in the center of each globular subunit of the junction (at arrows). This represents the pore or channel that traverses each connexon.

Figure 93. Gap junction from the saccular macula of a goldfish. (Micrograph courtesy of Kiyoshi Hama.)

Figure 94. Freeze-fracture preparation of a gap junction from a granulosa cell at the rat ovary. (Micrograph courtesy of David Albertini.)

Figure 95. Freeze-fracture replica of a gap junction from the saccular macula of the goldfish. (Micrograph courtesy of Kiyoshi Hama.)

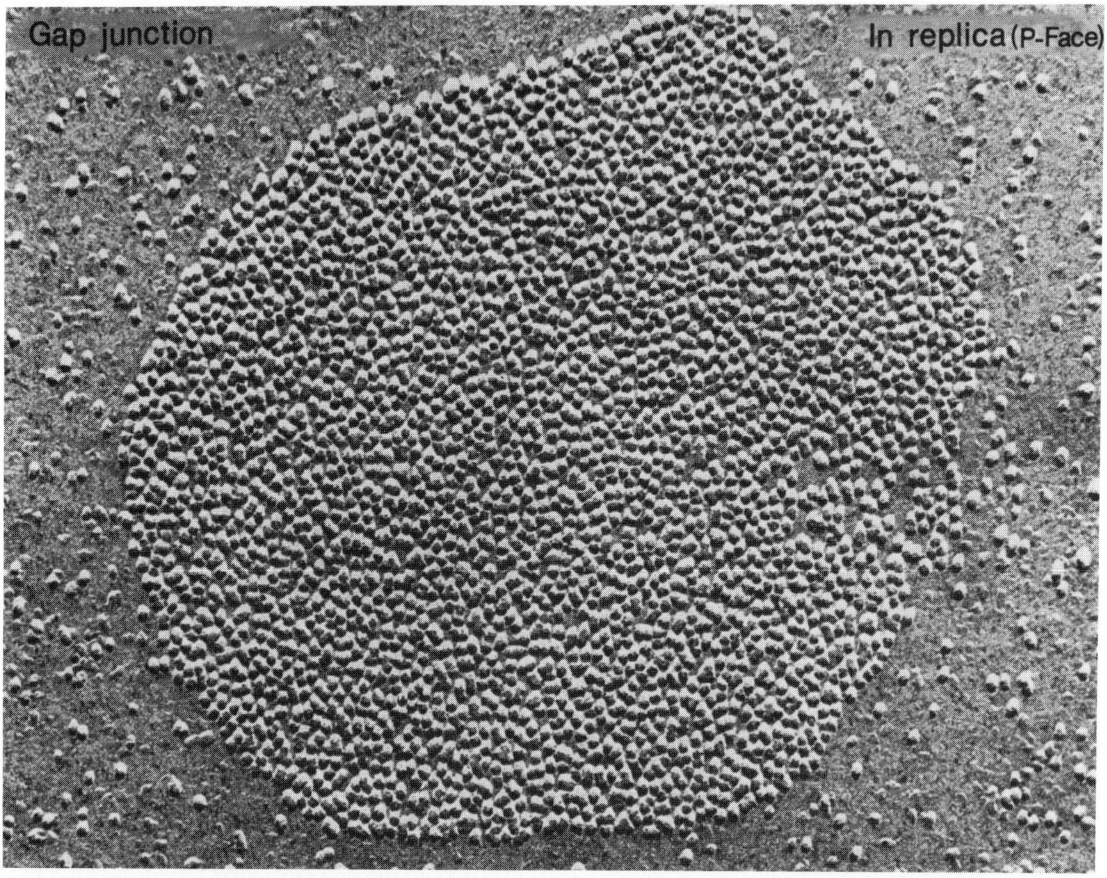

Figure 93, *upper* **Figure 94,** *center* **Figure 95,** *lower*

In the fracturing of a gap junction, the plane of cleavage often jumps from one membrane to the other. The resulting replica therefore presents in one area an image of the inner half-membrane (P-face) on one of the cells, and in another area the outer half-membrane (E-face) of the other cell. The lower half of the accompanying micrograph presents the P-face of the underlying cell which shows intramembrane particles randomly distributed in an unspecialized area of the membrane in the lower part of the figure. Above this are the closely packed 8 nm particles on the P-face of a gap junction. The irregular line crossing the field is where the fracture plane abruptly shifts to the membrane of the overlying cell, exposing the E-face of the gap junction which exhibits a pattern of shallow pits complementary to the junctional particles of the P-face of this membrane which has been cleaved away.

Figure 96. Replica of a freeze-fractured gap junction. (Micrograph courtesy of Daniel Friend.)

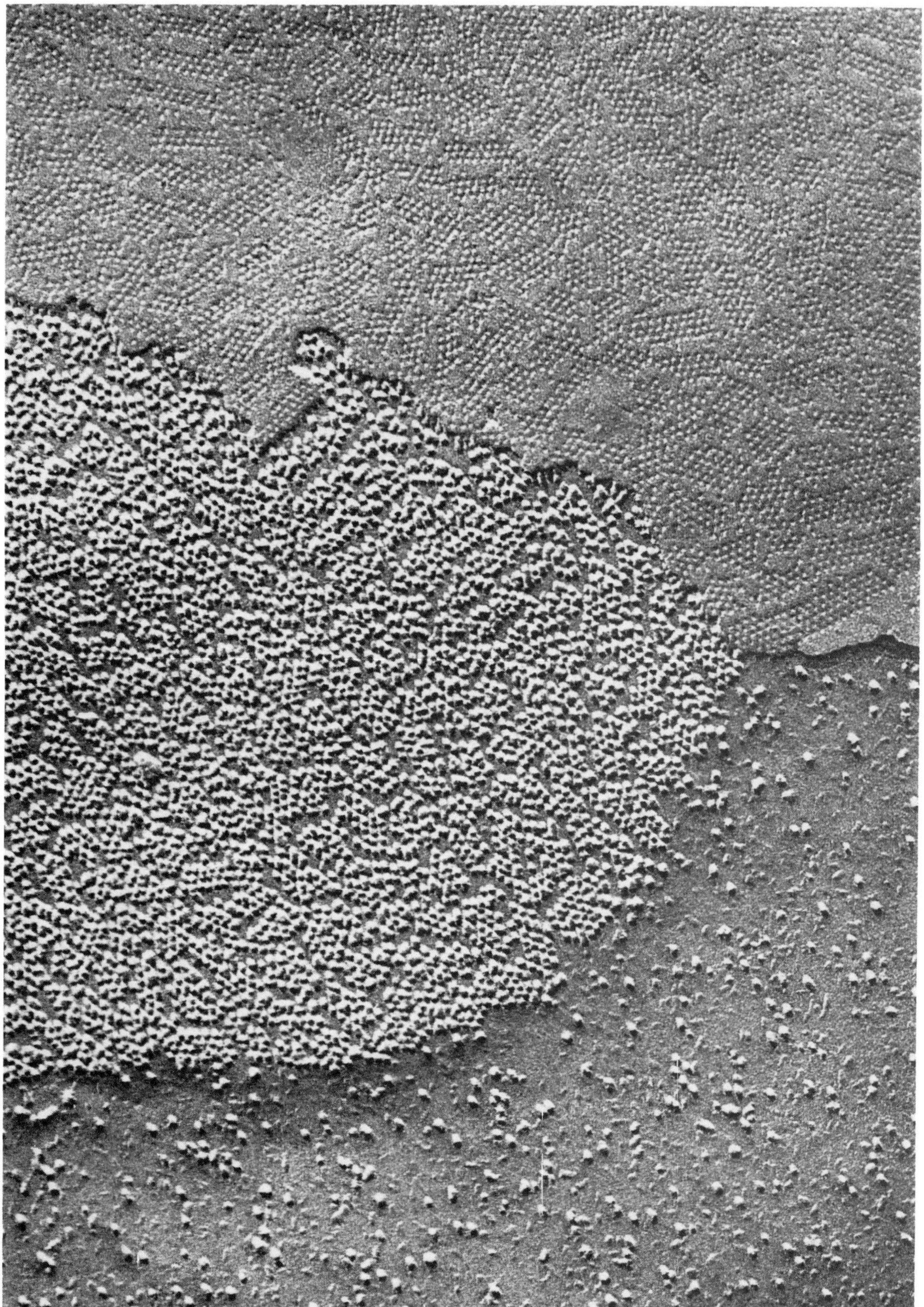

Figure 96

Gap junctions can be isolated from liver cells and examined by negative staining using phosphotungstate or uranyl formate as the contrast medium. The accompanying micrographs illustrate such a preparation. At high magnification in the lower figure, the 8 to 9 nm particles or connexons have a central 1.5 to 2 nm electron-dense region which is believed to represent the polar channel through which cell-to-cell communication takes place.

Figure 97 and **98.** Gap junctions isolated from mouse liver and negatively stained with uranyl formate. (Micrograph courtesy of Bernard Gilula.)

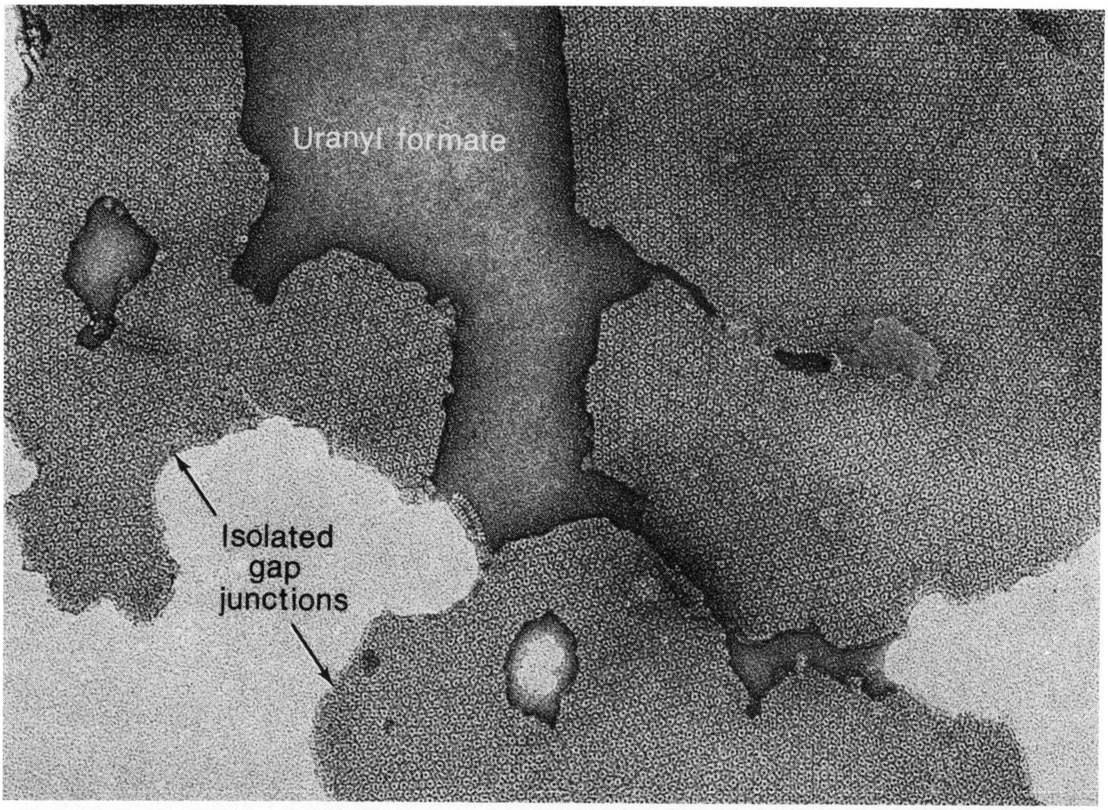

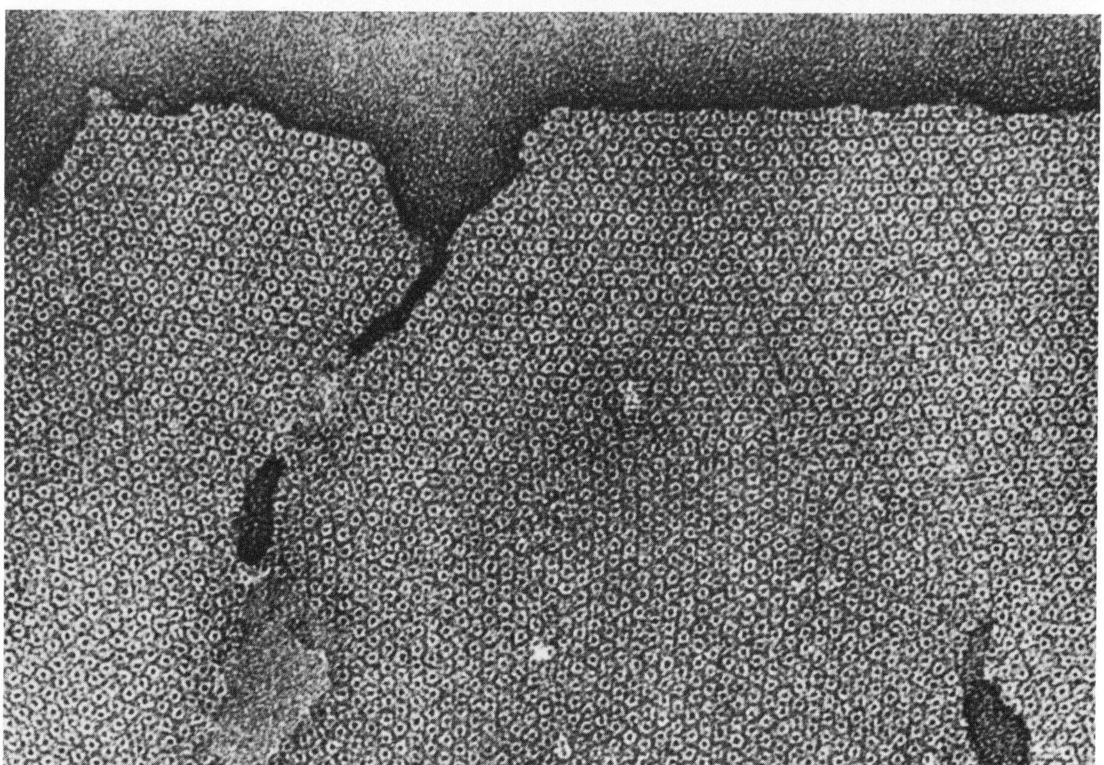

Figure 97, *upper* **Figure 98,** *lower*

Gap junctions vary greatly in size, ranging from aggregations of a few particles to large plaques up to 5 μm in diameter. Not infrequently in freeze-fracture preparations, the particles are packed in rectilinear arrays separated by particle-free aisles. Such patterns probably do not reflect the condition *in vivo,* but result from aggregation or crystallization of the particles as the junction is switched to the uncoupled high resistance state during excision and fixation of the tissue. Evidence for this interpretation is presented in the micrographs that follow.

Figure 99. Gap junctions from a granulosa cell of rat ovarian follicle.

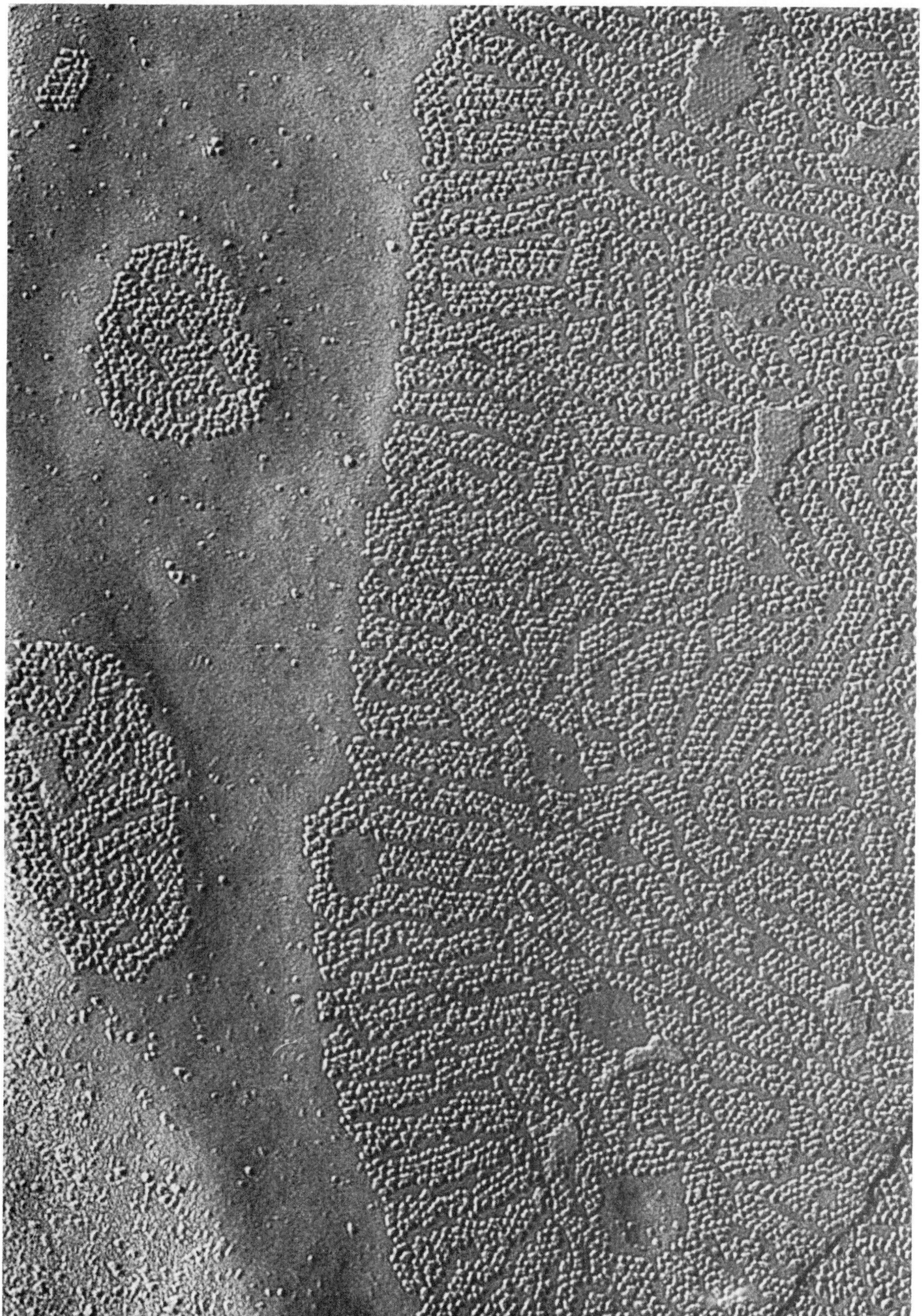

Figure 99

Replicas of gap junctions from tissue subjected to experimental conditions known to cause electrical uncoupling of the cells show a somewhat closer packing of the connexons than that normally observed (Peracchia, 1977). Since both hypoxia and aldehyde fixation are effective uncouplers, it seems likely that the images obtained with routine methods of specimen preparation may not accurately represent the functional state of those junctions *in vivo*. The development of a method for rapid freezing of tissue at the temperature of liquid helium without prior fixation or glycerination (Heuser et al., 1976) makes it possible to approximate more closely the disposition of connexons in their native state.

The upper figure on the facing page shows a gap junction from ciliary epithelium excised in oxygenated physiological salt solution and rapidly frozen with liquid helium. The connexons are spaced some distance apart and exhibit no ordered arrangement. The lower figure shows a junction from the same tissue subjected to anoxia by excision in physiological salt solution bubbled with nitrogen and carbon dioxide, and then rapidly frozen. Under these latter conditions, the gap junction consists of hexagonally packed connexons. Therefore it appears that in this tissue anoxia causes the connexons to aggregate into bidimensional crystals. Aldehyde fixation prior to freezing has a similar effect. In other organs such as stomach and liver, rapidly frozen gap junctions show less dispersion of the connexons and in these fixation produces relatively little change in the appearance of the junctions (Raviola, Goodenough and Raviola, 1980).

It seems clear that there are significant biochemical and physiological differences in the gap junctions of different tissues and in their sensitivity to anoxia and fixation, but in general a less highly ordered arrangement of connexons seems to be typical of their native state and crystalline order is characteristic of their high resistance state.

Figure 100. Gap junction from ciliary epithelium oxygenated and prepared by ultrarapid freezing.

Figure 101. Gap junction from the same tissue subjected to anoxia before ultrarapid freezing. (Both micrographs courtesy of Elio Raviola.)

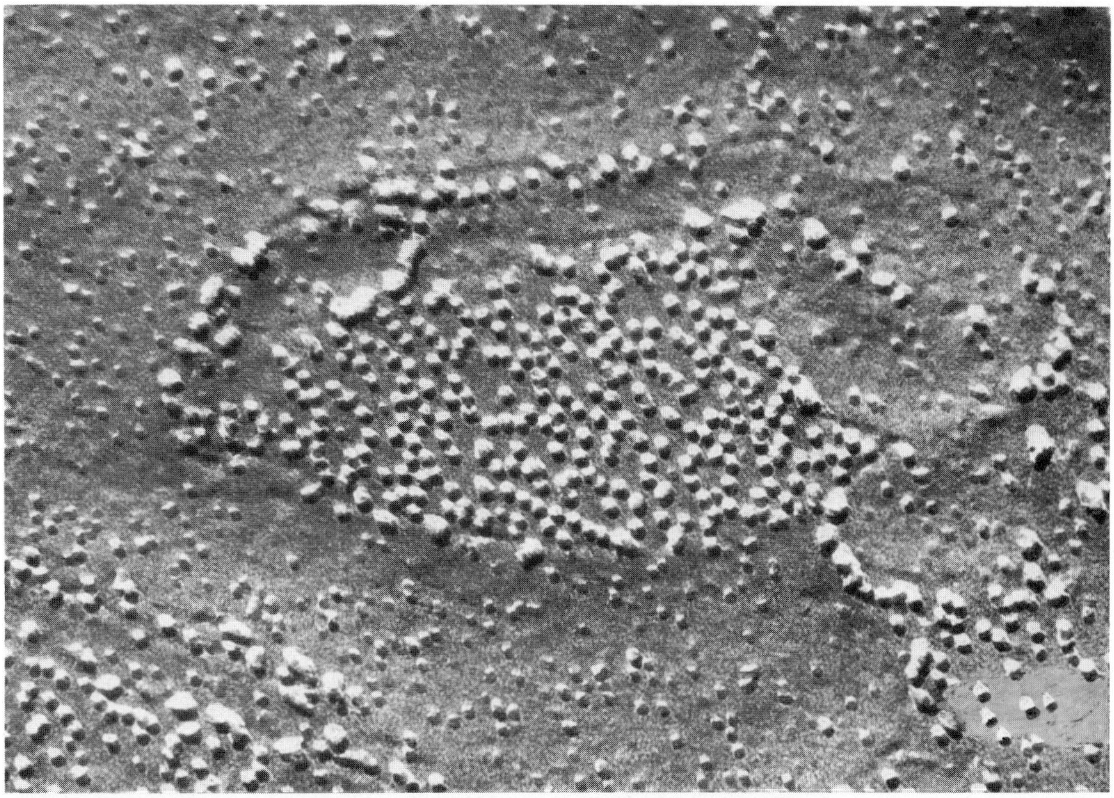

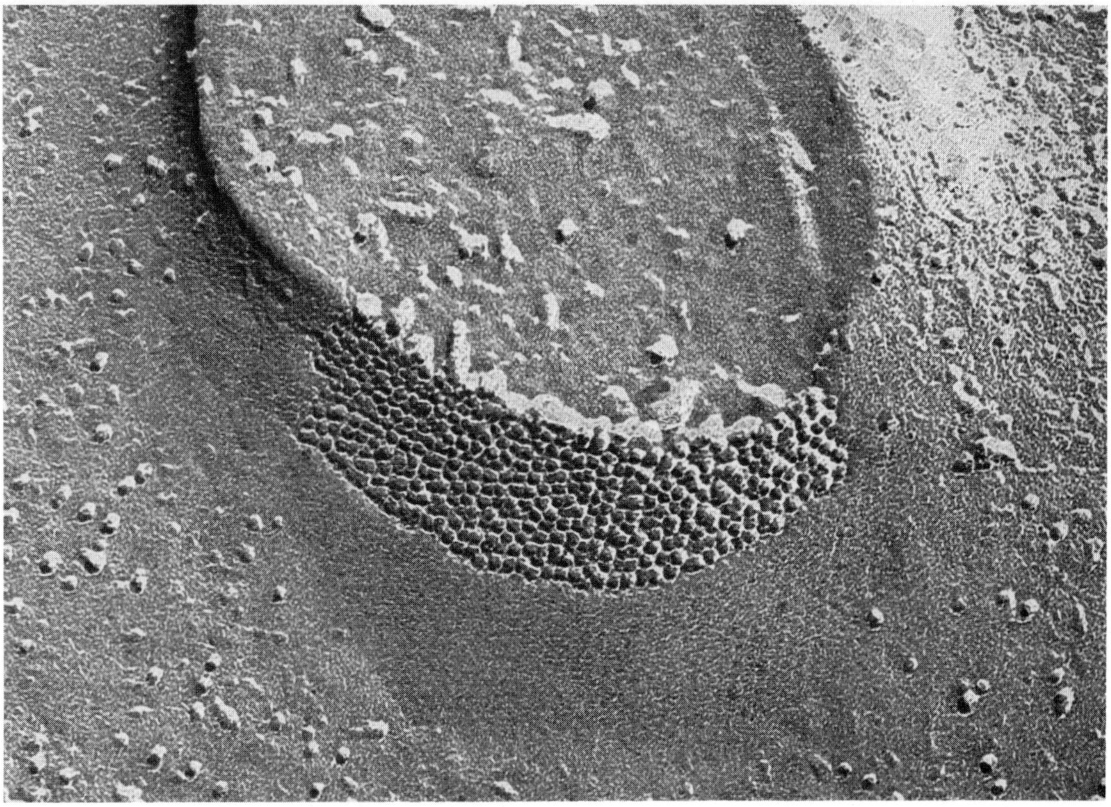

Figure 100, *upper* **Figure 101,** *lower*

Gap junctional areas are usually straight, as in the upper figure on the facing page. The normal intercellular space between cells abruptly narrows (at arrows). However, not infrequently, the junctional area is more or less curved, and in extreme examples such as that in the lower figure, a smooth contoured process limited by the gap junction projects into the adjacent cell. Some may detach and become spherical cytoplasmic inclusions. When these structures were first observed on the boundaries between liver cells (Fawcett, 1954), they were interpreted as mortise- and tenon-like devices for maintaining cell-to-cell cohesion. More recently they have been regarded as stages in a process of interiorization and intracellular degradation of gap junctions (Albertini et al., 1975). A variety of descriptive terms have been applied to them, including "annular gap junction" and "sphaera occlusa" (Espey and Stutts, 1972).

It has recently been found that junctions of this configuration may be artifacts. Rapid perfusion fixation, oxygenation during tissue processing, or rapid freezing in liquid helium eliminates such projections in epithelia where they are commonly observed after routine immersion fixation. It is suggested therefore that curvature of extensive gap junctions and their projection into the neighboring cell is probably an agonal event associated with postmortem uncoupling in the interval between interruption of the blood supply and arrival of aldehyde fixative at the site of the junction (Raviola and Raviola, 1978; Fawcett, 1978). This interpretation derives indirect support from the observation that experimentally induced uncoupling of large gap junctions results in their adopting a curved contour. This does not exclude the possibility that some of the intrusive junctions observed may be a consequence of physiological uncoupling.

Figure 102. Gap junction on the boundary between two hepatic cells illustrating their normal straight form.

Figure 103. A markedly curved gap junction projecting into one of the cells. (Both figures courtesy of Richard Wood.)

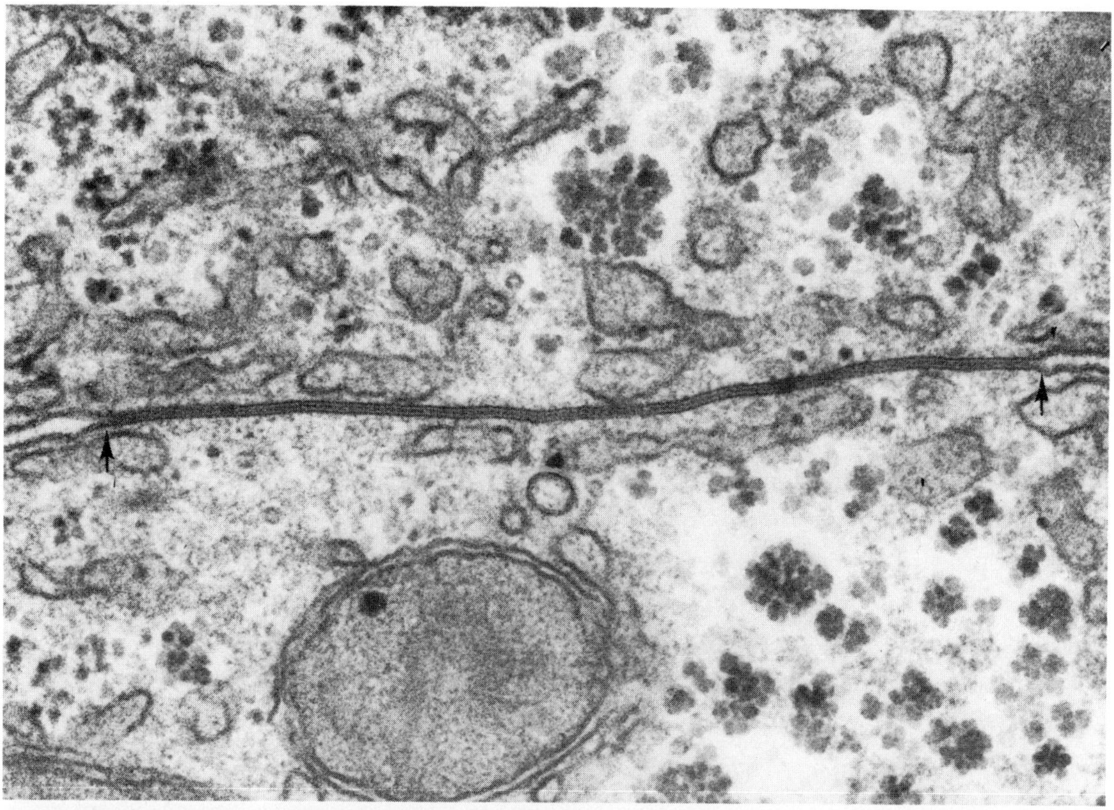

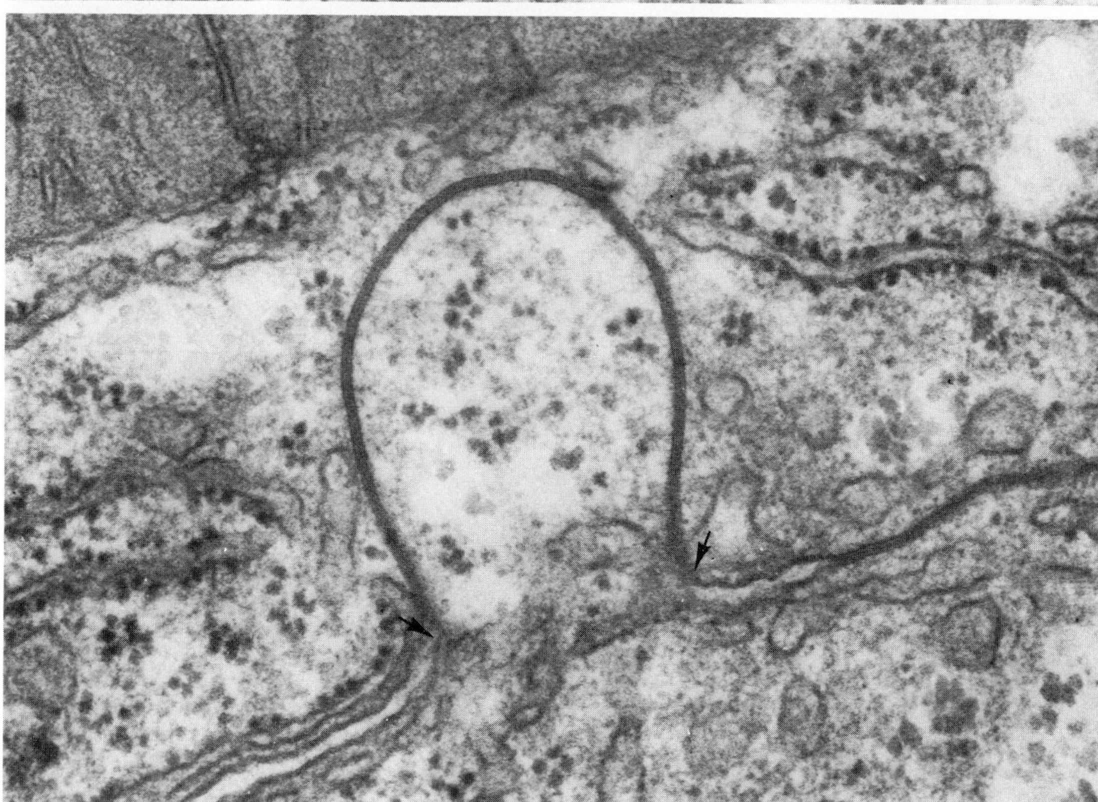

Figure 102, *upper* Figure 103, *lower*

Gap junctions occur in multicellular oganisms throughout the animal kingdom with relatively little morphological variation. When prepared by routine methods, the gap junctions of vertebrates consist of closely aggregated 8 to 9 nm particles on the P-face. The particles often exhibit hexagonal packing, and this tendency is sometimes expressed in a polygonal form of the gap junction as a whole. Examples are shown at the arrows in the upper figure.

The gap junctions of invertebrates, illustrated in the lower figure, consist of somewhat larger intramembrane particles. These are associated with the E-face and show little tendency to close packing. Because of the association of the particles with the E rather than the P-face, these are sometimes called *inverted gap junctions*.

Figure 104. Numerous gap junctions on cells from the ciliary epithelium of the eye in *Macaca mulatta*. (Courtesy of Giuseppina Raviola.)

Figure 105. Inverted gap junctions from cells of the mid-gut of the horseshoe crab, *Limulus polyphemus*. (From Lane, J. Cell Sci. 32: 293–305, 1978.)

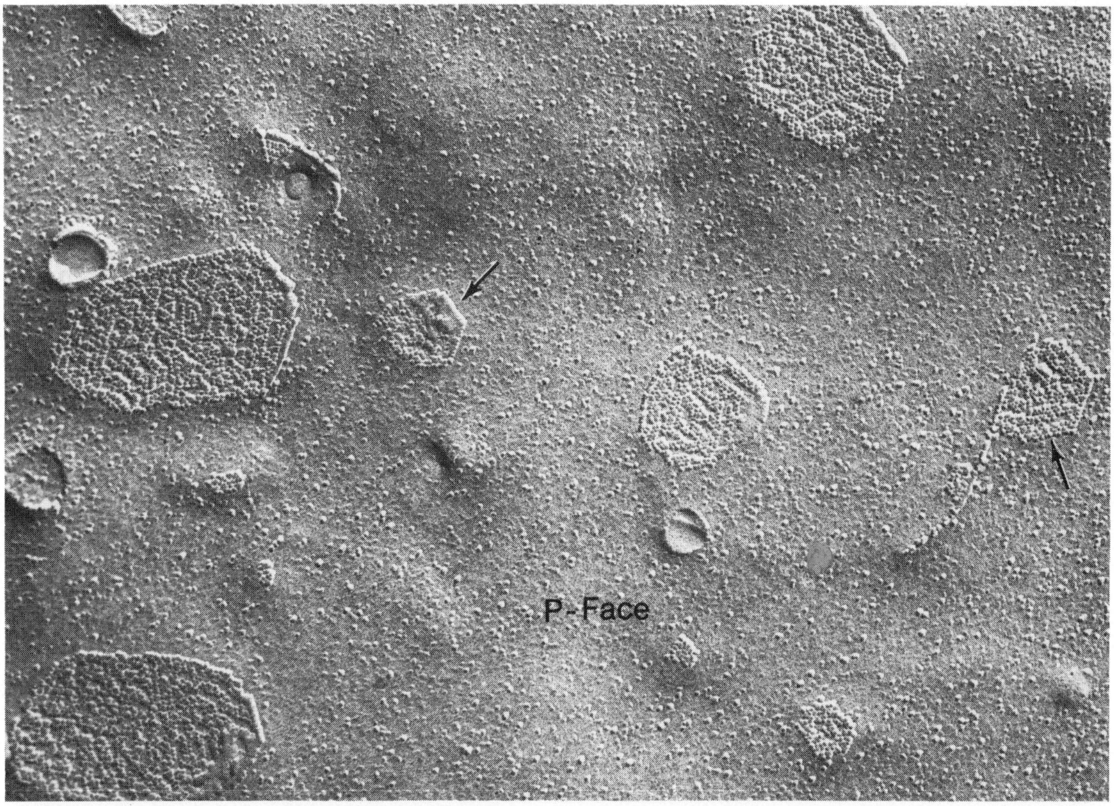

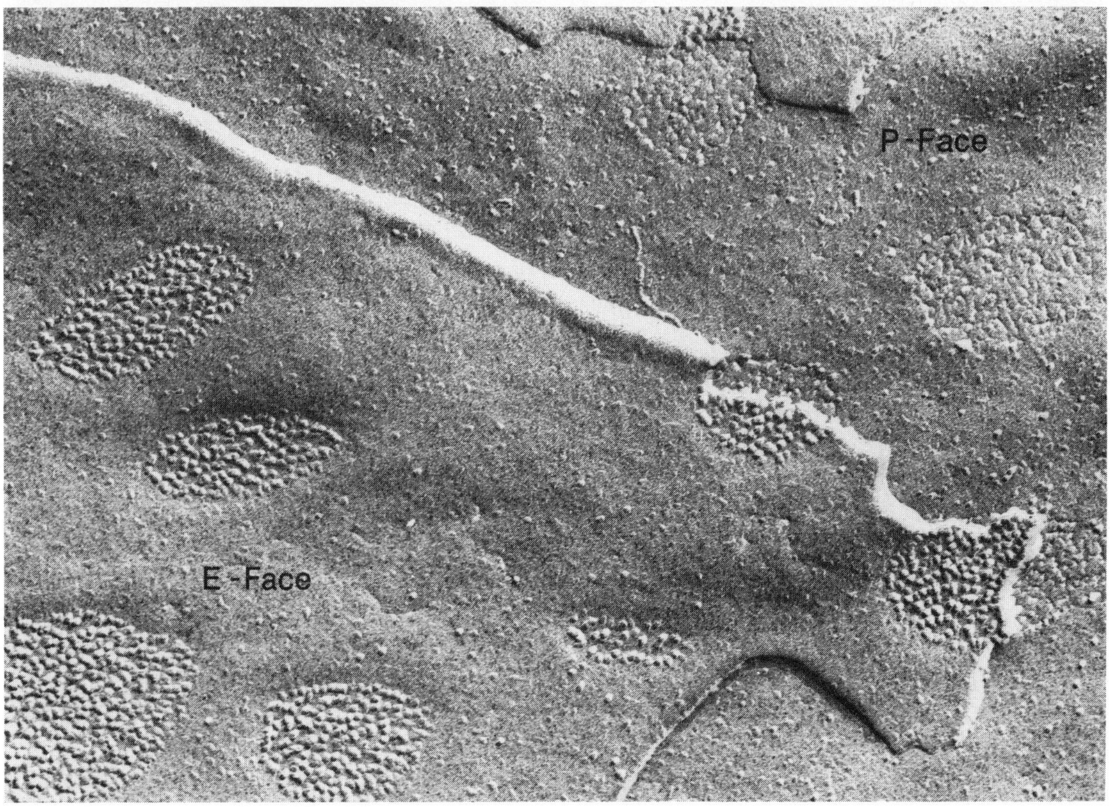

Figure 104, *upper* Figure 105, *lower*

INTERCALATED DISCS AND GAP JUNCTIONS OF CARDIAC MUSCLE

Histologists were unable to resolve cell boundaries in cardiac muscle with the light microscope and considered it to be a syncytium. Its coordinated contraction was believed to be a consequence of protoplasmic continuity throughout the myocardium. The transversely oriented, heavily stained *intercalated discs* were variously interpreted as irreversible contraction bands, sites of formation of new sarcomeres, or special devices for coordinating the contraction of the myofibrils.

The electron microscope set aside these erroneous earlier interpretations by clearly demonstrating the cellular nature of cardiac muscle. The intercalated discs were shown to be interdigitated end-to-end junctions of cellular units specialized to ensure firm cohesion and to provide for insertion of the myofibrils onto the ends of the cells. The ultrastructure of the discs at sites of myofibril attachment resembled that of the zonulae adherentes of epithelia. Typical desmosomes were also found on the apposed membranes between sites of myofibril termination. The longitudinal segments of the step-like cell boundaries, which had gone undetected with the light microscope, were straight and less obviously specialized. At high magnification, however, certain portions of the lateral membranes of adjoining cells were found to be in very close apposition. These areas of intimate contact were large plaques and not circumferential bands, but because of their superficial resemblance to the "tight junctions" of epithelia, they were referred to as "close junctions," "maculae occludentes," or "fasciae occludentes." These terms were rapidly abandoned when gap junctions were discovered and electron opaque tracers revealed in the intercellular space of the cardiac muscle junctions a hexagonal pattern comparable to that in the communicating junctions of epithelia. Freeze-fracture preparations later confirmed that these junctions contain globular particles in register in the opposing membranes and in contact across a 2 nm intercellular gap. It is now generally accepted that these are large gap junctions and that they mediate electrical coupling and coordination of the contractile activity of cardiac muscle by providing low resistance pathways for passage of ions from cell to cell.

The accompanying low power micrograph shows a step-like end-to-end junction of two cardiac muscle cells. The transverse segments of the cell boundary, corresponding to the intercalated discs of light microscopy, are elaborately interdigitated and the adjacent cytoplasm contains a dense matrix around the terminations of the myofibrils. The longitudinally oriented lateral cell boundaries are straight and exhibit extensive gap junctions (at stars).

Figure 106. Papillary muscle of cat heart.

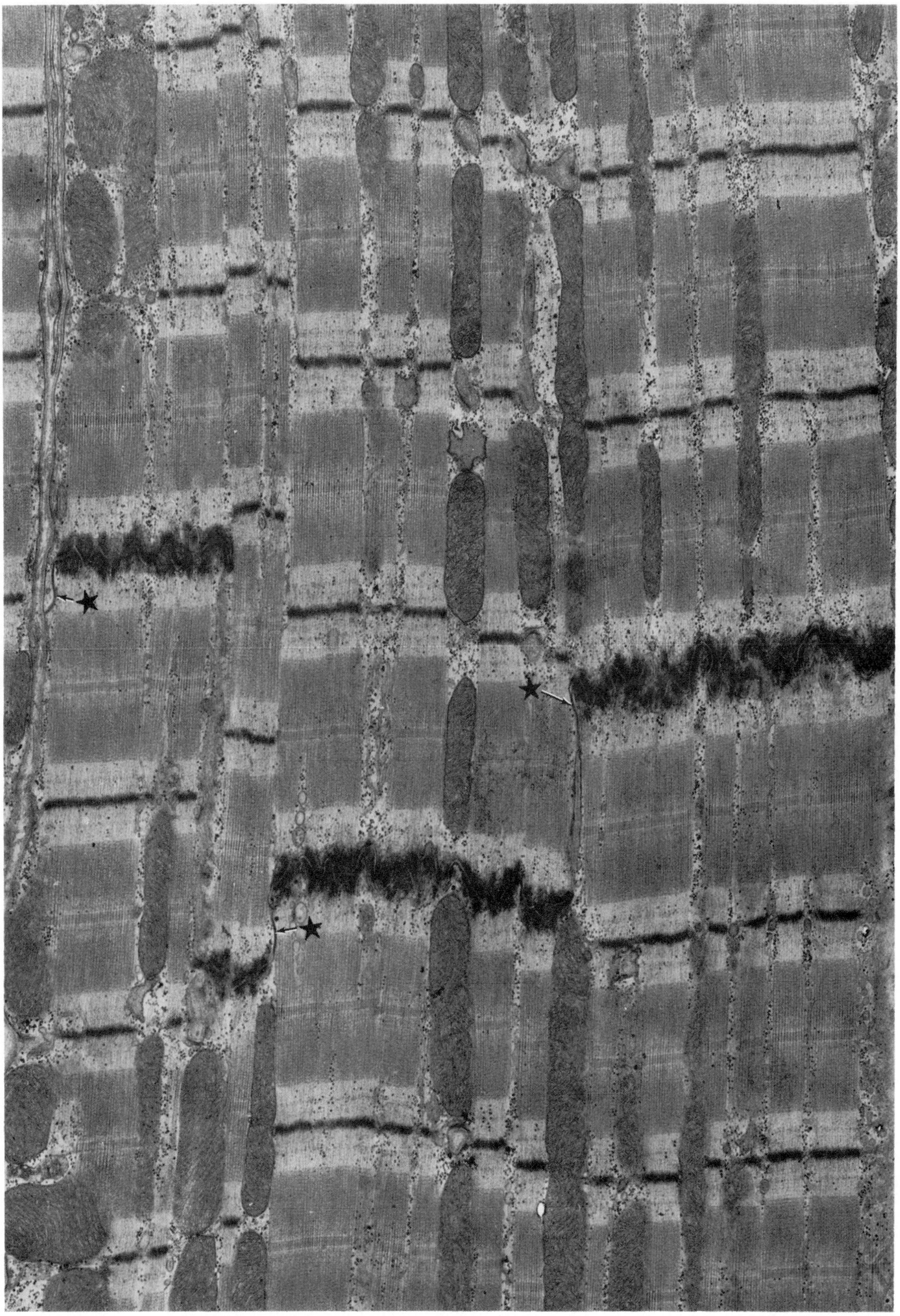

Figure 106

The gap junctions of cardiac muscle are seen to best advantage in transverse sections. The intercellular space is narrowed from the usual 25 to 30 nm to about 2 nm. A regular periodic structure can often be seen in the junctional interspace, representing the end-to-end abutment of the connexons.

Very small gap junctions are occasionally found in the intercalated discs, but in general the transverse segments of the cell surface are specialized for adhesion and transmission of force from the myofilaments of one cell to those of the next along the length of the fibers. For the most part, the communicating junctions responsible for spread of the wave of depolarization throughout the myocardium are located on the longitudinally oriented segments of the cell boundary. Desmosomes are also found on the lateral surfaces of cardiac muscle cells, often near the gap junctions, as in the examples shown on the facing page.

Figures 107 and **108.** Gap junctions in papillary muscle from cat heart.

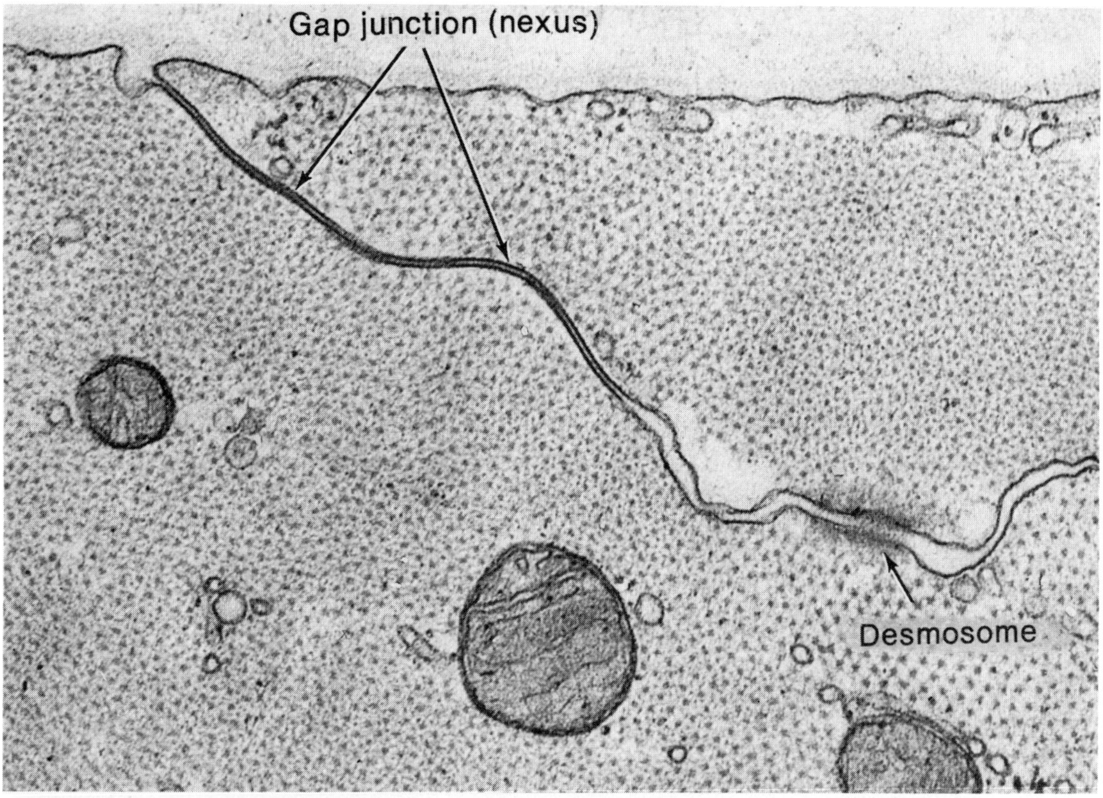

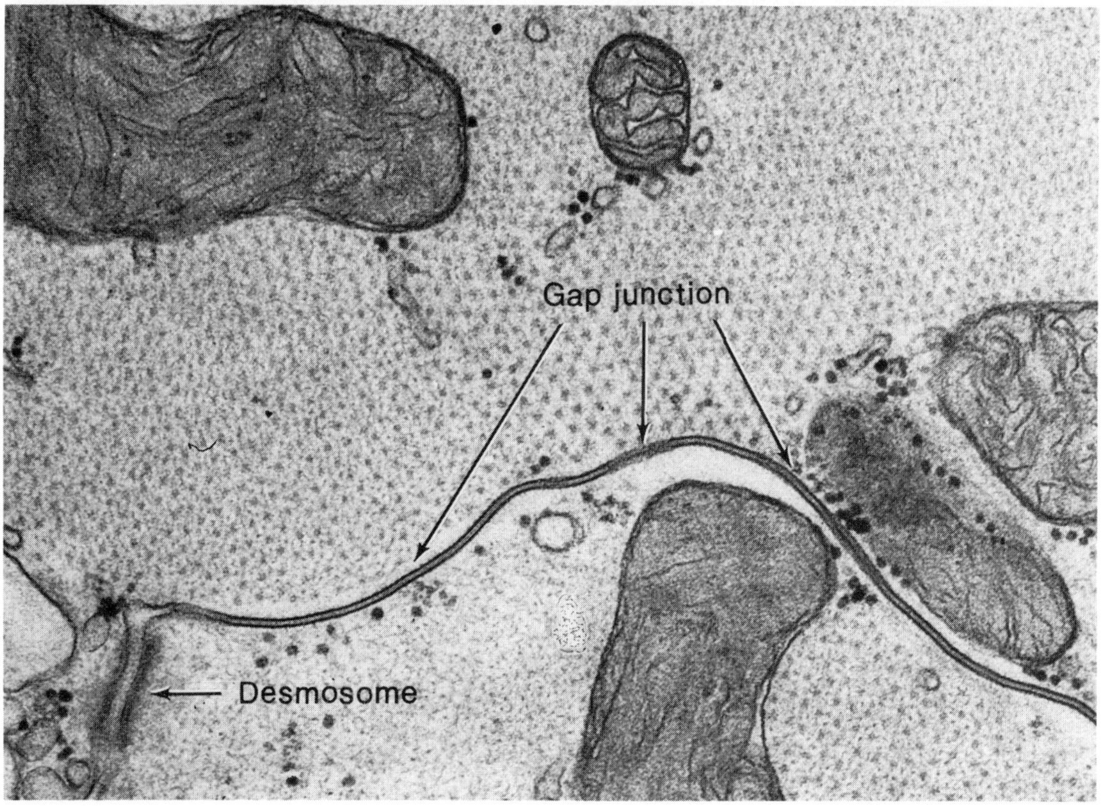

Figure 107, *upper* **Figure 108**, *lower*

REFERENCES

Tight Junction (Zonula Occludens)

Bullivant, S. Evidence that the tight junction sealing element consists of two side-by-side fibrils. J. Cell Biol. 70:35a, 1976.

Bullivant, S. The structure of tight junctions. Proc. International Congress on Electron Micrsocopy, Vol. III. State of the Art Symposia, pp. 659–671, Microscopical Society of Canada, 1978.

Chalcroft, J. P. and Bullivant, S. An interpretation of liver cell membrane and junction structure based on observation of freeze-fracture replicas of both sides of the fracture. J. Cell Biol. 47:49–60, 1970.

Claude, P., and Goodenough, D. A. Fracture faces of zolunae occludentes from "tight" and "leaky" epithelia. J. Cell Biol. 58:390–400, 1973.

Farquhar, M. G. and Palade, G. E. Junctional complexes in various epithelia. J. Cell Biol. 17:375–412, 1963.

Goodenough, D. A. and Revel, J. P. A fine structural analysis of intercellular junctions in the mouse liver. J. Cell Biol. 45:272–290, 1970.

Humbert, F., Grandchamp, A., Pricam, C., Perrelet, A. and Orci, L. Morphological changes in tight junctions of *Necturus maculosus* proximal tubules undergoing saline diuresis. J. Cell Biol. 69:90–96, 1976.

Kreutziger, G. O. Freeze-etching of intercellular junctions of mouse liver. Proc. 26th Annual Meeting, E.M.S.A. 234–235, 1968.

Reese, T. S. and Karnovsky, M. J. Fine structural localization of a blood-brain barrier to exogenous peroxidase. J. Cell Biol. 34:207–217, 1967.

Staehelin, L. A., Mukherjee, T. M. and Williams, A. W. Freeze-etch appearance of the tight junctions in the epithelium of small and large intestine of mice. Protoplasma 67:165–184, 1969.

Staehelin, L. A. Further observations on the fine structure of freeze-cleaved tight junctions. J. Cell Sci. 13:763–786, 1973.

Staehelin, L. A. Structure and functions of intercellular junctions. Int. Rev. Cytol. 39:191–283, 1974.

Wade, J. B., Revel, J. P. and Di Scala, V. Effect of osmotic gradients on intercellular junctions of the toad bladder. Am. J. Physiol. 224:407–415, 1973.

Wade, J. B. and Karnovsky, M. J. The structure of the zonula occludens. A single fibril model based on freeze-fracture. J. Cell Biol. 60:168–180, 1974a.

Wade, J. B. and Karnovsky, M. J. Fracture faces of osmotically disrupted zonulae occludentes. J. Cell Biol. 62:344–350, 1974b.

Zonula Continua and Septate Junction of Invertebrates

Filshie, B. K. and N. E. Flower. Junctional structures in Hydra. J. Cell Sci. 23:151–172, 1977.

Flower, N. E. Invertebrate gap junctions. J. Cell Sci. 25:163–171, 1977.

Flower, N. E. and B. K. Filshie. Junctional structures in the midgut cells of lepidopteran caterpillars. J. Cell Sci. 17:221–239, 1975.

Gilula, N. B. and P. Satir. Septate and gap junctions in molluscan gill epithelium. J. Cell Biol. 51:869–872, 1971.

Gouranton, J. Structure des desmosomes septaux. J. Microscopie 6:505–508, 1967.

Lane, N. J. Developmental stages in formation of inverted gap junctions during turnover in adult horseshoe crab, limulus. J. Cell Sci. 32:293–305, 1978.

Lane, N. J., H. LeB. Skaer and L. S. Swales. Intercellular junctions in the nervous system of insects. J. Cell Sci. 26:175–199, 1977.

Locke, M. The structure of septate desmosomes. J. Cell Biol. 25:166–169, 1965.

Noirot, C. and C. Noirot-Timothee. Un nouveau type de jonction intercellulaire (zonula continua) dans l'intestin moyen des insectes. C. R. Acad. Sci. 264:2796–2798, 1967.

Noirot-Timothee, C. and C. Noirot. Les jonctions "gap" chez les insectes. Structure, localisation, coexistence avec d'autres types de junctions. J. Microscopie 20:76a, 1974.

Satir, P. and N. B. Gilula. the fine structure of membranes and intercellular communication in insects. Ann. Rev. Entomol. 18:143–166, 1973.

Wood, R. Intercellular attachment in the epithelium of Hydra as revealed by electron microscopy. J. Biophys. Biochem. Cytol. 6:343–352, 1959.

Desmosomes

Bizzozero, G. Osservazione sulla structura degli epiteli pavimentosi stratificrati. Rend. r. 1st Lombardo 3:675, 1870.

Farquhar, M. G. and G. E. Palade. Junctional complexes in various epithelia. J. Cell Biol. 17:375, 1963.

Furshpan, E. J. and D. D. Potter. Mechanism of nerve impulse transmission at a crayfish synapse. Nature 180:342, 1957.

Furshpan, E. J. and D. D. Potter. Low-resistance junctions between cells in embryos and tissue culture. Curr. Top. Dev. Biol. 3:95–127, 1968.

Gilula, N. B. Isolation of rat liver gap junctions and characterization of the polypeptides. J. Cell Biol. 63:111a, 1974.

Gilula, N. B., M. L. Epstein and W. H. Beers. Cell-to-cell communication and ovulation. A study of the cumulus-oocyte complex. J. Cell Biol. 78:58–75, 1978.

Goodenough, D. A. and N. B. Gilula. Cell junctions and intercellular communication. *In* Membranes and Viruses in Immunopathology (S. B. Day and R. A. Good, eds.), pp. 155–168, Academic Press, New York, 1972.

Goodenough, D. A. Bulk isolation of mouse hepatic gap junctions. Characterization of the principal protein, connexin. J. Cell Biol. 61:557–563, 1974.

Goodenough, D. A. The structure and permeability of isolated hepatocyte gap junctions. Cold Spring Harbor Symposium on Quantitative Biology 40:37–43, 1976.

Karrer, H. E. Cell interconnections in normal human cervical epithelium. J. Biophys. Biochem. Cytol. 7:181, 1960.

Karrer, H. E. and J. Cox. The striated musculature of blood vessels. II. Cell interconnections and cell surface. J. Biophys. Biochem. Cytol. 8:135, 1960.

Lawrence, T. S., W. H. Beers and N. B. Gilula. Transmission of hormonal stimulation by cell-to-cell communication. Nature 272:501–506, 1978.

Loewenstein, W. R. Permeable junctions. Cold Spring Harbor Symposium on Quantitative Biology 40:49–63, 1976.

Makowski, L., D. L. D. Caspar, W. C. Phillips and D. A. Goodenough. Gap junction structures. II. Analysis of the x-ray diffraction data. J. Cell Biol. 74:629–645, 1977.

Muir, A. R. and A. Peters. Quintuple-layered membrane junctions at terminal bars between endothelial cells. J. Cell Biol. 12:443, 1962.

Pitts, J. D. and J. W. Simms. Permeability of junctions between animal cells. Intercellular transfer of nucleotides but not macromolecules. Exp. Cell Res. 104:153–163, 1977.

Prosser, C. L., N. Sperelakis and R. A. Bergmann. Conduction in intestinal circular muscle. Am. J. Physiol. 183:652, 1955.

Raviola, E., D. A. Goodenough and G. Raviola. The native structure of gap junctions rapidly frozen at 4° K. J. Cell Biol. 79: (abstract), 1978.

Raviola, G. and E. Raviola. Intercellular junctions in the ciliary epithelium. J. Invest. Ophthalmol. 17:958–981, 1978.

Revel, J. P. and M. J. Karnovsky. Hexagonal array of subunits in intercellular junctions of the mouse heart and liver. J. Cell Biol. 33:C7, 1967.

Robertson, J. D. New observations on the ultrastructure of the membranes of frog peripheral nerve fibers. J. Biophys. Biochem. Cytol. 3:1043–1047, 1957.

Robertson, J. D. Structural alterations in nerve fibers produced by hypotonic and hypertonic solutions. J. Biophys. Biochem. Cytol. 4:349–364, 1958.

Robertson, J. D. The occurrence of a subunit pattern in the unit membranes of club endings in the Mauthner cell synapses in goldfish brains. J. Cell. Biol. 19:201, 1963.

Rose, B., I. Simpson and W. R. Loewenstein. Calcium ion produces graded changes in permeability of membrane channels in cell junction. Nature 267:625–627, 1977.

Simpson, I., B. Rose and W. R. Loewenstein. Size limit of molecules permeating the junctional membrane channels. Science 195:294–297, 1977.

Turin, L. and A. Warner. Carbon dioxide reversibly abolishes ionic communication between cells of early amphibian embryo. Nature 270:56–57, 1977.

Ranvier, L. Nouvelles recherches sur le mode d'union des cellules du corps muqueux de Malpighi. Compt. Rendu. Acad. des Sci. 89:667, 1879.

Kolossow, A. Ueber die Struktur des Pleuroperitoneal — und Gefassepitheliels (Endothesis). Arch. Mikrskop. Anat. 42:318, 1893.

Carlier, E. W. On intercellular bridges in columnar epithelia. La Cellule 11:263, 1895.

Porter, K. R. Observations on the submicroscopic structure of animal epidermis. Anat. Rec. 118:433, 1954.

Fawcett, D. W. Intercellular bridges. Exp. Cell Res. Suppl. 8:174, 1961.

Odland, G. F. The fine structure of the interrelationship of cells in the human epidermis. J. Biophys. Biochem. Cytol. 4:529, 1958.

Kelly, D. E. The fine structure of desmosomes, hemidesmosomes and an adepidermal globular layer in developing newt epidermis. J. Cell Biol. 28:51, 1966.

Rayns, D. G., F. O. Simpson and J. M. Lidingham. Ultrastructure of desmosomes in the mammalian intercalated disc: Appearance after lanthanum treatment. J. Cell Biol. 42:322–325, 1969.

Staehelin, L. A. and B. E. Hull. Junctions between living cells. Sci. Am. 238:141–152, 1978.

Gap Junctions (Nexuses, Communicating Junctions)

Blackshaw, S. E. and A. E. Warner. Low resistance junctions between mesoderm cells during development of trunk muscles. J. Physiol. 255:209–230, 1976.

Bozler, E. Conduction automaticity and tonus of visceral muscle. Experientia 4:213–217, 1948.

Caspar, D. L. D., D. A. Goodenough, L. Makowski and W. C. Phillips. Gap junction structures. I. Correlated electron microscopy and x-ray diffraction. J. Cell Biol. 75:605–628, 1977.

Chalcroft, J. P. and S. Bullivant. An interpretation of liver cell membrane and junction structure based on observation of freeze-fracture replicas of both sides of the fracture. J. Cell Biol. 47:49–60, 1970.

Dewey, M. M. and L. Barr. A study of the structure and distribution of the nexus. J. Cell Biol. 23:553–560, 1964.

Devis, R. and D. W. James. Close assocation between adult guinea-pig fibroblasts in tissue culture, studied with the electron microscope. J. Anat. (London) 98:63–69, 1962.

Duguid, J. R. and J. P. Revel. The protein component of mouse hepatocyte gap junctions. FEBS Lett. 78:295–299, 1977.

NUCLEUS

In observing living blood cells of amphibians and birds, Leeuwenhoek in 1710 noted in each a centrally placed "clear area" which was almost certainly the structure we now recognize as the nucleus. But credit for discovery of the nucleus is usually given to Fontana (1781), who examined isolated epidermal cells from eel skin and observed in each an ovoid structure—no doubt the nucleus. Brown (1833) recognized the constant occurrence of a nucleus in botanical material and first enunciated the concept of the nucleated cell as the unit of structure in plants, a concept soon extended to animals.

The nucleus is now known to be the repository of the genome and the source of the informational macromolecules that control the synthetic activities of the cytoplasm. As such it is an essential organ present in nearly all eukaryotic cells. The few exceptional anucleate cells (mammalian erythrocytes, blood platelets, core fibers of the lens) are incapable of protein synthesis and are severely limited in their metabolic activities. After microsurgical removal of its nucleus, an amoeba survives for some time but stops moving, rounds up, and its protein synthesis ceases. Timely replacement of a nucleus reverses these changes and allows resumption of normal activity.

The nucleus is limited by a bilaminar *nuclear envelope* provided with *pore complexes* that permit transit of informational macromolecules to the surrounding cytoplasm. The genetic material of the nucleus, *chromatin,* takes various forms in different phases of the cell cycle. Between successive mitotic divisions, some of the chromatin remains condensed and is visible in microscopic preparations as heavily stained clumps of irregular outline, sometimes called *karyosomes.* This condensed and stainable form is referred to as *heterochromatin.* The rest of the chromatin is dispersed in the *nuclear matrix* in a form that is not identifiable with the light microscope. This invisible portion is called the *euchromatin.* During the early phases of cell division, it too becomes condensed and the entire complement of chromatin then becomes visible in the form of discrete elongated bodies, the *chromosomes.* The number and shape of the chromosomes is characteristic for each species.

In the interphase nucleus, much of the heterochromatin is situated in close association with the inner aspect of the nuclear envelope. In addition to these small peripheral clumps of deeply stained material, the nucleus contains a larger central or eccentrically placed *nucleolus* which also has an affinity for basic dyes. This conspicuous nuclear organelle is a site of processing of ribonucleoprotein for export to the cytoplasm.

These several structural components of the nucleus are discussed in greater detail in later pages.

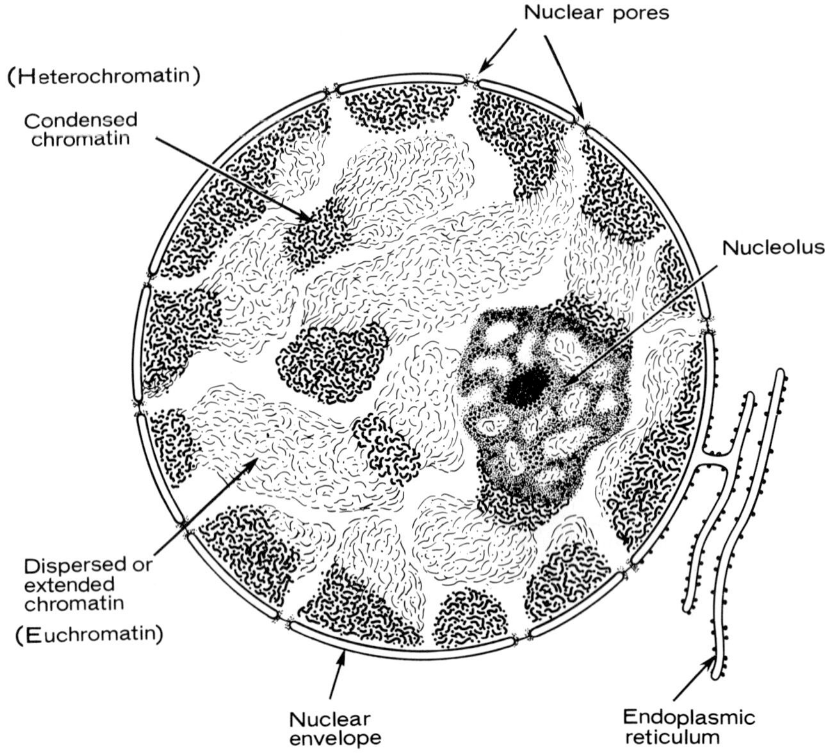

Schematic depiction of the principal structural components of the nucleus. The condensed heterochromatic regions of the chromosomes are shown in continuity with extended or euchromatic segments, which are assumed to be the regions active in transcription. The nucleolus is associated with a euchromatic segment of a chromosome carrying the gene for ribosomal RNA. Pores traverse the bilaminar nuclear envelope, which is continuous at one or more sites with the membranes of the endoplasmic reticulum. (From W. Bloom and D. W. Fawcett, Textbook of Histology, 10th Ed., W. B. Saunders Co., Philadelphia, 1975.)

NUCLEAR SIZE AND SHAPE

In the great majority of cells, the nucleus is spheroid or ovoid, a form that would provide minimal surface in relation to volume. Departures from these simple shapes may be imposed by external forces, as in actively amoeboid cells where the nucleus may be deformed incidental to changes of cell shape during locomotion, or in contractile cells such as smooth muscle where the ellipsoid nucleus in the relaxed cell takes on a helical configuration when the cell shortens during contraction. In such examples, the nucleus appears to be passively deformed.

In cells that are extremely active in protein synthesis such as the spinning glands of insects, the nucleus may be elaborately convoluted. This is interpreted as a device for increasing the surface area of the nucleus to meet the exceptional demand for interaction of nucleus and cytoplasm involved in synthesis of very large amounts of protein. But the metabolism of cells ultimately depends upon their genome and the nuclei may be altered to provide for increased functional demands in several different ways: *multinuclearity,* achieved by karyokinesis without cytokinesis; *polyploidy,* in which the number of chromosome sets per nucleus is increased; *polyteny*, in which the number of chromosomes remains the same but the amount of DNA per chromosome is multiplied; or by *gene amplification* as in the case of the multiple "nucleoli" in oocytes containing extra copies ribosomal DNA.

In general, there is a correspondence between the size of the nucleus and the amount of genetic material it contains. Thus when the number of chromosome sets increases in a polyploid series, there is a roughly proportional increase in the volume of the nucleus. Thus in the huge megakaryocytes of bone marrow, the large lobulated nucleus contains 16 or more sets of chromosomes.

A striking difference in nuclear size is illustrated in the accompanying micrograph of neighboring epithelial cells in tick salivary gland. The nucleus of the interstitial epithelial cell at the lower right is diploid, while the large nucleus of the glandular cell above is undoubtedly polyploid or polytene.

Figure 109. Adjacent cells in acinus type II from salivary gland of the tick *Rhipicephalus appendiculatus.*

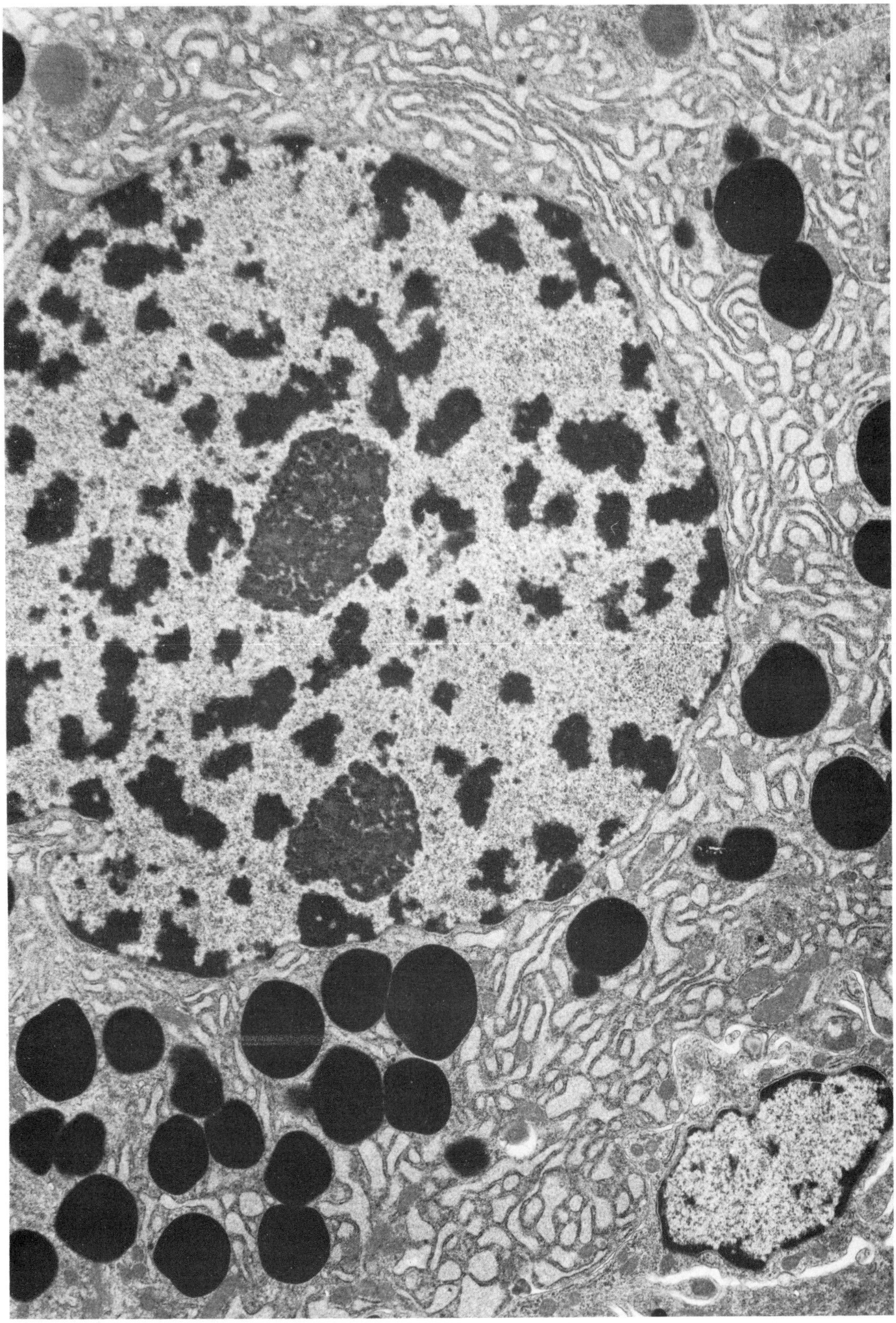

Figure 109

Nuclei of elaborate shape usually occur in long-lived fully differentiated cells that are metabolically very active, but nuclear lobulation may occur in end-stage cells in which protein synthesis is minimal. An example is the neutrophilic leucocyte, a cell of limited life span that exhibits a progressive increase in nuclear lobulation with time. The formation of lysosomes and specific granules in this cell line is greatest at the myelocyte stage when the nucleus is ovoid or reniform. Synthetic activity progressively declines as the cell ages. Therefore the significance of the increase in surface area of the nucleus in this cell type is obscure.

The accompanying micrograph shows a young neutrophil with a relatively simple nucleus. For comparison a photomicrograph of two older cells with more highly lobulated nuclei is presented in the inset.

Figure 110. Human polymorphonuclear neutrophil.

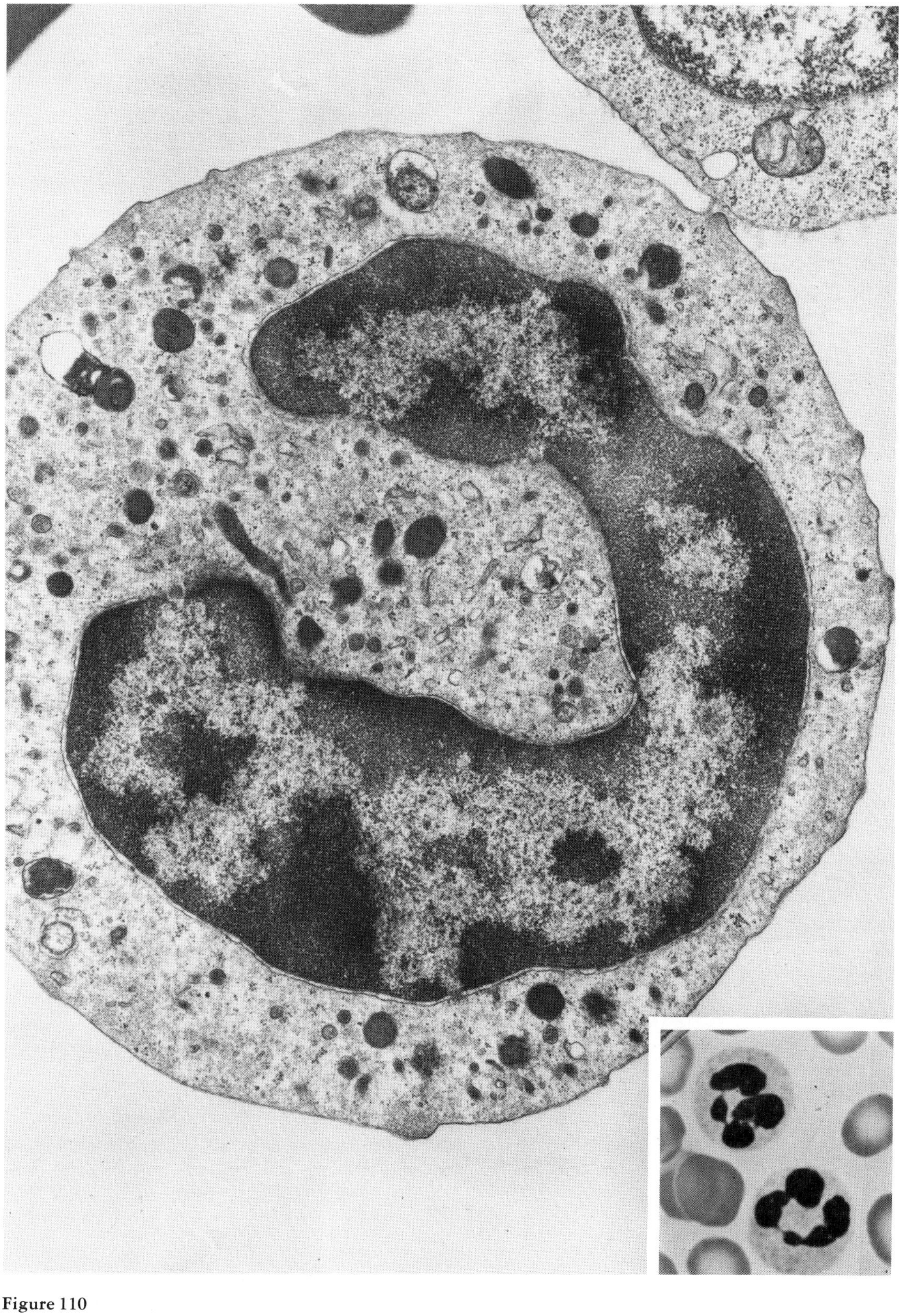

Figure 110

There are other puzzling examples of multilobulated nuclei. For example, the principal cells of the epithelium lining the epididymis in some mammalian species have an extraordinary lobulation of the nuclei. This is also seen in the epithelium of the human ductus deferens. Although our knowledge of the secretory products of these cells is still incomplete, there is little evidence at present that their metabolic and synthetic activity is great enough to require such extensive alteration of the normal surface-to-volume ratio. If this nuclear modification is a response to increased demand for nuclear cytoplasmic interaction, one might expect an associated increase in the number of nuclear pores. There is no evidence for a greater than average number of pores per unit area of nuclear envelope. The functional significance of the nuclear pleomorphism in these epithelia thus remains unexplained.

Figure 111. Principal cell from the epididymal epithelium of the chinchilla.

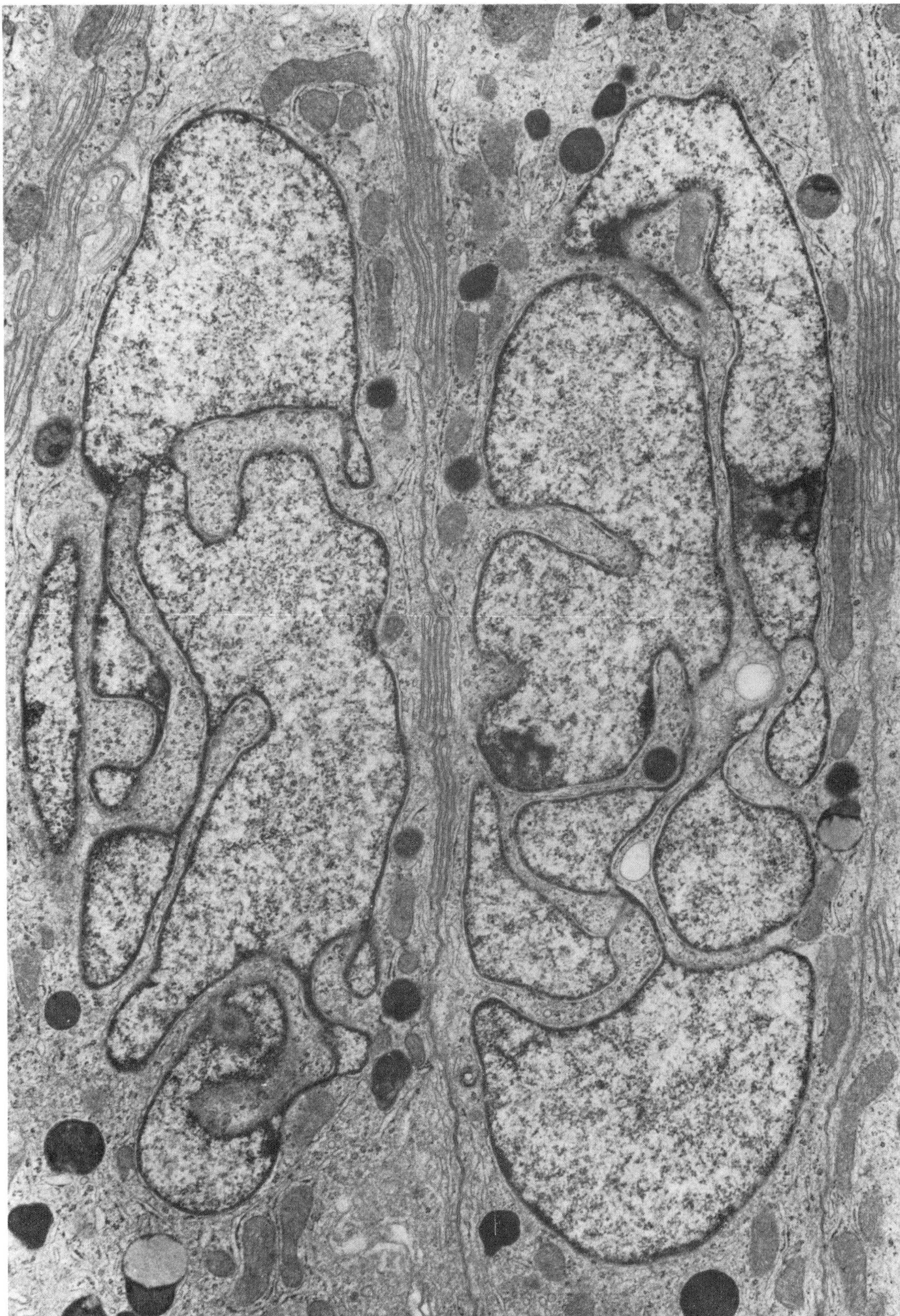

Figure 111

CHROMATIN

The earliest microscopic observations on the nucleus were made on free-hand slices or on cells teased from tissues. It is remarkable that using such simple preparative procedures, von Baer (1834) could suggest the possibility of cell division and Remak (1852) was able to conclude that division into two equal daughter cells was the general mechanism for increase in cell numbers – a concept which received enduring expression in 1855 in Virchow's aphorism "*omnis cellula e cellula.*" But nothing was known about the role of the nucleus in cell division until the latter half of the 19th century, when the introduction of chemical fixation, the development of simple microtomes, and the use of natural and synthetic dyes made it possible to demonstrate structures within the nucleoplasm. Flemming in 1882 applied the term *chromatin* to the stainable constituents of the nucleus. The term originally included the nucleolus but its distinctive tinctorial properties with certain combinations of dyes later led to its recognition as a separate organelle.

It was soon noted that the limits of the nucleus were not always discernible and that in some cells the discontinuous clumps of chromatin were transformed into long convoluted strands that condensed and shortened into intensely stained rods (batonets) of varying length. These structures were clearly described by Balbiani (1875) and were later called *chromosomes* by Waldeyer (1888). Flemming (1879) observed that these split longitudinally and it was reported by van Beneden (1884) that the longitudinal halves, later named *chromatids,* passed to the respective daughter cells where they were reconstituted into a nucleus. By the late 1880s the various stages of this process had been observed in favorable living cells of both plants (Strasburger, 1879) and animals (Flemming, 1882) and it was generally accepted that cells are formed from preexisting cells and division of the nucleus (*karyokinesis*) procedes that of the cytoplasm (*cytokinesis*).

The chromosomes, as discrete stainable entities, disappeared during reconstitution of the nucleus in the daughter cells and only the nucleolus and scattered clumps of chromatin remained. But in favorable material, the arrangement of chromosomes in early prophase of cell division appeared to be identical to that observed in telophase, and it was suggested that the chromosomes persisted during interphase and retained their relative positions even though they ceased to be visible (Rabl, 1885; Boveri, 1887). The persistence of chromosomes during interphase gradually gained acceptance, but its conclusive demonstration continued to be one of the most refractory problems in cytology.

Around the turn of the century, the number of chromosomes was found to be constant and characteristic for each species. It was discovered that this number was reduced by half during gametogenesis and the same number of chromosomes was contributed by egg and sperm at fertilization (van Beneden, 1887; Boveri, 1892; Hertwig, 1890). The hereditary and evolutionary implications of these discoveries were evident to Weismann, who by 1885 had developed the theory that heredity is a consequence of the continuity of cells through successive mitotic generations. The chromatin of the cell nucleus emerged from these morphological studies as the most likely physical basis of heredity. The term *euchromatin* was applied to that portion of chromosomes which disappears during interphase and *heterochromatin* to that which persists (Heitz, 1929). The euchromatin was believed to be uncoiled or extended and in this state was thought to be more active, whereas heterochromatin was considered inactive. Chromosomes became an absorbing subject of study after the rediscovery in 1900 of Mendel's principles of genetics, for it then seemed possible to derive from those principles and from microscopic observations on the nucleus a chromosomal theory of inheritance.

Some chemical investigations of cells had also been undertaken in the latter part of the 19th century, but little effort was made to relate the findings to the structures observed by microscopists. Miescher (1874) analyzed fish spermatozoa, and pus obtained from surgical bandages, both consisting of cells with relatively little cytoplasm. These studies established *nucleic acid* as a major constituent of the nucleus. *Histones* were discovered by Kossel (1884), who is also credited with the isolation and identification of most of the purines and pyrimidines of nucleic acids. He also recognized the existence of two types of nucleic acid, originally called "thymus nucleic acid" and "yeast nucleic acid." These differed in the kind of sugar that they contained. Many years later, the sugar in yeast nucleic acid was identified as ribose and that in thymus nucleic acid, as deoxyribose (Levine and London, 1929). From the 1930s onward the nucleic acids were designated *ribose nucleic acid* (RNA) and *deoxyribose nucleic acid* (DNA).

The discovery of the Feulgen reaction, a specific cytochemical stain for deoxyribose nucleoprotein (Feulgen and Rosenbeck, 1924), enabled microscopists to demonstrate that this form of nucleic acid occurs in the nuclei of all cells, whether obtained from plants or animals, and is always associated with ribose nucleic acid. The subsequent development of ultraviolet absorption microspectrophotometry for localization of nucleic acids (Caspersson, 1936) and the widespread use of selective staining methods for DNA and RNA made it possible not only to localize these substances in histological sections but also to achieve a rough quantitation by photometric methods. DNA was localized exclusively in the nucleus and was present in highest concentration in the interphase heterochromatin and in chromosomes of dividing cells, whereas RNA was abundant both in the nucleolus and in the cytoplasm (Caspersson and Schultz, 1940). The amount of DNA per haploid set of chromosomes was found to be constant (Boisvin, Vendreley and Vendreley, 1948) and the amount in diploid nuclei was shown to double just before cell division (Alpert and Swift, 1953). The constancy of the DNA content of cells and its precise relation to the number of chromosome sets was clearly consistent with the possible function of DNA as the carrier of genetic specificity, but proof was lacking and there was at that time no obvious mechanism for linking the genome to the physiological activities of the cell. It was noted, however, that RNA was most abundant in the nucleolus and cytoplasm of those cell types that were known to be active in the synthesis of protein and it could be shown to vary in amount with their degree of functional activity.

The emergence of DNA as the probable genetic material of the cell received a major impetus from classical studies on micro-organisms, which provided compelling experimental evidence that DNA was the agent which brought about the inheritable transformation of nonvirulent pneumococci to the virulent form of the organism (Avery, Macleod and McCarty, 1944). Not long thereafter it was shown in ingenious radiolabeling experiments that when a bacteriophage infects a bacterium only the DNA enters the cell, while the capsule remains outside. All the information needed by the host for synthesis of new virus particles was thus transmitted by the injected DNA molecule (Hershey and Chase, 1952). In the light of these experiments there was no longer any reason to doubt that in prokaryotes, and probably in eukaryotes as well, DNA was the molecular basis of inheritance.

The paramount biological importance of this molecule stimulated renewed efforts to determine its structure. It was known to be a long polymer of four kinds of nucleotides each consisting of a purine or pyrimidine base and deoxyribose, bearing a phosphate group. The purine bases are adenine and guanine and the pyrimidine bases are thymine and cytosine. The successive nucleotides in the polymer are linked together by phosphodiester bonds between the adjacent sugars. Relying upon x-ray diffraction data and modeling, Watson and Crick (1953) proposed that the DNA molecule is composed of two chains of phosphate-linked sugars arranged in a double helix of uniform diameter and pitch. The purines and pyrimidines project inward from the two sugar backbones in a plane perpendicular to the axis of the model and are arranged in *base pairs;* that is, with the purines of one chain in register with the pyrimidines of the other. The two chains are held together by hydrogen bonds between the apposed pairs of bases. Each turn of the double helix includes 10 base pairs and occupies 3.4 nm along the axis of the molecule. Steric conditions and the pattern of hydrogen bonding impose the restriction that a given sequence of bases in one chain is compatible with only one sequence in the opposite chain. This complementarity suggested the possibility that if the chains were to separate and each induce formation of the appropriate base pairs to make a complementary chain, the results would be two new double helices with a sequence of base pairs identical to the original. This proposed structure of the DNA molecule offered the possibility of storing an infinite variety of genetic information encoded in different sequences of nucleotide bases. This model is now generally accepted and has proved to be one of the most fundamental discoveries in the history of biology.

Analysis of the role of the histones of the nucleus and their relationship to the DNA has proved more difficult. More than 20 years ago, it was proposed that they formed complexes with certain parts of the DNA and that they might thus inhibit and control the activity of genes (Stedman and Stedman, 1945). Strong support for this hypothesis came from *in vitro* experiments on cell-free systems which demonstrated that naked DNA was five times as active in RNA synthesis as DNA complexed with histone (Bonner et al., 1968). Thus it seemed likely that the DNA of heterochromatin is present as nucleohistone and that this portion of the genome is not expressed, while the portion not complexed with histone (euchromatin) is active in transcription. There has since been considerable progress in sequencing and working out the chemistry of five major types of histone (H_1, H_2A, H_2B, H_3, and H_4) but much remains to be learned about the mode of association of histones with the template inactive regions of the chromosomes. The suggestion that histone might occupy one of the grooves in the DNA double helix has now been abandoned in the light of recent studies of the higher order structure of chromatin that will be described below.

The application of the electron microscope to the study of cells in the 1950s brought rapid advances in analysis of the organization of the cytoplasm but contributed relatively little to our understanding of the nucleoplasm. The membrane-limited cytoplasmic organelles were adequately preserved by the available osmic acid fixatives and presented in thin sections clearly defined profiles from which their three-dimensional configuration could easily be inferred. The macromolecular constituents of the nucleoplasm, on the other hand, are so small and so varied in their orientation that it was difficult to make valid inferences about their form or organization from the small sample included in an ultrathin section.

With primary osmic acid fixation, there was no obvious pattern of density variations in the nucleoplasm that corresponded to the chromatin pattern seen with the light microscope in stained preparations. In the accompanying micrograph of an osmium-fixed pancreatic acinar cell, there is a conspicuous nucleolus but areas of chromatin cannot easily be identified. This disappointing lack of visible structure led some to conclude that osmium is a poor fixative for nuclei, but when ultraviolet absorption images of living cells were compared with the same cells fixed in osmium, it was found that most of the nucleic acid remained after fixation. Therefore the chromatin is still present after osmium fixation. The difficulty encountered in identifying it in electron micrographs is attributable to the relatively poor osmium staining of nucleic acids and to the fact that the limits of the chromatin are obscured by intermingling with interchromosomal granular material of similar electron density.

Figure 112. Pancreatic acinar cell fixed with osmium and stained with lead hydroxide.

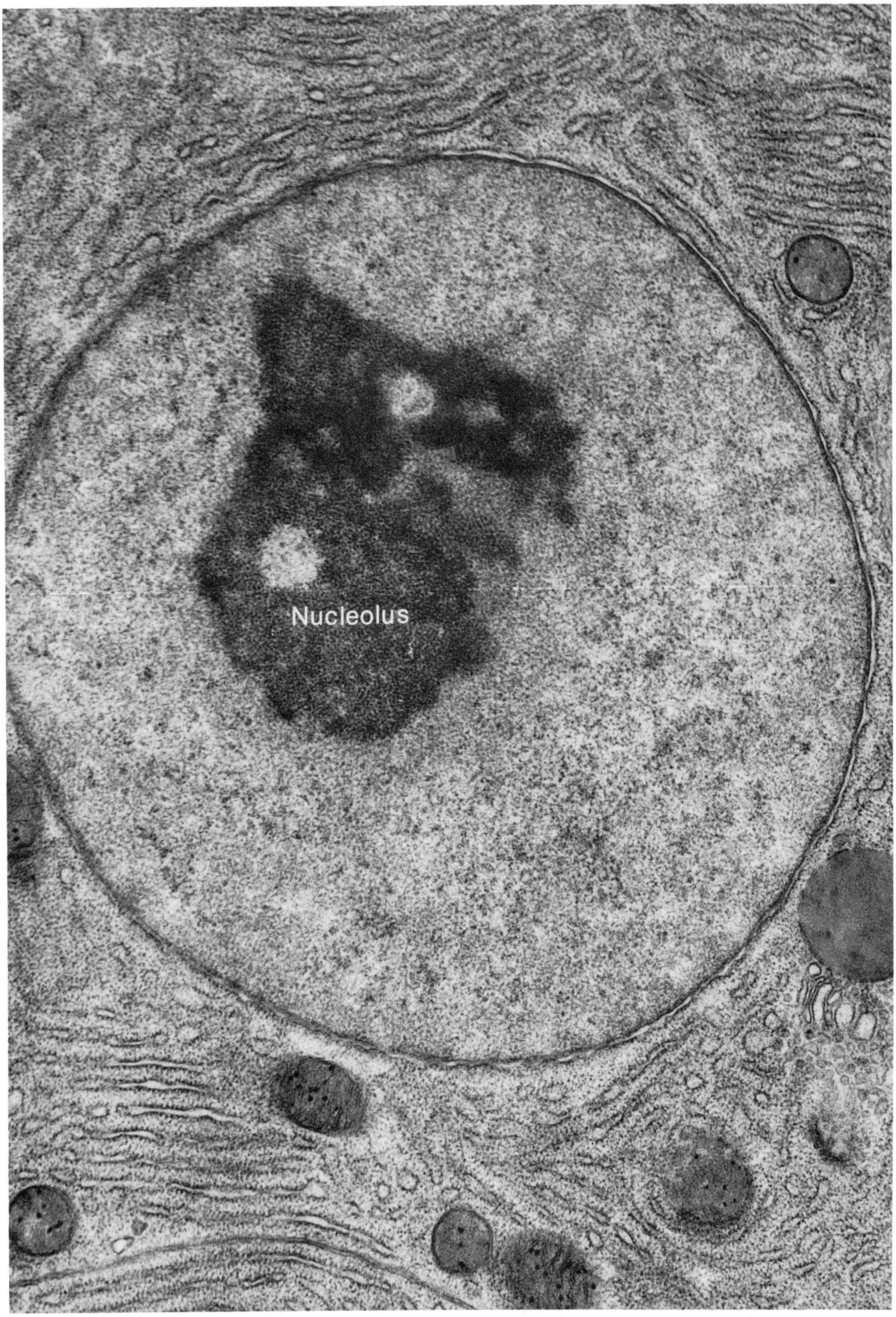

Figure 112

With the introduction of glutaraldehyde as a fixative for electron microscopy (Sabatini et al., 1962) interpretation of nuclear structure was greatly facilitated. In cells fixed in glutaraldehyde followed by osmium tetroxide, the chromatin was preserved in coarser, more discrete granules or filaments and when subsequently exposed to uranyl acetate and lead citrate it was deeply stained and stood out in bold contrast against a nuclear matrix of relatively low density. The pattern of chromatin after this method of specimen preparation corresponded closely to that seen with the light microscope in tissue stained with basic dyes or with the Feulgen reaction.

The nucleus on the facing page, fixed with glutaraldehyde, presents an appearance that is strikingly different from that in the previous micrograph of the same cell type after osmium fixation alone. With this method of specimen preparation, the chromatin appears as dense aggregations of small granules.

In a glandular cell such as that shown here, the heterochromatin typically occurs in irregular clumps adjacent to the nuclear envelope and around the nucleolus. The DNA carrying the genes for ribosomal RNA occupy the relatively unstained interstices of the nucleolus and the genes for messenger RNA and transfer RNA that are directing the synthesis of the cell's secretory product are believed to reside in the pale areas of the nucleoplasm.

Figure 113. Acinar cell from the pancreas of the bat *Myotis lucifugus*, fixed in glutaraldehyde followed by osmium tetroxide.

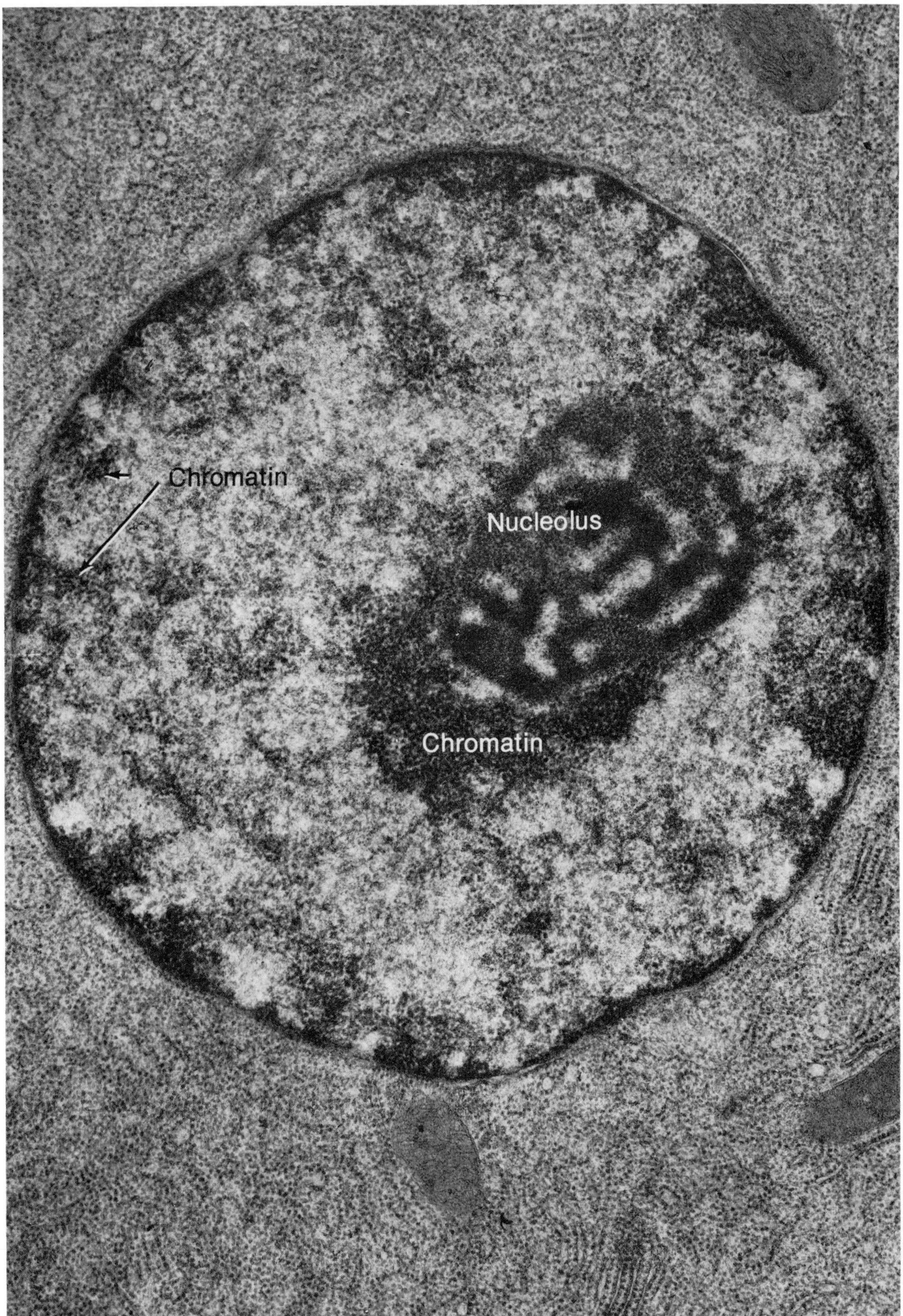

Figure 113

Variations in Amount of Heterochromatin

Although the nuclei of all somatic cells in a given animal species contain the same quantity of DNA, there are marked differences from tissue to tissue in the amount and distribution of visible chromatin. Since it is only the transcriptionally inactive, condensed form that is stained, cells with little demonstrable chromatin are usually considered to be more active in protein synthesis than are those with conspicuous masses of heterochromatin. Thus in neurons, which were traditionally described as having a "vesicular" nucleus containing very little basophilic material in the karyoplasm, a large proportion of their complement of chromatin is in the euchromatic form.

In contrast, the plasma cell shown here has abundant heterochromatin in a characteristic pattern of large blocks around the periphery and a large mass in the center. Such a coarse chromatin pattern might be unexpected in a cell actively synthesizing immunoglobulin. But the amount of protein synthesized is small relative to that of many glandular cells and it is likely that this highly specialized cell needs only a small fraction of its DNA in an active form to direct the narrow spectrum of synthetic activities carried out in its cytoplasm.

Figure 114. Plasma cell from guinea pig bone marrow.

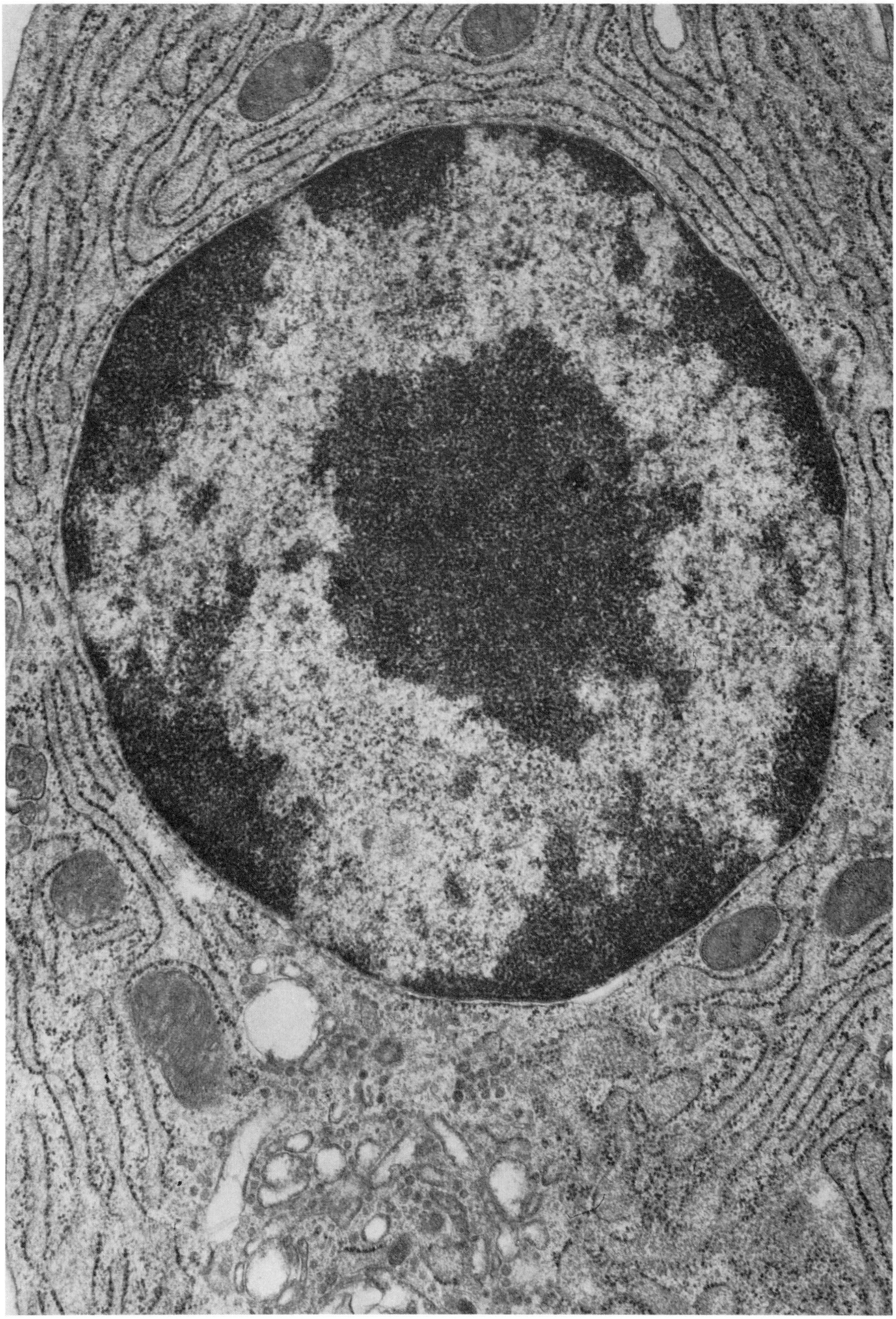

Figure 114

The chromatin pattern may undergo marked changes at successive stages in the course of differentiation of the same cell type. This is exemplified in erythropoiesis. Basophilic erythroblasts have a rather diffuse chromatin pattern and a prominent nucleolus. By the time the polychromatophilic erythroblast stage, illustrated in this electron micrograph, has been reached, the pattern has become much coarser, with large chromatin masses distributed throughout the nucleus. At this stage, synthesis of hemoglobin is still in progress. Numerous dense polyribosomes are present in the cytoplasm, and a profusion of less dense particles of hemoglobin are identifiable both in the cytoplasm and in the interchromosomal areas of nucleoplasm. With additional cell divisions and further differentiation, there is a diminution in the size of the nucleus and a progressive concentration of its chromatin, as illustrated in the next micrograph.

Figure 115. Polychromatophilic erythroblast from guinea pig bone marrow.

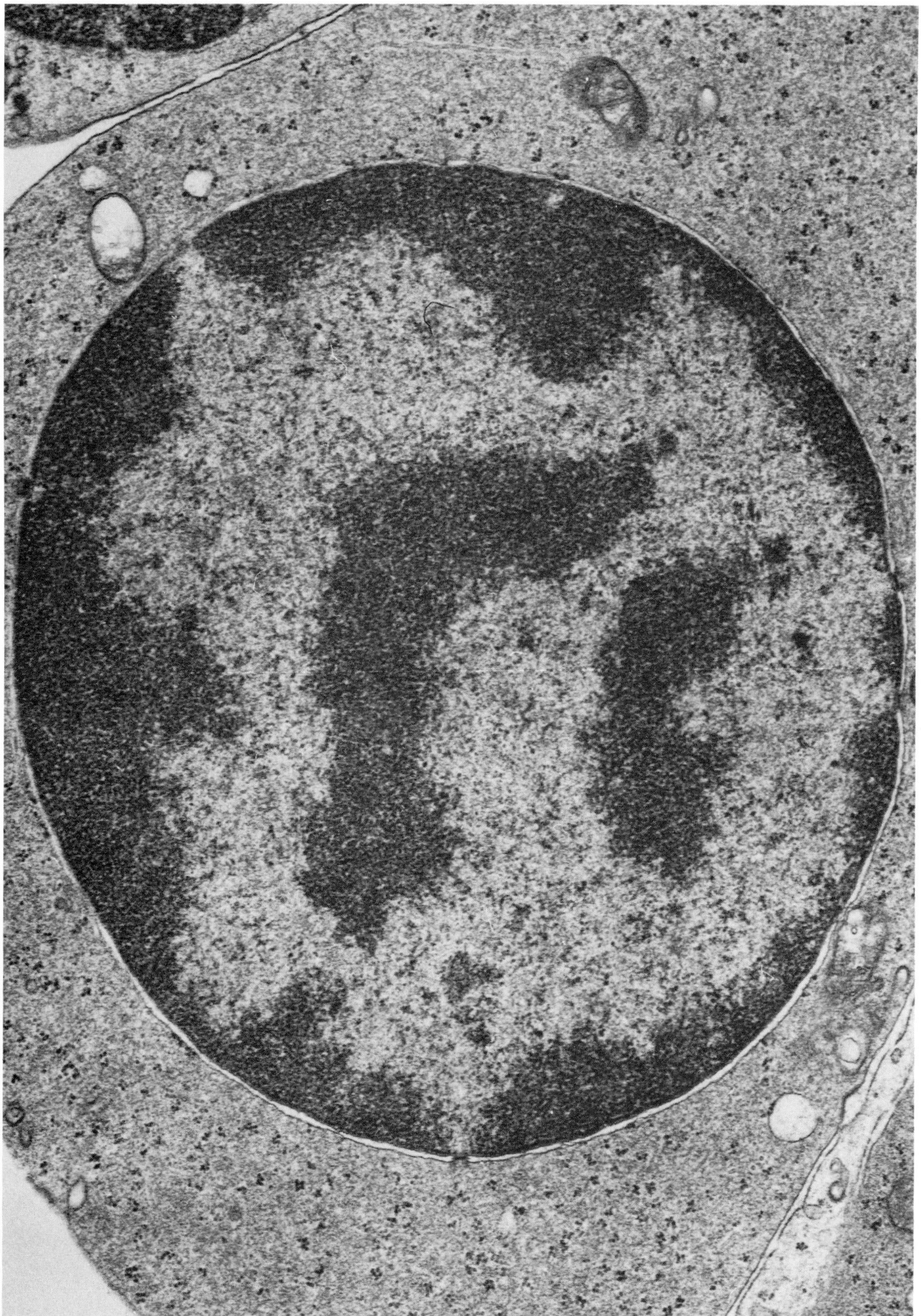

Figure 115

A more advanced stage of nuclear condensation is seen in the orthochromatic erythroblast shown here. Relatively little interchromosomal nucleoplasm remains. The condensed nucleus is now metabolically inert and would soon be extruded. Although hemoglobin synthesis will continue for some time in the cytoplasm of the reticulocyte, it gradually comes to a halt as the supply of messenger RNA is progressively depleted in this anucleate cell. It is evident in this micrograph that the number of polyribosomes has already diminished.

Figure 116. Orthochromatic erythroblast (normoblast) from guinea pig bone marrow.

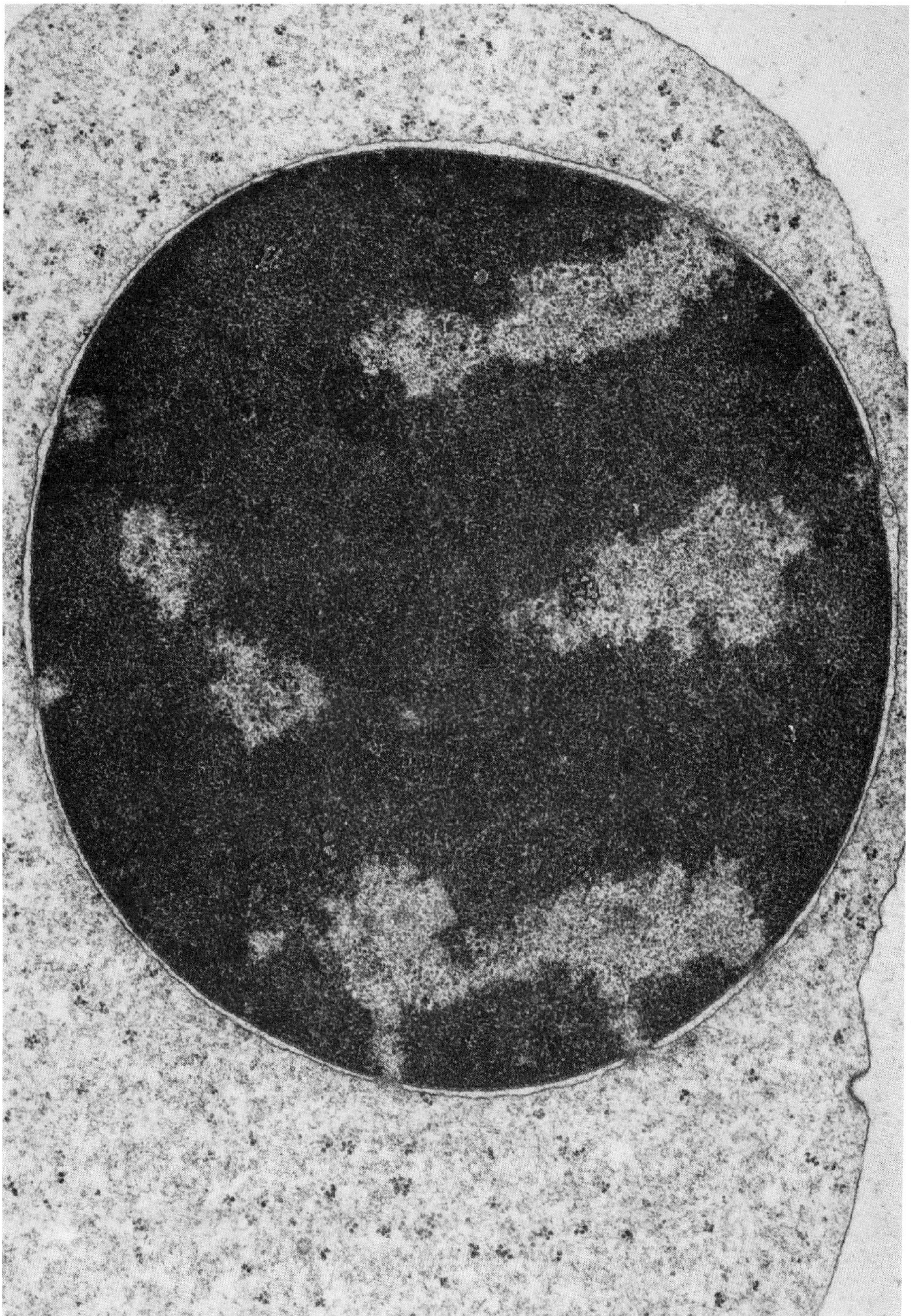

Figure 116

The most extreme example of chromatin condensation is found in the nucleus of a mature spermatozoon, such as that shown on the facing page. The developmental stages leading to this condition are quite different from those just described for the nucleus in the erythropoietic cell line.

Late in spermiogenesis the lysine-rich histones of the nucleus are replaced by arginine-rich histones. This is followed by a complete reorganization of the chromatin involving a sequence of morphological changes that is peculiar to spermiogenesis. Filaments appear in the nucleoplasm and these associate laterally to form coarser strands. These in turn shorten and become compacted into an extremely dense, homogeneous mass having a shape characteristic of the sperm head of the particular species. The shape of the sperm nucleus appears to be determined by a specific pattern of aggregation of the nucleohistones and proteins of the nucleoplasm. The chromosomes are assumed to retain their identity throughout this process of differentiation, but at no stage are there any morphological indications of their limits or their disposition within the condensed nucleus. The metabolically inert mature sperm head is incapable of incorporating labelled amino acids and is so extensively crosslinked that it is resistant to most chemical agents including deoxyribonuclease. The function of the sperm nucleus is solely to transmit genetic information from one generation to the next. Its condensation to an extremely unreactive, resistant state probably serves to protect the genome from damage during the journey from the male to the site of fertilization. Within the egg cytoplasm it rapidly decondenses to form a male pronucleus with an ultrastructure resembling that of somatic cell nuclei.

Figure 117. Mouse epididymal spermatozoon.

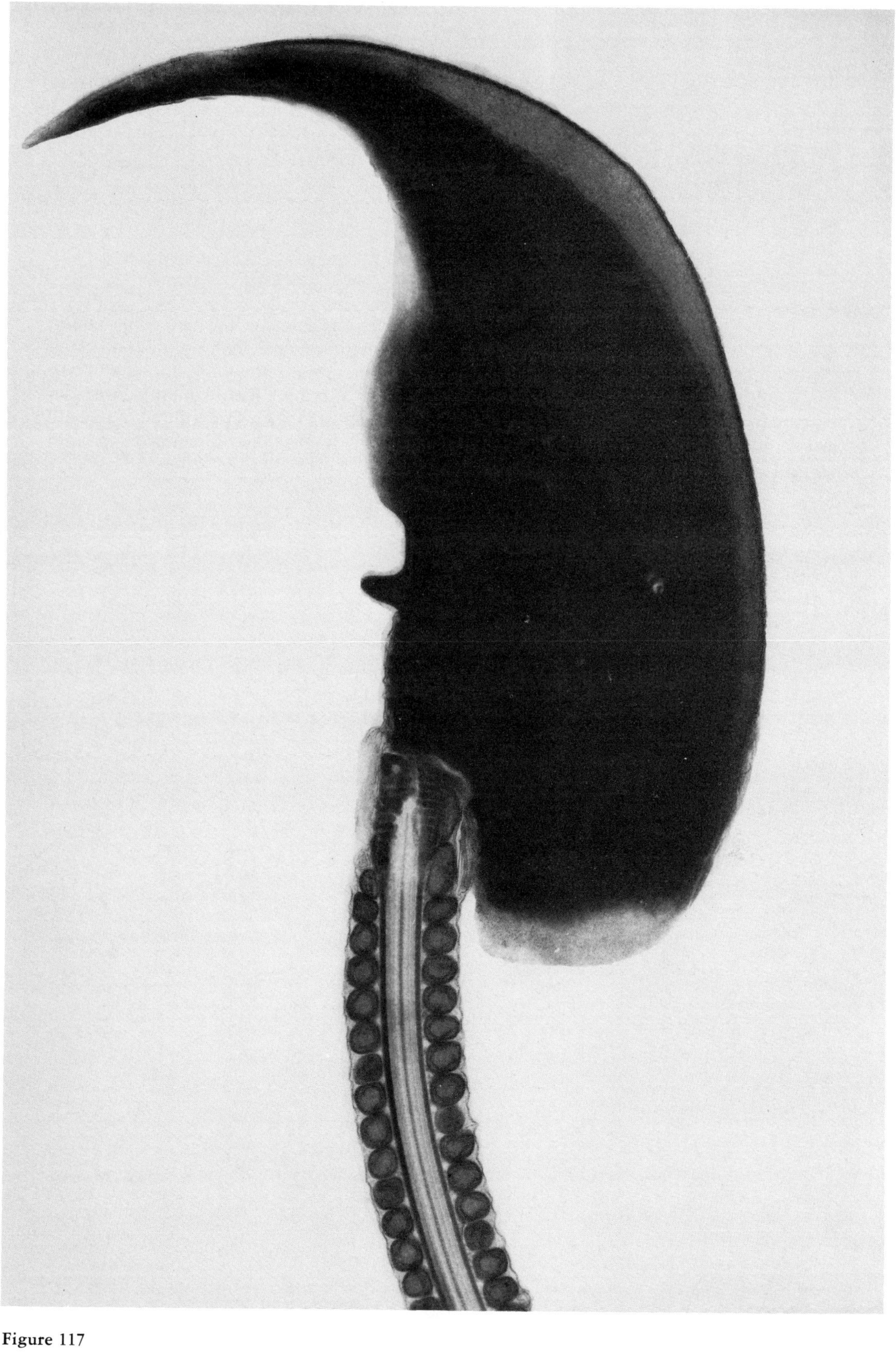

Figure 117

Higher Order Structure of Chromatin

When viewed at high magnification after aldehyde fixation the heterochromatin appears to be composed of 20 nm granules, only slightly larger than the ribosomes on the endoplasmic reticulum in the cytoplasm. Such images alone would not lead one to conclude that chromatin is composed of filamentous subunits. But filaments cut transversely or obliquely have punctate profiles, and if they are highly convoluted or closely interwoven, aggregations of filaments may present a granular appearance in thin sections.

Figure 118. Thin section of the nucleus and adjacent cytoplasm of an acinar cell from pancreas. (Micrograph courtesy of Susumu Ito.)

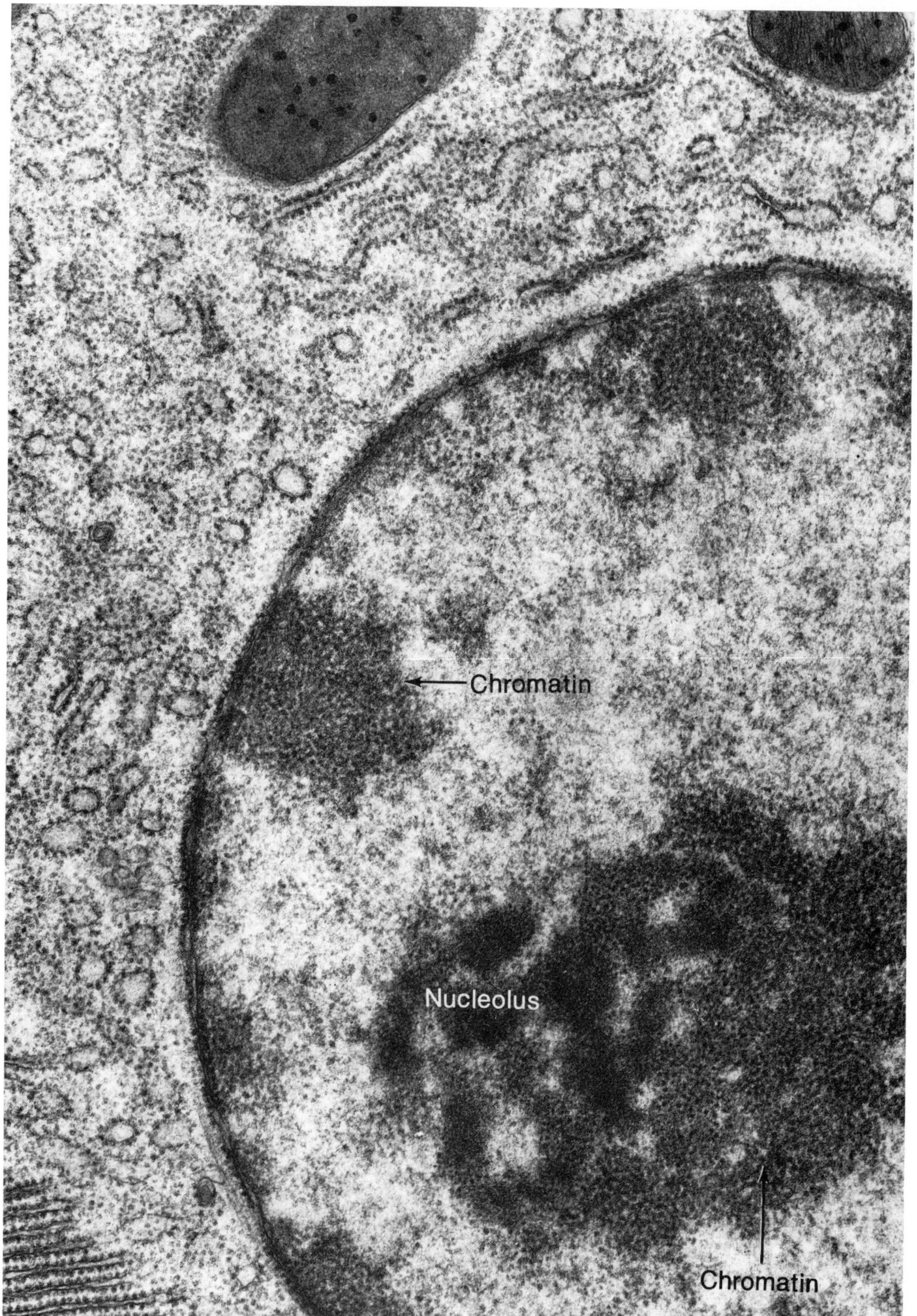

Figure 118

When some cells with abundant heterochromatin, such as the amphibian erythrocyte in the upper micrograph, were examined in very thin sections, occasional elongate profiles were observed (see at arrows), which suggested that the subunits of inactive chromatin were 20 nm filaments.

This interpretation was strongly supported by examination of insect spermatids in the early stages of nuclear condensation when the entire complement of chromatin loosens up and becomes uniformly distributed throughout the karyoplasm. In a thin section of such a nucleus in the lower micrograph on the facing page, the filamentous nature of chromatin becomes obvious.

Figure 119. Section of frog erythrocyte nucleus. Two per cent formaldehyde and 1 per cent osmium tetroxide fixation. (Micrograph courtesy of Hans Ris.)

Figure 120. Section of a spermatid nucleus from an insect. (Micrograph courtesy of David Phillips.)

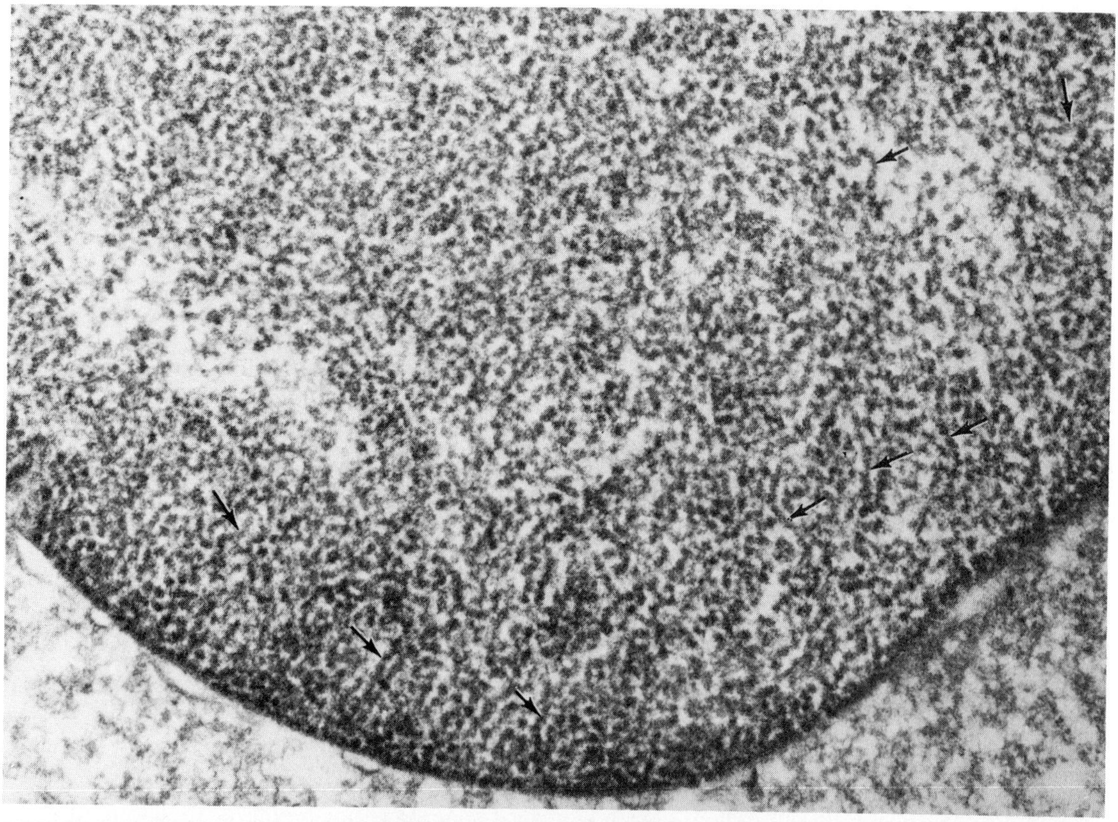

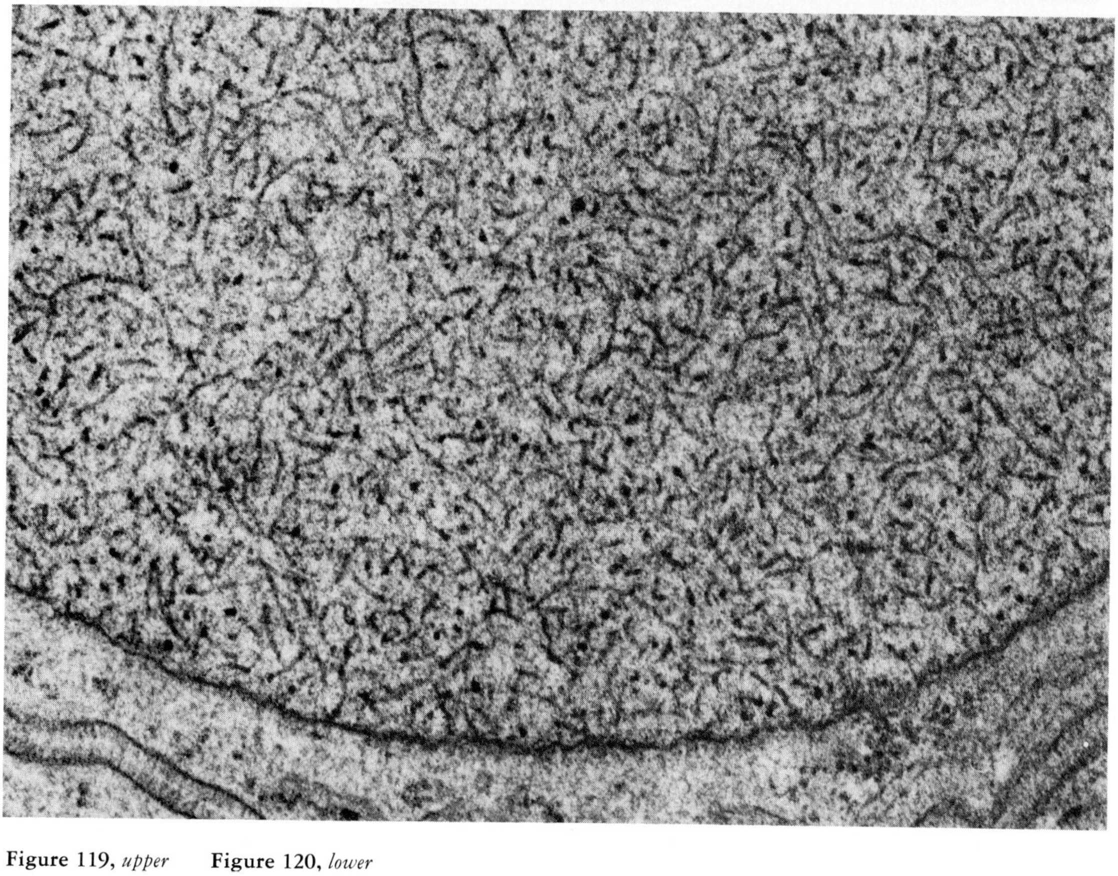

Figure 119, *upper* Figure 120, *lower*

When the contents of amphibian erythrocyte nuclei were prepared for electron microscopy by spreading on the surface of water, examination of whole mounts of the dissociated material revealed a tangle of 20 to 30 nm fibers (upper micrograph). After pretreatment with chelating agents, the chromatin in such preparations appeared as knobby 10 nm filaments (Ris, 1968). If the isolated chromatin was pretreated with urea, 2 to 4 nm filaments were observed.

For several years it remained unclear as to how these three categories of fibers were related to one another and which, if any, corresponded to the native state of the nucleohistone in the living nucleus, but a consensus gradually developed that heterochromatin of the interphase nucleus probably consisted of DNA-histone filaments about 10 nm in diameter folded or helically coiled into fibers 20 to 30 nm in diameter and that the "fibrogranular" appearance of chromatin in routine electron micrographs represented thin sections of a tangled mass of such fibers.

This interpretation has received strong reinforcement in more recent studies of dissociated chromatin. When nuclei are ruptured and spread upon the surface of low-salt buffers, the heterochromatin is dissociated into 10 nm filaments. If extended sufficiently by the shearing forces generated in outflow of the chromatin, these filaments have the appearance of delicate strings of beads composed of regularly repeating discoid subunits 11 nm in diameter and 5.5 nm thick. These subunits, which were originally termed nu bodies (Olins and Olins, 1973) but are now generally called *nucleosomes* (Oudet et al., 1975), are joined together by thin segments of double-stranded DNA, ~ 4 nm in diameter.

The morphological observations on chromatin structure have been greatly extended by biochemical analysis of isolated nucleosomes. Histone molecules have numerous basic amino acids concentrated in their amino-terminal end, while their midregion and the carboxyl-terminal portion form a compact globular structure. The core of the nucleosome is an octomer of two molecules each of histones H_4, H_3, H_2A, H_2B. A segment of DNA, 140 base pairs in length, is coiled around the outside of the histone core. It is therefore accessible to modification or cleavage by enzymes (Kornberg, 1974; Felsenfeld, 1975). The highly basic tails of the histones are believed to interact specifically with the DNA. Certain segments of the DNA are protected from DNAase digestion by their interaction with histone, whereas other exposed regions are susceptible to digestion.

Figure 121. Chromatin fibers from erythrocyte of the salamander *Notophthalmus viridescens* spread on water, fixed in 2 per cent formaldehyde, critical point dried, and shadowed with carbon-platinum. (Micrograph courtesy of Hans Ris.)

Figure 122. Nucleofilaments from chicken erythrocyte nuclei lysed on low ionic buffer in the absence of divalent metals. (Micrograph courtesy of Ada Olins.)

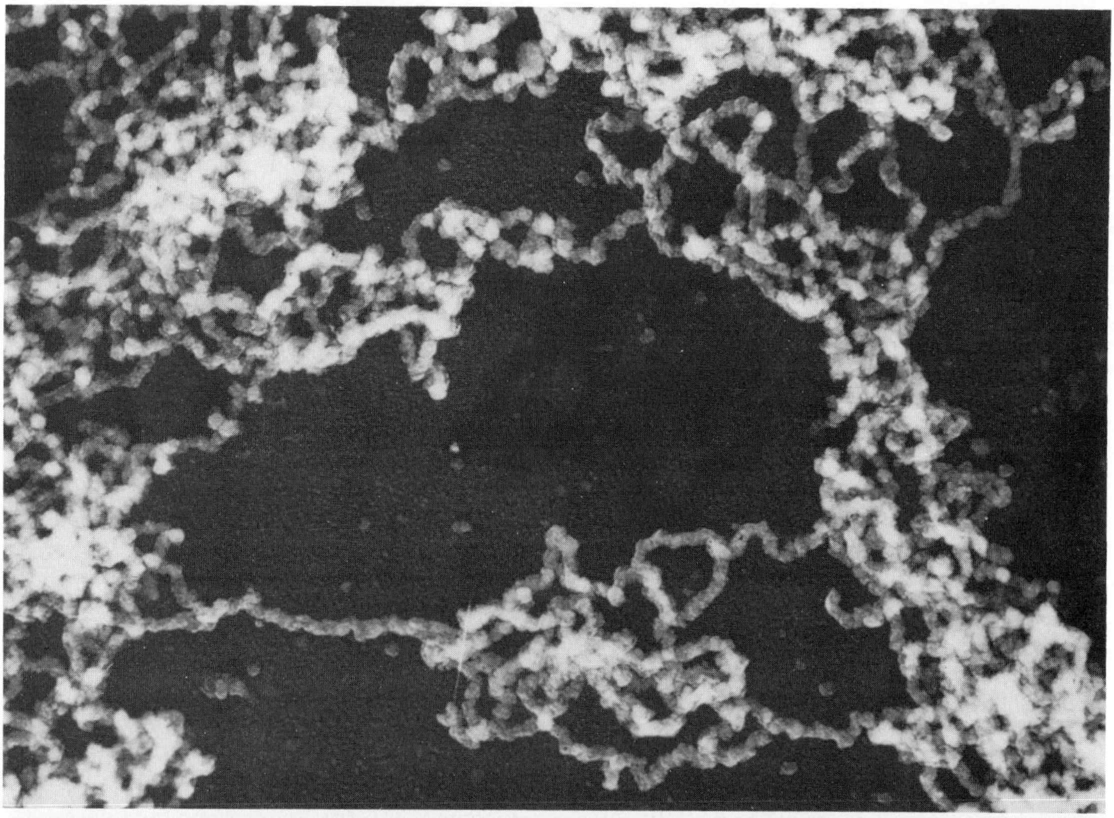

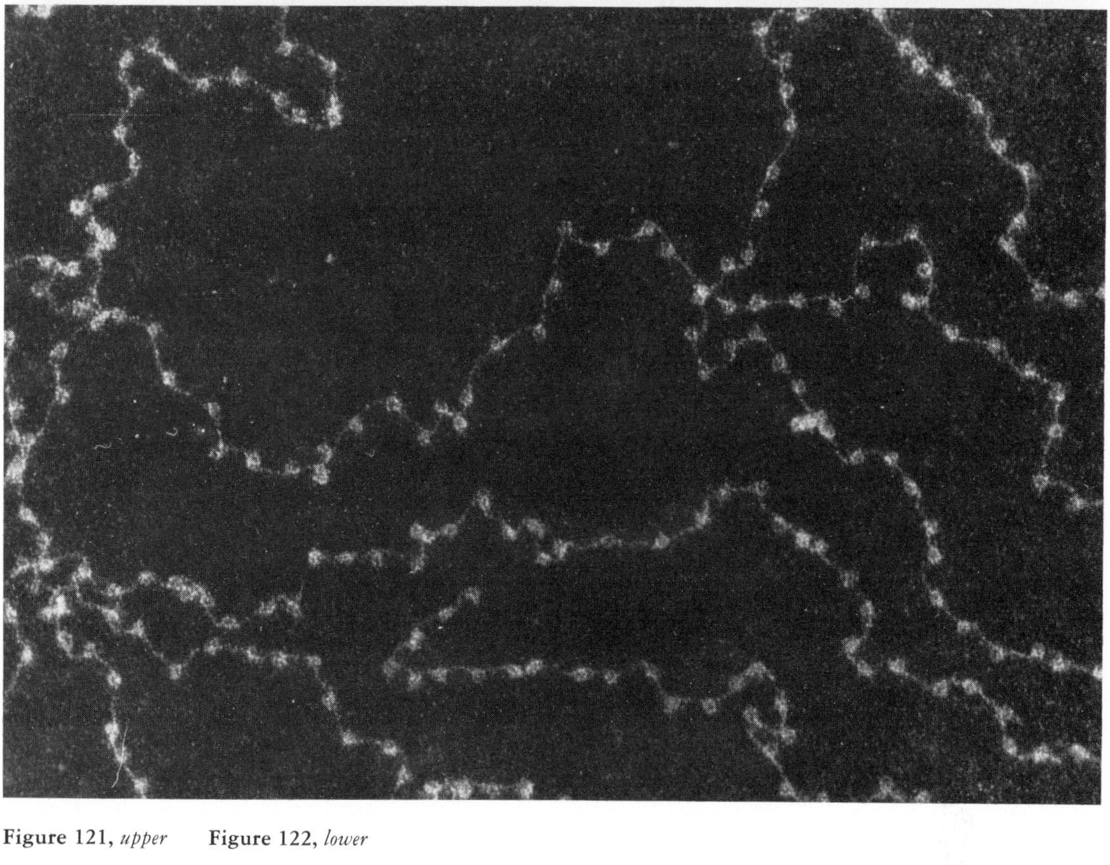

Figure 121, *upper* Figure 122, *lower*

The DNA of the spacer regions between nucleosomes is variable in length and consists of from 10 to 70 base pairs. The H_1 histone is associated with these segments. When not experimentally extended, the nucleosomes and spacer DNA are closely packed into *10 nm filaments* corresponding to the smallest elements normally observed in electron micrographs after removal of divalent metals (Ris, 1975). In the presence of magnesium ions, the 10 nm filaments condense into 20 to 30 nm *chromatin fibers*. The H_1 class of histones is implicated in the establishment and stabilization of this higher order structure. The exact arrangement of nucleosomes in the chromatin fibers is still a subject of dispute. Some investigators report that, in the presence of magnesium ions, the 10 nm filaments condense into a helix or solenoid with six nucleosomes per gyre arranged around a central channel (Finch and Klug, 1976). Other investigators favor a helical packing without a central channel, or alternative patterns of close packing.

The accompanying figure attempts to depict the salient features of the organization of heterochromatin. The DNA double helix is wrapped one and three quarters turns around a nucleosome core consisting of two molecules each of histones H_4, H_3, H_2A, and H_2B. When the chromatin fibers are maximally extended as indicated in the lower part of the drawing, they have a beads-on-a-string appearance. Histone H_1 is associated with spacer segments of varying length between successive nucleosomes. When not experimentally extended, the nucleosomes and spacer DNA are compacted into a more or less uniform 10 nm nucleoprotein filament. This in turn is believed to be helically coiled around a central channel with six nucleosomes per gyre. The resulting solenoid corresponds to the basic 20 to 30 nm nucleoprotein fiber of heterochromatin present in the intact nucleus.

A great deal of artistic license is inevitable at this stage of our knowledge of chromatin and the drawing has a number of shortcomings. The shape and scale of the nucleosomes is inaccurate. The discs have a diameter (11 nm) twice their thickness (5.5 nm) and thus are flatter than depicted here. As shown here the diameter of the double strand of DNA is small relative to the dimensions of the nucleosome. There is no evidence for the change of handedness shown in the DNA coil in the lower part of the figure as it goes from one nucleosome to the next. Moreover, it is not known precisely how the nucleosome discs are arranged in the 20 to 30 nm fiber. When the fiber is unwound at low ionic strength to the 10 nm filament and viewed in electron micrographs, the discs are oriented edge to edge and not face to face. If the 10 nm fiber coils with this orientation of the nucleosomes, the discs would have an orientation orthogonal to that shown.

Although x-ray diffraction of 20 to 30 nm fibers reconstituted *in vitro* (Finch and Klug, 1976) and high voltage electron microscopy of chromosomes (Ris and Korenberg, 1978) support a regular helical array of nucleosomes, other evidence favors their packing in clustered arrays called "super-beads" (Hozier et al., 1977). Parallel or staggered arrangements have also been suggested on the basis of electron microscopic studies (Rattner and Hamalko, 1978).

In this rapidly advancing area of cell biology, any attempt to depict current concepts of the higher order structure of heterochromatin in a simple drawing may be premature. Nevertheless it is hoped that with the stated caveats, this figure will be useful until more precise data are available.

Figure 123. Composite diagram of chromatin structure based upon drawings in Bradbury (La Recherche, 9:644–653, 1978); Worcel and Benyajati (The Cell, 12:83–100, 1977); and Olins and Olins (American Scientist, 66:704–771, 1978).

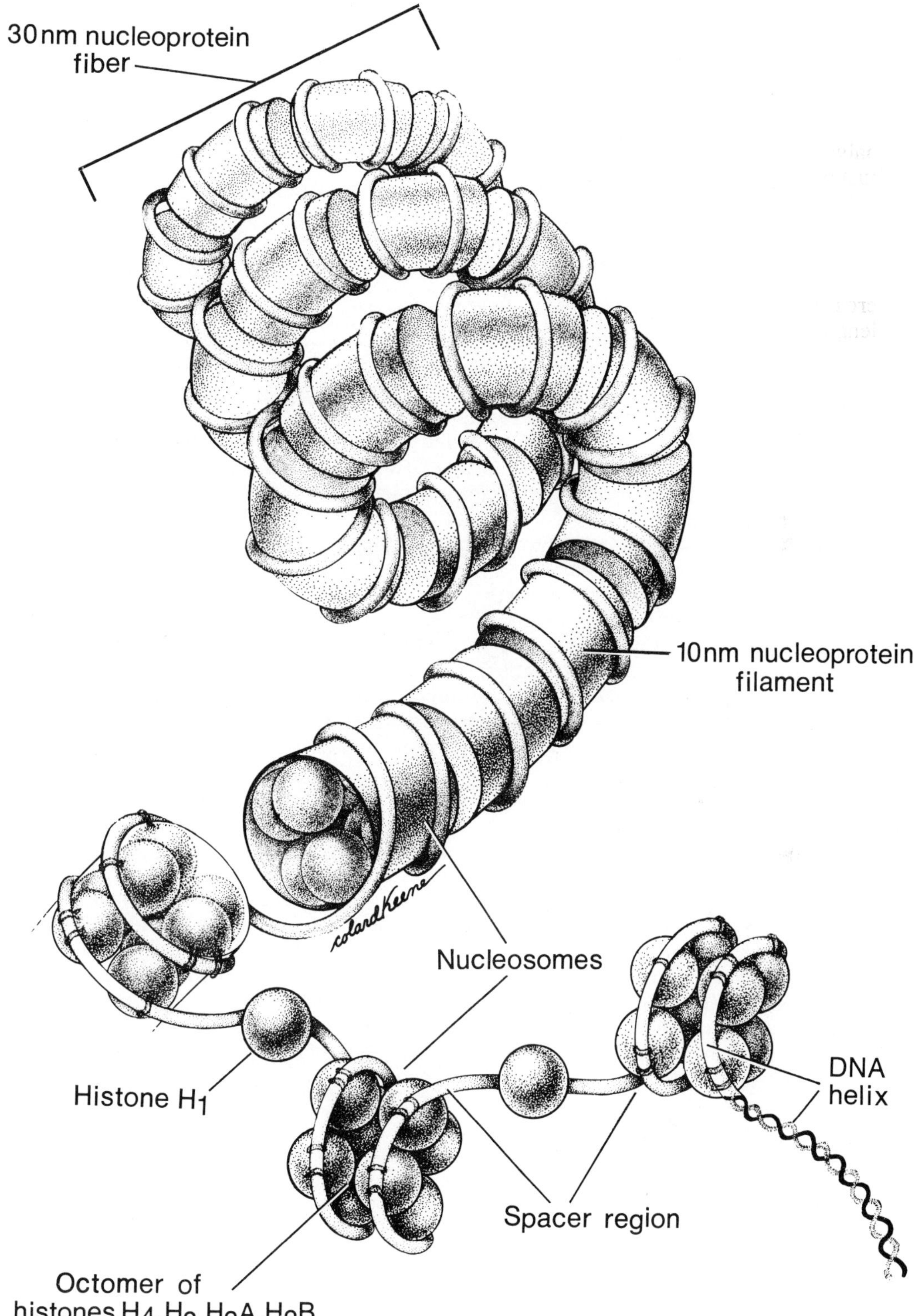
30 nm nucleoprotein
fiber
10 nm nucleoprotein
filament
Nucleosomes
Histone H_1
DNA
helix
Spacer region
Octomer of
histones H_4, H_3, H_2A, H_2B

Figure 123

MITOTIC CHROMOSOMES

One of the more challenging structural problems in cell biology has been the analysis of the organization of chromosomes. Any model of chromosome structure must be compatible with replication of the genome, gene transcription, sister chromatid exchanges, pairing and chiasma formation in meiosis, and other aspects of chromosome behavior.

The so-called lampbrush chromosomes of amphibian oocytes have been most successfully analyzed because of their large size and relative simplicity. They consist of two homologous chromatids in an extended state held together by one or more crossovers, or *chiasmata*. Each is made up of two long DNA molecules. Along their length, they are folded to form hundreds of symmetrical lateral loops that are the basis for their bottle-brush appearance. The loops are sites of active RNA synthesis and represent functional units of the DNA molecule called *replicons*. Other contracted segments of the DNA appear to constitute the central axis of the chromosome (Gall, 1956).

During prophase of the first meiotic division of many other vertebrates, there is a distinctive axial structure called the *synaptonemal complex* which is shared by the paired homologous chromosomes (Moses, 1956; Fawcett, 1956). In electron micrographs, it consists of three linear densities, a central element flanked by two parallel lateral elements and interconnected by regular periodic cross-bridges. Since the synaptonemal complex is resistant to deoxyribonuclease digestion but is degraded by proteases, its integrity depends upon proteins, presumably nonhistone proteins (Comings and Okada, 1970).

The presence of an axial structural component of mitotic chromosomes has been postulated by a number of investigators. In certain procedures for isolating chromosomes, the bulk of the DNA may be sheared off, leaving a ribbon-like 50 nm longitudinal "core" which is also believed to consist of DNA. The chromosome was therefore envisioned as composed of a DNA core to which are attached radially arranged loops of DNA. This proposed model therefore had some similarity to the lampbrush chromosomes of amphibian oocytes but distinguished between a structural category of DNA constituting the core and the associated loops that contain the genetic information (Stubblefield and Wray, 1971).

This interpretation has been challenged by electron microscopic observations on histone-depleted metaphase chromosomes. Extraction of histones results in dissolution of the nucleosome cores and uncoiling and extension of the associated filament of DNA into long loops that radiate from a loose network, or scaffolding, of nonhistone proteins that extends throughout most of the length of the chromatid (Paulson and Laemmli, 1977). According to this model, the radially distributed loops of DNA seen in these preparations are normally condensed by association with histones into a basic 20 to 30 nm nucleoprotein fiber that forms shorter, thicker loops arranged radially around a chromatid axis or core of nonhistones which holds the bases of the loops together (Mariden and Laemmli, 1979).

The morphological evidence for a radial loop organization of mammalian chromosomes is now compelling, but the existence of an axial protein framework remains a subject of controversy, with some workers suspecting that the scaffolding seen in histone-depleted chromosomes may be an artifact resulting from simple aggregation of nonhistone proteins released in the extraction procedure (Comings and Okada, 1979).

The concept of coiling in the structure of chromosomes has a long history in classical cytology (Manton, 1950) and continues to have its proponents among modern molecular biologists endeavoring to explain the high degree of compaction that occurs during formation of metaphase chromosomes. It has been suggested that the 10 nm beaded filament of nucleosomes and spacer DNA is wound helically to form a 30 nm tubular structure which in turn is coiled into a 200 nm tube. This is said to be coiled again to form a chromatid with a diameter of about 600 nm (Sadat and Manuelides, 1977). Others postulate a similar hierarchy of helices but with different dimensions (Bak, Zeuthen and Crick, 1977). The contraction ratios for each step of coiling in these schemes would explain the mass to unit length ratios involved in metaphase chromosome formation, but the morphological evidence for this model is less than convincing.

Whether the morphogenesis of metaphase chromosomes involves looping or coiling or a combination of the two, the degree of compaction of the genetic material that is achieved is quite remarkable. It can be calculated from the amount of DNA per human diploid cell (6.4×10^9 base pairs) that the total length of double helix would be 2.2 meters. Since this is distributed among 23 pairs of chromosomes, the average length of DNA per chromosome would be 4.8 cm but the average length of the chromosomes is only 6 μm. It follows then that in reorganization of the genetic material in preparation for mitotic division, there is an 8000-fold shortening or compaction (Hood, Wilson and Wood, 1975). How this orderly process is controlled at each division so as to maintain the same chromosomal form and linear sequence of genes at present defies understanding.

Classical accounts of chromosome structure based upon light microscopy described a number of parallel longitudinal subunits called *chromonemata*. These were believed to contain the genetic material and were usually depicted as coiled and embedded in an achromatic *matrix*. The matrix was said to be bounded at its outer limit by a membrane-like *sheath* or *pellicle* (Schrader, 1953). It was anticipated that with a little more magnification and resolution these components and others would be clearly delineated. Electron micrographs of dividing cells were therefore a great disappointment to cytogeneticists. No pellicle, chromonema, or matrix was discernible. The chromosomes in micrographs of thin sections appeared as irregularly shaped homogeneous masses of fibrogranular material with ill-defined boundaries. Owing to their varying orientation with respect to the plane of the thin section, only a portion of each chromosome was included and no identification of specific chromosomes on the basis of their length or form was possible.

The accompanying images of metaphase and anaphase chromosomes in dividing spermatogonia illustrate why routine electron microscopy of thin sections has contributed little to cytogenetics.

Figure 124. Metaphase chromosomes of a dividing spermatogonium from ram testis.

Figure 125. Separating sets of chromosomes in early anaphase of mitosis in a spermatogonium from Chinese hamster.

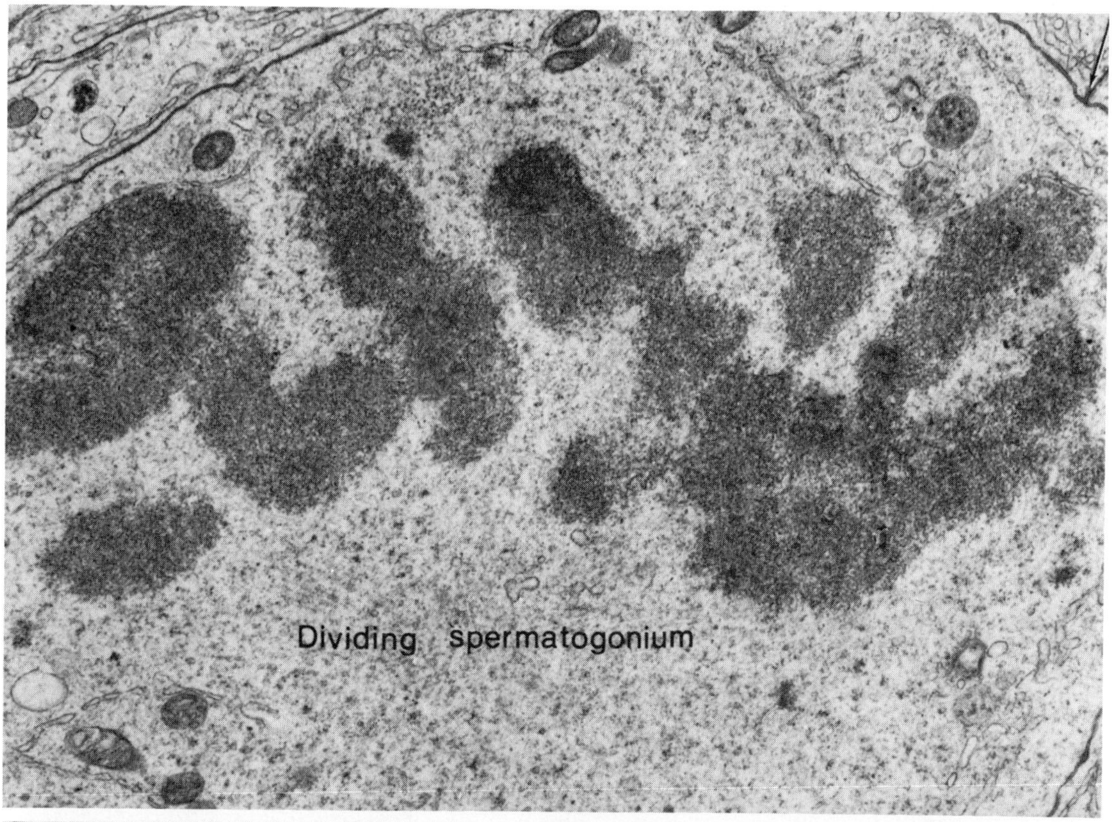

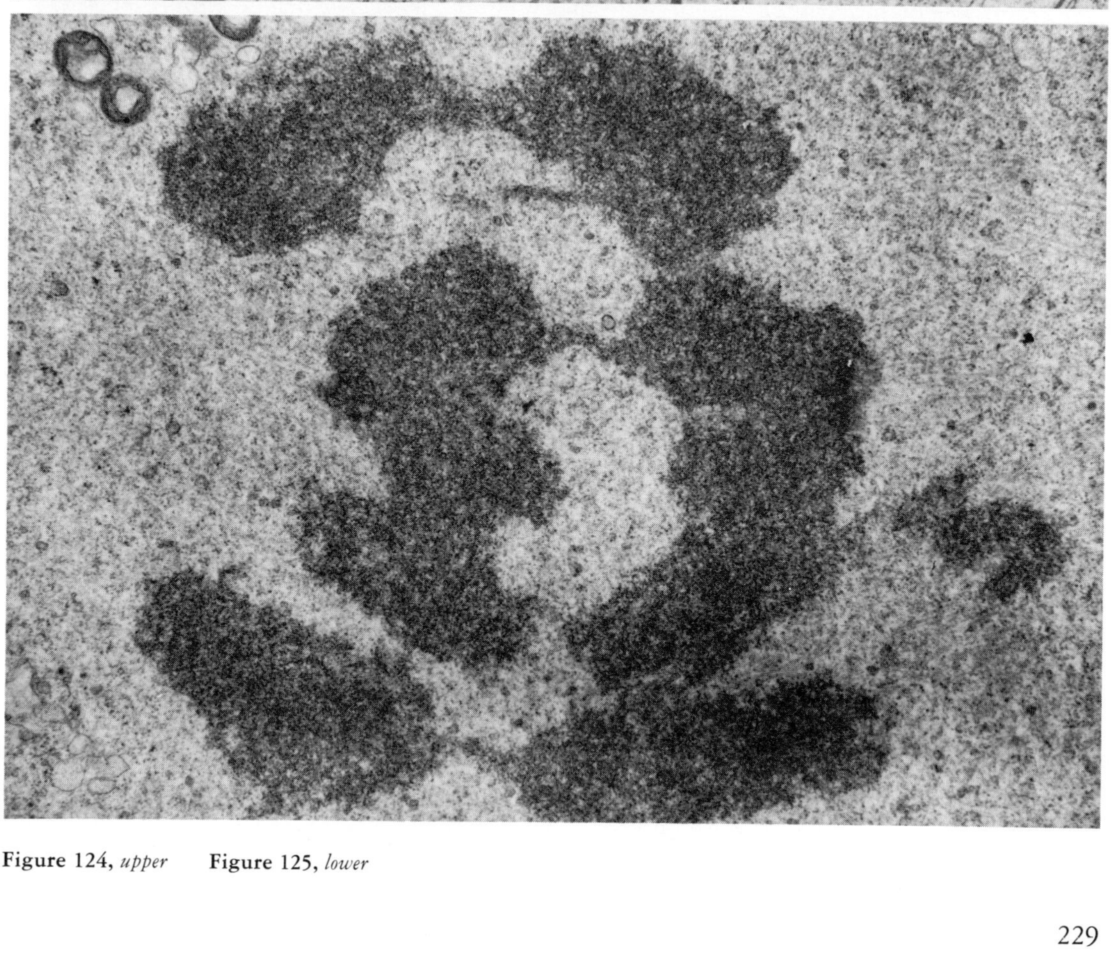

Figure 124, *upper* Figure 125, *lower*

Electron micrographs of chromosomes that are variously oriented with respect to the plane of the thin section provide no valid indication of their length or shape and yield little interpretable information as to their ultrastructural organization.

Figure 126. Polar view of metaphase chromosomes of a dividing myelocyte from the slender salamander, *Batrachoseps attenuatus.*

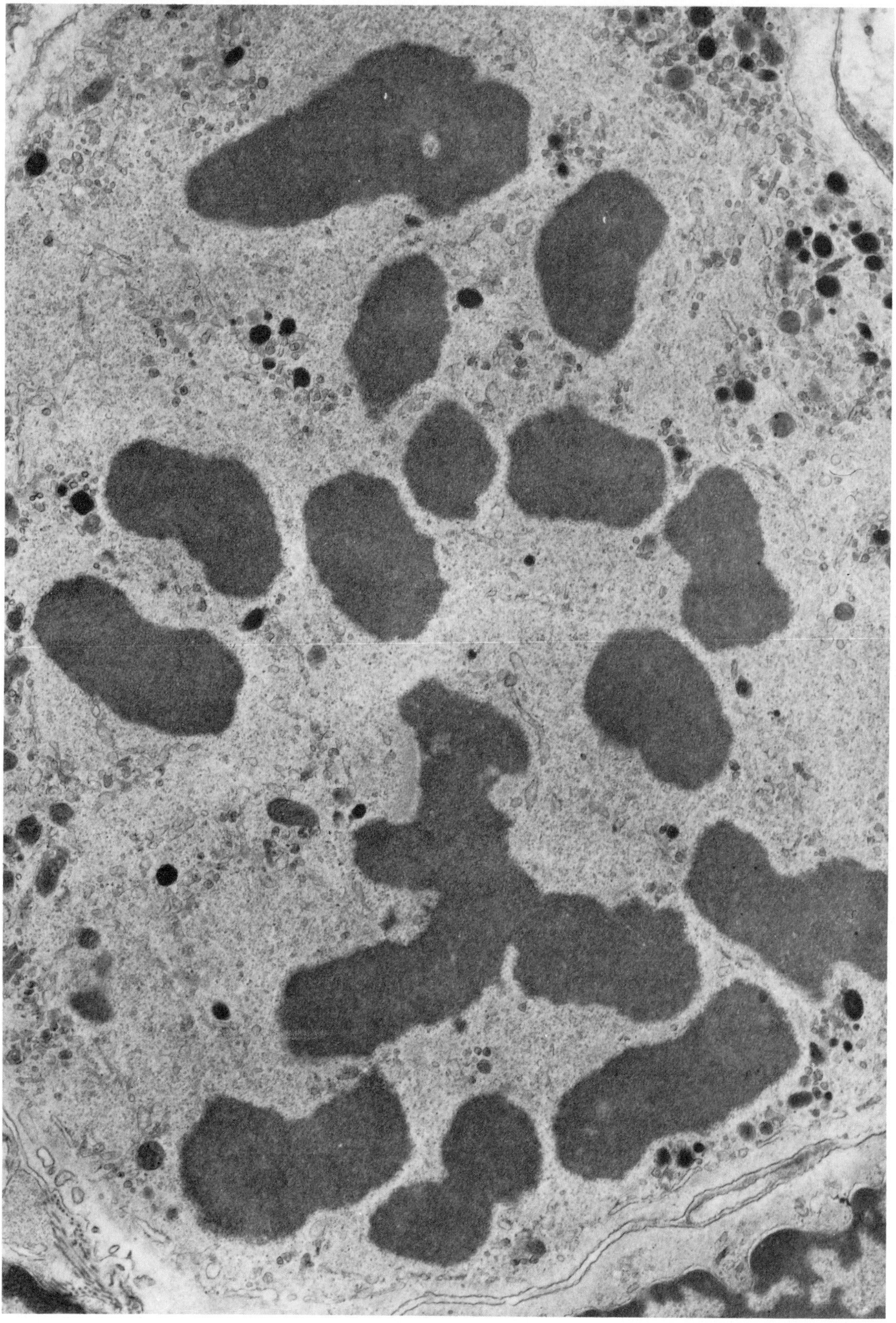

Figure 126

In the accompanying micrograph of a dividing insect cell in late anaphase, the homologous chromosomes can be identified in the daughter cells which are being separated by a cleavage furrow.

Figure 127. Dividing insect spermatocyte. (Micrograph courtesy of David Phillips.)

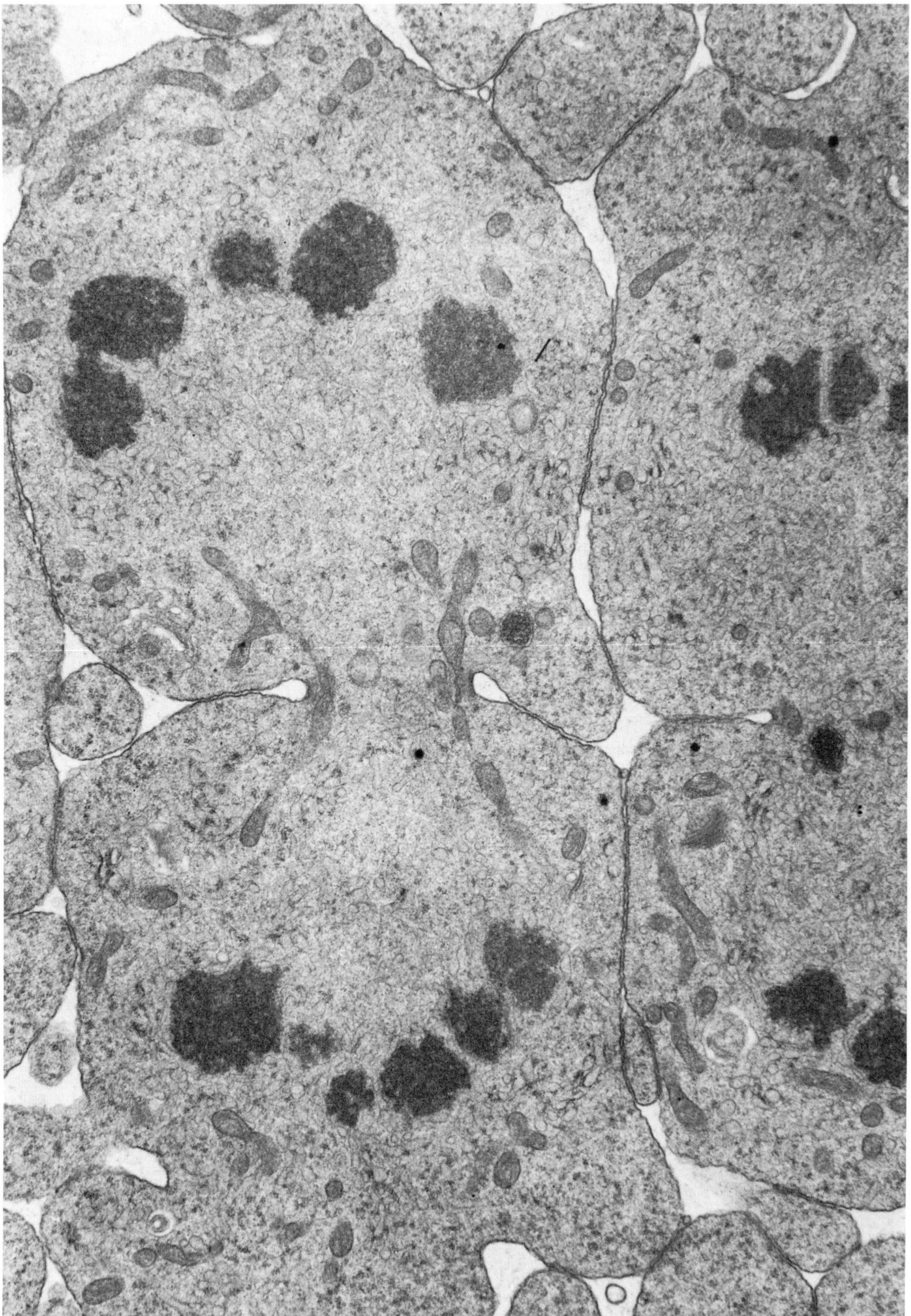

Figure 127

The study of isolated intact chromosomes by high voltage transmission electron microscopy has proved to be more informative. When stereo pairs of micrographs are examined, the isolated metaphase chromosomes appear to be made up of loops of 20 to 30 nm fibers arranged radially around a central axis (upper figure).

When prepared in the presence of calcium (lower figure), the fiber forming the loops is thicker (~50 nm) than in the absence of divalent cations and is believed to result from a higher order coiling of the 20 nm fiber. A faint suggestion of a 200 nm periodicity along the length of the chromatids in the upper figure is more evident when viewed in three dimensions. This may correspond to the regular scalloping attributed by light microscopists to the coiling of chromonemata.

Figure 128. Metaphase chromosomes of CHO cell treated 10 min with 0.075 M KCl, fixed in 3:1 methanol-acetic acid and squashed in 50 per cent acetic acid. (From Ris, Electron Microscopy, 1956.)

Figure 129. Metaphase chromosomes from Chinese hamster cell culture (CHO) isolated in Wray-Stubblefield isolation buffer. (From Ris and Korenberg in Cell Biology, Vol. 2 (Goldstein and Prescott, eds.), Academic Press, New York, 1979.)

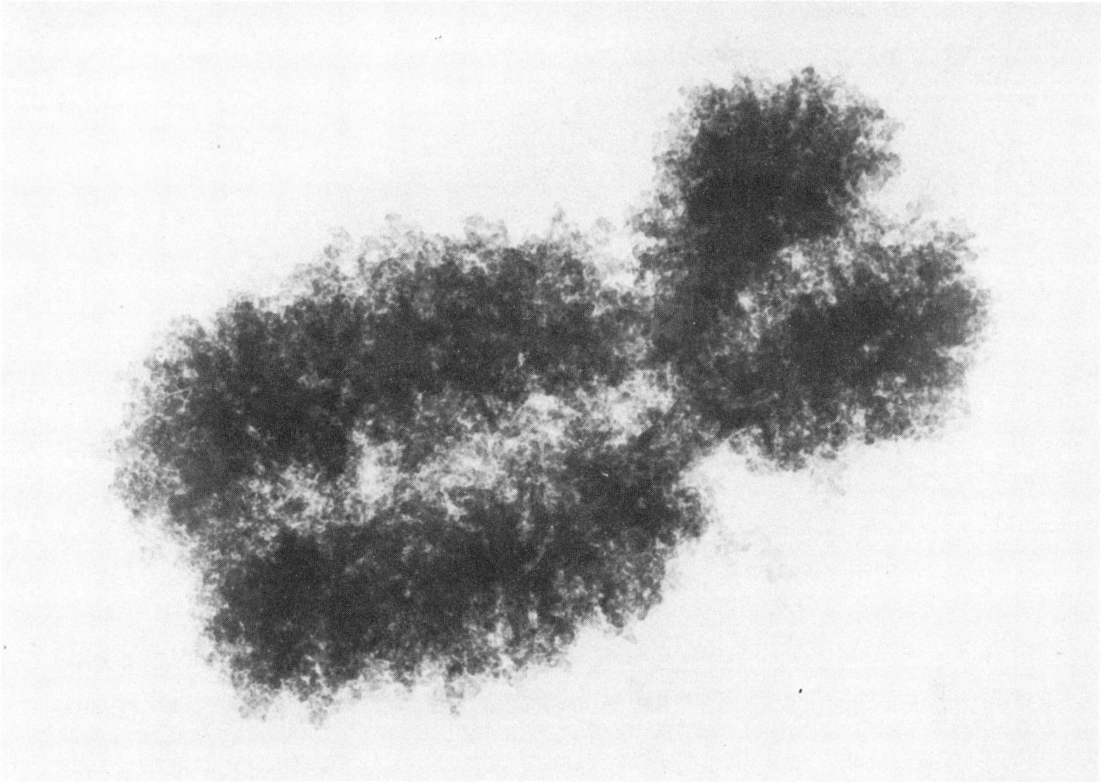

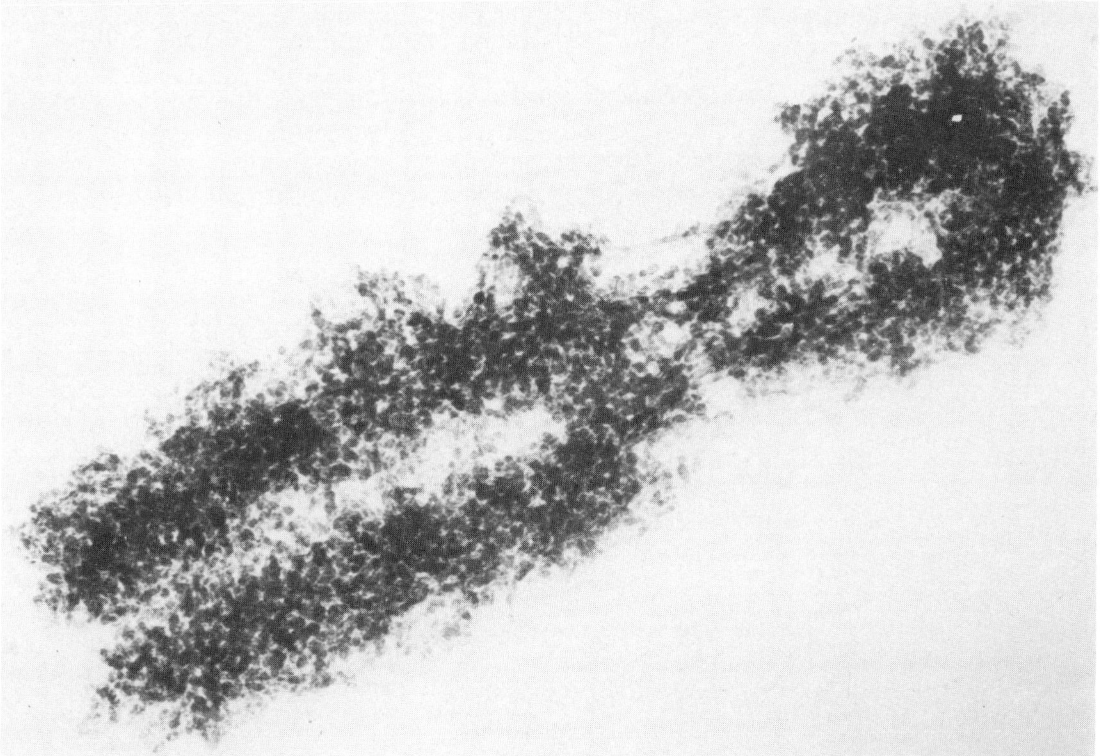

Figure 128, *upper* **Figure 129,** *lower*

A typical mammalian nucleus 4 to 5 μm in diameter contains a quantity of DNA which if extended would be nearly a meter in length. The orderly coiling and condensation of this great length of DNA into the small volume occupied by the chromosomes depends upon its highly specific association with histones and nonhistone proteins. By treatment of chromosomes with dextran sulfate and heparin, nearly all of the histones and many of the other proteins can be extracted. When such histone-depleted chromosomes are spread upon grids and examined with the electron microscope, a remarkable structural transformation is seen to have occurred. A dense framework of protein remains and retains the original form of the chromosome, but this is surrounded by a broad halo of DNA filaments which have uncoiled and spread outward. The uniform radius of the halo suggests that the DNA exists in long loops 20 to 24 μm in length with both ends anchored close to one another in a protein scaffold which maintains the general form of the chromosome.

An area such as that enclosed in the rectangle is shown at higher magnification in the following figure.

Figure 130. A spread preparation of a metaphase chromosome from a HeLa cell depleted of histone by treatment with dextran sulfate and heparin. (Micrograph from J. R. Paulson and U. K. Laemmli, Cell 12:817–828, 1977.)

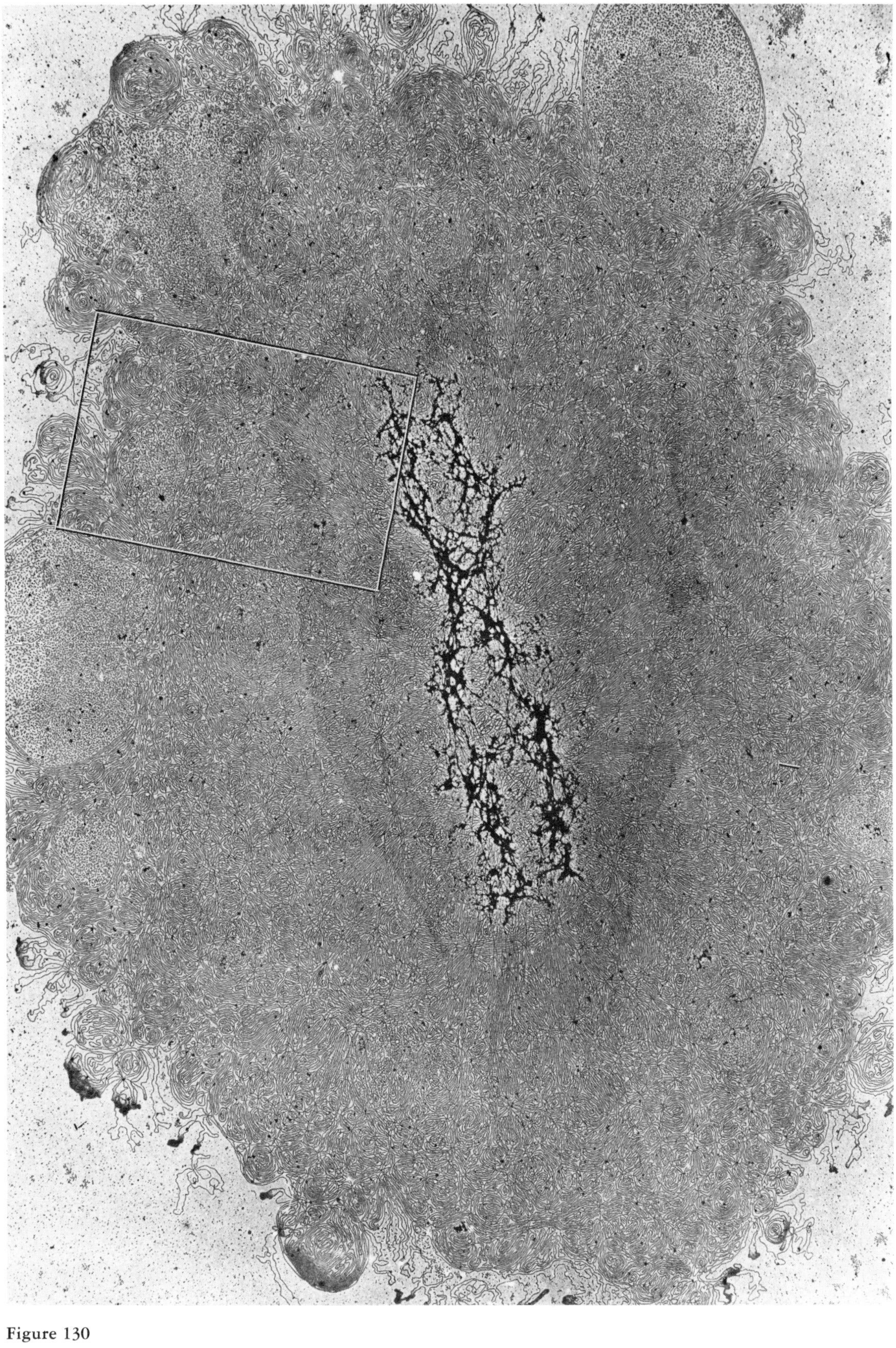

Figure 130

A more highly magnified area from the halo surrounding a histone-depleted metaphase chromosome. (For orientation see the previous figure.) A portion of the protein framework is seen at the bottom of the figure. Extending outward from this is a dense, convoluted pattern of DNA. Although it is not possible to follow individual strands through this complex labyrinth, examination of the periphery of the halo strongly suggests that the DNA filaments form long loops.

In the intact chromosome the DNA and associated histones are coiled to form 10 nm nucleofilaments which in turn are further compacted by supercoiling into 20 to 30 nm nucleohistone fibers that form short loops (0.5 to 2 μm). These loops are no doubt responsible for the appearance of dense bundles of radially arranged chromatin fibers illustrated in earlier micrographs of isolated chromosomes. These observations suggest that although mammalian chromosomes are more highly compacted and more difficult to study, their basic organization is not very different from that of lampbrush chromosomes in amphibian oocytes.

Figure 131. Part of the protein framework and surrounding DNA from a histone-depleted metaphase chromosome. (Micrograph courtesy of J. R. Paulson and U. K. Laemmli.)

Figure 131

REFERENCES

Chromatin Structure

Berkowitz, E. M. Chemical and physical properties of fractionated chromatin. Proc. Nat. Acad. Sci. 72:3328–3332, 1975.

Bradbury, E. M. La chromatine. La Recherche 9:644–653, 1978.

Felsenfeld, G. Chromatin. Nature 271:115–122, 1976.

Finch, J. T., M. Noll and R. D. Kornberg. Electron microscopy of defined lengths of chromatin. Proc. Nat. Acad. Sci. 72:3320–3322, 1975.

Finch, J. T., and A. Klug. Solenoid model for superstructure in chromatin. Proc. Nat. Acad. Sci. 73:1897–1901, 1976.

Hozier, J., M. Renz and R. Niels. The chromosome fiber: Evidence for an ordered superstructure of nucleosomes. Chromosoma 62:301–317, 1977.

Kornberg, R. Structure of chromatin. Ann. Rev. Biochem. 46:931–954, 1977. (Review)

Kornberg, R. D. Chromatin structure: A repeating unit of histone and DNA. Science 184:868, 1974.

Kornberg, R. D., and J. O. Thomas. Chromatin structure: Oligomers of the histones. Science 184:865–867, 1974.

Noll, M. Subunit structures of chromatin. Nature 251:249, 1974.

Olins, A. L., R. D. Carlson and D. E. Olins. Visualization of chromatin substructure: Nu bodies. J. Cell Biol. 64:528–537, 1975.

Olins, A. L., and D. E. Olins. Spheroid chromatin units (nu bodies). Science 183:330, 1974.

Olins, D. E., and A. L. Olins. Nucleosomes: The structural quantum in chromosomes. Am. Sci. 66:704–711, 1978. (Review)

Oudet, P., M. Gross-Bellard and P. Chambon. Electron microscopic and biochemical evidence that chromatin structure is a repeating unit. Cell 4:281, 1975.

Rattner, J. B., and B. A. Hamkalo. Higher order structure in metaphase chromosomes. I. The 250 Å fiber. Chromosoma 69:363–372, 1978.

Rattner, J. B., and B. A. Hamkalo. Higher order structure in metaphase chromosomes. II. The relationship between the 150 / fiber, superbeads and beads-on-a-string. Chromosoma 69:373–379, 1978.

Rattner, J. B., and B. A. Hamkalo. Nucleosome packing in interphase chromatin. J. Cell Biol. 81:453–457, 1979.

Scheer, U. Changes in nucleosome frequency in nucleolar and non-nucleolar chromatin as a function of transcription: An electron microscope study. Cell 13:535–549, 1979.

Senior, M. B., A. L. Olins and D. E. Olins. Chromatin fragments resembling nu bodies. Science 187:173, 1975.

Woodcock, C. L. F. Ultrastructure of inactive chromatin. J. Cell Biol. 59:368a, 1973.

Worcel, A., and C. Benyajati. Higher order coiling of DNA in chromatin. Cell 12:83–100, 1977.

Chromosome Structure

Bak, A. L., J. Zeuthen and F. H. C. Crick. Higher order structure of human mitotic chromosomes. Proc. Nat. Acad. Sci. 74:1595–1599, 1977.

Comings, D. E., and T. A. Okada. Whole mount electron microscopy of meiotic chromosomes and the synaptonemal complex. Chromosoma, 30:269, 1970.

Comings, D. E., and T. A. Okada. Some aspects of chromosome structure of eukaryotes. Cold Spring Harbor Symposium on Quantitative Biology 38:145–153, 1974.

Comings, D. E., and T. A. Okada. Chromosome scaffolding structure — real or artefact? J. Cell Biol. 83:150a (abstr.), 1979.

DuPraw, E. J. DNA and Chromosomes. Holt, Rinehart & Winston, New York, 1970.

Fawcett, D. W. The fine structure of chromosomes in meiotic prophase of vertebrate spermatocytes. J. Biophys. Biochem. Cytol. 2:403–408, 1956.

Gall, J. G. On the submicroscopic structure of chromosomes. Brookhaven Symposium in Biology 8:17–32, 1956.

Hood, L., J. Wilson and W. Wood. Structure and organization of eucaryotic chromosomes, pp. 30–85. In Molecular Biology of Eucaryotic Cells, W. A. Benjamin, New York, 1975.

Laemmli, U. K., S. M. Cheng, K. W. Adolph, J. R. Paulson, J. A. Brown and W. R. Baumbach. Metaphase chromosome structure: The role of non-histone proteins. Cold Spring Harbor Symposium on Quantitative Biology 42:351–360, 1977.

Manton, I. The spiral structure of chromosomes. Biol. Rev. 25:486–508, 1950.

Marsden, M. P. F., and U. K. Laemmli. Metaphase chromosome structure: Evidence for a radial loop model. Cell 17:849–858, 1979.

Moses, M. H. Chromosomal structures in crayfish spermatocytes. J. Biophys. Biochem. Cytol. 2:215–217, 1956.

Moses, M. J. Synaptonemal complex. Ann. Rev. Genet. 2:363, 1968.
Ohnuki, Y. Structure of chromosomes. I. Morphological studies of the spiral structure of human somatic chromosomes. Chromosoma 25:401–428, 1968.
Paulson, J. R., and U. K. Laemmli. The structure of histone-depleted metaphase chromosomes. Cell 12:817–828, 1977.
Ris, H., and D. Kubai. Chromosome structure. Ann. Rev. Genet. 4:263–294, 1970.
Ris, H. Chromosomal structure as seen by electron microscopy. In Structure and Function of Chromatin, Ciba Symposium, No. 28. pp. 7–28. Assoc. Sci. Publishers, Amsterdam, 1975.
Ris, H. Higher order structures in chromosomes, pp. 545–556. In Electron Microscopy 1978. Proc. IX International Congress of Electron Microscopy, Toronto.
Sadat, J. and L. Manuelidis. A direct approach to the structure of eukaryotic chromosomes. Cold Spring Harbor Symposium on Quantitative Biology 42:331–350, 1977.
Stubblefield, E. The structure of mammalian chromosomes. Int. Rev. Cytol. 35:1–60, 1973.
Stubblefield, E. and W. Wray. Architecture of the Chinese hamster metaphase chromosome. Chromosoma 32:262–294, 1971.
Wischnitzer, S. The lampbrush chromosomes: Their morphology and physiological importance. Endeavor 35:27, 1976.

NUCLEOLUS

The first observation of the nucleolus is attributed to Fontana (1781), but it was rarely described until after 1830 when the nucleus itself became widely accepted as a constituent of all cells. Initial confusion of the nucleolus with prominent clumps of chromatin persisted until the development of improved staining methods, which consistently showed that it differed from karyosomes in its staining affinities. Speculations as to the origin and function of the nucleolus were highly varied and understandably vague. The occasional observation of small "vacuoles" within it led to the suggestion that the nucleolus was an excretory organ for the nucleus, discharging its wastes into the cytoplasm (Lukjanow, 1888). More prevalent was the interpretation that it was involved in some way in the maintenance and growth of the nucleus (Montgomery, 1898). A prophetic suggestion that the size of the nucleolus might be related to the intensity of the interaction between the nucleus and cytoplasm (Hacker, 1895) gained little acceptance. There was general agreement on the disappearance of the nucleolus during mitosis, and there was much speculation on the distribution of its substance in cell division and the reformation of nucleoli in the daughter cells.

Little progress was made toward resolution of this problem until the 1930s. It was then noticed that in mitotic prophase of plant material when the chromosomes became visible, the nucleolus was always associated with a specific region on one of the chromosomes. After the nucleolus dispersed, the site that it had formerly occupied was visible as a constriction on one or more of the chromosomes. These were termed *secondary constrictions* to distinguish them from the sites of attachment of the spindle fibers which were called *primary constrictions*. The number and size of these secondary constrictions were observed to correspond to the size and number of nucleoli present in the interphase nucleus (Heitz, 1931; McClintock, 1934). When nucleoli reformed in the daughter cells at anaphase, they first appeared in relation to the secondary constrictions which therefore came to be regarded as *nucleolus-organizing centers*. In current usage the terms secondary constriction and nucleolus-organizing region are used interchangeably. The latter term is probably preferable since the so-called constrictions are, in fact, merely short segments that show little affinity for the stains commonly used to give contrast to the chromosomes. In the interphase nucleus, the chromosome bearing the organizing center follows a meandering course through the reconstituted nucleolus (Godward, 1950).

The configuration of the nucleolus in any particular cell type is relatively consistent. Although a plethora of minor variations were described in the tissues of plants and animals (Montgomery, 1898), no common organizational plan was discovered by classical cytologists. But with a special silver impregnation technique, Estable and Sotelo (1951) found that the nucleoli of many cell types appeared to consist of an amorphous inner region surrounded by a network of anastomosing strands which they called the *nucleolonema*. Around the periphery of the nucleolus, variable amounts of material staining with basic dyes were designated the *nucleolus-associated chromatin*.

With the advent of the electron microscope, the presence of a nucleolonema was confirmed, and it was found to consist of closely interwoven 5 nm filaments with interspersed small, dense granules 15 to 20 nm in diameter (Borysko and Bang, 1951; Bernhard, 1952; Horstmann and Knoop, 1957). Considerable variation in nucleolar organization was observed from one cell type to another. The nucleolonema in some formed an easily identifiable loose reticulum while in others its strands were so closely aggregated that the nucleolus resembled a tight ball of yarn with the interstices occupied by matrix material of nearly equal density. The filamentous and granular components were intermingled in some cell types and in others they were segregated into concentric zones. Whatever the topography of the nucleolus, the same four components always appeared to be present — delicate closely packed filaments, dense granules, an amorphous matrix, and associated chromatin. The latter was not confined to the periphery of the organelle but also extended into its interior (Marinozzi, 1963).

It was known from histochemical studies at the light microscope level that the RNA of cells was concentrated in the nucleolus and in certain basophilic components of the cytoplasm (Brachet, 1940; Caspersson and Schultz, 1940). Correlated biochemical and electron microscopic analysis had also established that the bulk of the cytoplasmic RNA resides in the 15 nm particles called *ribosomes*. The morphological resemblance of the granules of the nucleolus to the ribosomes of the cytoplasm encouraged the belief that they might also be ribonucleoprotein. For cytochemical validation of this assumption, aldehyde-fixed tissues were embedded in water-soluble plastics which permitted enzymatic digestion of specific substrates from the nucleolus. Subsequent staining of the undigested residues made it possible to identify in electron micrographs those components that had been extracted (Granboulan and Bernhard, 1964, 1965). It was hoped that by combining this approach with autoradiography of labeled precursors, the synthetic activity of the nucleolus could be localized to specific components. Exposure of the sections to pepsin digested the amorphous matrix of the nucleolus but left the fibrillar and granular components intact. These were removed, however, by digestion with ribonuclease, indicating that the granular and filamentous elements both contained RNA (Marinozzi and Bernhard, 1963). By autoradiography with tritiated thymidine, DNA could be localized within certain areas of the nucleolus as well as in the peripheral chromatin. The uptake of tritiated uridine in these regions demonstrated that this DNA served as a template for nucleolar RNA synthesis. The label appeared first in the fibrillar and later in the granular regions, suggesting the possibility that the nucleolar granules arose from a conformational change in the filaments.

An understanding of the relationship of the nucleolus to the cytoplasmic RNA and to protein synthesis gradually emerged in the 1960s from ingenious applications of autoradiography, biochemistry, and microsurgery. Isolation and analysis of nuclei and of nucleolar subfractions disclosed at least three types of RNA in the nucleus, one of which was in the nucleolus and the others in the extranucleolar nucleoplasm (Maggio, Siekewitz and Palade, 1963). Earlier autoradiographic studies had suggested that the RNA of the nucleolus was different from that associated with the chromatin and the cytoplasm. When tritiated cytidine was used as an RNA precursor, it was incorporated first into the RNA of the nucleolus and at longer time intervals label was found to be extranuclear, thus clearly demonstrating translocation of RNA from nucleolus to cytoplasm (McMaster and Kaye, 1958; Woods and Taylor, 1958). When it was subsequently shown that the RNA components of the ribosome (18s and 28s RNA) had a structural rather than an informational function (Jacob, Bremer and Meselson, 1961), it was suggested that the nucleolus was a site of assembly of the structural RNA of the ribosomes. Several additional lines of evidence provided strong support for this interpretation. The base composition of nucleolar and ribosomal RNA was found to be similar, and the composition of both differed from that of the extranucleolar RNA of the nucleoplasm (Edstrom and Beermann, 1962). The antibiotic actinomycin D inhibited the incorporation of labeled uridine into nucleolar and ribosomal RNA, without significantly affecting the synthesis of other RNAs (Perry, 1962).

The movement of RNA from the nucleus to the cytoplasm, first suggested by autoradiographic studies, was further established by microsurgical experiments on amoebae. When radioactively labeled nuclei were transplanted to enucleated amoebae, autoradiography demonstrated a progressive appearance of radioactivity in the previously unlabeled cytoplasm (Prescott, 1960). Further experimental evidence for an essential role of the nucleolus in the generation of ribosomal RNA was provided by its selective inactivation with microbeam irradiation. Destruction of the nucleoli of living cells in tissue culture by a collimated beam of ultraviolet light or by laser bombardment resulted in a striking reduction in the cell's production of ribosomal RNA (Perry, 1961; Rounds et al., 1968).

Perhaps the most compelling evidence for the essential role of the nucleolus came from an experiment of nature — a mutant of the South African clawed frog, *Xenopus laevis,* in which one fourth of the tadpole progeny lacked nucleoli (Fischberg and Wallace, 1960; Hay and Gurdon, 1967). The tadpoles without nucleoli were shown to be unable to synthesize 18s and 28s ribosomal RNA (Brown and Gurdon, 1964), even though messenger RNA (mRNA) and transfer RNA (tRNA) were synthesized at the normal rate. Without ribosomal RNA (rRNA), the tadpoles could not make ribosomes. Therefore protein synthesis and growth stopped and the animals died after exhausting the supply of ribosomes originally present in the egg cytoplasm. The molecular basis of this genetic defect was subsequently elucidated by the technique of RNA-DNA hybridization, in which DNA is first denatured by heat to unwind the double helix and make the nucleotides of the single strands accessible. Radio-labeled rRNA will then bind only to those complementary segments of the single strand DNA representing the rRNA gene. After washing away, the excess unbound rRNA, the amount hybridized can be measured by its radioactivity (Wallace and Bernstiel, 1966). Hybridization experiments with Xenopus showed that DNA of heterozygous tadpoles, possessing a single nucleolus per cell nucleus, bound half the amount of rRNA that was bound by normal binucleolate larvae, and the DNA of anucleolate tadpoles bound none (Brown and Weber, 1968).

Dramatic morphological localization of the genes for rRNA synthesis was achieved by hybridizing tritium-labeled rRNA to chromosomes denatured *in situ* and then identifying the site of its binding by autoradiography. When this was done, the silver grains were consistently found over the secondary constriction of the chromosome. Thus it was demonstrated beyond all reasonable doubt that the nucleolus-organizing region of the chromosome is the site of the genes coding for ribosomal RNA. There are several hundred such genes arranged in tandem and each codes for a 40s precursor molecule of RNA, which is subsequently cleaved to yield the 18s and 28s RNAs typical of ribosomes (Landsman and Gross, 1969).

The topographical relations between the organizer and the other structural components of the nucleolus are exceedingly difficult to study by electron microscopy of thin tissue sections. Much of our present understanding of the organization and function of ribosomal genes is therefore based upon work with amphibian oocytes which are exceptionally favorable material for such studies. During the early period of oocyte enlargement, the nucleolus-organizing regions of the chromosomes are reduplicated to form about a thousand extrachromosomal nucleoli (Gall, 1968). Like typical nucleoli, these have a central filamentous region and a dense granular peripheral zone. When these supernumerary nucleoli are isolated and dispersed on water, the central filamentous component becomes unwound and extended. The spread cores of the nucleoli can then be centrifuged onto specimen grids, dried, and examined with the electron microscope. They consist of an axial DNA fibril about 10 nm in diameter decorated in regularly recurring segments along its length with fine lateral filaments that radiate from the core fiber. At one end of each segment the radiating filaments are very short, but their length increases progressively toward the other end to give a tapering "bottle-brush" or "Christmas tree" configuration. Each of these decorated regions (2.5 to 3.5 μ long) is interpreted as a nucleolar gene being transcribed. Small granules or beads associated with the axial DNA molecule appear to be RNA polymerase, and the radiating fibrils are interpreted as molecules of ribosomal precursor RNA gradually lengthening as more of the gene is transcribed (Miller and Beatty, 1969). After attaining a certain length, the RNA fibrils acquire a small thickening at their free end that results from interaction of the molecule with protein and folding or coiling to form a terminal ribonucleoprotein (RNP) knob or granule. The genes and their associated transcription products are arranged in tandem separated by intergene spacer regions of DNA that are not transcribed. Some of these features of nucleolar structure are illustrated in the micrographs that follow.

The nucleolonema appears in electron micrographs as a dense strand that branches and anastomoses to form an irregular three-dimensional network. In the nucleoli illustrated here, the nucleolonema is rather loosely organized but in many cell types it may be so closely compacted that its reticular nature is obscured. As seen in the upper micrograph, the nucleolonema may be organized around one or more spherical regions composed of fine filaments or granules of lower density than the other regions of the reticulum. These areas appear to correspond to the *pars amorpha* described by light microscopists. The pale interstices in the meshes of the nucleolonema are occupied by fine filaments and granules indistinguishable from those in the ground substance of the surrounding nucleoplasm. Although not identifiable in routine micrographs, the presence of DNA in some of these interstices can be demonstrated in autoradiographs.

The nucleolonema consists of a mass of very fine filaments and 10 nm granules, both composed of ribonucleoprotein. There are local variations in the relative proportions of these two components. Regions of the nucleolonema consisting entirely of very closely compacted filaments are more electron dense than regions rich in granules.

Figures 132 and 133. Nucleoli from spermatogonia of opossum testis.

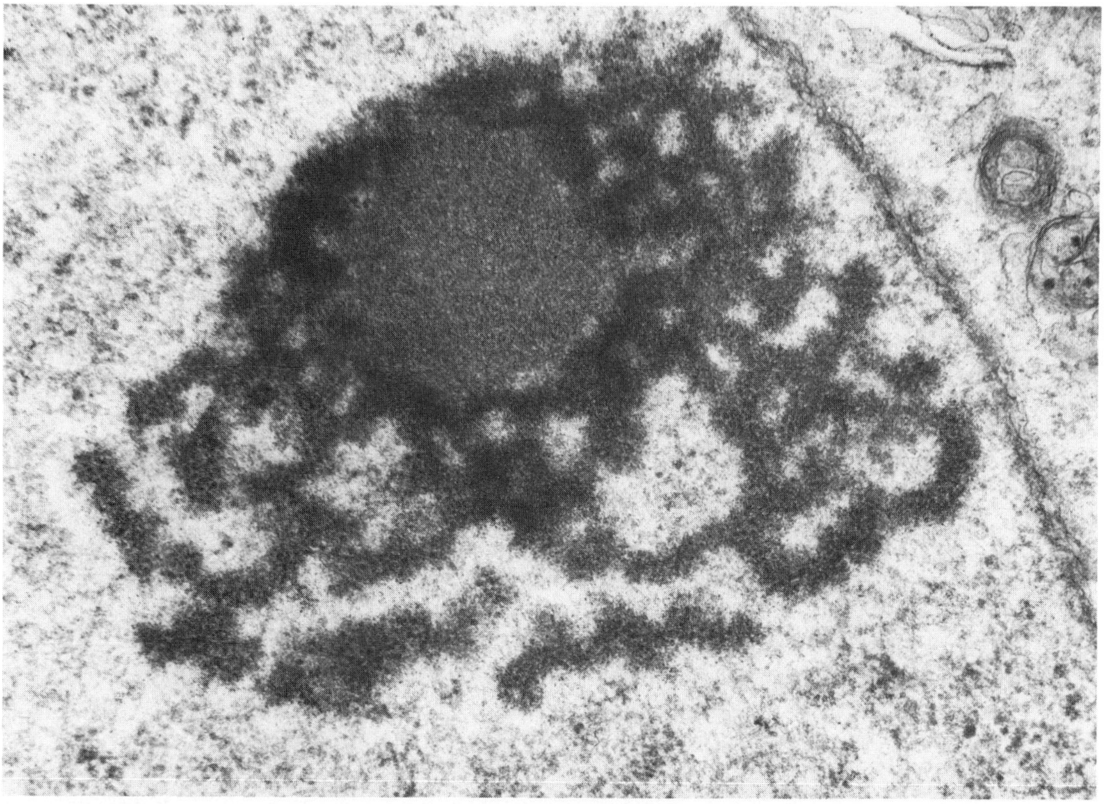

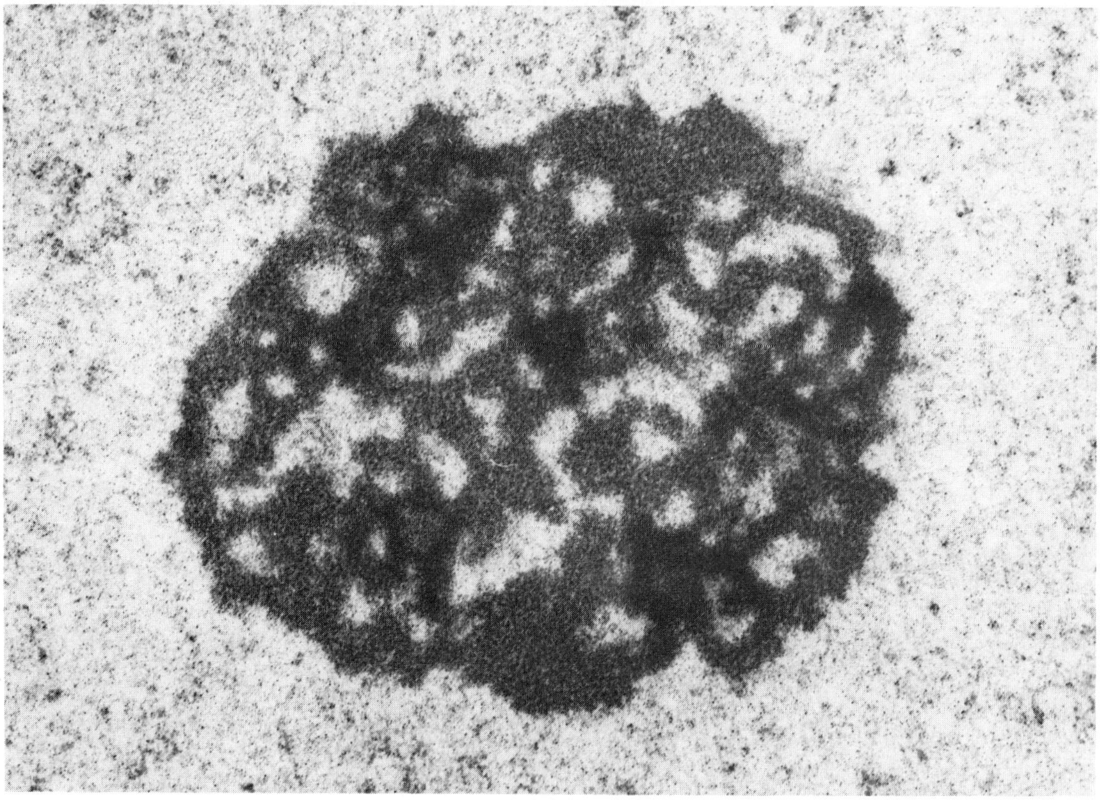

Figure 132, *upper* **Figure 133,** *lower*

In this typical nucleolus at high magnification, the nucleolonema is organized around several homogeneous spherical masses of relatively low density. These are interpreted by some investigators as the organizer region of the associated chromosome or chromosomes (Jordan and Chapman, 1971). The nucleolonema associated with those areas is very dense and appears to consist exclusively of closely aggregated filaments. The remainder of the nucleolonema is made up of granules in a more loosely organized matrix of filaments.

Figure 134. Nucleolus of spermatogonium from testis of Chinese hamster. (Micrograph courtesy of David Phillips.)

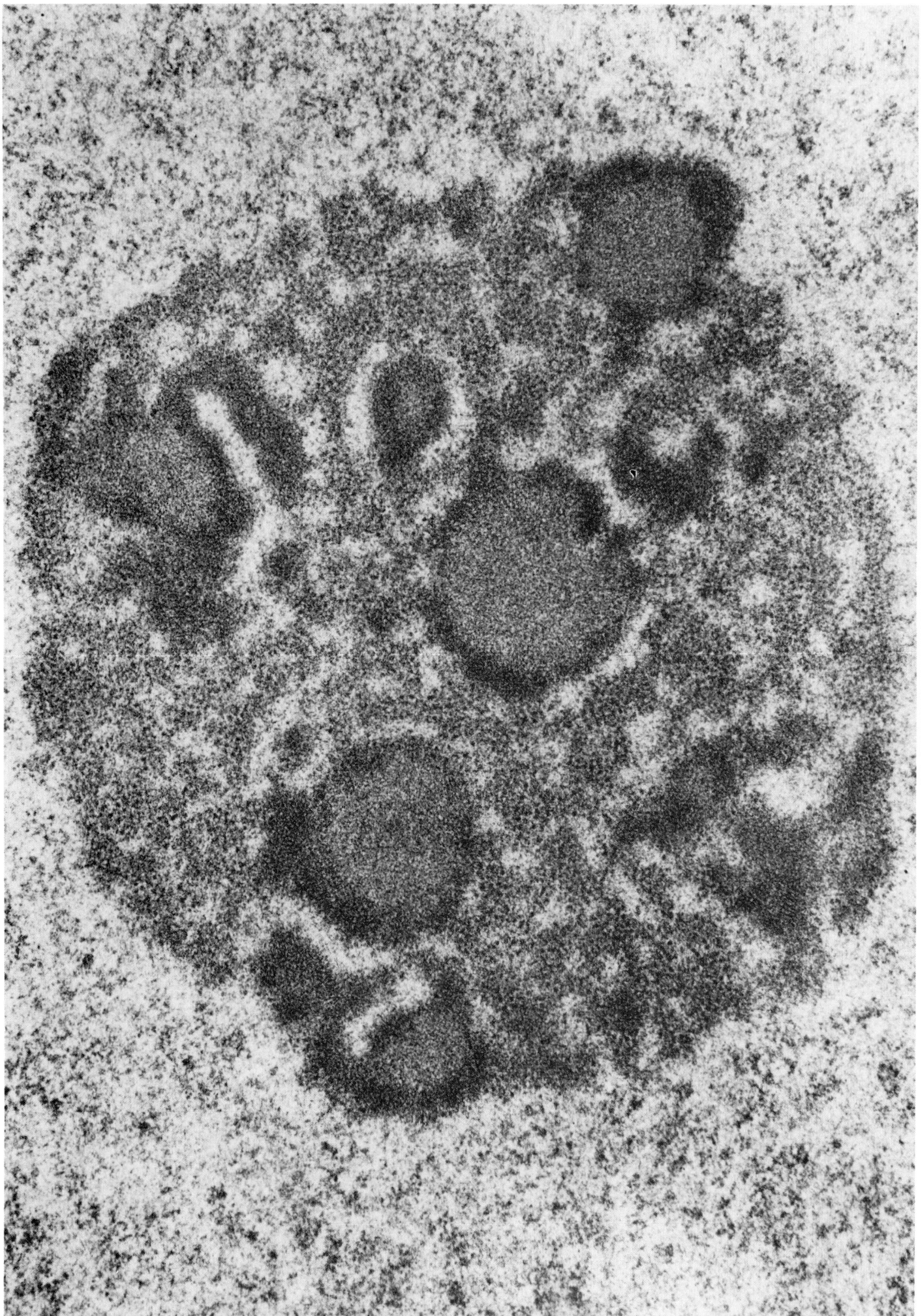

Figure 134

Some cell types have an extremely compact nucleolus. Such is the case in the salamander liver cell shown here. This dense, spherical nucleolus has a concentric organization with a central region apparently composed entirely of fine fibrils surrounded by a peripheral zone rich in granules. Although no interstices are evident in the outer zone, a faintly mottled pattern of granules suggests that it is composed of a closely compacted nucleolonema.

Although highly variable in appearance in different tissues, the nucleolus has certain basic organizational features that are common to all cells. The apparent variability in nucleolar form is attributable in part to different incident planes of section through an irregular structure with a complex internal architecture. In cells with a compact spherical nucleolus, its appearance is less affected by the plane of section. This suggests that there is a greater degree of uniformity in the topographical relations of nucleolar components in the same cell type than is generally appreciated.

A particular internal organization of the nucleolus is apparently related to its normal functional activity, for when rRNA synthesis is blocked by actinomycin D, this organization is lost and the components of the nucleolus are segregated into separate regions in an entirely unfamiliar pattern (Schoefl, 1964).

Figure 135. Portion of a liver cell nucleus from the salamander, *Batrachoceps attenuatus*.

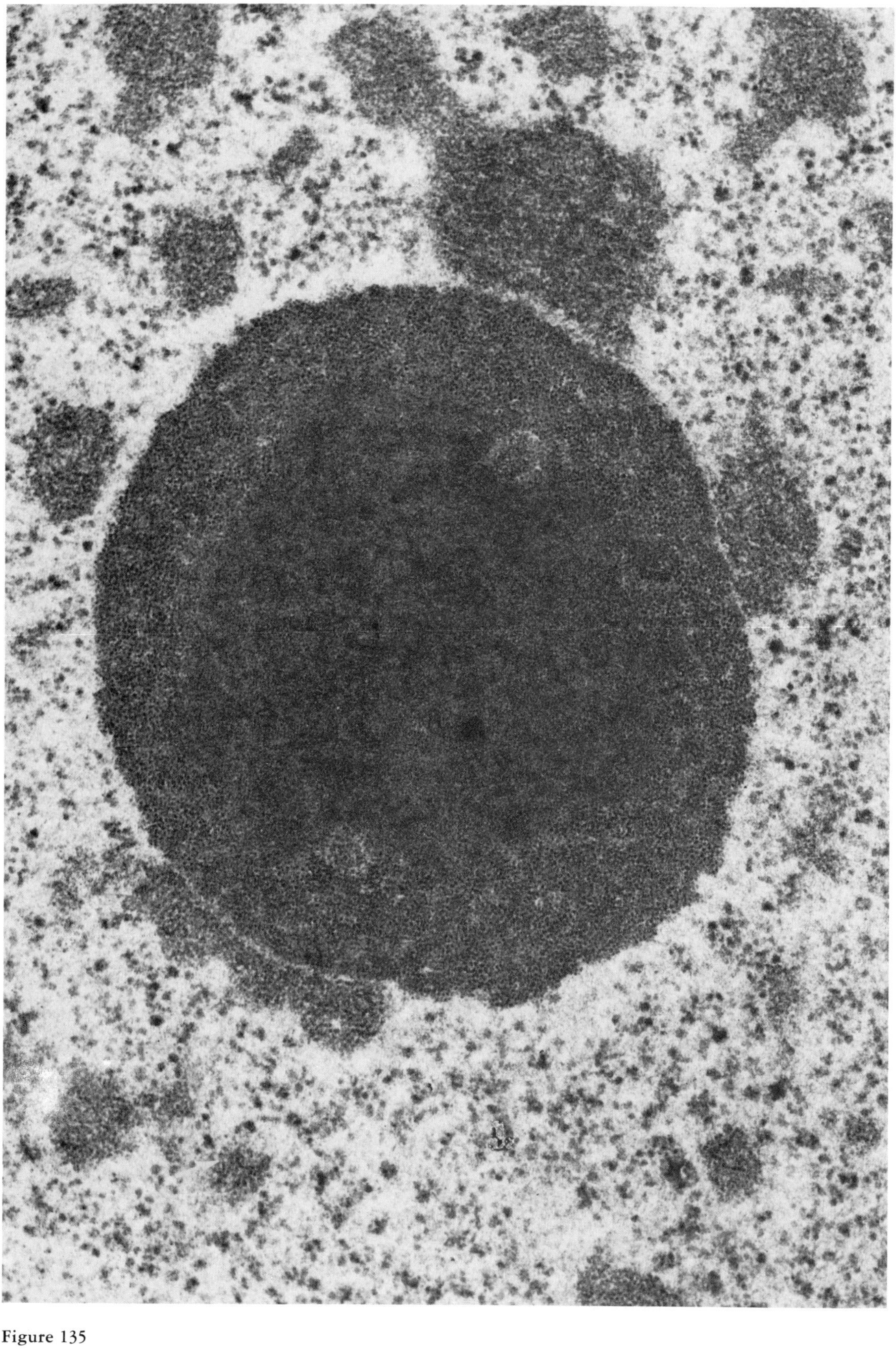

Figure 135

NUCLEOLUS-ASSOCIATED SEX CHROMATIN

Cajal (1909) described and illustrated a mass of chromatin consistently associated with the nucleolus in the neurones of several species and called it the *paranucleolus*. No particular significance was attached to this structure for the next half century. Then, in the course of cytological studies on the hypoglossal nucleus of the brain after stimulation of the hypoglossal nerve, Barr and Bertram (1949) noticed that a discrete mass of chromatin normally associated with the nucleolus became dissociated from it during the period of chromatolysis and early stages of recovery. However, this nucleolar satellite appeared to be lacking in some of their animals. When this inconsistency was investigated, it was found that it is present only in females. Sexual dimorphism discovered in cat neurones was rapidly extended to other cell types and confirmed in the majority of mammalian species studied, but not in birds, reptiles, amphibians, or molluscs. It was subsequently demonstrated that one of the two X−chromosomes in the somatic cells of females condenses early in embryonic development and remains heterochromatic throughout life. In the male, the single X chromosome is generally euchromatic.

The terms paranucleolus and nucleolar satellite were therefore replaced by *sex chromatin*. In many cell types, it is not located next to the nucleolus but is closely applied to the nuclear envelope. It is conspicuous in this position in cells scraped from the human buccal mucosa. In polymorphonuclear leucocytes, the sex chromatin is incorporated in a small appendage of the polymorphous nucleus, called the "drumstick." The ready availability of these two cell types for cytodiagnostic purposes has made the sex chromatin test a valuable clinical method for determining the genetic sex in patients with genital anomalies. Examination of cells exfoliated into the amniotic fluid makes possible the antenatal determination of the sex of the fetus by amniocentesis.

The accompanying micrograph shows the nucleolus of a neuron in the spinal cord of a female cat. The sex chromatin, closely applied to a dense fibrillar region of the nucleolonema is visible at the arrow. The sex chromatin appears to take the form of a convoluted dense fiber of somewhat different texture than heterochromatic regions of autosomes.

Figure 136. Micrograph of nucleolus from a neuron in the spinal cord of a female cat. (Micrograph courtesy of Sanford Palay from The Fine Structure of the Nervous System, W. B. Saunders Company, Philadelphia, 1976.)

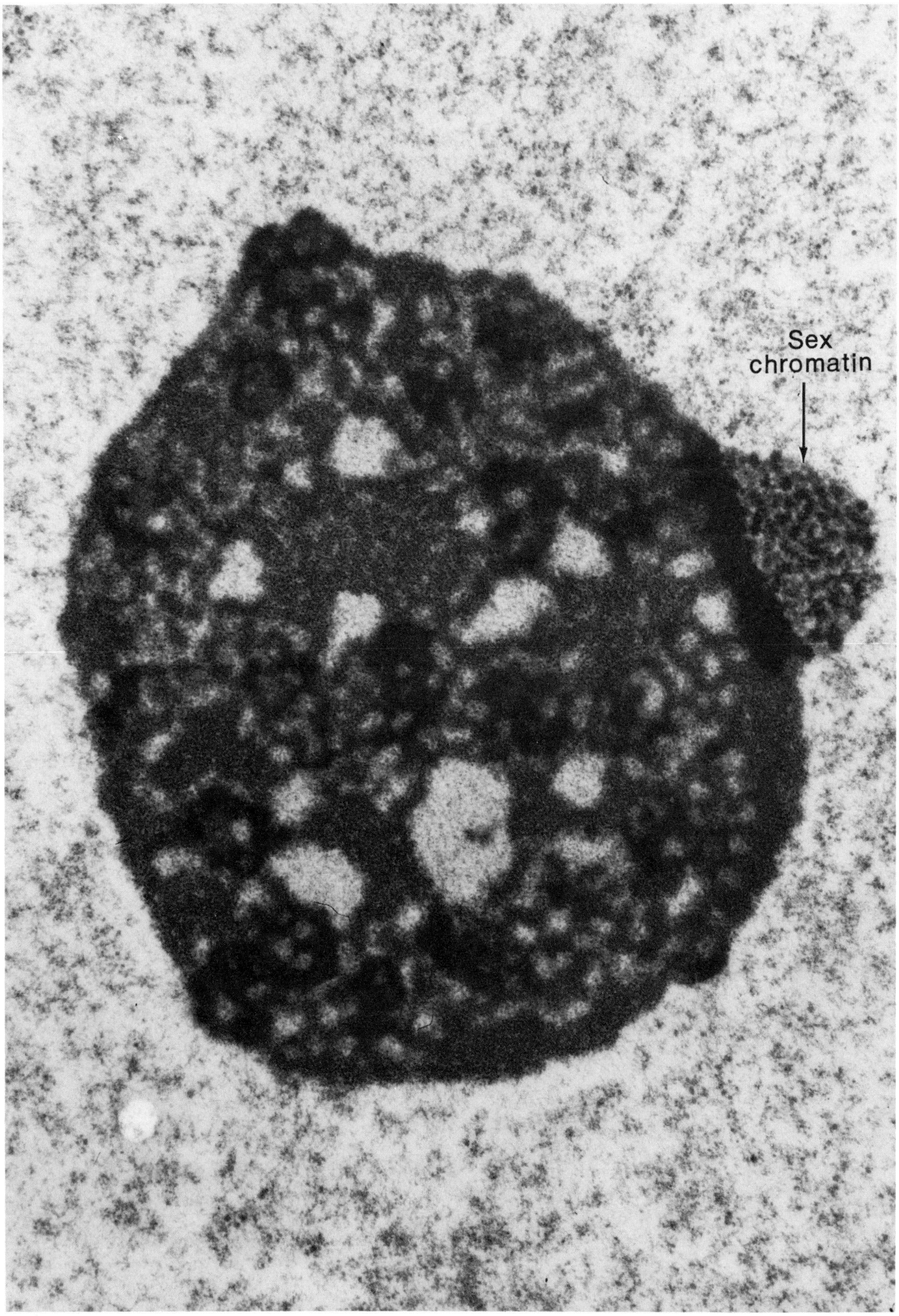

Figure 136

Structures comparable in ultrastructure to the sex chromatin are seldom found in males in the absence of chromosomal abnormalities. In the accompanying micrograph of a Leydig cell from boar testis, the dense structure associated with the nucleolus (at the arrow) bears a striking resemblance to sex chromatin of females in other species. With the light microscope, sex chromatin has been identified in the neurons of female swine but in other cell types an unusually dense chromatin pattern tends to obscure the sex chromatin. It is conceivable that in this species part of the single X chromosome of males is normally heterochromatic.

Figure 137. Leydig cell nucleus from testis of the domestic boar, *Sus scrofa*.

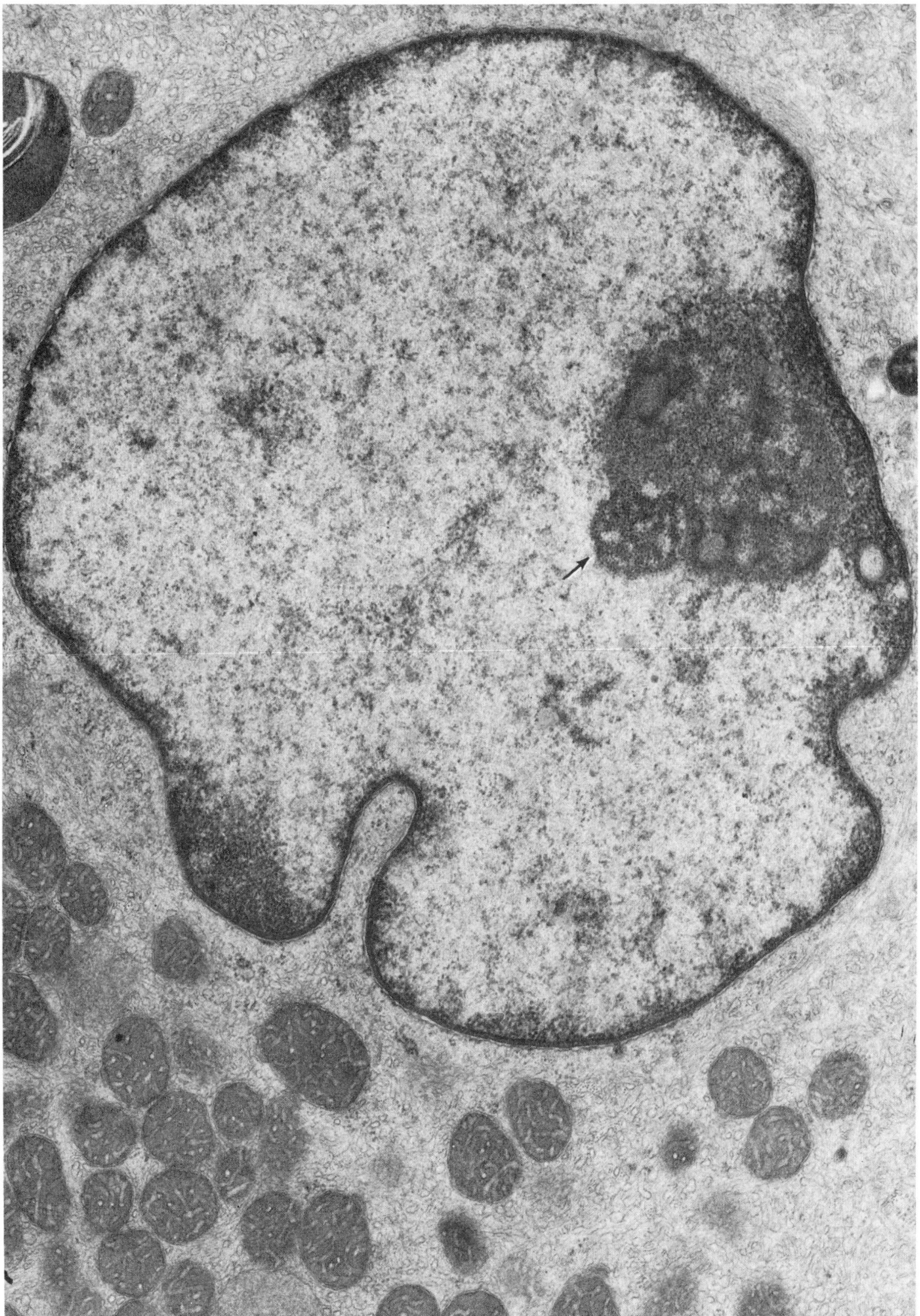

Figure 137

UNUSUAL NUCLEOLI

In an early study of the cytology of the testis, Hermann (1889) drew attention to the unusual configuration of the nucleolus in Sertoli cells. A large central body resembling a typical nucleolus was flanked by two smaller spherical bodies with different staining affinities. These juxtanucleolar dense bodies are Feulgen-positive and have been called *satellite karyosomes,* or *paranucleolar spheres.* They are usually the only stainable heterochromatin present in the Sertoli cell nucleus. It is likely that the highly repetitive DNA sequences called satellite DNA, which are associated with the centromeric regions of the chromosomes, are concentrated in this cell type in the two bodies adjacent to the nucleolus (Pardue and Gall, 1970). In other cell types, the centromeric regions tend to be associated with the nuclear envelope.

On the facing page are two favorable micrographs of Sertoli cell nucleoplasm in which the plane of section passed through the nucleolus and both satellite karyosomes.

Figures 138 and 139. Sections of the nucleolus and accessory structures in guinea pig and Chinese hamster Sertoli cells. (Courtesy of David Phillips.)

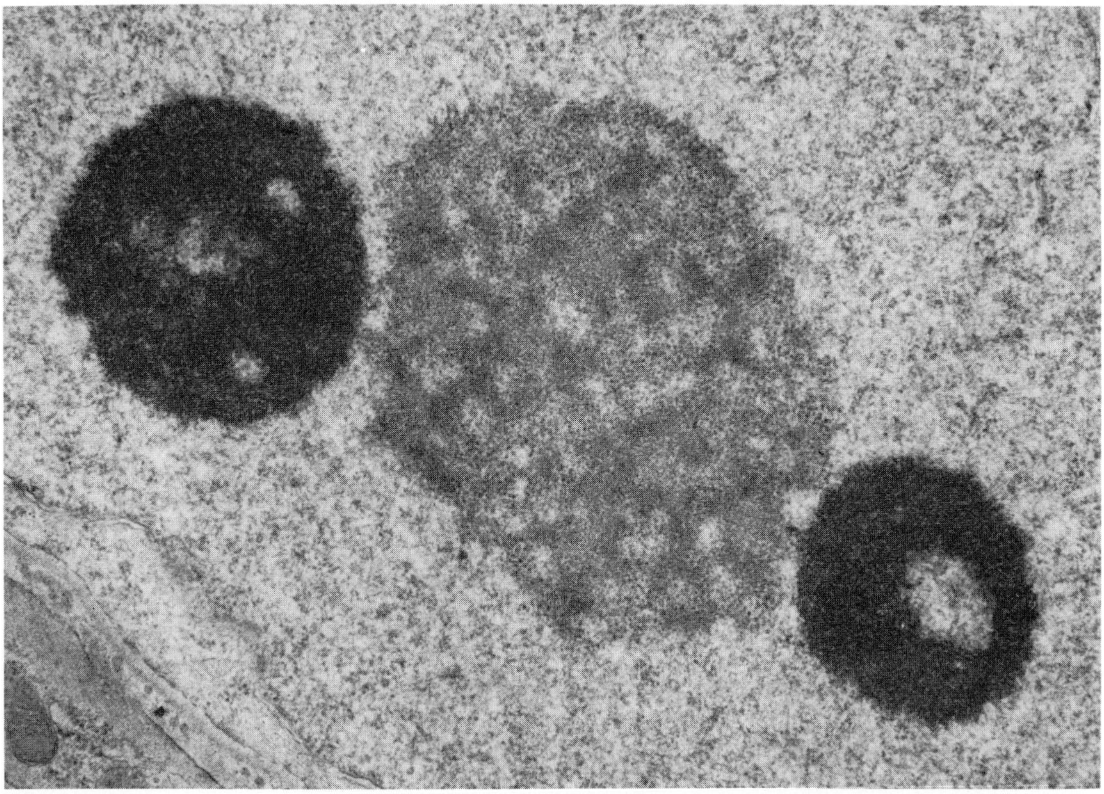

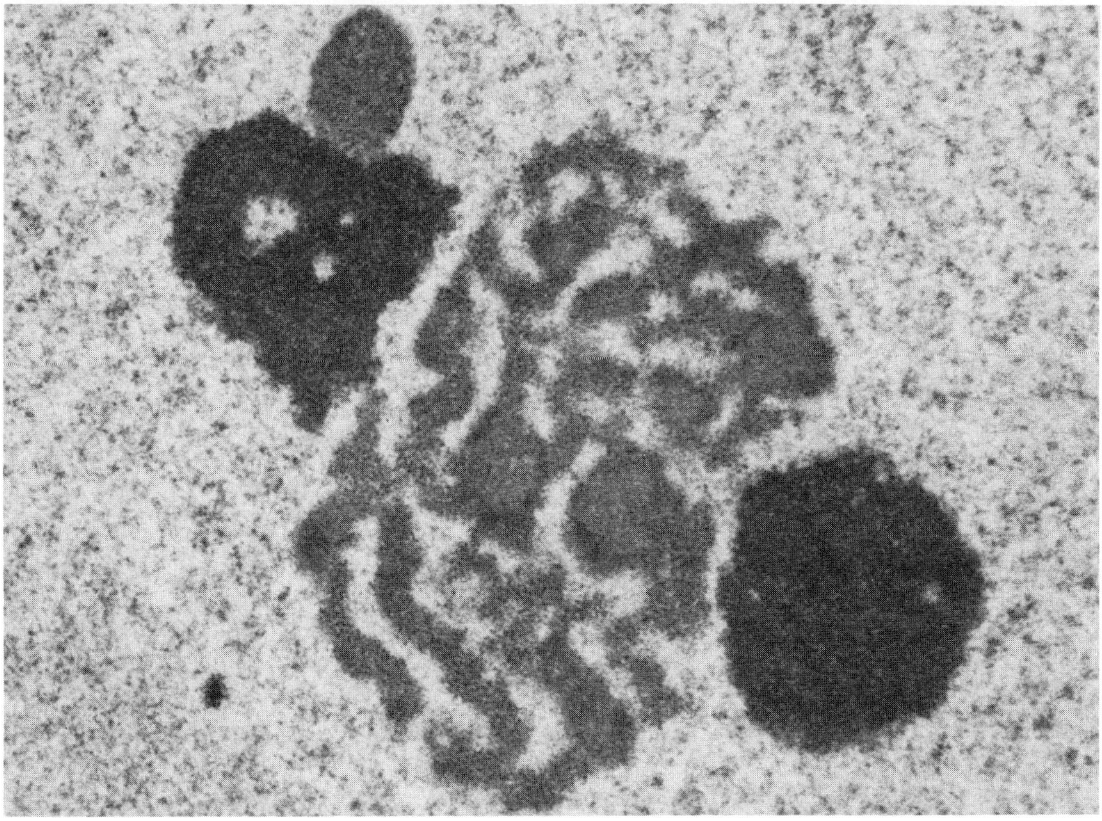

Figure 138, *upper* **Figure 139,** *lower*

Exceptional nucleoli are found in the Sertoli cells of the bull, ram, and certain other ruminants. In these species the nucleoli normally contain varying numbers of membrane-limited vesicles and tubules with dense granules associated with the outer surface of the membrane in much the same manner as ribosomes are associated with the membrane of the endoplasmic reticulum. The vesicles are occasionally seen at the periphery of the nucleus but there is no evidence that they discharge their contents into the perinuclear cistern or into the cytoplasm.

The only comparable nucleolar structure reported to date is in the cells of the human endometrium. A system of convoluted tubules called the *nucleolar channel system* is found within the nucleolus in the secretory phase of the menstrual cycle. Continuity of the tubules with the inner membrane of the nuclear envelope has been reported (Clyman, 1963; Tersakis, 1965). Unlike the vesicles in the Sertoli cells of ruminants, the nucleolar channel system is a transient structure reappearing cyclically and is apparently under endocrine control. It appears to be peculiar to the human uterine endometrium.

Figure 140. Portion of the nucleus of a Sertoli cell from the domestic bull, *Bos taurus.*

Figures 141, 142, and 143. Examples of nucleoli from human endometrium. (Micrographs courtesy of John Tersakis, J. Cell Biol. 27:158–161, 1965.)

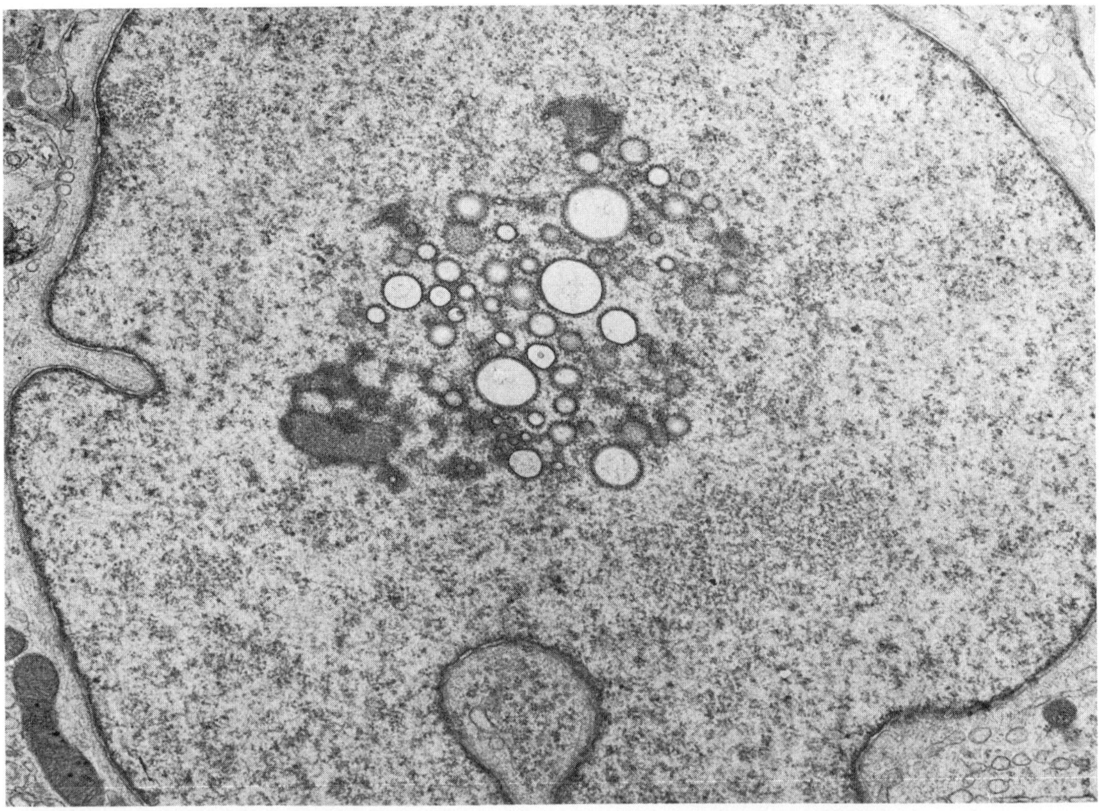

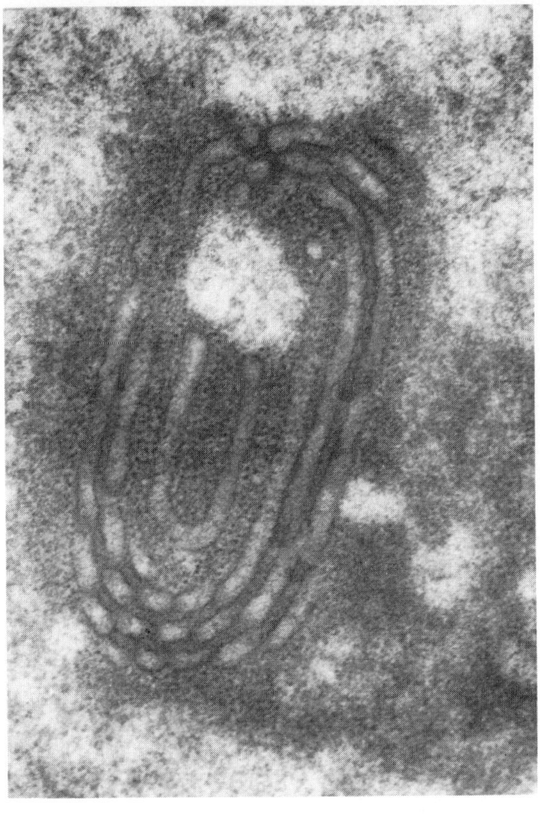

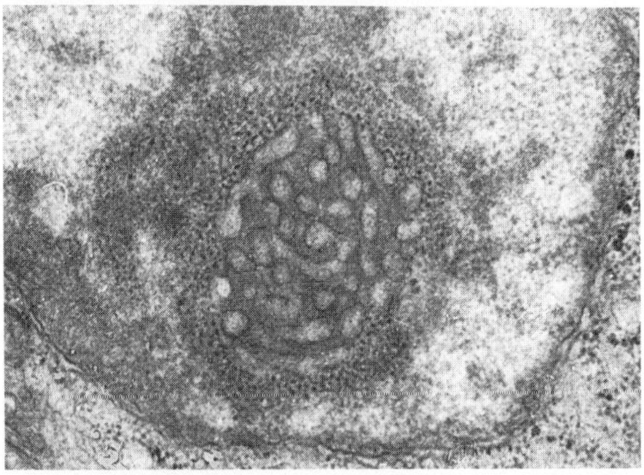

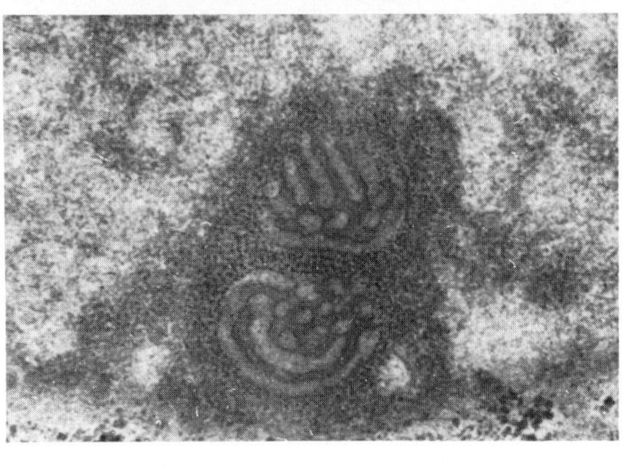

Figure 140, *upper* **Figure 141,** *lower left* **Figure 142,** *center right* **Figure 143,** *bottom right*

MULTIPLE NUCLEOLI OF AMPHIBIAN OOCYTES

In general all of the DNA of the nucleus is in the chromosomes, but an exception is found in the female germ cells of amphibia, which develop very large numbers of "nucleoli." Brachet (1940) and Painter and Taylor (1942) reported that a small Feulgen-positive mass was associated with each of the many nucleoli in amphibian oocytes. It was inferred from these observations that DNA could exist and function in the nucleus apart from chromosomes, and with admirable insight it was suggested that the association of chromatin with multiple nucleoli was part of a mechanism for production of the RNA which accumulates in the cytoplasm of amphibian eggs.

Some 20 years later, it was shown that the nucleoli of amphibian oocytes isolated in saline of low molarity can be dissociated into a cortex, rich in RNA and a circular core with a beaded appearance (Miller, 1964, 1966) which was unaffected by ribonuclease but fragmented upon exposure to deoxyribonuclease (Peacock, 1965). These observations not only confirmed the presence of DNA in these nucleoli but also showed that it occurred in circular form. In ingenious experiments, Gall (1967) was able to hybridize *in situ* ^{3}H-labeled rRNA that was complementary to ribosomal DNA. In autoradiographs a massive concentration of grains was found over the extrachromosomal Feulgen-positive areas of the nucleoli in young oocytes of *Xenopus laevis* (Gall, 1968). When nuclei of these oocytes were isolated and their DNA extracted, it was found by ultracentrifugation that they contained a very large amount of ribosomal DNA which could be identified by its characteristic buoyant density and by hybridization *in vitro* with ribosomal RNA (Brown and Dawid, 1968). Thus, it was clearly established that the multiple nucleoli of amphibian oocytes are a result of selective replication of the genes for ribosomal RNA — a phenomenon known as *gene amplification.* Much of what we now know about gene transcription was learned from combined morphological and biochemical studies on the nucleoli of amphibian oogonia and oocytes.

The gene complex that is amplified in Xenopus oocytes consists of about 450 tandem repeats of the segments of the DNA double helix that are complementary to 28s and 18s ribosomal RNA alternating with intervening spacer sequences. Gene amplification enables the Xenopus oocytes to synthesize as much rRNA in a few months as would be made by an ordinary diploid cell in four hundred years (Petrowska et al., 1968). The resulting ribosomes serve in the protein synthesis that occurs from fertilization of the egg to the feeding tadpole stage (Brown and Gurdon, 1964).

During early growth of amphibian oocytes, a thousand or more extrachromosomal nucleoli are formed in each nucleus. When these are isolated and dissociated on water, the fibrillar core of the nucleolus is unraveled. It consists of a thin circular filament about 3 nm in diameter believed to be a double helix of DNA, formed as a product of replication of the nucleolus-organizing region of a chromosome. At intervals along the length of the axial strand of DNA, there are 2 to 2.5 μm segments decorated by about 100 thin lateral filaments of progressively increasing length from one end of the unit to the other. These are interpreted as partially transcribed molecules of 45s RNA. The shorter filaments at one end are the product of recently initiated transcription. The longest filaments at the other end are RNA molecules nearly completed. The small knob or thickening at the end of the lateral filaments is believed to represent beginning folding of the molecule of ribosomal precursor RNA. These bottle-brush or Christmas tree–like segments each appear to represent one gene coding for the precursor of ribosomal RNA. It shows very clearly that many molecules are simultaneously synthesized on each gene. This remarkable micrograph was the first visual demonstration of an individual gene and its associated transcription product.

Figure 144. Electron micrograph of a portion of a dissociated nucleolar core isolated from an oocyte of *Notophthalmos viridescens.* (Micrograph courtesy of Oscar Miller, from Miller and Beatty, Science 164: 955–957, 1969.)

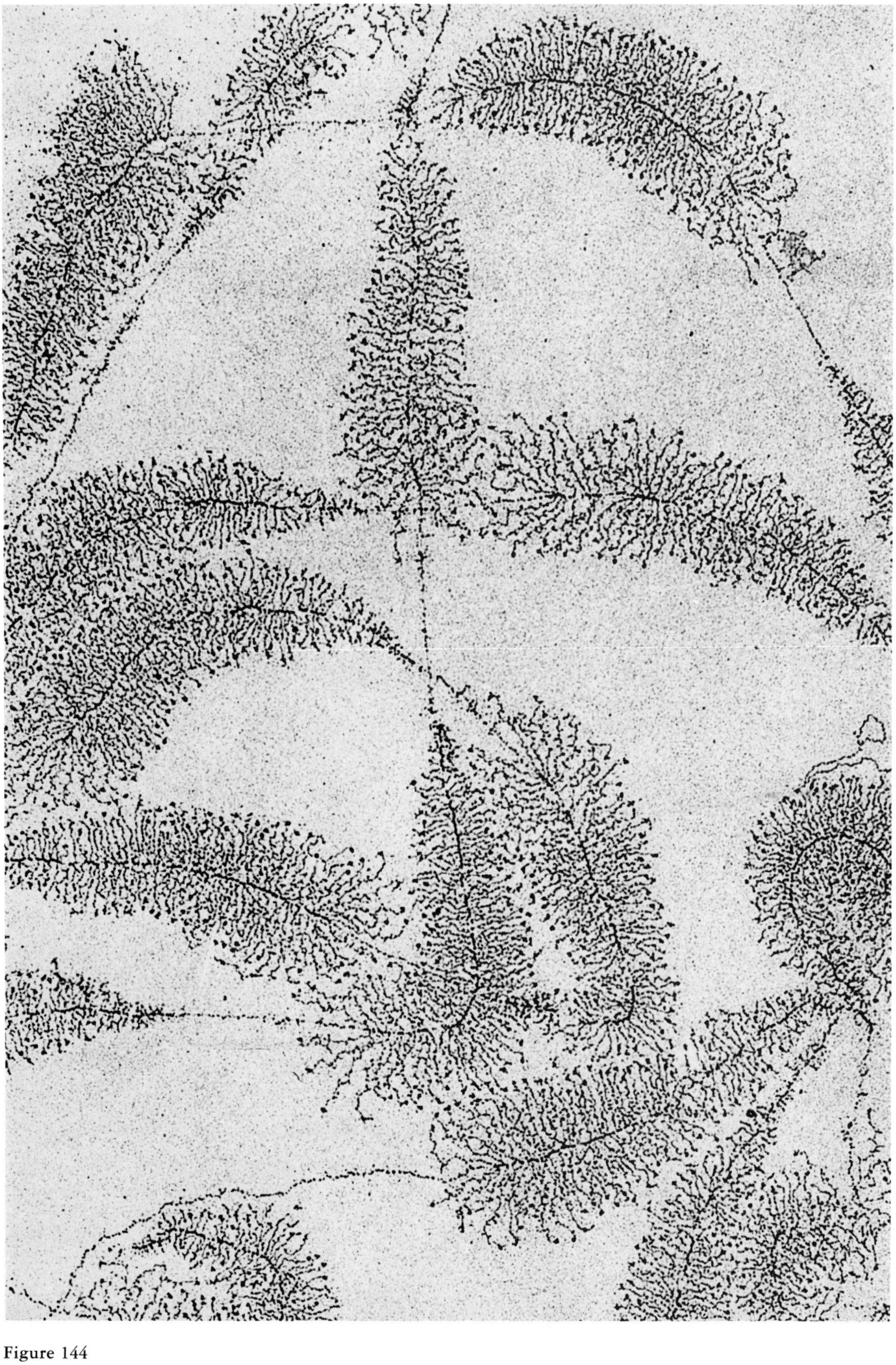

Figure 144

REFERENCES

Barr, M. L., and E. G. Bertram. A morphological distinction between neurones of the male and female, and the behavior of the nucleolar satellite during accelerated nucleoprotein synthesis. Nature 163:676, 1949.

Bernstiel, M. L., J. Speirs, I. Purdom, K. Jones and U. E. Loening. Properties and composition of the isolated ribosomal DNA satellite of *Xenopus laevis*. Nature 219:454–463, 1968.

Borysko, E., and F. B. Bang. Structure of the nucleolus as revealed by electron microscope: A preliminary report. Bull. Johns Hopkins Hosp. 89:468–473, 1951.

Brachet, J. La localisation des acides pentosenucléiques dans les tissues des animaux et les oeufs d'amphibiens en voie de développement. Arch. Biol. 53:207–257, 1942.

Brown, D. D., and I. B. Dawid. Specific gene amplification in oocytes: Oocyte nuclei contain extrachromosomal replicas of the genes for ribosomal RNA. Science 160:272–280, 1968.

Brown, D. D., and J. B. Gurdon. Absence of ribosomal RNA synthesis in the anucleoate mutant of *Xenopus laevis*. Proc. Nat. Acad. Sci. 51:139–146, 1964.

Brown, D. D., and C. S. Weber. Gene linkage by DNA-RNA hybridization. II. Arrangement of redundant gene sequences for 28s and 18s ribosomal RNA. J. Mol. Biol. 34:681–697, 1968.

Cajal, R. Y. Histologie du Système Nerveux de l'Homme et des Vèrtebres. Tome 1. A. Malone, Paris, 1909.

Clyman, M. J. A new structure observed in the nucleolus of the human endometrial epithelial cell. Am. J. Obstet. Gyncol. 86:430–432, 1963.

Edström, J. E., and W. Beermann. The base composition of nucleic acids in chromosomes, puffs, nucleoli, and cytoplasm of chrionomus salivary gland cells. J. Cell Biol. 14:371–380, 1962.

Estable, C., and J. R. Sotelo. Una neuva estructura celulare: El nucleolonema. Publ. Invest. Sci. Biol. 1:105–126, 1951.

Feulgen, R., and H. Rossenbeck, Mikroskopischchemischer Nachweis einer Nucleinsure von typus der Thymonucleinsäure und die darauf berühende elektive Färbung von Zellkernen in Mikroskopischen. Hoppe-Seyler's Z. Physiol. Chem. Präparaten 135:203–248, 1924.

Fischberg, M., and H. Wallace. A mutation which reduces nucleolar member in *Xenopus laevis*. In The Cell Nucleus (J. S. Mitchell, ed.), pp. 30–34, Butterworths, London, 1960.

Gall, J. G. Synthesis of nucleolar DNA in amphibian oocytes. J. Cell Biol. 35:43A, 1967.

Gall, J. G. Differential synthesis of the genes for ribosomal RNA during amphibian oogenesis. Proc. Nat. Acad. Sci. 60:553–560, 1968.

Gall, J. G. Early studies on gene amplification. In The Harvey Lectures. Series 71, pp. 55–70. Academic Press, New York, 1978.

Hay, E. D. The structure and function of the nucleolus in developing cells. In Ultrastructure in Biological Systems, Vol. 3, The Nucleus (A. J. Dalton and F. Hagenau, eds.), pp. 2–79, Academic Press, New York, 1968. (Review)

Hay, E. D., and J. B. Gurdon. Fine structure of the nucleolus in normal and mutant *Xenopus* embryos. J. Cell Sci. 2:151–162, 1967.

Hutz, E. Nukleolen und Chromosomen in der Gattung Vicia. Planta 15:495–505, 1931.

Hermann, F. Beiträge zur Histologie des Hodens. Arch. Mikrobiol. Anat. 34:58–106, 1889.

Jacob F., S. Brenner and F. Cuzin. On the regulation of DNA replication in bacteria. Cold Spring Harbor Symposium on Quantitative Biology 28:329–348, 1964.

Jordan, E. G., and J. M. Chapman. Ultrastructural changes in the nucleoli of Jerusalem artichokes (*Helianthus tuberosus*) tuber discs. J. Exp. Bot. 22:627–634, 1971.

King, H. D. The oogenesis of *Bufo lentiginosus*. J. Morphol. 19:369–438, 1908.

Maggio, R., P. Siekevitz and G. E. Palade. Studies on isolated nuclei. II. Isolation and chemical characterization of nucleolar and nucleoplasmic subfractions. J. Cell Biol. 18:293–313, 1963.

Marinozzi, V., and W. Bernhard. Présence dans le nucléole de deux types de ribonucléoproteines morphologiquément distinctes. Exp. Cell Res. 32:595–598, 1963.

McClintock, B.: The relation of a particular chromosomal element to the development of the nucleoli in *Zea mays*. Zellf. Mikr. Anat. 21:294–328, 1934.

Miller, O. L. Extrachromosomal nucleolar DNA in amphibian oocytes. J. Cell Biol. 23:604, 1964.

Miller, O. L. Structure and composition of peripheral nucleoli of salamander oocytes. Nat. Cancer Institute Monograph 23:53–66, 1966.

Miller, O. L., and B. R. Beatty. Visualization of nucleolar genes. Science 164:955–957, 1969.

Miller, O. L., and B. R. Beatty. Portrait of a gene. J. Cell Physiol. 74: Suppl. 1, 225–232, 1969.

Painter, T. S., and A. N. Taylor. Nucleic acid storage in toad's egg. Proc. Nat. Acad. Sci. 28:311–317, 1942.

Palade, G. E. A small particulate component of the cytoplasm. J. Biophys. Biochem. Cytol. 1:59–67, 1955.

Pardue, M. L., and L. G. Gall. Chromosomal localization of mouse satellite DNA. Science 168:1356–1358, 1970.

Peacock, W. J. Chromosome replication. Nat. Cancer Institute Monograph 18:101–131, 1965.

Perry, R. P. Role of the nucleolus in ribonucleic acid metabolism and other cellular processes. Nat. Cancer Institute Monograph 18:325–340, 1964.

Perry, R. P., A. Hill, and M. Errera. The role of the nucleolus in ribonucleic acid and protein synthesis. I.

Incorporation of cytidine into normal and nucleolar inactivated HeLa cells. Biochim. Biophys. Acta 49:47–57, 1961.
Phillips, S., and D. M. Phillips. Sites of nucleolus production in cultured Chinese hamster cells. J. Cell Biol. 40:248–268, 1969.
Prescott, D. M. Nuclear synthesis of cytoplasmic ribonucleic acid in *Amoeba proteus*. J. Biophys. Biochem. Cytol. 6:203–206, 1959.
Schoefl, G. I. The effect of actinomycin D on the fine structure of the nucleolus. J. Ultrastr. Res. 10:224–243, 1964.
Shea, J. R., and C. P. Leblond. Number of nucleoli in various cell types of the mouse. J. Morphol. 119:425, 1966.
Tersakis, J. A. The nucleolar channel system in human endometrium. J. Cell Biol. 27:293–304, 1965.
Wallace, H., and M. L. Birnstiel. Ribosomal cistrons and the nucleolar organizer. Biochim. Biophys. Acta 111:296–310, 1966.
Woods, P. S., and J. H. Taylor. Studies of ribonucleic acid metabolism with tritium-labeled cytidine. Lab. Invest. 8:309–318, 1959.

NUCLEAR ENVELOPE

The smooth contour and sharp discontinuity in density at the interface between the peripheral chromatin and the cytoplasm suggested to the early cytologists that the nucleus was probably limited by a membrane. This was later substantiated by micromanipulation which demonstrated that the nucleus could be moved around in the cytoplasm, that it was deformable, and that it appeared to be bounded by a membrane that offered considerable resistance to penetration by the microdissection needle. Disruption of the membrane led to death of the cell (Kite, 1913; Chambers, 1917).

Although osmotically induced volume changes suggested that the nuclear envelope had the properties of a semipermeable membrane, these observations posed a conceptual problem. The genetically controlled differentiation of cells clearly depended upon interchange of substances between the genes and the cytoplasm, but it was inconceivable that genetic expression could be mediated by ions and small molecules that could traverse a semipermeable membrane. Somehow informational macromolecules had to pass through the nuclear envelope.

Resolution of this problem had to await the development of the electron microscope. The large nuclei of amphibian oocytes were then isolated manually and the disrupted nuclear envelopes dried onto a Formvar-coated specimen grid. The resulting electron micrographs were initially interpreted as evidence for the presence of two membranes, one continuous and the other penetrated by pores (Callan and Tomlin, 1950). At that time, it was difficult to assess the degree to which the pores were artifacts of drying, but a few years later when it was possible to examine various cell types in ultrathin sections, a double nuclear membrane was consistently reported and ring-like structures were observed in it where the plane of section passed tangential to the nucleus (Bahr and Beerman, 1954; Rhodin, 1954; Palay and Palade, 1955). In the first detailed study of the nuclear envelope in thin sections, Watson (1954) reported that it consisted of inner and outer membranes which enclosed a *perinuclear space*. The two membranes were continuous with each other around circular pores through which the nucleoplasm appeared to be in communication with the cytoplasm. Pores were found in the interphase nucleus of all cell types examined and were assumed to be of general occurrence. The outer nuclear membrane was occasionally found to be continuous with tubular and saccular elements of the endoplasmic reticulum — a cytoplasmic organelle then newly discovered with the electron microscope. It was proposed, therefore, that two pathways existed for exchange between the nucleoplasm and cytoplasm. Small molecules and ions could diffuse across the inner membrane into the perinuclear space and thence throughout the cytoplasm via the endoplasmic reticulum, and large molecules could pass through the pores directly into the cytoplasmic matrix. Soon thereafter images of material in passage through the nuclear pores began to appear in the literature (Anderson and Beams, 1956).

The term *nuclear pore,* as originally defined, referred only to the membrane-bounded channels traversing the nuclear envelope, but it was soon noted that a ring or *annulus* of dense material was associated with the pores. This was interpreted by some as material deposited on the inner and outer lips of the pore (Afzelius, 1955) and by others as a cylindrical structure lining the pore and projecting a short distance into both the nucleoplasm and cytoplasm (Wischnitzer, 1958). As the terminology evolved, *pore* was used to describe the membrane-bounded passage through the envelope, *annulus* referred to the nonmembranous material associated with it, and the two together constituted the *pore complex*. Since the two never occur independently of each other, the term *nuclear pore* usually subsumes the annulus.

The annulus was later found to be composed, in part, of globular subunits 10 to 20 nm in diameter (Gall, 1956). These are 16 in number with eight on the nucleoplasmic and eight on the cytoplasmic side, arranged in radial symmetry around the rims of the membranous pore (Franke, 1966). The globular units at either end of the annulus are in register and probably connected by fine filaments that pass through the pore. The outlines of the membranous pores when seen in tangential sections of the nuclear envelope are usually circular but they may appear octagonal in negatively stained preparations of isolated nuclear envelopes.

In micrographs of thin sections, a linear density with a central thickening or granule often extends across the middle of the pore. This was originally interpreted as a *pore diaphragm*. However, some investigators insist that this transverse density is simply due to inclusion of a portion of the opposite side of the membranous pore in the thin section. Others consider the linear density to be too thin and too sharply defined to be the result of a tangential section of the opposing pore margin. Instead it is attributed to superimposition of images of very thin, radially arranged filaments that project inward from the lining of the pore toward a central granule (Barnes and Davis, 1959; Abelson and Smith, 1970). The existence of a continuous septum or diaphragm is no longer widely accepted, but there is some measure of agreement on the presence of a small granule somehow suspended in the center of the pore. Several elaborate and imaginative diagrams of the organization of the pore complex have been published. Although there is no doubt that it is a highly ordered and complex structure, some caution should be exercised in interpretation of its components that are near the limit of resolution of the electron microscope.

The occurrence of polyribosomes on the outer nuclear membrane and its continuity with the membranes of the cytoplasm led to the concept that the nuclear envelope is analogous to the membrane-bounded flat saccules or cisternae of the endoplasmic reticulum. This view is reinforced by the observation that during reconstruction of the nucleus after cell division the nuclear envelope appears to arise by extension and coalescence of discontinuous elements of the endoplasmic reticulum. The nuclear envelope is therefore regarded as a specialized region of the endoplasmic reticulum (Porter and Machado, 1960), and the *perinuclear cistern* is analogous to cisternae of the reticulum in the surrounding cytoplasm. There is abundant evidence that it shares some of the functional activities of the endoplasmic reticulum. When granular, crystalline, or amorphous products of protein synthesis accumulate in the lumen of the reticulum, they can occasionally be observed in the perinuclear space as well. Cytochemical and immunocytochemical studies on experimental induction of new enzymes (Brokelman and Fawcett, 1969) or synthesis of new antibodies (Leduc et al., 1968) have shown that these appear first in the perinuclear cistern and subsequently in the lumen of the reticulum throughout the cytoplasm. It seems likely therefore that the nuclear envelope participates early in the biosynthesis of a new product and that substances accumulating in its lumen have not simply refluxed from more peripheral sites of synthesis.

Around the periphery of the nucleus illustrated here, the two membranes forming the nuclear envelope are clearly seen. The outer membrane appears thick, for, like the membranes of the surrounding endoplasmic reticulum, it is studded with closely spaced ribosomes which are not individually resolved. At several points on the circumference of the nucleus, indicated by arrows, the perinuclear cistern is traversed by nuclear pores. The location of pores is betrayed by small areas of lower density that interrupt the otherwise continuous layer of heterochromatin. The invariable occurrence of such areas of rarefaction in the nucleoplasm adjacent to the pores is indirect evidence that they constitute pathways for exchange of materials between the nucleoplasm and cytoplasm.

Figure 145. Pancreatic acinar cell from the bat *Myotis lucifugus.*

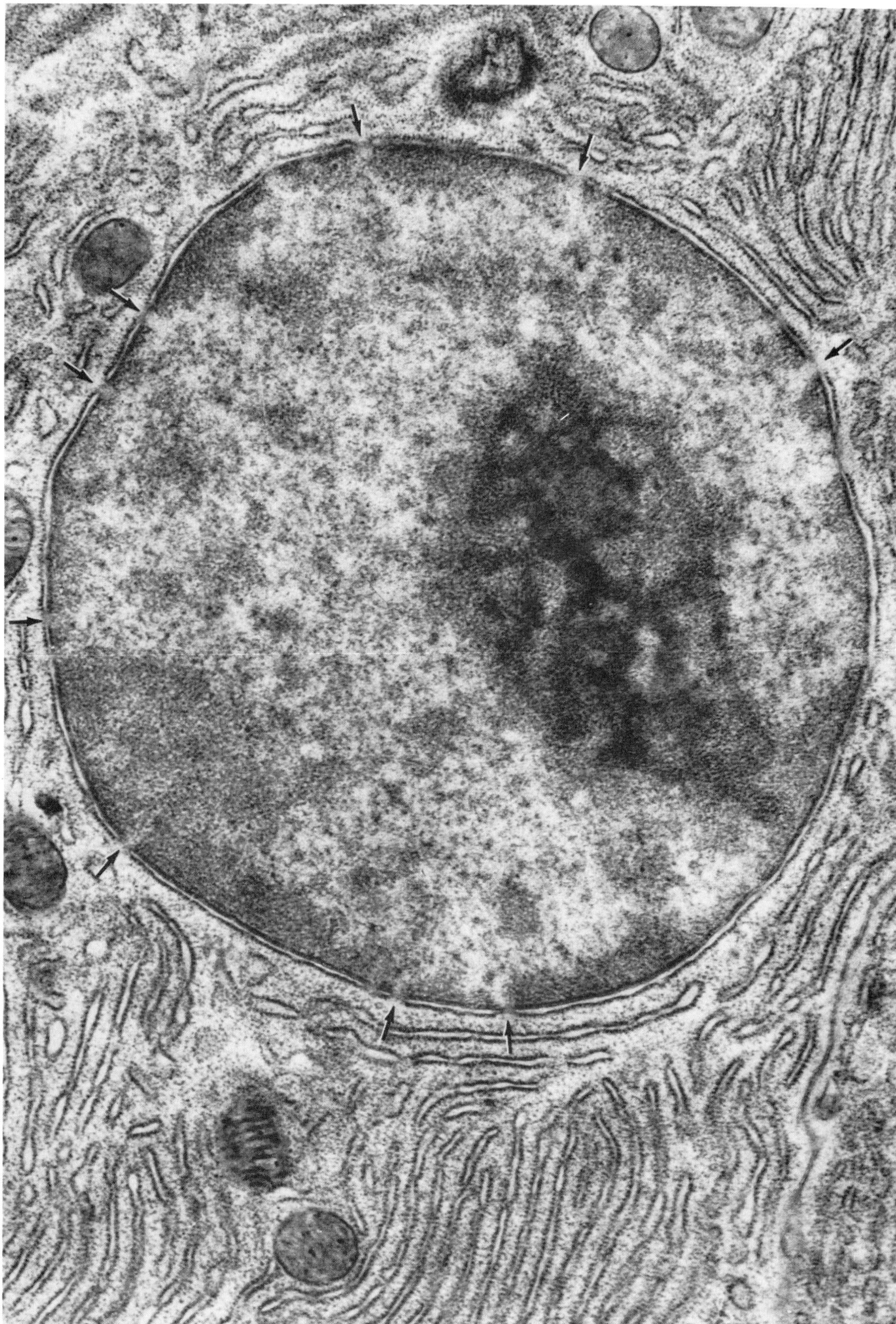

Figure 145

When examined at higher magnification, there is a strong suggestion of a transverse pore diaphragm. However, it is now generally agreed that this appearance is not due to the presence of a complete septum but is a consequence of superimposition of the images of eight small granules or radially arranged filaments attached to the inner aspect of the membranous pore. This linear density is especially prominent in erythroblasts.

An inconstant component of the pore complex is a thin ridge or flange encircling the membranous pore and projecting into the perinuclear cistern. This is discernible in the upper figure (at the arrow). This structure has received little attention in the literature but is frequently seen in some cell types.

Outlined by arrows in the lower figure is a typical channel or area of rarefaction in the peripheral chromatin commonly observed extending inward from a nuclear pore.

Figure 146. Polychromatophilic erythroblast from guinea pig bone marrow.

Figure 147. A portion of an endothelial cell nucleus.

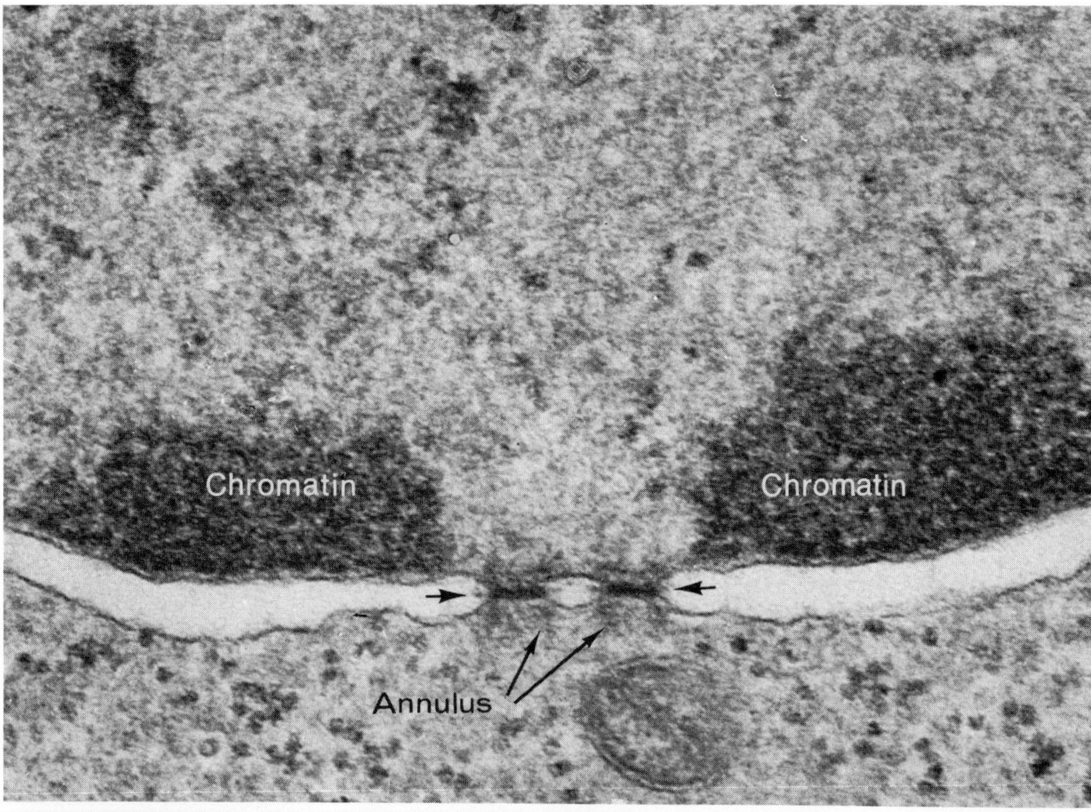

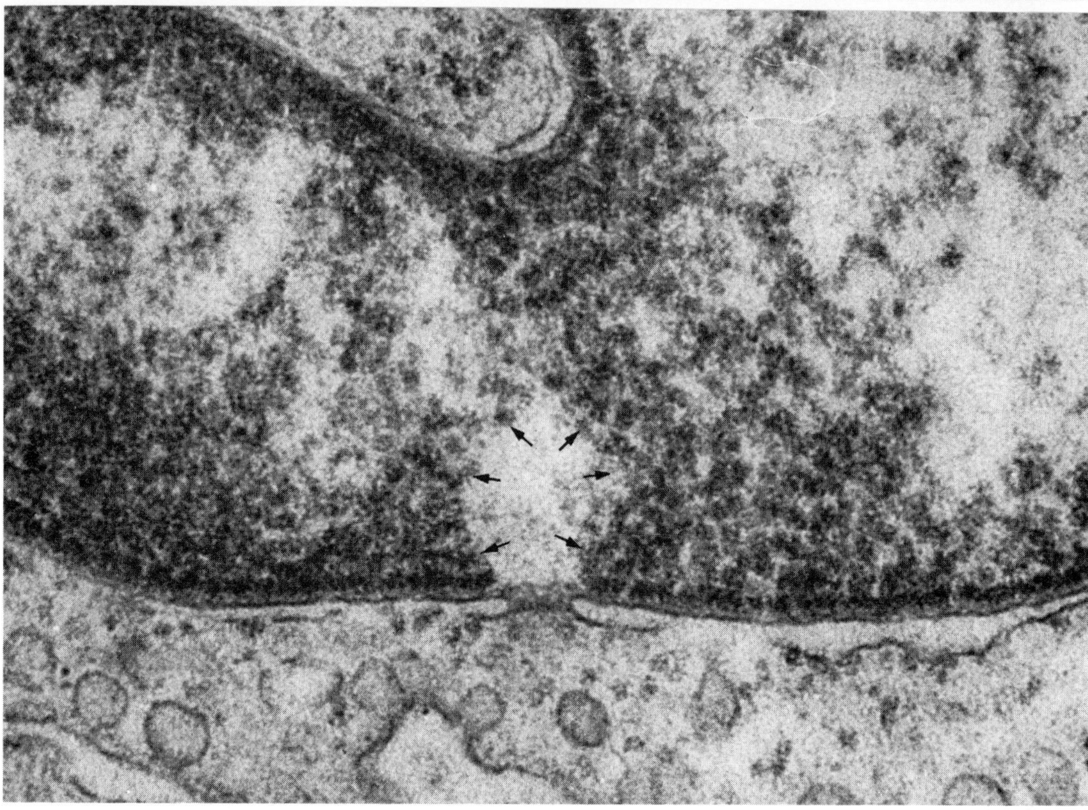

Figure 146, *upper* **Figure 147**, *lower*

Electron micrographs of thin sections provide little information about the pattern of distribution of the nuclear pores or their number per unit area. But in freeze-fracture preparations, the fracture path often follows the hydrophobic region of either the inner or the outer membrane for several micrometers, affording extensive *en face* views of the nuclear envelope. As seen in the upper micrograph, the pores are numerous and rather evenly distributed in somatic cell types. The pores are cross-fractured and appear as flat-topped elevations or shallow craters.

In the lower figure the plane of cleavage has followed the inner membrane and then has broken across the nuclear envelope and into the cytoplasm. The two membranes, the intervening perinuclear cistern, and several pores (at arrows) are therefore seen on edge. The other membrane-limited structures at the top of the figure are tubular and cisternal elements of the endoplasmic reticulum.

Figure 148. Freeze-fracture preparation of the nuclear envelope of a Sertoli cell from guinea pig testis. (From D. W. Fawcett and H. Chemes, Tissue Cell 11:147–162, 1979.)

Figure 149. Nuclear envelope and adjacent cytoplasm of an acinar cell from rat pancreas. (Micrograph courtesy of Bernard Gilula.)

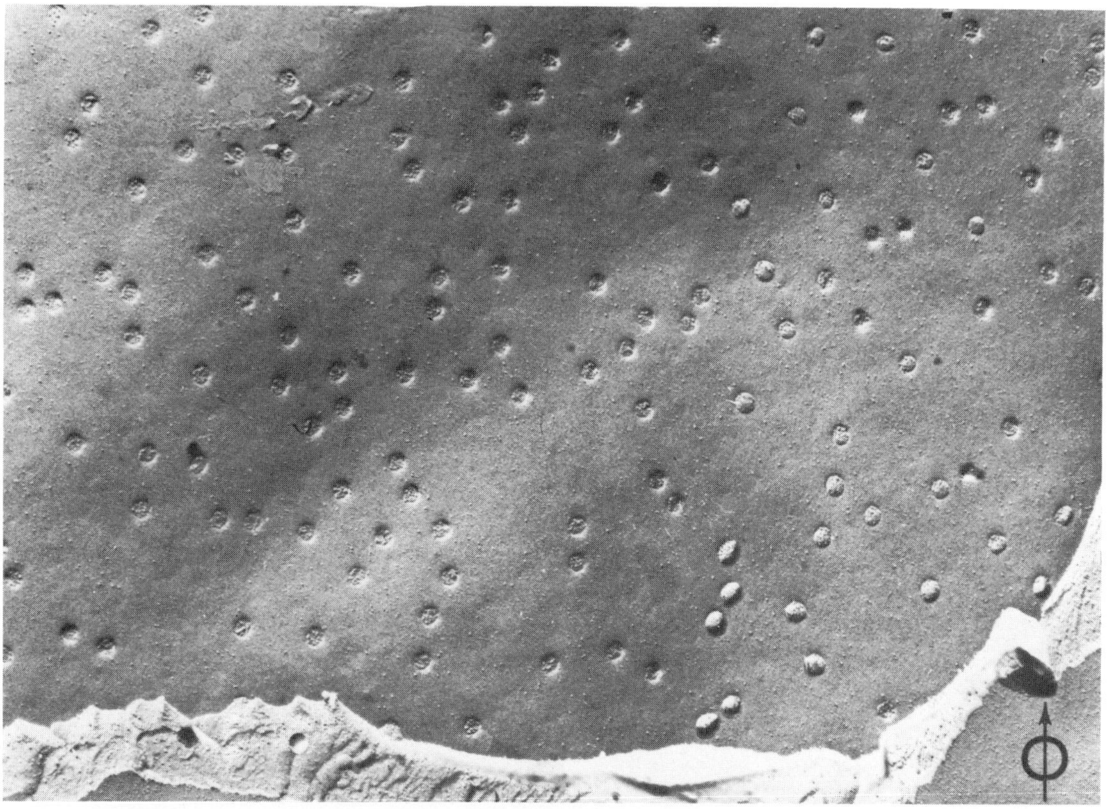

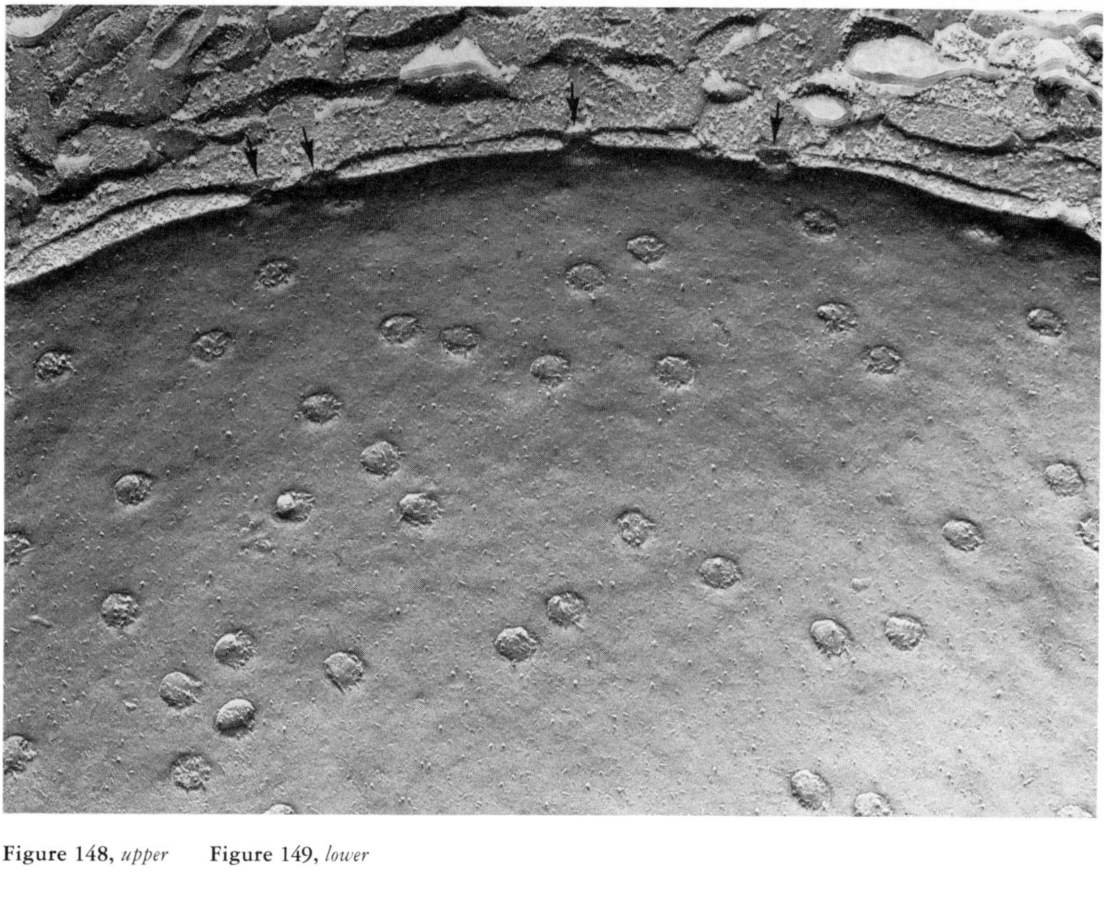

Figure 148, *upper* **Figure 149,** *lower*

In contrast to the random distribution of pore complexes in the somatic cells illustrated in the previous figures, the spermatocyte nucleus on the facing page exhibits large aggregations of closely packed pores and extensive pore-free areas. At the onset of meiotic prophase in many mammalian species, the chromosomes assume the so-called "bouquet" arrangement, forming long loops with their ends all clustered and attached to the nuclear envelope in a limited area at one pole of the nucleus. Concurrently with this chromosomal rearrangement, the great majority of the nuclear pores become concentrated in the region where the ends of the chromosomes are attached. The actual attachment sites are small, pore-free islands in a region of high pore density. In spermatocytes of some invertebrate species the Golgi complex, centrioles, and a mass of mitochondria are concentrated in the cytoplasm adjacent to the region of chromosomal attachment at the base of the "bouquet." It is not clear whether the conspicuous pore aggregation in this region is directly correlated with prophase chromosomal rearrangement or is functionally related to the concentration of metabolically active organelles in the adjacent cytoplasm.

Figure 150. Freeze-fracture replica of the nucleus of a guinea pig spermatocyte. (From D. W. Fawcett and H. Chemes, Tissue Cell 11:147–162, 1979.)

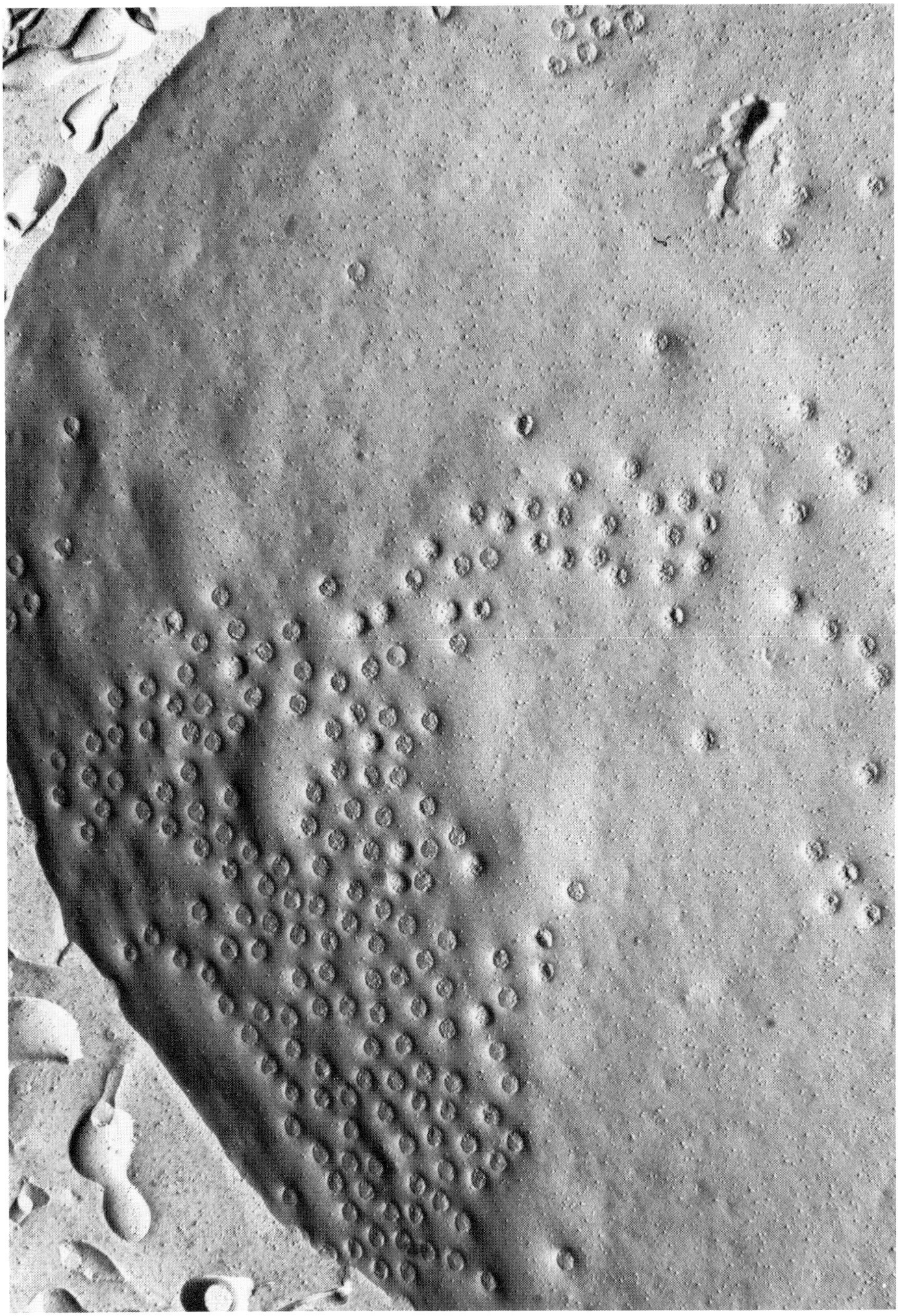

Figure 150

The membranous portion of the pore complex is clearly visualized in thin sections or in freeze-fracture replicas. The other constituents of the pore complex can be studied to advantage in isolated nuclear envelopes of amphibian oocytes negatively stained with phosphotungstate. Examples are shown in the accompanying micrographs. In the upper figure the contrast medium clearly outlines eight electron lucent areas around each pore. These correspond to the superimposed images of 16 granules, eight uniformly spaced around the rim of the pore on the inner aspect of the nuclear envelope, and eight in a corresponding location on its outer aspect. A central granule is also evident in many of the pores.

In the lower figure, the membranous component of the pores is seen in negative image. Their outline appears to be polygonal instead of circular as they are commonly depicted. When the negative image of a pore is subjected to the Markham multiple exposure rotational analysis, the outline is clearly octagonal (see inset). When this observation was first reported, it seemed likely that the eightfold symmetry of the globular constituents of the annulus imposed octagonality upon the membranous pore during the drying involved in specimen preparation. However, octagonal pores have since been observed in freeze-fracture replicas, and since no dehydration is involved in this procedure, these images are probably a true representation of their shape *in vivo*.

Figure 151. Negatively stained nuclear envelope from an oocyte of *Taricha granulosa*. (Micrograph courtesy of A. Fabergé, Z. Zellforsch. 136:183–190, 1973.)

Figure 152. Negatively stained nuclear envelope from an oocyte of the newt, *Notophthalmus viridescens*. (Micrograph from Joseph Gall, J. Cell Biol. 32:391, 1967.)

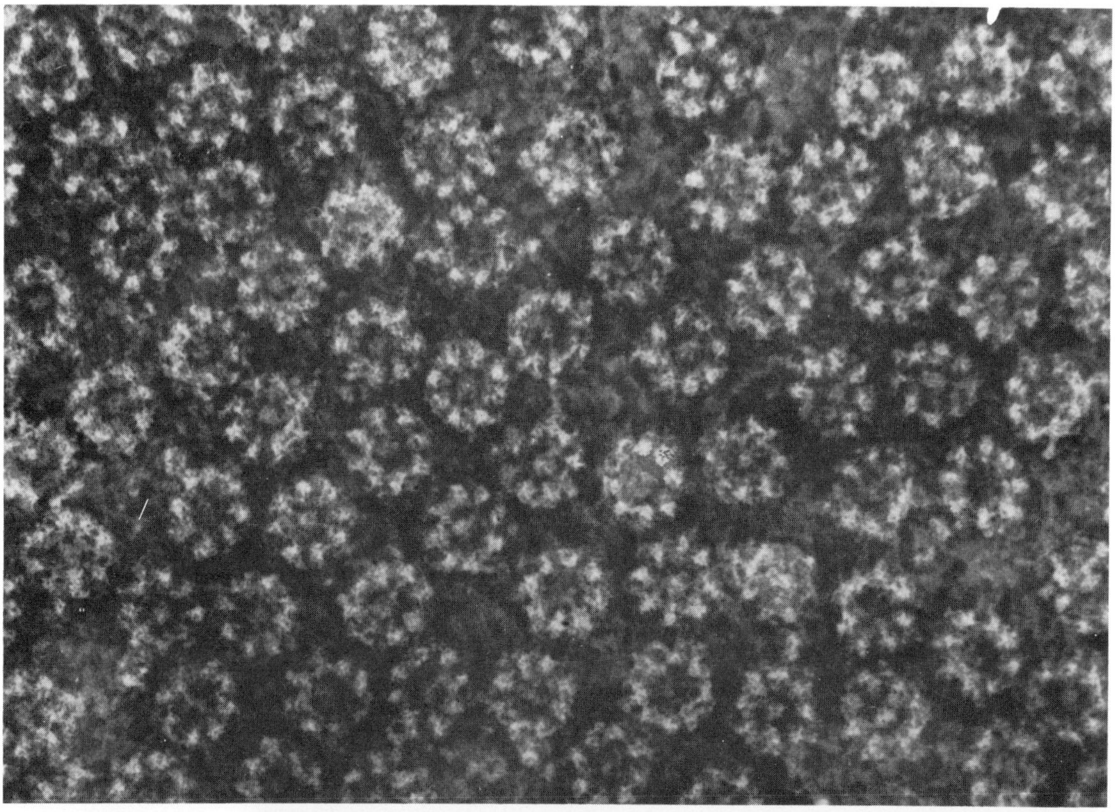

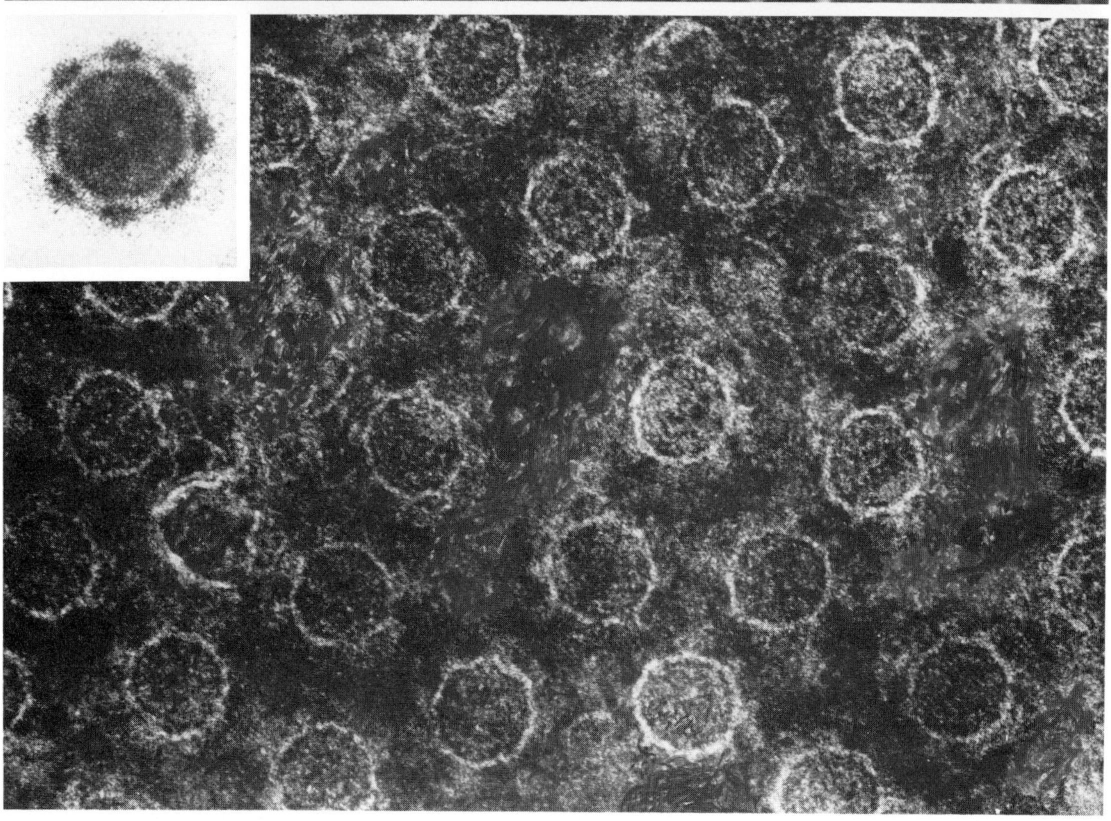

Figure 151, *upper* **Figure 152,** *lower*

It is generally accepted that the interaction of nucleus and cytoplasm is two-way traffic and involves informational macromolecules. Messenger RNA molecules move from the nucleus to cytoplasm and specific proteins move from the cytoplasm into the nucleus to influence gene expression. Evidence for this rests mainly on biochemical and cytochemical studies, but after transplantation of nuclei it can be shown that radioactively labeled nuclear components move into the cytoplasm and vice versa. However, the electron microscopic observation of substances in transit through the nuclear pores is relatively rare. The majority of published reports are based upon studies of oocytes which are extremely active in synthesis of mRNA and rRNA during the diplotene stage of meiosis. At this time, there are abundant perinuclear aggregations of dense material associated with the outer aspect of the nuclear envelope. Filaments can occasionally be seen in the nucleoplasm extending into a pore (see at arrows) and emerging to join one of the cytoplasmic masses of dense material. Such images are interpreted as gene products in transit to the cytoplasm. Their relatively low frequency suggests that passage through the pore is rapid. On the other hand, since the number of nuclear pores in amphibian oocytes is estimated to range from 1.7 million in early stages to 50 million in mature amphibian oocytes, only a small fraction need be involved at any one time to move a large amount of material from one compartment to the other.

Figures 153 and 154. Portions of the nucleus and adjacent cytoplasm of tadpole oocytes. (Micrographs courtesy of Susumu Ito and Edward Eddy.)

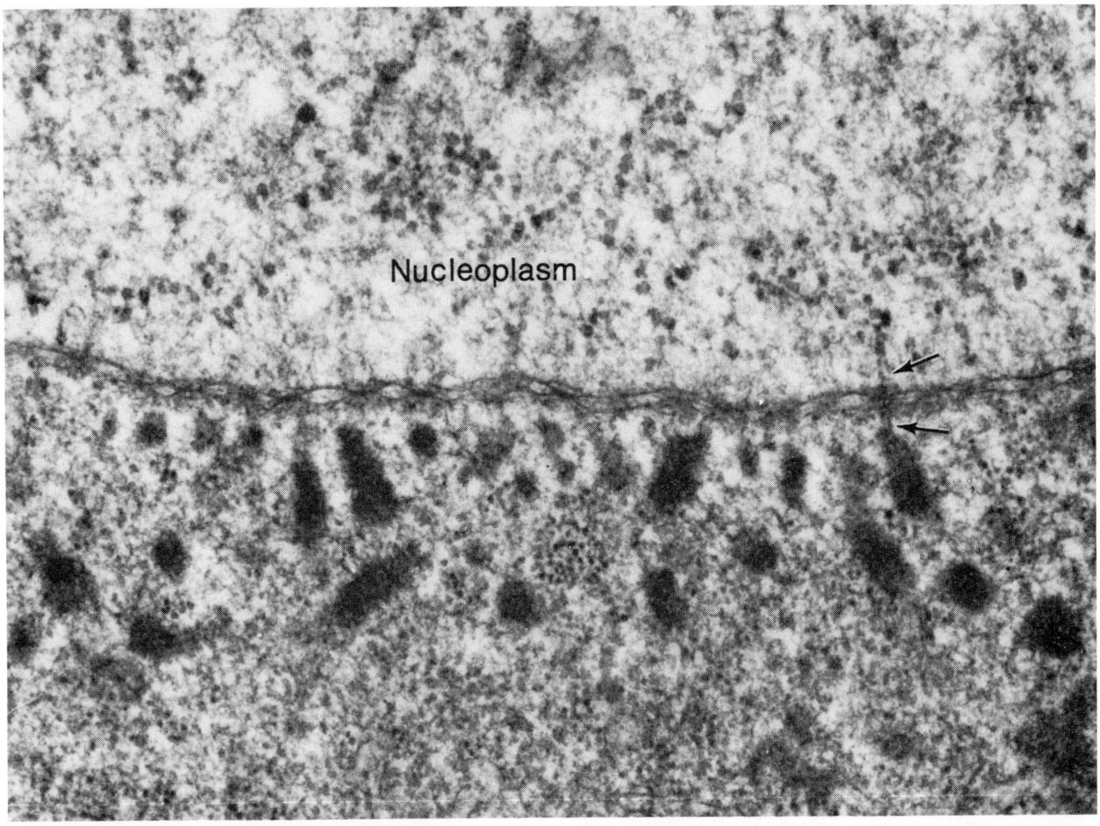

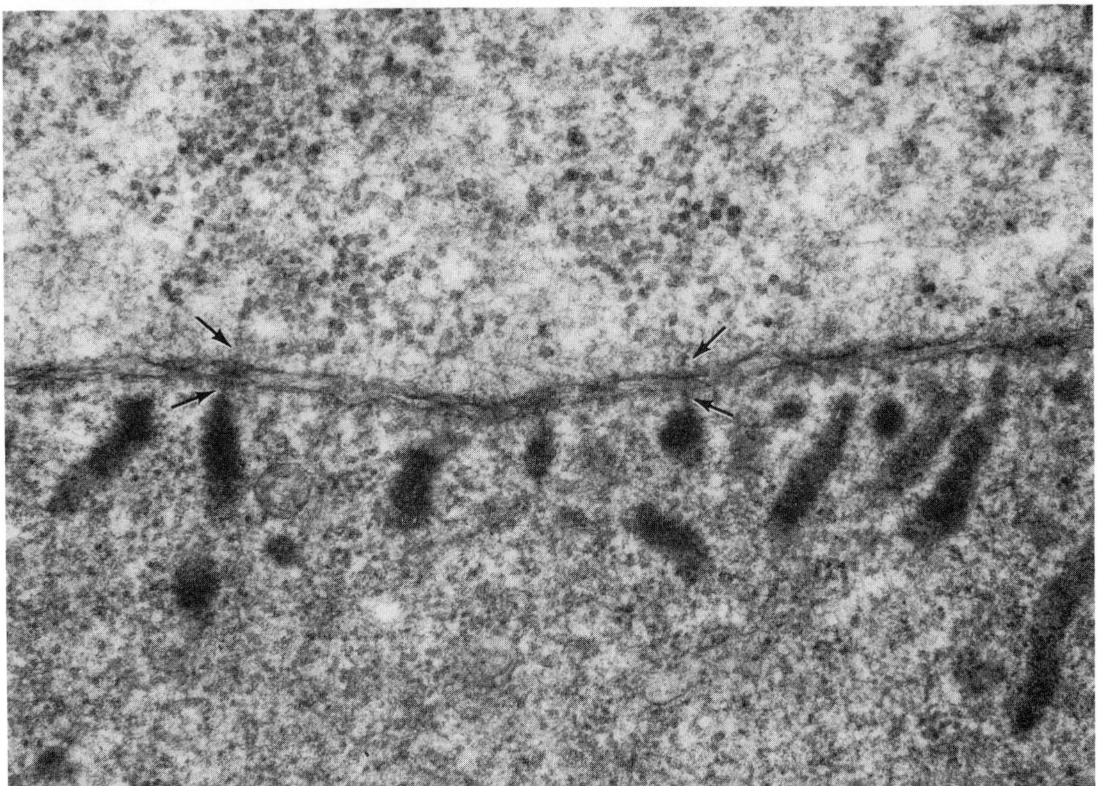

Figure 153, *upper* **Figure 154,** *lower*

FIBROUS LAMINA

In early electron microscopic studies of amoebae (Pappas, 1956) and gregarine protozoa (Beams et al., 1957), a complex supporting layer of fine filaments was observed on the inner aspect of the nuclear envelope. Its configuration in the amoeba was reminiscent of a single layer of cells in a honeycomb with a nuclear pore located at the base of each hexagonal cell. A similar layer was reported in nuclei of ganglion cells of the leech (Coggeshall and Fawcett, 1964). At the time these were described they were thought to be exceptional cases, but a comparable but thinner structure was later observed in many cell types of vertebrates and was called the *fibrous lamina* (Fawcett, 1966). Other terms applied to this structure were *zonula nucleum limitans* (Patrizi and Poger, 1967) and *dense lamina* (Kalifat et al., 1967; Andre and Stevens, 1969). To describe it as a dense lamina is not entirely appropriate since its staining intensity varies with the cell type and method of preparation. In fact, it is often detectable because it has much less affinity for the stain than the adjacent chromatin.

The fibrous lamina appears as a thin feltwork of interwoven fine filaments interposed between the inner nuclear membrane and the peripheral heterochromatin. Its thickness varies in different cell types from 30 to 80 nm, and its dimensions can change in the same cell type under different conditions (Ghadially et al., 1972; Oryschak et al., 1976). Although it is not always visualized in routine electron micrographs, it is now believed to be a ubiquitous component of the eukaryotic nuclear envelope. It evidently has a supporting or cytoskeletal function, stabilizing the inner nuclear membrane and providing sites of attachment for other structural components of the nucleoplasm.

A fibrous lamina is not ordinarily seen in thin sections of liver, but nonetheless it can be isolated as a thin, coherent lamina containing the annuli of the pore complex (Aaronson and Blobel, 1978). It is composed of three major polypeptides between 60,000 and 70,000 daltons. These have been localized immunocytochemically in tissue culture cells derived from several species. The staining is largely confined to the periphery of the nucleus, suggesting that at least a major portion of the lamina is biochemically distinct from any general internal nuclear matrix (Gerace et al., 1978). Localization of these proteins undergoes dramatic changes during mitosis. When the nuclear envelope is disassembled during prophase, the antigenic proteins of the fibrous lamina assume a diffuse localization throughout the cell, and in telophase they localize again at the surface of the aggregated chromosomes in the daughter cells. The polypeptides of the lamina are presumably in a soluble form dispersed throughout the cytoplasm during mitosis, and they polymerize as they become associated with the inner membrane of the reconstituting nuclear envelope in telophase. The substance of the fibrous lamina is thus regarded as a biological polymer which undergoes reversible depolymerization during cell division (Gerace et al., 1978). Its protein constituents can be regarded as peripheral proteins of the inner nuclear membrane. Their specific association with this surface implies significant biochemical differences between the outer and inner membranes of the nuclear envelope.

The fibrous lamina is well developed in certain cells of invertebrates. In the neuron from *Hirudo medicinalis* illustrated here, it is 150 to 200 nm thick. In equatorial sections, it has a scalloped appearance with thicker regions forming septa outlining compartments with a center-to-center distance of about 200 nm. A nuclear pore is situated at the bottom of each concavity in the fibrous lamina which appears to be discontinuous over the pore.

It can be seen in the inset that the lamina is composed of very thin filaments.

Figure 155. Leech ganglion cell. (From D. W. Fawcett, Am. J. Anat. 119:129–146, 1966.)

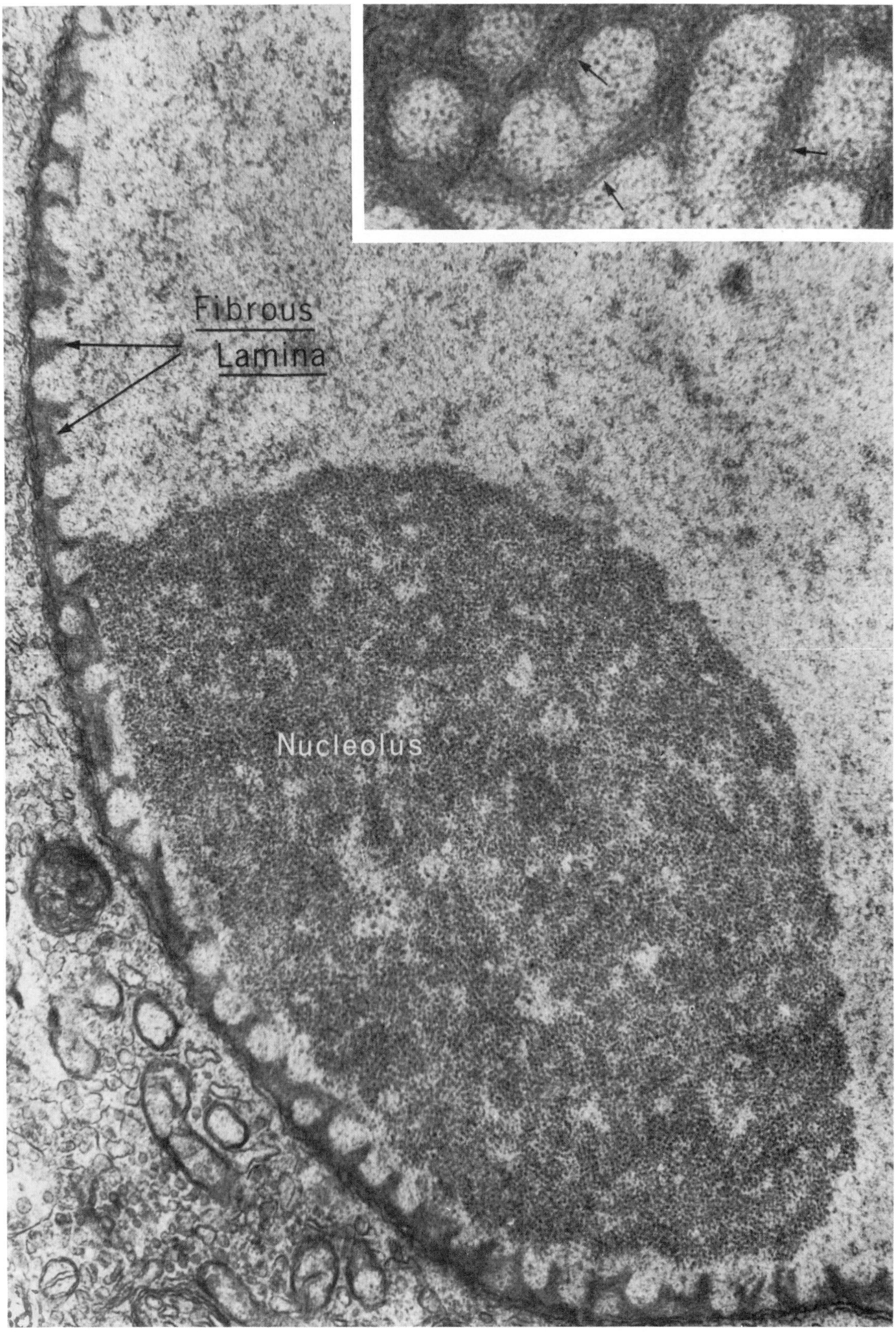

Figure 155

In the cells of vertebrates, the fibrous lamina is less conspicuous and is often overlooked, especially when it is heavily stained and tends to match the density of the peripheral clumps of heterochromatin. In the nucleus of an amphibian intestinal epithelial cell shown here, it can be seen as a continuous dense layer about 60 nm thick interposed between the inner nuclear membrane and the heterochromatin.

Figure 156. Intestinal epithelial cell of *Amphiuma tridactylum*.

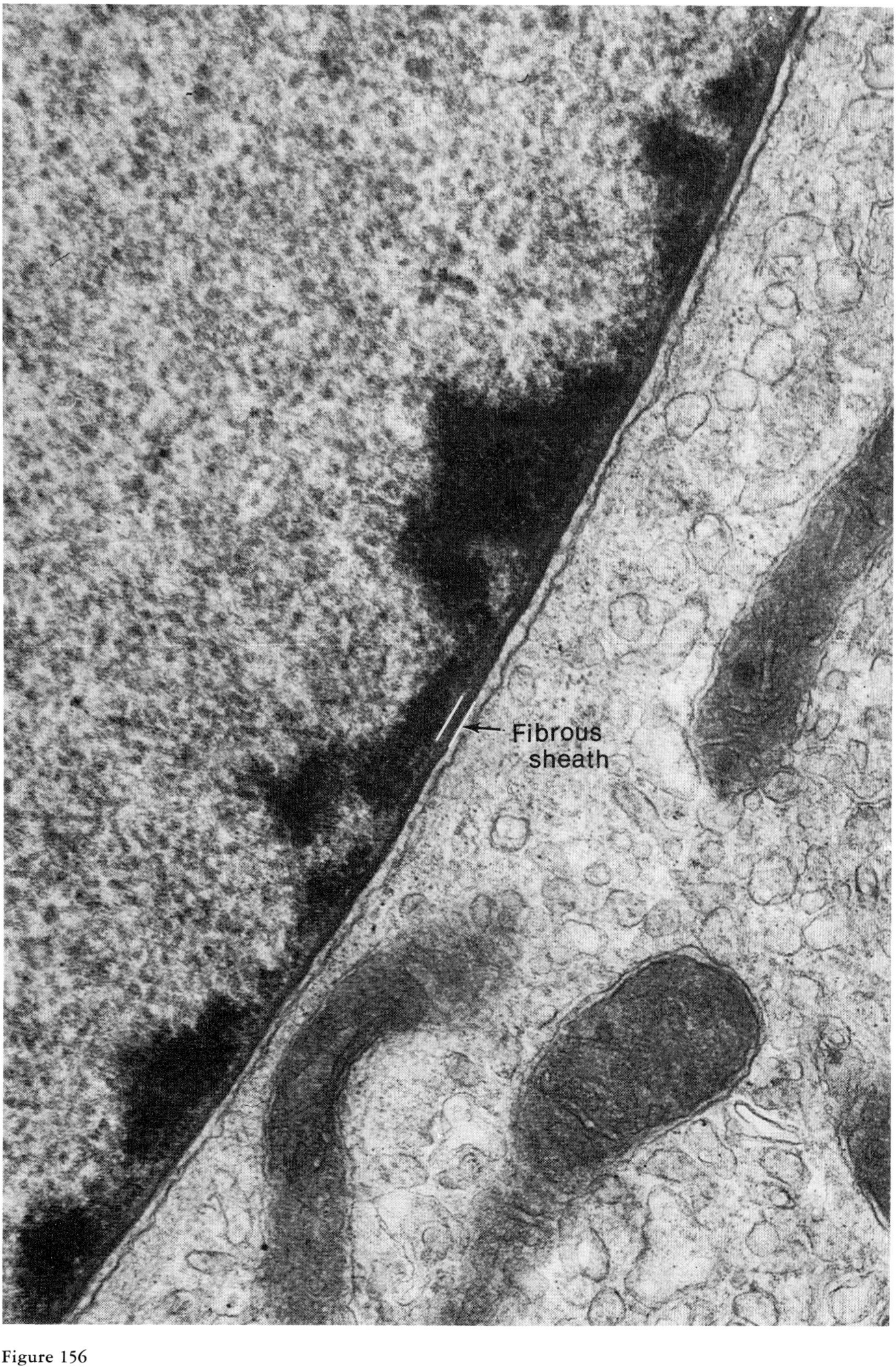

Figure 156

Some cell types of humans have a rather thick fibrous lamina. The upper figure on the facing page shows part of the nucleus and adjacent cytoplasm of an epidermal epithelial cell. The lamina is unusually conspicuous because it stains less intensely than the heterochromatin. Over the pore at the right of the figure, the lamina appears to be lacking.

In the lower figure, the fibrous lamina is more heavily stained but its inner limit is demarcated by an electron lucent line between it and the chromatin. At the arrows, a dense layer continues over the nuclear pores. This appearance is probably due to the fact that the section does not pass through the middle of the pores and includes a sector of the annulus which is of the same density as the adjacent fibrous lamina.

Figure 157. Portion of an epidermal cell from human skin. (Micrograph courtesy of George Szabo.)

Figure 158. Portion of a Leydig cell from rodent testis.

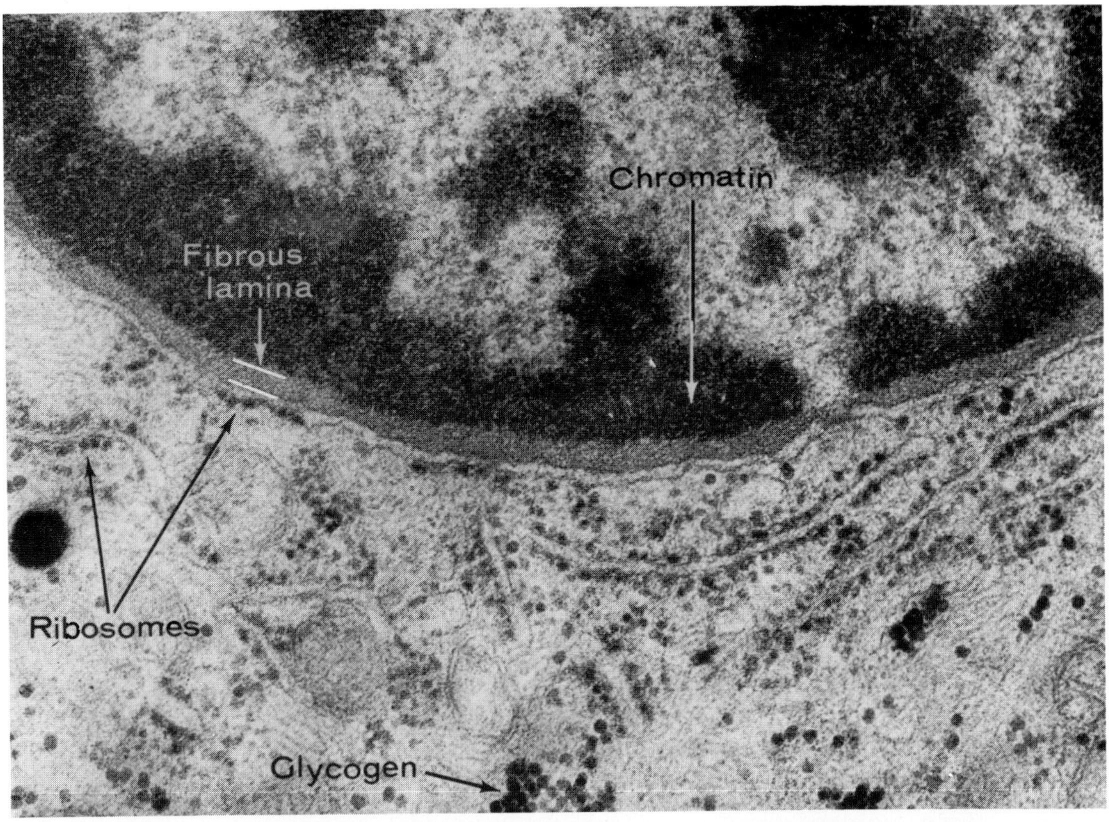

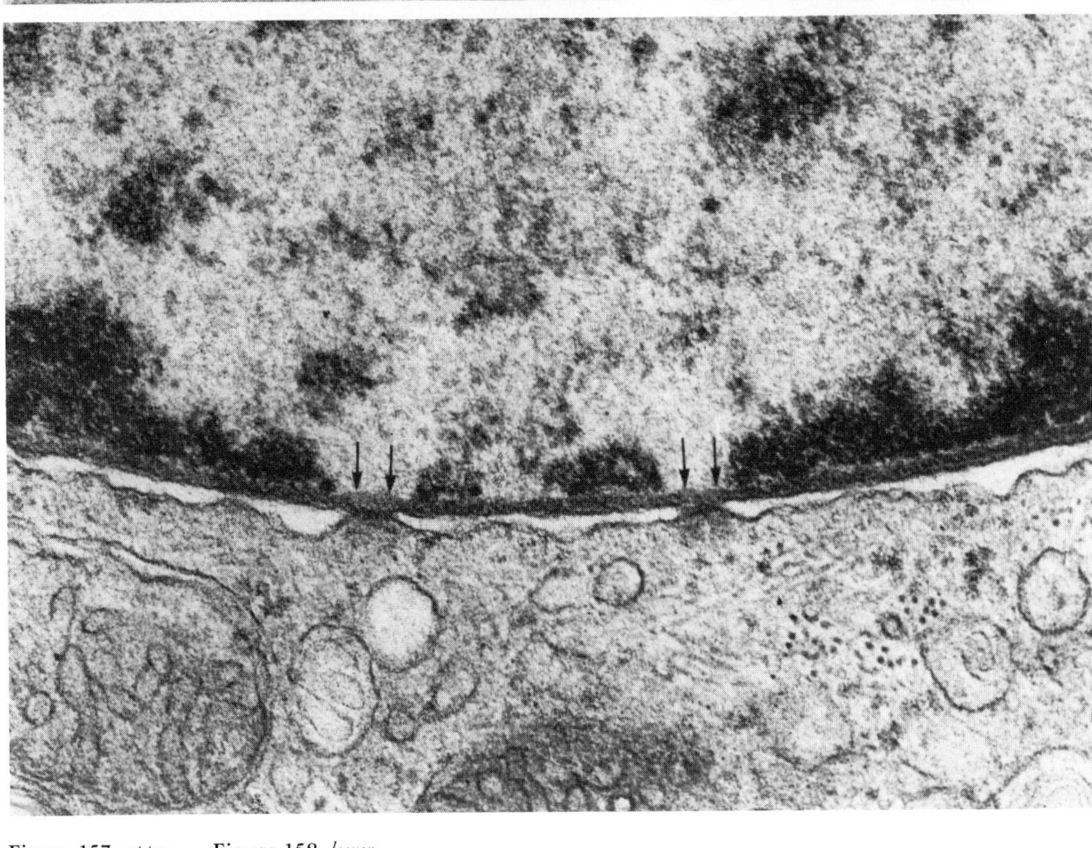

Figure 157, *upper* **Figure 158**, *lower*

The fibrous lamina and associated nuclear pore complexes have been isolated from hepatic cells and found to consist of three principal polypeptides that migrate on SDS gels in the range of 60,000 to 70,000 daltons (Aaronson and Blöbel, 1975). When antibodies raised to these polypeptides are used for immunohistochemical localization, the fluorescence is confined to the nuclei of interphase cells. The upper figure on the facing page is a phase contrast image of a confluent layer of cells in tissue culture, and the lower figure is the same field stained by indirect immunofluorescence with antibody to one of the principal polypeptides of the fibrous lamina-pore complex.

When ferritin-labeled antibody is applied to tissue sections, the label is bound only at the periphery of the nucleus. Therefore the fibrous lamina is not a portion of a general structural protein matrix extending throughout the nucleus but is a biochemically distinct component associated with the nuclear envelope.

Figure 159. Phase contrast photomicrograph of cultured K2 cells.

Figure 160. Same cells stained by fluorescent antibody against the fibrous lamina-pore complex. (From Gerace, Blum and Blöbel, J. Cell Biol. 79:546–566, 1978.)

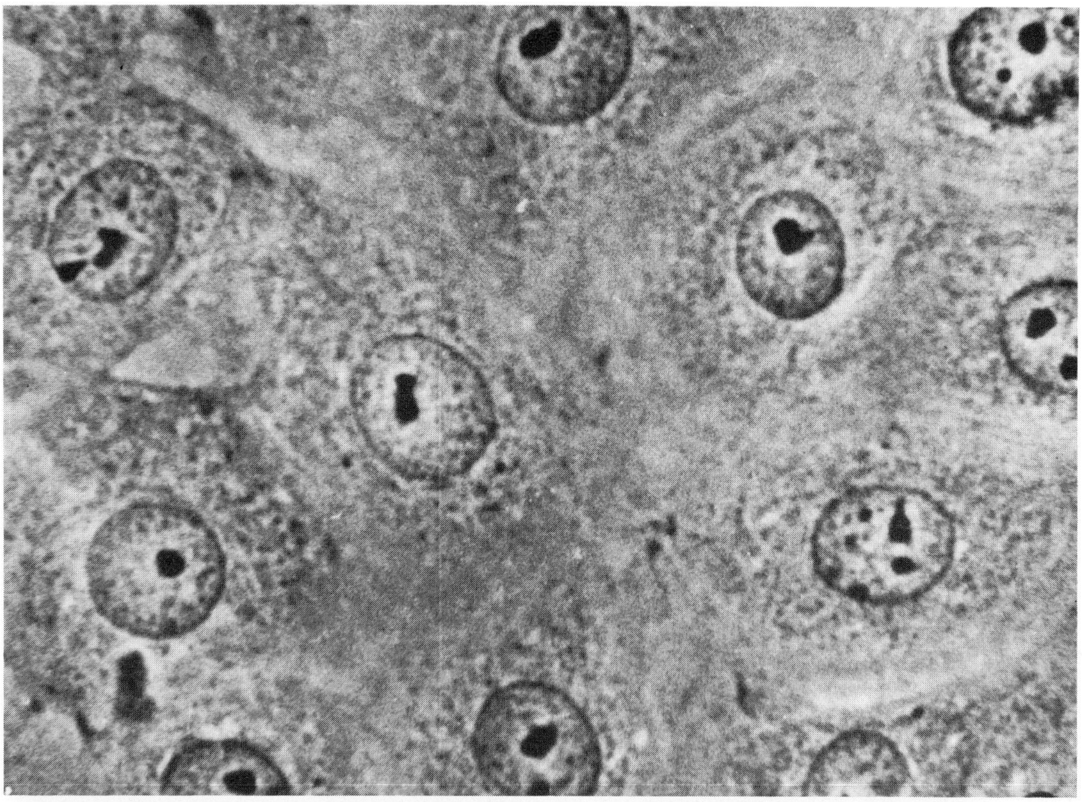

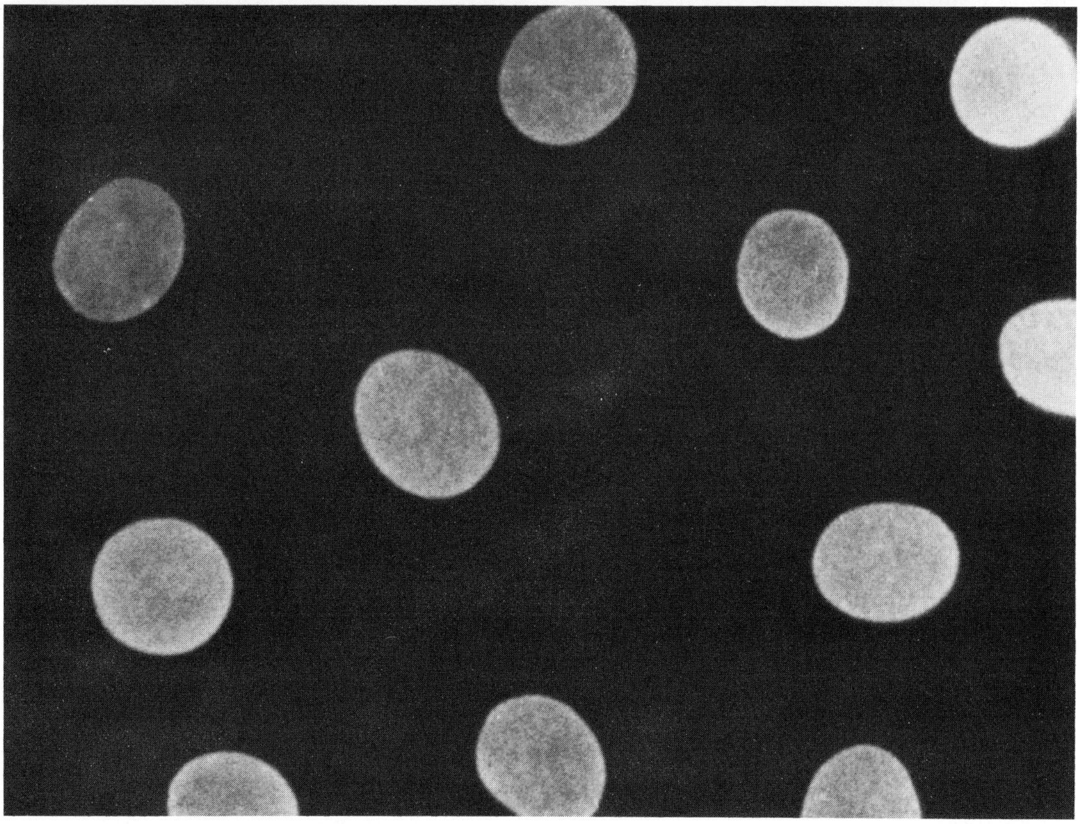

Figure 159, *upper* **Figure 160,** *lower*

REFERENCES

Nuclear Pore Complexes

Aaronson, R. P. and G. Blöbel. On the attachment of the nuclear pore complex. J. Cell Biol. 62:746–754, 1974.

Afzelius, B. A. The ultrastructure of the nuclear membrane of the sea urchin oocyte as studied with the electron microscope. Exp. Cell Res. 8:147, 1955.

Anderson, E. and H. W. Beams. Evidence from electron micrographs for the passage of material through pores of the nuclear membrane. J. Biophys. Biochem. Cytol. (Suppl. 2):439, 1956.

Barnes, B. G. and J. M. Davis. The structure of nuclear pores in mammalian tissues. J. Ultrastr. Res. 3:131, 1959.

Brokelman, J. and D. W. Fawcett. The localization of endogenous peroxidase in the rat uterus and its induction by estradiol. Biol. Reprod. 1:59–71, 1969.

Callan, S. G. and S. G. Tomlin. Experimental studies on amphibian oocyte nuclei. I. Investigation of the structure of the nuclear envelope by means of the electron microscope. Proc. Roy. Soc. Lond. 137B:367–378, 1950.

Eddy, E. M. and S. Ito. Fine structural and radioautographic observations on dense cytoplasmic material in tadpole oocytes. J. Cell Biol. 49:90–109, 1971.

Faberge, A. C. Direct demonstration of eightfold symmetry in nuclear pores. Z. Zellforsch. 136:183–190, 1973.

Fawcett, D. W. and H. Chemes. Changes in distribution of nuclear pores during differentiation of the male germ cells. Tissue Cell 11:147–162, 1979.

Feldherr, C. M. Evidence for changes in nuclear permeability during different functional states. Tissue Cell 3:1, 1971.

Franke, W. W. Zur Feinstruktur isolierter Kernmembranen aus tierischen Zellen. Z. Zellforsch. Mikros. Anat. 80:585, 1967.

Franke, W. W. On the universality of nuclear pore complex structure. Z. Zellforsch. 105:405–429, 1970.

Franke, W. W. Structure, biochemistry and functions of the nuclear envelope. Int. Rev. Cytol. (Suppl.) pp. 72–236, 1974. (Review)

Gall, J. G. Octagonal nuclear pores. J. Cell Biol. 32:291–400, 1967.

Kessel, R. G. Structure and function of the nuclear envelope and related cytomembranes. Progress in Surface and Membrane Science 6:243–329, 1973. (Review)

Leduc, E. H., S. Avrameus and M. Bouteille. Ultrastructural localization of antibody in differentiating plasma cells. J. Exp. Med. 127:109, 1968.

Markovics, J., L. Glass and G. G. Maul. Pore patterns on nuclear membranes. Exp. Cell Res. 85:443–451, 1974.

Maul, G. C. On the octagonality of the nuclear pore complex. J. Cell Biol. 51:558, 1971.

Maul, G. C. The nuclear and cytoplasmic pore complex structure dynamics, distribution and evolution. Int. Rev. Cytol. (Suppl. 6):76–186, 1977. (Review)

Maul, G. G., H. M. Maul, J. E. Scogna, M. W. Lieberman, G. S. Stein, B. Y. Hsu and T. W. Borun. Time sequence of nuclear pore formation in PHA stimulated lymphocytes and in HeLa cells during the cell cycle. J. Cell Biol. 55:433, 1972.

Maul, G. G., J. W. Price and M. W. Lieberman. Formation and distribution of nuclear pore complexes in interphase. J. Cell Biol. 51:405–418, 1971.

Maul, H. M., B. Y. L. Hsu, T. M. Borun and G. G. Maul. Effect of metabolic inhibitors in nuclear pore formation during the HeLa S_3 cell cycle. J. Cell Biol. 59:669–676, 1968.

Moses, M. Breakdown and reformation of the nuclear envelope at cell division. In Internat. Conf. on Electron Microscopy, Berlin, 1958, pp. 230–233, Springer-Verlag, Berlin, 1960.

Porter, K. R. and R. D. Machado. Studies on the endoplasmic reticulum. IV. Its form and distribution during mitosis in cells of onion root tip. J. Biophys. Biochem. Cytol. 7:167, 1960.

Watson, M. L. The nuclear envelope. Its structure and relation to cytoplasmic membranes. J. Biophys. Biochem. Cytol. 1:257, 1955.

Watson, M. L. Further observations on the nuclear envelope of the animal cell. J. Biophys. Biochem. Cytol. 6:147, 1959.

Weiner, J., D. Spiro and W. R. Louwenstein. Ultrastructure and permeability of nuclear membranes. J. Cell Biol. 27:107, 1965.

Fibrous Lamina

Aaronson, R. P. and G. Blöbel. On the attachment of the nuclear pore complex. J. Cell Biol. 62:746–754, 1974.

Aaronson, R. P. and G. Blöbel. Isolation of nuclear complexes in association with a lamina. Proc. Nat. Acad. Sci. 721:1007–1011, 1975.

Coggeshall, R. E. and D. W. Fawcett. The fine structure of the central nervous system of the leech, *Hirudo medicinalis*. J. Neurophysiol. 27:229–289, 1964.

Fawcett, D. W. On the occurrence of a fibrous lamina on the inner aspect of the nuclear envelope in certain cells of vertebrates. Am. J. Anat. 119:129–146, 1966.

Gerace, L., A. Blum and G. Blöbel. Immunocytochemical localization of the major polypeptides of the nuclear pore complex-lamina fraction. Interphase and mitotic distribution. J. Cell Biol. 79:546–566, 1978.

Ghadially, F. N., N. Bhatnager and J. Fuller. Waxing and waning of the nuclear fibrous lamina. Arch. Pathol. 94:303–307, 1972.

Kalifat, S. R., M. Bouteille and J. Delarue. Étude ultrastructurale de la lamelle dense observée au contact de la membrane nucléaire interne. J. de Microscopie 6:1019–1026, 1967.

Marzanic, K. Présence de la "zonula nucleum limitans" dans quelques cellules humaines. J. de Microscopie 6:1027–1032, 1967.

Pappas, G. D. The fine structure of the nuclear envelope of *Amoeba proteus*. J. Biophys. Biochem. Cytol. 2 (Suppl.):431–434, 1956.

Patrizi, G. and M. Poger. The ultrastructure of the nuclear periphery. J. Ultrastr. Res. 17:127–136, 1967.

ANNULATE LAMELLAE

The structures now called annulate lamellae may have been detected by light microscopists as birefringent elements in the cytoplasm of Arbacia eggs (McCullock, 1952). They were first recognized as parallel arrays of double membranes in electron micrographs by Lansing et al. (1952). The striking resemblance of these lamellae to the nuclear envelope was pointed out by Afzelius (1955), who considered it possible that they were fragments that remained in the cytoplasm following breakdown of the nuclear envelope during cell division. They were studied in some detail by Swift (1956), who noted that in some of the oocytes where these structures were observed several months had elapsed since the last oogonial division, and he considered it unlikely that they were persisting residues of those divisions. Swift favored their origin as a budding off from the nuclear envelope, and it was he who introduced the descriptive term *annulate lamellae*.

They consist of pairs of parallel membranes enclosing a 30 to 50 nm space. At intervals of 100 to 200 nm, the membranes cross this space to bound circular fenestrations, or pores, about 50 nm in diameter. These openings are lined by material of appreciable density, forming a cylindrical annulus within the pore and projecting a short distance beyond its limits on either side of the pair of parallel membranes. The pore may seem to be closed by a transverse pore diaphragm, but this appearance may be due to inclusion of the rim of the pore in the thin section. Thus annulate lamellae are membrane-bounded cisternae traversed by uniformly spaced pore complexes indistinguishable from those of the nuclear envelope. They are frequently stacked in parallel array and are often continuous at their margins with cisternae of granular endoplasmic reticulum. The pores in neighboring lamellae may either be offset or in precise register.

Annulate lamellae were first described in developing oocytes and are most consistently encountered in this cell type in both invertebrates and vertebrates. But in the past 20 years there have been innumerable reports of their sporadic occurrence in a great variety of normal embryonic and adult tissues and in many tumors. The majority of authors have assumed that they are derived from the nuclear envelope and secondarily migrate to other locations in the cytoplasm. Some have speculated that they have a messenger function, carrying information from the nucleus to other regions of the cell (Swift, 1956; Kessel, 1968) but there is no substantial evidence to support this thesis. Indeed it is by no means established that they arise exclusively at the nuclear membrane. If one subscribes to the view that the nuclear envelope itself is a specialized portion of the endoplasmic reticulum and that the latter contributes to reconstitution of the nuclear envelope in telophase of each mitotic division, then it is not unreasonable to believe that annulate lamellae may arise anomalously anywhere in the cytoplasm by local modification of the endoplasmic reticulum. Although this specialization of the reticulum is normally induced by proximity to the chromatin of the reconstituting nucleus, it is possible that it may occur as an inappropriate response to some nonspecific ectopic stimulus. Pore complexes in the nuclear envelope are clearly essential for interchange between the nuclear and cytoplasmic compartments but pores in cisternae that do not delimit separate compartments would seem to serve no useful purpose. The infrequency of occurrence of annulate lamellae in most cell types argues against their designation as organelles possessing an important function.

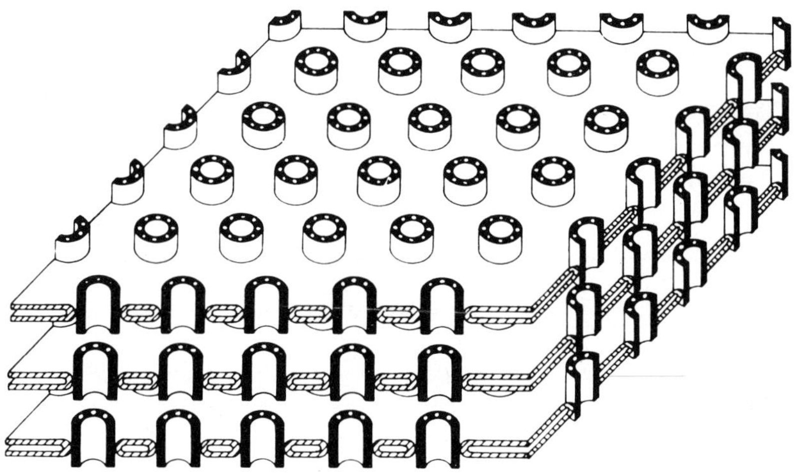

Diagram of annulate lamellae. The annuli are depicted here as simple cylindrical elements. They are now known to have a more complex ultrastructure consisting of granular and filamentous subunits arranged in eight-part radial symmetry within the membranous pore. (From R. G. Kessel, J. Ultrastr. Res. Suppl. 10:1–82, 1968.)

Annulate lamellae consist of membrane-bounded cisternae in the cytoplasm traversed by pore complexes identical to those in the nuclear envelope. The upper figure includes a portion of the nuclear envelope and in the adjacent cytoplasm a stack of parallel annulate lamellae. Their frequent proximity to the nucleus and the similarity of their annuli to nuclear pores has led to the suggestion that they may form by delamination from the nuclear envelope. However, they may also occur at the periphery of the cell near the cell membrane. The lamellae are usually continuous with tubular or cisternal elements of the endoplasmic reticulum, and it seems likely that they may arise anywhere in the cytoplasm as local specializations of this organelle. What triggers their differentiation or what function they serve, if any, is unknown.

Figures 161 and 162. Annulate lamellae in oocytes of sea urchin, Arbacia. (Micrographs courtesy of Susumu Ito.)

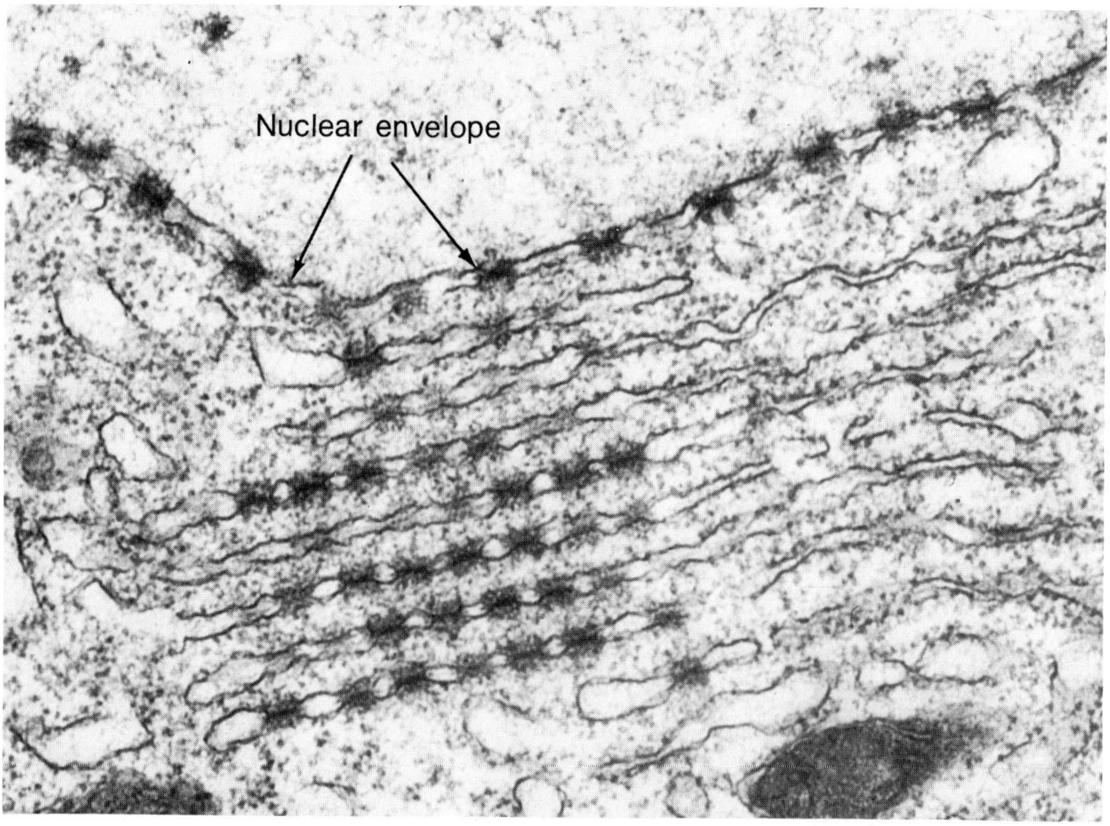

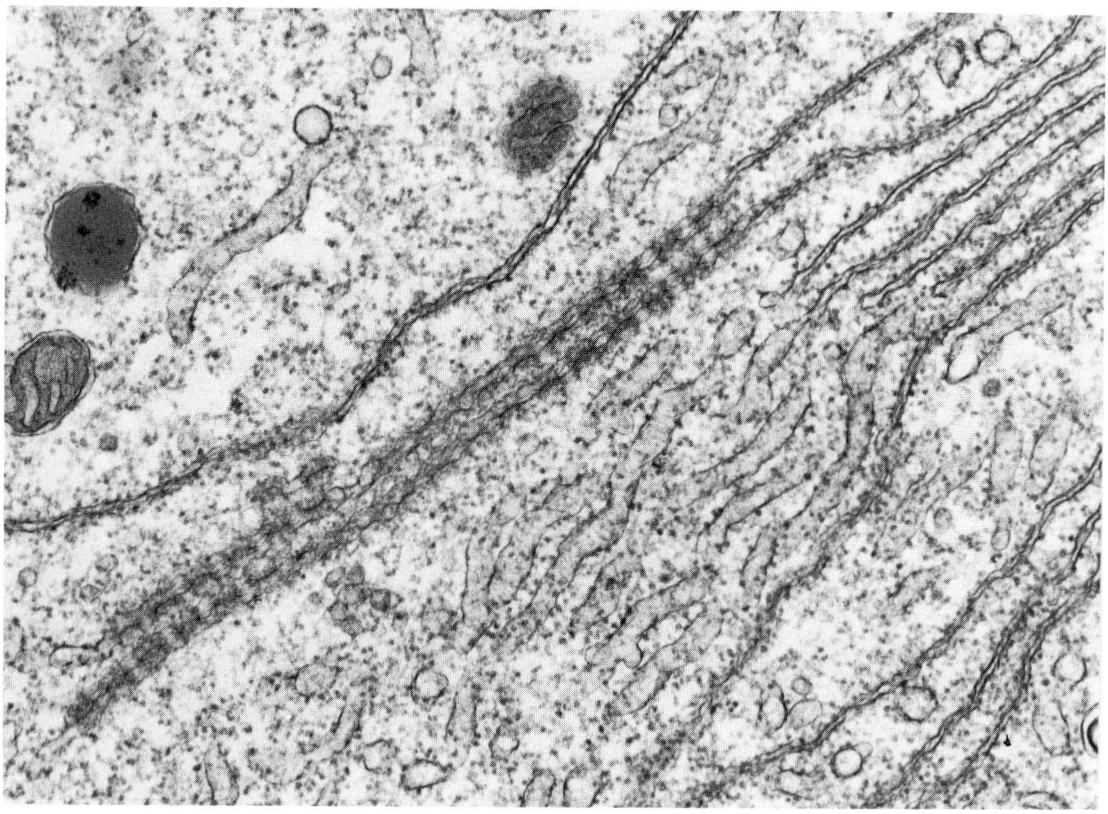

Figure 161, *upper* Figure 162, *lower*

The occasional occurrence of annulate lamellae has been reported in a great many cell types, but they are a constant constituent of very few. The accompanying micrograph illustrates a typical aggregation of annulate lamellae in a Sertoli cell from human testis. The organelle is seldom seen in this epithelium. The arrows draw attention to sites of continuity between the annulate lamellae and ribosome-bearing cisternae of the endoplasmic reticulum.

Figure 163. Annulate lamellae in a Sertoli cell of human seminiferous epithelium. (Micrograph courtesy of Hector Chemes.)

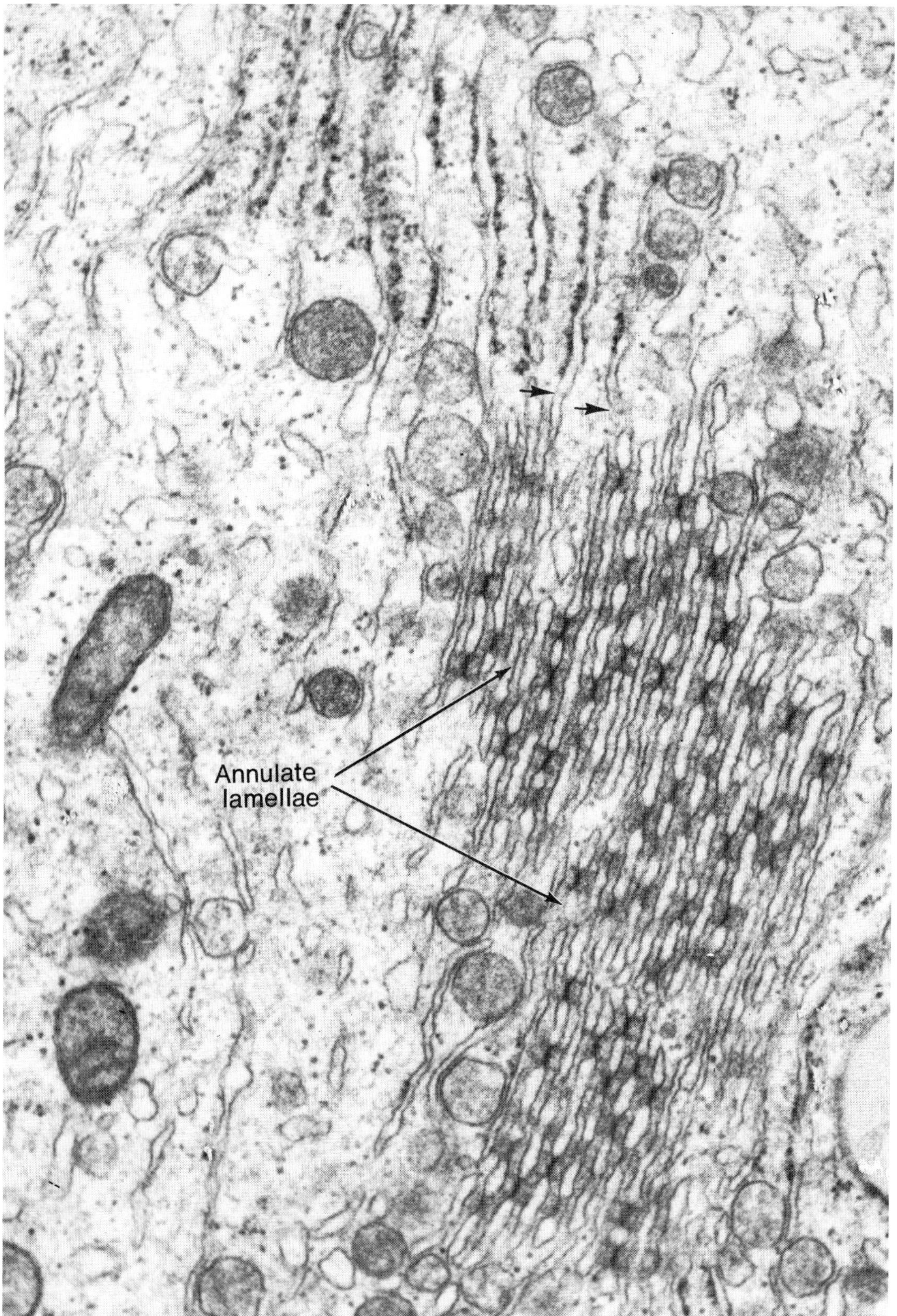

Figure 163

Annulate lamellae are observed most consistently in oocytes. The micrographs presented here show conspicuous examples of large stacks of annulate lamellae in the peripheral cytoplasm of a mature frog oocyte.

Figures 164 and 165. Annulate lamellae in an oocyte of *Rana pipiens*. (Micrographs courtesy of Richard Kessel.)

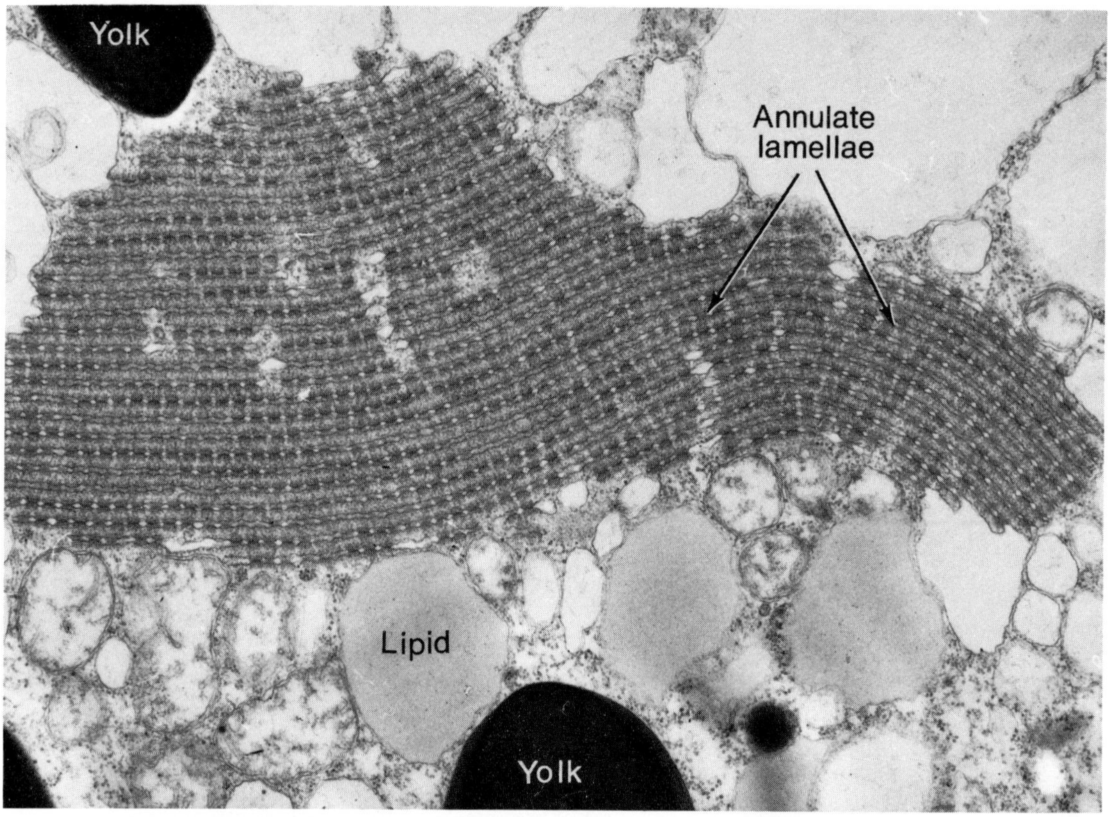

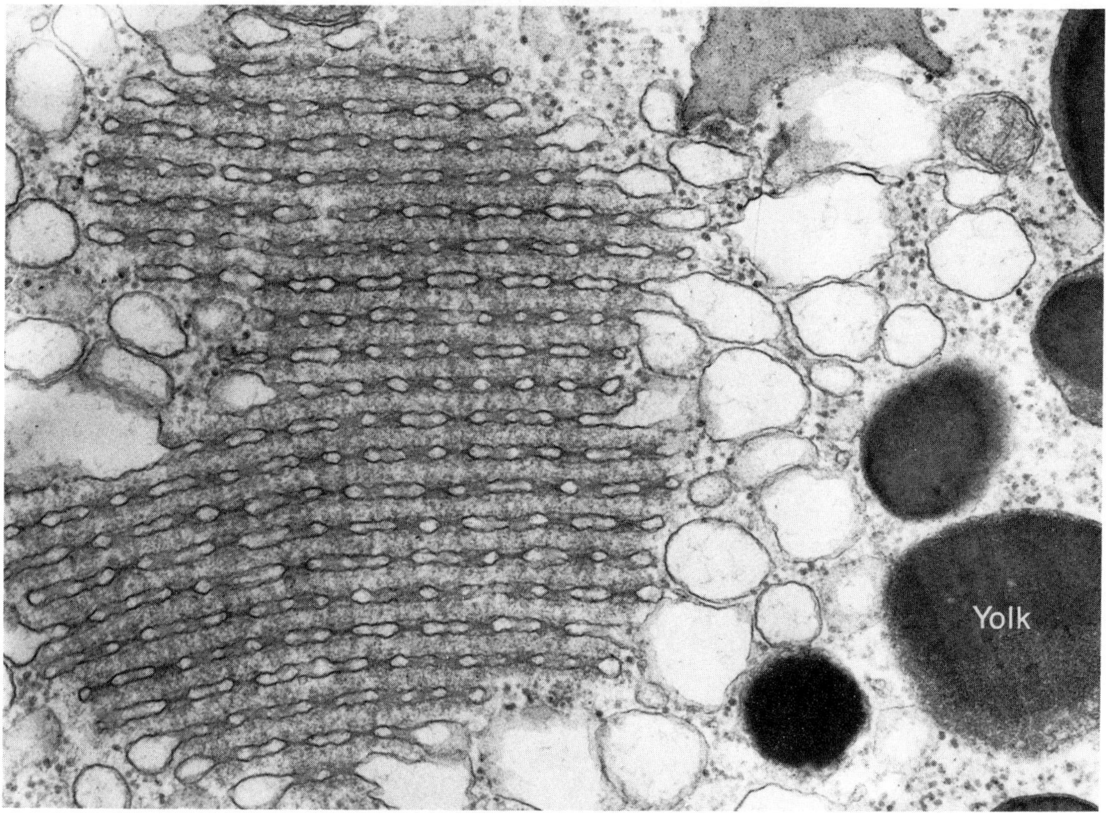

Figure 164, *upper* **Figure 165,** *lower*

Annulate lamellae vary greatly in linear extent from short profiles with one or two pore complexes to cisternae many microns long. They may occur singly as in the upper figure on the facing page or in parallel stacks of 30 or more cisternae as illustrated in the lower figure. The disposition of pore complexes in parallel arrays of cisternae is usually random, but in some instances (lower figure) they may be in register, creating channels extending through the entire stack. In the lower figure, the nuclear envelope is at the right and from the pores at the arrow, the apposed annuli of the associated lamellae form channels that can be traced deep into the array of parallel lamellae.

Figure 166. Nuclear envelope and single annulate lamellae of varying length in the neighboring cytoplasm of an Arbacia oocyte. (Micrograph courtesy of Susumu Ito.)

Figure 167. Stack of annulate lamellae associated with the nuclear envelope of a human oocyte. (Micrograph courtesy of Arthur Hertig.)

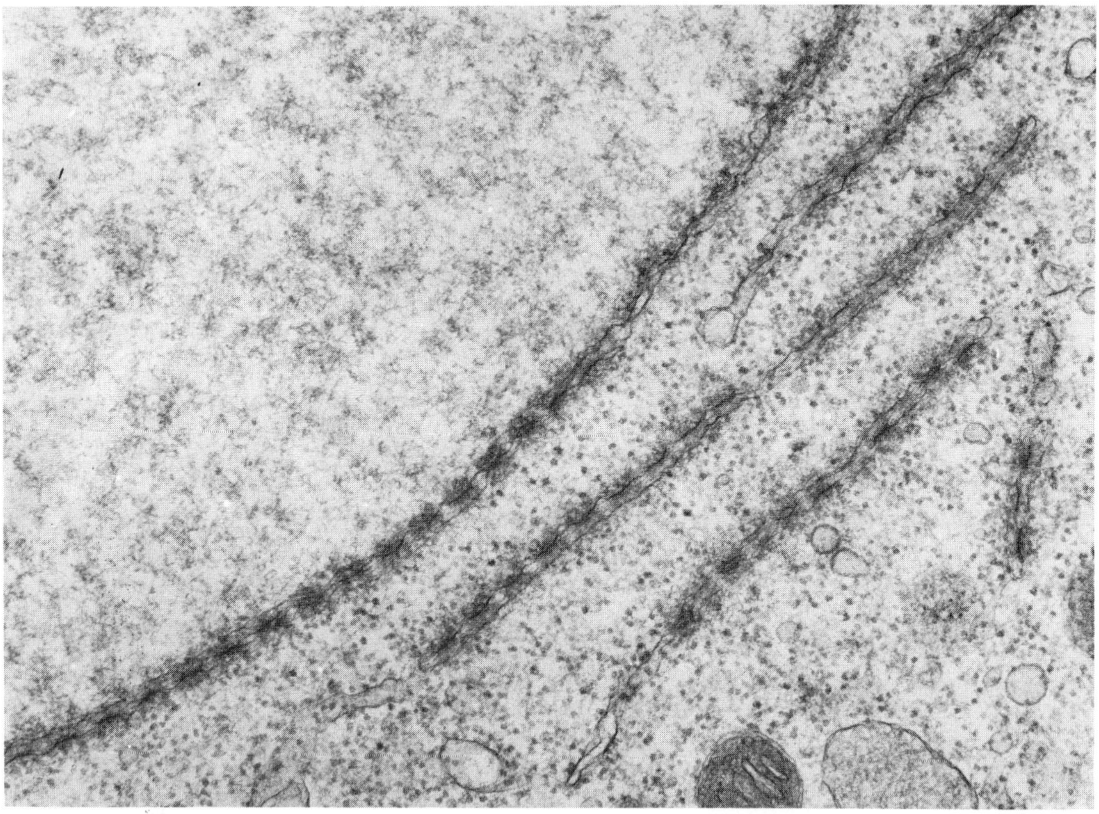

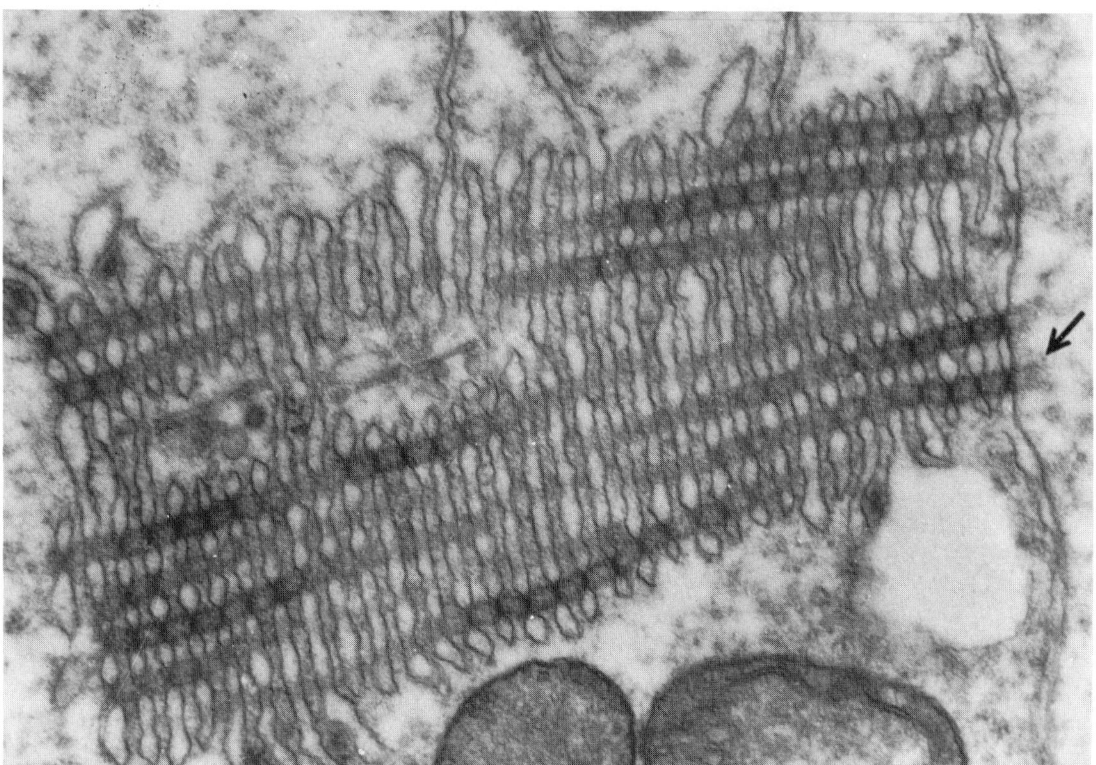

Figure 166, *upper* **Figure 167**, *lower*

REFERENCES

Afzelius, B. A. The ultrastructure of the nuclear membrane of the sea urchin oocyte as studied with the electron microscope. Exp. Cell Res. 8:147, 1955.

Gross, B. G. Annulate lamellae in the axillary apocrine glands of adult man. J. Ultrastr. Res. 14:64, 1966.

Harrison, G. A. Some observations on the presence of annulate lamellae in alligator and seagull adrenal cortical cells. J. Ultrastr. Res. 14:158, 1966.

Hertig, A. T. and E. C. Adams. Studies on the human oocyte and its follicle. I. Ultrastructural and histochemical observations on the primordial follicle stage. J. Cell Biol. 34:647–675, 1967.

Kessel, R. G. Electron microscope studies on the origin of annulate lamellae in oocytes of Necturus. J. Cell Biol. 19:291–414, 1963.

Kessel, R. G. Intranuclear and cytoplasmic annulate lamellae in tunicate oocytes. J. Cell Biol. 24:471–488, 1965.

Kessel, R. G. Annulate lamellae. J. Ultrastr. Res. Suppl. 10:1–82, 1968. (Review)

Lansing, A. I., J. Hillier and T. B. Rosenthal. Electron microscopy of some marine egg inclusions. Biol. Bull. 103:295 (Abstr.), 1952.

McCulloch, D. Fibrous structures in the ground cytoplasm of Arbacia eggs. J. Exp. Zool. 119:47, 1952.

Merkow, L. and J. Leighton. Increased numbers of annulate lamellae in myocardium of chick embryos incubated at abnormal temperatures. J. Cell Biol. 28:127–137, 1966.

Rebhun, L. I. Electron microscopy of basophilic structures of some invertebrate oocytes. J. Biophys. Biochem. Cytol. 2:93–103, 1956.

Sakai, A. and M. Shigenaga. The annulate lamellae in spermatogonia of the grasshopper, *Atractomorpha bedeli*. Cytologia (Japan) 33:34–45, 1968.

Swift, H. The fine structure of annulate lamellae. J. Biophys. Biochem. Cytol. 2 Suppl.:415–418, 1956.

Wischnitzer, S. Observations on the annulate lamellae of immature amphibian oocytes. J. Biophys. Biochem. Cytol. 8:558–563, 1960.

ENDOPLASMIC RETICULUM

ROUGH ENDOPLASMIC RETICULUM

In the closing years of the last century, the French cytologist Garnier described filamentous structures in the basal cytoplasm of cells in the pancreas and salivary glands which stained intensely with basic dyes. He observed that this material, which he called *ergastoplasm,* varied in form and quantity in different phases of the secretory cycle. He considered it to be a fundamental constituent of glandular cells, involved in their synthetic functions. The basophilic masses of ergastoplasm were regarded as local concentrations of a fibrillar network that extended throughout the cytoplasm. Similar observations were made independently by Bensley (1898) and Mathews (1899). Throughout the next 50 years the existence of ergastoplasm was widely accepted; however, the question of whether its apparent fibrillar structure was an artifact of fixation continued to be a subject of controversy. Interest in the basophilic component of the cytoplasm was revived in the 1940s by the development of staining and ultraviolet absorption methods that identified it as ribonucleoprotein.

The concurrent development of methods for differential centrifugation of liver homogenates led to identification of a cell fraction consisting of 50 to 300 nm particles that were called *microsomes* (Claude, 1943). This fraction rapidly became a focus of investigative attention because it was found to contain the bulk of the cytoplasmic ribonucleoprotein and therefore was thought to be involved in protein synthesis.

In his pioneering application of the electron microscope to the study of thinly spread tissue culture cells, Porter (1945) observed a system of delicate branching and anastomosing strands that formed a lace-like network throughout the cytoplasm. This *endoplasmic reticulum* was considered to be a new cell organelle (Porter and Thompson, 1947). Also noted in these preparations were small vesicles 50 to 200 nm in diameter, sometimes connected in rows and sometimes entirely separate. It was suggested that this vesicular component corresponded to Claude's "microsomes."

The earliest electron micrographs of thin sections of liver cells provided nebulous images of "lamellar" or "filamentous" structures with a distribution comparable to that of the basophilic bodies (ergastoplasm) seen with the light microscope (Dalton et al., 1950; Bernhard et al., 1951, 1952). Better methods of fixation and microtomy soon yielded micrographs of greatly improved quality, and Porter and Palade clearly demonstrated in a series of classic papers in 1953 and 1954 that the "fibrillae" and "lamellae" previously interpreted as the ergastoplasm were, in fact, sections of membrane-limited tubules and cisternae of the endoplasmic reticulum which they had described earlier in intact tissue culture cells. Thus it was established that this continuous canalicular system corresponded in its distribution to the basophilic bodies seen by light microscopy.

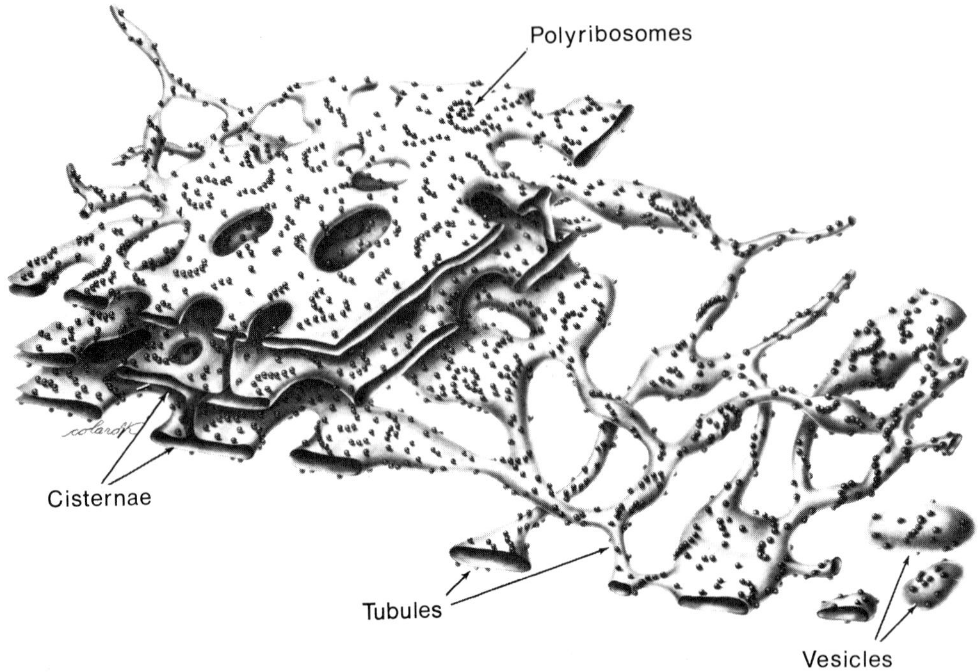

Drawing of the three-dimensional configuration of the rough endoplasmic reticulum occurring in fenestrated cisternae or anastomosing tubules. Isolated vesicles are also observed, but these are often a result of less-than-ideal fixation.

It soon became apparent, however, that the basophilia did not reside in the membranes or contents of the reticulum. Palade (1953) described a small particulate component consisting of 15 to 20 nm dense granules that were attached in large numbers to the outer surface of the membranes bounding the endoplasmic reticulum. They also occurred singly or in small aggregates scattered throughout the cytoplasmic matrix. Where the light microscope showed basophilia localized in discrete masses near the base of secretory cells, the electron microscope revealed cisternae of the endoplasmic reticulum stacked in parallel array. On the other hand, in rapidly proliferating embryonic cells where the cytoplasm was diffusely basophilic, the endoplasmic reticulum was often poorly developed but small dense particles free in the cytoplasmic matrix were very abundant. It was suggested therefore that cytoplasmic basophilia did not reside in the reticulum *per se*, but in the small particulate component adherent to its surface.

The nature of the microsomes and their relation to the endoplasmic reticulum was established by Palade and Siekevitz (1956) in correlated biochemical and ultrastructural studies of cell fractions. In micrographs of thin sections, the microsomes appeared as spherical vesicles 80 to 200 nm in diameter and limited by a membrane bearing 15 to 20 nm dense particles on its outer surface. It was evident therefore that they were formed during tissue homogenization by fragmentation of the endoplasmic reticulum and resealing of the fragments into closed vesicles. Upon solubilization of the membranes and recentrifugation at higher speed, a pellet was obtained that consisted entirely of the small particulate component and contained all of the RNA of the microsome fraction. The dense granules were thus identified as *ribonucleoprotein particles* and later called *ribosomes*.

These discoveries in the 1950s with the electron microscope, together with parallel advances in cell fractionation and the use of radio-labeled compounds, stimulated intense biochemical interest in the mechanisms of protein synthesis. Zamecnik and Keller (1954) developed a cell-free system consisting of liver homogenate, amino acids, and ATP, which could synthesize small amounts of protein *in vitro*. By a process of successive elimination of particular organelles from the system they established that the microsomes were essential for protein synthesis. When radioactively labeled amino acids became available, it was shown that after administration of C^{14}-leucine to animals, the protein most rapidly labeled in their liver was in the microsome fraction, and the label was still more highly concentrated in the RNP particles that remained after solubilization of the microsomal membranes (Littlefield et al., 1955). Thus it was concluded that the ribosomal particles were the site of incorporation of amino acids into protein.

To obtain synthesis of protein in significant amounts in a cell-free system, it was found to be necessary to add a soluble fraction of RNA, which served as a carrier of amino acids to the ribosomes (Hoagland, Zamecnik and Stephenson, 1957). This low-molecular weight species was later called *transfer RNA* (tRNA). Nirenberg and Matthei (1961), working with a cell-free bacterial system, discovered that it was necessary to add a third type of RNA, called *messenger RNA* (mRNA). Once a complete and efficient cell-free system had been assembled, the way was clear to study in greater detail the functions of its components.

Messenger RNA had a base ratio similar to that of DNA and it was ultimately shown to be a product of *transcription* of genes coding for the structural, catalytic, or secretory proteins of the cell. Its role is to carry information from the genome to the ribosomes in the cytoplasm, where its encoded message is *translated* into the amino acid sequence of specific proteins. When mRNA was added to a mixture of amino acids, tRNA, ribosomes, and cofactors, it could be shown to attach to the ribosomes and to determine a specific sequence of amino acids in the polypeptides synthesized (Brenner, Jacob and Meselsohn, 1961). It was evident from correlated biochemical and ultrastructural studies that protein synthesis did not take place on single ribosomes but in groups called *polyribosomes (polysomes)* joined together by a slender 2 nm strand. Since this tenuous connecting strand could be cleaved by ribonuclease, it was concluded that it consisted of a long molecule of mRNA spanning several ribosomes. It was known that the lengthening polypeptide chains were attached to the ribosomes and since each ribosome needed the information encoded in the full length of the mRNA molecule, it seemed likely that the mRNA strand had to move relative to the ribosomes for the entire message to be read (Warner, Knopf and Rich, 1963). When the ribosomes reached the end, they were believed to release the completed polypeptide and detach from the messenger RNA. Although the ribosomes themselves seemed to have no template function, they appeared nevertheless to play a coordinating structural role in translation of the information encoded in the mRNA molecule to which they were attached.

Systematic investigation of ribosomes revealed a surprisingly complex organization. They consisted of two subunits of unequal size with the smaller attached to a flattened area on the larger subunit. Each was found to contain characteristic RNA and constituent protein molecules. Estimates of the total number of ribosomal proteins range from 50 to 60. The two subunits can be dissociated by altering the pH and Mg^{++} ion concentration. Specific binding sites for tRNA and mRNA have been demonstrated. In negatively stained preparations, the messenger RNA appears to be associated with the smaller subunit at or near the groove between the subunits. While it is not possible to visualize the nascent polypeptide chains with the electron microscope, biochemical evidence indicates that they are always fixed to the large subunit (Sabatini, Tashiro and Palade, 1966). Where polysomes are associated with the endoplasmic reticulum, the ribosomes are invariably attached to its membrane by their larger subunit.

The mechanism of protein synthesis has been worked out in great detail by molecular biologists during the past 20 years. It is beyond the scope of this book to go into this complex process in any depth. It will suffice to point out that the information required for synthesis of specific proteins resides in the sequence of nucleotides in DNA. This sequence is *transcribed* in the nucleus to produce a messenger RNA which carries the information from the genome to a site of protein synthesis in the cytoplasm.

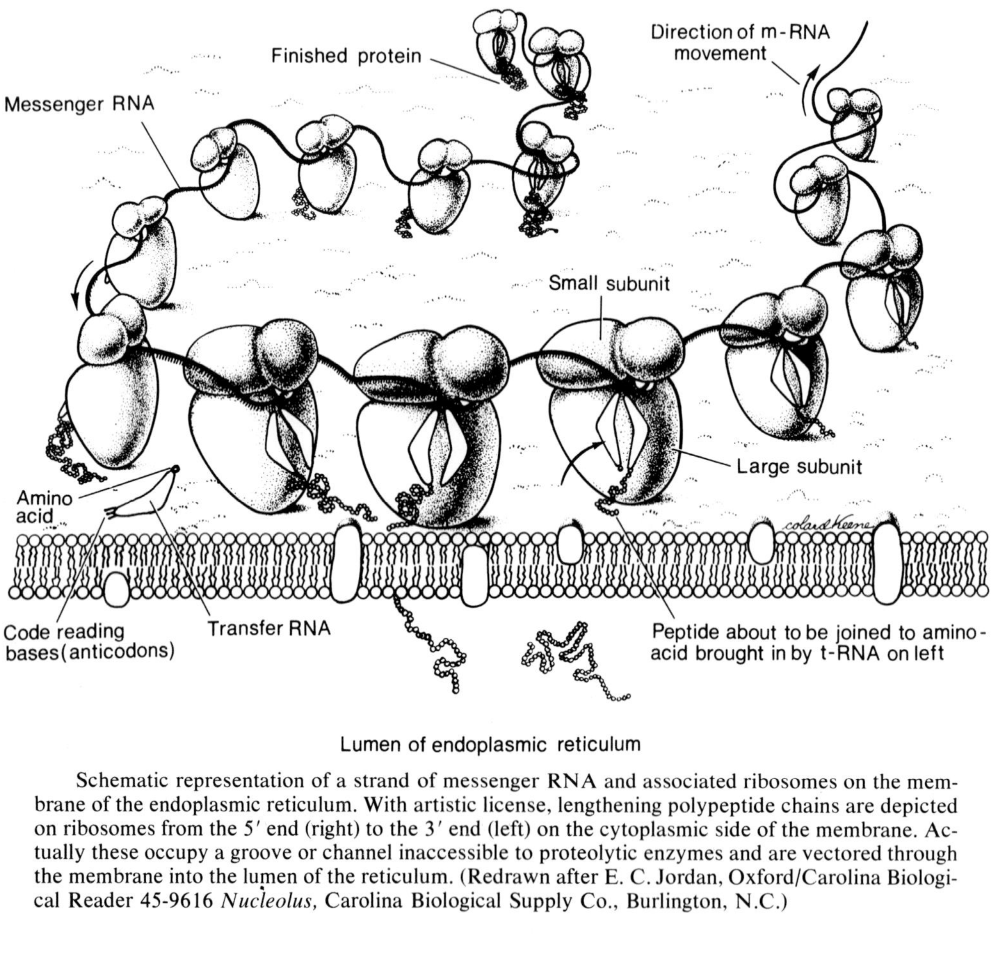

Schematic representation of a strand of messenger RNA and associated ribosomes on the membrane of the endoplasmic reticulum. With artistic license, lengthening polypeptide chains are depicted on ribosomes from the 5′ end (right) to the 3′ end (left) on the cytoplasmic side of the membrane. Actually these occupy a groove or channel inaccessible to proteolytic enzymes and are vectored through the membrane into the lumen of the reticulum. (Redrawn after E. C. Jordan, Oxford/Carolina Biological Reader 45-9616 *Nucleolus,* Carolina Biological Supply Co., Burlington, N.C.)

An essential initial event in translation of the message is the binding of the 5′ end of the mRNA to a site on the smaller subunit of a ribosome. In the surrounding cytoplasm, there are separate tRNAs for all 20 amino acids and each tRNA molecule can carry only one kind of amino acid to the ribosome. Also present in the cytoplasm are at least 10 specific enzymes called *aminoacyl-tRNA-synthetases*. These are each capable of activating one kind of amino acid and joining it to the appropriate tRNA to form an *aminoacyl-tRNA*. The various aminoacyl-tRNA molecules needed in synthesis of the polypeptide are selected one at a time by attraction to specific trinucleotide sequences (*codons*) in the mRNA molecule. There are two binding sites for tRNA molecules on the ribosome, the acceptor (A) site and the peptidyl (P) site. The tRNA bound initially to the A site moves to the P site, carrying the messenger with it. The second aminoacyl-tRNA then binds to the vacated A site. A peptide bond is formed between the first amino acid and the second with release of the initial tRNA. The tRNA at the A site, now linked to the peptide chain, moves to the P site, again vacating the A site. At the same time the mRNA molecule is advanced, exposing the next codon. A third amino acid occupies the A site and this cycle is repeated until the *termination codon* at the 3′ end of the mRNA is reached. Thus any given aminoacyl-tRNA is transiently bound to the ribosome-mRNA complex until the amino acid is incorporated into the nascent polypeptide chain. The tRNA is then replaced at site A by the next aminoacyl-tRNA. The nascent protein is at all times coupled to the tRNA that has just brought a new amino acid to the ribosome. Each step in the lengthening of the polypeptide chain is accompanied by a movement of the translation apparatus along the messenger one codon at a time. The formation of the peptide linkage between amino acids is catalyzed by an active site on one of the ribosomal proteins. Therefore the ribosome plays an active role in the synthetic process as well as providing by its binding sites the proper steric relations for the tRNA-mRNA interaction (Kurland, 1970).

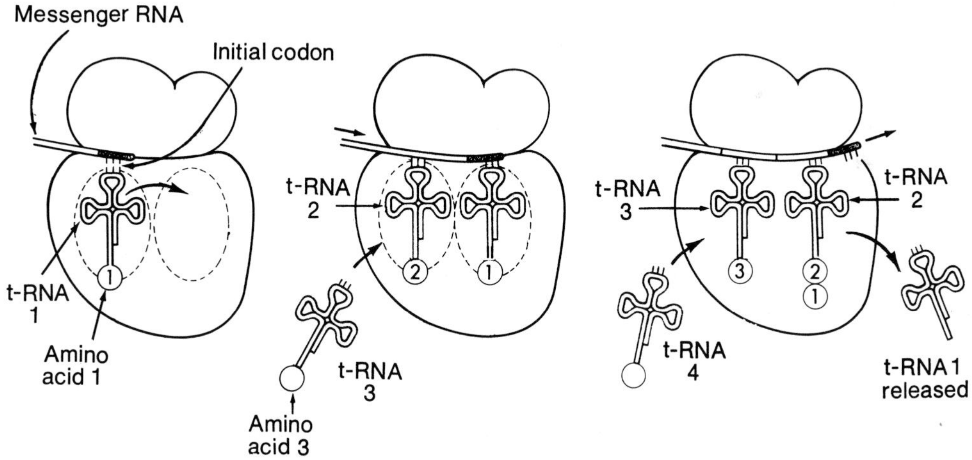

Diagrammatic representation of successive steps in attachment of tRNA to the acceptor site on the ribosome; translocation to the peptidyl site with creation of a peptide bond to the previous amino acid; and release of the unoccupied tRNA. For details see text.

The precise topographical relationship of the nascent polypeptide chain to the ribosome remains somewhat unclear. It was found in biochemical studies that the greater part of the nascent chains of protein associated with free ribosomes could be degraded by proteolytic enzymes, but that a segment consisting of 30 to 40 amino acids at the carboxyl end resisted degradation. This led to the suggestion that the developing polypeptide chain occupies a groove or channel in the large ribosomal subunit to which proteolytic enzymes are denied access (Blobel and Sabatini, 1970). When ribosomes were separated from rough microsomes, they were found to contain a polypeptide segment of similar length, whereas the remainder of the protein resided in the interior of the microsomal vesicle. It was therefore proposed that the hypothetical tunnel in the large ribosomal subunit is linked to a channel in the underlying membrane, and that the nascent chains grow vectorially along this path into the lumen of the endoplasmic reticulum (Sabatini and Blobel, 1970). Studies involving *in vitro* disassembly of ribosomes from their membranes suggested that the nascent polypeptide chains themselves played some role in maintaining ribosome attachment, but this did not explain the process by which mRNAs for secretory proteins become attached to the endoplasmic reticulum while those for integral proteins remain associated with free polysomes.

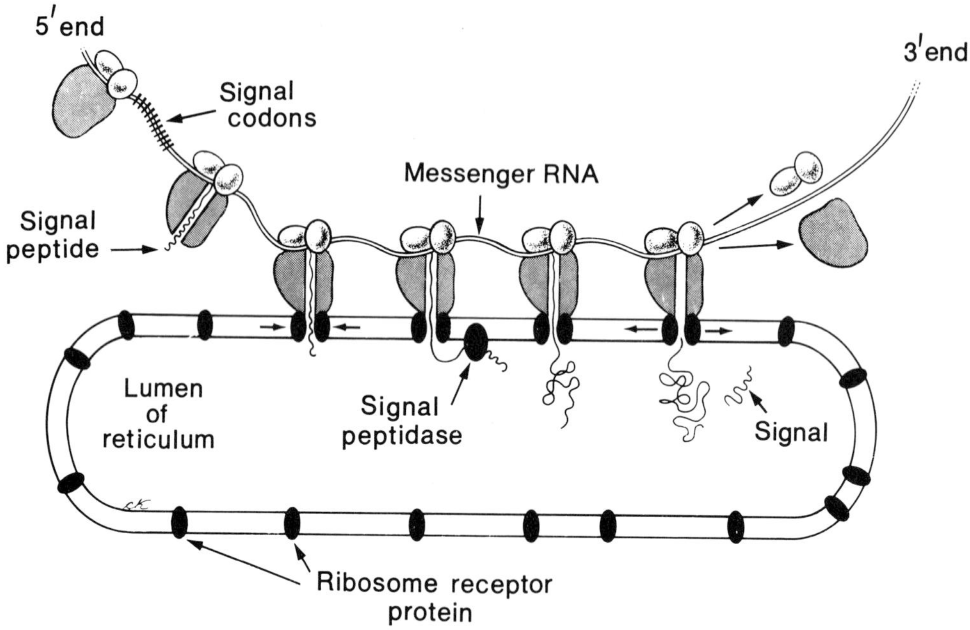

Schematic representation of the signal hypothesis for the transfer of proteins across the membrane of the endoplasmic reticulum. The signal sequence of the nascent chain is indicated by the wavy line. Endoproteolytic removal of the signal sequence from peptides by a signal peptidase is indicated. For details see text. (Redrawn after G. Blobel in *Cell Biology 1976–1977,* Rockefeller University Press, New York.)

To provide an explanation, Blobel and Sabatini (1971) advanced the *signal hypothesis,* which postulates that mRNAs for secretory proteins contain a sequence of *signal codons* and that translation of the signal codons to initiate protein synthesis takes place on ribosomes free in the cytoplasm. When the sequence of amino acids emerges from the channel in the large ribosomal subunit, it is believed to interact with the membrane of the endoplasmic reticulum in such a way as to cause aggregation of *ribosome receptor proteins* to form a channel through the membrane. The receptor proteins in turn bind to sites on the large ribosomal subunit and connect it to the membrane with its tunnel in register with the transmembrane channel. When the progressive lengthening of the polypeptide chain has moved the signal sequence into the interior of the cistern, a hypothetical *signal peptidase* on the inner aspect of the membrane cleaves the signal sequence from the end of the nascent chain. After chain termination, the ribosome dissociates from the membrane and it is assumed that the channel in the membrane is obliterated by dissociation and lateral diffusion of the ribosome receptor proteins (Blobel and Dobberstein, 1975; Blobel, 1977).

Evidence in support of this attractive hypothesis continues to accumulate. A number of familiar protein secretory products synthesized in *in vitro* systems in the absence of membranes are distinguishable from secretory proteins by the continued presence of the signal sequence (Kemper et al., 1974). Investigations on reconstitution of microsomes using heterologous ribosomes, messengers, and membranes from unrelated species have supported the signal hypothesis and shown that the recognition and attachment sites and signal sequences have been highly conserved in evolution (Blobel, 1977).

Two integral membrane proteins, designated *ribophorin I* and *ribophorin II,* are present in rough but not in smooth endoplasmic reticulum (Kreibich, Ulrich and Sabatini, 1978). These are thought to be responsible for the binding of ribosomes and therefore correspond to the ribosome receptors postulated in the signal hypothesis. Complex though this hypothesis is, persuasive evidence for several of its predictions has been obtained and it has gained widespread acceptance.

As seen in electron micrographs of intact tissue culture cells, the endoplasmic reticulum is a continuous network of membrane-bounded channels. However, the continuity of the system is not evident in thin sections, where it appears as separate circular or elongate profiles such as those seen in the accompanying micrograph. The circular and elliptical profiles studded with small dense ribosomes are transverse or oblique sections of meandering tubular elements and the elongate profiles are sections through broad, flat dilatations of the system called *cisternae*. These latter tend to be arranged in conspicuous parallel arrays as seen in the lower half of this figure. This type of reticulum with adherent ribosomes is designated *granular* or *rough endoplasmic reticulum* to distinguish it from areas that lack adherent ribosomes and are therefore called *agranular* or *smooth endoplasmic reticulum*.

The large number of associated ribonucleoprotein particles (ribosomes) makes aggregations of parallel cisternae intensely basophilic in stained histological preparations. In the classical literature of light microscopy, they were referred to as *basophilic bodies, chromidia, chromophil substance,* or *ergastoplasm.* These terms are now rarely used.

Figure 168. Electron micrograph of acinar cell from the pancreas of the bat, *Myotis lucifugus*.

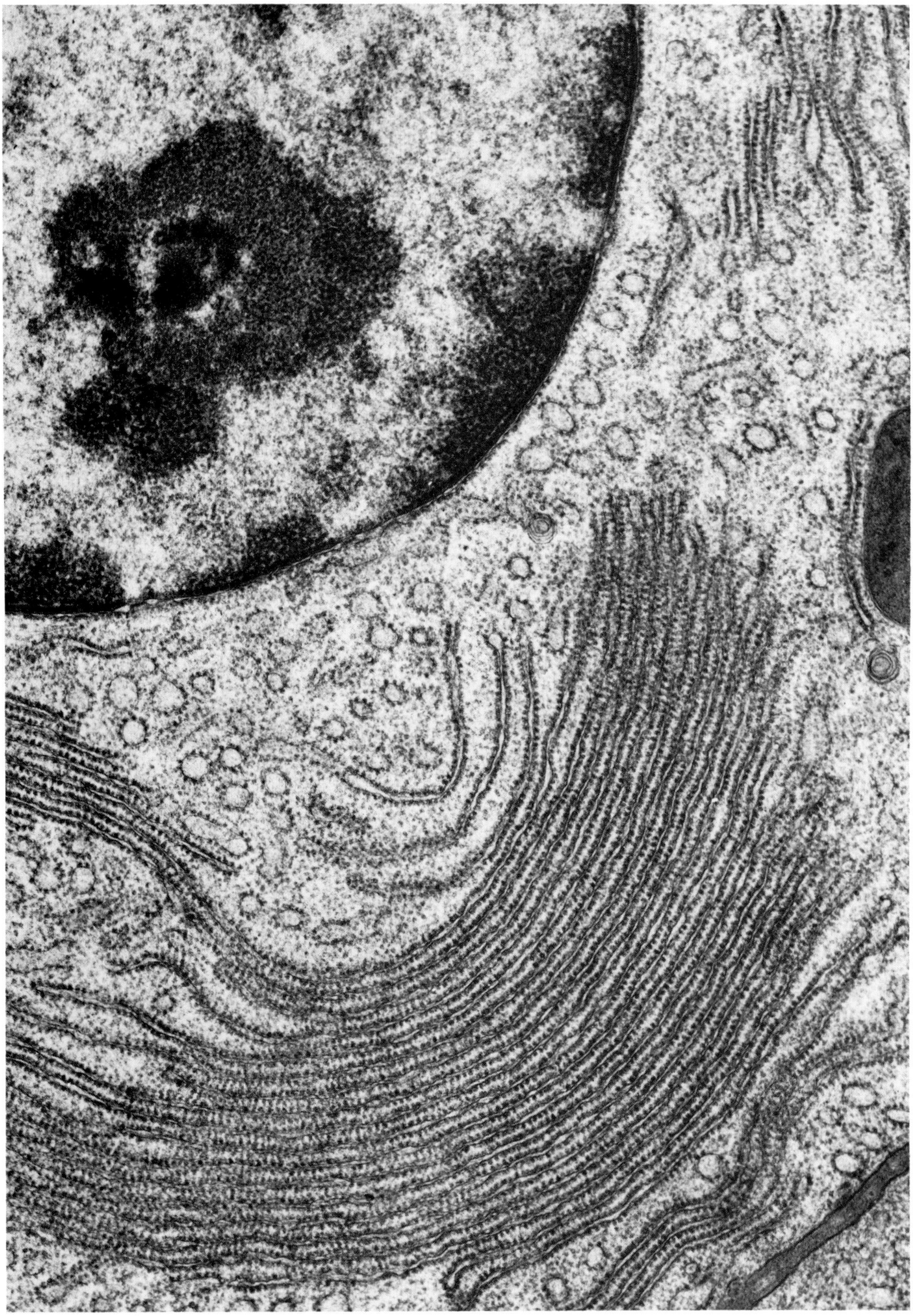

Figure 168

Cisternae of the reticulum are most abundant in cells synthesizing large amounts of protein for export as a secretory product. They may be arranged parallel in stacks 2 to 4 μ in length, or when very long they may form extensive concentric systems such as those shown on the facing page. In such parallel arrays, the lumen of the cisternae is quite narrow and the successive layers only about 35 nm apart. When the cisternae are continuous at their margins with tubular elements of the reticulum (see at arrows), the lumen widens.

Figure 169. Rough endoplasmic reticulum in the basal cytoplasm of a pancreatic acinar cell from the bat, *Myotis lucifugus.*

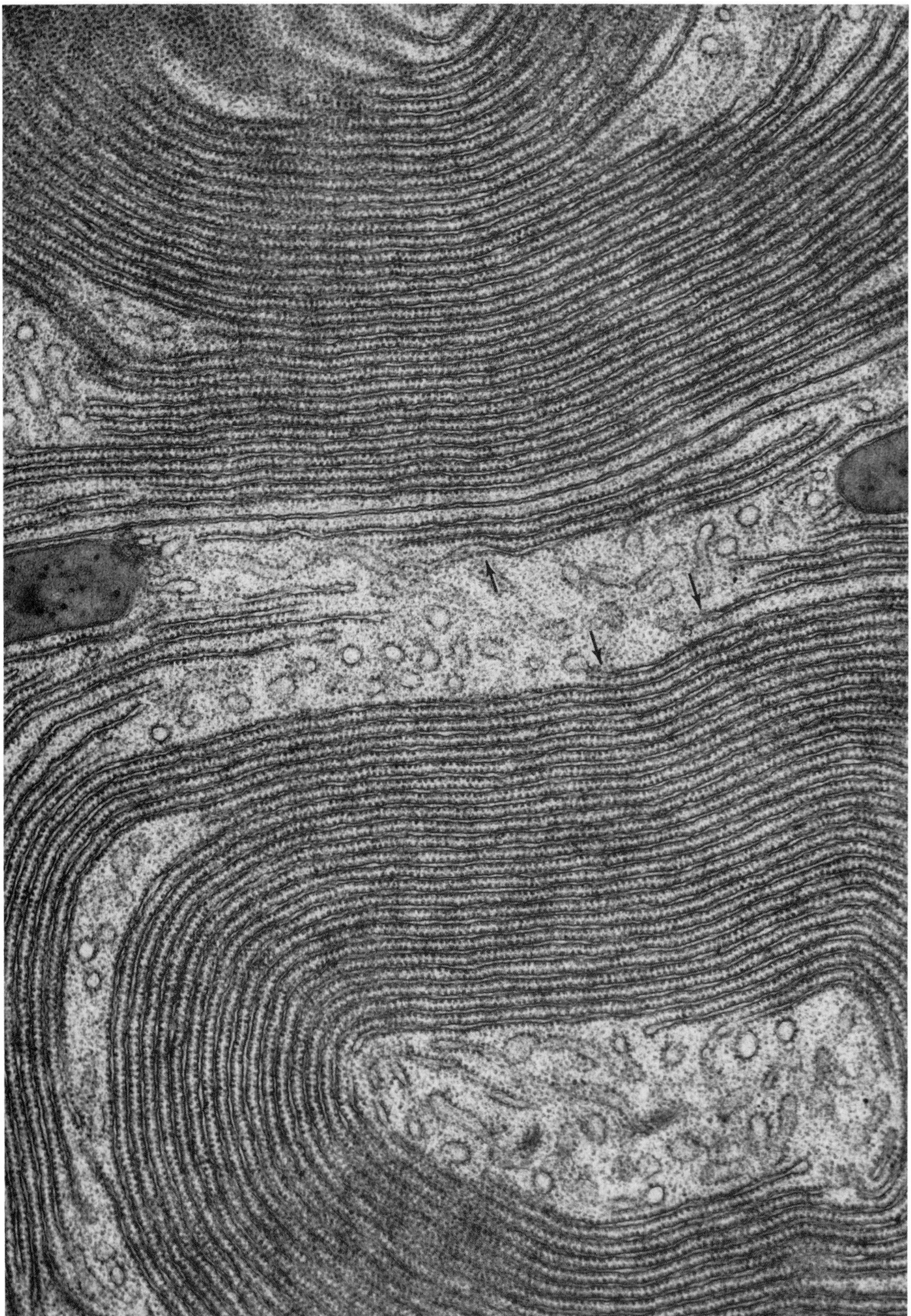

Figure 169

The contents of the rough endoplasmic reticulum are usually extracted during specimen preparation so that its lumen is of lower density than the cytoplasmic matrix between cisternae. With optimal fixation, a pale flocculent precipitate may be preserved in the interior of the reticulum. This represents a dilute solution of newly synthesized protein which will subsequently be transported to the Golgi complex for concentration and packaging into secretory granules.

Figure 170. Rough endoplasmic reticulum in pancreatic acinar cells of human pancreas. (Micrograph courtesy of Susumo Ito and Arthur Like.)

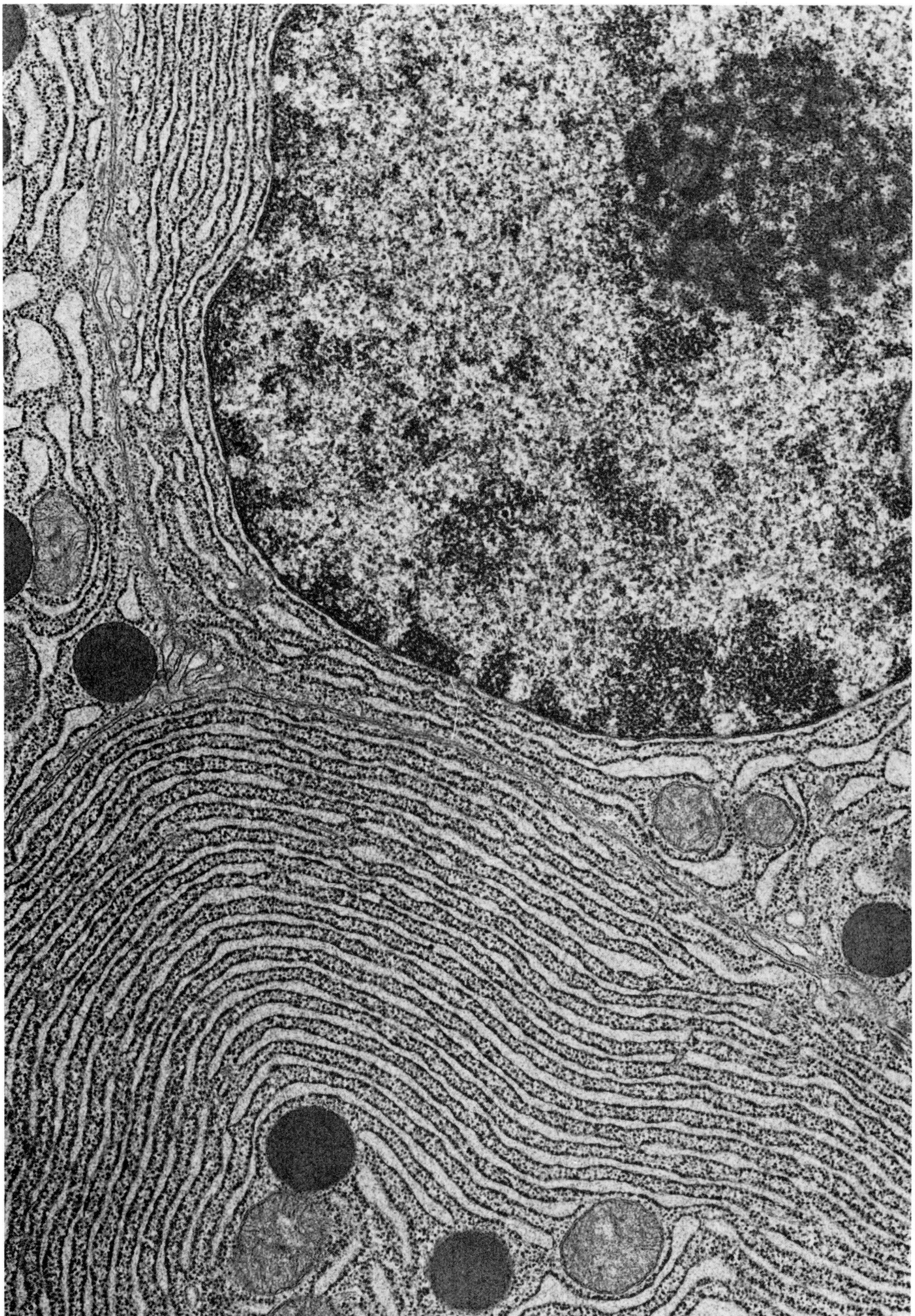

Figure 170

A more highly magnified micrograph of the periphery of two neighboring cells permits a comparison of the plasma membranes with those bounding the cisternae of the rough endoplasmic reticulum. The cell surface membranes are slightly thicker and although there are free ribosomes in the adjacent cytoplasm, none of these is in contact with the inner surface of the surface membrane. They tend to be aligned about 10 nm away from the membrane (see at arrows), suggesting that they are kept away from it by an intervening thin ectoplasmic layer. The membranes of the reticulum on the other hand are studded with numerous adherent ribosomes. At still higher magnifications, these can be shown to consist of two subunits, one larger than the other, and it is always the larger subunit that is bound to the membrane.

Figure 171. Surface membranes and adjacent endoplasmic reticulum of two pancreatic acinar cells from the bat, *Myotis lucifugus*.

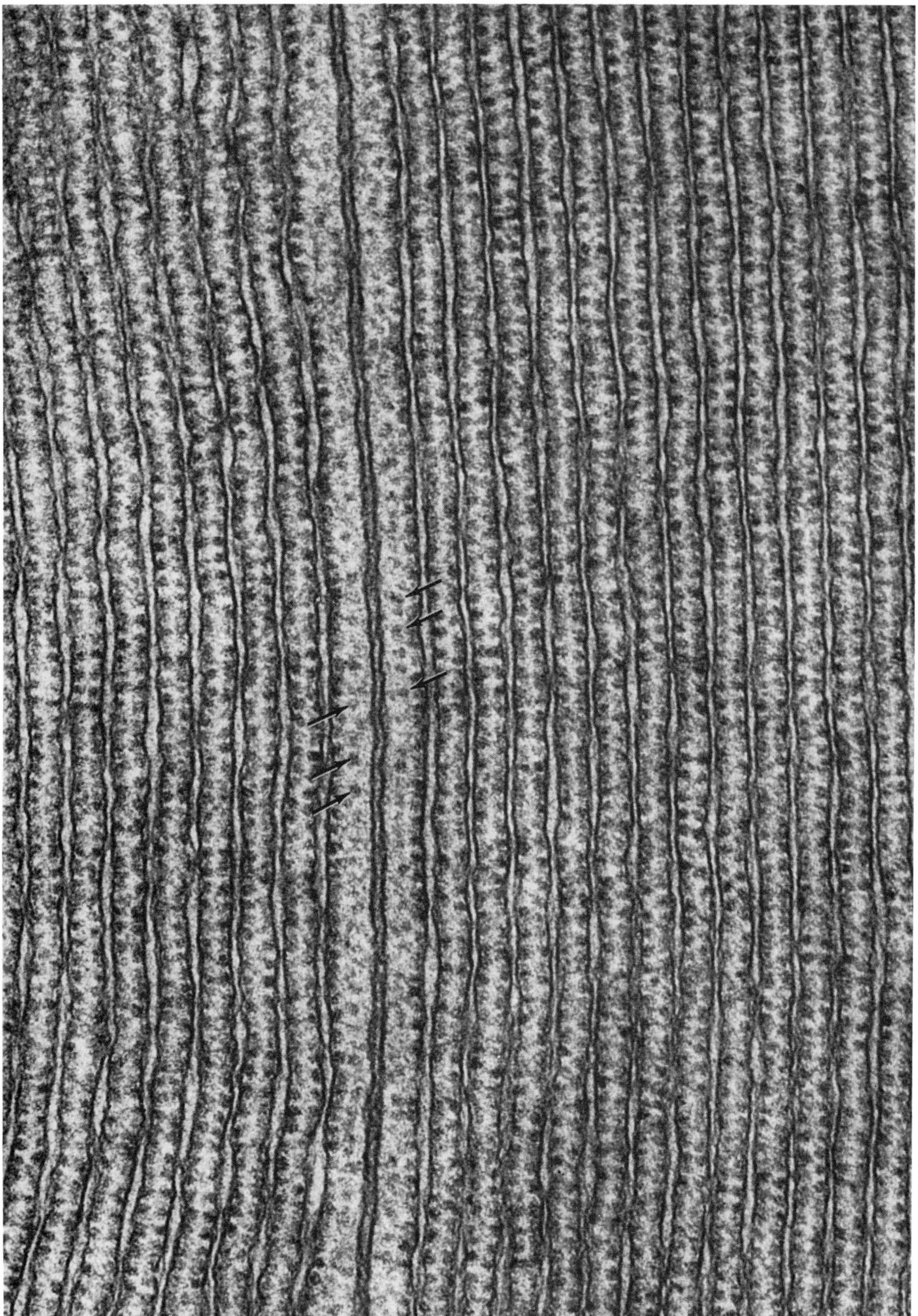

Figure 171

In light microscopic preparations, motor neurons are characterized by the presence of large angular blocks of basophilic material called *Nissl bodies*. When studied with the electron microscope, these are found to be orderly arrays of cisternae of the endoplasmic reticulum. They have a distinctive appearance unlike that of the stacks of reticulum in glandular cells. The fenestrated cisternae are separated by intervals of 0.2 to 0.5 μm and these interspaces are crowded with free polyribosomes. In addition, there are numerous polyribosomes attached to the membranes of the cisternae. The Nissl substance of neurons thus differs from the basophilic bodies of glandular cells in the greater separation of the parallel cisternae and the very large number of associated free polysomes. While both are centers of protein synthesis, the differences in architecture of the endoplasmic reticulum are no doubt related to the differing fate of the protein synthesized. In glandular cells concerned with synthesis of protein for export, the great majority of ribosomes are associated with the membranes of the reticulum. In neurons nearly all of the protein synthesized is for internal use in maintenance and renewal of the protoplasm in the perikaryon, dendrites, and the long axon. It is estimated that a large neuron may renew as much as a third of its protein content daily. It is the free polyribosomes that are mainly involved in this process. Relatively little product accumulates in the cisternae of the endoplasmic reticulum.

Figure 172. Nissl body of a large neuron. (Micrograph courtesy of Sanford Palay.)

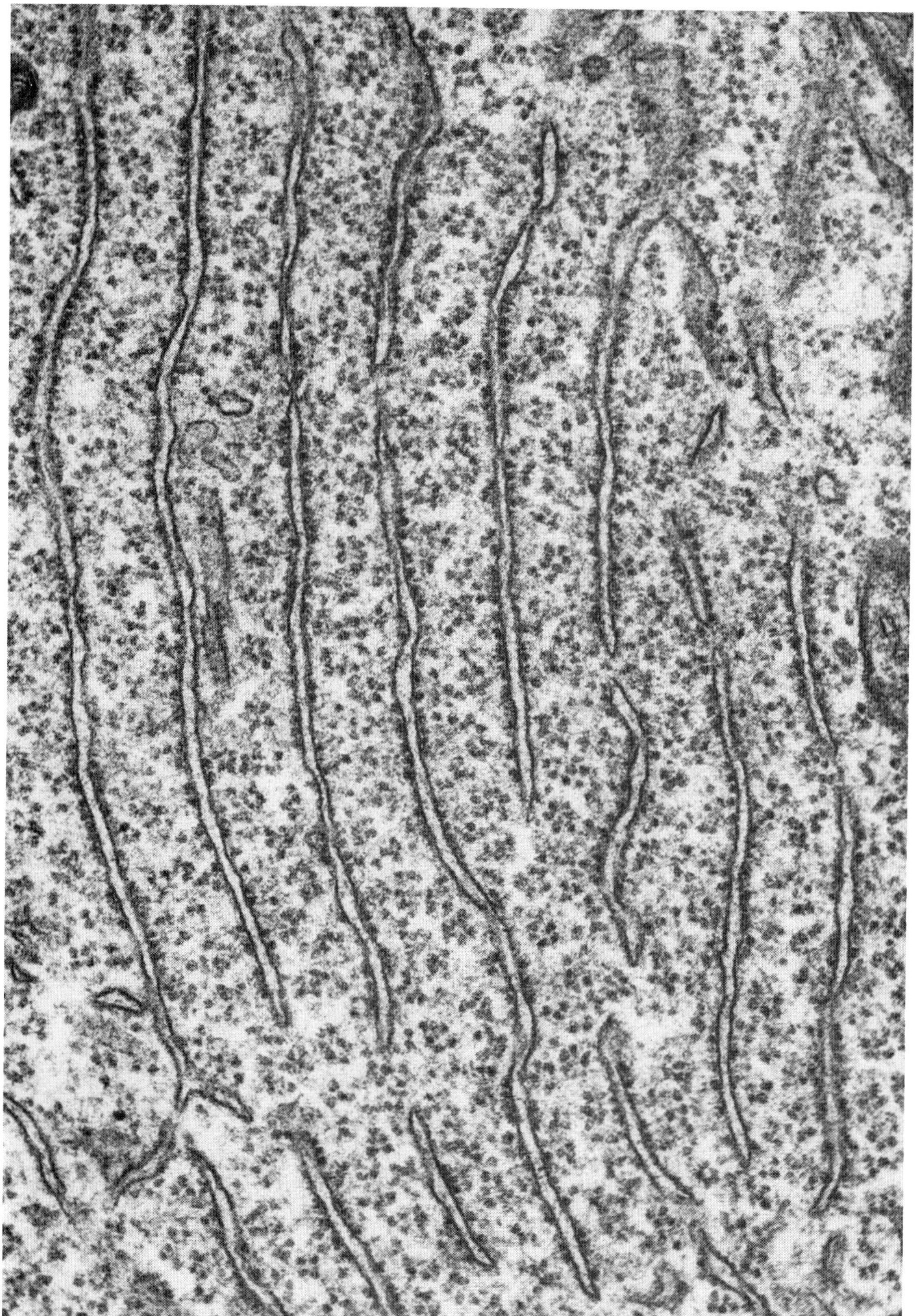

Figure 172

In random thin sections, it is not possible to resolve the thin messenger RNA molecule or to ascertain the number of ribosomes in any given polysomal cluster. However, when the plane of section is parallel to a cistern of the endoplasmic reticulum, one can observe *en face* views of limited areas of its surface. The number and arrangement of ribosomes in each polysome is then visible. This is illustrated in the micrographs on the facing page. The upper figure shows numerous free polysomes associated with a Nissl body, and in two areas the section passes tangential to cisternae, showing ribosomes in rows, spirals and rosettes against the grey background of the grazing section of membrane (see at arrows).

In the lower figure, from a glandular cell, the slightly oblique section passes tangential to three cisternae. In each there are long loops and spirals consisting of multiple ribosomes aligned on the same molecule of messenger RNA. The length of the message and number of ribosomes translating it are related to the size of the polypeptide being synthesized. As many as 32 ribosomes have been observed on the same messenger RNA molecule in neurons (Peters, Palay and Webster, 1976).

Figure 173. Section of a Nissl body. (Micrograph courtesy of Sanford Palay.)

Figure 174. Section of reticulum in an adrenocortical cell of a human fetus of 27 weeks. (Micrograph courtesy of Eichi Yamada.)

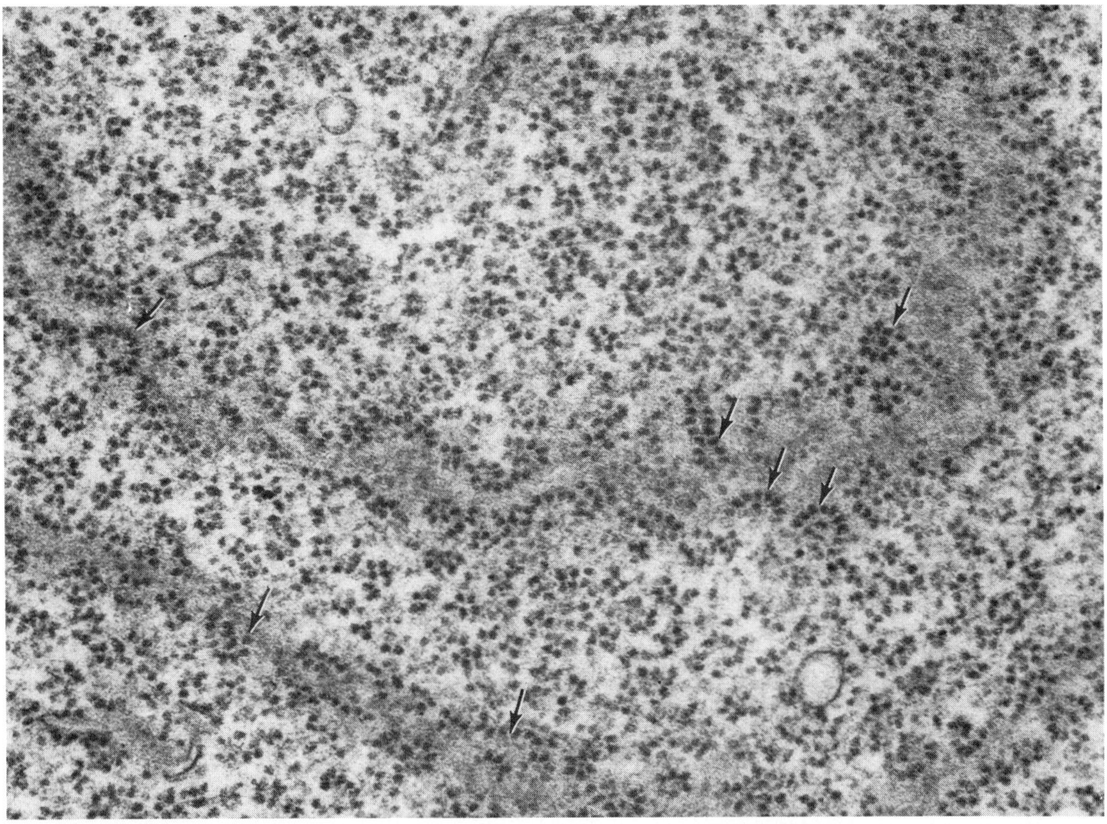

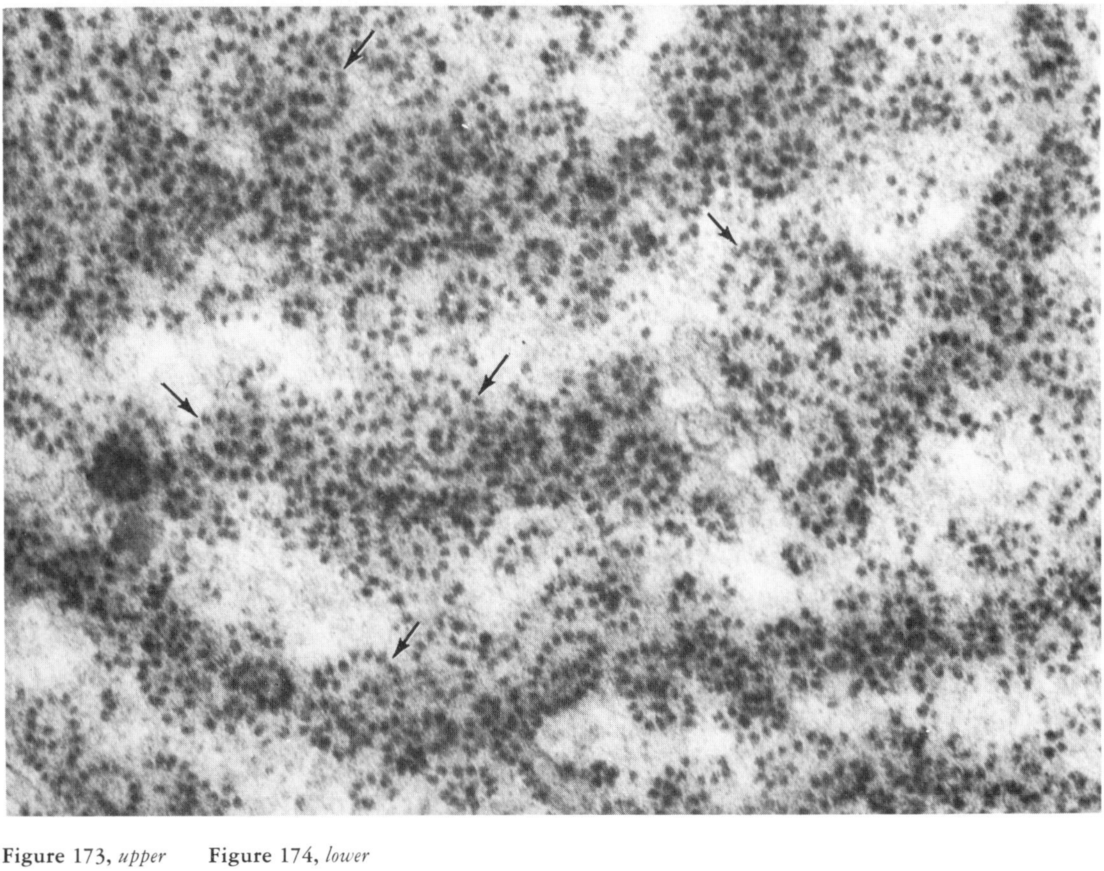

Figure 173, *upper* **Figure** 174, *lower*

The plasma cell shown here is actively engaged in the synthesis of immunoglobulin and therefore has an extensive rough endoplasmic reticulum. The product accumulates in the lumen of the reticulum where it appears as a grey precipitate of protein. Unlike in glandular cells, the product is not concentrated in the Golgi complex into secretory granules.

Figure 175. Plasma cell from guinea pig bone marrow.

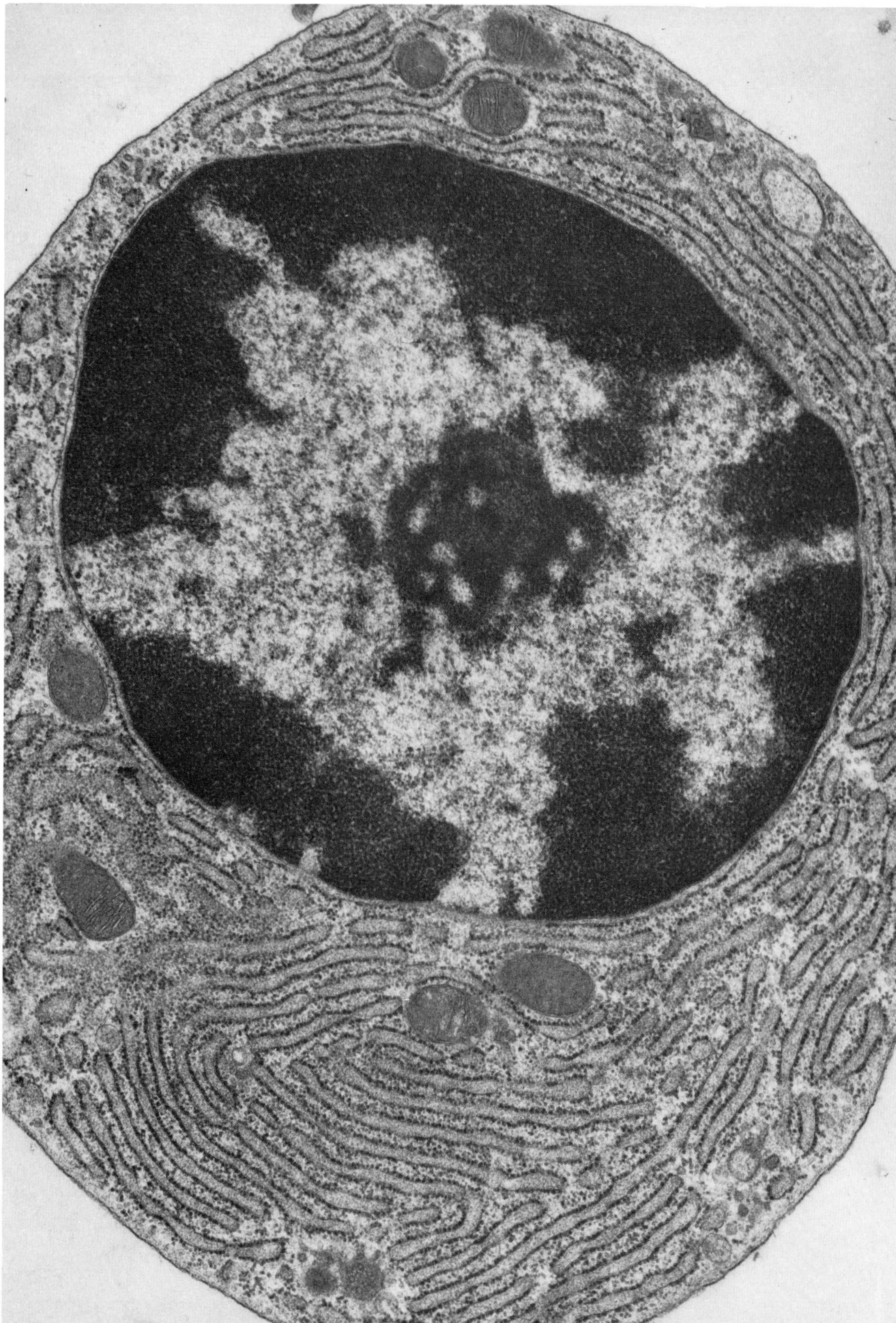

Figure 175

In the plasma cell which does not concentrate its product in discrete secretory granules, storage takes place within the cisternae of the endoplasmic reticulum, which become greatly distended. In the cell illustrated here, the cisternae distended with protein form a labyrinthine system of intercommunicating spaces, separated by narrow septa of cytoplasm containing mitochondria and free ribosomes.

Figure 176. Plasma cell from guinea pig bone marrow.

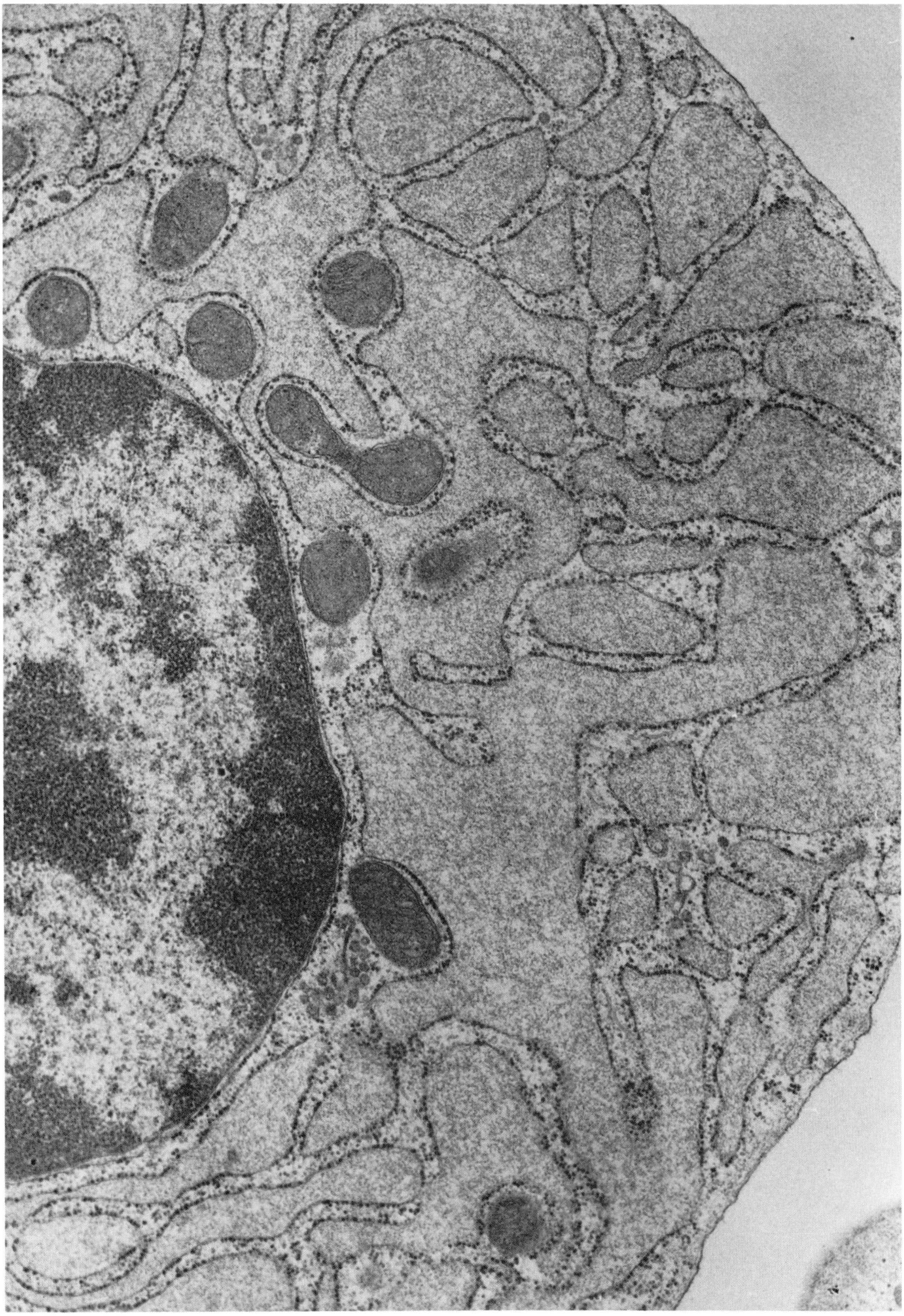

Figure 176

In some cell types, the products of protein synthesis are preserved in specimen preparation and appear as a flocculent precipitate in the lumen of the endoplasmic reticulum. An example is shown in the accompanying micrograph of the supranuclear region of an odontoblast actively secreting dentin matrix.

Figure 177. Micrograph of an odontoblast from a rat incisor. (Micrograph courtesy of Melvyn Weinstock.)

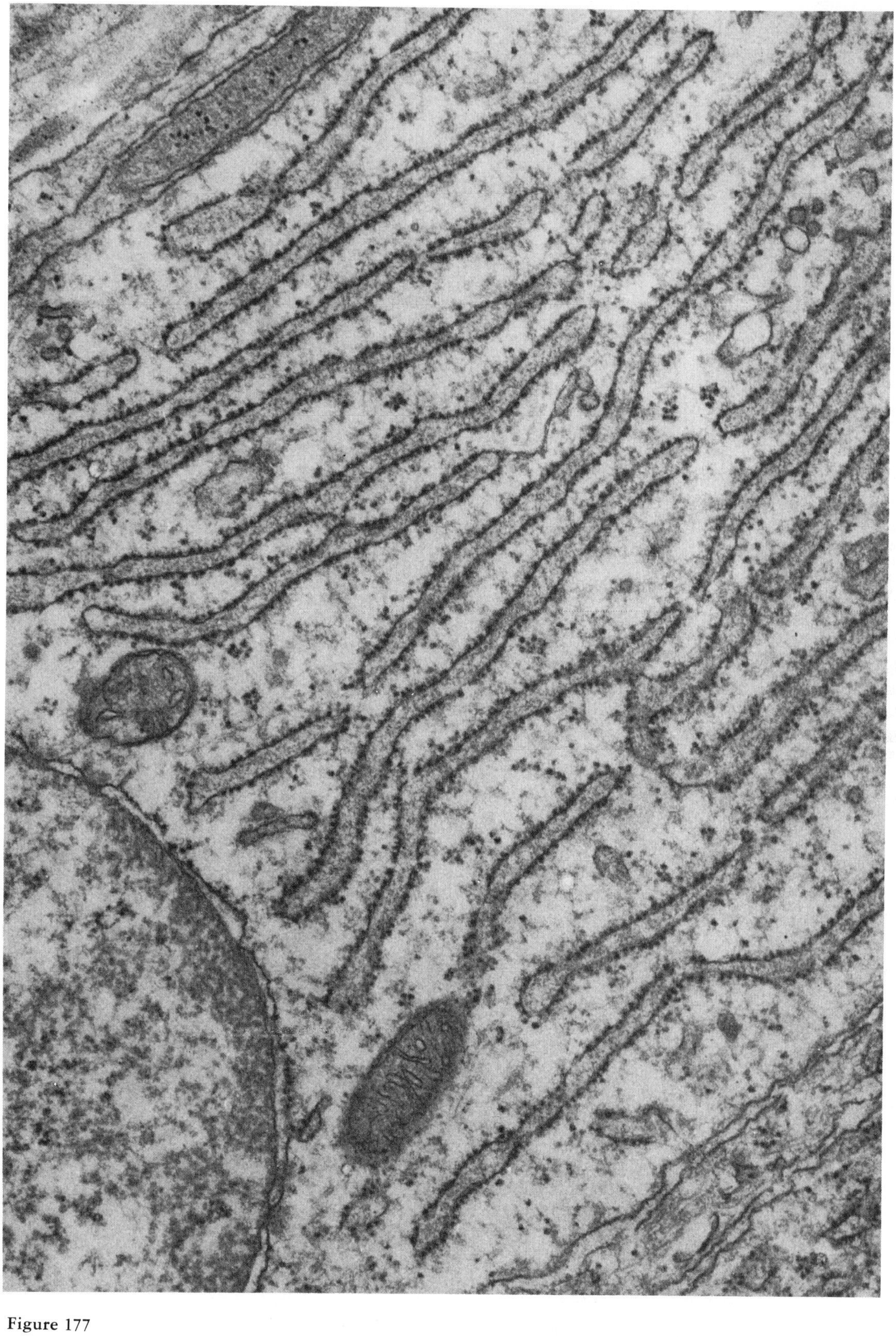

Figure 177

In rare instances, the product of protein synthesis may accumulate in high enough concentration to crystallize in the lumen of the endoplasmic reticulum. An example is shown in the accompanying micrograph from the liver of the slender salamander, *Batrachoseps*. The crystals are never found in the cytoplasmic matrix nor in the smooth endoplasmic reticulum. The nature of the crystals is not known.

Figure 178. Liver cell from the slender salamander, *Batrachoseps attenuatus.*

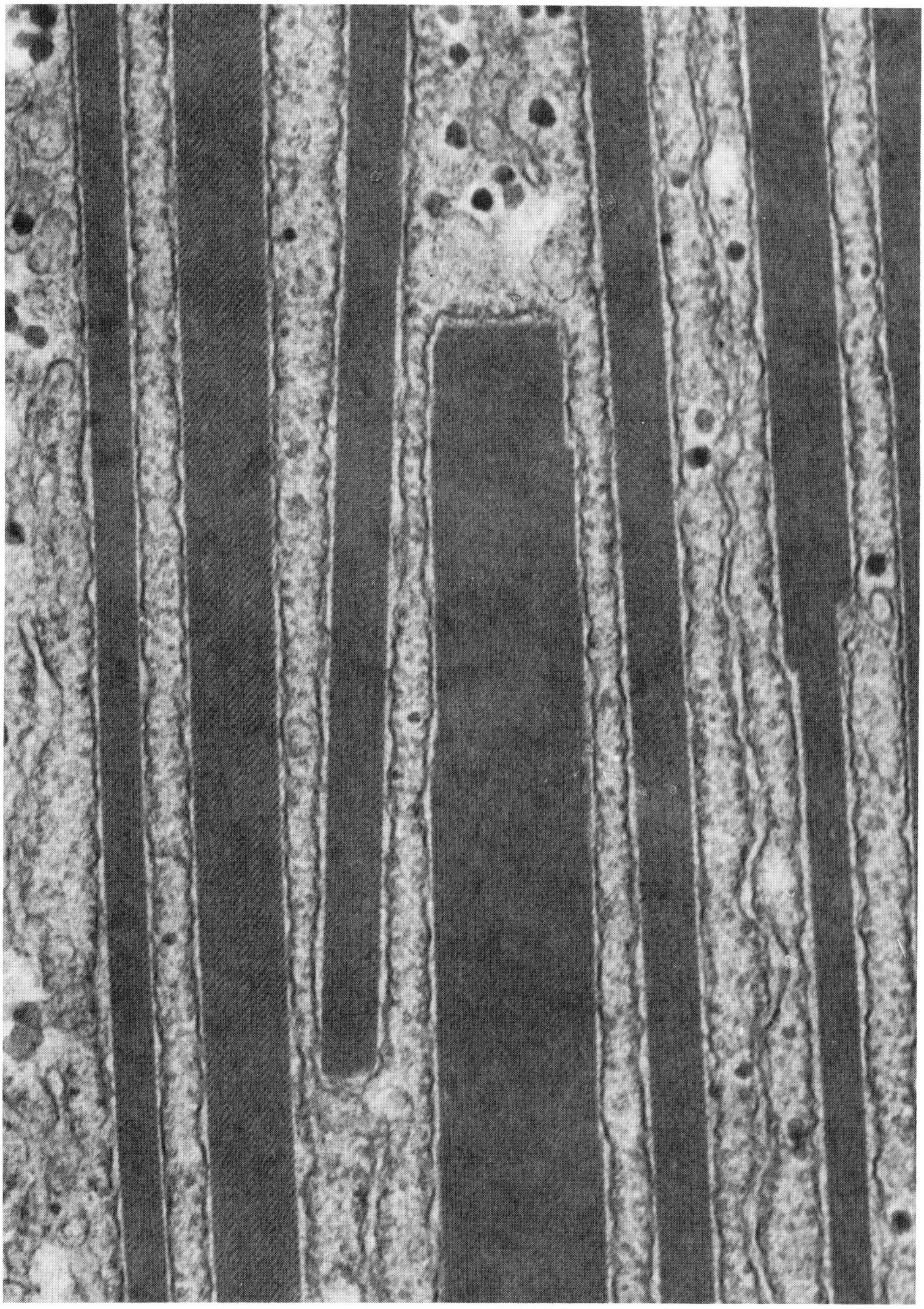

Figure 178

SMOOTH ENDOPLASMIC RETICULUM

The early studies on the endoplasmic reticulum concentrated on the granular- or rough-surfaced form of the organelle and its correspondence to the ergastoplasm of classical cytology. However, it was soon recognized that membrane-limited tubules without associated ribonucleoprotein particles were a conspicuous feature of some cell types. Palay and Palade (1954) described as "agranular reticulum" the closely apposed flattened cisternae and associated vesicles which were subsequently identified as the Golgi complex. These smooth membranes were initially interpreted as a ribosome-free local differentiation of the endoplasmic reticulum but were later shown to be a distinct organelle only in intermittent communication with the reticulum by vesicular transport. Compact masses of smooth-surfaced tubules distinct from the Golgi complex were described by Fawcett (1954) in hepatic cells of animals refed after a period of fasting. These were often continuous at their periphery with profiles of rough endoplasmic reticulum, and it was suggested that they might represent sites of regeneration of rough endoplasmic reticulum by neogenesis of smooth membranes followed by binding of free ribosomes. It later became evident that less compact networks of smooth-surfaced tubules were present in the liver cells of normally nourished animals.

A network of smooth-surfaced tubules around the myofibrils of striated muscle was described by Porter and Palade (1957) and interpreted as a specialized form of the endoplasmic reticulum. As other cell types were examined with the electron microscope, many were found to contain branching and anastomosing tubules devoid of ribosomes in addition to the familiar rough endoplasmic reticulum. These came to be interpreted as integral parts of the same continuous system of tubules and cisternae, but since they lacked associated ribonucleoprotein particles, it was assumed that they subserved functions other than protein synthesis. Morphological observations correlating differences in degree of development of smooth endoplasmic reticulum with well-established functions of specific cell types supported the concept of separate functions for the two kinds of reticulum. The finding of a well-developed smooth reticulum in meibomian glands (Palay, 1958) and in intestinal epithelium during fat absorption (Palay and Karlin, 1959) suggested a role in lipid metabolism. Very extensive development of smooth membranes in the adrenal cortex (Ross et al., 1958), in interstitial cells of the testis (Christensen and Fawcett, 1960) and in the corpus luteum of the ovary (Yamada and Ichikawa, 1960; Enders, 1962) strongly suggested its involvement in biosynthesis of steroid hormones. A function in glycogenesis or glycogenolysis was proposed on the basis of a close association of glycogen with areas of smooth reticulum in the liver (Porter and Bruni, 1959). The observation of a striking hypertrophy of the smooth reticulum and a concomitant increase in certain mixed function oxidases in the liver implied an important function in drug metabolism (Remmer and Merker, 1963; Jones and Fawcett, 1966).

These morphological observations encouraged the speculation that the composition and enzymatic properties of the smooth membranes were different from those of the rough endoplasmic reticulum. When the technical difficulties involved in separating the microsome fraction into subfractions of rough and smooth vesicles were overcome (Rothschild, 1963; Dallner and Ernster, 1968), this expectation was not borne out. Apart from the disparity in ribonucleoprotein content, the biochemical properties of the two subfractions were very similar. The smooth microsomes have a lower net negative surface charge, a somewhat higher concentration of cholesterol and galactose sialic acid, but to date no enzymatic activity has been localized exclusively in rough or smooth microsomes (De Pierre and Dallner, 1975). The possible existence of quantitative difference in enzyme activity has not been ruled out.

The endoplasmic reticulum is a highly versatile organelle involved in synthesis and transport of proteins, glycoproteins, and lipoproteins; synthesis of cholesterol, steroids, phospholipids, and triglycerides; it participates in degradation of glycogen, and in the metabolism of xenobiotics. The extent of development of rough and smooth reticulum in any given cell type appears to depend upon the importance of protein synthesis on ribosomes relative to other functions carried out by the enzymatic constituents of the membrane.

In an earlier section on the cell surface, we noted that one of the major conceptual advances in histology and cell biology in the past 30 years has been a greater awareness of the structural devices for gaining efficiency by increasing the area of physiologically important interfaces. The *convolution* of tubular organs, the *plication* of their mucosa, and the *formation of villi* were cited as examples. These same devices are repeated at the intracellular level in the convolution of the endoplasmic reticulum, the plication of the internal membrane of the mitochondria to form the cristae, the projection of inner membrane subunits from the cristae, and so on. Long before the advent of the electron microscope, the British physiologist Sherrington wrote, "Part of the secret of life is the immense internal surface of the cell." Even he would have been astonished by the extraordinary degree of compartmentation of the cell by membrane-bounded cell organelles. The application of stereological and morphometric methods to electron micrographs yields estimates of over 8 square meters of membrane per cubic centimeter of liver, and the endoplasmic reticulum makes a major contribution to this impressive total.

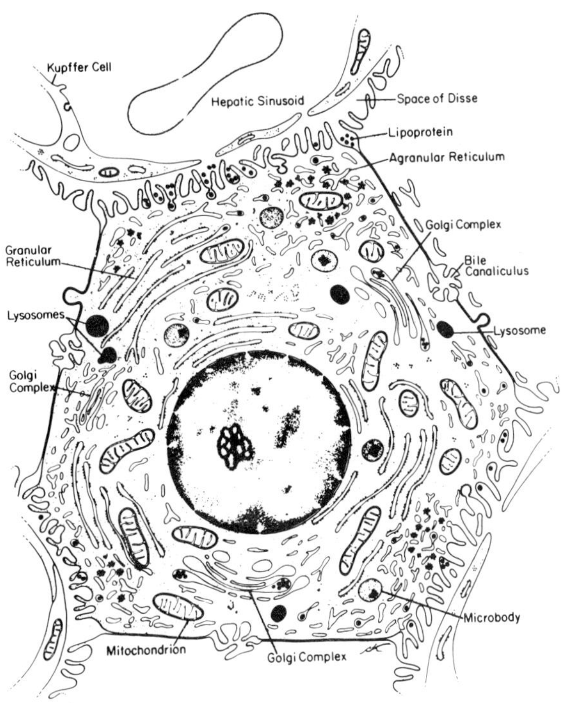

Membrane Surface Area
m^2/cm^3

Membrane	m^2/cm^3
Plasma membrane	0.56
Endoplasmic reticulum	3.80
(Rough	2.41)
(Smooth	1.39)
Mitochondria	3.83
Golgi apparatus	0.17
Lysosomes	0.07
Pinocytosis vesicles	0.01
Total membranes	8.44

(Data of Blouin, Bolender, and Weibel)

Illustration of the extensive compartmentation of a liver cell by membrane-limited organelles, together with stereological estimates of the surface area of membrane contributed by the various cell components.

In addition to the cisternae of rough endoplasmic reticulum, liver cells contain branching and anastomosing tubules that are devoid of ribosomes. These form close meshed networks of the kind illustrated at the left upper quadrant of the accompanying micrograph. The tubular elements of this smooth endoplasmic reticulum are continuous at many sites with cisternae of the rough endoplasmic reticulum (see at arrows). The reticulum is therefore a continuous membrane-limited system of tubules and cisternae, some regions of which have associated ribosomes and are involved in sequestration and transport of the products of protein synthesis, while other regions, lacking ribosomes, are specialized for the other functions specified above.

Figure 179. Electron micrograph of a portion of the cytoplasm of a liver cell. (Micrograph courtesy of Robert Bolender.)

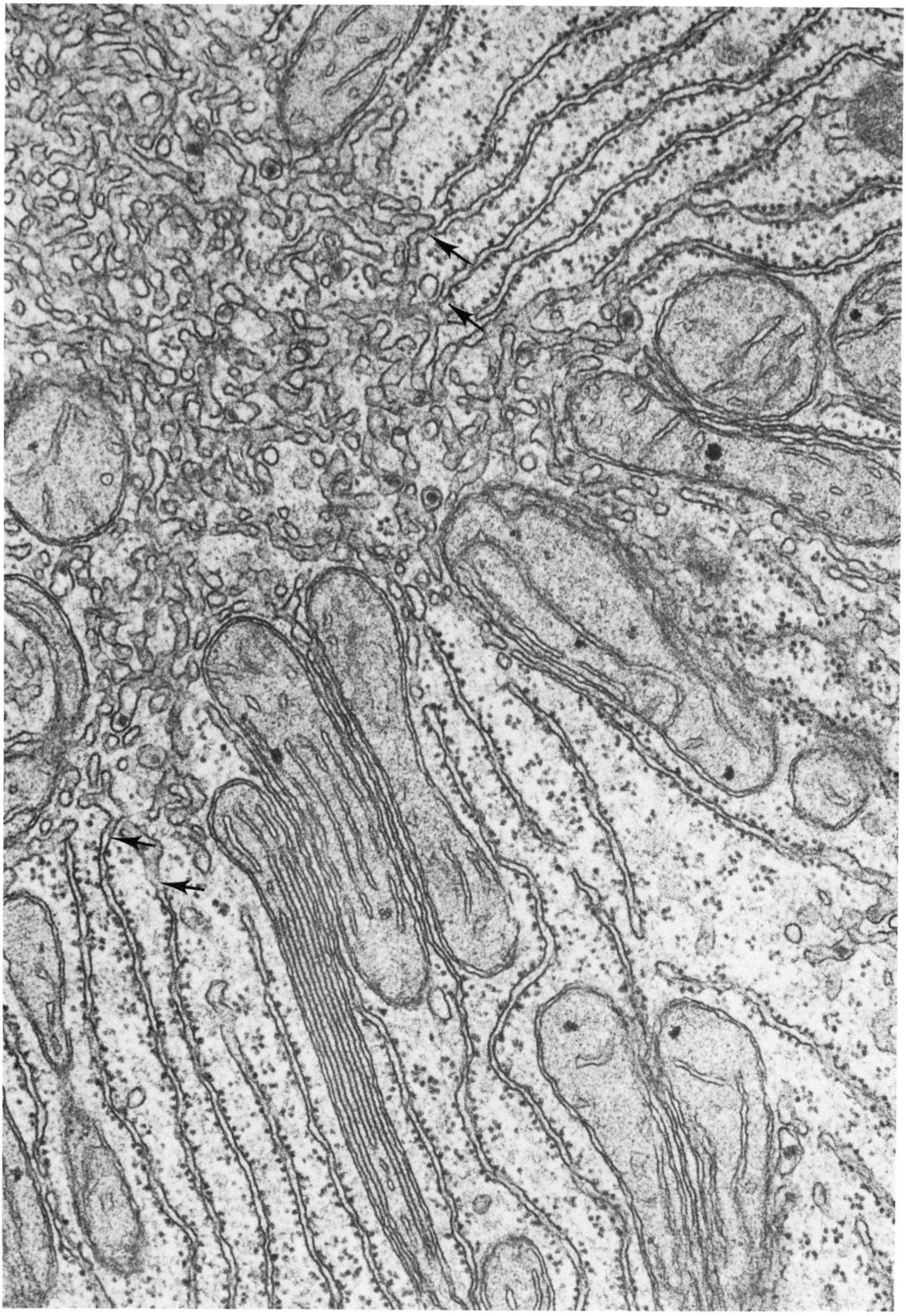

Figure 179

The smooth endoplasmic reticulum is often well developed in cell types that are active in the metabolism of lipids. This is exemplified in the accompanying micrograph of an intestinal epithelial cell fixed during absorption of fat. After ingestion of fat the lumen of the intestine contains an emulsion of di- and triglycerides stabilized by conjugation with bile salts. The triglycerides are hydrolyzed by pancreatic lipase in the lumen to free fatty acids and monoglycerides. These are able to diffuse through the membrane of the microvilli and into the apical cytoplasm of the absorptive cells. There the enzyme fatty acid:Co-A ligase in the membrane of the smooth endoplasmic reticulum catalyzes the reaction of fatty acids with Co-A to yield highly reactive thiolesters. These in turn react with monoglycerides and are transformed by other microsomal enzymes into triglycerides which accumulate in the form of osmiophilic droplets of lipid in the lumen of the smooth endoplasmic reticulum (Senior, 1964; Cardell et al., 1967). Several such droplets are indicated by arrows in the accompanying micrograph. These droplets containing triglyceride, phospholipid, cholesterol, and stabilizing protein are transported through the lumen of the reticulum to the lateral aspect of the cell, where they are released into the intercellular cleft as chylomicrons. These ultimately enter the lymphatics of the intestinal villi and are carried in the lymph to the bloodstream.

Figure 180. Electron micrograph of a rat intestinal epithelial cell during absorption of a fatty meal. (Micrograph courtesy of Sanford Palay.)

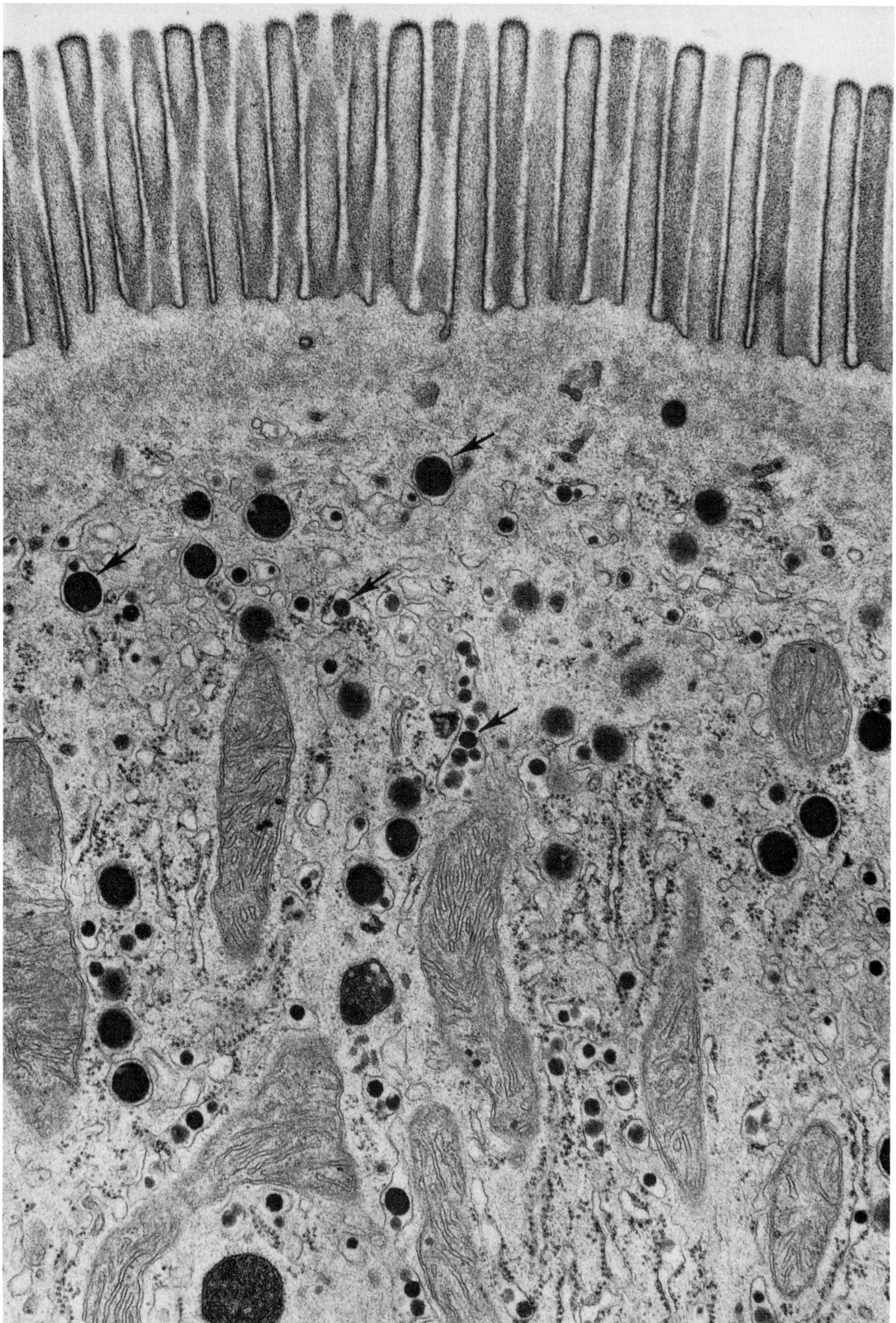

Figure 180

An important function of the liver is the metabolism of dietary lipid and maintenance of lipid levels in the circulating blood. The vehicle for the transport of lipid from the liver to the fat depots is the plasma lipoprotein. Ingested fat absorbed in the intestine reaches the bloodstream via the lymphatics and is taken up by the hepatic cells, which transform it into 30 to 100 nm low-density lipoprotein particles. These consist of a core of triglyceride and a surface coat of phospholipid, cholesterol, and protein. The formation of these particles in the liver cells appears to be a cooperative effort of the rough and smooth reticulum. The protein is synthesized in the rough reticulum and combined with triglycerides that are formed from fatty acids and monoglycerides by enzymes in the smooth reticulum. The particles appear in the lumen of the smooth reticulum and are transported to the cell surface via the Golgi apparatus and are released into the space of Disse and thence into the sinusoids (Jones et al., 1967).

Figure 181. Electron micrograph of rat liver. (Micrograph courtesy of Robert Bolender.)

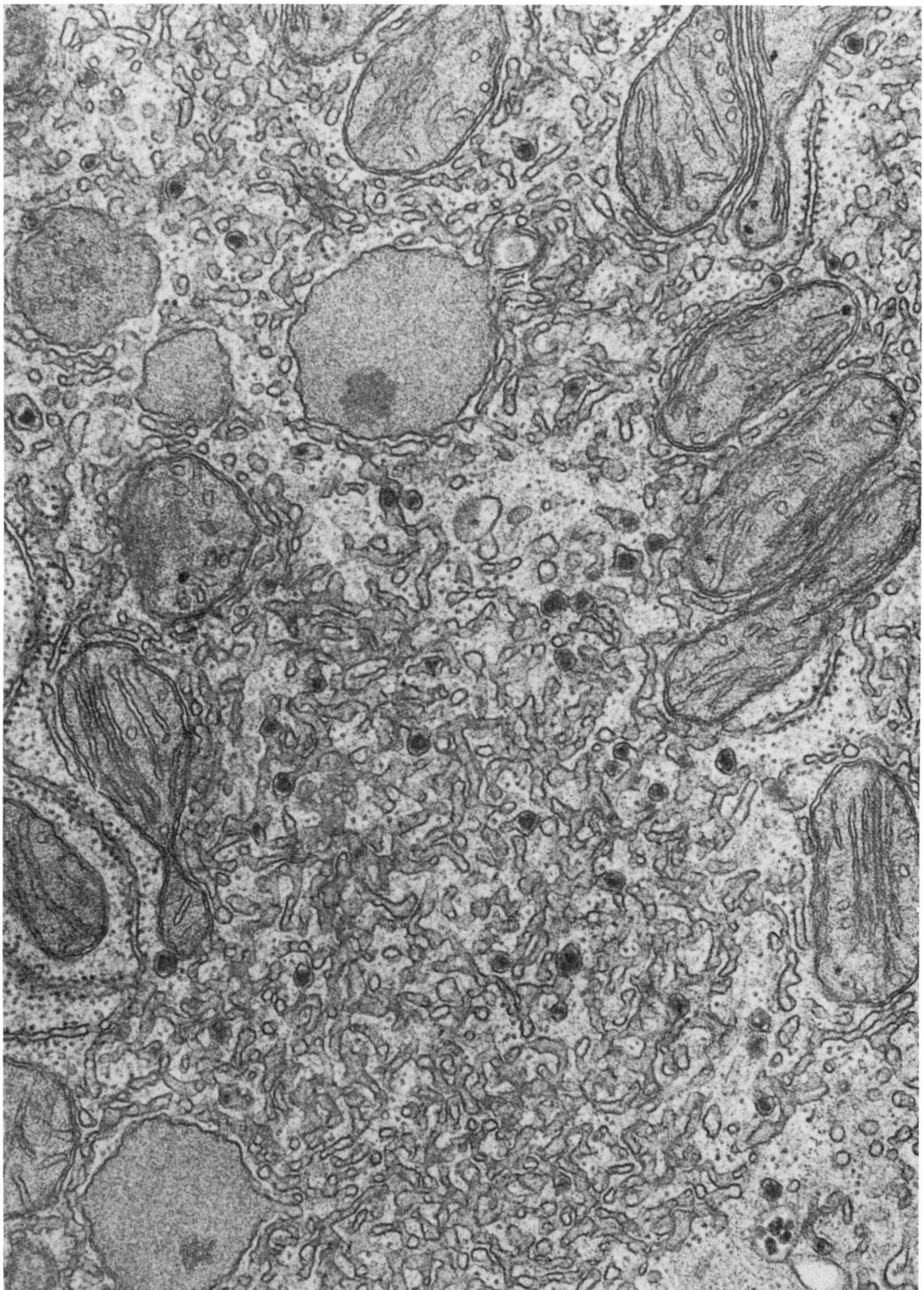

Figure 181

The smooth endoplasmic reticulum is a dynamic organelle responsive to a variety of endogenous and exogenous stimuli. Remmer and Merker (1963) observed that administration of phenobarbital to rats resulted in a selective hypertrophy of the smooth endoplasmic reticulum in the liver and a corresponding increase in the activity of several microsomal enzymes involved in the metabolism of drugs. It was subsequently found that other barbiturates, halogenated hydrocarbon insecticides, carcinogens, and a great variety of other compounds that are metabolized in the liver are also powerful inducers of drug-metabolizing enzymes and of hypertrophy of the smooth reticulum. This adaptive response of the liver cells enhancing their capacity to metabolize and eliminate drugs is the basis for drug tolerance observed in humans.

The accompanying micrograph is from the liver of a hamster that received injections of phenobarbital on four successive days. The treatment has resulted in a remarkable hypertrophy of the smooth reticulum, which now occupies nearly all available space in the cytoplasm. Although this is the only obvious morphological response of the liver cell, the synthesis of additional drug-metabolizing enzymes and their incorporation into the newly formed membranes also involves enhanced protein synthesis in the rough endoplasmic reticulum. These findings clearly establish the metabolism of exogenous compounds as one of the major functions of the endoplasmic reticulum in the liver. Some of the same oxidases are normally involved in metabolism of lipid-soluble endogenous compounds such as steroid hormones.

Figure 182. Liver of a hamster treated four days with phenobarbital.

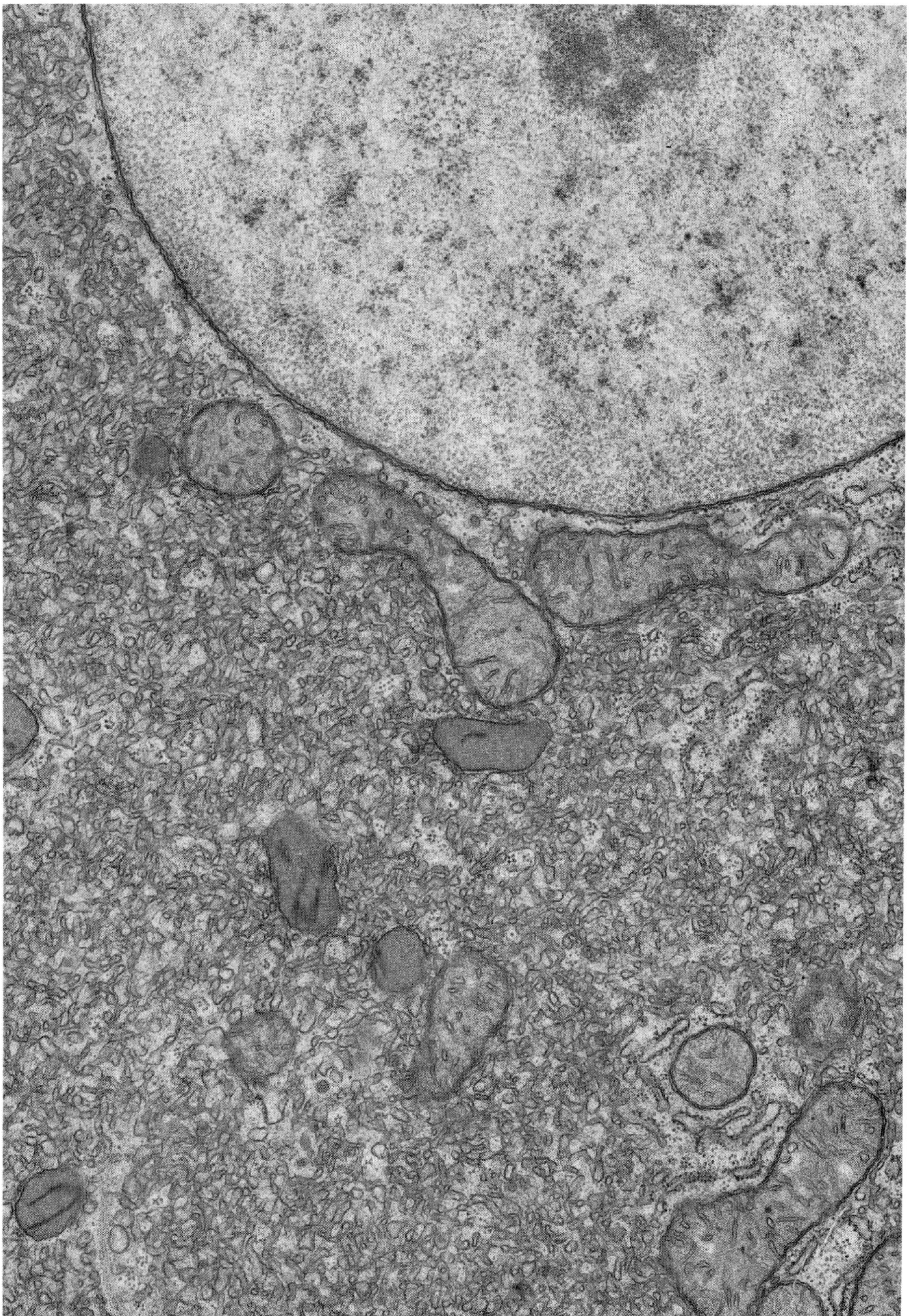

Figure 182

The glycogen in liver cells is not uniformly distributed throughout the cytoplasm but tends to be concentrated in areas of smooth endoplasmic reticulum. This is illustrated in the micrograph on the facing page, where numerous alpha particles of glycogen are found in the interstices between profiles of smooth reticulum.

This close topographical relationship led Porter and Bruni (1959) to suggest that this form of the reticulum might play a role in glycogen synthesis or glycogenolysis. Upon separation of a smooth membrane fraction from liver glycogen pellets, no UDPG-glycogen transferase activity was found (Luck, 1961) and it was concluded therefore that the smooth endoplasmic reticulum does not participate in glycogen synthesis. The glucose-6-phosphatase known to be present in the microsome fraction may, however, be involved in the degradation of glycogen and return of glucose to the blood. However, this does not explain the selective association of glycogen with the smooth reticulum, for this enzyme is also a constituent of the rough reticulum.

Figure 183. Glycogen in an area of a hamster hepatic cell rich in smooth endoplasmic reticulum.

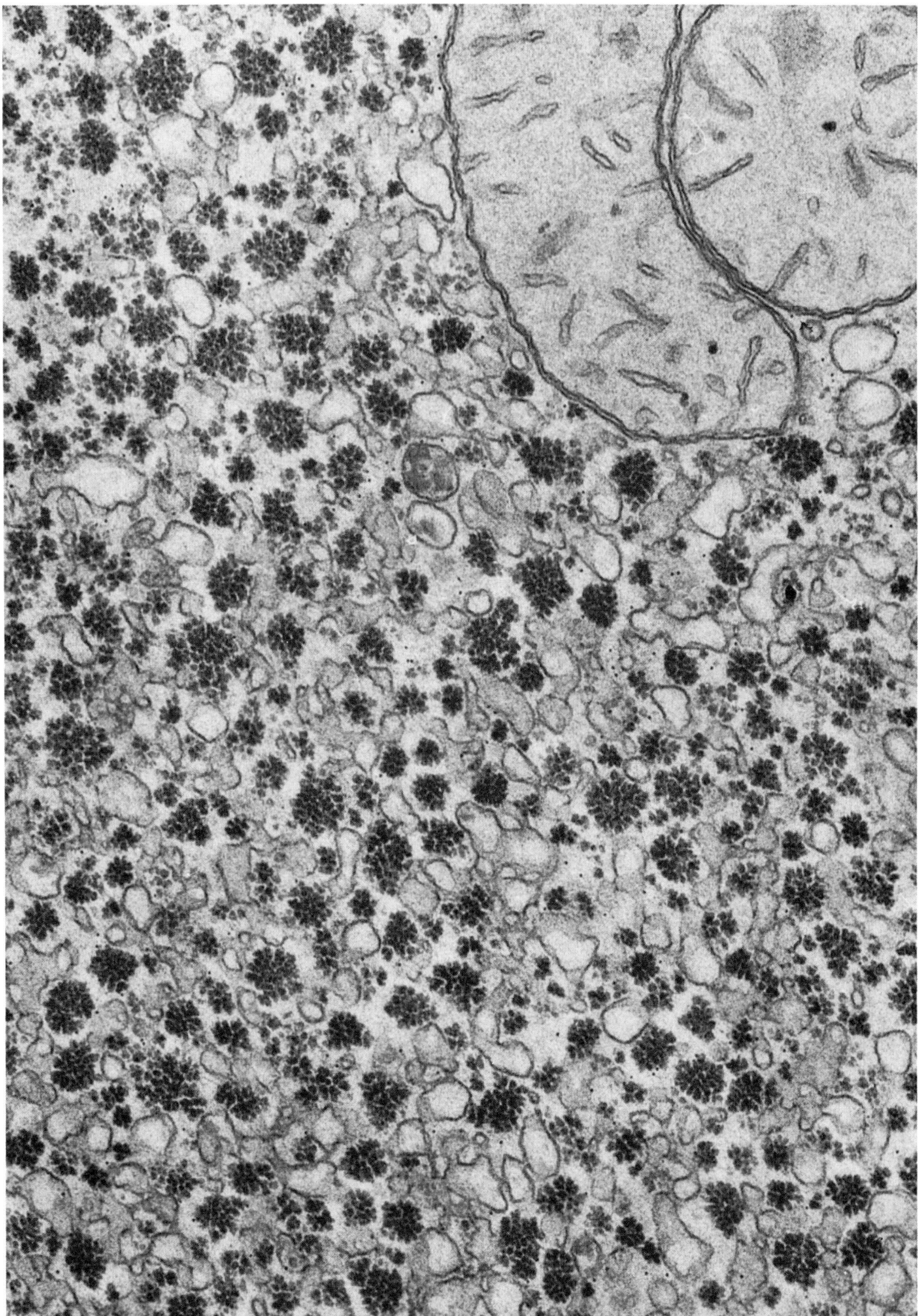

Figure 183

The close association of glycogen with smooth endoplasmic reticulum is not confined to the liver. More highly ordered parallel arrays of membranes alternating with layers of glycogen particles are seen in the paraboloid of the photoreceptor cells of the reptilian retina (Yamada, 1960), occasionally in normal and abnormal muscle (Miledi and Slater, 1969), in sensory nerves of muscle spindles (Corvaja et al., 1971) and in the extraocular muscles (Davidowitz et al., 1975a, 1975b).

The accompanying micrographs illustrate some variations in these membrane-glycogen complexes in the extraocular muscles of the rabbit. At the upper left is an example of alternation of layers of glycogen beta particles with concentric or spirally arranged cisternae of the smooth sarcoplasmic reticulum. At the upper right and below are similar systems of membranes with little associated glycogen but situated adjacent to arrays of alternating membranes and layers of glycogen (at arrows). In some instances the cisternae are distended with a flocculent material of low density. These glycogen bodies occur in extraocular muscle fibers with multiple innervation, but not in those singly innervated. Their functional significance remains obscure.

Figures 184, 185, and 186. Particle-free and glycogen-bearing arrays of smooth membranes from extraocular muscles of rabbit. (Micrographs courtesy of J. Davidowitz, Fig. 185 from Am. J. Pathol. 78:191, 1975 and Proc. Electron Microscope Society of America, 1976.)

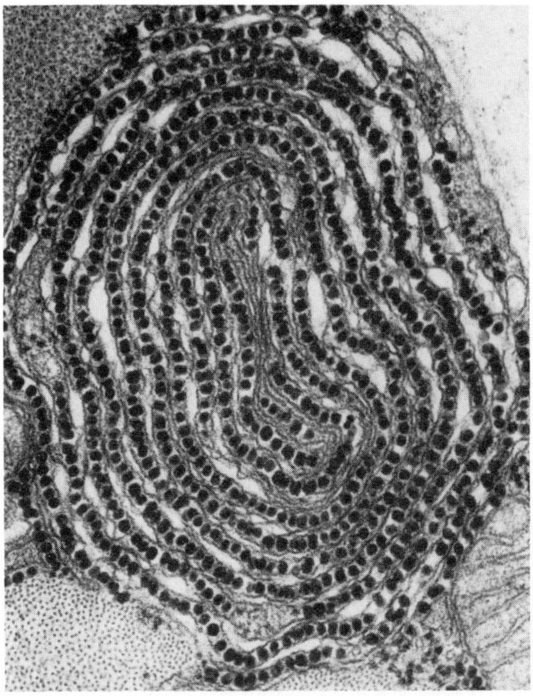

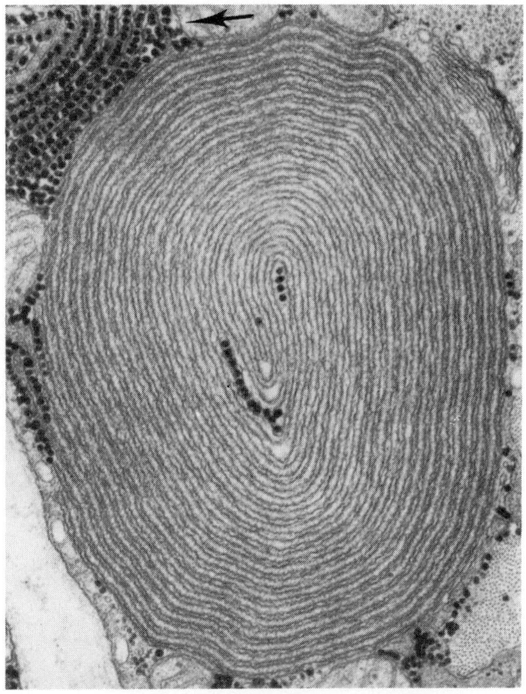

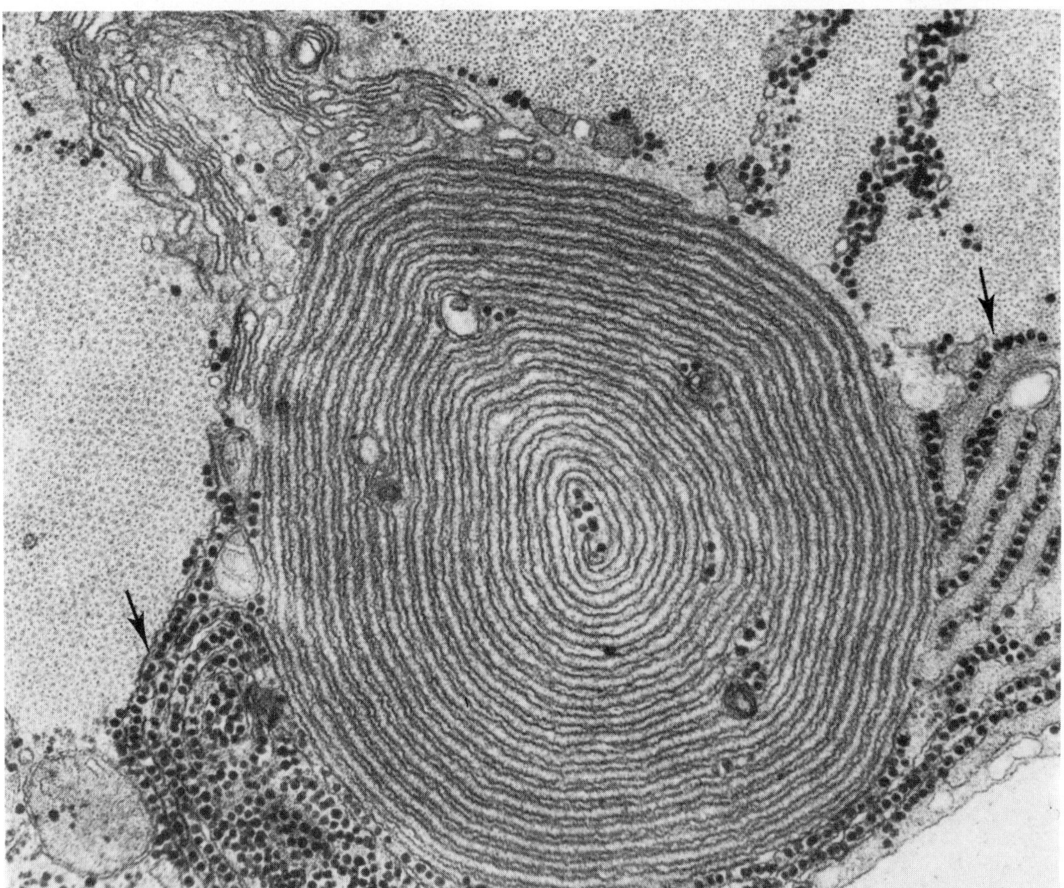

Figure 184, *upper left* **Figure 185,** *upper right* **Figure 186,** *lower*

The most prominent cytoplasmic organelle in all steroid-secreting endocrine glands is the smooth endoplasmic reticulum. The accompanying micrograph of an interstitial cell of the testis illustrates its typical appearance. It consists of an interconnected meshwork of 80 to 120 nm tubules extending throughout most of the cytoplasm.

Before two categories of endoplasmic reticulum were recognized, Bucher and McGarrahan showed that the enzymes necessary for *in vitro* synthesis of cholesterol from acetate were present in the microsome fraction of liver. Although steroid-secreting cells contain some rough endoplasmic reticulum, the microsome fraction of such cells consists predominantly of vesicular fragments of the smooth reticulum. There is evidence that the abundance of smooth reticulum in steroid-secreting cells varies with the extent to which the cell synthesizes its own cholesterol as a precursor in the biogenesis of steroids (Christensen and Gillim, 1969). With the exception of the enzymatic cleavage of the side chain of cholesterol in the mitochondria, the isomerase, hydroxylase, and hydroxysteroid dehydrogenases involved in the further steps of the biosynthesis of steroid hormones are located in the smooth endoplasmic reticulum (Christensen, 1975). The abundance of smooth endoplasmic reticulum is responsive to gonadotrophic hormones. Experimental stimulation or suppression of steroid secretion produces corresponding changes in the amount of smooth reticulum (Aoki, 1970; Neaves, 1973, 1976.)

Figure 187. Leydig cell from interstitial tissue of opossum testes (Christensen and Fawcett, J. Biophys. Biochem. Cytol. 9:653, 1961).

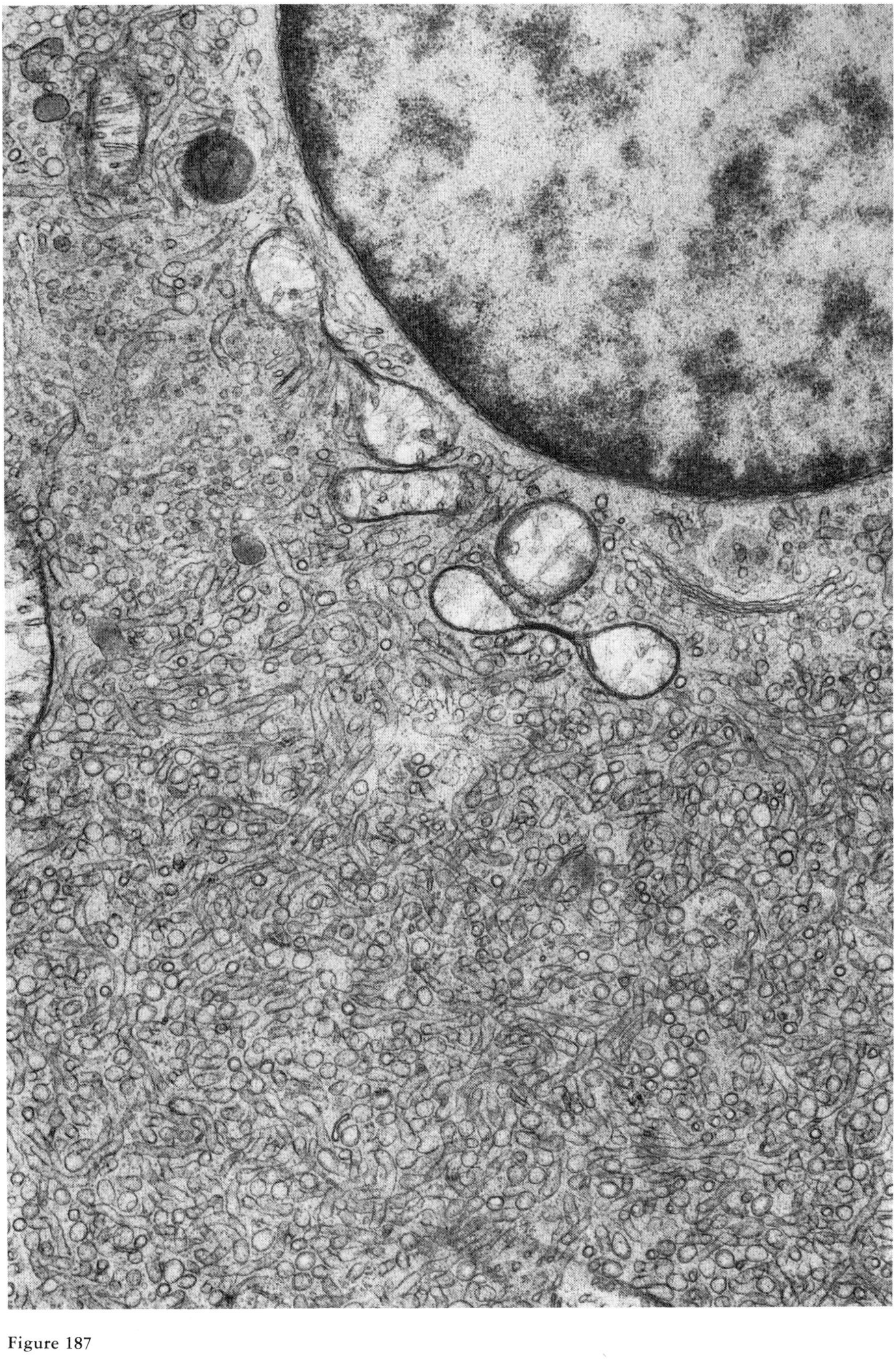

Figure 187

The general configuration of the smooth reticulum is much the same in all steroid-secreting cells. The accompanying micrograph of an adrenal cortical cell would be difficult to distinguish from a Leydig cell of the testis or a cell from the corpus luteum of the ovary.

Figure 188. Cell from the zona fasciculata of guinea pig adrenal cortex. (Micrograph courtesy of Daniel Friend.)

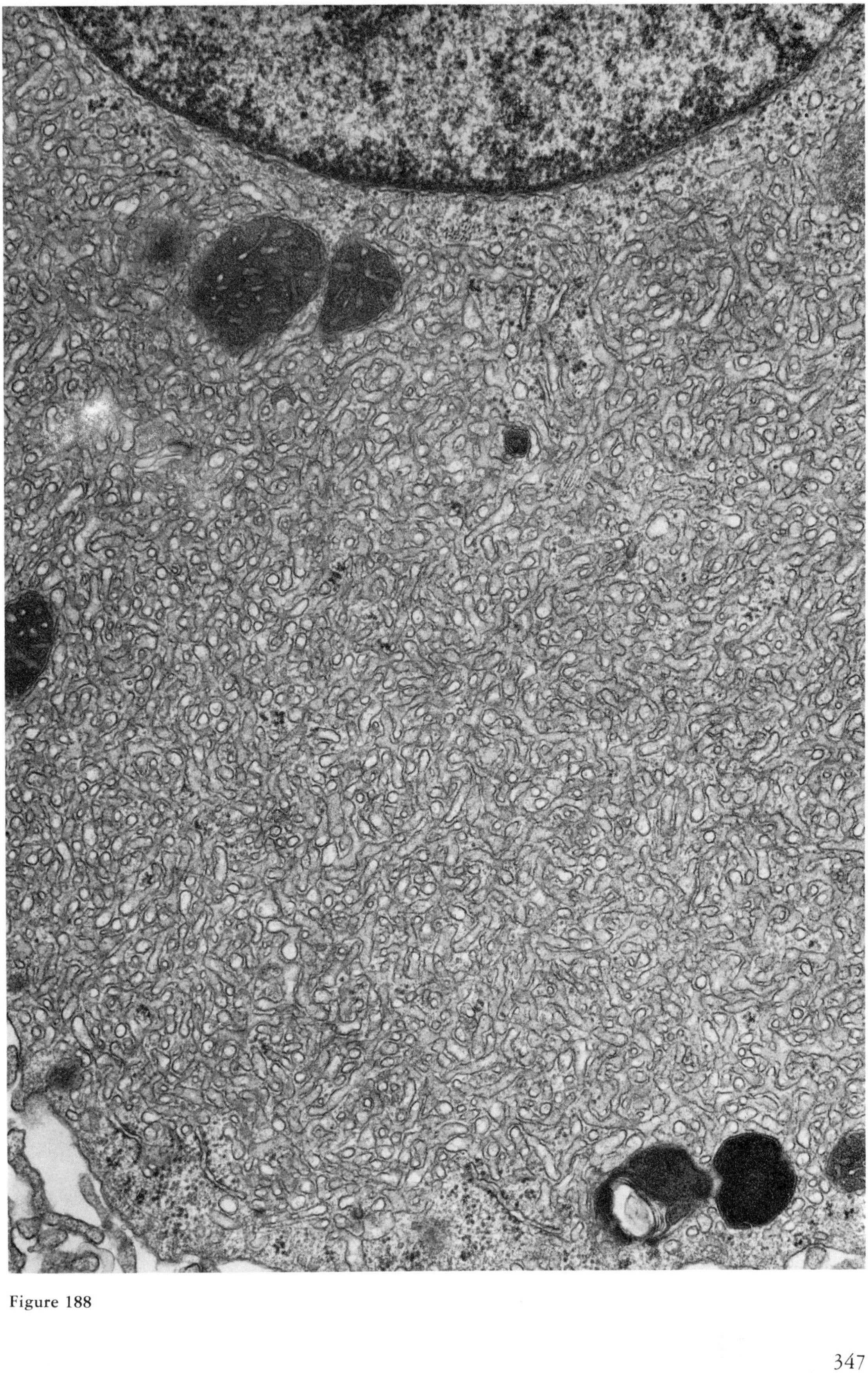

Figure 188

At high magnifications the geometry of the smooth reticulum is quite different from that of the rough reticulum. The tubules are more variable in diameter along their length and they branch and anastomose frequently to form a very compact three-dimensional reticulum. Although this is its most common configuration, it may also take the form of concentric or spiral arrays of highly fenestrated cisternae, as illustrated in the next figure.

Figure 189. Smooth endoplasmic reticulum in an adrenal cortical cell. (Micrograph courtesy of Daniel Friend.)

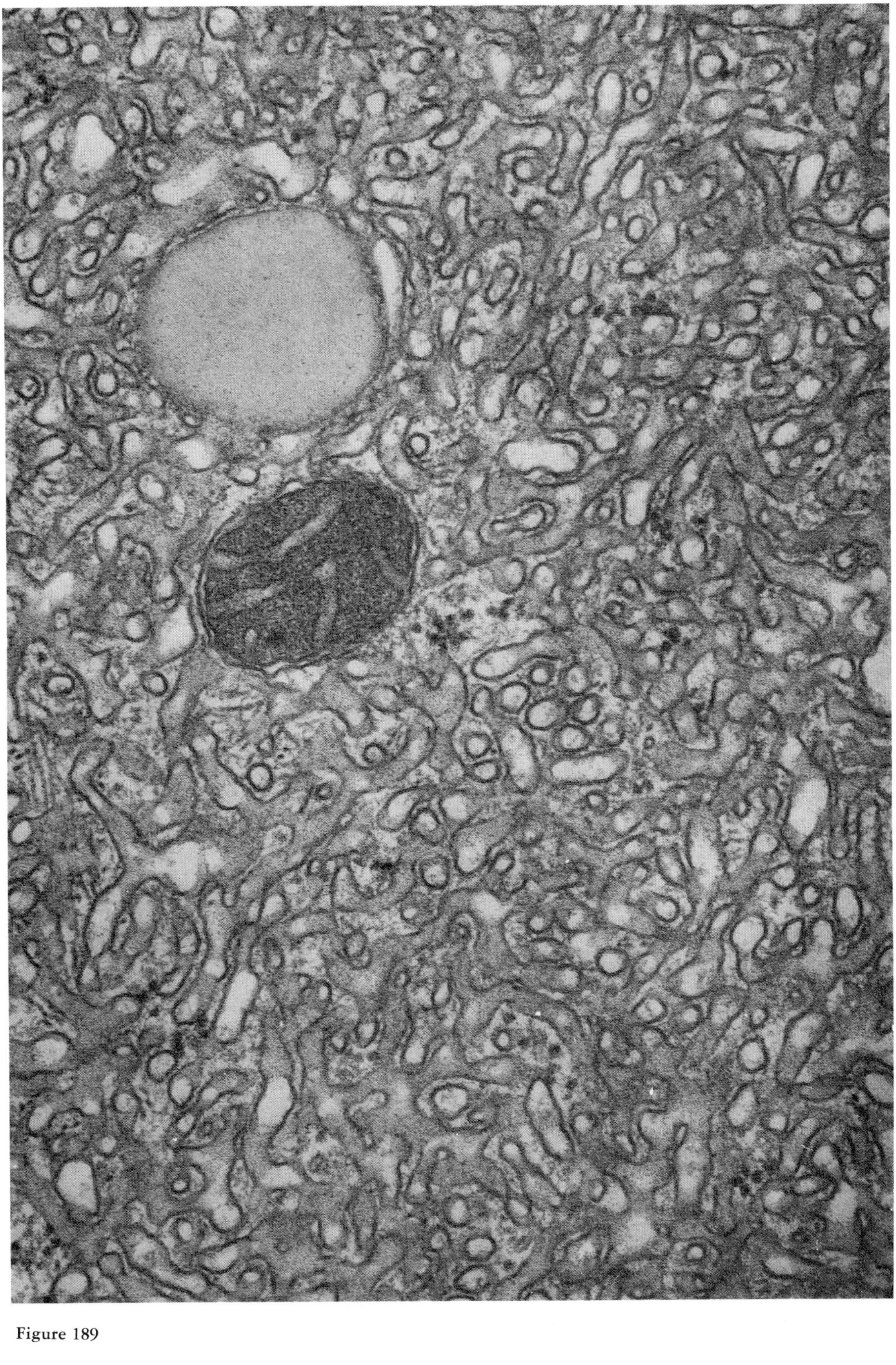

Figure 189

In steroid-secreting cells, the major part of the smooth reticulum may be in the form of extensive spiral or concentric arrays of fenestrated cisternae which in section resemble dermatoglyphic patterns. Not infrequently, the systems of cisternae are organized around lipid droplets, as in the upper part of the accompanying micrograph.

Figure 190. Fenestrated cisternae of smooth reticulum in a guinea pig Leydig cell.

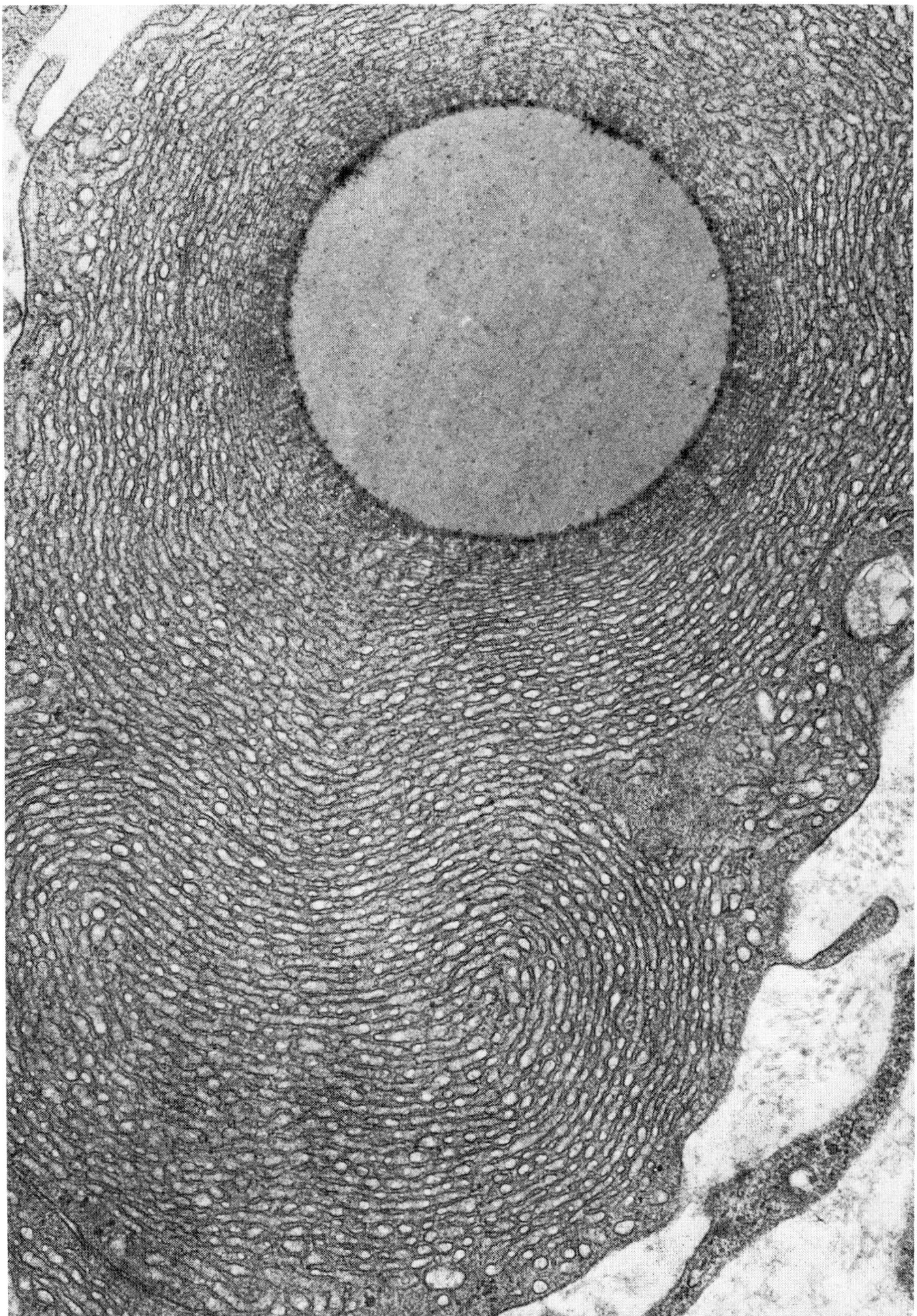

Figure 190

SARCOPLASMIC RETICULUM

The early microscopic studies of striated muscle by 19th century cytologists led to a lively controversy over widely divergent interpretations. Bowman (1840) and Rollett (1888) among others correctly concluded that muscle fibers consist of contractile fibrils embedded in a semifluid cytoplasmic matrix. However, many of their most influential contemporaries, including Cajal (1888), Carnoy (1884), Van Gehuchten (1888), and Retzius (1881), considered the fibrils to be artifacts resulting from coagulation of the protoplasm by the fixative and insisted that the contractile element was a reticulum embedded in an elastic cytoplasmic matrix. This interpretation was no doubt influenced by the prevailing view of the alveolar or reticular nature of all protoplasm (Bütschli, 1892, 1901). The universal properties of contractility and irritability of protoplasm were thought to reside in a reticulum suspended in a more or less fluid matrix. Consistent with this interpretation was the view that the muscle fiber was simply a cell in which the longitudinal and transverse elements of the protoplasmic reticulum were more highly ordered and arranged so as to result in longitudinal contraction (Carnoy, 1884). This interpretation rested heavily upon results of gold impregnation methods that showed a relatively coarse reticulum in muscle fibers. Rollett (1888) correctly insisted that these methods merely impregnated the entire system of sarcoplasmic septa between unstained myofibrils and therefore delineated a honeycomb-like pattern that was imposed upon the sarcoplasm by the disposition of contractile fibrils embedded within it. Although he rejected the idea that this coarse reticulum was the contractile component of muscle, he conceded that the sarcoplasm might ultimately prove to have finer structural differentiations not revealed by the gold impregnation methods of his contemporaries.

This was borne out in a remarkable comparative study of muscle by Veratti (1902). Applying the so-called black reaction of Golgi, he described in a wide variety of species a delicate reticulum throughout the sarcoplasm, consisting of mutually anastomosing strands with small enlargements at nodal points. The reticulum surrounded all of the myofibrils and its transverse elements appeared to occupy a fixed position relative to the pattern of striation in the muscle fiber. Surprisingly, Veratti's *reticular apparatus of the sarcoplasm* went virtually unnoticed for a half century.

Soon after the development of adequate preparative procedures for biological electron microscopy, Bennett and Porter (1953) rediscovered a reticular structure in the sarcoplasm distributed in a repeating pattern related to the sarcomeres of the myofibrils. It was composed of anastomosing tubules resembling those of the recently described endoplasmic reticulum but without associated ribonucleoprotein particles. These formed lace-like sleeves of tubules around the myofibrils with prominent transverse elements consistently associated with particular bands in the myofibrils. The finding of a system of channels throughout the muscle fiber suggested that this *sarcoplasmic reticulum* might distribute essential metabolites to the contractile elements. Furthermore, the occurrence of a repeating pattern of transverse specializations in relation to specific bands in each sarcomere and their periodic extension from peripheral fibrils to the sarcolemma suggested the possibility that the limiting membrane of the reticulum might conduct excitatory stimuli from the cell surface to responsive regions of the myofibrils (Porter and Palade, 1961). This speculation gained additional credence from the observation that the system was extremely well developed in fast-acting muscles (Peachey and Porter, 1959; Fawcett and Revel, 1961).

In ingenious physiological experiments, Huxley and Taylor (1955) had shown that when a microelectrode was applied at various points along the length of isolated muscle fibers, the signal for contraction was transmitted into the interior of the fiber at certain levels of the banding pattern and not at others. In frog muscle, the sensitive points were at the Z-line and in reptilian muscle at the A-I junction. Of special interest in relation to these experiments was the morphological observation that the longitudinal *sarcotubules* of the reticulum were confluent at regular intervals with a pair of parallel transverse elements of larger diameter, called *terminal cisternae,* situated on either side of a slender intermediate tubule. These triads of transverse elements were located at the Z-line in amphibian muscle and at the A-I junction in mammals, birds, and reptiles (Edwards et al., 1956; Robertson, 1956). The intermediate element of the triad appeared as a row of vesicles in the early electron microscopic studies (Porter and Palade, 1957), but with improved fixation it proved to be a continuous tubule that could be traced to the cell surface. With electron opaque probes, its lumen was shown to be open to the extracellular space (Huxley, 1964; Endo, 1964).

The coincidence in location of the triads of the sarcoplasmic reticulum with the sites on the surface of the muscle fiber where stimulation resulted in inward spread of excitation, and the demonstration of continuity of the limiting membrane of the intermediate tubule with the sarcolemma led to general acceptance of the concept that this transverse element (*T-tubule*) was involved in the coupling of excitation to contraction. How the signal that was propagated inward traversed the specialized areas of contact between the T-tubule and the terminal cisternae of the sarcoplasmic reticulum and how it triggered contraction was not immediately apparent. Nor was the mechanism of muscle relaxation.

A particulate centrifugal fraction prepared from muscle had been shown by Ebashi (1958) to have adenosine triphosphatase activity and to be involved in muscle relaxation. This fraction was capable of ATP-dependent thousand-fold concentration of calcium (Hasselbach and Makinose, 1961). When examined in electron micrographs, this fraction was found to consist of vesicular fragments of the sarcoplasmic reticulum (Ebashi and Lipmann, 1962). This convergence of biochemical and morphological observations thus led to the current concept that depolarization of the muscle cell surface spreads inward along the T-tubules and current flowing across the serrated junctions from the T-tubule to the terminal cisternae of the reticulum results in release of calcium, which triggers muscle contraction. Calcium is then sequestered by the sarcotubules of the reticulum, causing relaxation and reaccumulation of calcium in the terminal cisternae (Hasselbach, 1964; Constantin et al., 1965).

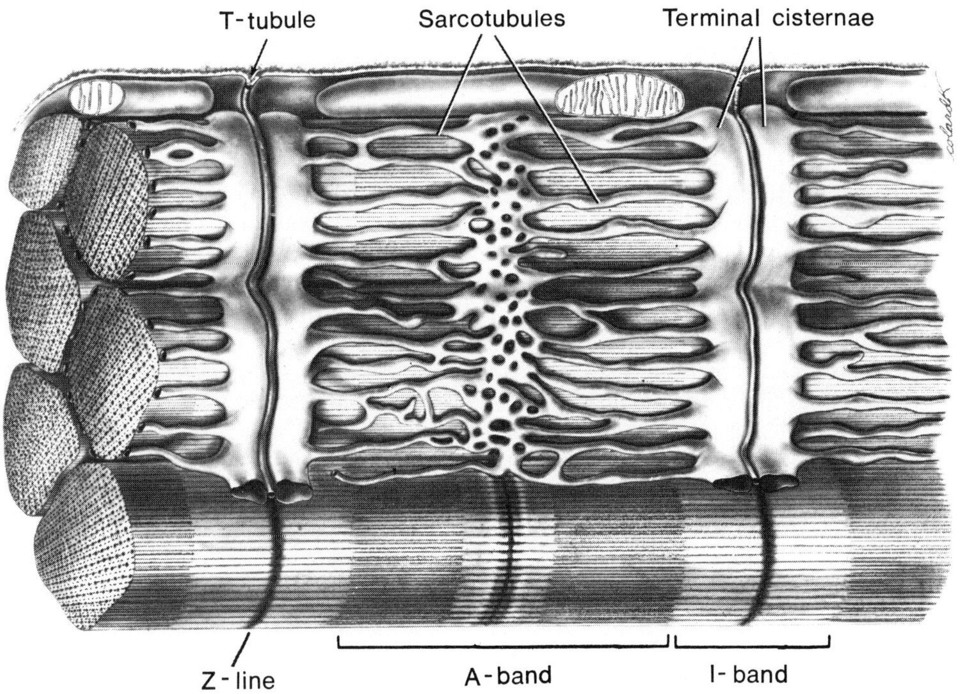

Drawing of the distribution of the sarcoplasmic reticulum around myofibrils of amphibian skeletal muscle. In mammalian muscle there are two triads in each sarcomere located at the A-I junctions. (From D. W. Fawcett and S. M. McNutt, J. Cell Biol. 42:1, 1969.)

The organization of the sarcoplasmic reticulum is somewhat simpler in amphibian than in mammalian muscle. The accompanying micrograph of frog sartorious muscle illustrates well the differentiations of the system in relation to a single sarcomere. Seen in the vertical strip at the center of the figure are the components of the reticulum included in the thin layer of sarcoplasm between two myofibrils. On the left side, the section passes through a myofibril, by reference to which the specializations of the reticulum can be related to specific bands. The myofibril on the right is slightly out of register for technical reasons.

The triads of the reticulum at the top and bottom of the figure are seen to be opposite the Z-lines. Over the middle of the A-band numerous lateral anastomoses between longitudinal sarcotubules form a fenestrated collar around the middle of the sarcomere. The reticulum passes out of the plane of the thin section at either end of the A band so that the continuity of the sarcotubules with the terminal cisternae is not shown. The dense granules are particles of glycogen in the sarcoplasm between the tubules of the reticulum.

Figure 191. A thin section of frog sartorius. (Micrograph courtesy of Lee Peachey.)

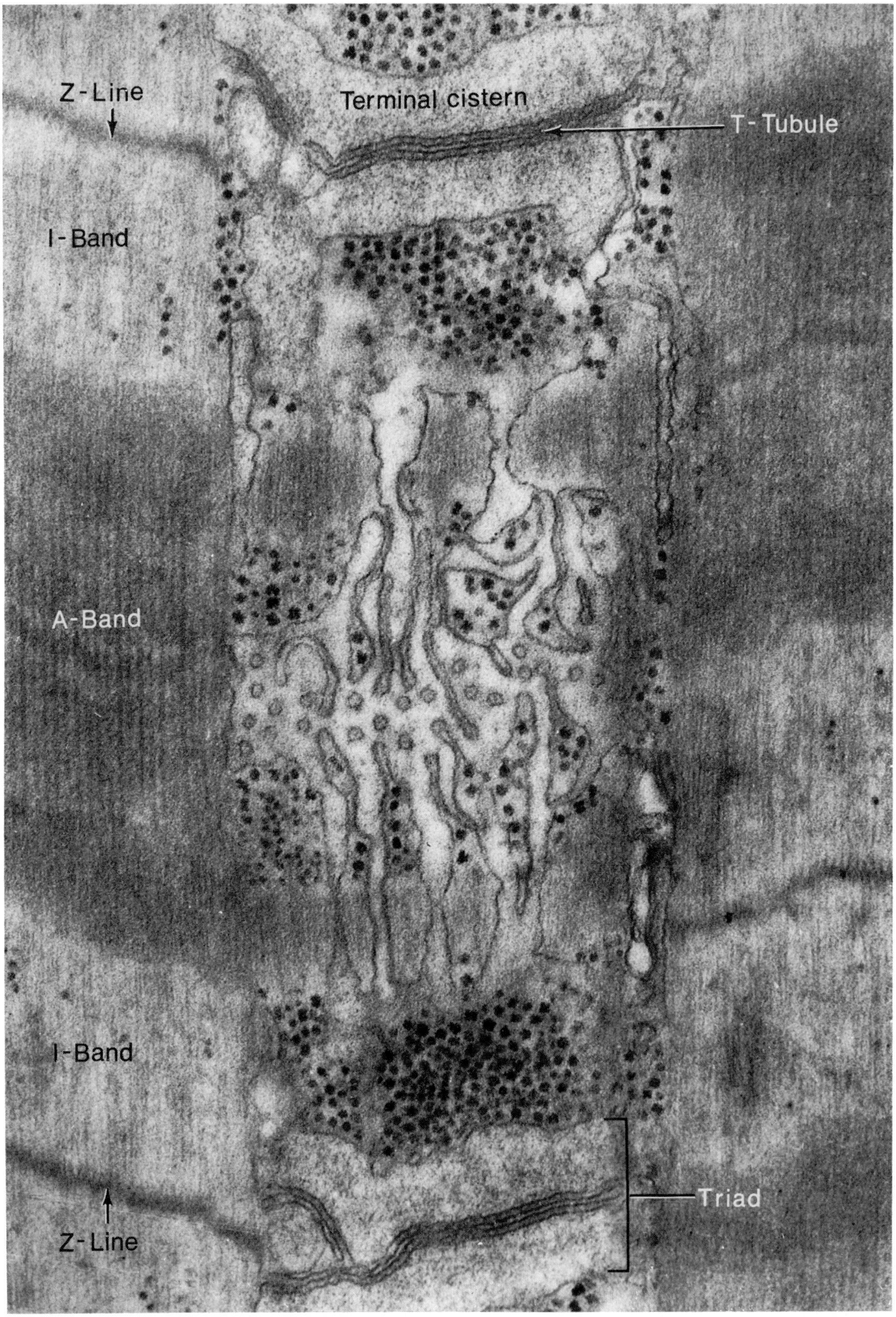
Z-Line
Terminal cistern
T-Tubule
I-Band
A-Band
I-Band
Triad
Z-Line

Figure 191

The highly ordered arrangement of the sarcoplasmic reticulum is clearly seen in a fast-acting fish muscle. In the accompanying low-power micrograph, the intermyofibrillar clefts are filled with membrane-limited canaliculi of an extensive reticulum in which the terminal cisternae and T-tubules are located at the A-I junctions.

Rapid contractions of this intrinsic striated muscle in the wall of the gas-filled swim bladder set up vibrations that produce the sounds emitted by toadfish, croakers, and drums. In this unusually fast-acting muscle, peak contraction is attained in 5 to 8 milliseconds and relaxation is complete 5 to 7 milliseconds later. It requires a stimulation frequency as high as 300 cycles per second to tetanize (Skoglund, 1959).

Figure 192. Muscle of the swim bladder of the toadfish, *Opsanus tau*.

Figure 192

In most striated muscles, a single layer of sarcotubules is interposed between adjacent myofibrils. In the fast-acting fish muscle illustrated here, multiple layers of sarcotubules converge upon the terminal cisternae. With the method of preparation used, the banding pattern of the myofibrils is not prominently displayed. However, with reference to the Z-lines that are visible, it can be seen that there are two triads to each sarcomere and these are located at the A-I junctions.

The lower figure shows a one-sarcomere length of sarcoplasmic reticulum in three intermyofibrillar clefts. The slender T-tubules and adjacent terminal cisternae forming the triads are seen in transverse section. The multiple layers of longitudinal sarcotubules are evident. This extraordinary development of the reticulum is no doubt related to the very rapid contraction and relaxation cycle of this muscle.

Figures 193 and 194. Intrinsic muscle from the swim bladder of the toadfish, *Opsanus tau*. (From Fawcett and Revel. J. Cell Biol. 10 (Suppl) 89–109, 1961.)

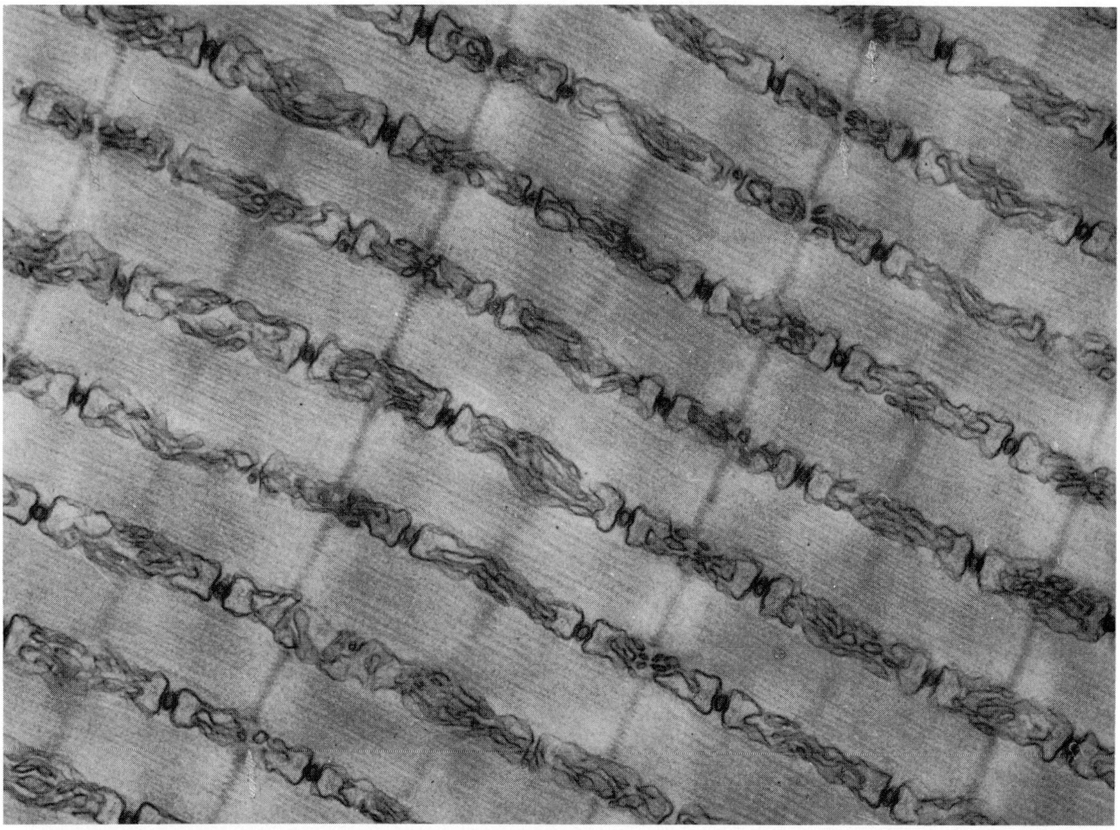

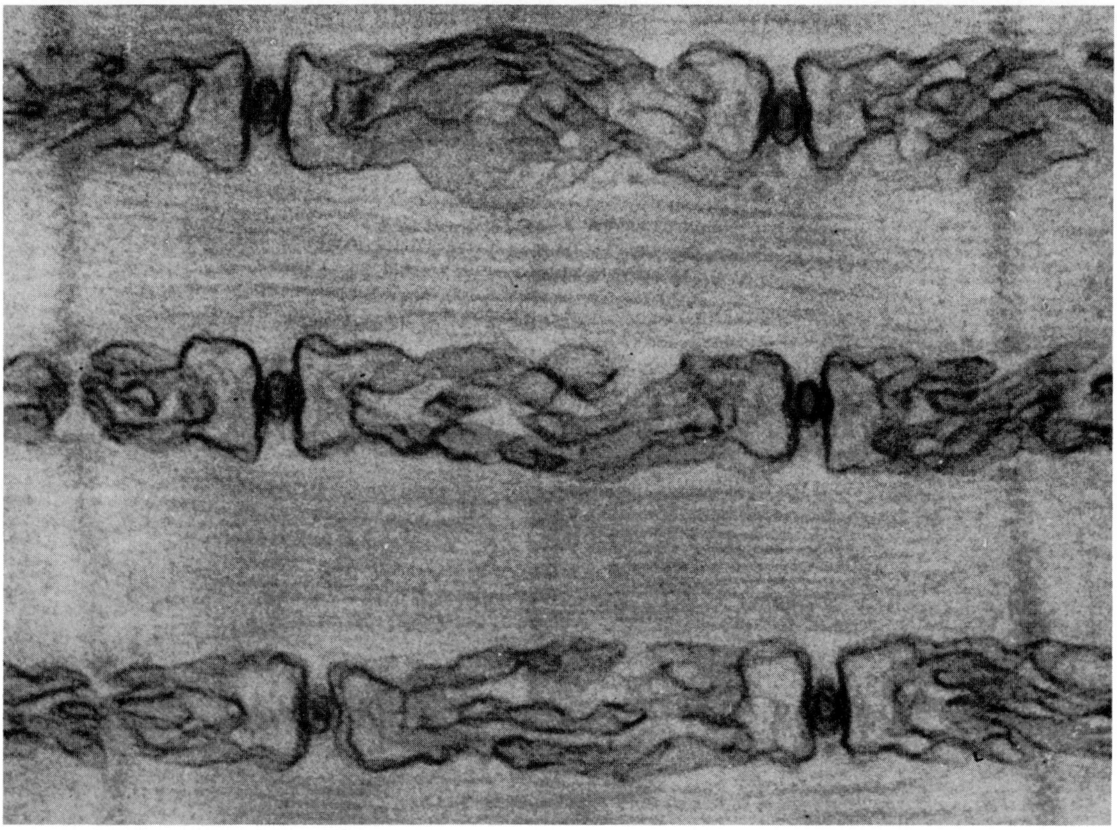

Figure 193, *upper* **Figure 194,** *lower*

Study of the disposition of the sarcoplasmic reticulum and the system of associated tubular invaginations of the sarcolemma has been facilitated by the development of a method for their selective staining. After fixation in a glutaraldehyde solution containing calcium, postfixation in osmium-ferrocyanide solution results in staining of the T-tubules and sarcotubules of the reticulum (Forbes et al., 1977). The exact mechanism of staining is poorly understood, but it appears to have some features in common with the black reaction of Golgi (based upon successive actions of osmium-bichromate mixture and a solution of silver nitrate). This enabled Veratti to visualize the reticular apparatus of the sarcoplasm with the light microscope 75 years earlier.

The accompanying micrograph shows the results of applying this method to mammalian skeletal muscle. The segments of T-tubules included in the section are deeply stained, while the adjacent terminal cisternae and longitudinal sarcotubules are less intensely stained. The triads are at the A-I junctions. Continuity of the reticulum over the I-band is obscured by mitochondria preferentially located on either side of the Z-line.

Figure 195. Micrograph of the tibialis anterior of the mouse. (Micrograph courtesy of Michael Forbes.)

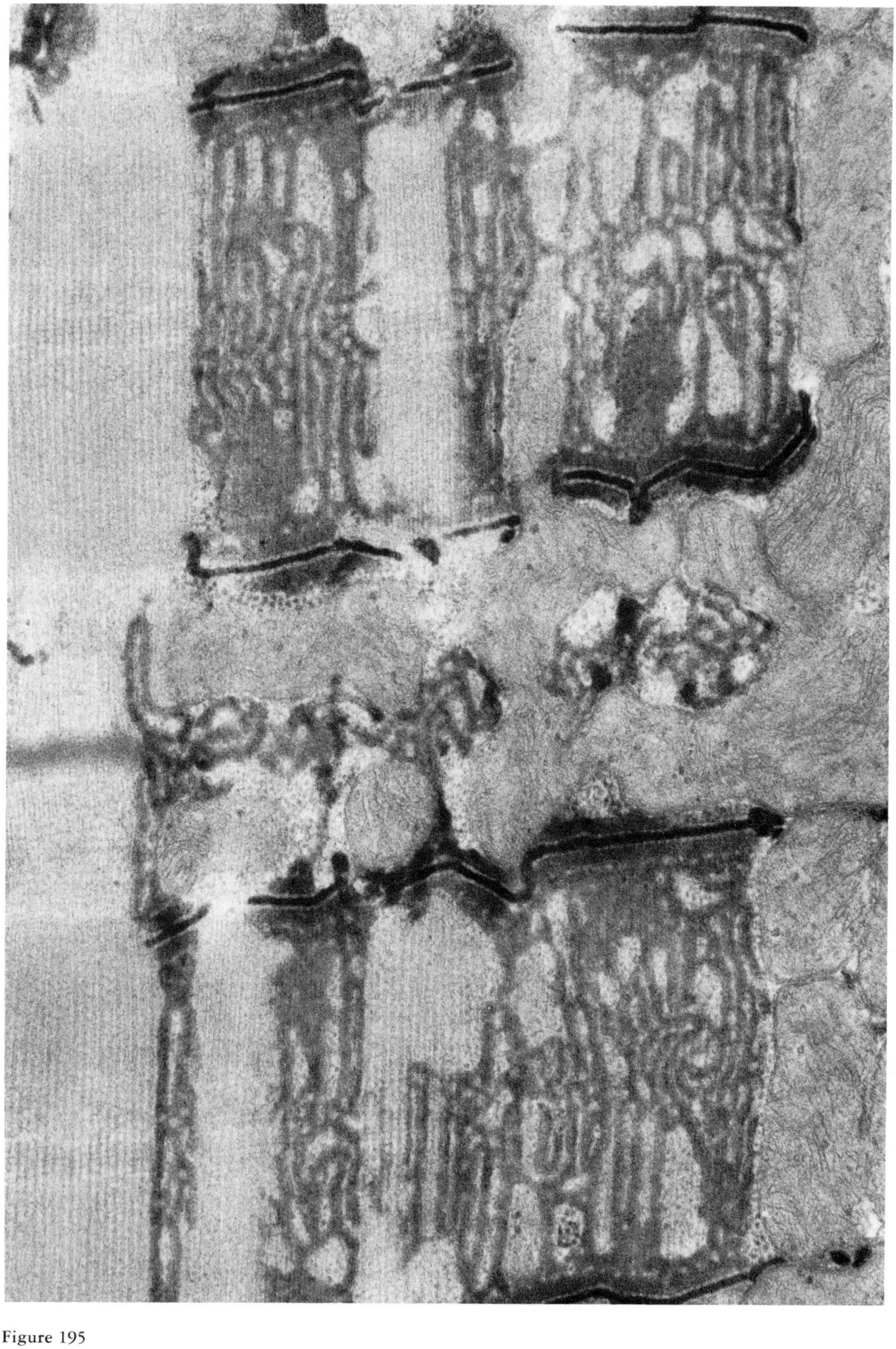

Figure 195

When the same method is applied after fixation in a glutaraldehyde solution lacking calcium or containing phosphate, it results in staining of the extracellular space and T-tubules but not of the sarcoplasmic reticulum. In the accompanying micrograph, two T-tubules are seen in each sarcomere at the junctions of the A- and I-bands. The sarcoplasmic reticulum is unstained.

Figure 196. Micrograph of tibialis anterior of the mouse. (Micrograph courtesy of Michael Forbes.)

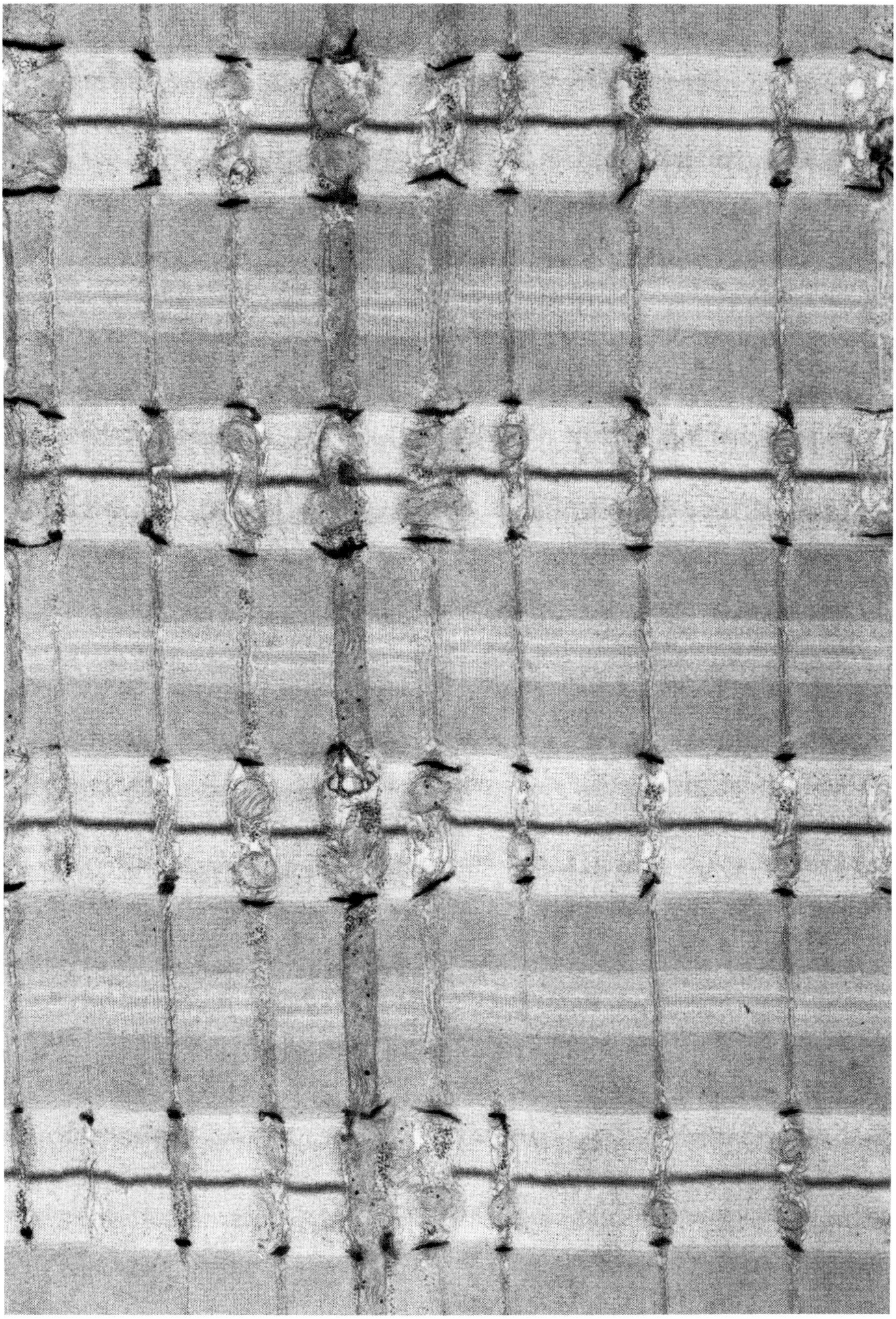

Figure 196

REFERENCES

Rough Endoplasmic Reticulum

Bensley, R. R. Studies on the pancreas of the guinea pig. Am. J. Anat. 12:297–389, 1911.

Bernhard, W., A. Gautier and Ch. Oberling. Eléments fibrailaires de nature probablement ergastoplasmique dans le cytoplasme de la cellule hépatique révélés au microscope électronique. Compt. Rendu. Soc. Biol. 145:566–571, 1951.

Bernhard, W., F. Haguenau, A. Gautier and Ch. Oberling. 1952. La structure submicroscopique des éléments basophiles cytoplasmiques dans le foie, le pancréas et les glandes salivaires. (Étude de coupes ultrafines au microscope electronique.) Z. Zellforsch. 37:281, 1952.

Blobel, G. and D. D. Sabatini. Controlled proteolysis of nascent peptides in rat liver cell fractions. I. Location of polypeptides within ribosomes. J. Cell Biol. 48:130–145, 1970.

Blobel, G. and B. Dobberstein. Transfer of proteins across membranes. I. Presence of proteolytically processed and unprocessed nascent immunoglobulin light chains on membrane-bound ribosomes of murine myeloma. J. Cell Biol. 67:835–851, 1975.

Blobel, G. and B. Dobberstein. Transfer of proteins across membranes. II. Reconstitution of functional rough microsomes from heterologous components. J. Cell Biol. 67:852–862, 1975.

Brachet, J. Biochemical Cytology. Academic Press, New York, 1957.

Caspersson, T. Cell Growth and Cell Function. Norton, New York, 1950.

Claude, A. The constitution of protoplasm. Science 97:451–456, 1943.

Dalton, A. J., H. Kohler, M. J. Striebick and B. Lloyd. Finer structure of hepatic, intestinal and renal cells of the mouse as revealed by the electron microscope. J. Nat. Cancer Inst. 11:439–461, 1950.

Garnier, Ch. Contribution a l'étude de la structure et du functionnement des cellule glandulaires séreuses. Du rôle de l'ergastroplasme dans la secretion. Thesis, Nancy, No. 50, 1899.

Hoagland, M. B., M. L. Stephenson, J. F. Scott, L. Hecht and P. Zamecnik. A soluble ribonucleic acid intermediate in protein synthesis. J. Biol. Chem. 231:241–258, 1958.

Holley, R. W. The nucleic acid sequences of a nucleic acid. Sci. Am. 214:30–39, 1966.

Kreibich, G., B. L. Ulrich and D. D. Sabatini. Proteins of rough microsomal membranes related to ribosome binding. I. Identification of ribophorin I and II. Membrane proteins characteristic of rough microsomes. J. Cell Biol. 77:464–487, 1978.

Littlefield, J. W., E. Keller, J. Gross and P. C. Zamecnik. Studies on cytoplasmic ribonucleoprotein particles from the liver of the rat. J. Biol. Chem. 217:111–123, 1955.

Mathews, A. The changes in structure of the pancreas cell. A consideration of some aspects of cell metabolism. J. Morphol. (Suppl.) 15:171–222, 1899.

Nirenberg, M. W. and J. H. Matthai. The dependence of cell-free protein synthesis in *E. coli* upon naturally occurring or synthetic polyribonucleotides. Proc. Soc. Acad. Sci. 47:1588–1602, 1961.

Nirenberg, M. and P. Leder. RNA codewords and protein synthesis. The effect of trinucleotides upon the binding of sRNA to ribosomes. Science 145:1399–1407, 1964.

Palade, G. E. A small particulate component of the cytoplasm. J. Biophys. Biochem. Cytol. 1:59–68, 1955.

Palade, G. E. and K. R. Porter. Studies on the endoplasmic reticulum. I. Its identification in cells in situ. J. Exp. Med. 100:641–656, 1954.

Porter, K. R. Observations on a submicroscopic basophilic component of the cytoplasm. J. Exp. Med. 97:727–750, 1953.

Porter, K. R., A. Claude and E. Fullam. A study of tissue culture cells by electron microscopy. Methods and preliminary observations. J. Exp. Med. 81:233–241, 1945.

Porter, K. R. and H. P. Thompson. Some morphological features of cultured rat sarcoma cells as revealed by the electron microscope. Cancer Res. 7:431–438, 1947.

Sabatini, D. D. and G. Blobel. Controlled proteolysis of nascent polypeptides in rat liver cell fractions. II. Localization of the polypeptides in rough microsomes. J. Cell Biol. 45:146–157, 1970.

Warner, J. R., P. M. Knopf and A. Rich. A multiple ribosomal structure in protein synthesis. Proc. Nat. Acad. Sci. 49:122–129, 1963.

Zachaus, H. G., D. Dutting, H. Feldmann, F. Melchers and W. Karau. Serine specific transfer RNAs. XIV. Comparison of nucleotide sequences and secondary structure models. Cold Spring Harbor Symposium on Quantitative Biology 31:417–424, 1966.

Zamecnik, P. C. and E. B. Keller. Relation between phosphate energy donors and incorporation of labeled amino acids into proteins. J. Biol. Chem. 209:337–354, 1954.

Smooth Endoplasmic Reticulum

Aoki, A. Hormonal control of Leydig cell differentiation. Protoplasma 71:209–225, 1970.

Bucher, N. L. R. and K. McGarrahan. The biosynthesis of cholesterol from acetate–1–C^{14} by cellular fractions of rat liver. J. Biol. Chem. 222:1, 1956.

Cardell, R. R. S. Badenhausen and K. R. Porter. Intestinal triglyceride absorption in the rat. An electron microscopical study. J. Cell Biol. 34:123–255, 1967.
Christensen, A. K. and D. W. Fawcett. The fine structure of the interstitial tissue of opossum testis. J. Biophys. Biochem. Cytol. 9:653–670, 1961.
Christensen, A. K. and G. W. Gillim. Correlation of fine structure and function in steroid secreting cells with emphasis on those of the gonads. In The Gonads (K. W. McKerns, ed.), pp. 415–488, Appleton-Century-Crofts, New York, 1969.
Christensen, A. K. Leydig cells. In Handbook of Physiology, Section 7, Endocrinology, Vol. V, Male Reproductive System, pp. 57–94, (D. W. Hamilton and R. O. Greep, eds.), American Physiological Society, Washington, 1975.
Corvaja, N., P. C. Magherini and O. Pompeiano. Ultrastructure of glycogen membrane complexes in sensory nerve fibers of cat muscle spindles. Zeitsch. Zellforsch. Mikr. Anat. 121:199–217, 1971.
Dallner, G. and L. Ernster. Subfractionation and composition of microsomal membranes. A review. J. Histochem. Cytochem. 16:611–632, 1968.
De Pierre, J. W. and G. Dallner. Structural aspects of the membrane of the endoplasmic reticulum. Biochim. Biophys. Acta 415:411–472, 1975.
Davidowitz, J., G. H. Philips, B. R. Pachter and G. M. Breinin. Particle free and glycogen bearing double membrane arrays in extraocular muscles of rabbit. Am. J. Pathol. 78:191–195, 1975a.
Davidowitz, J., G. H. Philips, B. P. Pachter and G. M. Breinin. Cisternal distension in membrane-glycogen complexes of rabbit extraocular muscle. J. Ultrastr. Res. 51:301–313, 1975b.
Enders, A. C. Observations on the fine structure of lutein cells. J. Cell Biol. 12:101, 1962.
Fawcett, D. W. Observations on the cytology and electron microscopy of hepatic cells. J. Nat. Cancer Inst. 15:1475, 1955.
Jones, A. L. and D. W. Fawcett. Hypertrophy of the agranular endoplasmic reticulum in hamster liver induced by phenobarbital. (With a review on the functions of this organelle in liver.) J. Histochem. Cytochem. 14:251–232, 1966.
Jones, A. L., N. B. Ruderman and M. G. Herrera. Electron microscopic and biochemical study of lipoprotein synthesis in the isolated perfused rat liver. J. Lipid Res. 8:429, 1967.
Luck, D. J. L. Glycogen synthesis from uridine diphosphate glucose. J. Biophys. Biochem. Cytol. 10:195, 1961.
Miledi, R. and C. R. Slater. Electron microscopic structure of denervated skeletal muscle. Proc. Roy. Soc. Lond. B 174:253–269, 1969.
Neaves, W. B. Changes in testicular Leydig cells and in plasma testosterone levels among seasonally breeding rock Hyrax. Biol. Reprod. 8:451–466, 1973.
Palay, S. L. and G. E. Palade. The fine structure of neurons. J. Biophys. Biochem. Cytol. 1:69–88, 1954.
Palay, S. L. The morphology of secretion. In Frontiers in Cytology (S. L. Palay, ed.), pp. 305–342, Yale University Press, New Haven, 1958.
Palay, S. L. and L. J. Karlin. An electron microscopic study of the intestinal villus. II. The pathway of fat absorption. J. Biophys. Biochem. Cytol. 5:373, 1959.
Porter, K. R. and G. E. Palade. Studies on the endoplasmic reticulum. III. Its form and distribution in striated muscle cells. J. Biophys. Biochem. Cytol. 3:269, 1957.
Porter, K. R. and C. Bruni. An electron microscopic study of the early effects of 3′–Me–DAB on rat liver cells. Cancer Res. 19:997, 1959.
Porter, K. R. and E. Yamada. Studies on the endoplasmic reticulum. V. Its form and differentiation in pigment epithelial cells of the frog retina. J. Biophys. Biochem. Cytol. 8:161–205, 1960.
Remmer, H. and J. J. Merker. Drug induced changes in the liver endoplasmic reticulum: Association with drug metabolizing enzymes. Science 142:1657, 1963.
Ross, M. H., G. Pappas and J. Lanman. Electron microscope observations on the endoplasmic reticulum in the human fetal adrenal. J. Biophys. Biochem. Cytol. 4:659, 1958.
Rothschild, J. Isolation of microsomal membranes. Biochem. Soc. Symp. 22:4–28, 1962.
Senior, J. R. Intestinal absorption of fats. J. Lipid Res. 4:495, 1964.
Yamada, E. The fine structure of the parabaloid in the turtle retina as revealed by electron microscopy. Anat. Rec. 137:172, 1960.
Yamada, E. and T. M. Ichikawa. The fine structure of the corpus luteum in the mouse ovary as revealed by electron microscopy. Kyushu J. Med. Sci. 11:235, 1960.

Sarcoplasmic Reticulum

Bennett, H. S. and K. R. Porter. An electron microscope study of sectioned heart muscle of the domestic fowl. Am. J. Anat. 93:61–105, 1953.
Bowmann, W. On the minute structure and movements of voluntary muscle. Phil. Trans. Roy. Soc. Lond. 130:457–501, 1840.
Bütschli, O. Untersuchungen über mickorskopische Schäume und das Protoplasma. Leipzig, 1892. Trans. E. A. Minchin, 1894.
Bütschli, O. Meine Ansicht über die Struktur des Protoplasmas und einige ihrer Kritiker. Arch. Entwicklungsmechanik 11, 1901.

Cajal, S. R. Observations sur la texture des fibres musculaires des pattes et des ailes des insectes. Internat. Monatschr. Anat. Physiol. 5:205–232, 253–275, 1888.

Carnoy, J. B. La biologie cellulaire, étude comparée de la cellule dans les deux règnes. Lierre, Von In., 1884.

Constantin, L. L., C. Franzini-Armstrong, R. J. Podolsky. Localization of calcium accumulating structures in striated muscle fibers. Science, 147:158, 1965.

Ebashi, S. A granule-bound relaxation factor in skeletal muscle. Arch. Biochem. Biophys. 76:410–423, 1958.

Ebashi, S., F. Takeda, M. Otsuke and H. Kumagai. Enzymic nature of relaxing factors in skeletal muscle. Symp. on Enzyme Chem. (Japan) 11:11, 1956.

Ebashi, S. and F. Lipmann. Adenosine triphosphate-linked concentration of calcium ions in a particulate fraction of rabbit muscle. J. Cell Biol. 14:389–400, 1962.

Edwards, G. A., H. Ruska, P. de Souza-Santos and A. Vallejo-Freire. Comparative cytophysiology of striated muscle with special reference to the role of the endoplasmic reticulum. J. Biophys. Biochem. Cytol. 2(Suppl).:143, 1956.

Endo, M. Entry of a dye into the sarcotubular system of muscle. Nature 202:1115, 1964.

Fawcett, D. W. and J. P. Revel. The sarcoplasmic reticulum of a fast-acting fish muscle. J. Biophys. Biochem. Cytol. 10 (Suppl):89, 1961.

Hasselbach, W. Relaxation and the sarcotubular pump system. Fed. Proc. 23:909, 1964.

Hasselbach, W. and M. Makinose. Die Calciumpumpe der Erschlaffungsgrana des Muskets und ihre Abhängigkeit von der ATP–spaltung. Biochem. Z. 333:518, 1961.

Huxley, A. F. Local activation of muscle. Ann. N.Y. Acad. Sci. 81:446, 1959.

Huxley, A. F. and R. E. Taylor. Local activation of striated muscle fibres. J. Physiol. 144:426, 1958.

Huxley, H. E. Evidence for continuity between the central elements of the triads and extracellular space in frog sartorius muscle. Nature 202:1067, 1964.

Peachey, L. D. The sarcoplasmic reticulum and transverse tubules of the frog's sartorius. J. Cell Biol. 25:209–231, 1965.

Peachey, L. D. and K. R. Porter. Intracellular impulse conduction in muscle cells. Science 129:721, 1959.

Porter, K. R. and G. E. Palade. Studies on the endoplasmic reticulum. III Its form and distribution in striated muscle cells. J. Biophys. Biochem. Cytol. 3:269, 1958.

Retzius, G. Zur Kenntniss der quergestrelften Muskelfaser. Biol. Untersuch. 1:1–26, 1881.

Revel, J. P. The sarcoplasmic reticulum of bat cricothyroid muscle. J. Cell Biol. 12:571, 1962.

Rollett, A. Uber die Flossenmuskeln des Supferdchensf(Hippocampus antiquorum) und ueber Muskel Structur in Allgemeinen. Arch. Mikr. Anat. 32:233–266, 1888.

Skoglund, C. R. Neuromuscular mechanism of sound production in Opsanus tau. Biol. Bull. 117:438, 1959.

Van Gehuchten, A. Étude sur la structure intime de la cellule musculaire striée chez les vertébrés. Cellule 4:245–318, 1888.

Verrati, E. Richerche sulle fine struttura della fibra muscolare striata. Mem. Ist. Lomb. Classe Sc. Mat. Nat. 19: 87–133, 1902 (Translation J. Biophys. Biochem. Cytol. 10:(Suppl.) 1–60, 1961.)

GOLGI APPARATUS

The Golgi apparatus is a pleomorphic organelle present in nearly all eukaryotic cells. Its principal function is in the secretory activities of the cell and it is subject to changes in position, size, and configuration in different states of physiological activity. Although its role is best understood in glandular cells, it may well have other functions unrelated to secretion.

The organelle was probably observed by LaValette St. George (1865) and by Platner in 1889 in studies on spermatogenesis. It was Golgi who developed a method that stained it intensely and made possible the demonstration of its occurrence in a wide variety of cell types. He worked at the Institute of General Pathology and Histology in Pavia under the directorship of Giulio Bizzozero at a time when biomedical studies were flourishing in Italy. Golgi developed an intense interest in the nervous system. In the course of devising a series of modifications of existing neurocytological procedures, he developed the so-called black reaction (la reazione nera), which later became known as the Golgi method. It consisted of preliminary fixation of tissues in potassium bichromate followed by immersion in silver nitrate. This resulted in a blackening of the elements of what he described as the *internal reticular apparatus* and which was later to become known as the *Golgi apparatus.*

There was little agreement as to the form of the Golgi apparatus, owing in part to its considerable variability from one cell type to another, and the distortions resulting from different methods of specimen preparation. Golgi considered it a reticulum, others interpreted it as canalicular, but Hirschler (1918) insisted that its elements were lamellar or membranous structures that appeared in optical section as half-circles or rings. Its lamellar form was supported by Nassonov (1923), Bowen (1926), and Pollister (1957). The latter, attempting to reconstruct its shape from sections in different planes, concluded that it was a ''plate-work'' perforated, branched, or convoluted in a complex manner.

In studying dividing cells, Perroncito (1910) observed that the Golgi apparatus dissociated into a number of smaller arciform structures which he called *dictyosomes.* These were scattered in the cytoplasm and were distributed to the daughter cells, where they aggregated to reconstitute a typical Golgi complex. This process of division was called *dictyokinesis.* The term *dictyosome* has persisted and is still used by some authors to describe the multiple dispersed Golgi bodies that occur in the cells of invertebrates and in the eggs and embryonic tissues of vertebrates. There is some basis for considering these as the primitive form of the organelle and as subunits of the aggregated Golgi apparatus found in most animal cells.

Nearly all structures that were proposed by light microscopists as new organelles were soon challenged as alternative forms of other organelles or artifacts of specimen preparation. The internal reticular apparatus of Golgi was no exception. From the outset, it was a subject of lively controversy. Holmgren (1902) described a system of canaliculi within the cytoplasm of certain cells which seemed to open onto the cell surface. This was designated the *trophospongium* and was assumed to play a role in the access of nutrients to the interior of the cell. For a short period early in this century, such leading cytologists as Cajal (1908) and Bensley (1910) considered it likely that the canalicular system described by Holmgren was comparable to the internal reticular apparatus reported in cells of the nervous system by Golgi.

The French school of cytologists represented by Parat and Painlevé (1924) recognized the existence of a system of cavities in the cytoplasm which they designated the *vacuome*. The cytologists thought that, under the influence of Golgi's fixative, these coalesced to form a reticulum. Baker (1944) at Oxford also rejected the concept of a reticulum and believed that many cell types contained a juxtanuclear aggregation of "liposomes" stainable with the lipid soluble stain Sudan black. He contended that these lipid droplets and other components of the cell reduced osmium or became a site of deposition of silver nitrate in Golgi's method and that the various structures impregnated could not be considered a single organelle of general occurrence in cells. Palade and Claude (1949) reported experiments which seemed to show that intracellular structures resembling the Golgi apparatus could be produced in fresh tissue by treating them with 50 per cent ethanol and then staining them with Sudan black. Similar structures could also be produced by treating tissues with conventional Golgi fixatives. They concentrated Neutral Red and blackened with osmium. It was concluded from these studies that the Golgi apparatus was a gross artifact — a myelin figure or complex of myelin figures that developed in cells during fixation and then blackened with silver or osmium during later steps of the classical methods of specimen preparation. This work, intended to dispose of the Golgi controversy, did have a profound effect. But other investigators, led by the volatile Irish cytologist Gatenby (1959), rejected these studies on tissue fragments and homogenates as just another example of "mash cytology" by what was then beginning to be called by traditionalists the "grind and find" school of cell biology.

The inability to see the Golgi apparatus in living cells had contributed to doubts about its reality, but Dalton (1952) reported that it was visible by phase-contrast microscopy in the first few minutes after removal of the tissue from the body and was recognizable in electron micrographs as osmiophilic lamellae or vacuoles. Through examination with the electron microscope of cells that had been fixed and then postosmicated as in the classical Golgi methods, the membranous lamellae which he had identified as Golgi components were found to be sites of selective deposition of granular masses of reduced osmium. The Golgi apparatus was thus reinstated as a true organelle of widespread occurrence and with a distinctive membranous structure. In the next two decades, Palade and Claude, who had once considered the Golgi apparatus to be an artifact, were to become major contributors to cell biology in general, and especially to the analysis of the function of this important organelle in the secretory process. They were appropriately recognized for this and other fundamental studies by award of the Nobel prize in 1974, as Golgi had been in 1906.

The earliest accounts of the Golgi apparatus were purely descriptive and made little effort to establish its function. Nassonov (1923) is credited with presenting the first convincing evidence for its role in secretion. In a series of studies on various glandular cells, he observed that the droplets, or granules, of secretory product first appeared in close association with the Golgi apparatus and later separated from it to accumulate in the apex of the cell. These studies on secretory cells were confirmed and extended by Bowen (1929), who concluded that other components of the cell contributed to the process but that the Golgi apparatus played an essential role in the process of "accumulation and final synthesis" of the products of secretion. This perceptive interpretation has been substantiated by ultrastructural and biochemical investigations.

When examined in electron micrographs of thin tissue sections, the "lamellae" that had been seen by Nassonov and Bowen with the light microscope proved in turn to have a lamellar substructure. At the resolution of the electron microscope, the "lamellae" are membrane-limited flattened saccules, or cisternae, closely stacked in parallel array. The membranes are smooth contoured, and the number of associated cisternae in the stack varies with the cell type and its physiological state. Two to eight are common but much larger numbers are seen, especially in invertebrates. When viewed in section, the lumen of each cistern is relatively narrow in its central portion but slightly expanded at its ends. Observed *en face*, the discoid cisternae are uninterrupted in their central region but highly fenestrated peripherally and in this outer zone may present a very regular reticular pattern. The cisternae are usually slightly curved so that the stack as a whole has a convex and a concave surface. These curved assemblages of parallel cisternae correspond to the dictyosomes observed by classical cytologists in stained preparations. In many cell types, the Golgi apparatus consists of several such stacks of cisternae arranged to form a discontinuous hollow sphere or hemisphere around the centrosome.

In glandular cells, where it has been most thoroughly studied, the Golgi apparatus is located between the apical pole of the nucleus and the lumenal surface. In free cells, the centrioles and associated Golgi apparatus occupy a shallow concavity in the nucleus. In neurons, where the organelle is very extensive, multiple stacks of cisternae are interconnected to form a continuous reticular system throughout the perikaryon.

There is evidence of a functional polarity in the organization of the Golgi apparatus. The lumen of the cisternae is usually quite narrow at the convex side of the stack and becomes wider in successive cisternae toward the concave side. Moreover, in some actively secreting cells, the density of the content of the cisternae increases progressively from the outer convex to the inner concave surface (Rambourg et al., 1969). When subjected to prolonged postosmication, the metal is deposited preferentially in the outermost cisternae (Friend and Murray, 1965). Similarly, in tissue stained by the histochemical method for thiamine pyrophosphatase, the reaction product is confined to the inner cisternae and is not found in the outer (Novikoff and Essner, 1962; Wise and Flickinger, 1970). On the other hand, the innermost cistern and associated tubules and vacuoles give a positive reaction for acid phosphatase. Based in part upon their distinctive cytochemical staining, these elements have been interpreted by Novikoff and coworkers (1971) as a functionally distinct organelle described by the acronym GERL. They consider these cisternal and tubular elements to be a specialized region of the endoplasmic reticulum in close topographical relation to the Golgi apparatus but not an integral part of it.

ROLE IN SECRETION

The interaction of the endoplasmic reticulum and the Golgi apparatus in the elaboration of secretory products remains controversial. It is agreed that in glandular cells, protein synthesis takes place on ribosomes associated with the endoplasmic reticulum. The product is segregated in the lumen of the reticulum and transported through this canalicular system to the Golgi region. Cisternae of the reticulum that are closely associated with the outer aspect of the Golgi apparatus are usually devoid of ribosomes on the side facing this organelle. Numerous small evaginations of this ribosome-free surface bud off to form *transport vesicles* that carry quanta of the protein-rich product. There is less agreement as to the destination of these intermediary vesicles. According to the simplest interpretation of the morphological evidence, these fuse with the outermost Golgi cistern, thus transporting the secretory product from the endoplasmic reticulum to the Golgi apparatus. Then the product is said to move through the stack of cisternae toward the concave inner surface, undergoing progressive chemical modification. The product accumulates in expansions of the innermost cistern, which round up and separate off as sizable *condensing vacuoles*. By progressive concentration of their content, these become transformed into *secretory granules* that move into the apical cytoplasm and ultimately discharge by fusion of their Golgi-derived membrane with the plasmalemma.

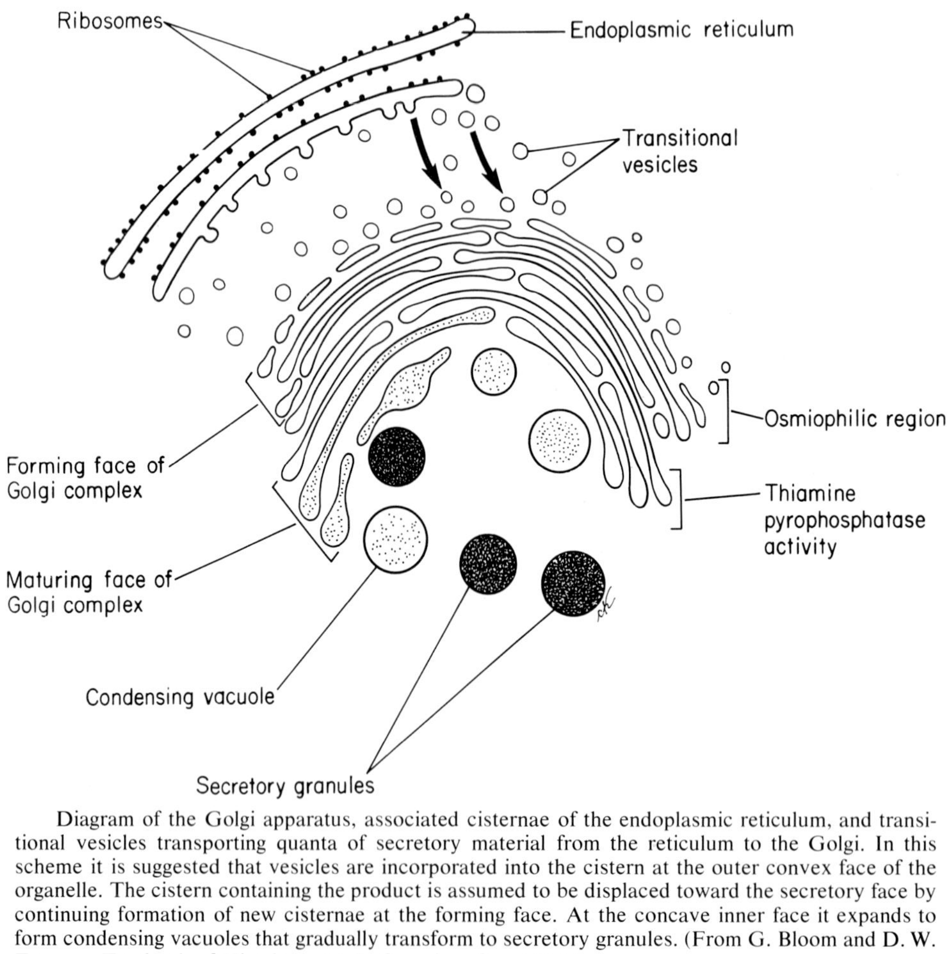

Diagram of the Golgi apparatus, associated cisternae of the endoplasmic reticulum, and transitional vesicles transporting quanta of secretory material from the reticulum to the Golgi. In this scheme it is suggested that vesicles are incorporated into the cistern at the outer convex face of the organelle. The cistern containing the product is assumed to be displaced toward the secretory face by continuing formation of new cisternae at the forming face. At the concave inner face it expands to form condensing vacuoles that gradually transform to secretory granules. (From G. Bloom and D. W. Fawcett, Textbook of Histology. W. B. Saunders Co., 1975.)

This simplistic interpretation has a number of shortcomings, not the least of which is its failure to explain how the secretory product moves through a static stack of cisternae in the absence of any visible connections between them. It is also evident that formation of condensing vacuoles at the inner face without a mechanism for replacement would soon result in disappearance of the organelle. A more dynamic view of the Golgi apparatus has evolved which assumes that new cisternae are continuously formed at the convex face by coalescence of the transport vesicles from the reticulum. This cistern then moves through the stack as new cisternae are formed above it. During this passage there is believed to be a progressive modification in the properties of its membrane and in the composition and concentration of its content. When it reaches the inner aspect of the stack, its content has attained its definitive composition and its membrane has acquired the properties necessary to permit fusion of the secretory granule with the plasmalemma. The outer convex aspect of the Golgi stack is therefore commonly called the *forming face* and the inner concave side the *maturing face*. According to this dynamic view of the organelle, the secretory pathway traverses the entire stack as each cistern is displaced from the forming to the maturing face, and there is no need for transfer of the secretory product from cistern to cistern.

Although this interpretation of the Golgi apparatus is now widely accepted, it involves a number of assumptions that have yet to be validated. If the organelle is being continually renewed by contributions of membrane from the endoplasmic reticulum, Golgi-specific enzymes and other integral proteins must be synthesized in the reticulum and segregated in the smooth membrane that buds off to form the transport vesicles, otherwise the properties of the Golgi membrane would become identical to those of the reticulum. Biochemical evidence indicates that there is little or no mixing of either the lipid or protein components of the membranes in the two compartments (Keenan and Morré, 1970; Bergeron, Ehrenreich, Siekevitz and Palade, 1973). By immunocytochemical localization of cytochrome P-450, a marker enzyme for the endoplasmic reticulum, it has been shown that the membranes of the vesicles that transport low-density lipoprotein particles from the reticulum to the Golgi apparatus in liver do not contain this enzyme (Matsuura and Tashiro, 1979). It seems likely therefore that membrane proteins characteristic of the reticulum are excluded and those destined for incorporation in the Golgi are clustered in the transitional region from which the transport vesicles arise by budding.

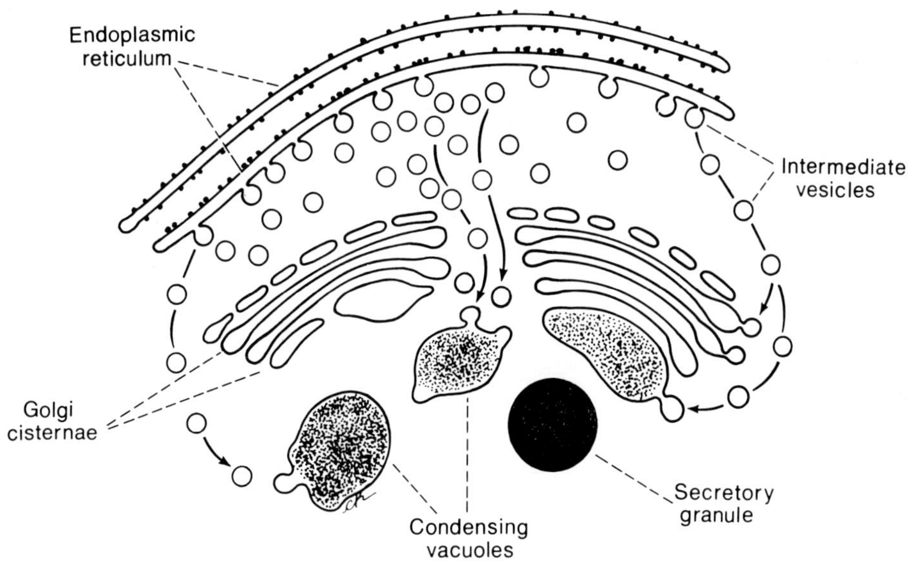

An alternative interpretation of the secretory path through the Golgi region. According to this view the transport vesicles fuse directly with the condensing vacuoles, bypassing the Golgi cisternae. In the hyperstimulated cell vesicles may fuse with the periphery of some cisternae as shown at the right of the figure.

The concept of delivery of the product of protein synthesis to cisternae at the forming face of the Golgi and its progressive modification during the transit of those cisternae to the secretory face has been questioned by Palade and coworkers on the basis of their extensive studies on the guinea pig pancreas. These studies suggest that the transport vesicles go directly from the transitional elements of the endoplasmic reticulum to the condensing vacuoles. In support of this interpretation, they cite autoradiographic studies in which tritiated leucine incorporated into the secretory product of acinar cells was localized over the condensing vacuoles but not over the stacks of Golgi cisternae (Caro and Palade, 1964). Similarly, when slices of pancreas incorporated labeled leucine and were fractionated after various time intervals, the radioactivity was detected first in rough microsomes, then in smooth microsomes consisting in part of transitional vesicles, and then in condensing vacuoles and zymogen granules, apparently bypassing the Golgi cisternae (Jamieson and Palade, 1967). They concede, however, that in overstimulated guinea pig pancreas and in other secretory cells, transport vesicles do fuse with the expanded rims of the Golgi cisternae and that these participate in product modification and condensation. While it is not unreasonable to expect some variation in Golgi function among glandular cells having different products and rates of secretion, it would be surprising if the stacks of cisternae which comprise the bulk of the Golgi apparatus did not have a dominant role in the secretory pathway of most cell types.

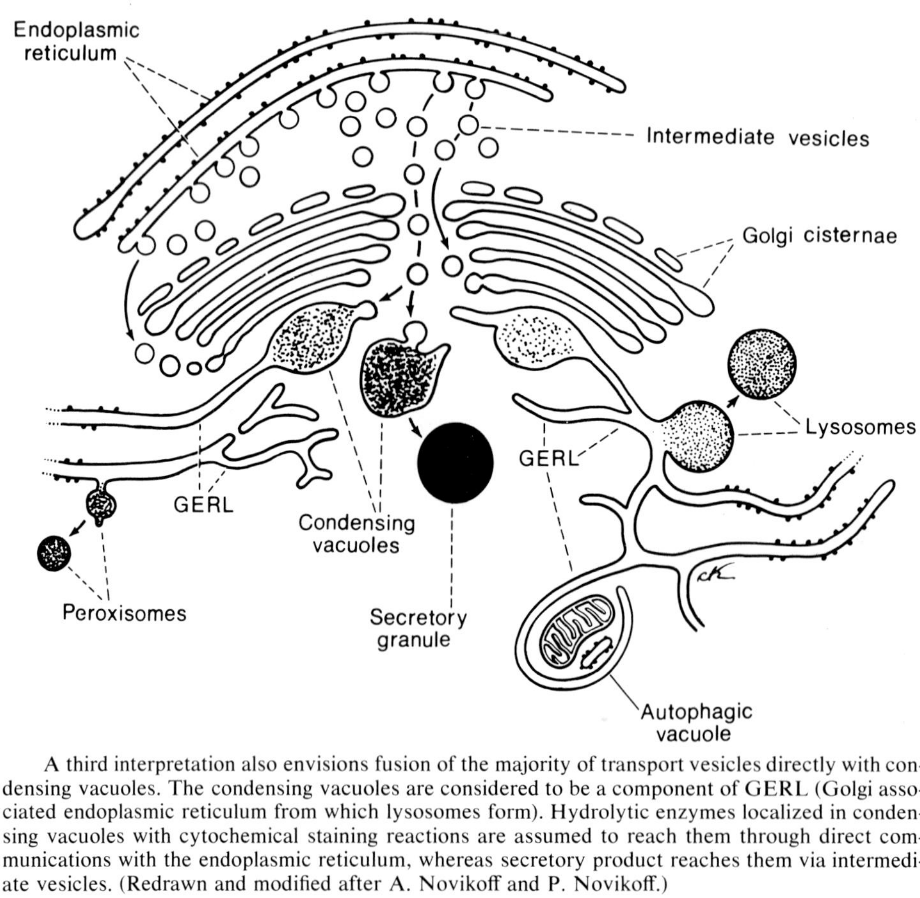

A third interpretation also envisions fusion of the majority of transport vesicles directly with condensing vacuoles. The condensing vacuoles are considered to be a component of GERL (Golgi associated endoplasmic reticulum from which lysosomes form). Hydrolytic enzymes localized in condensing vacuoles with cytochemical staining reactions are assumed to reach them through direct communications with the endoplasmic reticulum, whereas secretory product reaches them via intermediate vesicles. (Redrawn and modified after A. Novikoff and P. Novikoff.)

Another interpretation that assigns a subsidiary role to the Golgi cisternae has gained some measure of acceptance in recent years. It relies heavily upon ultrastructural cytochemistry to distinguish between elements of the Golgi apparatus and other membrane-limited cytoplasmic organelles. Novikoff and coworkers observed that condensing vacuoles, neighboring smooth-surfaced tubules, and a cistern near the secretory face of the Golgi consistently exhibit a positive staining reaction for acid phosphatase. These structures, which had generally been considered by others to be components of the Golgi complex, were shown by Novikoff to be continuous with elements of the endoplasmic reticulum. Observing that lysosomes arise in this region, investigators concluded that the acid phosphatase–positive tubules and cisternae constitute a specialized route for transfer of hydrolases directly from their site of synthesis in the endoplasmic reticulum to lysosomes forming near the inner face of the Golgi apparatus.

To describe this system, Novikoff proposed the term GERL, an acronym for *Golgi-associated endoplasmic reticulum from which lysosomes form* (Novikoff, 1964; Novikoff et al., 1971). When preparations of secretory cells are stained in parallel for thiamine pyrophosphatase and for acid phosphatase activity, the thiamine pyrophosphatase is localized in one or two of the inner cisternae of the Golgi stack, while acid phosphatase stains neighboring tubules and dilated cisternae which are interpreted as GERL. Since condensing vacuoles and immature secretory granules also exhibit acid phosphatase activity, it is argued that these arise from GERL and not from the thiamine phosphatase–positive cisternae at the maturing face of the Golgi stacks (Novikoff et al., 1977). Thus, in addition to formation of lysosomes and autophagic vacuoles, Novikoff and coworkers attribute to GERL the origin of condensing vacuoles. The system could therefore bypass the Golgi apparatus insofar as it may receive both acid hydrolases and secretory proteins directly from the endoplasmic reticulum. The involvement of GERL in secretory processes has been reported in a number of cell types, but its functional relationship to the Golgi cisternae remains unclear. Hand and Oliver (1977) took advantage of the secretory protein peroxidase in the lacrimal gland as a natural marker for membrane-limited elements containing secretory product and concluded that the Golgi cisternae do participate in processing and transport of secretory product, but that GERL plays an important role in the formation of the secretory granules. In the absence of an unambiguous demonstration of continuity between Golgi cisternae and GERL or of vesicular transport between them, inclusion of GERL in the normal secretory pathway remains unconvincing.

A number of common features of the Golgi membranes and the cell membrane have been recorded. A major function of the Golgi apparatus in secretory cells is believed to be the packaging of the product in a membrane capable of fusing with the plasmalemma in exocytosis (Grove et al., 1968). The suggestion that condensing vacuoles are derived from GERL, a specialized region of the endoplasmic reticulum is difficult to bring in accord with this widely accepted concept.

ROLE IN CARBOHYDRATE AND GLYCOPROTEIN SYNTHESIS

Soon after the introduction of the periodic–acid Schiff reaction for staining carbohydrates in tissue sections, the Golgi complex of many secretory cell types was found to be capable of being stained and it was suggested that it might be involved in secretion of carbohydrate-containing materials (Leblond, 1950). Cytochemical studies localized nucleoside diphosphatase and thiamine pyrophosphatase and identified these as useful morphological markers for the Golgi apparatus (Novikoff and Goldfischer, 1961) but did not implicate the organelle in carbohydrate synthesis.

In autoradiographic studies on the secretion of mucus by goblet cells, Neutra and Leblond (1966, 1969) showed that labeled glucose and galactose were rapidly incorporated into the Golgi apparatus and suggested that this organelle might be a site of polysaccharide synthesis. Similar observations followed on a variety of cells secreting glycoproteins. It was considered likely that the carbohydrate moieties of these secretory products might be linked to protein during their passage through the Golgi apparatus. Convincing evidence of similar nature was obtained in studies of plant cells which secrete a variety of pectins and carbohydrate-rich components of the cell wall (Northcote and Pickett-Heaps, 1966). An especially striking example was in certain algae that are covered with scales consisting of complex carbohydrates; these structures could be seen in developmental stages within the cisternae of the Golgi complex (Manton and Leesdale, 1969; Brown et al., 1970).

Biochemical support for the morphological evidence of a role in carbohydrate synthesis was delayed by the technical difficulty of isolating relatively pure Golgi fractions. The initial efforts by Schneider and Kuff (1954) to isolate the large Golgi apparatus of the epididymal epithelium had simply verified the presence of phospholipid, protein, and some acid phosphatase activity in a relatively impure fraction. Later attempts to isolate dictyosomes from plant material (Morré and Möllenhauer, 1964) were more successful in maintaining the morphological integrity of the organelle but contributed little biochemical information. A method for isolation of Golgi membranes from liver (Chatham et al., 1970) yielded a fraction enriched in nucleoside diphosphatase and thiamine pyrophosphatase activity, substantiating earlier cytochemical localization of these enzymes. Although thiamine pyrophosphatase was considerably concentrated in this fraction of liver, it was not confined to this organelle (Cheetham et al., 1970). The first enzymes that appeared to fulfill all of the criteria of a "marker enzyme" for the Golgi apparatus were glycosyltransferases, enzymes that add sugar to a molecule in synthesis (Moore et al., 1969; Fleischer, Fleischer and Ozawa, 1969).

In animal cells in studies of the synthesis of glycoprotein secretory products such as thyroglobulin (Haddad et al., 1971) and immunoglobulin (Schenkein and Uhr, 1970), it has been shown that carbohydrates involved in linkage to the polypeptide are incorporated early in the endoplasmic reticulum together with amino acids, while those sugars added later are rapidly taken up into the Golgi apparatus. It is now generally accepted that some oligosaccharides are incorporated into polypeptides as they are synthesized on the ribosomes of the endoplasmic reticulum, but others are later added to the completed polypeptide chains when they reach the Golgi apparatus. The organelle is therefore responsible for addition of terminal sugar sequences to many glycoprotein secretory products.

Many cell types possess a carbohydrate-rich cell coat, or *glycocalyx*. In addition to its participation in elaboration of the carbohydrate moieties of secretory glycoproteins in glandular cells, the Golgi apparatus appears to play an important role in synthesis of plasma membrane glycoproteins. Autoradiographic studies with labeled galactose, fucose, and N-acetylmannosamine show early localization of these sugars in the Golgi apparatus and later over the plasmalemma. Thus terminal glycosylation of membrane proteins also appears to take place in this organelle. The products transported in small vesicles from the maturing face of the Golgi to the cell surface become incorporated in the membrane (Wise and Flickinger, 1970; Michaels and Leblond, 1976; Leblond and Bennett, 1977).

Secretory and plasma membrane glycoproteins seem to be produced in much the same manner. The polypeptide backbone of the molecule is synthesized on ribosomes associated with the endoplasmic reticulum and addition of carbohydrate begins there, while the near terminal and terminal sugars of long side chains are added to the glycoprotein molecule in the cisternae of the Golgi apparatus.

The cisternae that comprise the Golgi complex are thin in their central portions but expanded at their periphery. The parallel arrays or stacks of cisternae are usually curved so that a convex (forming) face is distinguishable from the concave (maturing or secretory) face. When the organelle is relatively inactive, the cisternae are uninterrupted, closely spaced, and tend to be of uniform thickness throughout the stack. In actively secreting cells, the cisternal profiles are shorter, often fenestrated, and show a progressive increase in width from the convex toward the concave face of the organelle. The upper figure on the facing page is an example of a relatively inactive Golgi complex from a late spermatid after completion of acrosome formation.

The lower figure presents the appearance of a somewhat more active Golgi in a freeze-fracture preparation. Cross fractures of the expanded peripheral portions of the cisternae are indicated by arrows. *En face* views of some of the cisternae show a regular pattern of circular fenestrae and dimples that may represent formative stages of new fenestrations.

Figure 197. Golgi complex in the caudal cytoplasm of a ram spermatid.

Figure 198. Freeze-fracture replica of a Golgi complex from a guinea pig spermatocyte.

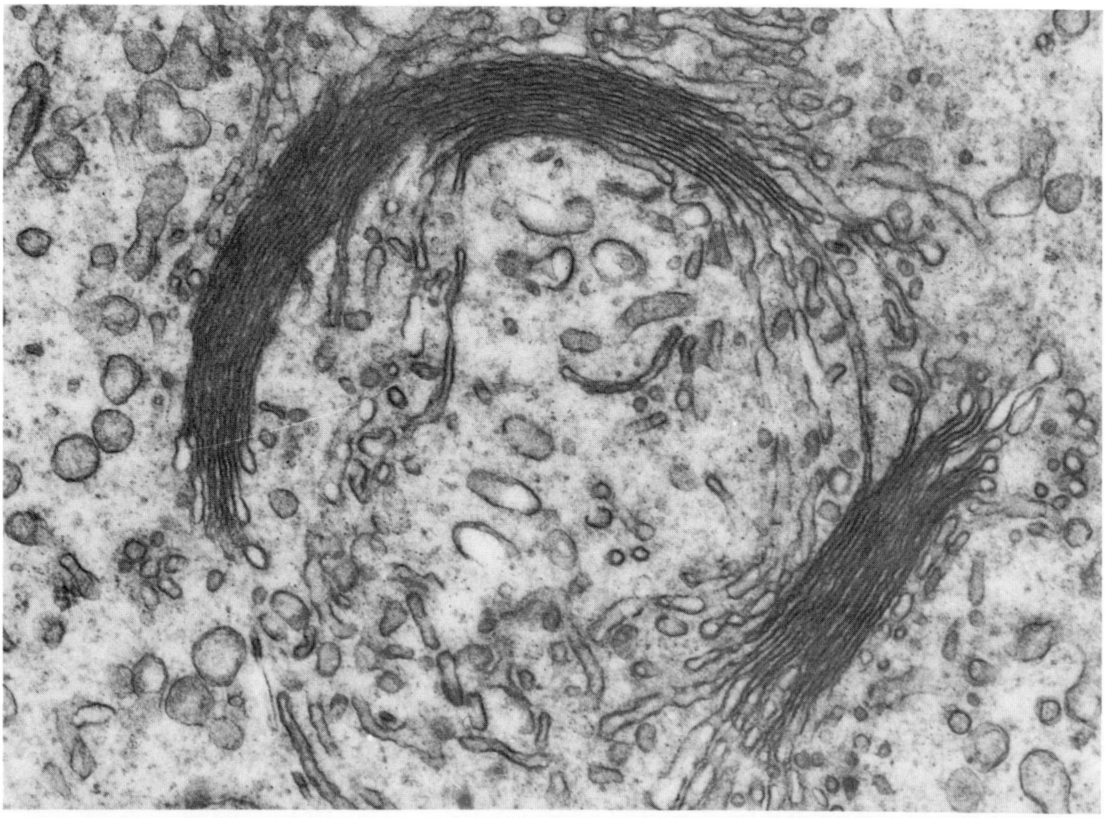

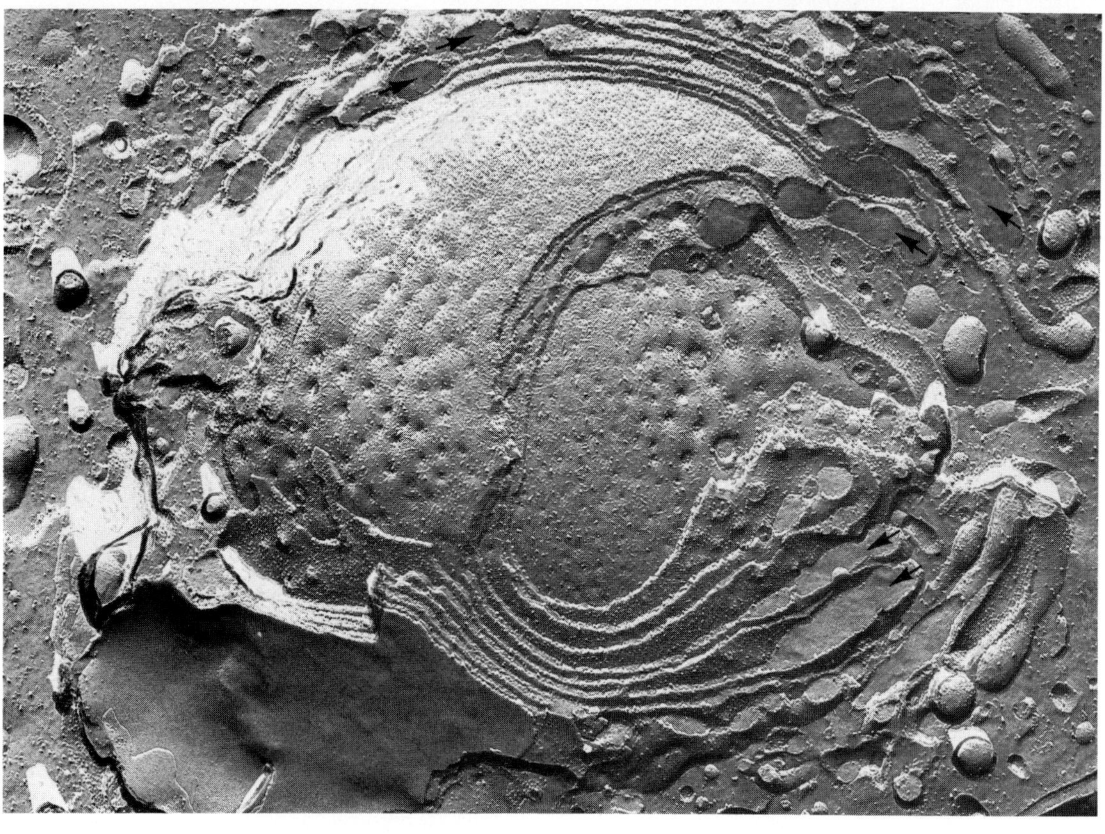

Figure 197, *upper* **Figure 198,** *lower*

When epithelial tissues are subjected to prolonged impregnation with osmium, as in some of the classical staining procedures for demonstration of the Golgi apparatus, electron micrographs show a selective deposition of osmium on or in the outermost profiles on the convex side of each stack of cisternae. The shorter cisternae near the concave or maturing face of the organelle do not become sites of osmium deposition even after very long periods of postosmication. The chemical basis for this reaction is not understood, but its reproducibility and its selectivity for the outermost elements is indicative of a cytochemical and functional polarity within the stacks of Golgi cisternae.

Figure 199. Mouse epididymis. Collidine-buffered osmium fixation with 40 hours postosmication at 37° C. (Micrograph courtesy of Daniel Friend.)

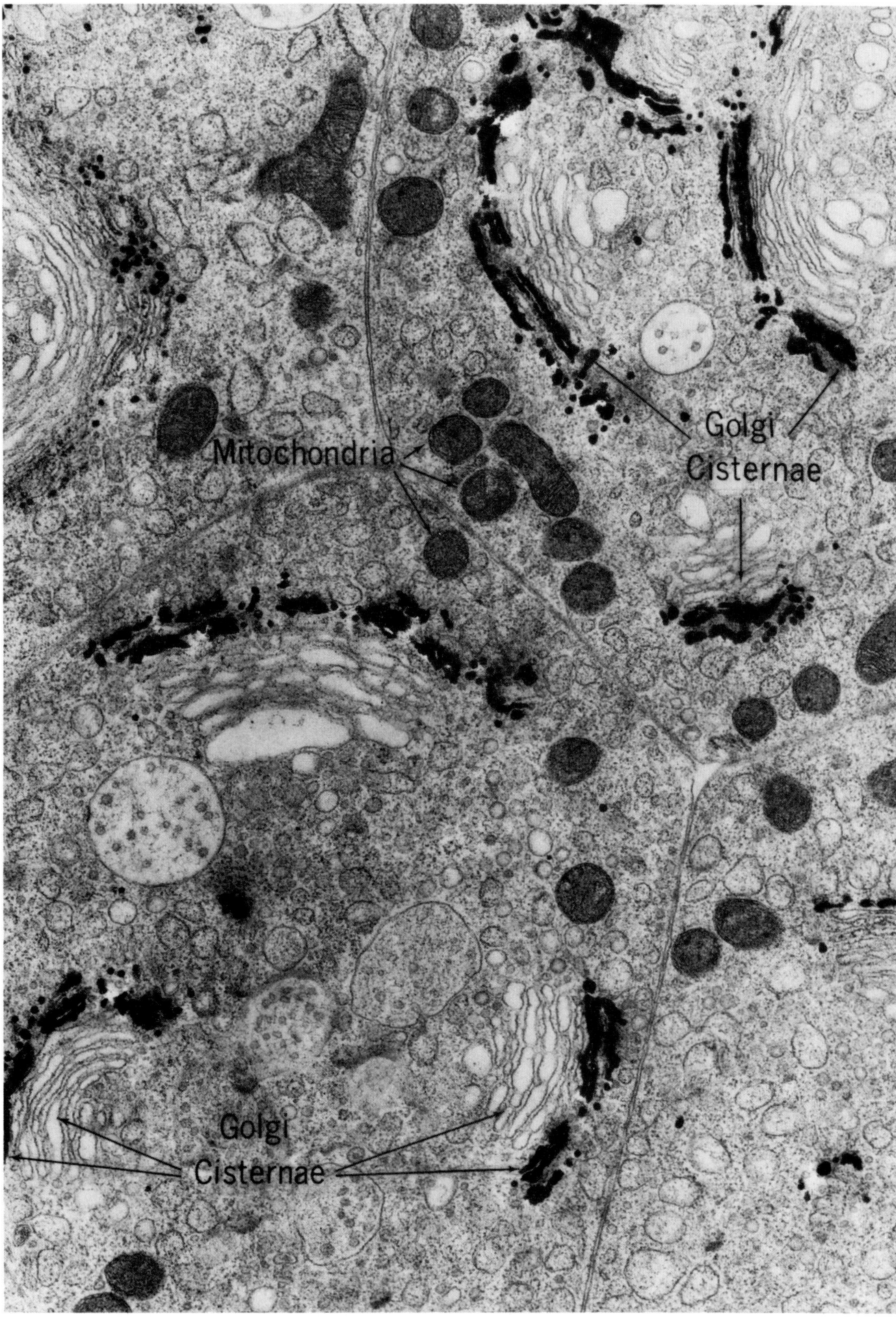

Figure 199

When stained by the histochemical method for thiamine pyrophosphatase, the reaction product is localized in the innermost cisternae and associated vesicles at the maturing face of the Golgi complex. This is in striking contrast to the localization of osmium staining on the opposite face illustrated in the previous figure. It provides further evidence of functionally distinct regions within the organelle.

Figure 200. Thiamine pyrophosphatase reaction of the Golgi apparatus in rat epididymis. (Micrograph courtesy of Daniel Friend.)

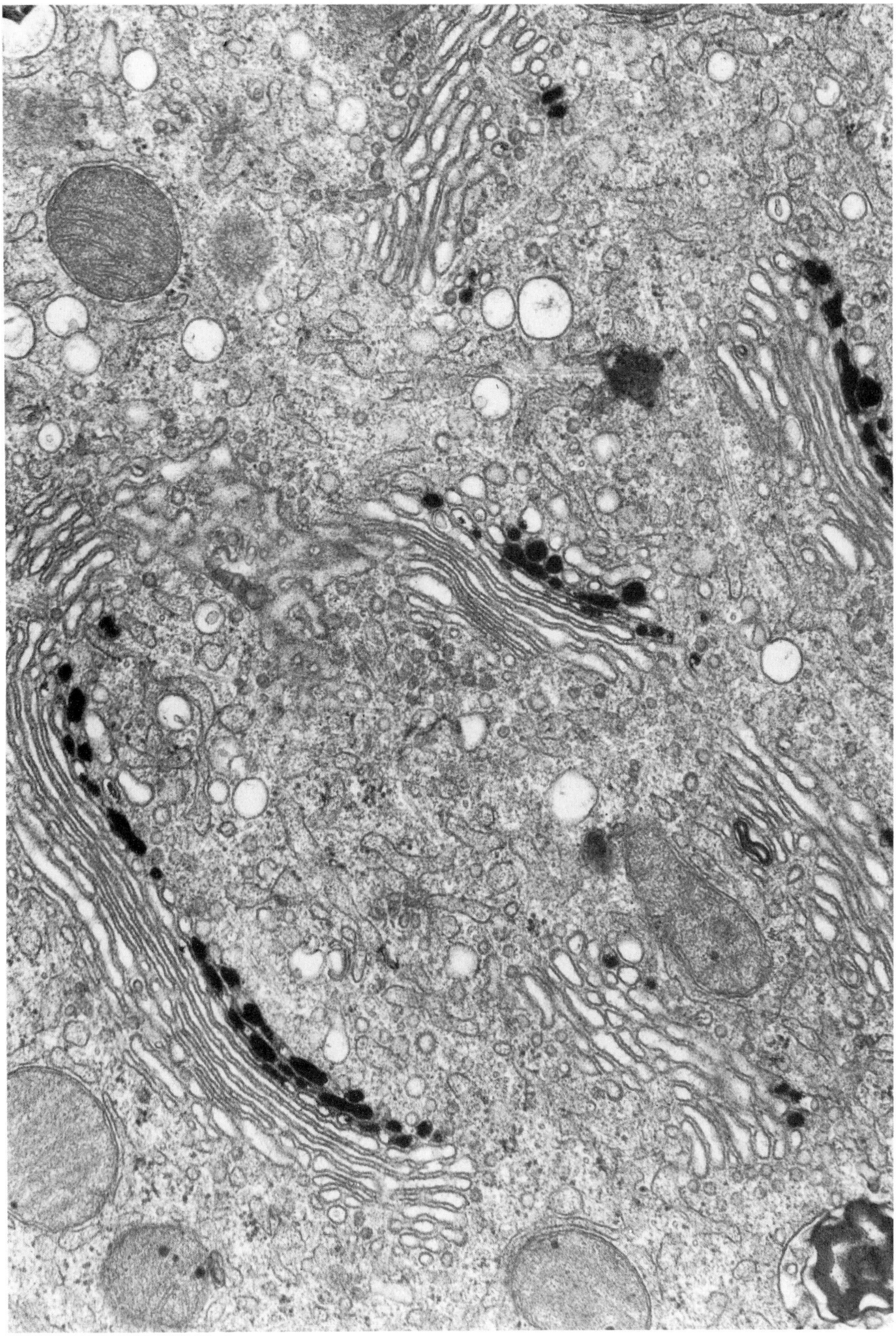

Figure 200

The content of the Golgi cisternae is usually extracted in the course of specimen preparation for electron microscopy, but in favorable material it may be preserved. Under these conditions, there is an obvious gradient in cisternal contents from the convex to the concave face of the organelle. In the accompanying micrograph, the fenestrated outermost cistern appears relatively empty, but there is a progressive increase in the density of the succeeding cisternae, with those near the inner face exceedingly dense. This observation is consistent with the interpretation that the cell product is concentrated and modified in its passage through the Golgi apparatus.

Figures 201 and 202. Golgi apparatus of nurse cells from the testis of the insect *Oniscus*. (Micrograph courtesy of David Phillips.)

Figure 201, *upper* **Figure 202**, *lower*

Direct continuity of the endoplasmic reticulum with the Golgi complex is rarely if ever observed. Communication between the two is maintained by intermediate or transport vesicles that bud off from a transitional region of the reticulum associated with the forming face of the Golgi complex.

In the upper figure vesicles can be seen budding from a ribosome-free region of the cistern of endoplasmic reticulum immediately below the mitochondrion. These small smooth-surfaced vesicles transport quanta of the secretory product to the Golgi cisternae or to condensing vacuoles on the concave face of this organelle.

The lower figure illustrates the same process in an alga, but here the vesicles are budding from the perinuclear cistern. Such images are further evidence that the nuclear envelope is functionally as well as morphologically similar to a cistern of the endoplasmic reticulum.

Figure 203. Transitional zone of endoplasmic reticulum and Golgi region from a cell of Brunner's gland. (Micrograph courtesy of Daniel Friend.)

Figure 204. Nuclear envelope–Golgi relationship in an alga. (From Massalski and Leedale, Br. Phycol. J. 4:159–180, 1969.)

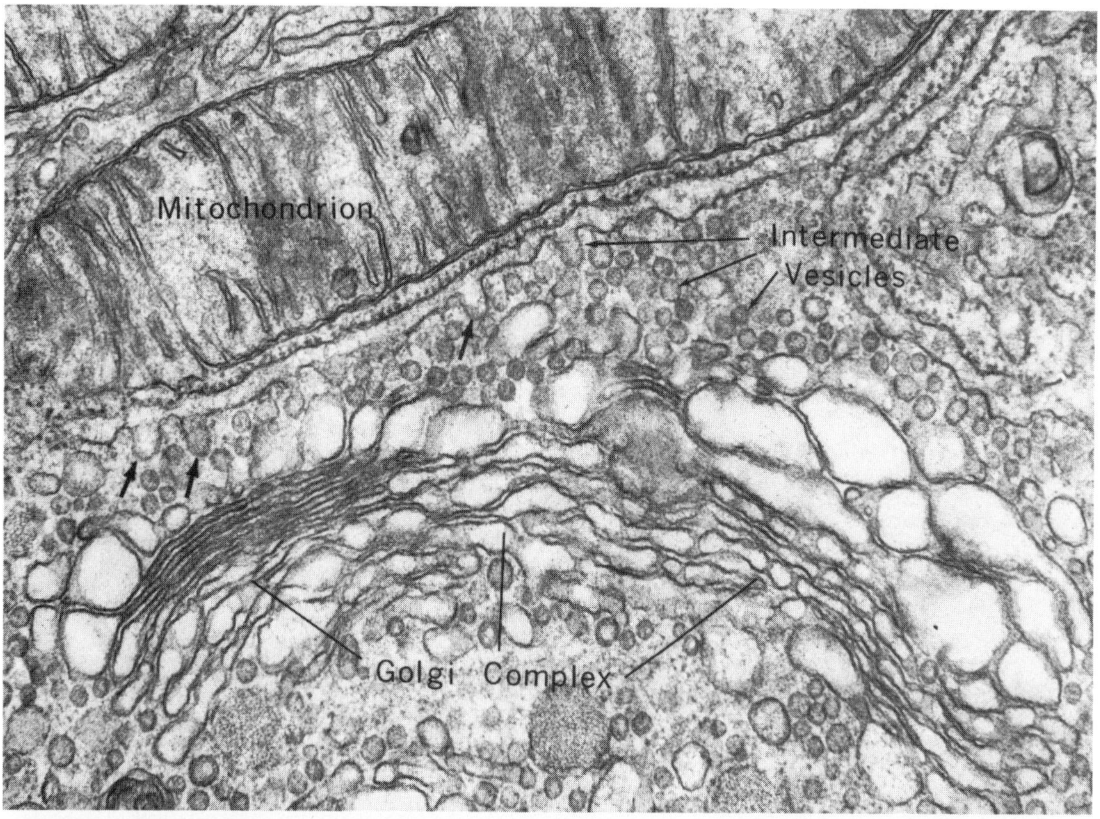

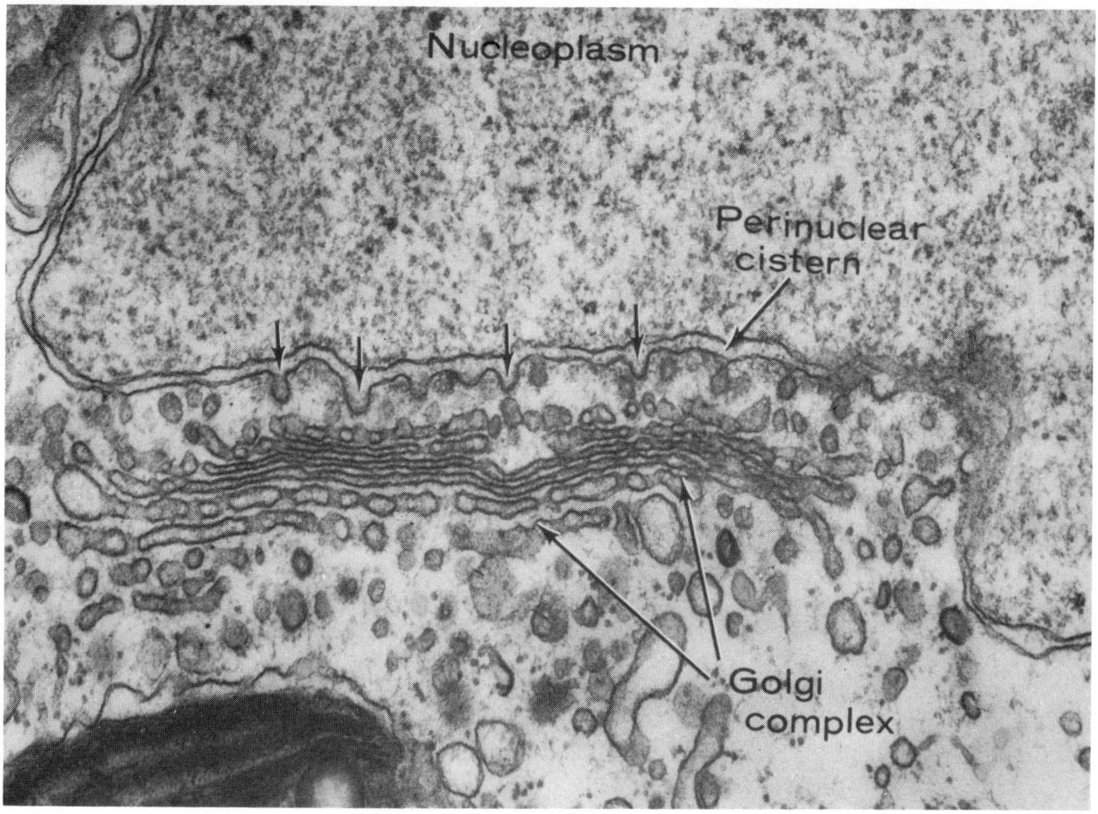

Figure 203, *upper* **Figure 204**, *lower*

Following Nassanov's classical investigations (1923) of the role of the Golgi apparatus in secretion, Bowen's studies of spermiogenesis (1923) led him to conclude that the sperm acrosome was generated by the Golgi apparatus in a manner analogous to the formation of secretory granules in glandular cells. Since the advent of the electron microscope, detailed studies of this process have contributed significantly to our understanding of the function of this organelle.

A number of small membrane-bounded granules first appear in the interior of the Golgi (*A*). These coalesce to give rise to a few larger proacrosomal granules (*B*). One of these attaches to the nuclear envelope (*C*) and those remaining coalesce with it to form a single large acrosomal granule (*D*). The acrosome is further enlarged by addition of material of lower density transported to it in small vesicles that appear to arise from the innermost cistern of the Golgi apparatus. With the progressive accretion of acrosomal contents and additional membrane, the acrosomal vesicle flattens and spreads over the entire anterior hemisphere of the spermatid nucleus to form the acrosomal cap (*E,F,G*). When the acrosome has attained its definitive volume, the Golgi apparatus leaves its surface and migrates into the caudal cytoplasm. Subsequent development of the acrosome involves changes in its shape and a redistribution of the substance comprising the acrosomal granule, but there is no further increase in its volume after dissociation of the Golgi complex.

The relationships of the endoplasmic reticulum and Golgi apparatus to the developing acrosome are shown in greater detail in the figures that follow.

Figure 205. Electron micrographs illustrating successive stages (*A–G*) of development of the acrosomal cap in spermatids from guinea pig testis.

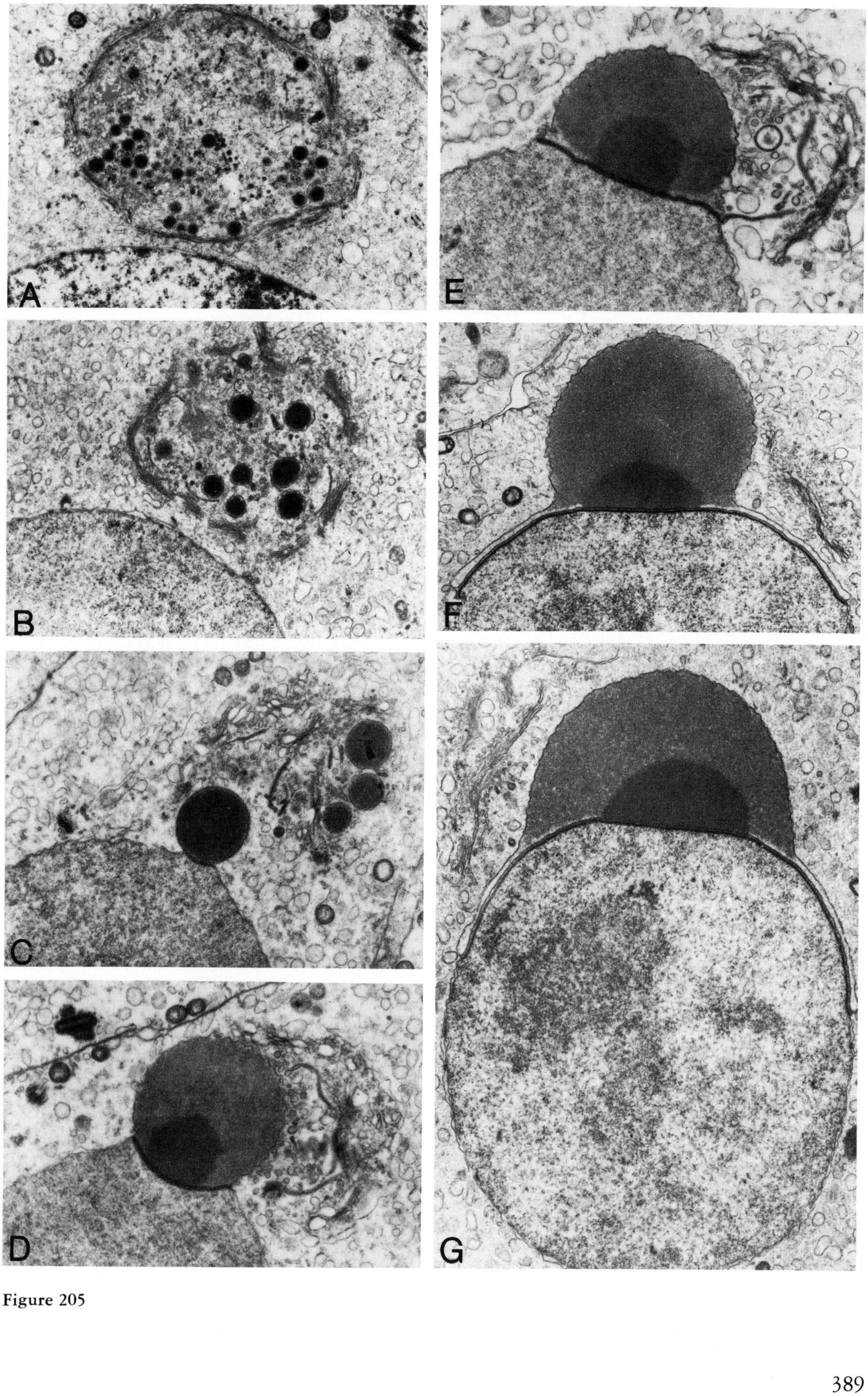

Figure 205

The transitional zone between the endoplasmic reticulum and the Golgi is seen with exceptional clarity in the spermatids of some species during formation of the acrosome. A curving cistern of endoplasmic reticulum parallels the convex outer face of the Golgi. Between it and the forming face of the Golgi are large numbers of small smooth-surfaced vesicles. These can be seen budding off the smooth inner aspect of the solitary cistern of endoplasmic reticulum (at arrows).

In addition to the smooth vesicles, coated vesicles are commonly associated with the Golgi complex, where they may be seen arising from, or, more likely, fusing with, the cisternae (see at stars). Their functional significance is not known, but in secretory cells it is suggested that they may be involved in recirculation of membrane from the plasmalemma back to the Golgi apparatus.

Figures 206 and 207. Golgi complex associated with the developing acrosome in chinchilla spermatids.

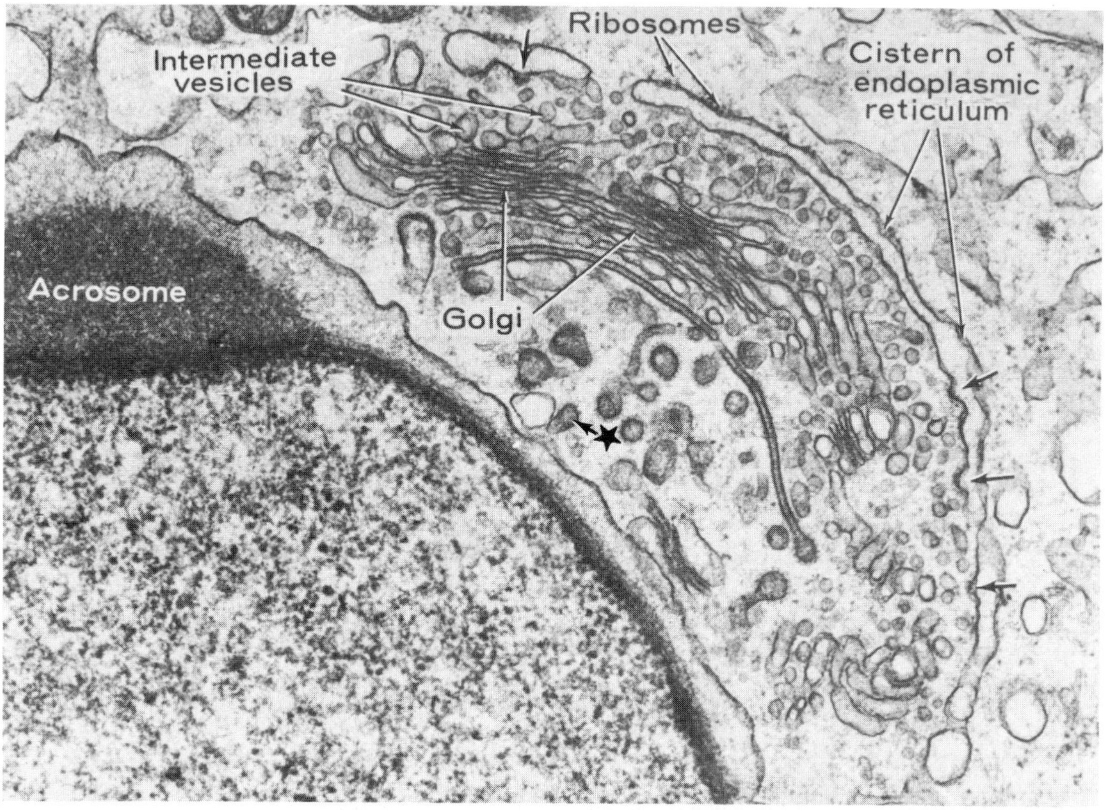

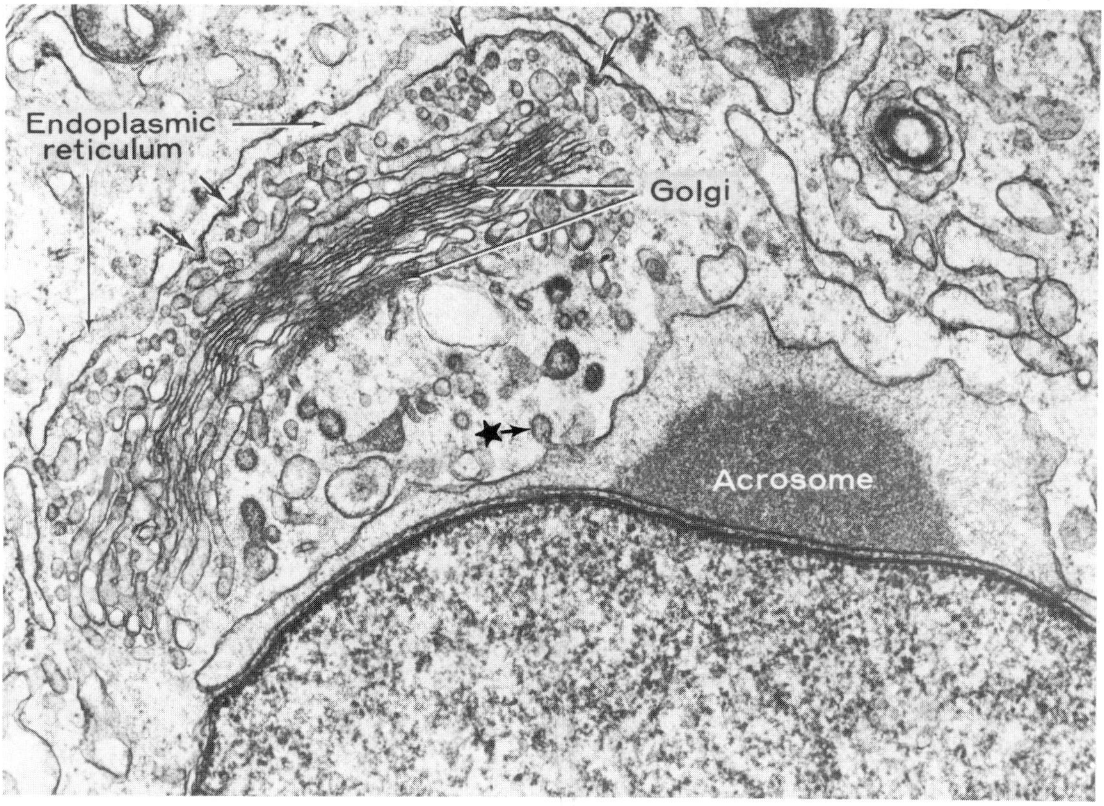

Figure 206, *upper* **Figure 207**, *lower*

Liver cells have multiple small Golgi complexes situated between the centrally placed nucleus and the intercellular bile canaliculi. One such is shown in the micrograph on the facing page. Very low density lipoprotein particles synthesized in the rough and smooth endoplasmic reticulum are carried to the Golgi cisternae in transport vesicles. A site of continuity between such a vesicle and the end of one of the Golgi cisternae is shown at the star. Other lipoprotein particles in the innermost cistern and associated vesicles are indicated by arrows. Such images suggest that the cistern at the forming face is not the only site of entry of secretory product into the Golgi apparatus. Transport vesicles apparently can fuse with the periphery of any of the cisternae or directly with condensing vacuoles associated with the concave face of the Golgi.

Figure 208. Golgi apparatus and associated organelles of a cell from rat liver. (Micrograph courtesy of Robert Bolender.)

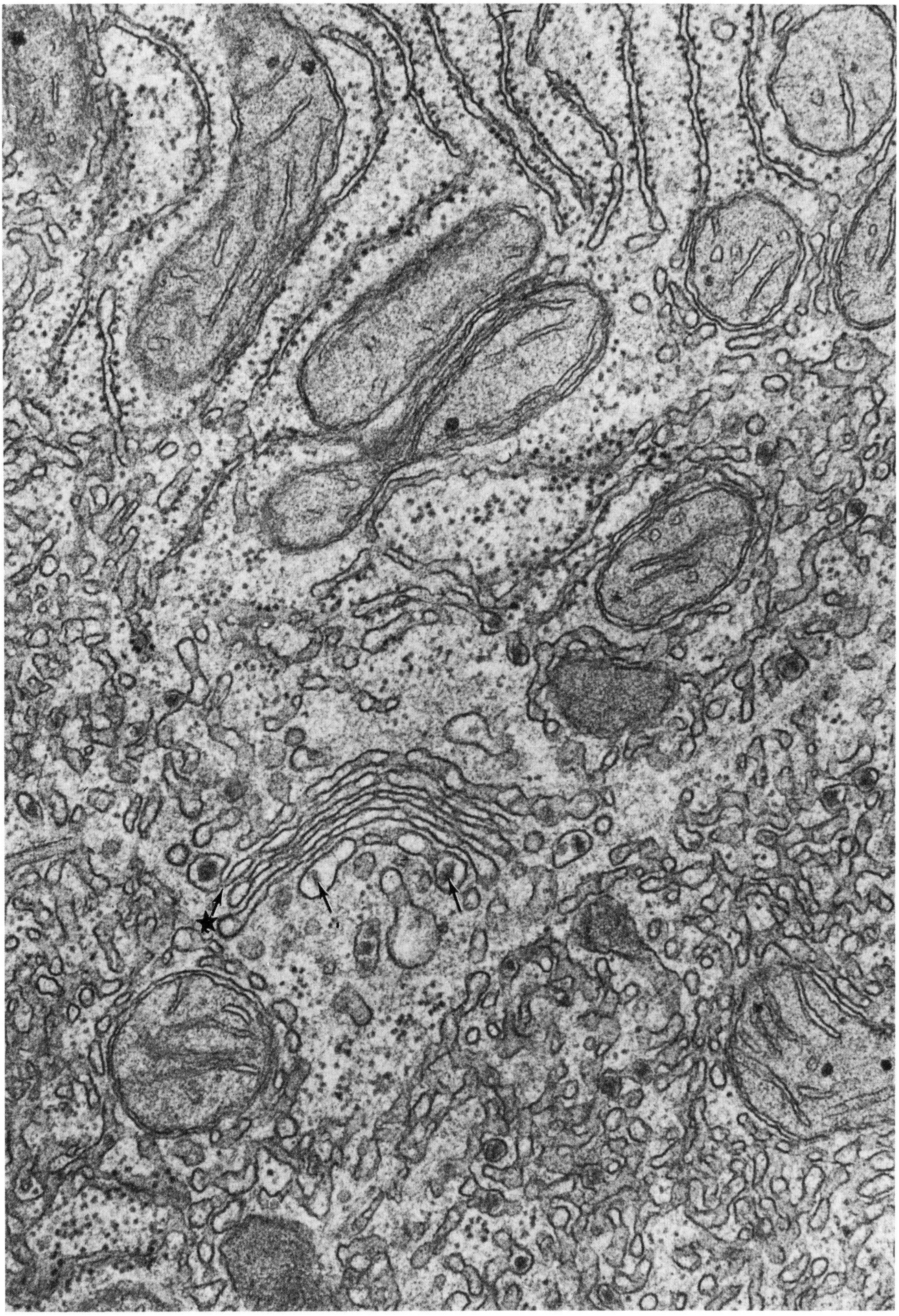

Figure 208

This micrograph of rat liver shows two stacks of Golgi cisternae in the vicinity of a bile canaliculus. Small vacuoles at the secretory face of both contain multiple very low density lipoprotein particles (at arrows). This secretory product requires no concentration in the Golgi apparatus but packaging in membrane of Golgi origin may be essential for its release by exocytosis.

Figure 209. Portions of two hepatic cells and the intervening bile canaliculus from rat liver. (Micrograph courtesy of Robert Bolender.)

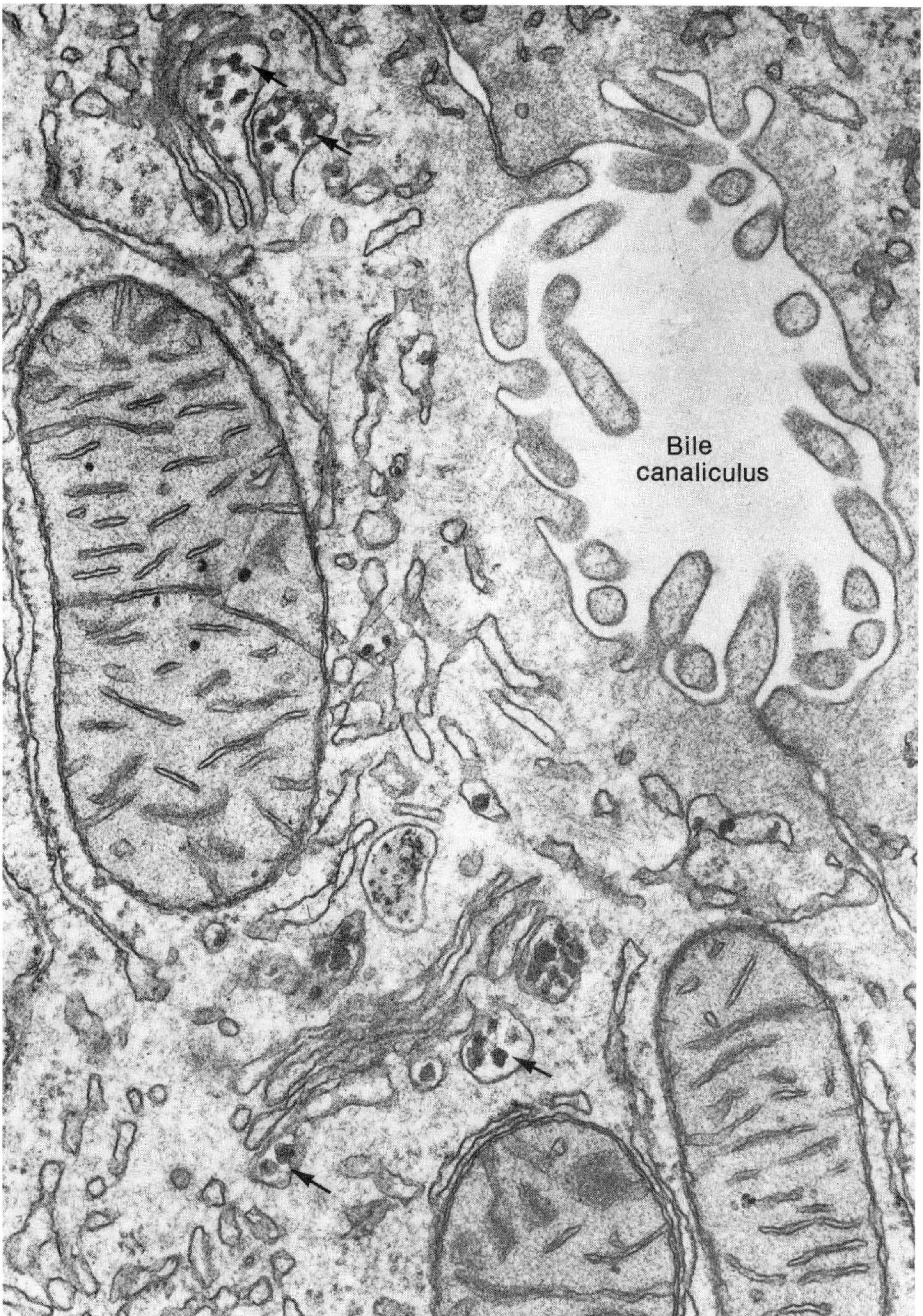

Figure 209

The cells of the mammalian epididymis have an exceptionally large Golgi apparatus. No satisfactory explanation has been advanced for the unusual size of the organelle in this cell type. The epididymal epithelium is active in absorption of fluid from the lumen by micropinocytosis. It is also known to secrete glycerophosphorylcholine and an acid glycoprotein. The glycosyl transferases of the Golgi membranes might be expected to play an important role in synthesis of glycoprotein, but, oddly enough, no typical secretory granules are formed and there is no morphological evidence of participation of the Golgi apparatus in the secretory process. The large dense granules in the accompanying micrograph are identified as lysosomes by their histochemical reaction for acid phosphatase. The numerous small circular profiles in the surrounding cytoplasm are sections of an extensive sparsely granulated endoplasmic reticulum.

Figure 210. Supranuclear region of a cell from rabbit epididymis.

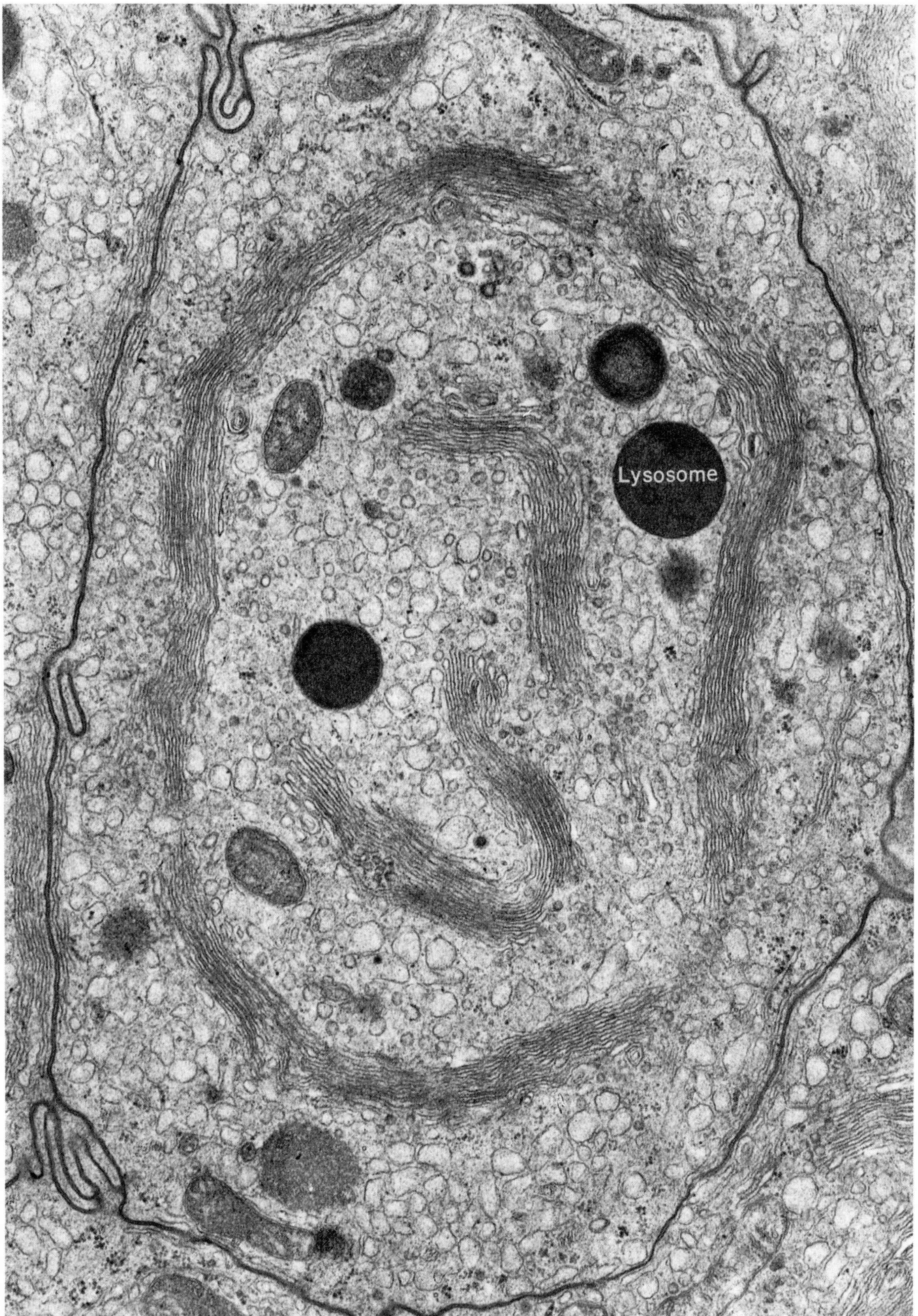

Figure 210

A well-developed supranuclear Golgi apparatus is present in the nonpigmented cells of the ciliary epithelium. This epithelium affords attachment for the fibers of the ciliary zonule and is responsible for secretion of the aqueous humor. The role of the Golgi apparatus in this secretory activity has not been defined. As in many other cell types, coated vesicles are abundant in the Golgi region (at arrows).

Figure 211. Golgi apparatus in a nonpigmented cell of the ciliary epithelium in *Macaca mulatta*. (Micrograph courtesy of Giuseppina Raviola.)

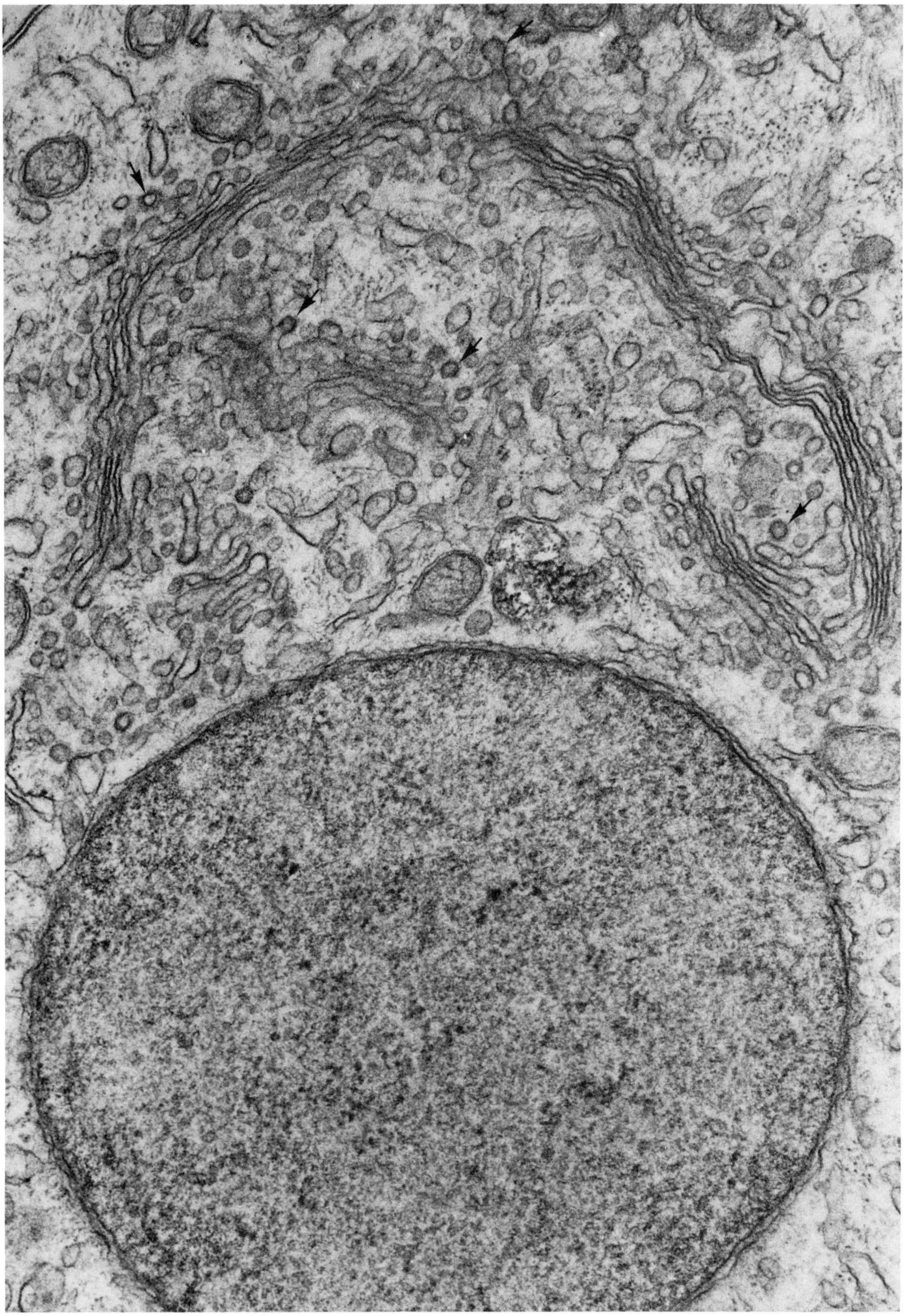

Figure 211

In very actively secreting glandular cells such as that shown in the accompanying micrograph from Brunner's duodenal gland, the Golgi complex is very extensive and the newly formed secretory granules are located near the inner aspect of the multiple stacks of cisternae.

Figure 212. Secretory granules associated with an extensive Golgi complex in a cell from Brunner's gland of mouse. (Micrograph courtesy of Daniel Friend.)

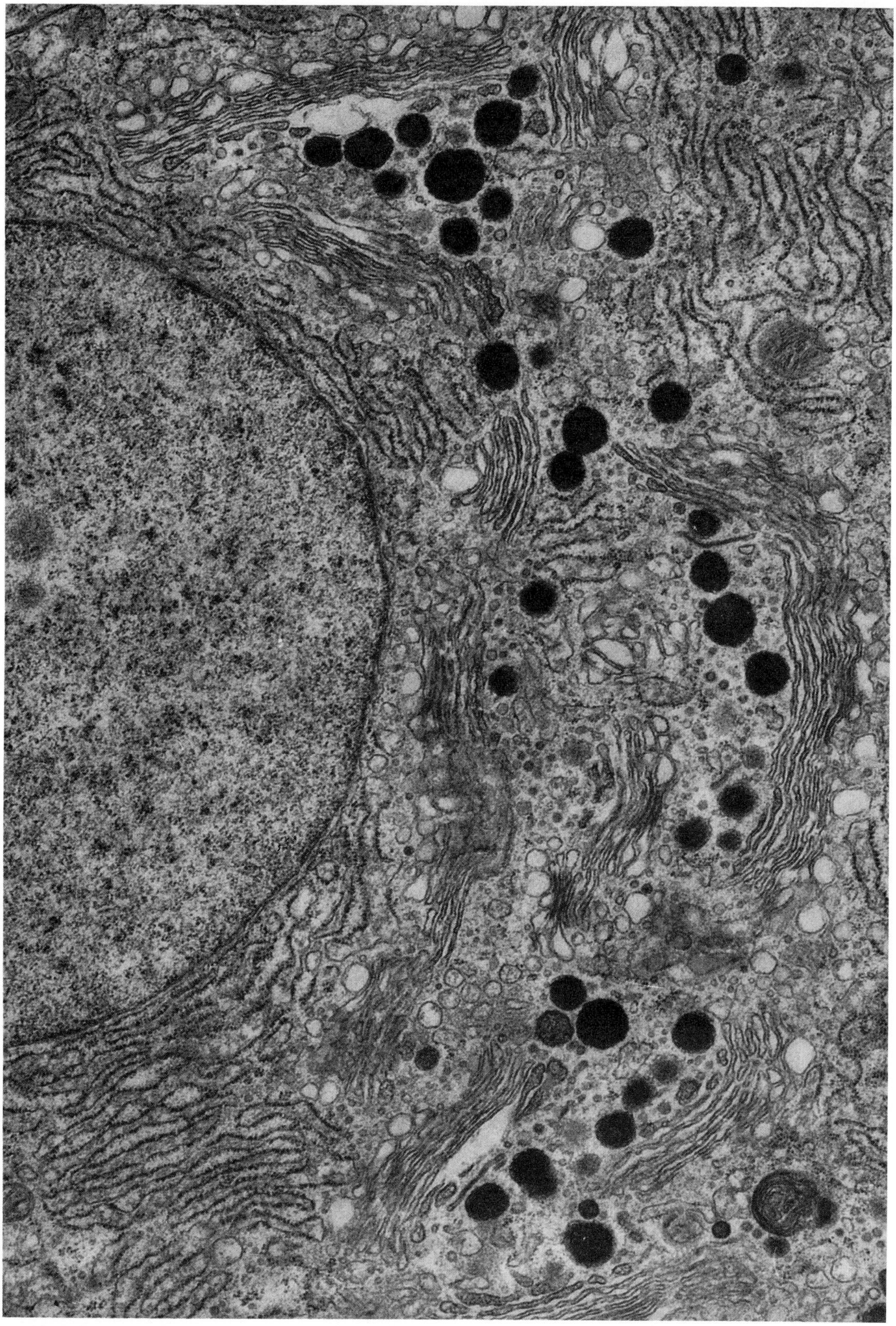

Figure 212

In a majority of cell types, the lumen of the Golgi cisternae appears empty, owing to extraction of their content during routine specimen preparation. Thus in the upper figure of a steroid-secreting cell, the highly fenestrated cisternae of the Golgi complex have no visible content.

In the protein-secreting cell in the lower figure, on the other hand, the fixative has preserved the protein content of the cisternae as a moderately dense, finely granular precipitate. It is evident from images such as this that the Golgi participates in concentration of the secretory product. The lumen of the tubular elements of the neighboring endoplasmic reticulum contains only a very sparse flocculent precipitate.

Figure 213. Golgi complex of a Leydig cell from guinea pig testis.

Figure 214. Golgi complex of an acinar cell from rat pancreas. (Micrograph courtesy of Daniel Friend.)

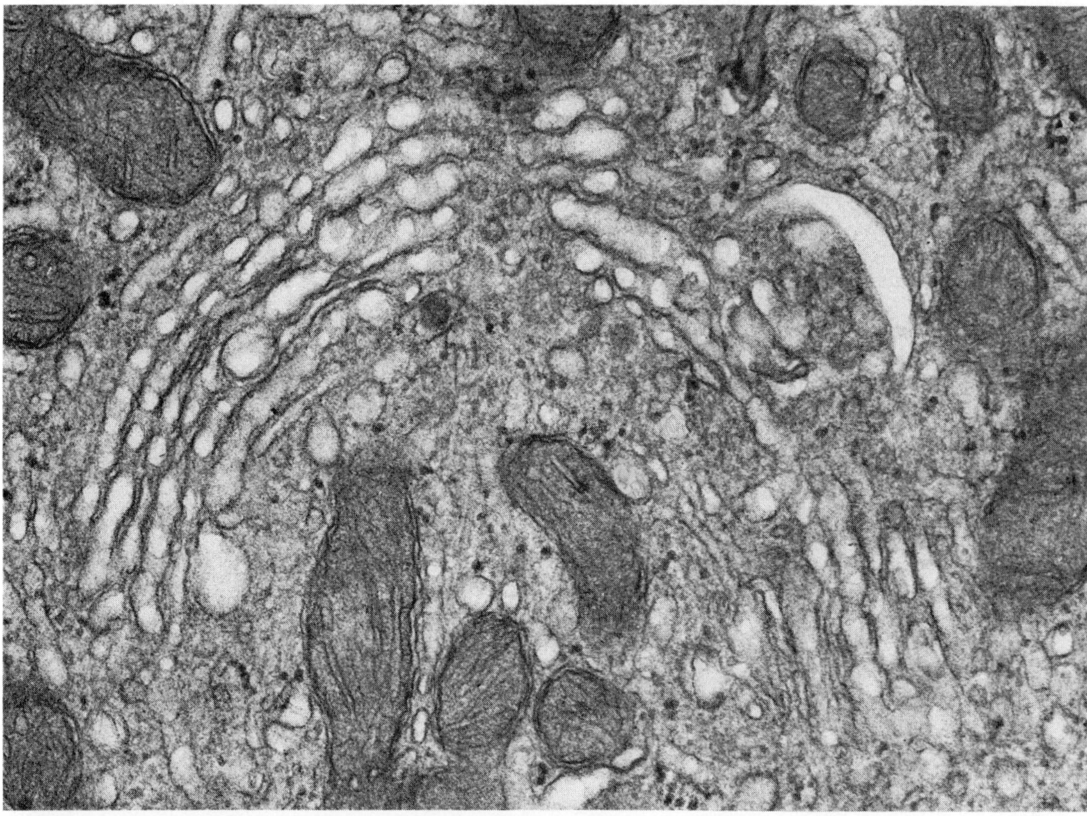

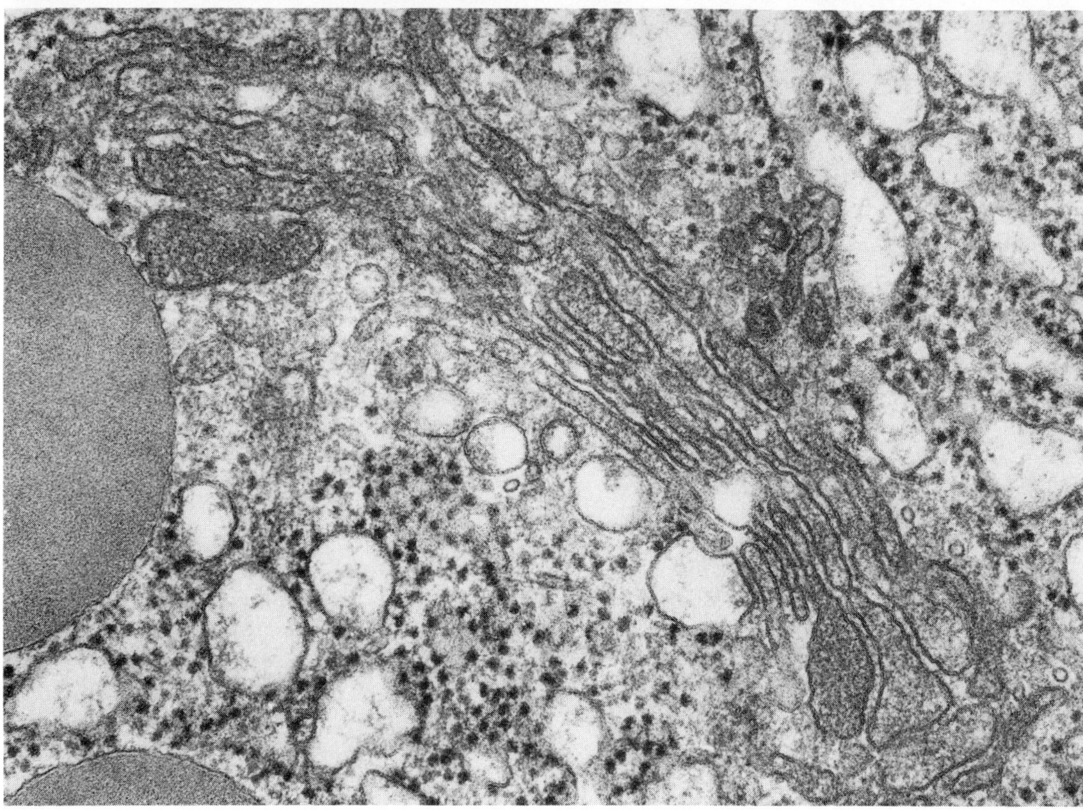

Figure 213, *upper* **Figure 214**, *lower*

The specific granules of eosinophil leucocytes stain positively with the histochemical reaction for peroxidase. This makes it possible to localize this product in myelocytes during the development of specific granules. In the upper figure, the dense reaction product is seen in the perinuclear cistern, throughout the endoplasmic reticulum, in the Golgi complex, and in the mature granules.

The lower figure shows the Golgi region of the same cell with reaction product in all of the cisternae, as well as in condensing vacuoles and immature granules. This distribution casts doubt on the validity of those interpretations of the secretory pathway that envision the product in intermediary vesicles bypassing the cisternae to fuse directly with condensing vacuoles. Similarly, such intense staining of the cisternae would be unexpected if the product reached the condensing vacuoles from the reticulum via direct communication through the GERL.

Figures 215 and 216. Eosinophilic myelocyte from rat bone marrow stained by the cytochemical reaction for peroxidase. (Micrograph from D. Bainton and M. Farquhar, J. Cell Biol. 45:54, 1970.)

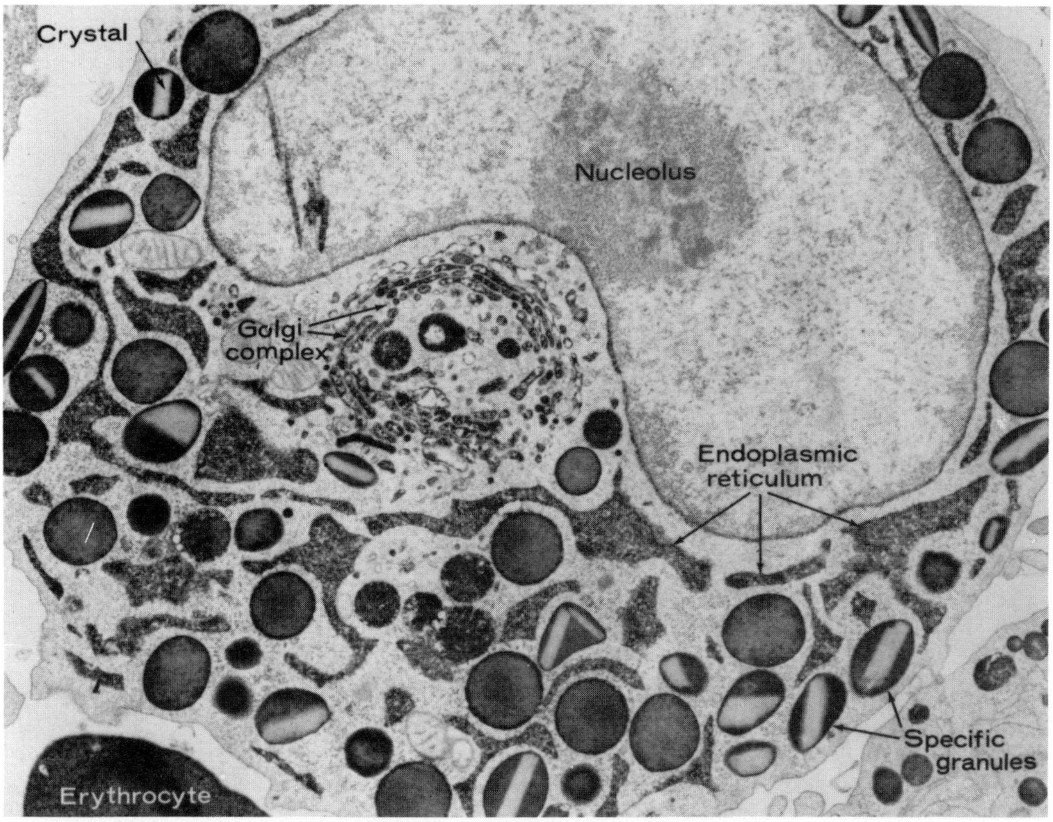

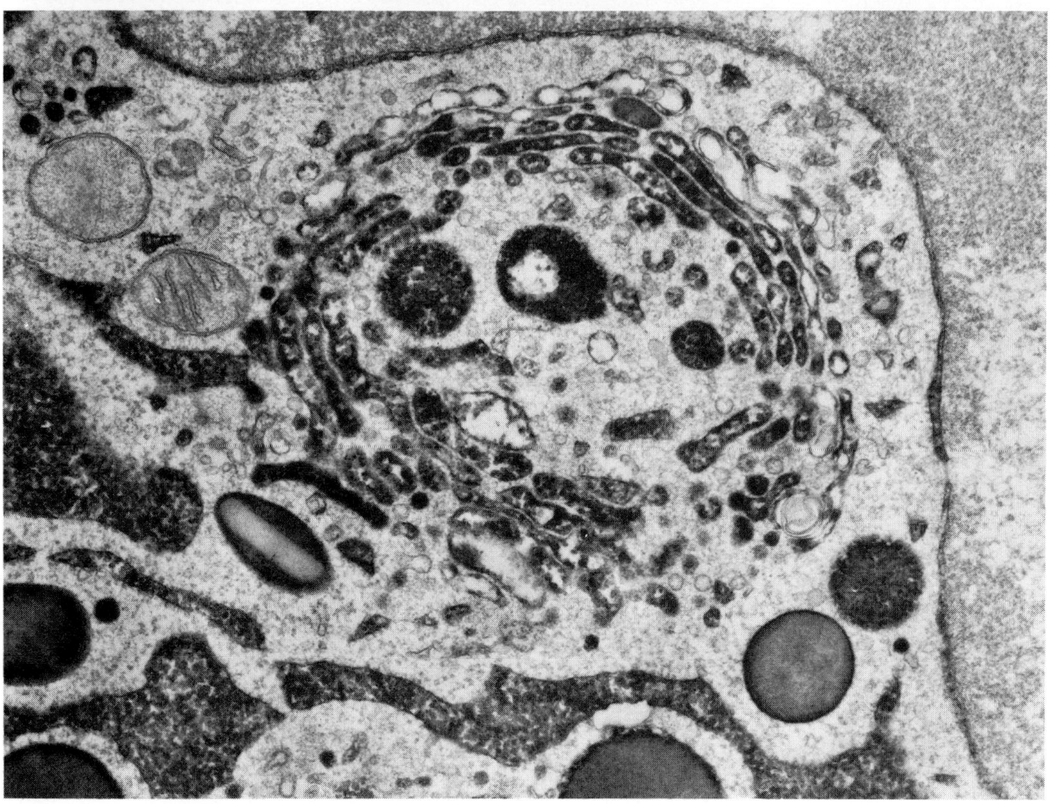

Figure 215, *upper* Figure 216, *lower*

CONTRIBUTIONS TO THE CELL MEMBRANE

The addition of Golgi-derived membrane to the free surface of secretory cells during exocytosis has already been discussed. The organelle also may play an important role in renewal of the surface membrane in cells that have no glandular function. We have previously referred to the synthesis of pectins and cell wall constituents in higher plants and to the synthesis and assembly in Golgi cisternae of carbohydrate-rich scales destined to cover the surface of certain algae. In mammals, the superficial cells of the bladder epithelium present one of the more dramatic examples of Golgi contribution to relatively stable differentiations of the plasmalemma.

The freeze-fracture replica presented here shows several discoidal vesicles in the cytoplasm exhibiting different stages in the assembly of plaques of intramembrane particles that are characteristic of the lumenal membrane of the urinary bladder. A fully formed plaque in the surface membrane is illustrated in the inset. When the discoidal vesicles seen here in the cytoplasm arise from Golgi cisternae, they contain very few intramembrane particles. After their dissociation from the Golgi apparatus, particles appear, increase progressively in number, and become closely packed in hexagonal array. The mature vesicles are ultimately incorporated in the surface membrane.

This sequence of events suggests that integral proteins are inserted into the membranes of the discoidal vesicles after their detachment from the Golgi apparatus.

Figure 217. Freeze-fracture replica of a surface epithelial cell of rat bladder showing discoidal vesicles destined to be incorporated in the surface membrane. (Micrographs courtesy of Nicholas Severs and Marian Hicks.)

Figure 217

REFERENCES

Golgi Apparatus

Baker, J. R. Structure and chemical composition of the Golgi element. Quart. J. Micro. Sci. 85:1–71, 1944.

Beams, H. W. and R. G. Kessel. The Golgi apparatus: Structure and Function. Int. Rev. Cytol. 23:209–267, 1968. (Review)

Bensley, R. R. On the nature of the canalicular apparatus of animal cells. Biol. Bull. 19:179–194, 1910.

Bergeron, J. J., J. H. Ehrenreich, P. Siekevitz and G. E. Palade. Golgi fractions prepared from rat liver homogenates. II. Biochemical characterization. J. Cell Biol. 59:73–88, 1973.

Bowen, R. H. The cytology of glandular secretion. Quart. Rev. Biol. 4:299–324; 484–519, 1929.

Brown, R. M., W. W. Franke, H. Kleinig, H. Falk and P. Sette. Scale formation in chrysophycean algae. I. Cellulosic and noncellulosic wall components made by the Golgi apparatus. J. Cell Biol. 45:246–271, 1970.

Caro, L. G. and G. E. Palade. Protein synthesis storage and discharge in the pancreatic exocrine cell. An autoradiographic study. J. Cell Biol. 20:473–495, 1964.

Cheetham, R. D., D. J. Morré and W. N. Younghaus. Isolation of a Golgi apparatus rich fraction from rat liver. II. Enzymatic characterization and comparison with other fractions. J. Cell Biol. 4:492, 1970.

Dalton, A. J. A study of the Golgi material of hepatic and intestinal epithelial cells with the electron microscope. Z. Zellforsch. 36:522–540, 1952.

Dalton, A. J. Golgi apparatus and secretion granules. In The Cell, Vol. II, pp. 603–618, Academic Press, New York, 1961. (Review)

Dalton, A. J. and M. D. Felix. Cytologic and cytochemical characteristics of the Golgi substance of epithelial cells of the epididymis — *in situ*, in homogenates and after isolation. Am. J. Anat. 94:171–207, 1954.

Dauwalder, M., W. G. Whaley and J. E. Kephart. Functional aspects of the Golgi apparatus. Sub-cell. Biochem. 1:225–275, 1972. (Review)

Favard, P. The Golgi apparatus. In Handbook of Molecular Cytology (A. Lima-de-Faria, ed.), North Holland Publishing Company, Amsterdam, 1969. (Review)

Fleischer, B., S. Fleischer and H. Ozawa. Isolation and characterization of Golgi membranes of bovine liver. J. Cell Biol. 43:59–79, 1969.

Friend, D. S. and M. J. Murray. Osmium impregnation of the Golgi apparatus. Am. J. Anat. 117:135–149, 1965.

Gatenby, J. B. The Golgi apparatus. J. Roy. Micr. Soc. 74:134–161, 1955.

Golgi, O. Sur la structure des cellules nerveuses. Arch. Ital. Biol. 30:60–71, 1898.

Grove, S. N., C. E. Bracker and D. J. Morré. Cytomembrane differentiation in the endoplasmic reticulum-Golgi apparatus-vesicle complex. Science 161:171–173, 1968.

Haddad, A., M. D. Smith, A. Herscovics, N. J. Nadler and C. P. Leblond. Radioautographic study of in vivo and in vitro incorporation of fucose ^{3}H into thyroglobulin by rat thyroid follicular cells. J. Cell Biol. 49:856–882, 1971.

Hand, A. H. and C. Oliver. Relationship between the Golgi apparatus, GERL, and secretory granules in acinar cells of the rate exorbital lacrimal gland. J. Cell Biol. 74:399–413, 1977.

Hibbard, H. Current status of our knowledge of the Golgi apparatus in the animal cell. Quart. Rev. Biol. 20:1–19, 1945. (Review)

Hirschler, J. Ueber den Golgischen Apparat embryonaler Zellen. Arch. Mikr. Anat. 91:140–181, 1918.

Holtzman, E., A. B. Novikoff and H. Villaverde. Lysosomes and GERL in normal and chromatolytic neurons of the rat ganglion nodosum. J. Cell Biol. 33:419, 1967.

Holmgren, E. Studien an der feineren Anatomie der Nervenzellen. Anat. Hefte 15:3–89, 1900.

Jamieson, J. D. and G. E. Palade. Intracellular transport of secretory proteins in the pancreatic exocrine cell. II. Transport to condensing vacuoles and zymogen granules. J. Cell Bio. 34:597–615, 1967.

Jaeken, L., D. Thines-Sempoux, and F. Verkeyen. A three-dimensional study of organelle interrelationships in regenerating rat liver. I. The GERL system. Cell Biol. Int. Rep. 2:501–503, 1978.

Keenan, T. W. and D. J. Morré. Phospholipid class and fatty-acid composition of Golgi apparatus isolated from rat liver and comparison with other cell fractions. Biochemistry 9:19–25, 1970.

Kirkman, H. and A. E. Severinghaus. A review of the Golgi apparatus. Anat. Rec. 70:413–481; 557–573; 71:79–102, 1938. (Review)

LaVallette St. George, A. J. H. Uber die Genese der Samenkörper. Arch. Mikrosh. Anat. 1:403–414, 1865.

Leblond, C. P. Distribution of periodic-acid reactive carbohydrates in the adult rat. Am. J. Anat. 86:1–49, 1950.

Leblond, C. P. and G. Bennett. Role of the Golgi apparatus in terminal glycosylation, pp. 326–336. In Cell Biology 1976–1977, Rockefeller University Press, New York, 1977.

Manton, I. Further observations on scale formation in Chrysochromulina chiton. J. Cell Sci. 2:411, 1967.

Manton, I. and G. F. Leedale. Observations on the microanatomy of Coccolithus pelagicus and Cre-

cosphaera carterae, with special reference to the origin and nature of the coccoliths and scales. J. Mar. Biol. Ass. 49:1, 1969.
Matsuura, S. and Y. Tashiro. Immunoelectron-microscopic studies of endoplasmic reticulum-Golgi relationships in the intracellular transport process of lipoprotein particles in rat hepatocytes. J. Cell Sci. 39:273–290, 1979.
Michaels, J. and C. P. Leblond. Transport of glycoprotein from Golgi apparatus to cell surface by means of carrier vesicles as shown by radioautography of mouse colonic epithelium after injection of tritiated fucose. J. Micros. Biol. Cell 25:243–248, 1976.
Morré, D. J. and H. H. Möllenhauer. Isolation of the Golgi apparatus from plant cells. J. Cell Biol. 23:295, 1964.
Morré, D. J., H. H. Möllenhauer and C. E. Bracket. Origin and continuity of the Golgi apparatus. In Origin and Continuity of Cell Organelles (J. Reinert and H. Ursprung, eds.), Springer-Verlag, New York, 1971. (Review)
Nassanov, D. Das Golgische Binnennetz und seine Beziehungen zu der Sekretion. Arch. Mikr. Anat. 100:433–471, 1924.
Neutra, M. and C. P. Leblond. Synthesis of the carbohydrate of mucus in the Golgi complex as shown by electron microscope radioautography of goblet cells from rats injected with H^3 glucose. J. Cell Biol. 30:119, 1966a.
Neutra, M. and C. P. Leblond. Radioautographic comparison of the uptake of H^3-galactose and H^3-glucose in the Golgi region of various cells secreting glycoproteins or mucopolysaccharides. J. Cell Biol. 30:137, 1966b.
Northcote, D. H. and J. D. Pickett-Heaps. A function of the Golgi apparatus in polysaccharide synthesis and transport in root-cap cells of wheat. Biochem. J. 98:159, 1966.
Novikoff, A. B. GERL, its form and function in neurons of rat spinal ganglia. Biol. Bull. 127:358, 1964.
Novikoff, A. B. The endoplasmic reticulum: A cytochemist's view (a review). Proc. Nat. Acad. Sci. 73:2781–2787, 1976.
Novikoff, A. B., E. Essner and N. Quintana. Golgi apparatus and lysosomes. Fed. Proc. 23:1010–1022, 1964.
Novikoff, A. B., M. Mori, N. Quintana and A. Yam. Studies of the secretory process in the mammalian exocrine pancreas. I. The condensing vacuole. J. Cell Biol. 75:148–165, 1977.
Novikoff, P. M., A. B. Novikoff, N. Quintana and J. J. Hauw. Golgi apparatus, GERL and lysosomes of neurons in rat dorsal root ganglia studied by thick section and thin section cytochemistry. J. Cell Biol. 50:859–886, 1971.
Novikoff, A. B. and W. Y. Shin. The endoplasmic reticulum in the Golgi zone and its relation to microbodies, Golgi apparatus and autophagic vacuoles in rat liver cells. J. Micros. 3:187, 1964.
Palade, G. E. and A. Claude. The nature of the Golgi apparatus. I. Parallelism between Golgi apparatus and intracellular myelin figures. J. Morphol. 85:35–70, 1949.
Palade, G. E. and A. Claude. The nature of the Golgi apparatus. II. Identification of the Golgi apparatus with a complex of myelin figures. J. Morphol. 85:71–112, 1949.
Parat, M. and J. Painlevé. Appareil réticulaire interne de Golgi, trophosponge de Holmgren et vacuome. C. R. Acad. Sci. (Paris), 179:844–846, 1924.
Perroncity, A. Contribution à l'étude de la biologic cellulaire. Mitochondres, chromidies et appareil réticulaire interne dans les cellules spermatiques. Le phénomène de la dictyokinèse. Arch. Ital. Biol. 54:307–345, 1910.
Platner, G. Beitrage zur Kenntinis der Zelle und ihrer Teilung. Arch. Mikrosk. Anat. 33:180–216, 1889.
Pollister, A. W. and P. F. Pollister. The structure of the Golgi apparatus. Int. Rev. Cytol. 6:85–106, 1957. (Review)
Rambourg, A., Y. Clermont and A. Marrand. Three-dimensional structure of the osmium impregnated Golgi apparatus as seen in the high voltage electron microscope. Am. J. Anat. 140:27–46, 1974.
Schenkein, I. and U. W. Uhr. Immunoglobulin synthesis. I. Biosynthetic studies on the addition of carbohydrate moieties. J. Cell Biol. 46:42–51, 1970.
Schneider, W. C. and E. L. Kuff. On the isolation and some biochemical properties of the Golgi substance. Am. J. Anat. 94:209–224, 1954.
Wagner, R. R. and M. A. Cynkin. Enzymatic transfer of ^{14}C-glucosamine from UDP-N-acetyl-^{14}C-glucosamine to endogenous acceptors in a Golgi apparatus-rich fraction from liver. Biochem. Biophys. Res. Comm. 35:139–143, 1969.
Whaley, W. G. The Golgi Apparatus. Springer-Verlag, New York, 1975. (Review)
Wise, G. E. and C. J. Flickinger. Relation of the Golgi apparatus to the cell coat in amoebae. Exp. Cell Res. 61:13–23, 1970.

MITOCHONDRIA

Mitochondria are ubiquitous organelles found in nearly all eukaryotic cells. Their major function is to provide the chemical energy necessary for the biosynthetic and motor activities of the cell. This is accomplished by a complex series of integrated chemical reactions. Carbohydrates, fatty acids, and amino acids from food are oxidized in the mitochondria to carbon dioxide and water and the free energy released is used to convert adenosine diphosphate (ADP) and inorganic phosphate to adenosine triphosphate (ATP) — a remarkable molecule responsible for most of the energy transfer involved in living processes. ATP is exported from the mitochondria into the surrounding cytoplasm. Wherever the energy stored in this molecule is required to drive synthetic or contractile processes, it is made available by splitting ATP to ADP and phosphate. These are then taken up again by the mitochondria, where ATP is again generated. To carry out this most important activity and a number of secondary functions, mitochondria contain 50 or more enzymes strategically located and precisely ordered in the various structural components of the organelle which are described in this chapter.

The assignment of priority for the observation of mitochondria is not possible. Many cytologists in the late 1800s observed granules, rodlets, and filaments in the cytoplasm, and surely some of these were mitochondria. Flemming was certainly among the first to recognize them as distinct cell organelles. He described granules and thread-like structures, which he called *fila*, in many cell types and devoted special attention to those of muscle. Indeed, in 1888 he teased some of these out of insect muscle, where they are unusually large, and observed that they swelled in water. He inferred from this that they were limited by a membrane — a suggestion that was not validated until some 75 years later.

Systematic study of this organelle did not begin until relatively selective staining methods were developed using acid fuchsin (Altmann, 1890) or crystal violet (Benda, 1898). Their form, relative abundance, and distribution were then described for many cell types. The term *mitochondrion*, from Greek roots meaning "thread" and "grain," was introduced by Benda but was not widely accepted at the time. A profusion of other terms appeared, of which the most enduring was *chondriosome*. This term persisted until very recent times, especially in the European literature, but it is now less commonly used.

Altmann looked upon the mitochondria as elementary organisms, or "bioblasts," and considered them the essential living units of protoplasm from which all other parts of the cell were derived. This extreme view was not widely shared. However, other leading cytologists of the time — including Benda (1903), Meves (1907), and Duesberg (1907) — noted that mitochondria increase before cell division, are distributed equally to the daughter cells, and are transferred by the sperm to the egg at fertilization. Their passage from cell to daughter cells and from generation to generation suggested to these authors that they might play a role in inheritance. Although erroneous, this conclusion does not seem unreasonable when we recall that the chromosomal theory of inheritance was originally based upon similar considerations of organelle continuity.

The literature of the early part of this century abounds with claims that mitochondria gave rise during differentiation to myofibrils, yolk, pigment, secretory granules, and many other cell components. These interpretations were based entirely upon circumstantial evidence such as the proximity of mitochondria to sites of formation of these structures and identification of elements which appear to be intermediate stages in transformation of mitochondria.

The development of methods for growing cells *in vitro* provided a new approach to the study of this organelle. Cells flattened against the glass substrate of the culture vessel offered ideal conditions for observation on the form and behavior of mitochondria in the living state (Lewis and Lewis, 1915–1920; Strangeways, 1924). They usually appeared as short rods or slender filaments but were capable of changing their shape under altered environmental conditions. Direct observations and lapse-time cinematography showed that they are constantly executing slow flexuous movements as they move about in the cytoplasm. It could not be ascertained whether they possessed an intrinsic capacity for locomotion or were passively transported by streaming of the cytoplasm. As mitochondria move, they usually maintain the shape characteristic for that cell type, but filamentous forms may be observed to branch or to break up into short rods. Such observations on living cells established that mitochondria are dynamic organelles displaying considerable plasticity and responding to changing physiological and environmental conditions.

Study of mitochondria and speculation as to their probable function were almost exclusively the prerogative of the morphologist, but there were two biochemical observations which later proved to be significant. Michaelis (1898) demonstrated that mitochondria of living cells reduced Janus green, an oxidation-reduction indicator dye, and Warburg (1913) showed that the capacity to consume oxygen resided in particulate components of the cell. The majority of biochemists that followed had little interest in the cell organelles, and they tended to regard particulate components of cells as an annoyance and obstacle to their efforts to isolate respiratory and other enzymes. The concept had evolved that oxygen is involved in cell metabolism through the action of an iron-containing catalyst, later called *cytochrome* (Warburg, 1925). The pyridine nucleotides were isolated and their structure determined and the principle was developed that oxidations in cell respiration take place by removal of hydrogen from the substrate. In the 1930s flavoproteins were found to be the link between dehydrogenases and the cytochromes in the flow of hydrogen in cell respiration.

A number of the dehydrogenation reactions by which oxidation of foodstuffs is accomplished were worked out in the years that followed. It was proposed that oxidation of carbohydrates took place in a cycle of reactions in which short carbon chains are continuously broken down to yield CO_2 — the *Krebs cycle*. It was also discovered that the oxidation of glyceraldehyde phosphate was somehow coupled to the synthesis of ATP (Warburg, 1938) and that generation of additional ATP is linked to the respiratory sequence of reactions that reduce oxygen to water. It became apparent that the coupling of ATP synthesis to respiration, called oxidative phosphorylation, is the principal source of ATP in the cell.

The cytologists Bensley and Hoerr (1934) first attempted to isolate mitochondria by differential centrifugation from homogenates of liver. This initial effort met with little success. However, when this approach was taken up again nearly a decade later (Claude, 1946), it was found that when 0.88 M sucrose was used as the medium, mitochondria could be successfully isolated. They retained near normal form and stained supravitally with Janus green (Hogeboom, Schneider and Palade, 1948). This extraordinarily important technical advance led to rapid expansion of knowledge on the function of mitochondria and other subcellular organelles. Mitochondria were found to contain nearly all of the cytochrome oxidase (Hogeboom et al., 1948) and fatty acid oxidase (Kennedy and Lehninger, 1948) of the homogenate from which they were isolated. They could carry out *in vitro* the oxidation of all the tricarboxylic acid cycle intermediates and generate ATP by oxidative phosphorylation (Kennedy and Lehninger, 1949). Thereafter mitochondria were regarded as the principal site for generation of energy for cellular activities.

The advent of the electron microscope and methods for preparation of ultrathin tissue sections permitted, for the first time, an examination of the interior of these cell organelles. Their typical internal structure was described independently by Palade (1953) and Sjöstrand (1953). The mitochondrion was found to be limited by a continuous outer membrane and lined by an inner membrane that formed narrow folds, or *cristae,* projecting inward. The interior of the organelle was occupied by a dense matrix that appeared homogenous except for occasional dense *matrix granules* 25 to 35 nm in diameter. Although subsequent examination of a great many cell types revealed considerable variation in size and shape of mitochondria and in the form and number of cristae, there was no significant departure from the basic structural plan described by Palade.

The fixatives routinely used for electron microscopy did not preserve all of their structural components. Examining fragments of unfixed mitochondria by the negative staining technique, Fernandez-Moran (1963) was able to visualize subunits of the inner membrane that had previously gone undetected. These were spherical particles 9 nm in diameter connected to the inner membrane by a slender stalk 3 to 4 nm wide and about 5 nm long. They were found only on the side of the membrane toward the matrix, and it was immediately suspected that they might consist of some of the membrane-associated mitochondrial enzymes. The subunits could be separated from the membranes of lysed mitochondria by sonication and isolated by centrifugation in a sucrose gradient. They did not contain cytochromes (Williams and Parsons, 1964), but they did catalyze oxidative phosphorylation and ATP hydrolysis (Racker et al., 1964). Moreover, when mitochondrial ATPase, purified by chemical means, was examined with the electron microscope in negatively stained preparations, it consisted of 9 nm particles similar to the isolated *inner membrane subunits*. Thus the units projecting into the matrix from the inner membrane appear to contain the complement of enzymes responsible for most of the energy production in animal cells.

The energy-generating machinery of mitochondria is activated when the products of digestion of food are delivered to them in the form of pyruvate (from carbohydrate), amino acids (from protein), and fatty acids (from fats). These substrates are oxidized in the respiratory chain, a multienzyme system within the inner mitochondrial membrane. The members of this system are not randomly distributed in the membrane but are believed to be specifically arranged in assemblies that facilitate their functional interaction. These assemblies recur throughout the membrane (Lehninger, 1964). Their arrangement moves protons (H^+) vectorially from the matrix into the space between the inner and outer membranes, creating a concentration gradient and therefore a membrane potential. The ATPase associated with the inner membrane subunits then pumps protons back through the membrane into the matrix, resulting in a proton current that drives the formation of ATP (Mitchell, 1966). Thus the powerful combination of ultrastructural and biochemical analysis has given us considerable insight into the functions of this remarkably complex and precisely organized organelle.

An unexpected development in the 1960s was the discovery that mitochondria possess their own DNA and RNA and have the capacity to synthesize some of their own proteins. Early morphological studies on trypanosomes had identified a structure near the base of the flagellum called the *kinetoplast*. This stained with the Feulgen reaction and therefore seemed to contain DNA. When it became possible to examine this organelle with the electron microscope, it was found to have outer and inner membranes and a limited number of cristae and therefore had to be interpreted as a specialized mitochondrion. The dense mass of filaments in its matrix was further identified as DNA by the demonstration that it incorporated tritiated thymidine before cell division (Steinert, 1960). The kinetoplast was considered an exceptional case with little bearing upon mitochondria in general, until mitochondria and chloroplasts of certain unicellular algae were also found to contain a few 4 nm filaments (Ris, 1962). Simultaneously and independently, similar filaments were observed in pale areas of matrix in mitochondria of chick embryos and soon thereafter in many other cell types (Nass and Nass, 1962; Nass, Nass and Afzelius, 1963). In biochemical studies, amphibian eggs were shown to contain 300 to 500 times the amount of DNA present in other cells of the same organisms and this extra DNA was found in the mitochondrial fraction prepared from homogenates of eggs (Dawid, 1966).

When mitochondria were disrupted and spread upon a water surface, the liberated DNA appeared in electron micrographs as filaments about 4 nm in diameter, forming circles about 5 μm in circumference. Denaturation of the DNA with heat separated the filament into single nucleotide chains, and it was concluded that *in vivo* it occurs as a double helix like that of the nucleus. The circular DNA strands closely resemble those of viruses and bacteria.

Also present in the mitochondrial matrix are small (10 to 15 nm) dense granules interpreted as ribonucleoprotein particles comparable to ribosomes of the general cytoplasm (André and Marinozzi, 1965). In size, they resemble the ribosomes of bacteria. It has now been shown that mitochondria contain all of the factors necessary for incorporation of nucleotides into DNA and RNA and for synthesis of proteins from amino acids. However, the limited information content of mitochondrial genomes makes it unlikely that they synthesize all of their own components. Thus, mitochondrial DNA probably codes for their DNA, RNA, and the protein of their ribosomes, for transfer RNA and for certain of their structural elements. However, mitochondria seem to be dependent upon the nuclear genome for synthesis of many, if not most, of their enzymes.

STRUCTURE OF MITOCHONDRIA

A mitochondrion in thin section exhibits the characteristic internal organization illustrated in the accompanying electron micrograph. It is bounded by a smooth-contoured outer membrane about 7 nm thick. Within this limiting membrane and separated from it by a space 8 to 10 nm wide is an inner membrane. The inner membrane has numerous infoldings, the *cristae mitochondriales*, that project into the interior of the organelle. These are of variable length and form a series of incomplete transverse septa. The membranes of the mitochondrion delimit two compartments. The narrow cleft between the outer and inner membrane and extending inward between the leaves of the cristae is called the *membrane space*, or *intracristal space*. Its content is of low density. The large cavity bounded by the inner membrane and its cristae is occupied by a homogenous or finely granular matrix, which varies in density from one cell type to another and with different degrees of extraction during specimen preparation. After freeze-substitution it is invariably quite dense. In most cell types dense *matrix granules* 30 to 50 nm in diameter are scattered at random in the inner compartment.

This pattern of internal structure with thin *lamellar* or *foliate* cristae is described as the typical, or orthodox, configuration of the mitochondrion and is the one found in micrographs of the great majority of cells after routine specimen preparation. However, numerous modifications of this basic architectural plan are encountered in specific cell types and in different physiological states of the same cell type, as will be illustrated in later pages.

Figure 218. A longitudinal section of a mitochondrion and surrounding cytoplasm from pancreas of the bat, *Myotis lucifugus*. (Micrograph courtesy of Keith Porter.)

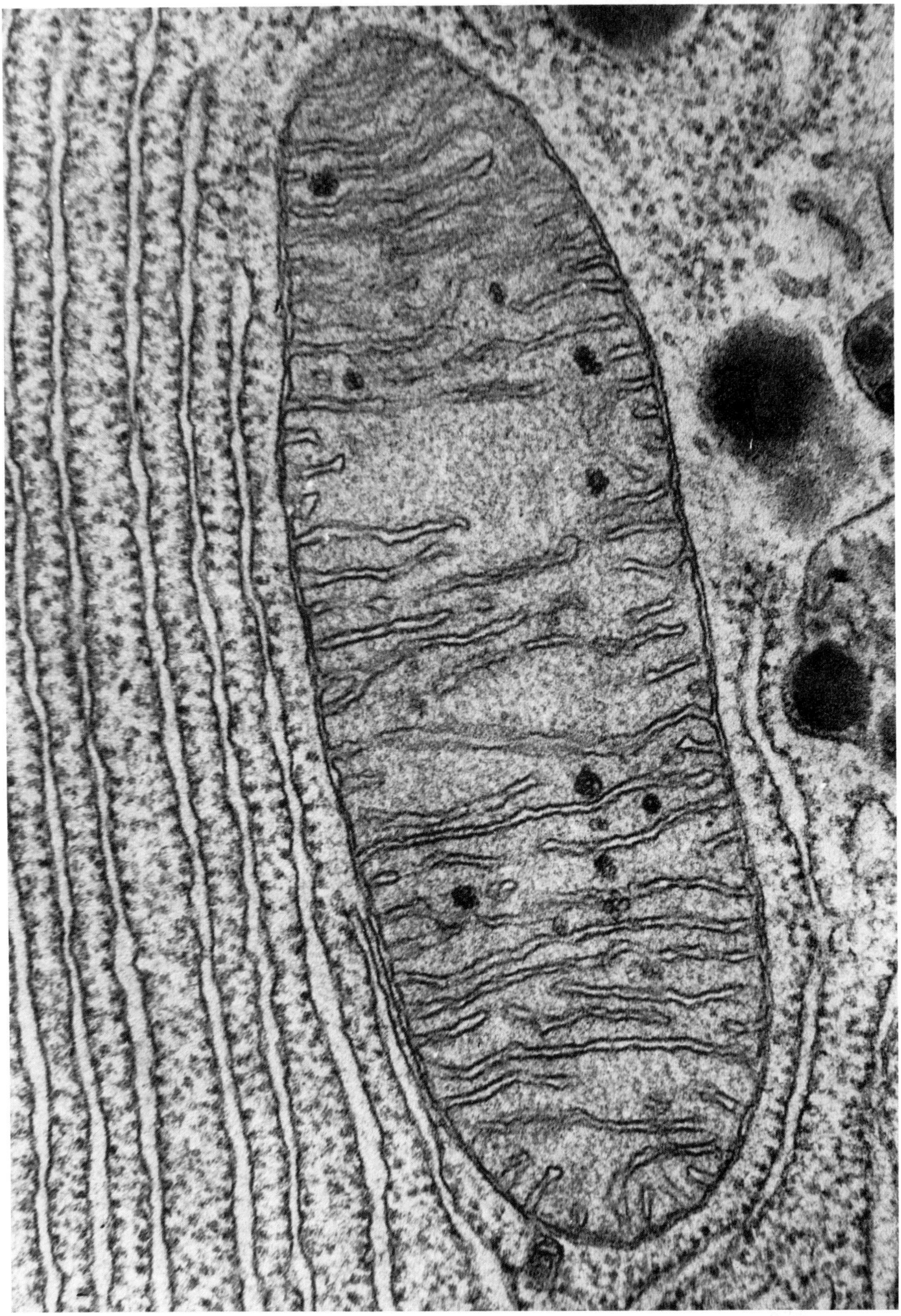

Figure 218

Electron micrographs of thin tissue sections are a poor basis for description of the external form of mitochondria. Their shape can be seen to better advantage in selectively stained histological sections or in living cells viewed with phase contrast microscopy. Such preparations are favorable because they are thick enough to include whole mitochondria. Electron micrographs, on the other hand, give a misleading impression of their shape because the section includes only thin slices through cell organelles. Randomly oriented mitochondria of uniform shape therefore present various round, elliptical, or elongated profiles, depending upon the orientation of the organelles in relation to the plane of the section.

In the micrograph shown here the majority of the mitochondrial profiles are round cross-sections; only a few are elongated, oblique, or longitudinal sections. In fact, all of the mitochondria of this cell type are long flexible rods, but this is not apparent in thin sections.

Figure 219. Micrograph of a cell from the ductulus efferens of the ground squirrel, *Citellus lateralis*. (Micrograph courtesy of Jeffrey Pudney.)

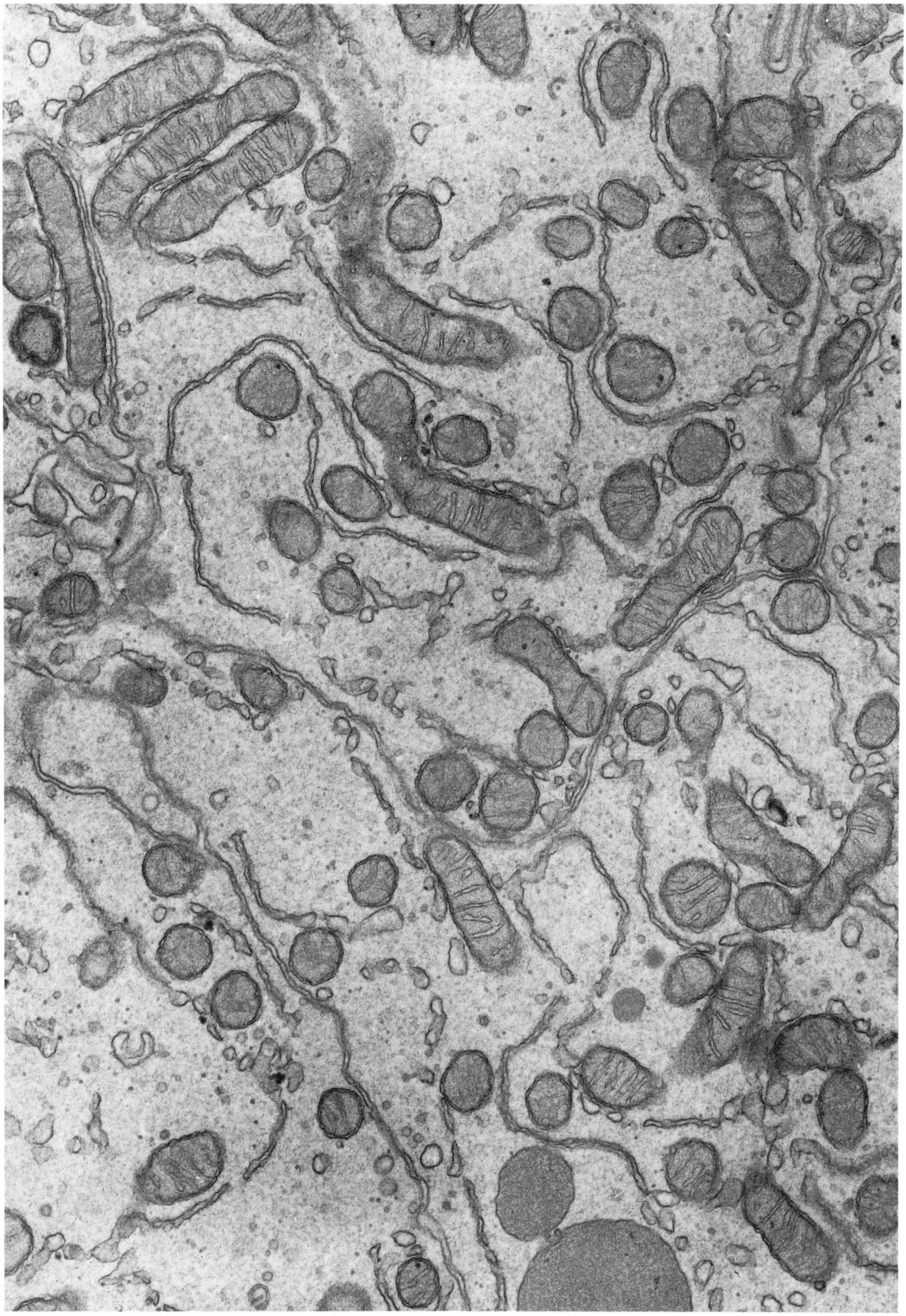

Figure 219

In general, the number of mitochondria per cell and the number of cristae per mitochondrion are related to the energy requirement for the function carried out by that cell type. The mitochondrion in the upper figure from moderately active epididymal epithelium has an average number of lamellar cristae. In the lower figure, the mitochondrion, from a very rapidly contracting skeletal muscle with high energy requirements, contains many long and closely packed cristae.

Figure 220. Mitochondrion from mouse epididymis.

Figure 221. Mitochondrion from cricothyroid muscle of the bat, *Myotis lucifugus.*

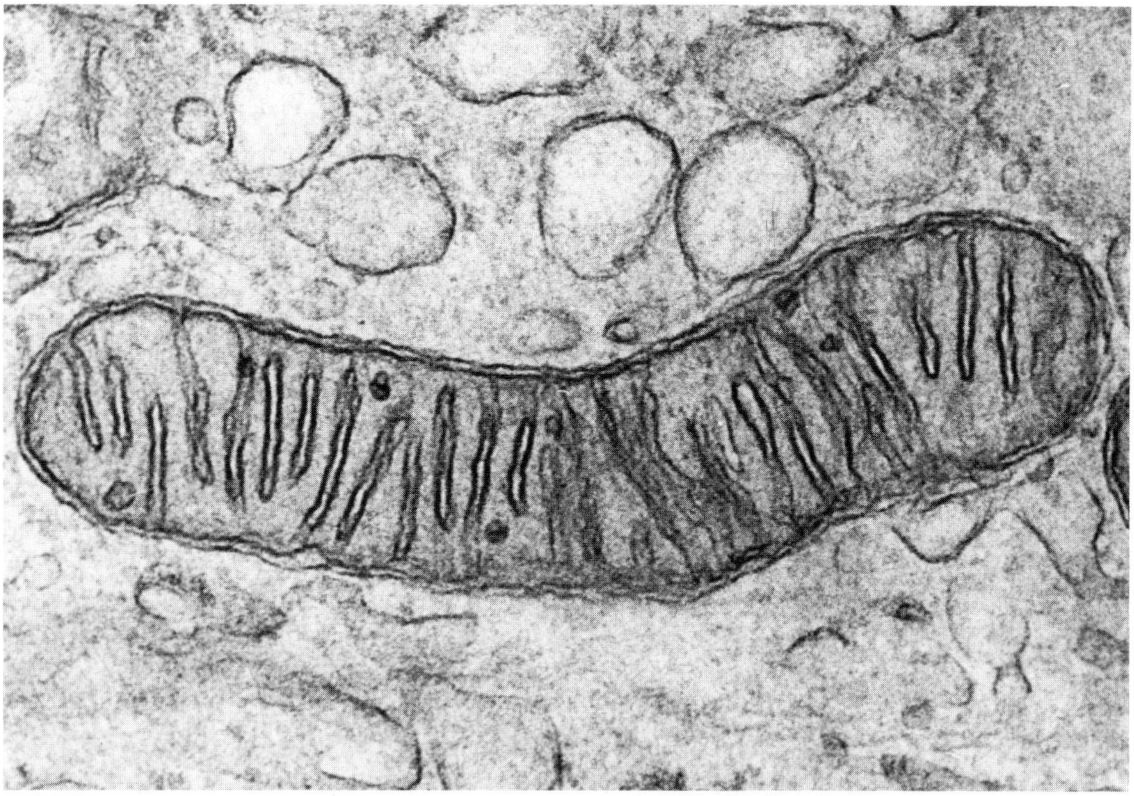

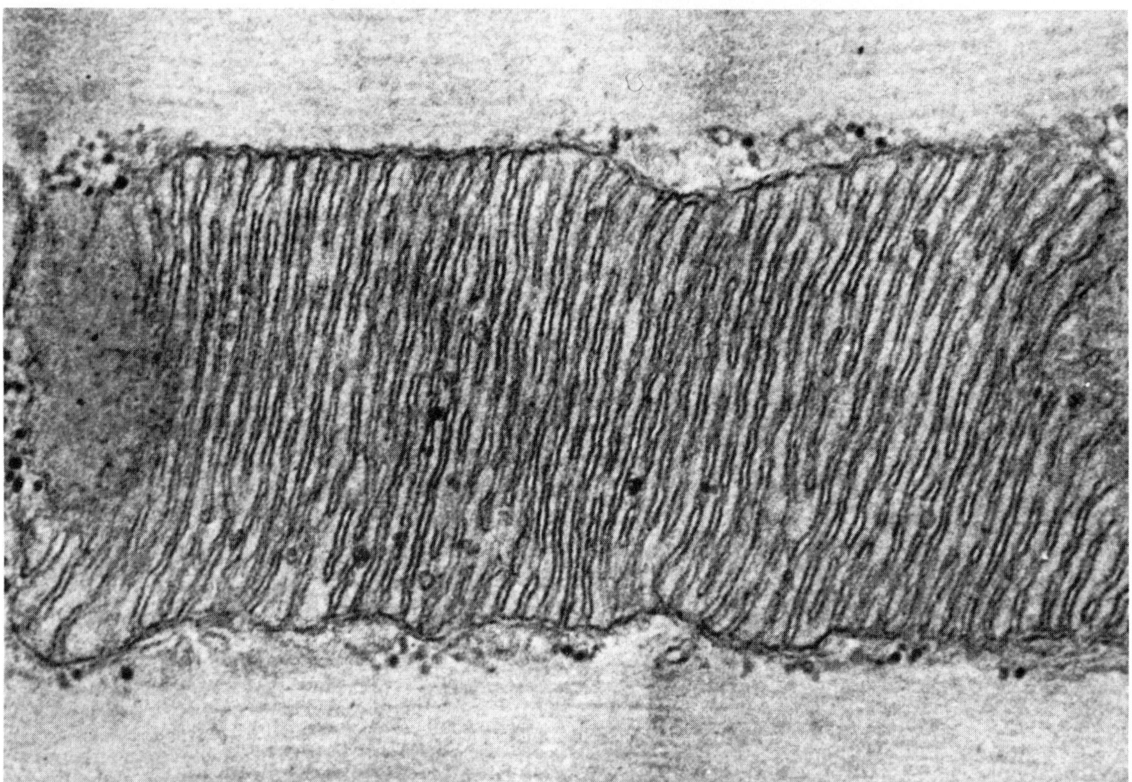

Figure 220, *upper* Figure 221, *lower*

MATRIX GRANULES

The matrix granules are approximately spherical, osmiophilic inclusions located within the space bounded by the inner mitochondrial membrane. They may be free in the matrix but are often closely associated with the membranes of the cristae. Their density usually obscures any internal structure, but in very thin sections viewed at high magnification they appear to be subdivided by thin membrane-like septa into minute loculi or compartments. These membrane-like profiles may also assume concentric configurations.

The composition and function of the matrix granules continue to be subjects of controversy. When calcium or other divalent cations are present in high concentration in the fluid bathing isolated mitochondria, large dense granules are found in their matrix. It seemed likely that these conspicuous densities were not inorganic precipitates formed *de novo* but that cations had accumulated in the preexisting matrix granules. It was suggested therefore that these granules probably functioned in the regulation of the internal ionic environment of the mitochondrion (Peachey, 1964).

This interpretation was strengthened by the detection of calcium phosphate in the large matrix granules of cells in calcifying cartilage and bone by x-ray microprobe analysis (Sutfin et al., 1971). Although it was assumed from these observations that sequestration of calcium was probably a general property of mitochondrial matrix granules, it was difficult to exclude the possibility that the calcium detected had diffused into the mitochondria from the calcium-rich environment of these calcifying tissues during specimen preparation.

When mitochondria from several noncalcifying tissues were subjected to x-ray microanalysis after glutaraldehyde fixation, no calcium was found in the matrix granules. But when fixed in osmium supplemented with 5 mM Ca^{2+} they were found to bind detectable amounts of this element (Barnard and Ruusa, 1979). In most cell types the granules have no inherent electron density but only become dense after exposure to osmium-containing fixatives, indicating that their composition is mainly organic. Their affinity for osmium and their membrane-like ultrastructure in cells other than bone suggests that their major component is lipid (Barnard and Afzelius, 1972). If they do seed the deposition of calcium and other cations these evidently make up only a small proportion of their mass. Efforts to isolate a matrix granule fraction for biochemical analysis have met with some success and indicate that they are composed of a phospholipoprotein (Brdiczka and Barnard, 1980). The lipoprotein binds calcium under some *in vitro* conditions, but there is no convincing evidence that the matrix granules in nonmineralized tissues are involved in mitochondrial calcium sequestration *in vivo*. Their close association with the cristae and their membrane-like composition suggests that they may instead be involved in assembly of the mitochondrial inner membrane.

Figure 222. Mitochondria from cat ventricular cardiac muscle with numerous matrix granules.

Figure 223. Mitochondrion of bat pancreas.

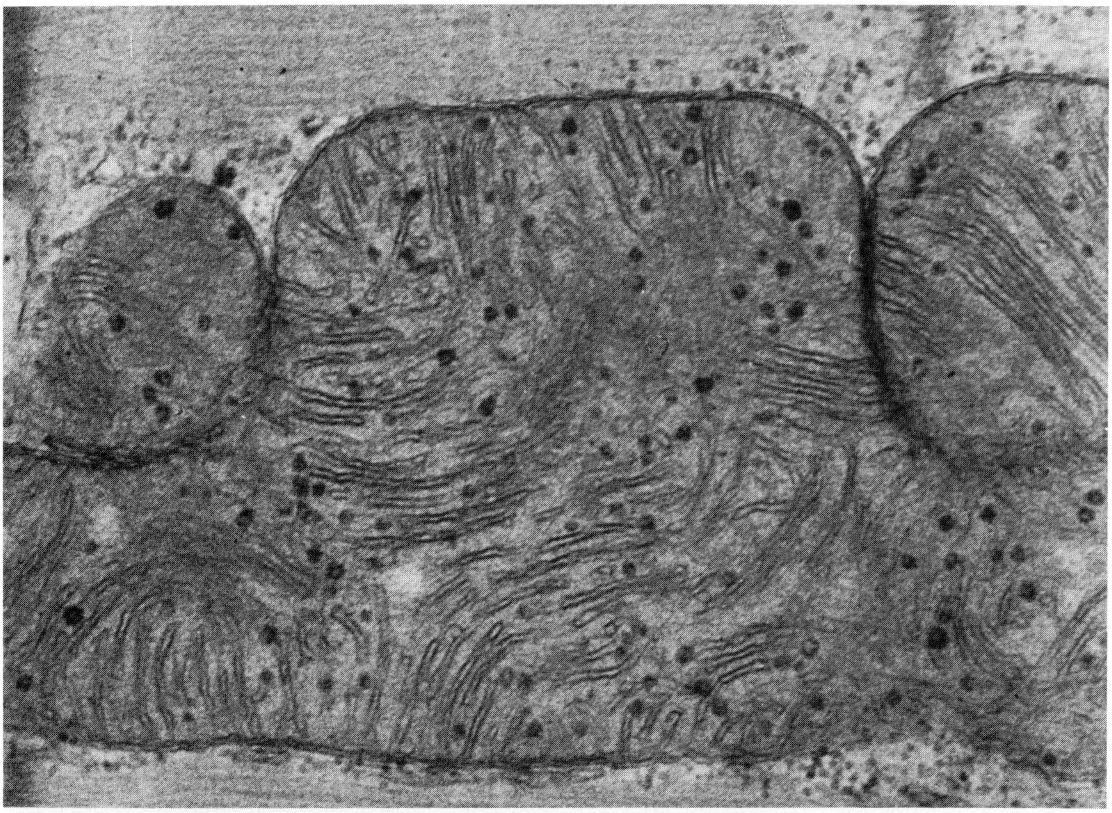

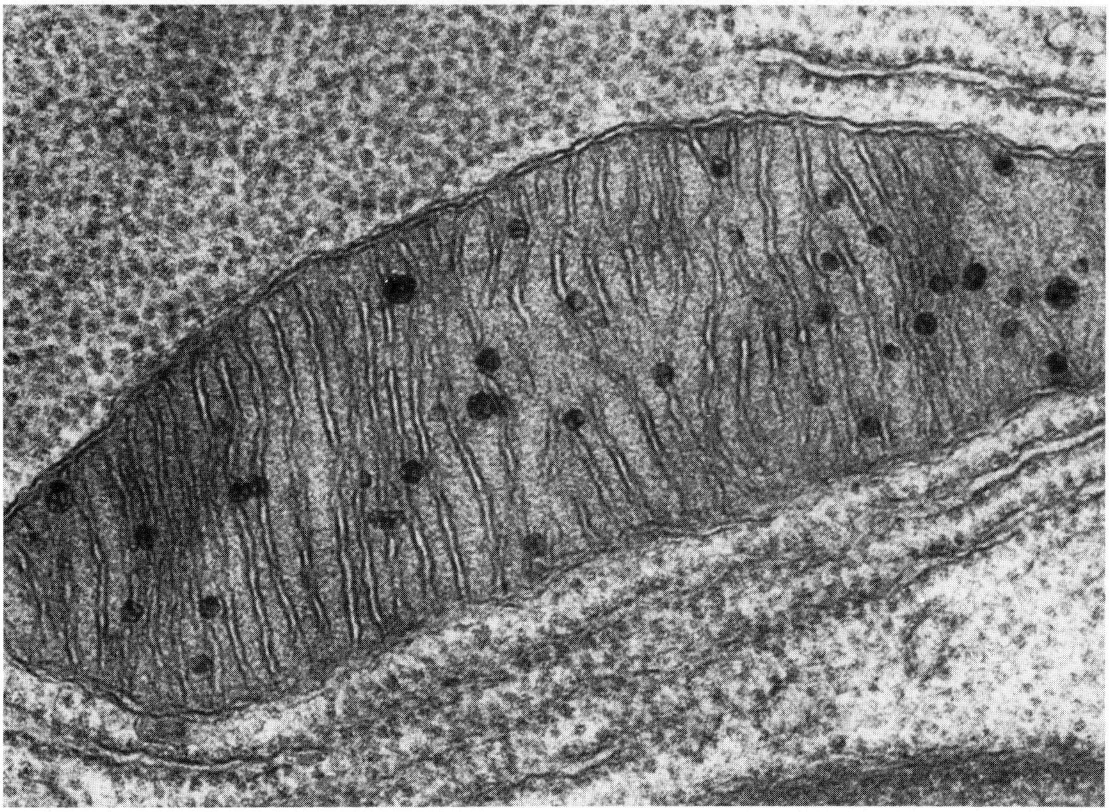

Figure 222, *upper* **Figure 223**, *lower*

Matrix granules have been reported to be sensitive to a great variety of physiological stimuli and to toxic agents, but from such observations no meaningful pattern has emerged that sheds light upon their origin or function.

Mitochondrial granules range in size from 25 nm to 120 nm and vary greatly in number in different tissues and in different physiological states of the same tissue. The upper figure shows the typical small granules in liver mitochondria. The lower figure illustrates the less common, large matrix granules occasionally encountered. Those illustrated here are from certain cells in proximal convoluted tubule of an apparently normal guinea pig kidney.

The mitochondria of osteoclasts active in bone resorption often contain clusters of large matrix granules which exhibit small dense granules in their interior (Gonzales and Karnovsky, 1961). The presence of calcium and phosphorus in granular deposits within mitochondria of bone cells has been confirmed by electron probe microanalysis (Sutfin et al., 1971). Matrix granules of bone cells and duodenal cells have been shown to sequester calcium from the fixative in the presence of osmium (Martin and Mathews, 1969; Sampson et al. 1970). Thus matrix granules in these cell types appear to have the capacity to sequester calcium by a nonenergy-requiring mechanism. Whether this is their major role *in vivo* or whether their organic constituents have other important functions remains unsettled.

Figure 224. Mitochondria from bat liver. (Micrograph courtesy of Susumu Ito.)

Figure 225. Mitochondria from guinea pig proximal convoluted tubule.

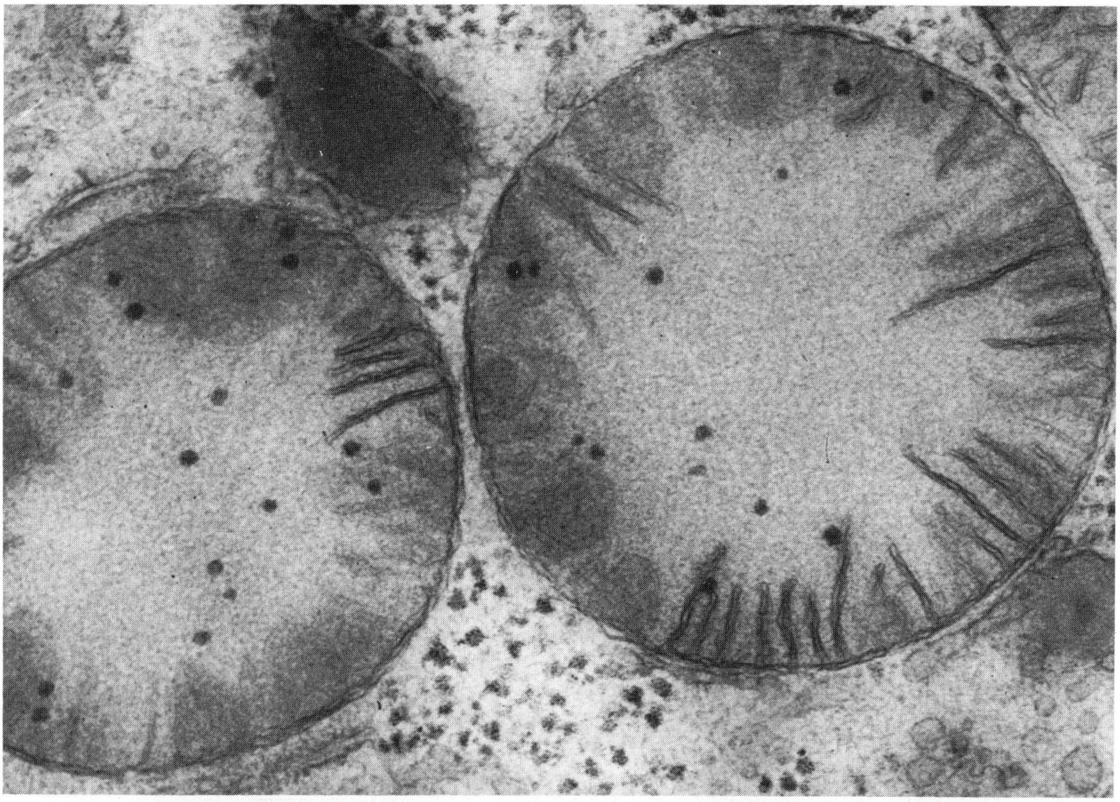

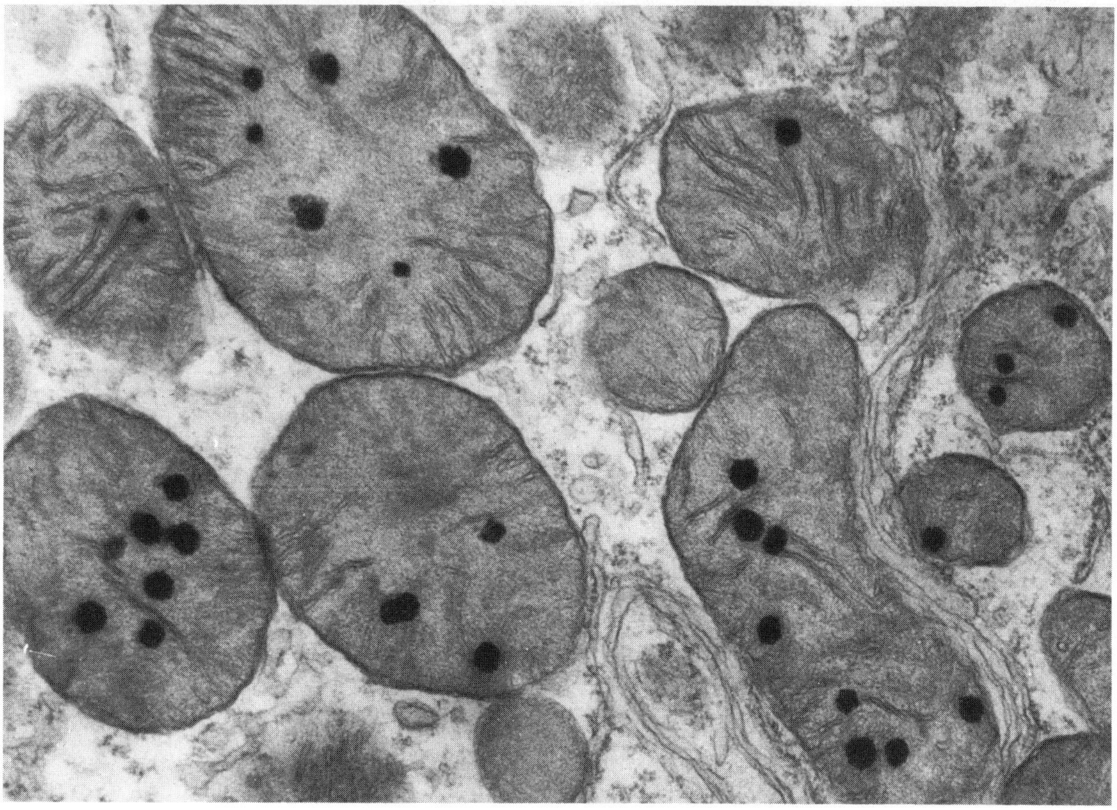

Figure 224, *upper* **Figure 225**, *lower*

MITOCHONDRIAL DNA AND RNA

DNA has been demonstrated in the mitochondria of many plant and animal cells. It is found in the form of filaments of varying thickness located in the matrix in circumscribed regions of lower density. Initially no significance was attached to these small pale areas of matrix because larger but otherwise similar areas occur as a common artefact of glutaraldehyde fixation. Microscopists have now learned to look for DNA in such areas. The thicker strands of DNA frequently appear to branch into more slender filaments. It is likely that the thicker strands are formed by lateral association of thin DNA filaments.

A mitochondrion from a plant cell is shown here in the upper figure. In addition to conspicuous DNA filaments in the matrix, there are a number of small dense particles of uniform size. These are RNP particles comparable in composition and function to cytoplasmic ribosomes. In mitochondria of animal cells these are usually obscured by a relatively dense matrix. A mitochondrion from adrenal cortex, shown in the lower figure, is crowded with tubular cristae, but short segments of DNA can be identified in two or three small areas of low density.

Figure 226. Mitochondrion from the plant *Vica fabia.* (Micrograph courtesy of Hewson Swift.)

Figure 227. Mitochondria from snake adrenal adjacent to a highly ordered array of endoplasmic reticulum. (Micrograph courtesy of John Long.)

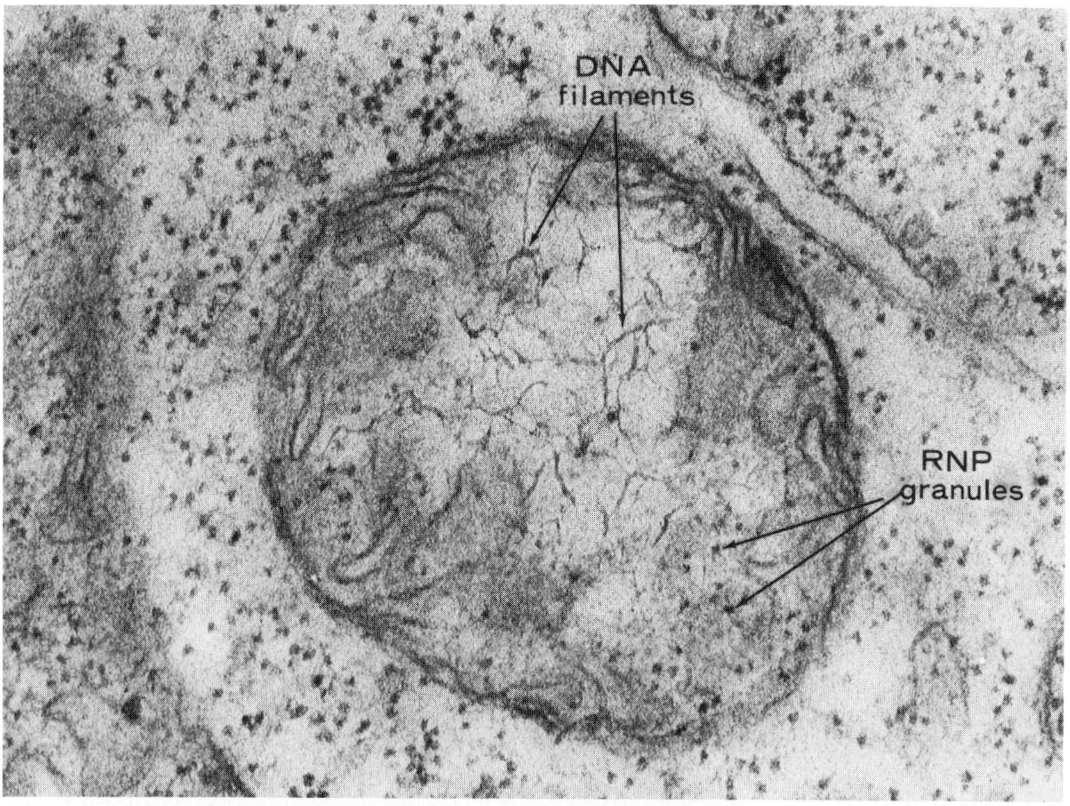

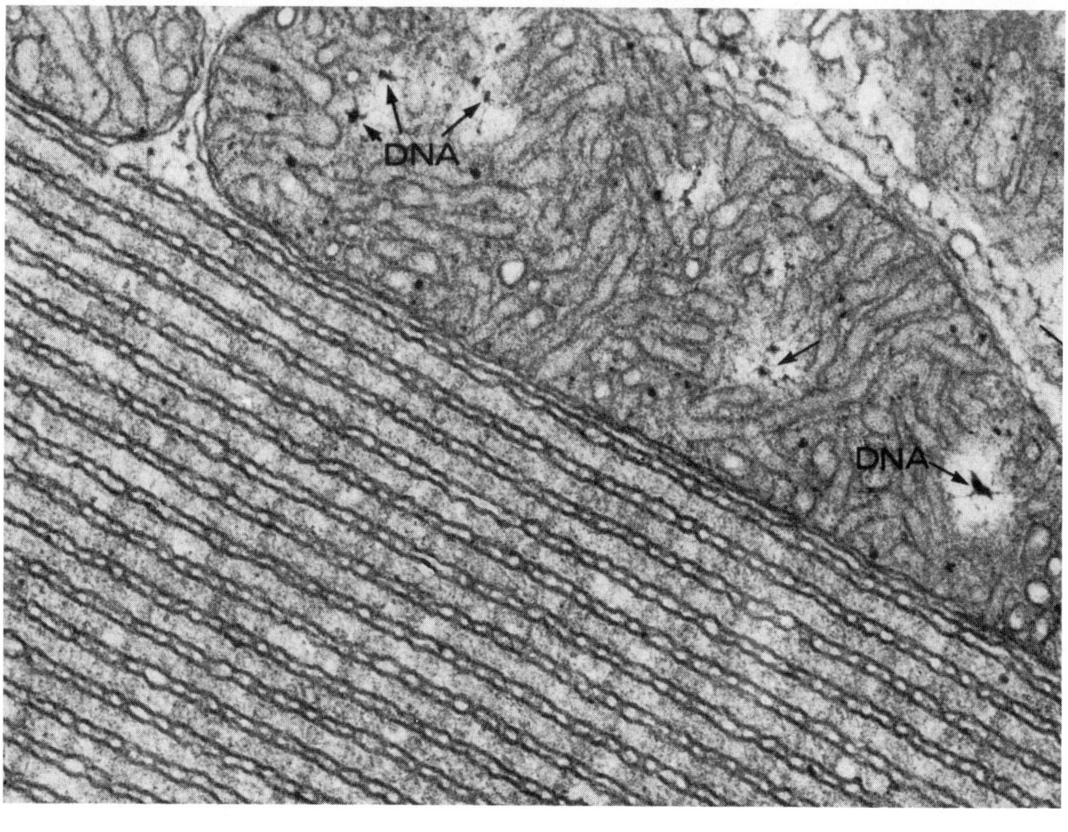

Figure 226, *upper* **Figure 227**, *lower*

It is rare to observe more than two or three sites of DNA in a single thin section of a mitochondrion. The example on the facing page, from cultured adrenal, is unusually large and shows a greater than normal number of DNA filaments. The abnormal environment *in vitro* may have resulted in growth of the organelle and DNA replication without subsequent division.

When disrupted mitochondria are spread by surface tension on a film of water and the liberated DNA is collected on an electron microscope grid, it appears in the form of circular filaments such as that shown in the lower figure. These circular DNA molecules are 5 to 6 μm long, have a molecular weight of about 10 million and consist of 15 to 17 kilobase pairs. There are normally from 3 to 6 per mitochondrion. The maximum coding capacity of this relatively small amount of DNA is estimated to be only about 5000 amino acids (Borst et al., 1967). Mitochondrial DNA codes for two ribosomal RNAs and a set of transfer RNAs. Evidence for mitochondrial messenger RNAs is still indirect but it seems clear that each mitochondrion has all of the ingredients necessary for transcription of the genetic information carried in its DNA. However, it is apparent from the limited coding capacity of its small genome, that the mitochondrion must be dependent upon the nuclear genome to direct the synthesis of many of its constituents.

Figure 228. Mitochondrion from an organ culture of rat adrenal. (Micrograph courtesy of Arvi Kahri.)

Figure 229. Mitochondrial DNA molecule from an oocyte of *Xenopus laevis*. (Micrograph courtesy of Igor Dawid.)

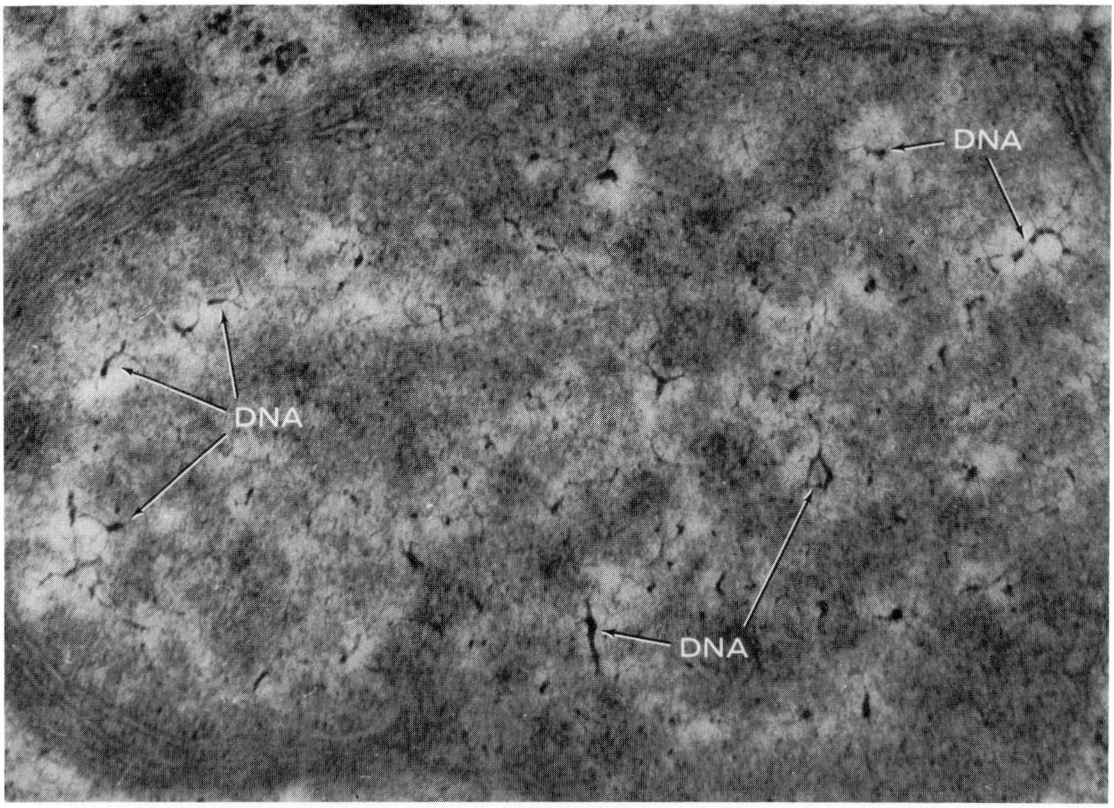

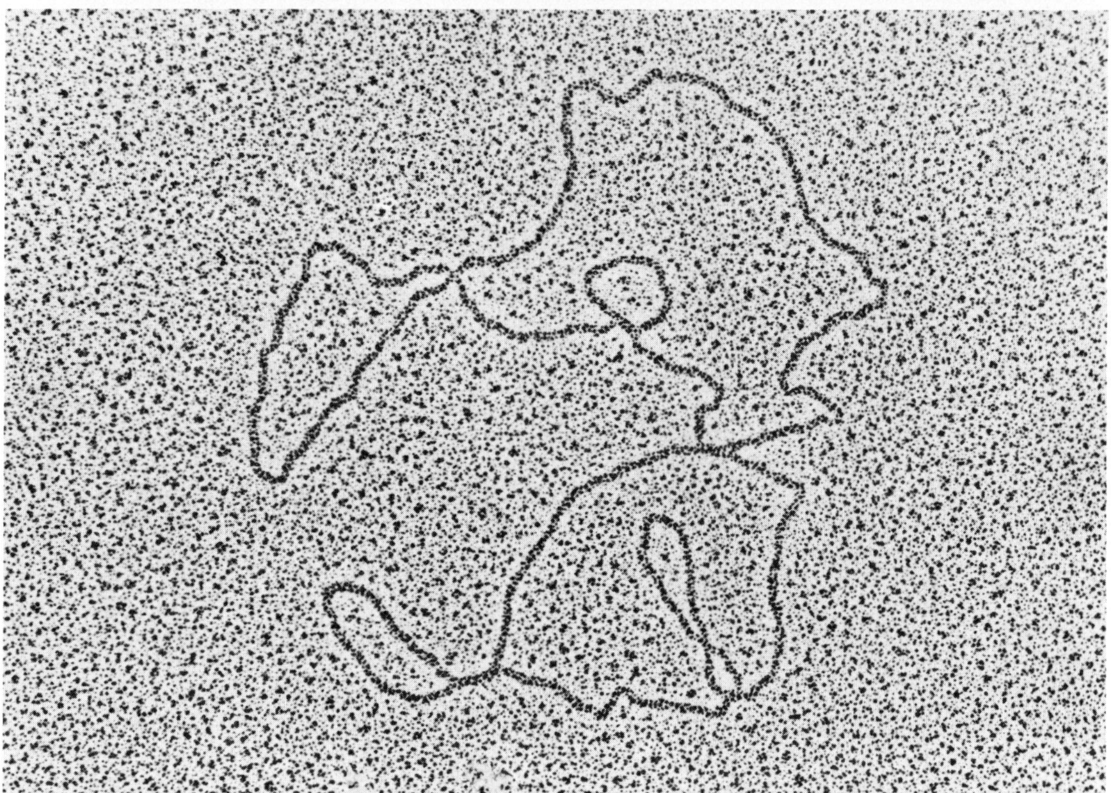

Figure 228, *upper* **Figure 229**, *lower*

The first suggestive evidence of mitochondrial DNA came from morphological and histochemical observations on trypanosomes over 50 years ago. In these organisms a unique organelle, the *kinetoplast,* is located near the base of the flagellum. Light microscopists suspected that this structure was a component of a modified mitochondrion because the organelle containing it could be stained with Janus green (Shipley, 1916). When the kinetosome was subsequently shown to give a positive Feulgen reaction (Bresslan and Scremin, 1924), the term *kinetonucleus* was preferred by some investigators. The paradoxical occurrence of Janus green and Feulgen staining of the same organelle was considered an exceptional situation limited to trypanosomes and therefore of little general interest.

The nature of the kinetoplast was clarified nearly 40 years later by the electron microscope, which showed that the Feulgen reactive component is a long rod or disc made up of intensely staining DNA filaments oriented transversely to its long axis, as illustrated in the lower illustration on the facing page. This structure is located in the matrix of a modified mitochondrion with outer and inner membranes and a few typical cristae (Steinert, 1960). The kinetoplast replicates and divides prior to cell division. Interference with DNA synthesis by treatment with acridine dyes results in a marked decrease in oxidative metabolism, which suggests that normal mitochondrial function is dependent upon kinetoplast DNA (Trager and Rudzinska, 1964).

In some stages of the life cycle of parasitic trypanosomes, the Krebs cycle enzymes appear to be absent but they become active again when the organism enters a more virulent phase. This transformation is accompanied by striking changes in size, shape and internal configuration of the trypanosome's single mitochondrion (Vickermann, 1962, 1965; Hecker et al., 1972).

Figure 230. Kinetoplast region of the *Trypanosoma mega.* (Micrographs courtesy of Winston Anderson.)

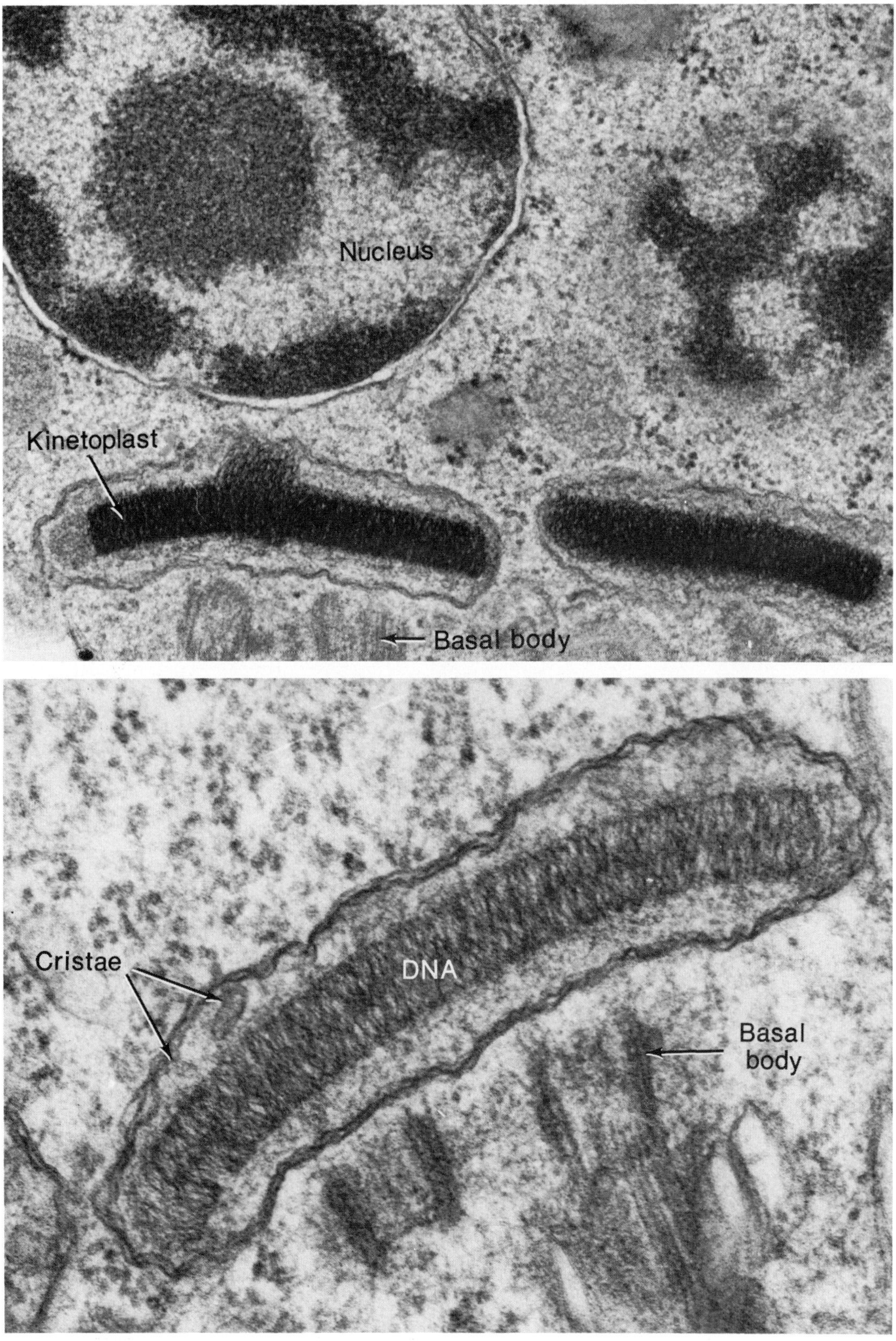

Figure 230, *upper and lower*

DIVISION OF MITOCHONDRIA

In the 1930s, time lapse cinematography of cells maintained in culture recorded frequent examples of apparent division or fragmentation of mitochondria. Some 20 years later, when the electron microscope was applied to the study of ultrathin sections of liver (Fawcett, 1955), mitochondria were occasionally observed in which a pair of membranes formed a partition extending across the organelle from the inner membrane on one side to that on the other. It was suggested that these probably represented stages in a process of mitochondrial division. However, from study of static images, the possibility that they were stages of mitochondrial fusion could not be excluded. This latter interpretation was rendered highly improbable by the subsequent observation of an increased frequency of septa under conditions in which rapid increase in mitochondrial number could be induced experimentally or was known to occur naturally. Hepatic cells of riboflavin-deficient mice have reduced numbers of exceptionally large mitochondria, but when riboflavin is added to the diet, their normal number and size are restored within two days. Electron micrographs of the liver during this period of rapid proliferation show large numbers of mitochondria with transverse partitions and associated constriction of the limiting membrane (Tandler et al., 1969). Many of the mitochondria of the fat body in certain insects undergo autolysis before pupation and then in a 12-hour period after emergence of the adult, there is a sevenfold increase in the number of mitochondria. During this period, partitioned and constricted mitochondria are very abundant (Larsen, 1970).

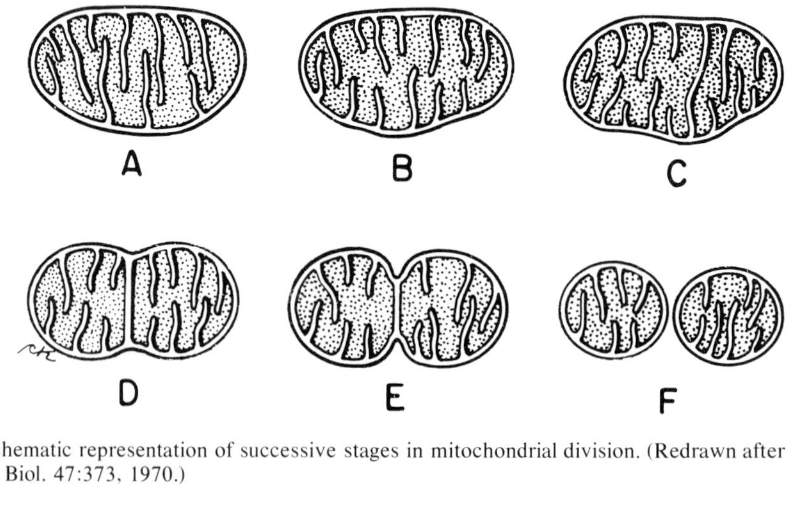

Schematic representation of successive stages in mitochondrial division. (Redrawn after Larsen, J. Cell Biol. 47:373, 1970.)

The most compelling evidence for the origin of new mitochondria from preexisting mitochondria came from autoradiographic studies on the distribution of isotopically labeled membrane precursors among mitochondria isolated from cultures of a strain of *Neurospora* requiring choline (Luck, 1963). Labeled choline was supplied to cultures in a logarithmic phase of growth, and at suitable intervals after transfer to nonradioactive medium, mitochondria were isolated and spread upon slides for autoradiography. If *de novo* formation of mitochondria had occurred under these conditions, there should have been two populations, one labeled and the other unlabeled. Instead the label was found to be uniformly distributed among all mitochondria. In comparable studies on *Tetrahymena*, mitochondrial DNA was labeled with tritiated thymidine (Parsons and Rustad, 1968). After transfer to unlabeled medium, the label remained evenly distributed among all mitochondria for four generations of logarithmic growth. Therefore it was concluded that the increase in mitochondrial mass in the course of the experiment had been due to growth and division of the preexisting labeled mitochondria.

As in the proliferation of bacteria, mitochondrial division involves more than simple fission. Filaments of DNA are seen most frequently in mitochondria of rapidly growing cells. When isolated, some of the circular molecules exhibit a segment that branches and rejoins in a configuration suggesting DNA replication by a mechanism comparable to that postulated for the circular DNA of bacteria. It is believed that in the initial stage of mitochondrial multiplication the genome is replicated and this is followed by DNA-directed synthesis of components of mitochondrial ribosomes and some structural proteins. However, since the coding potential of mitochondrial DNA is quite limited, most of its structural protein, its complement of enzymes, and its lipid components are probably synthesized in the cytoplasm under the direction of nuclear genes. Little is known about how these components enter the growing organelle or what controls their ordered assembly to create the specific properties of the outer and inner membranes.

Mitochondria are occasionally seen in the process of degradation within autophagic vacuoles and they apparently have a limited life span. Investigations of the turnover rate of mitochondrial DNA in liver indicate that they have a half-life of about 8 days compared to estimates of 350 days or more for the nucleus. Similar analyses on other tissues yield values ranging from 6 days in heart muscle to 31 days in brain (Gross et al., 1969; Fresco and Bendick, 1960).

The septum of dividing mitochondria consists of two membranes separated by a narrow interspace with dimensions similar to those of an intracristal space. This might suggest that partitioning occurs by extension of one of the preexisting cristae across the organelle and fusion of its edge with the inner membrane on the opposite side. However, the membranes of the septa are slightly thinner than those of the cristae and in freeze-fracture preparations, they have fewer intramembrane particles (Kanaseki, 1975). Therefore the septum is not simply a modified crista but a distinctive structure formed anew from the inner mitochondrial membrane for its specific partitioning role in division of the organelle. A circumferential fold of the outer membrane then invades the space between the septal membranes and when its advancing edges meet and fuse, the separation of the daughter mitochondria is completed.

The micrograph on the facing page shows an early stage of invasion of the septum by the outer mitochondrial membrane. The paucity of such images suggests that this event proceeds more rapidly than the earlier formation of the septum.

Figure 231. Dividing mitochondrion in rat liver. (Micrograph courtesy of Daniel Friend.)

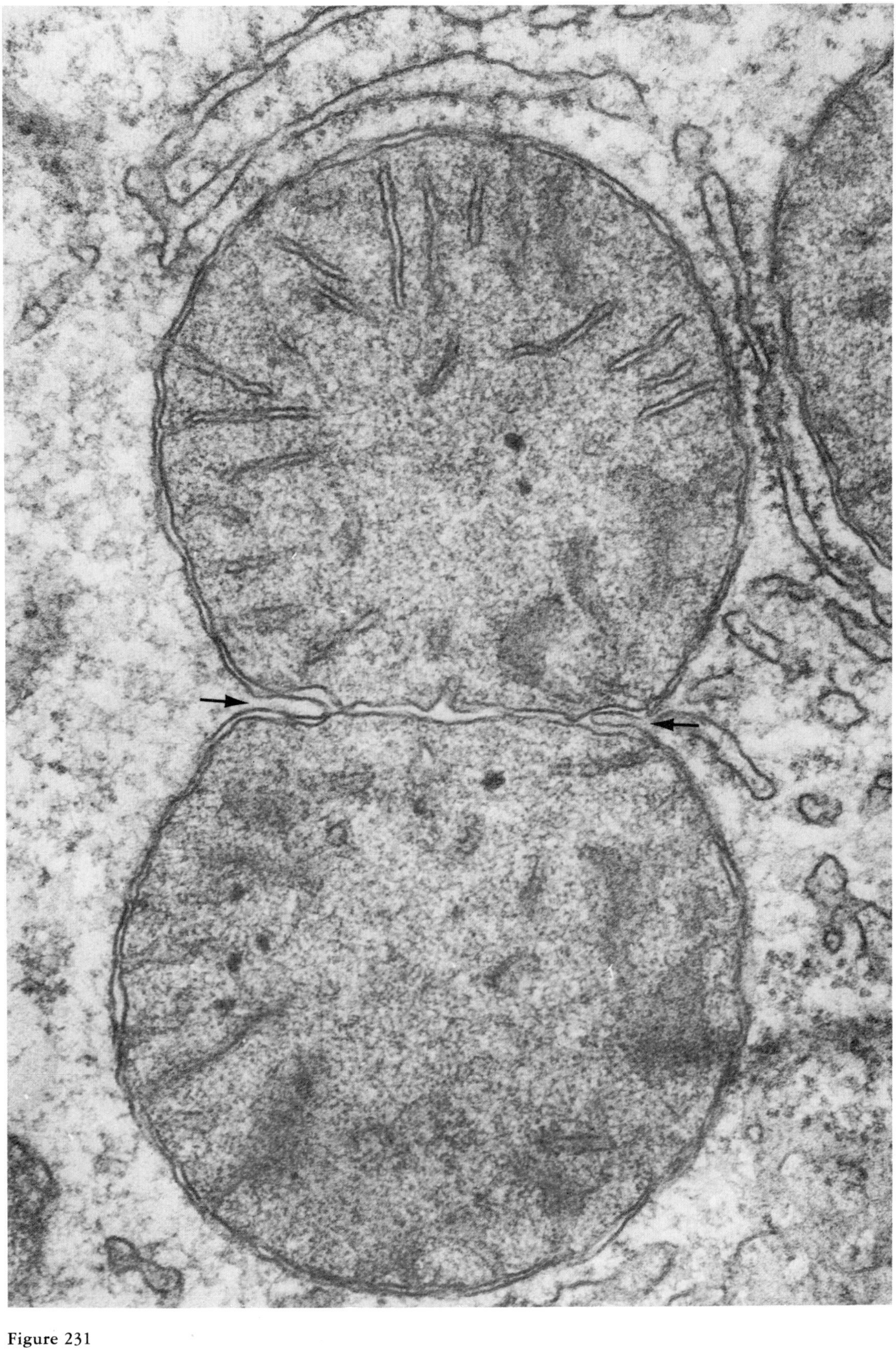

Figure 231

In the upper figure of a dividing mitochondrion, the bilaminar septum has been formed but separation of the two compartments by invasion of a fold of the outer membrane has not yet begun. It is evident from examination of the lower figure that the outer membrane does not always form a circumferential invagination of uniform width. In the central portion of the septum (at star), the membranes are separated by a deep and slender invagination from a sector outside the plane of section, while the outer mitochondrial membrane in this plane (at arrows) has advanced only a short distance into the cleft.

Figure 232. Dividing mitochondria in gastric mucosa of a mole. (Micrographs courtesy of Toku Kanaseki.)

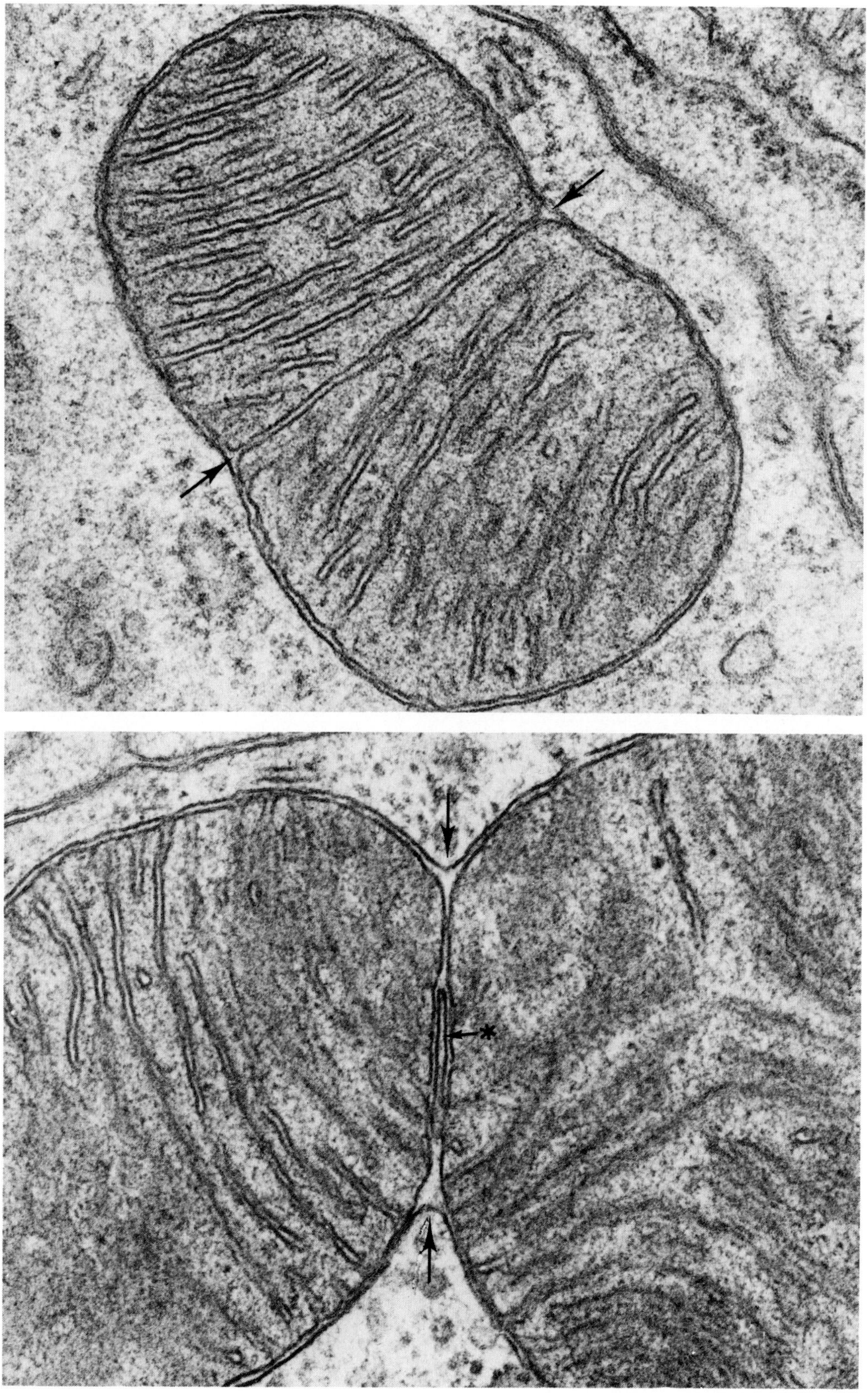

Figure 232, *upper and lower*

The accompanying micrographs represent three of the stages in mitochondrial division. In *A*, the free edge of the developing septum has not yet reached the opposite side of the organelle. In *B*, the asymmetrical penetration of the fold of outer membrane is evident. In *C*, the separation of the daughter mitochondria is complete.

Figure 233. Dividing mitochondria from gastric mucosa of a mole. (Micrographs courtesy of Toku Kanaseki.)

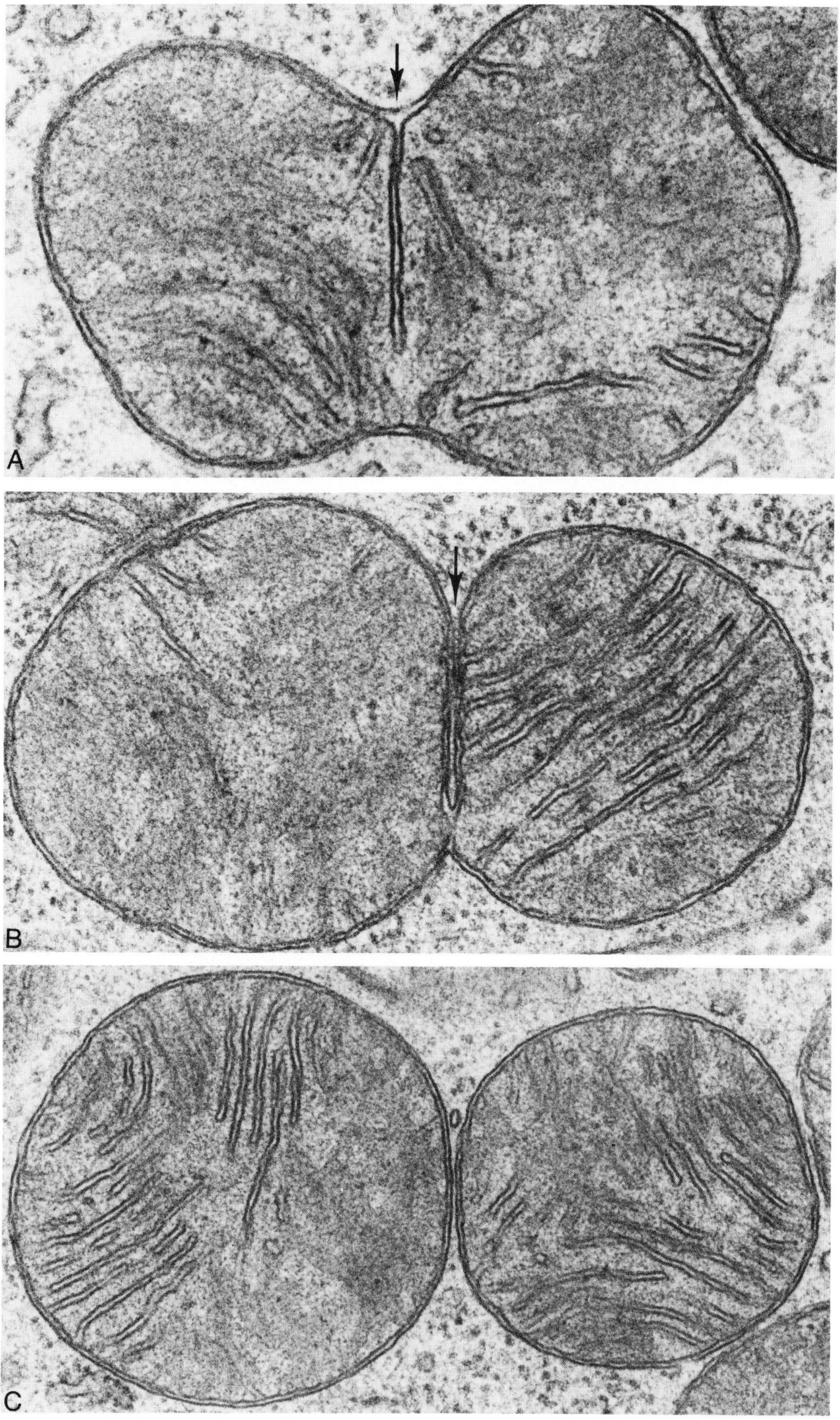

Figure 233

FUSION OF MITOCHONDRIA

No doubt in many cell types from time to time two mitochondria may make contact and fuse into a single mitochondrion, but these are isolated events of little physiological significance. The most dramatic examples of mitochondrial fusion are those that occur in insect spermiogenesis. Soon after the second meiotic division, spermatid mitochondria aggregate, as shown in the upper figure on the opposite page. They then begin a process of fusion that results in the formation of a large spherical mass traditionally called the *nebenkern.* As shown in the lower figure the progressive fusion of mitochondria results in longer and thicker organelles that are interwoven in a remarkably complex pattern.

Figures 234 and 235. Successive stages in the formation of the mitochondrial nebenkern in insect spermiogenesis. (Micrograph by David Phillips.)

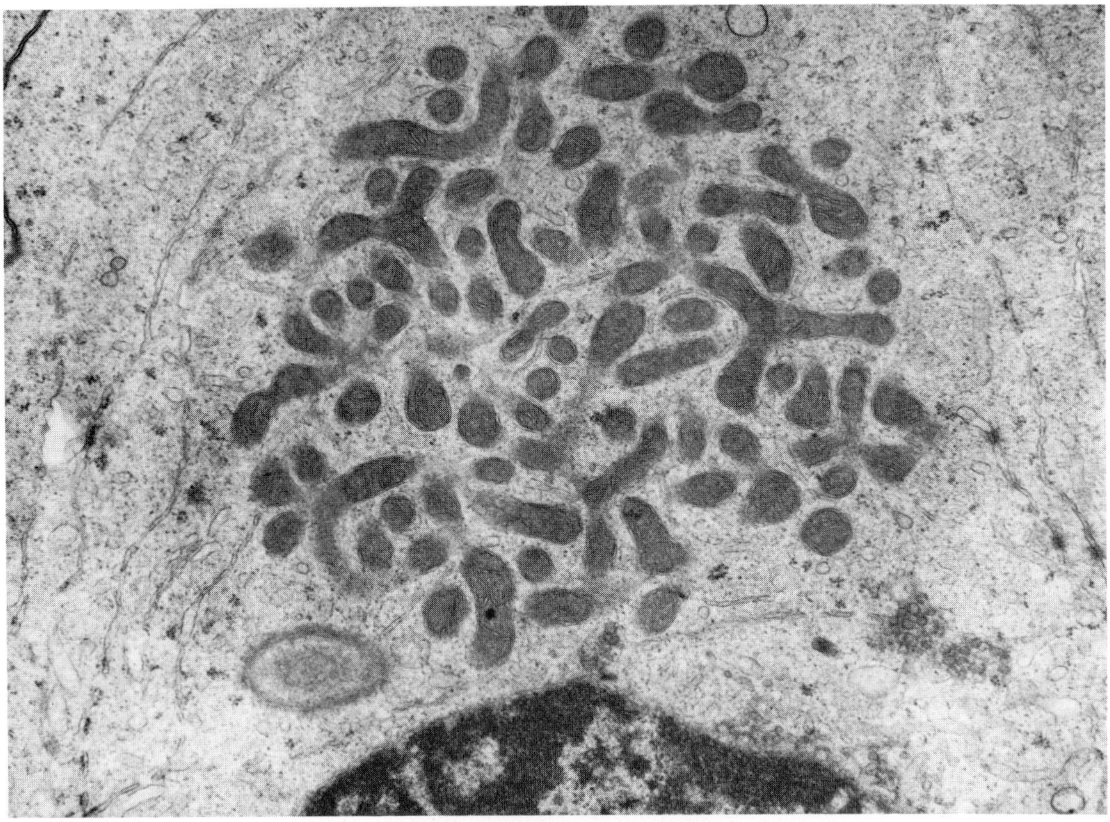

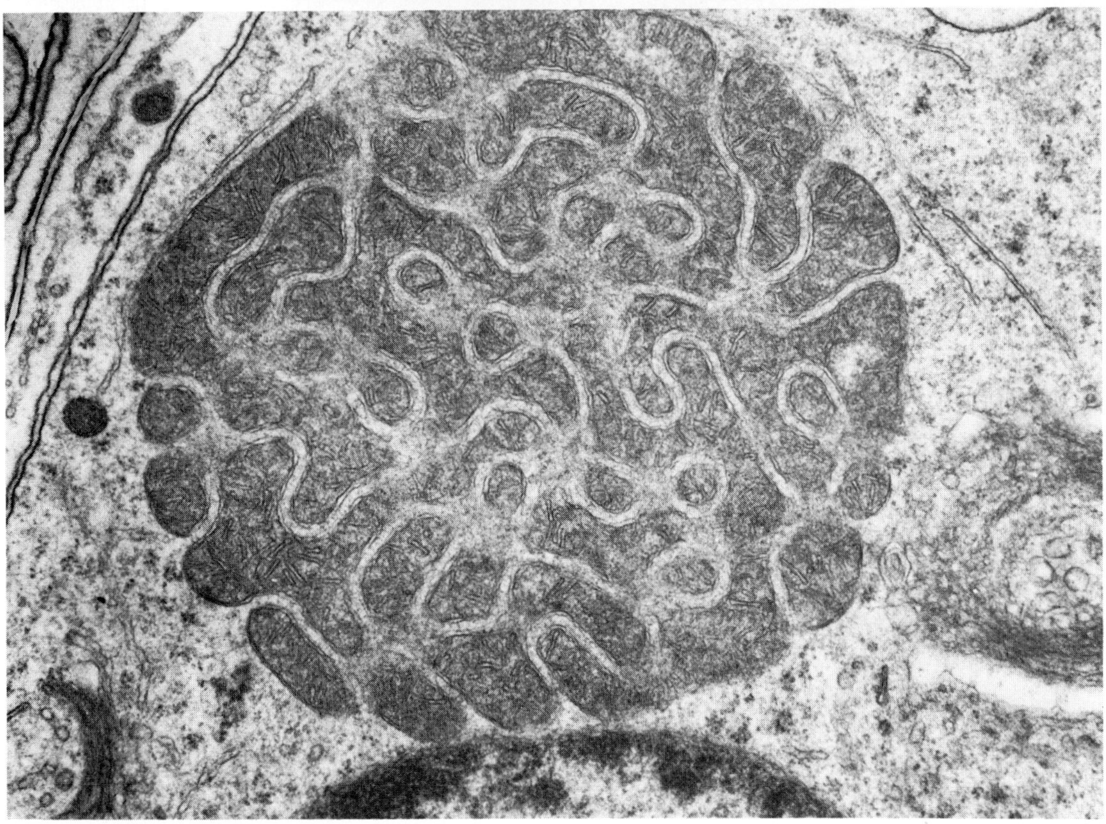

Figure 234, *upper* **Figure 235**, *lower*

In the further development of the nebenkern beyond the stage illustrated here, the initial complexity of its labyrinthine pattern is reduced by continuing fusion (Pratt, 1968). Internal reorganization within the nebenkern then results in formation of two hemispherical masses that are situated on either side of the flagellum. These elongate to form two mitochondrial derivatives that may course parallel to the axoneme for nearly the full length of the sperm tail. In the process of elongation and internal differentiation, the mitochondria become extensively modified, with a loss of cristae and formation of one or more long crystalline inclusions (Phillips, 1970).

No counterpart of the nebenkern is observed in mammalian spermiogenesis. The mitochondria assemble directly around the proximal portion of the flagellum, where they form a helical sheath in which the mitochondria usually retain their individuality. However, in some reptiles and birds they may fuse end-to-end in the mitochondrial helix.

Figure 236. Fully formed nebenkern from an insect spermatid. (Micrograph courtesy of David Phillips.)

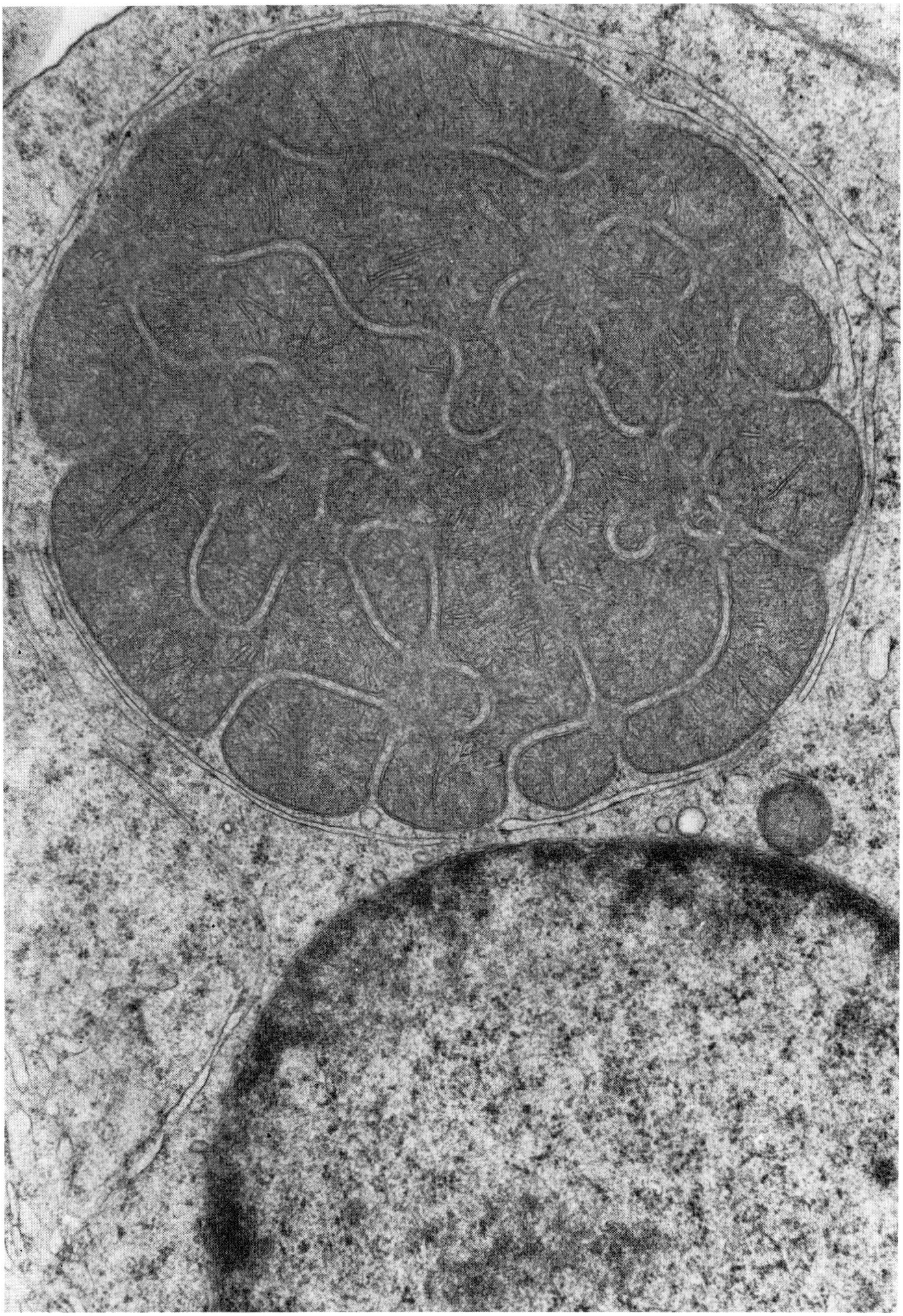

Figure 236

VARIATIONS IN INTERNAL STRUCTURE

The mitochondria of brown adipose tissue are exceedingly numerous and have an extensive internal membrane surface. The accompanying micrograph shows part of an adipose cell from the interscapular region of a bat collected near the end of its period of hibernation. The cell is depleted of lipid and most of the cytoplasm is occupied by large spherical mitochondria. The unusual number of mitochondria in this tissue and their abundant cristae are believed to be related to the high energy requirements for the synthesis of fat from carbohydrate. In the mitochondria of this organ, oxidation is incompletely coupled to phosphorylation. Therefore during arousal from hibernation, much of the energy from oxidation of fat is dissipated in the form of heat which helps raise the body temperature to normal.

Figure 237. Interscapular brown adipose cell from the bat, *Eptesicus fuscus*.

Figure 237

Mitochondria of brown adipose tissue may attain a diameter of 5 to 7 μm. The large size of the three shown here can be appreciated by comparison with a sector of the cell nucleus at the left of the figure. Despite the large size of the mitochondria, the cristae traverse the entire width of the organelle. At high magnification it can be shown that the cristae have numerous small fenestrations near their junction with the inner membrane. These serve to prevent separation of the matrix into multiple isolated compartments.

Figure 238. Mitochondria of brown adipose cell from a bat, *Myotis lucifugus.*

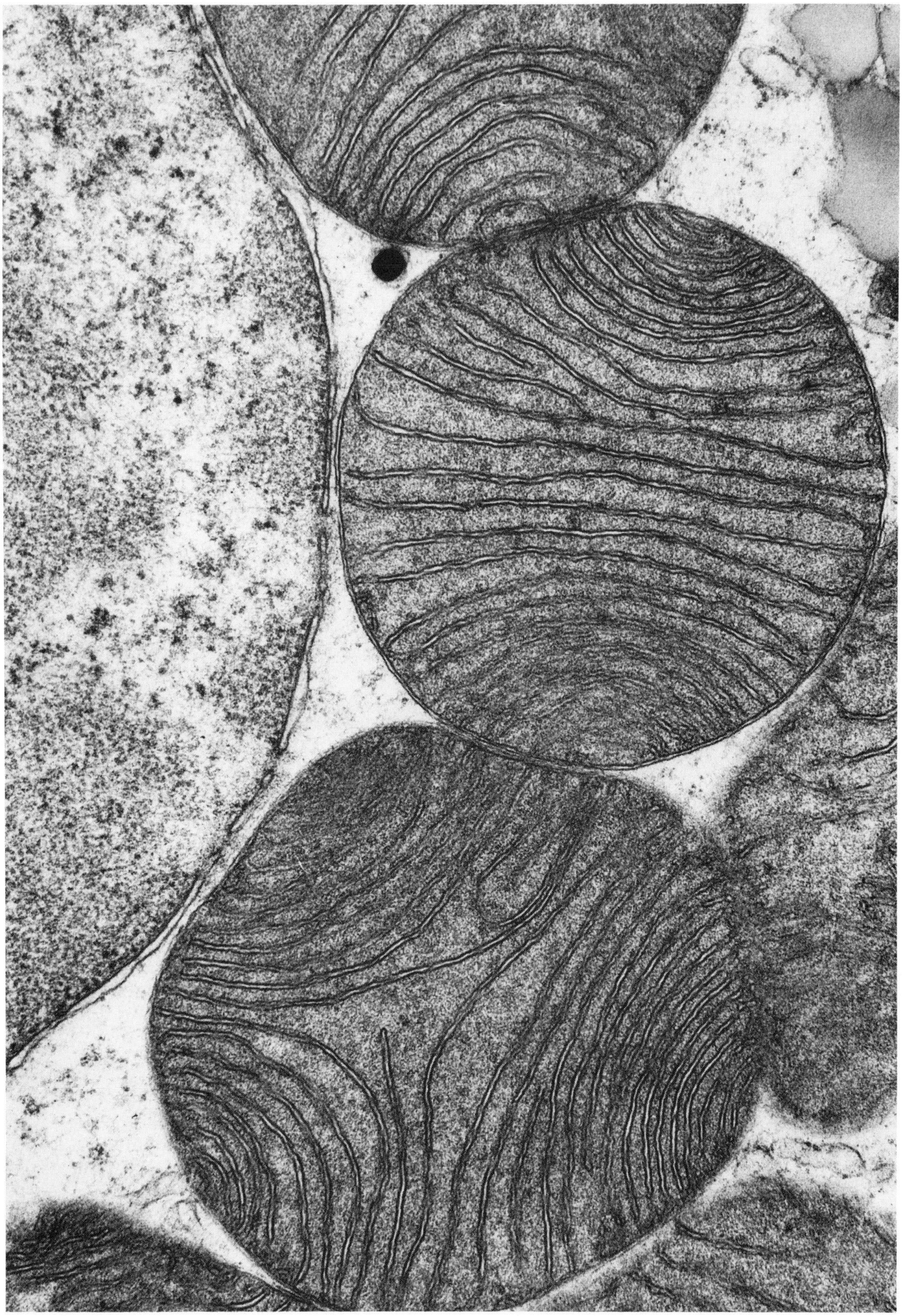

Figure 238

In mitochondria of well-preserved cells the cristae often exhibit sharp angulations of the membranes recurring at regular intervals along their length. The angulations frequently alternate in direction, occurring first on one leaf of the crista and then on the other. This gives the crista a zigzag course. Cristae with this angular configuration are commonly found in active tissues such as skeletal and cardiac muscle, but they have also been observed in a great variety of other cell types and in a wide range of species.

The examples illustrated in the accompanying figures are from cardiac muscle. The angular ridges on one crista are often opposite the concavities on the adjacent crista, but where this pattern is perturbed so that the angulations of neighboring cristae make contact they may coalesce to produce a honeycomb configuration of cristae outlining parallel cylindrical areas of matrix (see at arrows).

Figure 239. Mitochondria of cat ventricular papillary muscle.

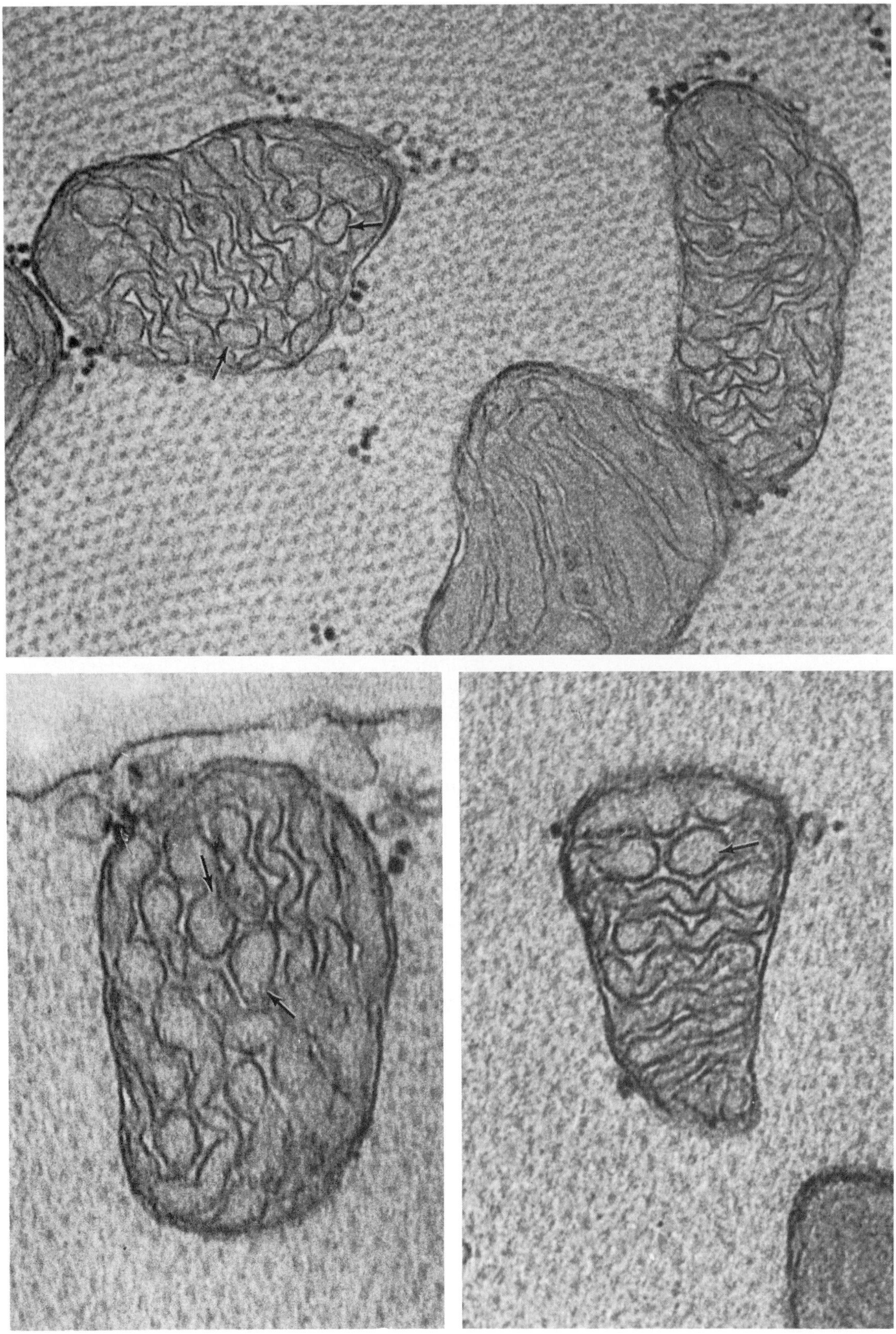

Figure 239

A thin fold of the inner membrane is the commonest configuration of the cristae mitochondriales. An alternative form, found in many protozoa and in certain organs of metazoa, is a slender villus or blind-ending tubule. Among mammalian tissues mitochondria with tubular cristae are frequently observed in steroid-secreting cells such as the Leydig cells of the testis, cells of the corpus luteum, and cells of the adrenal cortex. They are exceptionally well developed in the mitochondria of the adrenal cortex illustrated here. The interior of these long mitochondria is crowded with unbranched tubular cristae of uniform diameter, extending nearly the full width of the organelle. Their tubular form is evident in several places, indicated by arrows, where they have been cut transversely and present circular profiles.

This shape of the cristae is evidently specifically related to the biochemical function of the cells, for in fetal and neonatal life the mitochondria have the usual lamellar cristae but then develop tubular cristae as the cells become active in steroidogenesis. Lamellar cristae may persist in limited areas of the mitochondrion in the mature gland. One such area is marked by an X at the lower edge of the figure.

Figure 240. A portion of a cell from hamster adrenal cortex.

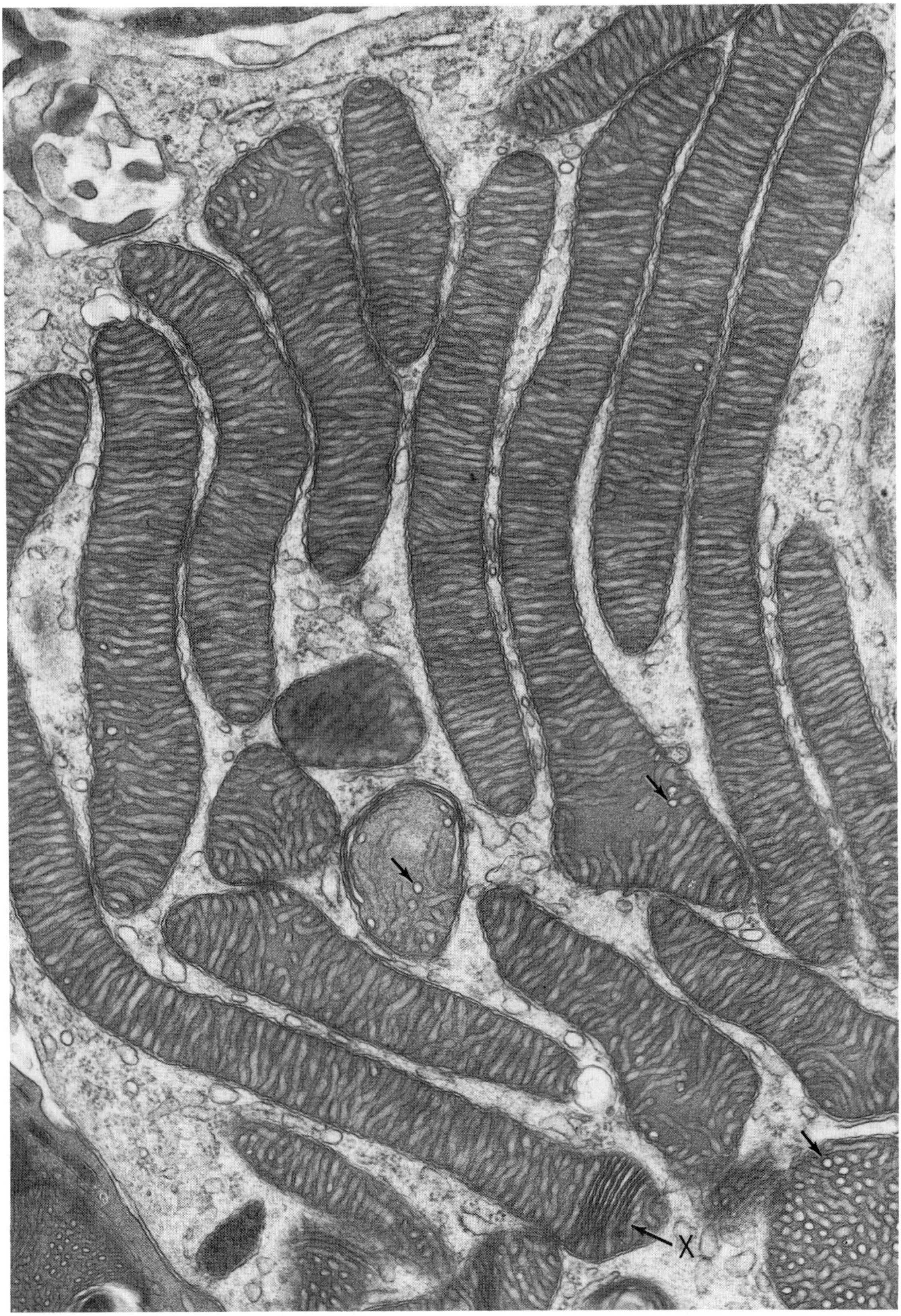

Figure 240

Tubular cristae of protozoan mitochondria are usually fewer in number than those shown on the previous page but are otherwise very similar. Another form which they may take is illustrated by the mitochondria of the Singh amoeba on the facing page. Instead of behaving as blind-ending simple tubules, they branch and anastomose to form an irregular three-dimensional network.

Figure 241. Mitochondria of Singh amoeba. (Micrograph courtesy of Tom Pollard.)

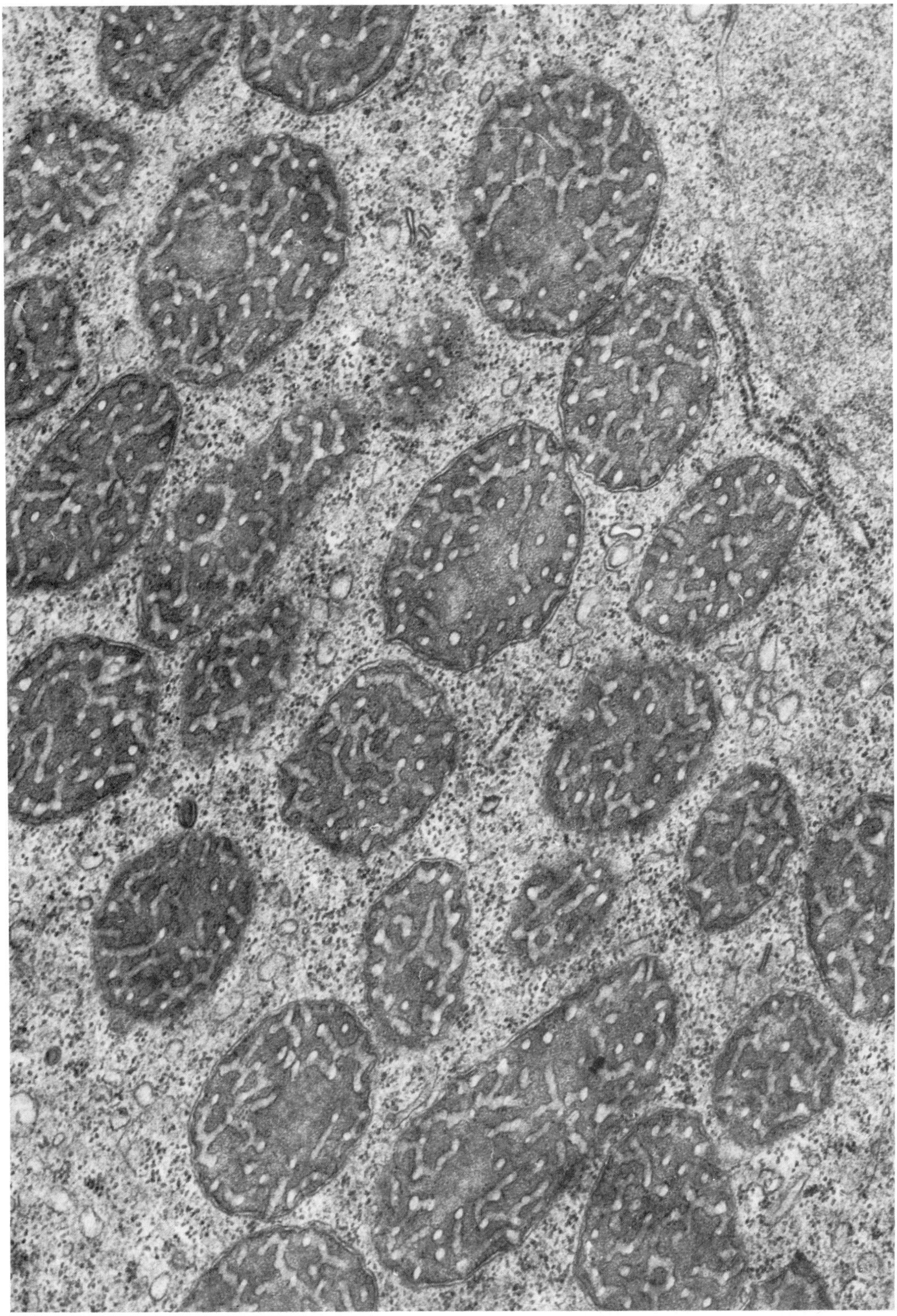

Figure 241

Two forms of tubular cristae are presented here for comparison at higher magnification. The small granules in the dense matrix in the lower figure are believed to be mitochondrial ribosomes. This micrograph shows that they are smaller than the ribosomes in the surrounding cytoplasm.

Figure 242. Mitochondria from the zona fasciculata of hamster adrenal cortex.

Figure 243. Mitochondria from Singh amoeba. (Courtesy of Tom Pollard.)

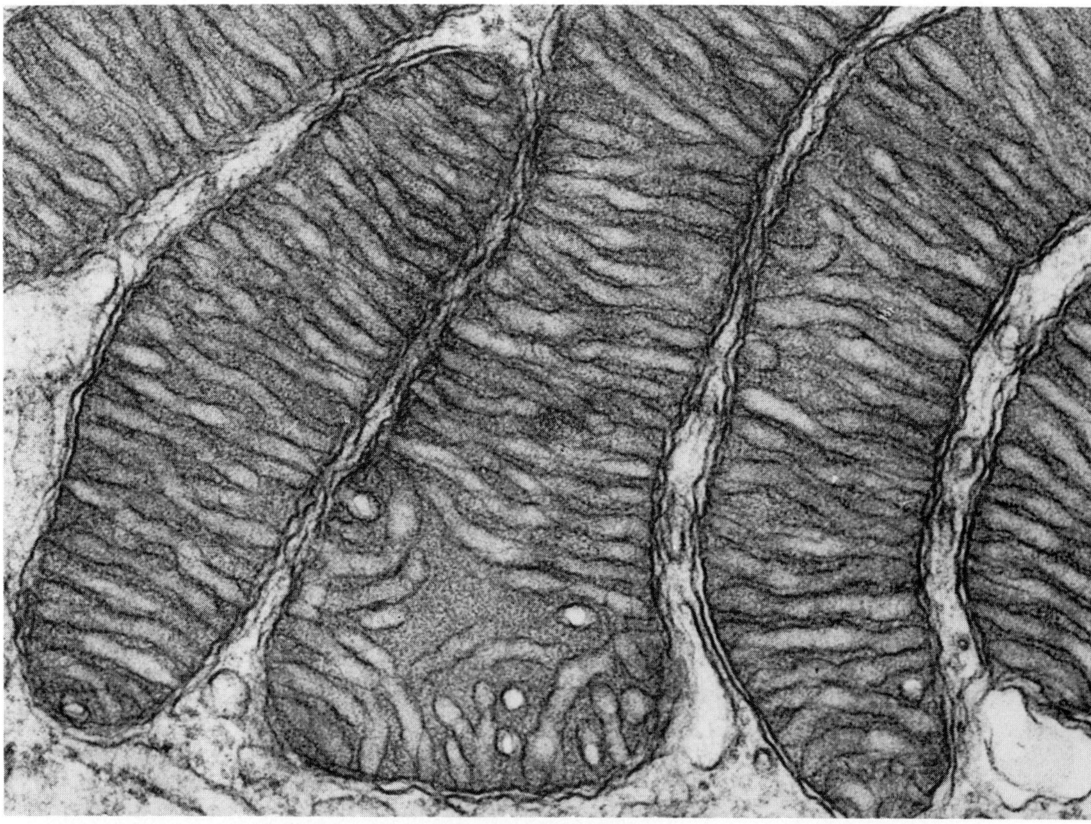

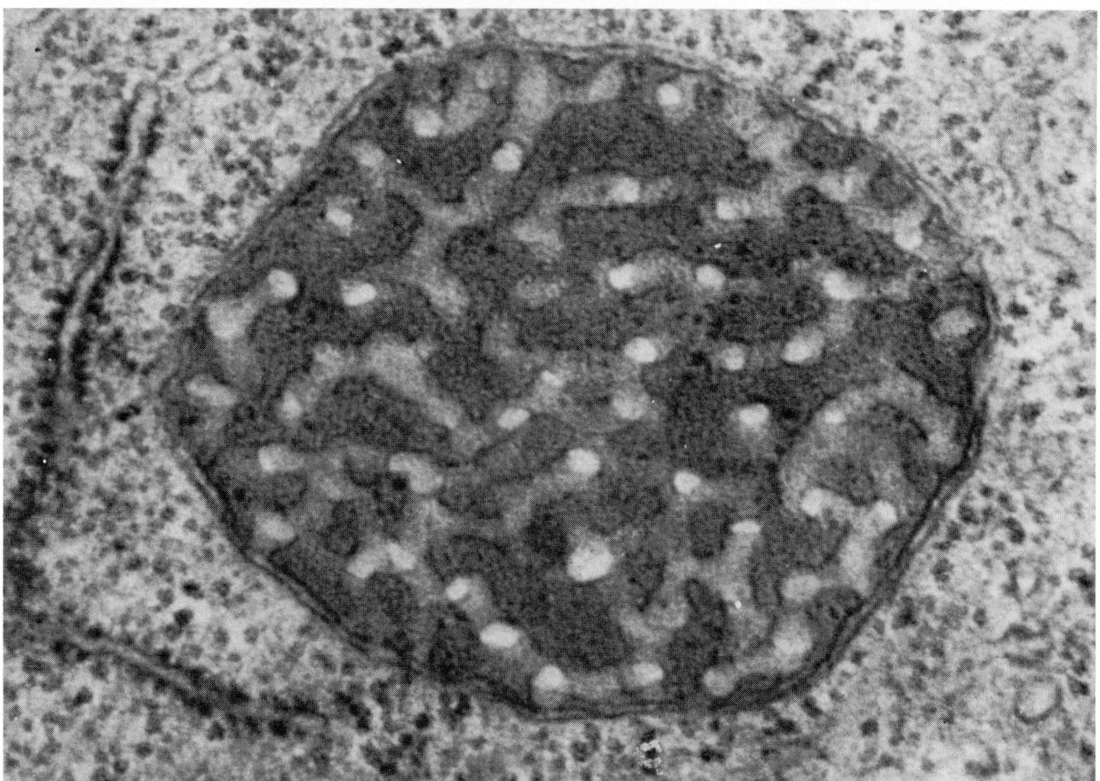

Figure 242, *upper* **Figure 243,** *lower*

In contrast to the irregular network of anastomosing tubular cristae previously illustrated, highly ordered three-dimensional lattices of interconnected tubules are regularly observed in mitochondria of *Amoeba proteus* and rarely in certain tissues of vertebrates as illustrated here.

As a rule, differences in size, external form, and internal organization are relatively slight among mitochondria of the same cell. There are, however, examples of marked changes in structure during functional differentiation of some cell types, and not all mitochondria change their form at the same rate. In the adrenal cortex of fetal or neonatal rats the mitochondria have lamellar cristae, but as the cells differentiate either *in vivo* or in organ culture the prevailing form of the cristae becomes tubular (Kahri, 1970). Similarly, immature supporting cells of *Xenopus* testis have mitochondria with conventional internal structure, but as they become mature Sertoli cells, the number of cristae increases and some of the mitochondria acquire the unusual internal structure which is depicted in the accompanying figures (Kalt, 1974). There are multiple parallel fenestrated layers of cristae that are interconnected by regularly arranged tubular elements. The resulting three-dimensional lattice presents varying patterns in different planes of section. These peculiar mitochondria occur together with others of orthodox configuration in the same cell.

Figure 244. Mitochondria from Sertoli cells of *Xenopus* in various planes of section. (Micrograph courtesy of Marvin Kalt, Anat. Rec. 182: 53, 1974.)

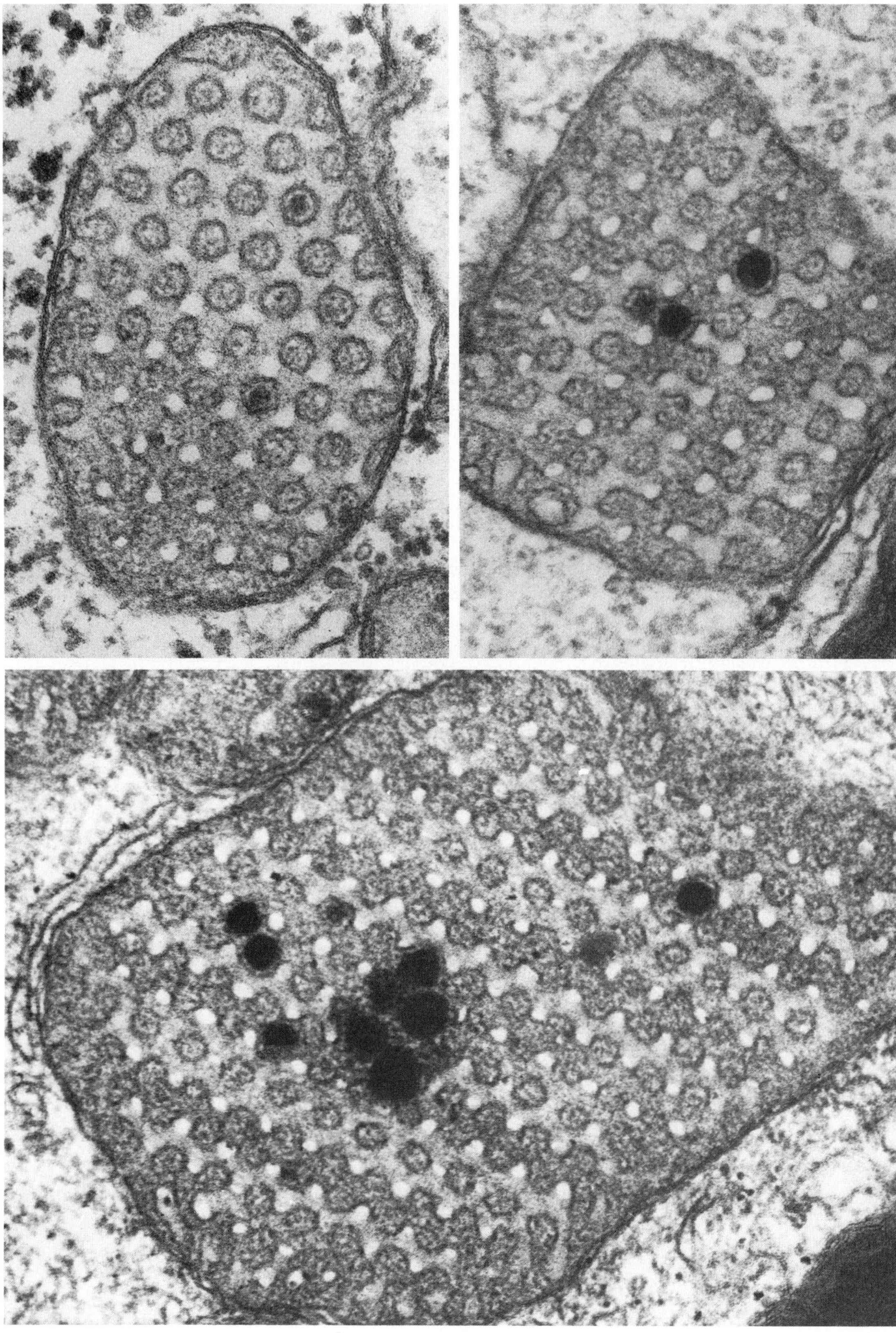

Figure 244

One of the most striking and puzzling configurations of mitochondrial cristae is prismatic tubules that are invariably triangular in cross section. These were first described in muscle (Revel et al., 1963) but have since been found sporadically in many tissues and many species. They were originally interpreted as an unusual and extreme expression of the tendency of the membranes of lamellar cristae to develop angulations at regular intervals along their length. It was later pointed out that in mitochondria with prismatic cristae in the mammalian nervous system the intervening matrix shows a very regular pattern of dots with a center-to-center space of 15 nm. It was suggested therefore that the triangular shape might be imposed upon tubular cristae by hexagonal packing of filamentous elements in a paracrystalline matrix. The densities in the matrix of the mitochondrion of the facing page are seen only faintly, but the evenly spaced triangular profiles of the cristae are well shown and some at the periphery are seen to be in continuity with the inner membrane.

The large spherical mitochondria of rat adrenal cortex, shown in the lower figure, are often described as having vesicular cristae but closer examinaton reveals that they are probably tubules with alveolar branches (see at arrows). In section, both the tubules and their branches have circular profiles that have been erroneously interpreted as vesicles floating free in the matrix.

Figure 245. Mitochondrion from an astrocyte of hamster brain. (Micrograph courtesy of K. Blinzinger, J. Cell Biol. 25: 293, 1965.)

Figure 246. Mitochondrion from rat adrenal cortex. (Micrograph courtesy of Daniel Friend.)

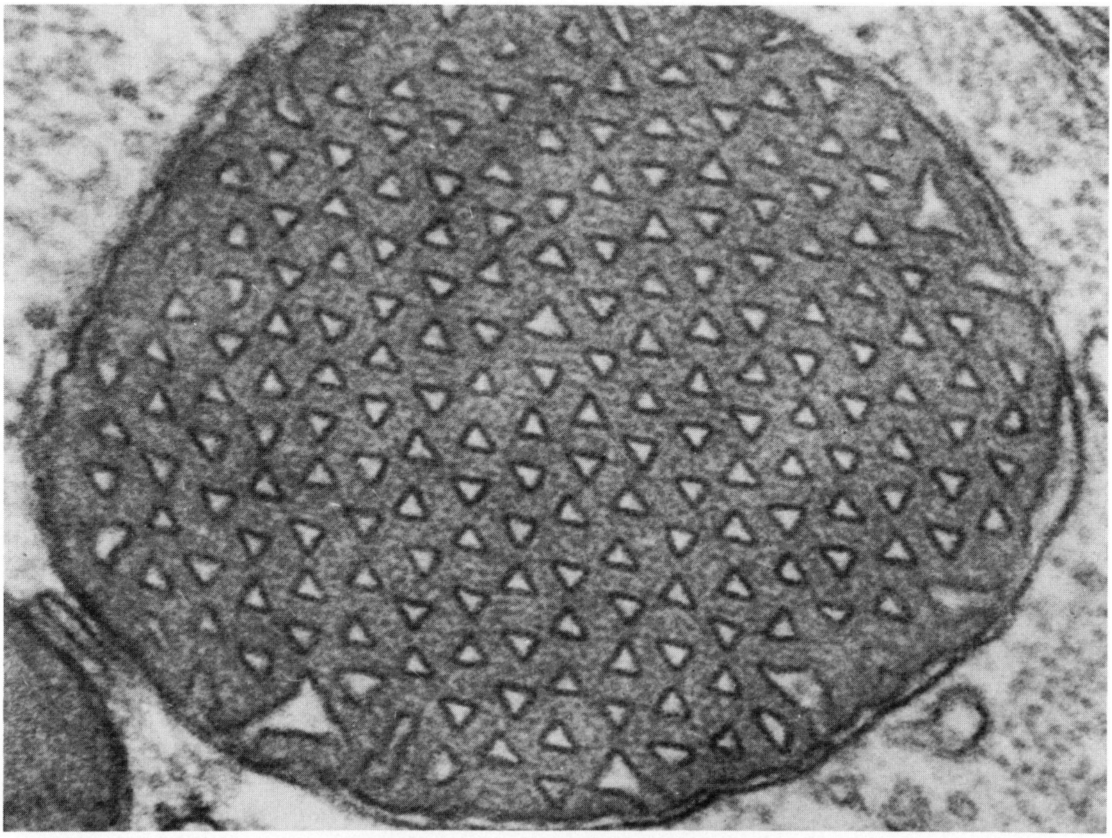

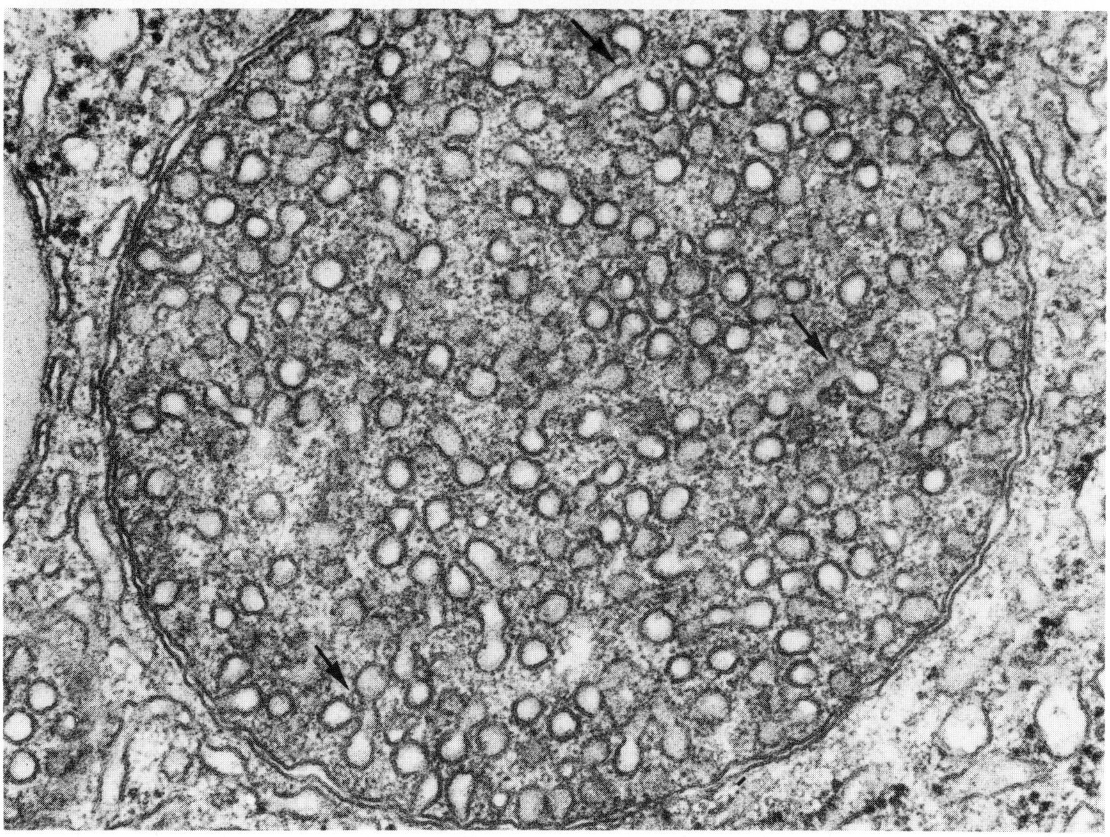

Figure 245, *upper* **Figure 246,** *lower*

In the majority of cell types the range of variation in mitochondrial size and shape is relatively narrow, but in a few cell types they exhibit an unusual degree of pleomorphism. Prominent among these are steroid-secreting cells such as those in the interstitial tissue of the testis. Reproduced here is a Leydig cell showing considerable variation in mitochondrial shape. Much more extreme examples of size difference are not infrequent, with huge spherical mitochondria 6 to 7 μm in diameter adjacent to small elongate mitochondria 0.5 μm wide.

Figure 247. Leydig cell from the testis of *Citellus lateralis*. (Micrograph courtesy of Jeffrey Pudney.)

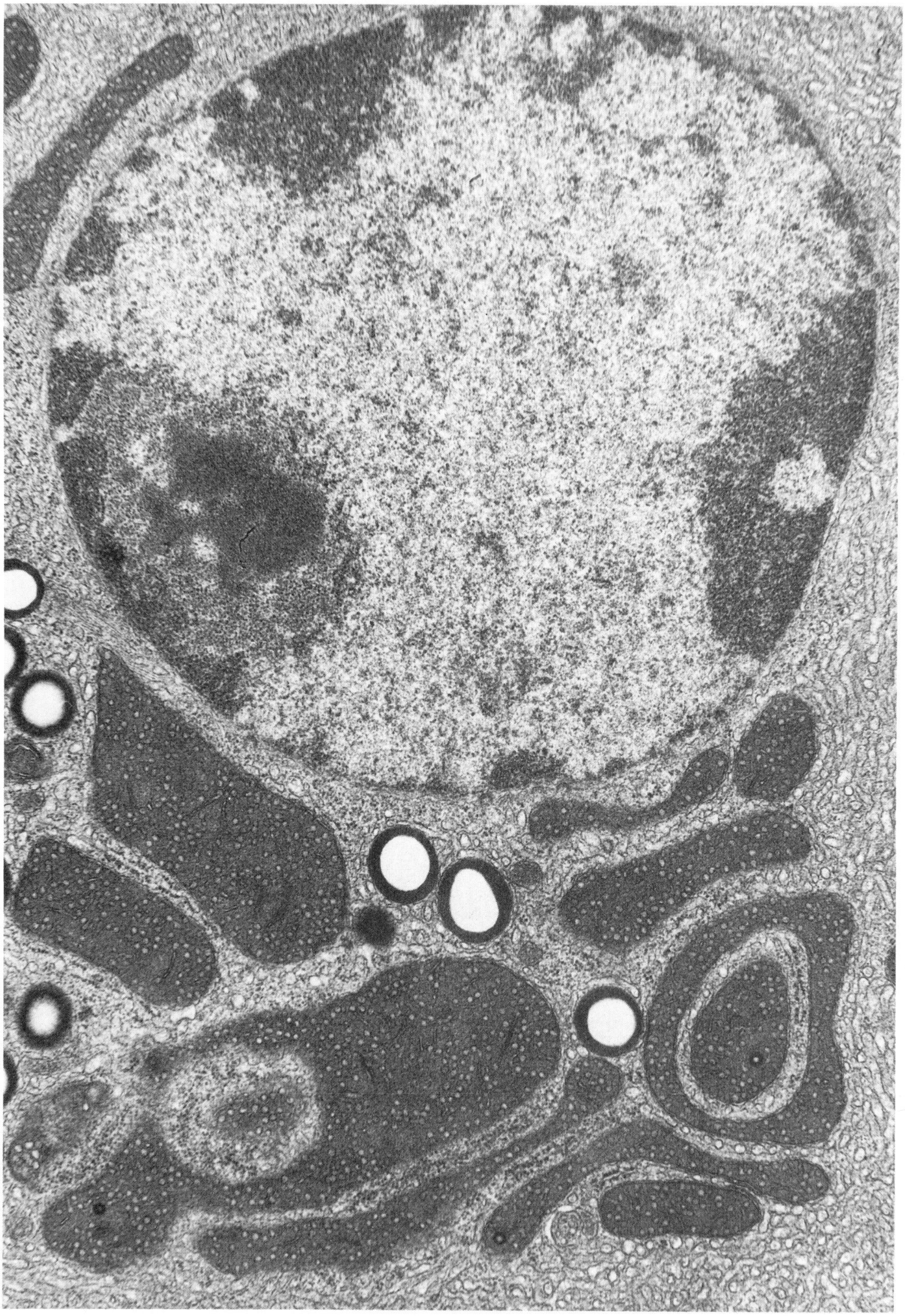

Figure 247

Steroid-secreting cells also exhibit unusual diversity in the form of the cristae in their mitochondria. Those shown here, for example, have small alveolar cristae along the inner membrane, longitudinally oriented lamellar cristae (single arrows), and bundles of parallel tubules coursing in various directions in the matrix (double arrows).

Figure 248. Mitochondria from testicular Leydig cells of the ground squirrel, *Citellus lateralis.* (Micrograph courtesy of Jeffrey Pudney.)

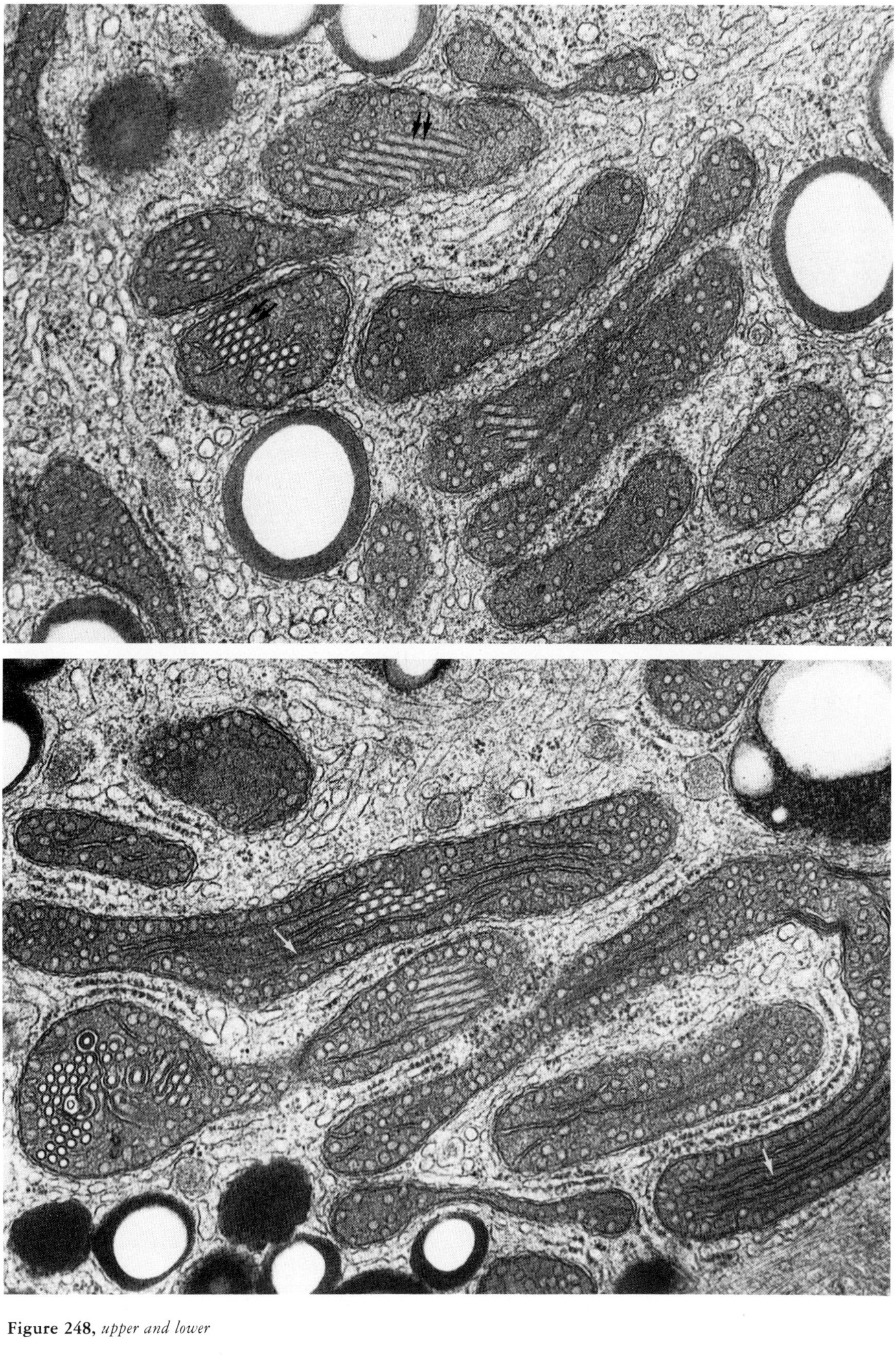

Figure 248, *upper and lower*

The ability of mitochondria to change their internal structure in response to altered physiological demands is exemplified in the pseudobranch of fish. This organ is believed to have a functional relationship with the eye. In fish maintained under normal illumination or exposed to brief periods of darkness, the great majority of the mitochondria have the usual foliate cristae. However, when fish are kept in constant light for three to four weeks, a second category of tubular cristae appears in 70 to 90 per cent of the mitochondria. These are arranged in rows between the longitudinally oriented foliate cristae. When viewed in longitudinal section (see inset), they have the appearance of helically twisted tubules.

Figure 249. Mitochondria from the pseudobranch of *Cyprinoden variegatus* exposed to constant light for four weeks. (Micrograph courtesy of Joseph Harb and Eugene Copeland.)

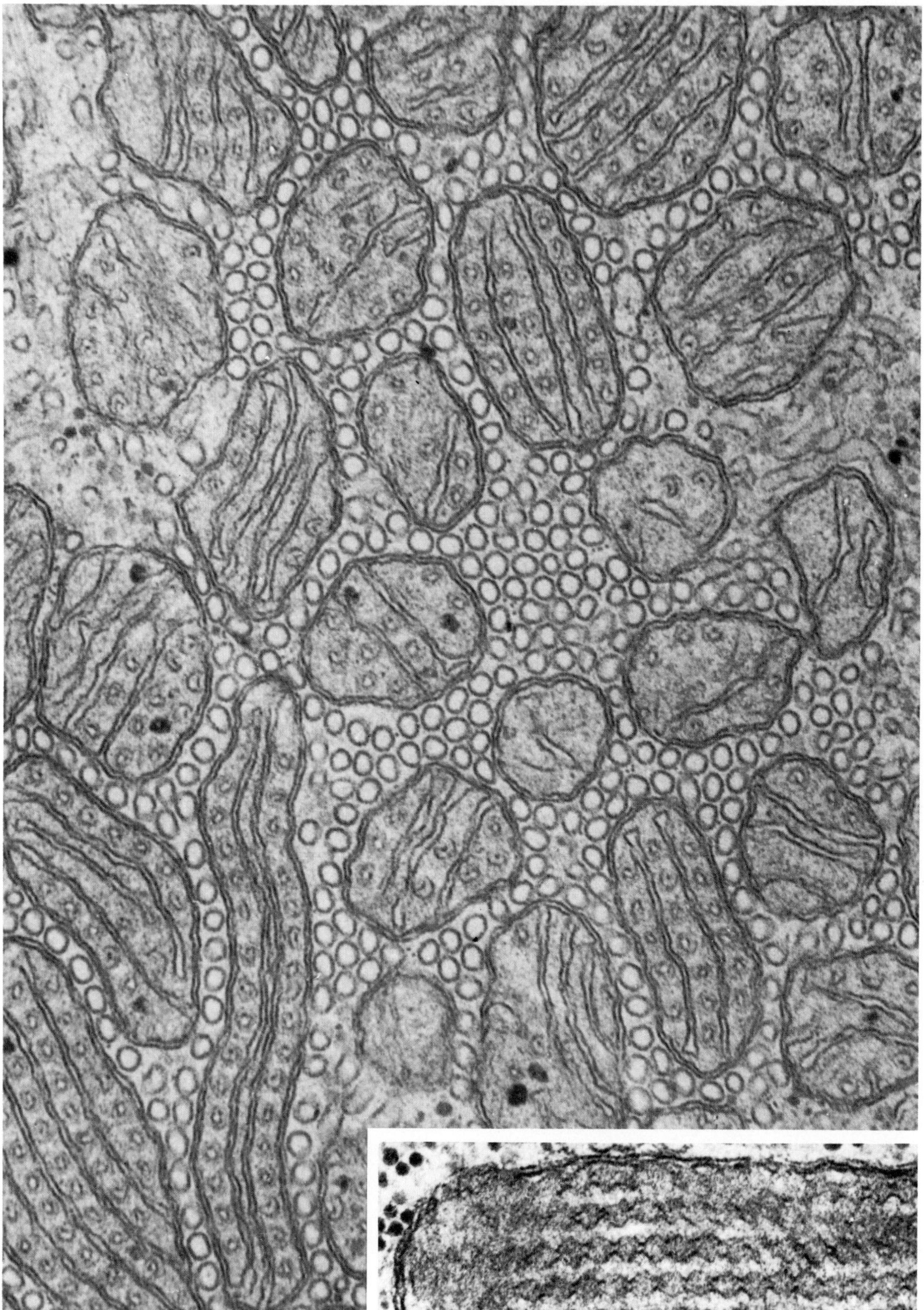

Figure 249

MITOCHONDRIAL INCLUSIONS

In addition to the standard structural components of mitochondria, a great variety of inclusions have been described. These may be located within the matrix or within an expanded intracristal compartment. Some are of frequent occurrence and must be regarded as normal for the particular cell type, while others occur more rarely and may be manifestations of disordered mitochondrial metabolism or disease states.

Two examples of stored material accumulating in an expanded intracristal space are shown in the micrographs on the facing page. The upper figure is a mitochondrion from a frog oocyte containing a hexagonal crystal with a 7.4 nm periodicity. These intramitochondrial crystals are similar in form and lattice dimensions to yolk crystals in the oocyte cytoplasm. It is postulated that these form within mitochondria and are subsequently extruded in an outpocketing of the mitochondrial membranes which pinches off to leave an intact mitochondrion and a membrane-limited yolk crystal in the cytoplasm (Massover, 1971). These have been observed in several species of frog and are considered to be part of the normal process of vitellogenesis. To what extent, if any, the synthesis of yolk lipoprotein is directed by the mitochondrial genome is unknown.

The lower figure is a mitochondrion from a Purkinje fiber of cat heart which has accumulated a large amount of glycogen. The glycogen may either be in the form of alpha or beta particles. Such mitochondria are relatively uncommon, but various stages in glycogen accumulation can be found, from a few particles in one or two dilated cristae to large masses such as illustrated here. This is an unexpected finding in that the enzymes for glycogen production are not thought to be present in mitochondria.

Figure 250. Mitochondrion from an ovarian oocyte of the bullfrog *Rana catesbiana*. (Micrograph courtesy of William Massover, J. Ultrastruct. Res. 36: 603–620, 1971.)

Figure 251. Mitochondria from conduction tissue of the cat myocardium.

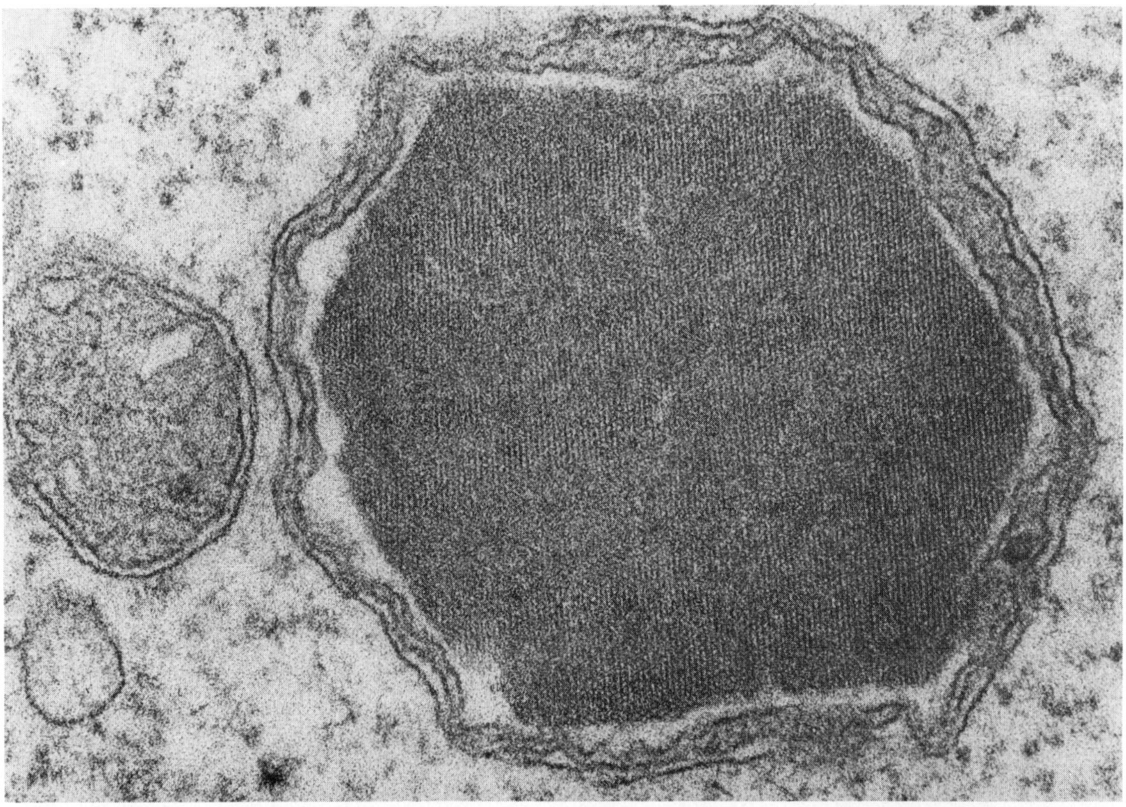

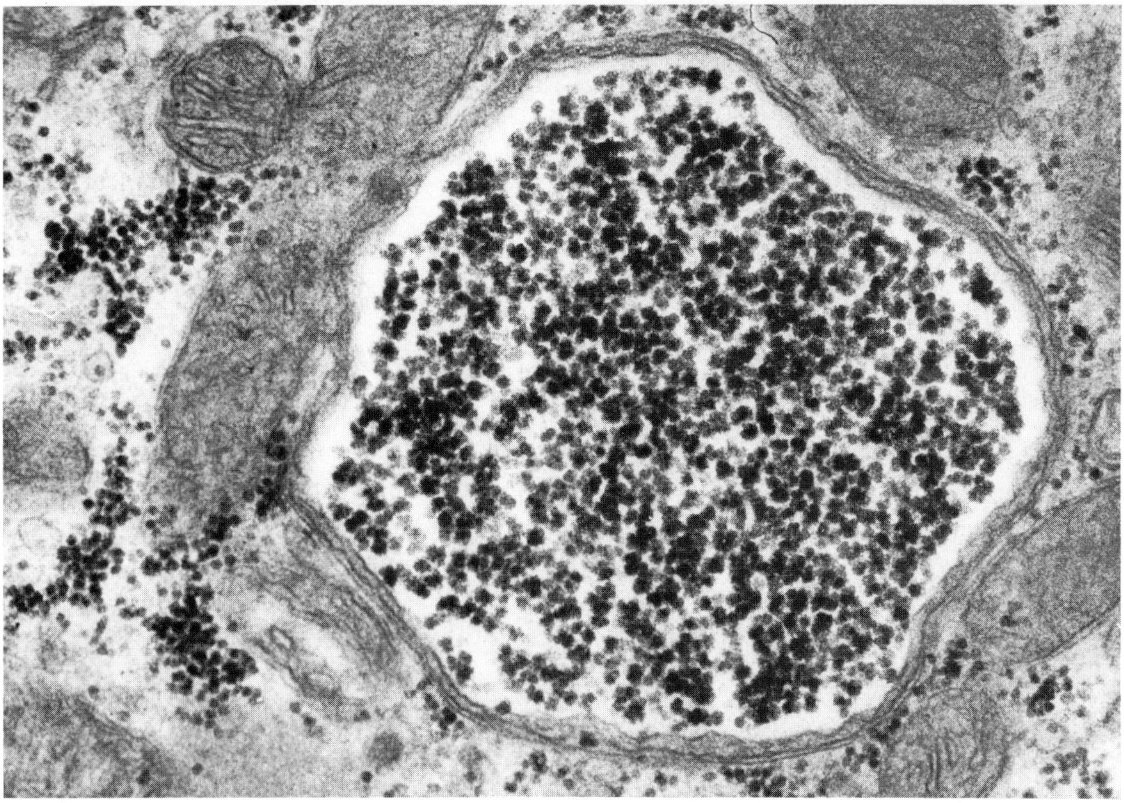

Figure 250, *upper* **Figure 251,** *lower*

During spermiogenesis in insects, the cristae, originally present in the individual mitochondria and later in the mitochondrial nebenkern, disappear and proteinaceous material is deposited in filaments that become closely packed in paracrystalline arrays. Crystals are a conspicuous feature of the mitochondrial derivatives of most insect sperm, but their size and shape vary greatly from species to species. In the accompanying micrograph, crystals are present in each of the two mitochondrial derivatives that flank the axoneme.

Figure 252. Mitochondrial derivatives in spermatozoa of the stilt bug, *Jalysus*. (Micrograph courtesy of David Phillips.)

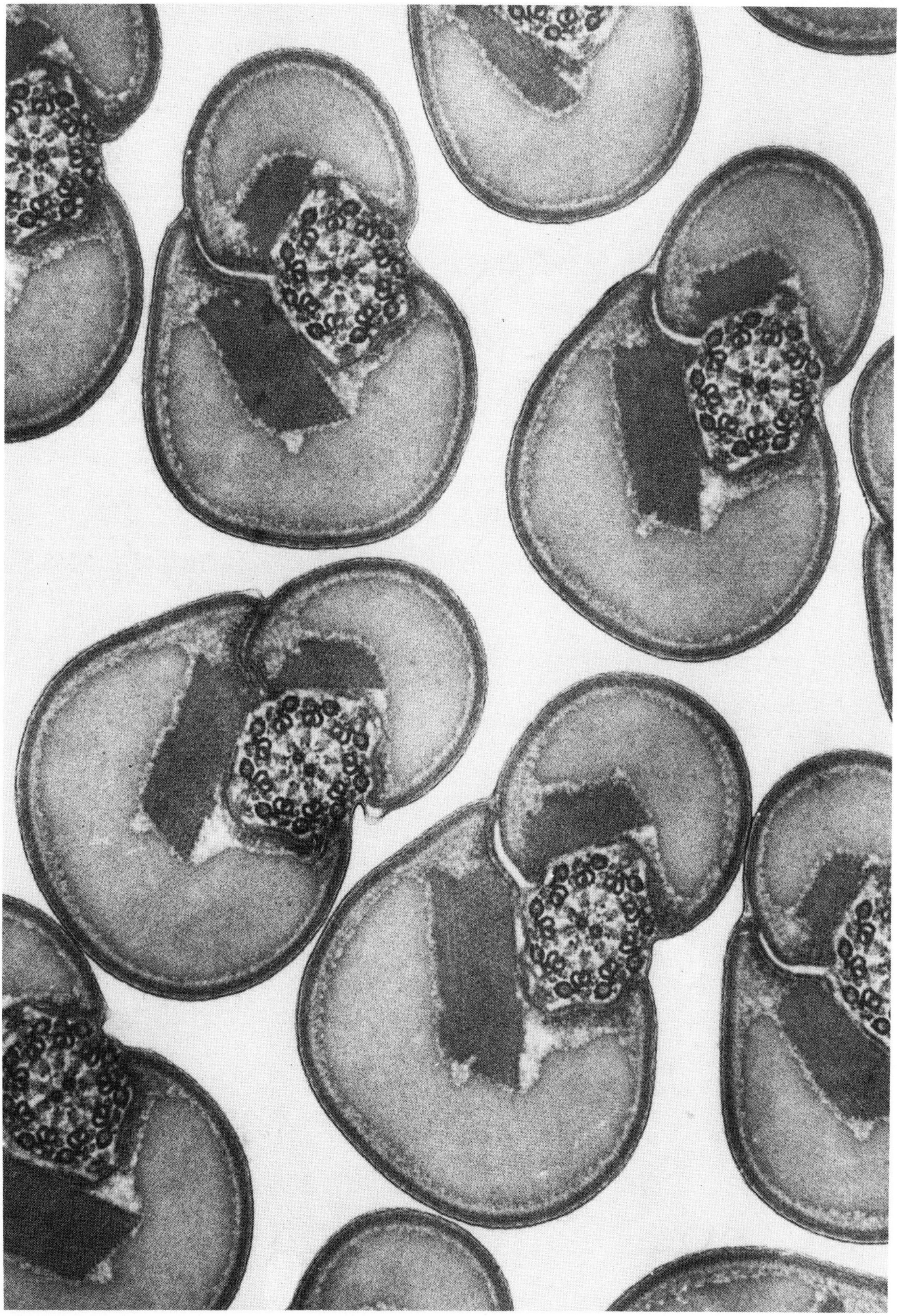

Figure 252

NUMBERS AND DISTRIBUTION

The numbers of mitochondria vary over a wide range depending upon the size of the cell and its energy requirements. Reliable estimates have been made for only a few cell types. Values for the liver are 800 to 2000; oocytes of echinoderms 150,000; and in the giant amoeba *Chaos chaos* as many as half a million.

They may be randomly dispersed in the cytoplasm or concentrated around the cell center as in the micrograph on the facing page. They exhibit a tendency to aggregate around deposits of substrate such as lipid droplets. They are often specifically located in very close proximity to sites of high energy utilization. A number of examples of such localization will be illustrated in the pages that follow.

The transport and control mechanisms responsible for mobilizing mitochondria at particular sites in the cell are not understood, but there are studies that seem to implicate microtubules in directing their movements. In nerve axons the proximity of mitochondria to microtubules is greater than would be expected from a random distribution and small cross-links between them have been described. It was suggested that the organelles might be moved along the tubules by a sequential making and breaking of such linkages (Smith, Jarlfors and Cameron, 1977). In studies on several cell types in tissue culture by immunofluorescence techniques, mitochondria were shown to be arranged along bundles of microtubules. Disruption of microtubules with colchimid caused a redistribution of mitochondria (Heggeness et al., 1978). These observations are suggestive of direct or indirect involvement of microtubules in mitochondrial movements but shed little light upon the mechanism.

Figure 253. Prominent clustering of mitochondria in the Golgi region of a Leydig cell from the domestic boar, *Sus scrofa*.

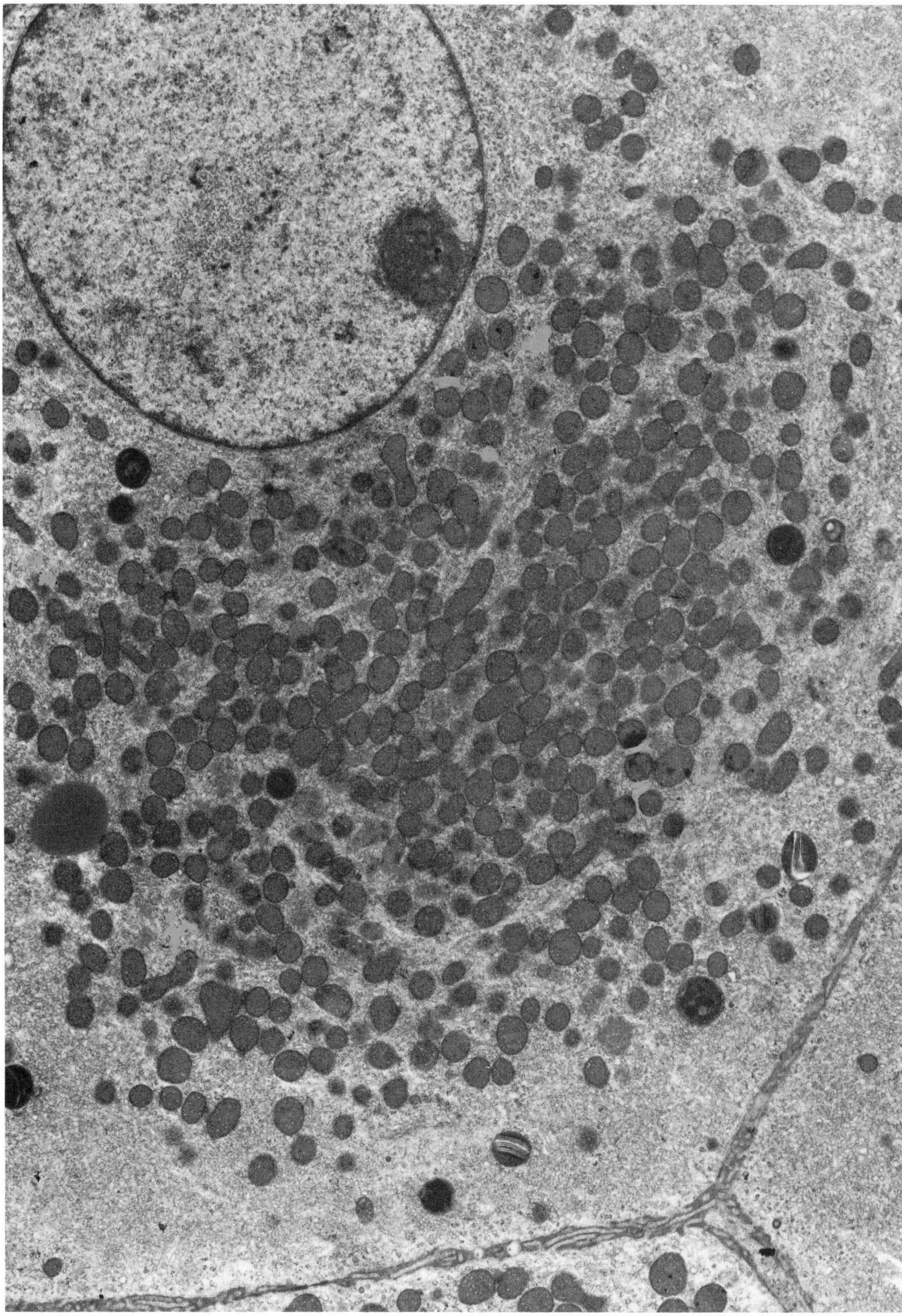

Figure 253

An example of localization of mitochondria in close proximity to a source of substrate is seen in the relationship of mitochondria to lipid droplets. They may partially surround small lipid droplets and when the droplets are very large, as in this liver cell from a fasted hamster, several mitochondria may be closely applied to its surface. The fatty acid oxidases of the mitochondria are involved in the metabolism of triglycerides, and this intimate relationship no doubt facilitates transport of substrate to the enzymes.

Figure 254. Portion of a liver cell from a hamster fasted for four days.

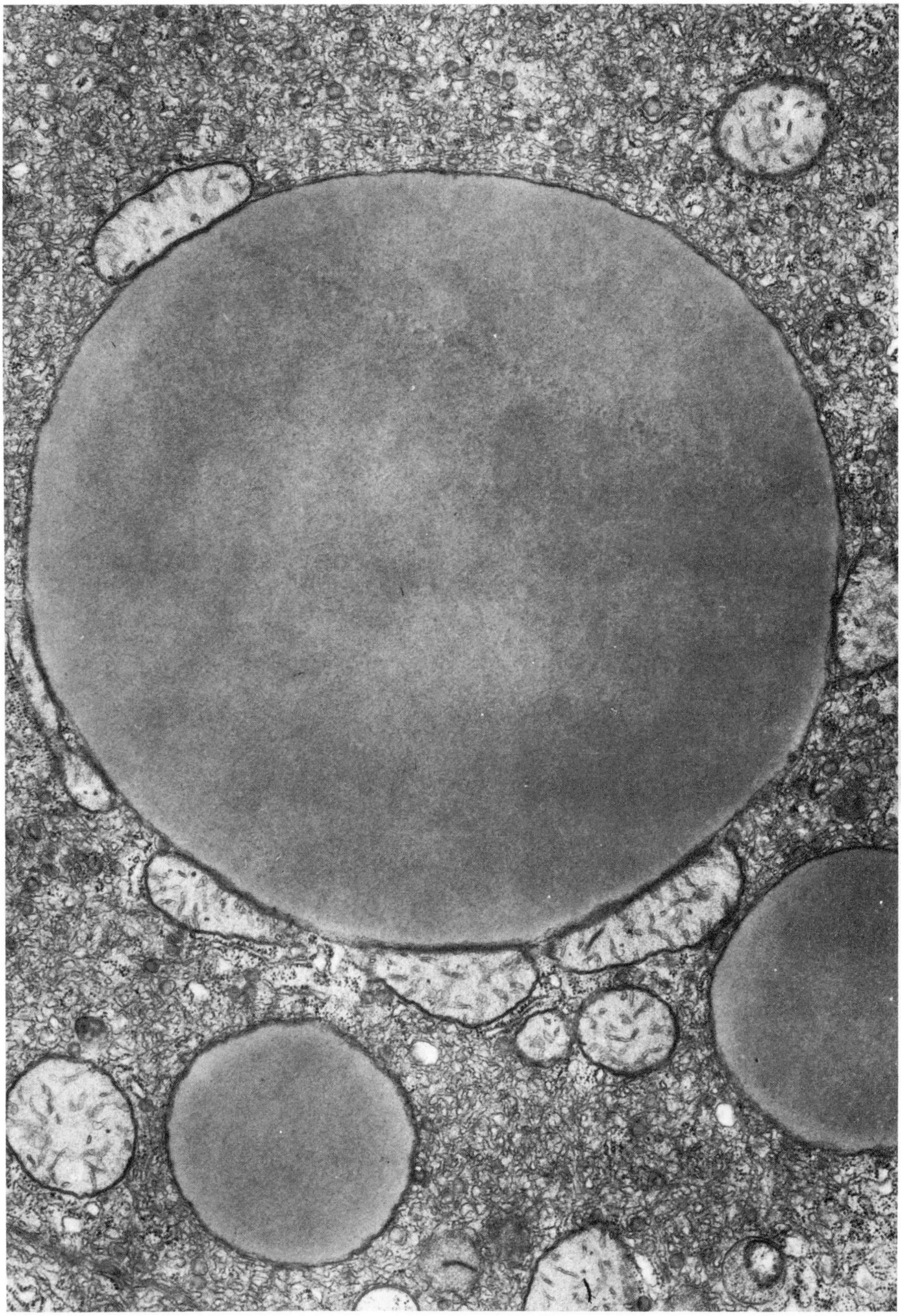

Figure 254

Energy for muscle contraction is derived mainly from metabolism of carbohydrate, but under certain conditions lipid also serves as an energy source. In cardiac muscle, lipid droplets are often partially enveloped by mitochondria in the interstices between myofibrils.

Figures 255 and 256. Mitochondria-lipid relationships in papillary muscle of the cat heart.

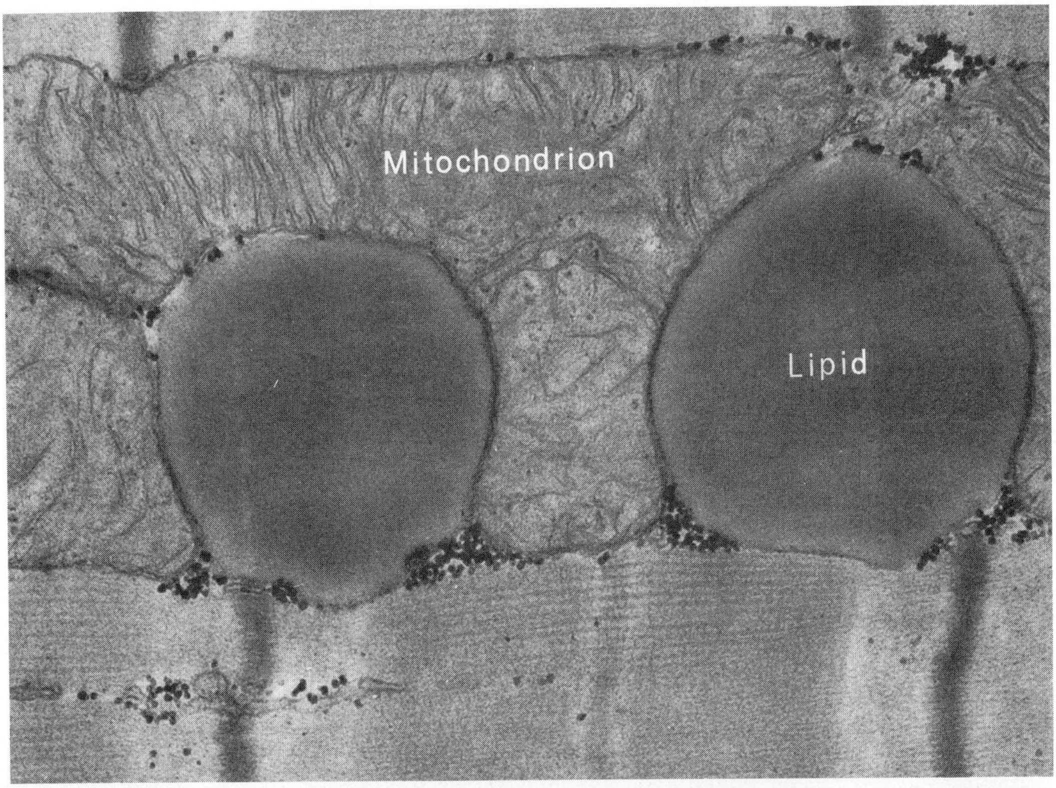

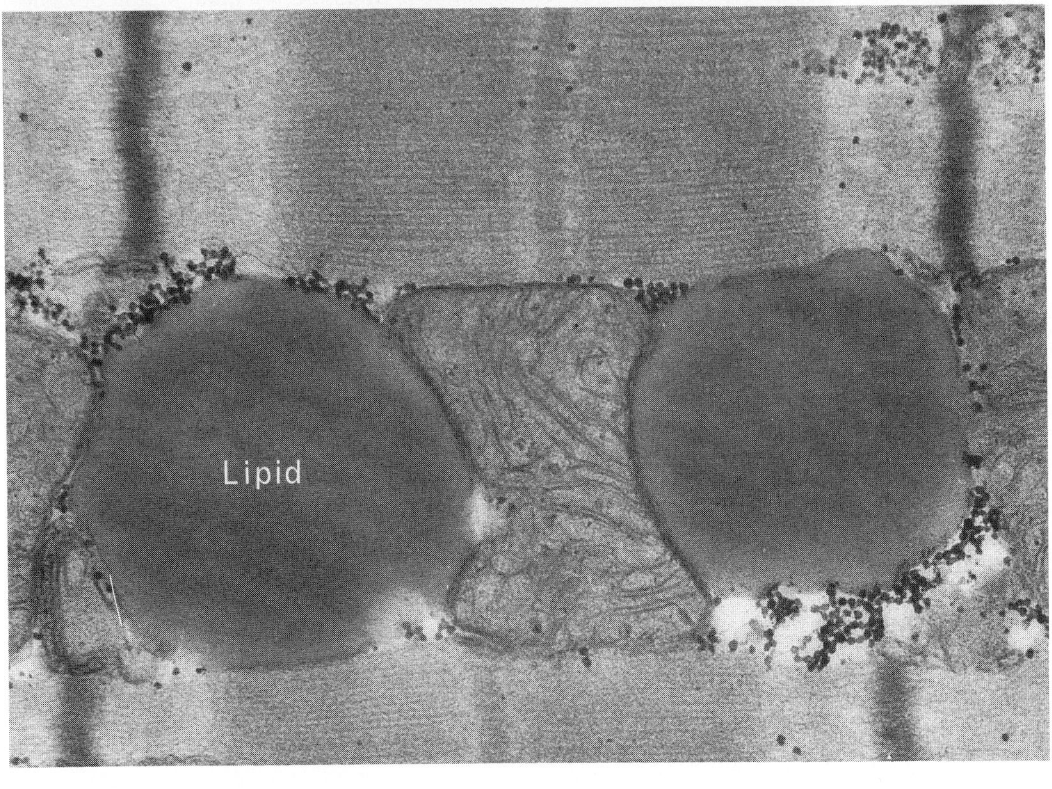

Figure 255, *upper* **Figure 256,** *lower*

One of the most striking local accumulations of mitochondria in the body is in the inner segment of the rods and cones of the retina. The closely packed mitochondria are very long and have a complex pattern of curved cristae. The accompanying micrograph shows the inner segment of a cone, flanked by the inner segments of two rods. The mitochondria are several micrometers in length and those of the cone appear to have a denser matrix. The molecular events related to energy transduction and amplification in the photoreceptors are still poorly understood, but the remarkable concentration of mitochondria near the photosensitive region of the cell indicates that the process has high energy requirements.

Figure 257. Electron micrograph of photoreceptors in human retina. (Micrograph courtesy of Toichiro Kuwabara.)

Figure 257

In epithelia engaged in active transport of ions, numerous mitochondria at the cell base are lodged in narrow compartments or interdigitating processes of neighboring cells in such a way that the mitochondria are brought into very close relationship to a large area of the cell surface. Details of the process are not well understood, but it is assumed that this arrangement provides a favorable juxtaposition of the energy source to the elaborately convoluted cell surface in which the active transport mechanism resides.

The micrograph on the facing page includes a vertical section of the base of the epithelium of the distal convoluted tubule in the kidney and a section of the subjacent peritubular capillary.

Figure 258. Distal convoluted tubules of guinea pig kidney. (Micrograph courtesy of Atsushi Ichikawa.)

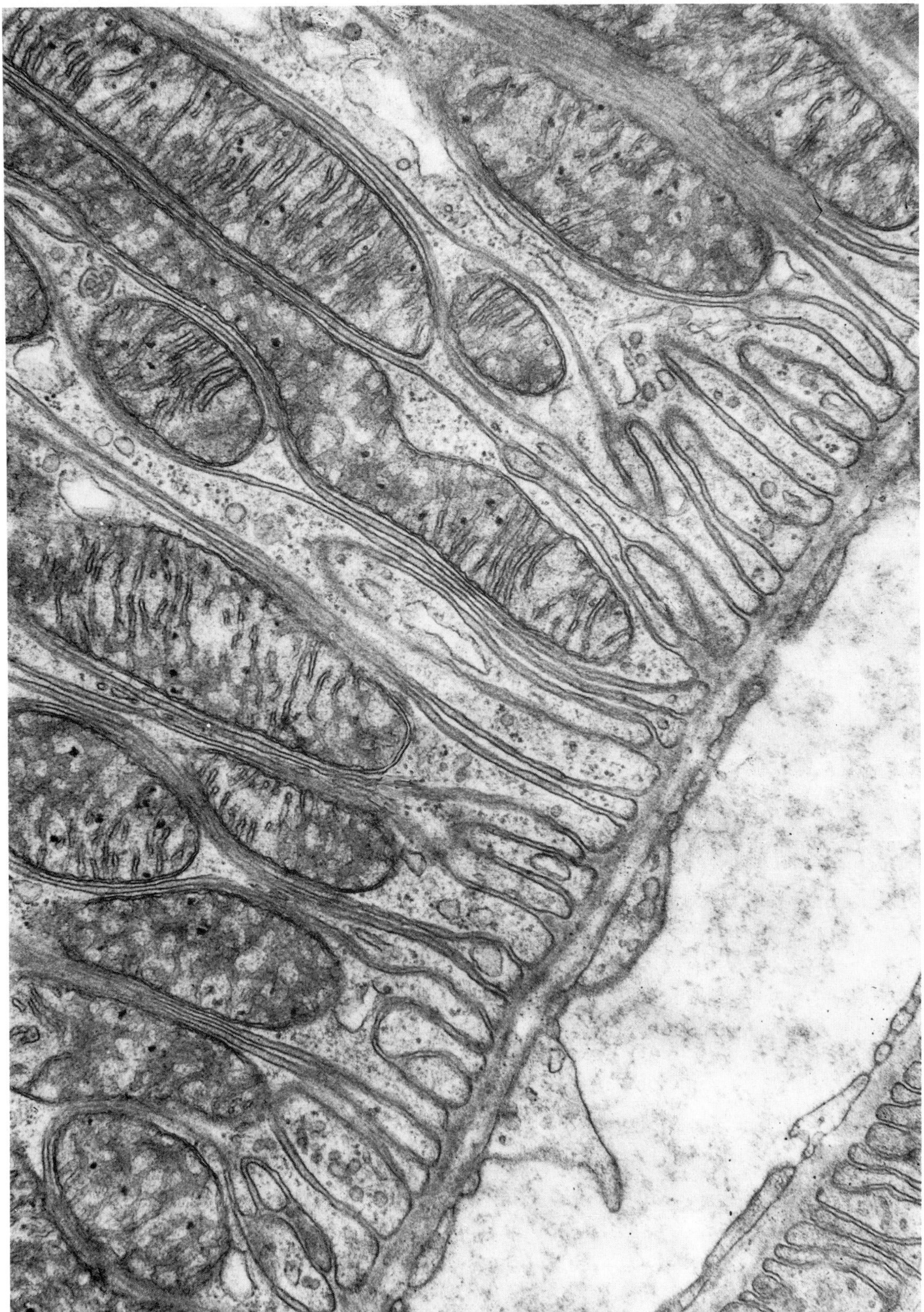

Figure 258

The intimacy of the association of mitochondria with the contractile elements of muscle is well illustrated in transverse sections such as that in the micrograph on the facing page. The boundaries of myofibrils are ill defined in cardiac muscle. The mitochondria are uniformly distributed in a more-or-less continuous field of myofilaments. The diffusion distance for ATP is thus minimized.

Figure 259. Right ventricular papillary muscle of cat heart.

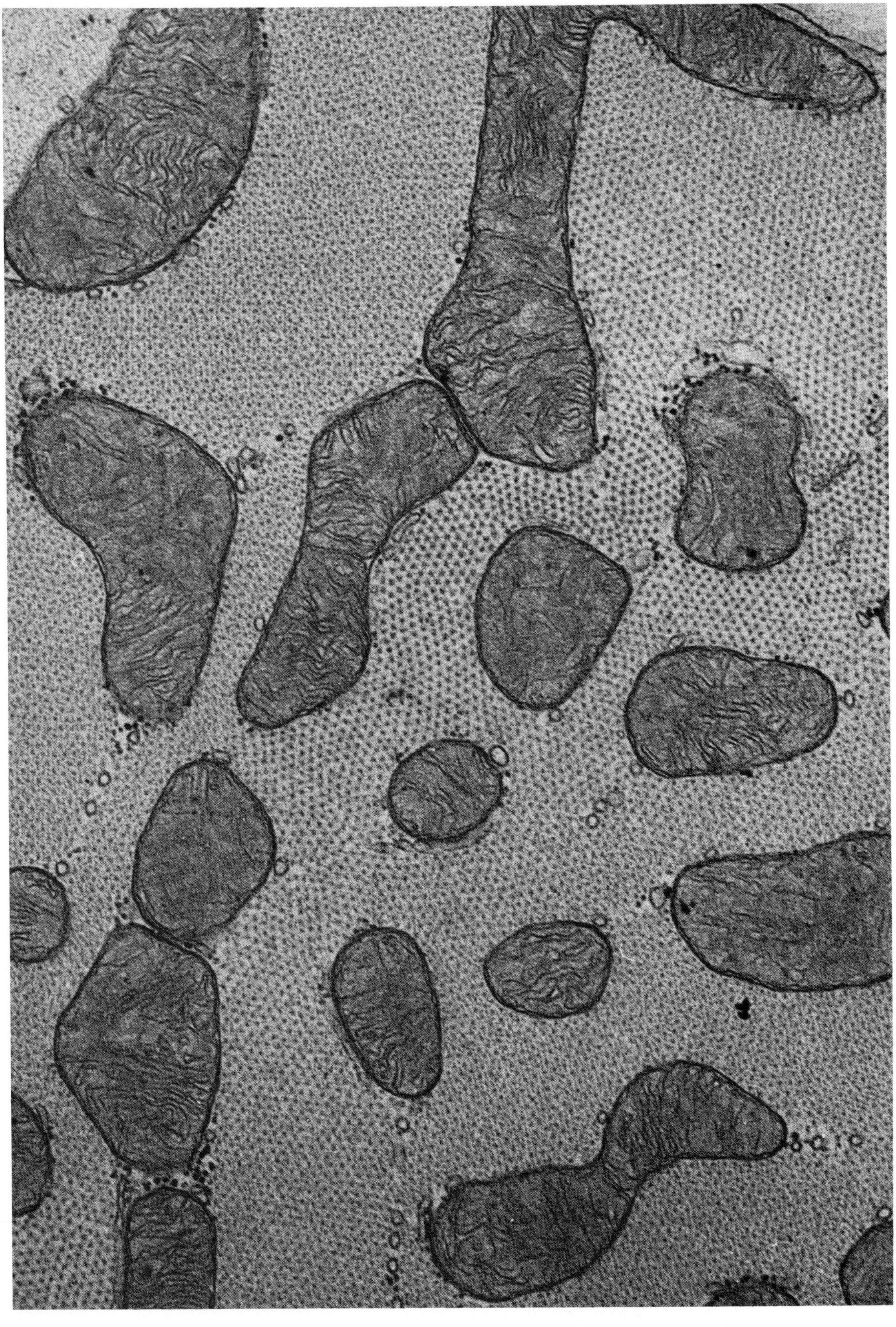

Figure 259

Insect flight muscle with a beat frequency of 50 per second exhibits discrete myofibrils of uniform size surrounded by large mitochondria (sarcosomes), which often assume bizarre shapes. In this species a very dense mitochondrial matrix obscures the numerous closely packed cristae.

Figure 260. Flight muscle of the moth, *Phytometra gamma*. (Micrograph courtesy of J. Auber, C. R. Acad. Sci. (Paris) 264: 621, 1967.)

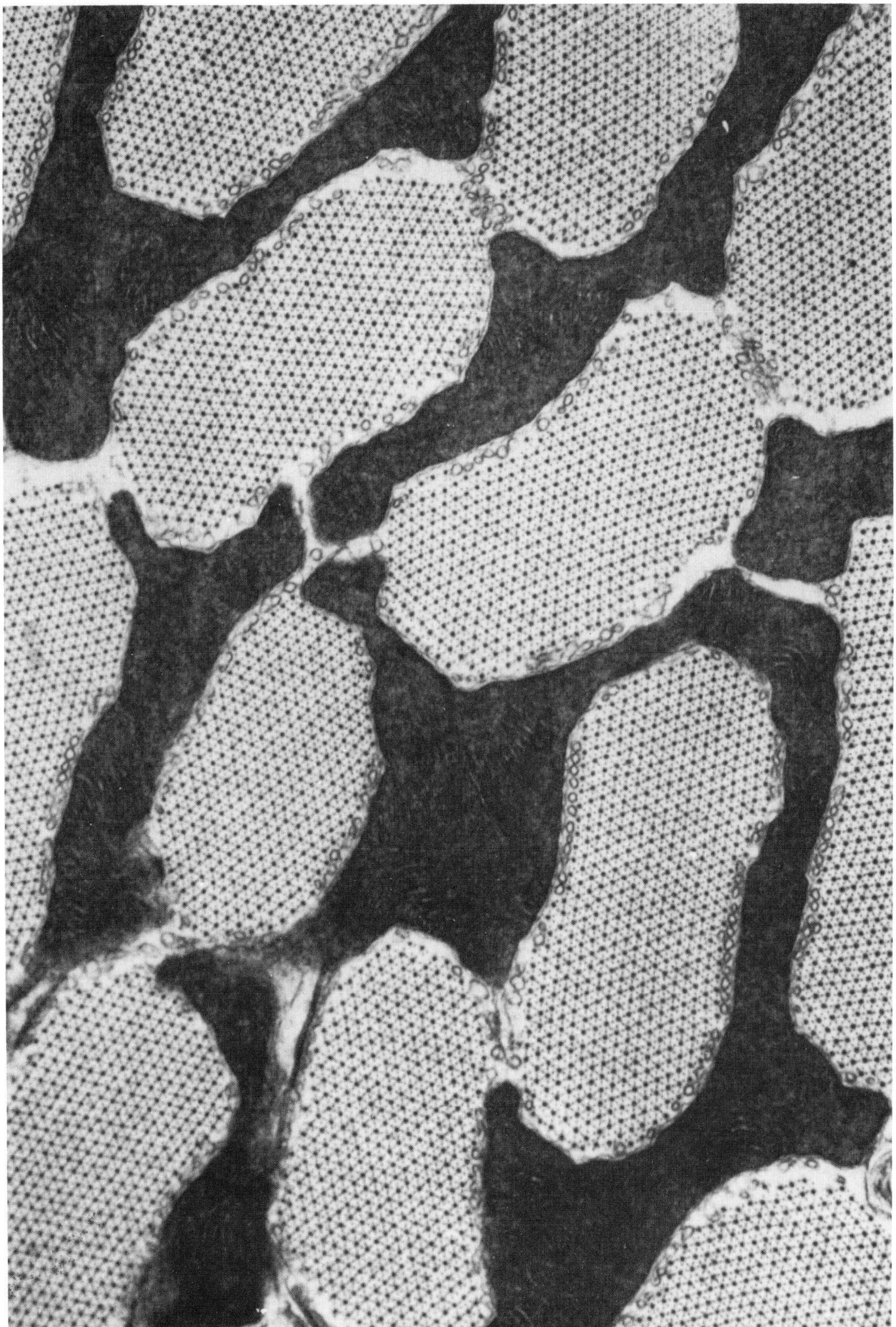

Figure 260

In the midpiece of the spermatozoon of nearly all vertebrates the axoneme is enclosed in a sheath of mitochondria. In mammals, such as the dormouse illustrated here, the elongate mitochondria are arranged end to end in a helix. Thus in longitudinal sections of the sperm tail the mitochondria are seen in cross-section. Their intimate relationship to the outer dense fibers and axoneme of the sperm tail facilitates diffusion of ATP, which is essential for motility.

Figure 261. Midpiece of spermatozoa from the dormouse, *Glis glis*, in longitudinal section. (Micrograph courtesy of David Phillips.)

Figure 261

REFERENCES

Mitochondria

André, J. Quelques données recentes sur la structure et la physiologie des mitochondries: glycogen, particules élémentaires, acides nucléiques. Archives de Biologie 76:277–304, 1965.

André, J. and V. Marinozzi. Presence, dans les mitochondries, de particules resemblant aux ribosomes. J. Microscopie 4:615–626, 1965.

Altmann, R. Die Elementärorganismen und ihre Beziehungen zu den Zellen. Viet Co., Leipzig, 1890.

Barnard, T. and J. Ruusa. Mitochondrial matrix granules in soft tissues I. Elemental composition by x-ray microanalysis. Exp. Cell Res. 124:339–347, 1979.

Benda, C. Weitere Mitteilungen über die Mitochondria. Verh. Otsch. Physiol. Ges., pp. 376–383, 1899.

Bensley, R. R. and N. Hoerr. Preparation and properties of mitochondria. Anat. Rec. 60:449–455, 1934.

Borst, P., E. F. J. Van Bruggen and G. J. C. M. Ruttenberg. Size and structure of mitochondrial DNA. In Biochemical Aspects of Biogenesis of Mitochondria (E. C. Slater, J. M. Tager, S. Papa and E. Quagliarello, eds.), pp. 51–59, Adriatica Editrice, Bari, 1968.

Brdiczka, D. and T. Barnard. Mitochondrial matrix granules in soft tissues. II Isolation and initial characterization of a calcium precipitable, soluble lipoprotein subfraction from brown fat and liver mitochondria. Exp. Cell Res. 126:127–136, 1980.

Claude, A. The morphology and significance of dumbbell-shaped mitochondria in early stages of regenerating liver. J. Cell Biol. 27:146A, 1965.

Cowdry, E. V. The mitochondrial constituents of protoplasm. Carnegie Contribution to Embryology 8:29, 1918. (Review)

Dawid, I. B. and D. R. Wolstenholme. The structure of frog oocyte mitochondrial DNA, pp. 83–89. In Biochemical Aspects of Biogenesis of Mitochondria. (E. C. Slater, J. M. Tager, S. Papa and E. Quagliarello, eds.), Adriatica Editrice, Bari, 1968.

Fawcett, D. W. Observations on the cytology and electron microscopy of hepatic cells. J. Nat. Cancer Inst. 15:1475, 1955.

Fernandez-Moran, H. Subunit organization of mitochondrial membranes. Science 140:381, 1963.

Gibor, A. and S. Granick. Plastids and mitochondria: inheritable systems. Science 145:890–897, 1964.

Gross, N. J., G. S. Getz and M. Rabinowitz. Turnover of mitochondrial DNA and phospholipid in rat tissues. *In* Biochemical Aspects of Biogenesis of Mitochondria (E. C. Slater, J. M. Tager, S. Papa and E. Quagliarello, eds.), Adriatica Editrice, Bari, 1968.

Hackenbrock, C. R. Ultrastructural bases for metabolically linked mechanical activity in mitochondria. I. Reversible ultrastructural changes with changes in metabolic state in isolated liver mitochondria. J. Cell Biol. 30:269–297, 1966.

Harris, R. A., C. H. Williams, M. Caldwell, D. E. Green and E. Valdivia. Energized configurations of heart mitochondria in situ. Science 165:700, 1969.

Harris, R. A., M. A. Asbell and D. E. Green. Conformational basis of energy transduction in membrane systems. II. Measurement of configurational change by light scattering. Arch. Biochem. Biophys. 132:545–560, 1969.

Hawley, E. S. and R. P. Wagner. Synchronous mitochondrial division in *Neurospora crassa*. J. Cell Biol. 35:489–499, 1967.

Hogeboom, G. H., W. C. Schneider and G. H. Palade. Isolation of intact mitochondria from rat liver; some biochemical properties of mitochondria and submicroscopic particulate material. J. Biol. Chem. 172:619, 1948.

Kahri, A. I. Effects of actinomycin D and puromycin on the ACTH-induced transformation of mitochondria of cortical cells of rat adrenals in tissue culture. J. Cell Biol. 36:181–195, 1968.

Kalf, G. F. Deoxyribonucleic acid in mitochondria and its role in protein synthesis. Biochemistry 3:1702–1706, 1964.

Kalf, G. F. and M. V. Simpson. The incorporation of valine-1-C^{14} into the protein of submitochondrial fractions. J. Biol. Chem. 234:2943–2947, 1959.

Kroon, A. M. Inhibitors of mitochondrial protein synthesis. Biochim. Biophys. Acta 76:165–167, 1963.

Lehninger, A. L. The Mitochondrion. W. A. Benjamin, Inc., New York, 1964.

Larsen, W. J. Genesis of mitochondria in insect fat body. J. Cell Biol. 47:373–383, 1970.

Luck, D. J. L. DNA in mitochondria of *Neurospora crassa*. Proc. Nat. Acad. Sci. 52:931–938, 1964.

McLean, J. R. et al. Incorporation of labeled amino acids into the protein of muscle and liver mitochondria. J. Biol. Chem. 233:657, 1958.

Nass, M. M. K. and S. Nass. Intramitochondrial fibers with DNA characteristics. I. Fixation and electron staining reactions. J. Cell Biol. 19:593–612, 1963.

Nass, M. M. K. and S. Nass. Intramitochondrial fibers with DNA characteristics. II. Enzymatic and other hydrolytic treatments. J. Cell Biol. 19:613–629, 1963.

Novikoff, A. B. Mitochondria. In The Cell (J. Brachet and A. E. Mirsky, eds.), Vol. II, Academic Press, New York, 1961. (Review)

Palade, G. An electron microscope study of mitochondrial structure. J. Histochem. Cytochem. 1:188–211, 1953.

Parsons, D. F. Mitochondrial structure: two types of subunits on negatively stained mitochondrial membranes. Science 140:985, 1963.

Parsons, D. F. Recent advances correlating structure and function in mitochondria. Int. Rev. Exp. Pathol. 4:1–54, 1965. (Review)

Parsons, J. A. Mitochrondrial incorporation of tritiated thymidine in *Tetrahymena pyriformis*. J. Cell Biol. 25:641–646, 1965.

Peachey, L. D. Electron microscopic observations on the accumulation of divalent cations in intramitochondrial granules. J. Cell Biol. 20:95–111, 1964.

Rabinowitz, M., J. Sinclair, L. DeSalle, R. Haselhorn and H. H. Swift. Isolation of deoxyribonucleic acid from mitochondria of chick embryo heart and liver. Proc. Nat. Acad. Sci. 53:1126–1133, 1964.

Reich, E. and D. J. L. Luck. Replication and inheritance of mitochondrial DNA. Proc. Nat. Acad. Sci. 55:1600, 1966.

Schuster, F. L. A deoxyribose nucleic acid component in mitochondria of *Didymium nigripes,* a slime mold. Exp. Cell Res. 39:329–345, 1965.

Sheridan, M. Fine structural changes seen in the livers of scorbutic and partially starved guinea pigs. Anat. Rec. 136:276, 1960.

Sinclair, J. H. and B. V. Stevens. Circular DNA filaments from mouse mitochondria. Proc. Nat. Acad. Sci. 56:508–514, 1966.

Smith, D. S. The structure of flight muscle sarcosomes in the blowfly, *Calliphora erythrocephala*. J. Cell Biol. 19:115, 1963.

Steinert, M. Mitochondria associated with the kinetonucleus of *Trypanosoma mega*. J. Cell Biol. 8:542, 1960.

Stoeckenius, W. Some observations on negatively stained mitochondria. J. Cell Biol. 17:443–454, 1963.

Stone, G. E. and O. L. Miller. Incorporation of tritiated thymidine into deoxyribonuclease sensitive material in mitochondria of tetrahymena. J. Cell Biol. 23:89A, 1964.

Sutfior, L. V., M. E. Holtrop and R. Ogilvie. Microanalysis of individual mitochondrial granules with diameters less than 1000 angstroms. Science 174:947, 1971.

Swift, H., N. Kislev and L. Bogorad. Evidence for DNA and RNA in mitochondria and chloroplasts. J. Cell Biol. 23:91A, 1964.

Swift, H., M. Rabinowitz and G. Getz. Cytochemical studies on mitochondrial nucleic acids. In Biochemical Aspects of Biogenesis of Mitochondria (E. C. Slater, J. M. Tager, S. Papa and E. Quagliarello, eds.), pp. 3–36. Adriatica Editrice, Bari, 1968.

Tandler, B. and C. L. Hoppel. Division of giant mitochondria during recovery from cuprizone intoxication. J. Cell Biol. 56:266–272, 1973.

Van Bruggen, E. F. J., P. Borst, G. J. C. M. Ruttenberg, M. Gruber and A. M. Kroon. Circular mitochondrial DNA. Biochim. Biophys. Acta 119:437, 1966.

Vogel, F. S. Structural and functional characteristics of DNA rich mitochondria of the common meadow mushroom *Agaricus campestris*. I. Intracellular environment. Lab. Invest. 14:1849, 1965.

Vogel, F. S. and L. Kemper. Structural and functional characteristics of DNA-rich mitochondria of the common meadow mushroom, *Agaricus campestris*. II. Extracellular cultures. Lab. Invest. 14:1868, 1965.

LYSOSOMES

The history of most cell organelles has been early description by microscopists followed many years later by isolation and biochemical characterization. The lysosome is an exception in that it originated as a biochemical concept and morphological identification followed. Among investigators isolating mitochondria and microsomes by differential centrifugation of cell homogenates, there was some disagreement as to which fraction contained the enzyme acid phosphatase. Some found it in the mitochondrial fraction (Palade, 1951; Schneider and Hogeboom, 1952), while others, using somewhat lower centrifugal force, found it in their microsome fraction (Novikoff et al., 1953). This problem was resolved by DeDuve and his associates, who were able to separate the classical mitochondrial fraction into two subfractions. The lighter (L) fraction consisted of particles rich in acid phosphatase but lacking the mitochondrial enzyme cytochrome oxidase (DeDuve et al., 1953). These particles, which also contained cathepsin, ribonuclease, deoxyribonuclease, and β-glucuronidase, were recognized as a new particulate component of cells distinct from mitochondria and were given the name *lysosomes* to draw attention to their richness in hydrolytic enzymes (DeDuve et al., 1955).

Another defining characteristic of lysosomes was the fact that they were impermeable to their substrates and were enzymatically active *in vitro* only after disruption or treatment with a surface active agent. It was thus inferred that they must be enclosed by a membrane-like barrier. When centrifugal pellets of fractions enriched in lysosomes were examined in thin sections with the electron microscope, a high proportion of the granules were obviously different from mitochondria. They had a dense heterogeneous content and, as postulated, they were enclosed by a membrane (Novikoff, Beaufay and DeDuve, 1956).

Fortunately a dependable cytochemical method was available for the lysosomal enzyme acid phosphatase (Gomori, 1952). Cytochemical observations at the light and electron microscope level confirmed that the acid phosphatase of liver was not localized in mitochondria but in membrane-bounded dense bodies in the vicinity of the bile canaliculi (Novikoff, 1959; Holt, 1959; Barka, 1960). Unlike mitochondria and other cell organelles which have a consistent, clearly defined and easily recognizable structure, the granules exhibiting acid phosphatase activity varied in size and were highly heterogeneous in their internal structure. Some were spherical with a uniform content of moderate density; others were irregular in outline and contained aggregations of very dense granules in a less dense matrix. Others contained myelin figures or crystalline inclusions. It was this extraordinary pleomorphism that had prevented cytologists from recognizing lysosomes as a distinct entity. This was the first instance of an organelle that could not be identified with confidence solely by morphological criteria. Biochemical or cytochemical demonstration of acid hydrolase activity was required. By 1962, the number of hydrolytic enzymes identified had increased to 10. It was apparent that while most of these were present in all lysosomes, their proportions probably varied considerably from tissue to tissue.

The assembly of multiple acid hydrolases in the same organelle led naturally to the speculation that lysosomes functioned as an intracellular digestive system and that the substances degraded were either *exogenous* — brought into the cell by phagocytosis — or *endogenous,* involving the controlled autolytic elimination of other organelles and inclusions in the course of normal renewal or in response to an altered state of physiological activity. An essential feature of this concept was the belief that in their lysosomes cells contained the seeds of their own destruction. In normal function, it was assumed that the digestive processes had to be sequestered within membranes to protect the surrounding cytoplasm, but in pathological conditions autolysis of cells was a consequence of breakdown of this protective barrier and release of digestive enzymes into the cytoplasm (DeDuve, 1963).

Evidence for the physiological degradation of cell components, called *autophagy,* rested upon numerous electron microscopic observations of mitochondria, profiles of endoplasmic reticulum or aggregations of glycogen particles enclosed in membrane-limited *autophagic vacuoles* (Ashford and Porter, 1962; Novikoff and Essner, 1962). It is believed that the organelles to be degraded are first surrounded by a membrane of uncertain provenance and that lysosomes then coalesce with the autophagic vacuole so formed, adding their lytic enzymes to its contents.

The relationship of lysosomes to exogenous material incorporated in *heterophagic vacuoles* was demonstrated experimentally in many cell types. In the kidney, intravenously injected peroxidase is taken up by endocytosis in the proximal convoluted tubules in phagosomes which have the morphological and biochemical properties of lysosomes (Strauss, 1959; DeDuve, 1961). The phagocytic Kupffer cells of hepatic sinusoids are rich in lysosomes and after administration of peroxidase or particulate markers, these substances are found sequestered in acid phosphatase–positive vacuoles or secondary lysosomes (Novikoff et al., 1960). Observations of phagocytosis by living leucocytes provided convincing support for the postulated mechanism involved in heterophagocytosis. The granules of these cells contain lysosomal hydrolases and antibacterial agents. When observed with the light microscope, the granules can be seen to coalesce with the phagocytic vacuoles, resulting in a progressive degranulation of the cells (Cohn and Hirsch, 1960; Hirsch, 1962).

The morphological heterogeneity of lysosomes is attributable in part to the fact that they represent various stages in a dynamic process of intracellular digestion. Small granules with relatively homogeneous content are called *primary lysosomes,* an inactive storage form or ready reserve of hydrolytic enzymes not yet involved in digestive events. Somewhat larger bodies with a heterogeneous content that may include recognizable residues of ingested material are *heterophagic vacuoles (phagosomes,* or *secondary lysosomes)* formed by fusion of primary lysosomes with endocytosis vacuoles. As digestion of the contents of the heterophagic vacuoles progresses, they are reduced in size and become *residual bodies* filled with granules of varying size and density in an amorphous matrix.

The fate of undigestable residues is not entirely clear. In amoebae they appear to be eliminated by exocytosis, but unambiguous evidence for such a process in cells of higher forms is lacking. Instead, residual bodies appear to coalesce and become consolidated into the large, irregularly shaped aggregations of dense material traditionally designated *lipofuscin,* or *wear-and-tear pigment.*

Primary lysosomes are generally believed to be formed in much the same manner as secretory granules, with synthesis of the enzymes in the rough endoplasmic reticulum and segregation and packaging in the Golgi apparatus. Some investigators, however, believe they originate in the GERL, a differentiated region of the endoplasmic reticulum associated with the secretory face of the Golgi.

Some 50 lysosomal enzymes have now been identified. These include several proteolytic enzymes, glycosidases, nucleases, phosphatases, phospholipases, and sulfatases. The pH optima of nearly all of these are in the acid range. This battery of enzymes can break down most constituents of cells into small molecules that can pass through the lysosome membrane into the cytoplasm. An inherited defect in the synthesis of one or more of these is the basis for 20 or more rare storage diseases of children (Neufeld et al., 1975). When a lysosomal enzyme is deficient, its normal substrate accumulates in large membrane-bounded structures in the cytoplasm which are, in effect, autophagic vacuoles that are unable to digest their contents.

While the main site of lysosomal function is intracellular, evidence has accumulated in recent years indicating that under certain conditions phagocytic cells can secrete their lysosomal enzymes. The secretion of collagenase and other proteases by osteoclasts is involved in the resorption of matrix in the normal remodeling of bone. The binding of complement or antigen-antibody complexes to leucocytes or macrophages stimulates their secretion of hydrolases that may destroy basal laminae, cartilage matrix, collagen or elastin and result in damage to kidneys, joints, or lungs (Davies and Allison, 1976; Weissmann et al., 1971).

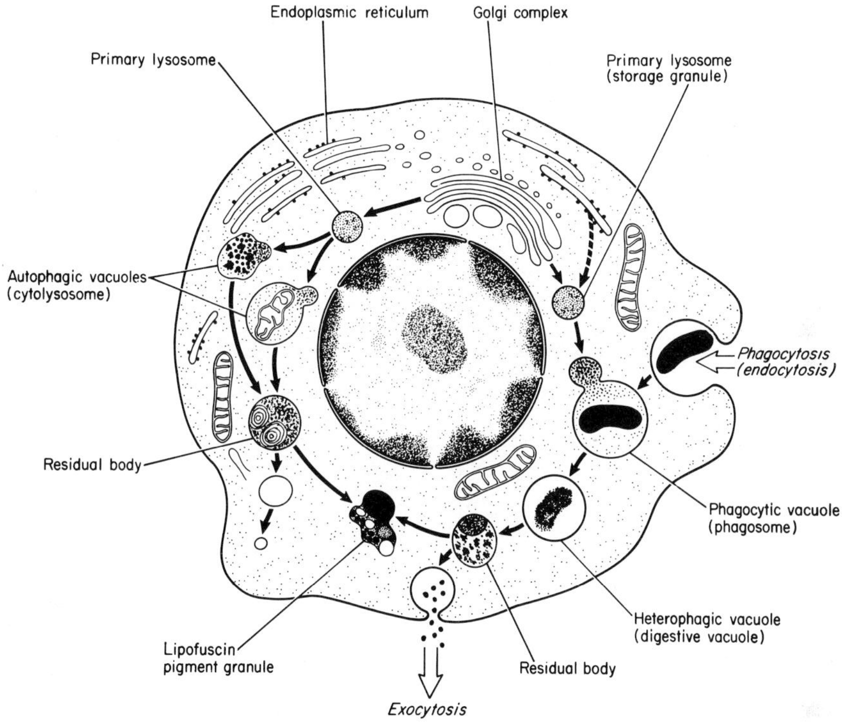

Schematic representation of the role of lysosomes in heterophagy (lower right) and in autophagy (upper left). Bacteria or other foreign matter may be taken into the cell by phagocytosis. Primary lysosomes fuse with the heterophagic vacuole and their enzymes degrade the foreign matter. Effete or supernumerary organelles may be enclosed by membranes to form autophagic vacuoles. Subsequent fusion of primary lysosomes with these vacuoles results in degradation of their contents. End products of either autophagy or heterophagy can be recognized in cells as residual bodies and lipofuscin pigment. (From G. Bloom and D. W. Fawcett, *Textbook of Histology,* 10th Ed., W. B. Saunders Co., 1975.)

The aggregation of primary lysosomes in the accompanying micrograph shows some of the variability that makes their microscopic identification difficult. They may be round, elliptical, or highly irregular in shape. They are invariably limited by a membrane but their content may be of homogeneous density or highly heterogeneous. Other variations in their appearance are recorded in the micrographs that follow.

Figure 262. Lysosomes in the supranuclear region of an epithelial cell from the epididymis of the bandicoot, *Parameles nasuta.*

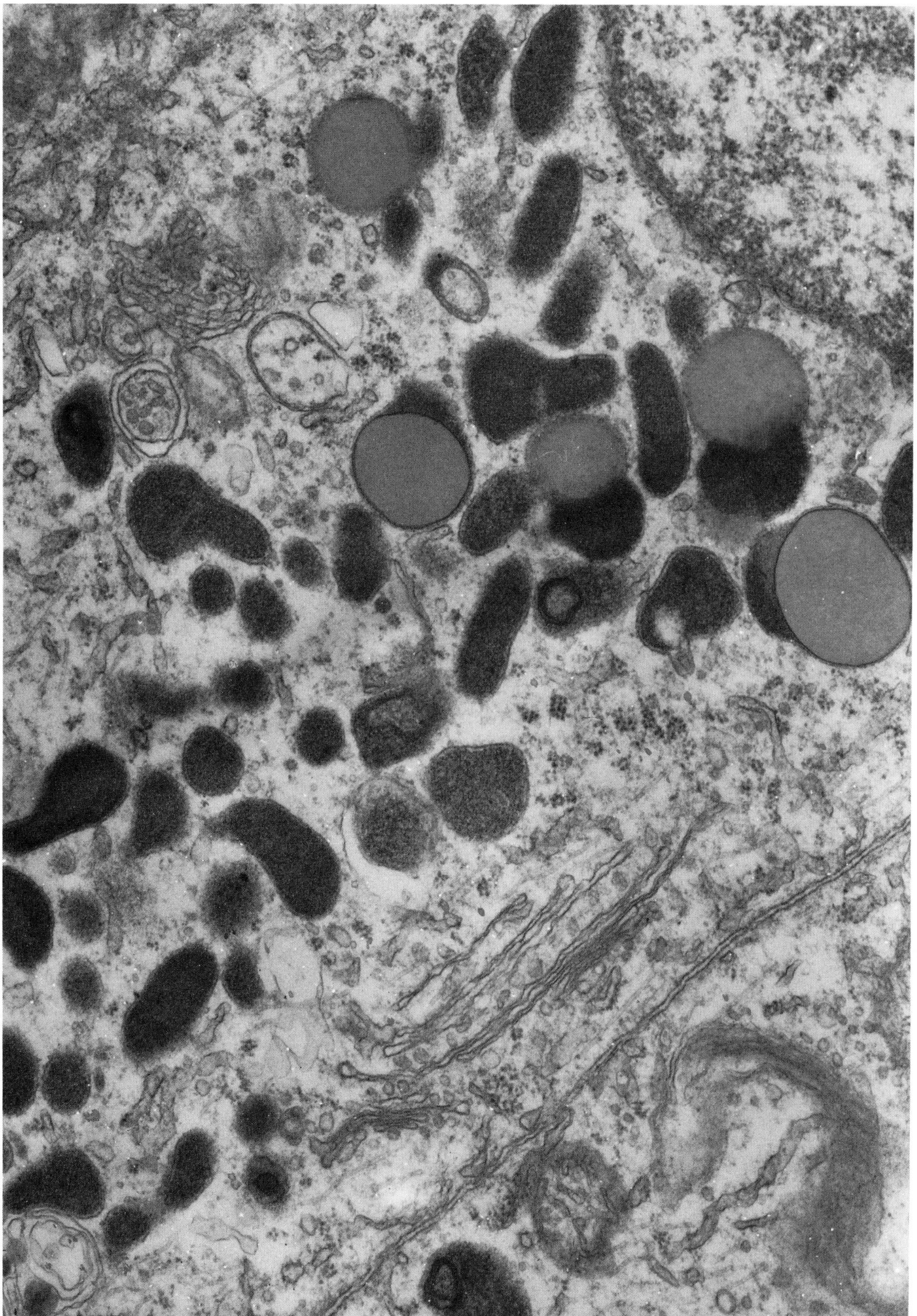

Figure 262

Another example of lysosomes clustered in the Golgi region. On these the limiting membrane, which is one of their defining characteristics, is clearly visible. It is often very closely applied to contents of similar density and is therefore difficult to resolve.

Figure 263. Region of the cell center from hamster suprarenal cortex.

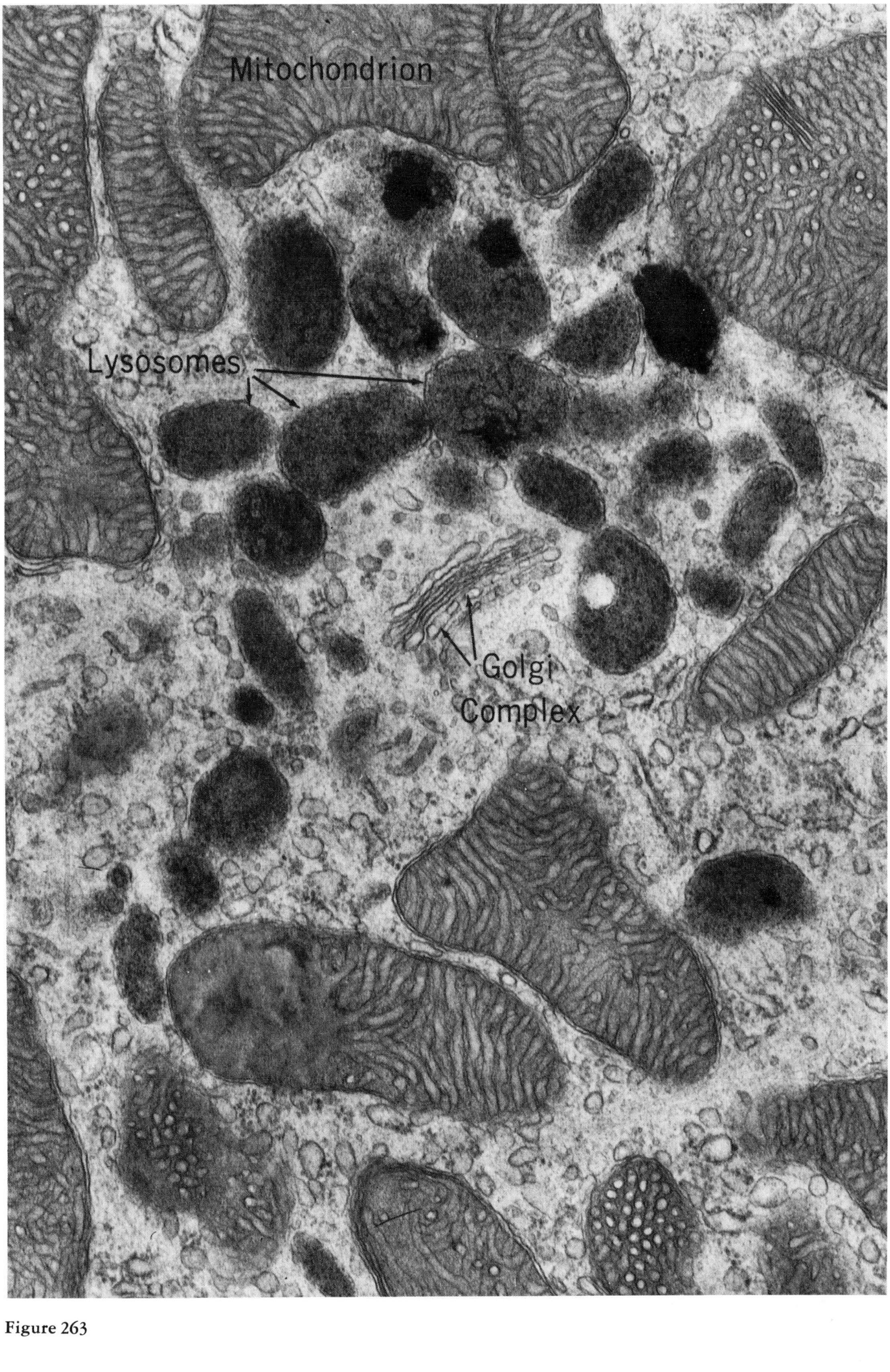

Figure 263

The lysosomes that have been most thoroughly studied are those of hepatic cells which were first centrifugally isolated from the mitochondrial fraction of liver homogenates. The primary lysosomes tend to have a dense homogeneous content but the secondary lysosomes may contain a variety of products of hydrolysis of differing density and texture. The majority of the membrane-bounded dense bodies in the accompanying figures are secondary lysosomes.

In the absence of a universally applicable set of morphological criteria, identification of lysosomes rests upon a general resemblance to the dense particles comprising the "lysosome fraction" of liver homogenates. Since one of their defining characteristics is the presence of hydrolytic enzymes, verification of the lysosomal nature of cytoplasmic particles requires the cytochemical demonstration of their acid phosphatase or beta glucoronidase activity.

Figures 264 and 265. Liver of phenobarbital-treated hamster showing a variety of lysosomes.

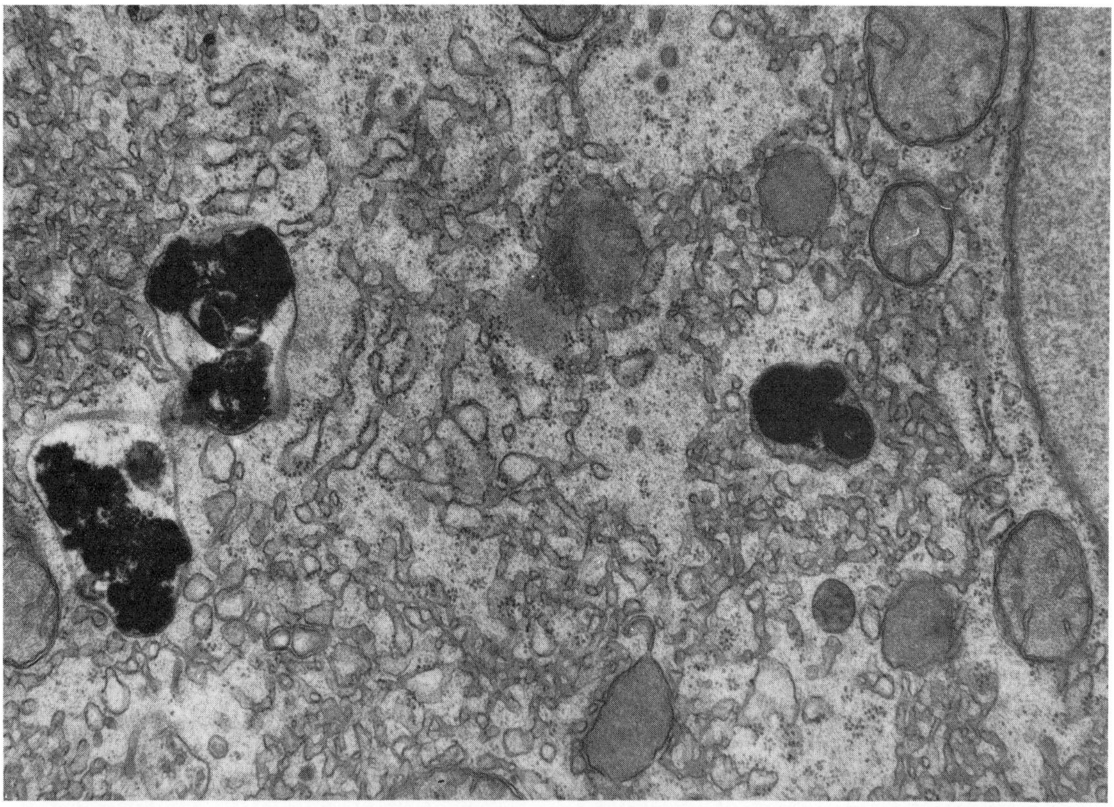

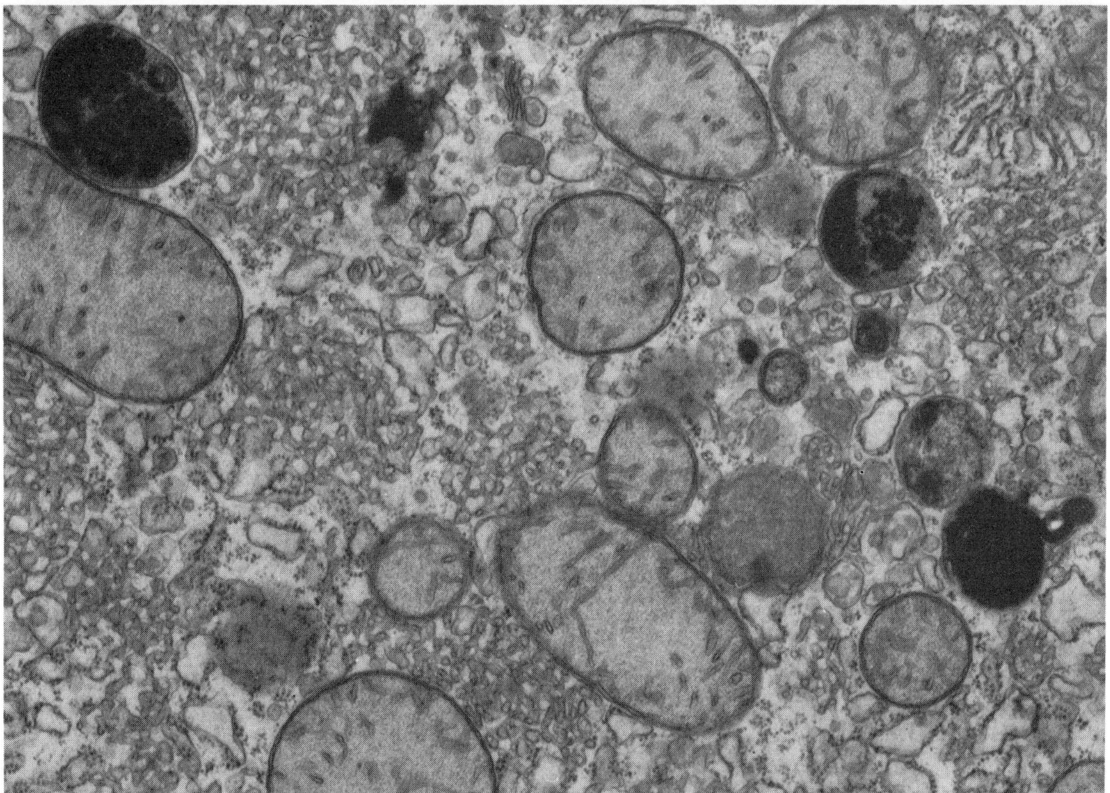

Figure 264, *upper* **Figure 265**, *lower*

In the internal remodeling of cells associated with diminished physiological activity, excess organelles are eliminated by autophagy. The organelle or inclusion to be destroyed is first surrounded by a membrane to form an autophagic vacuole. Lysosomes then fuse with it, discharging their hydrolytic enzymes into its interior. The origin of the membrane surrounding the material to be digested has not been established, but the fact that there are often two parallel limiting membranes suggests the possibility that a cistern of the smooth endoplasmic reticulum or GERL is involved.

The accompanying micrographs present typical examples of autophagic vacuoles. The upper figures are of liver from an animal in the period of recovery after experimental induction of drug-metabolizing enzymes. The content of the autophagic vacuoles is mainly smooth endoplasmic reticulum and free ribosomes. In the figure at the lower left, a mitochondrion and a peroxisome have been sequestered, and the autophagic vacuole at the lower right contains a mitochondrion in a more advanced state of dissolution but with cristae still identifiable.

Figures 266, 267, and 269. Autophagic vacuoles from liver several days after withdrawal of phenobarbital. (Micrographs courtesy of Robert Bolender.)

Figure 268. Autophagic vacuole from normal rat liver. (Micrograph courtesy of Daniel Friend.)

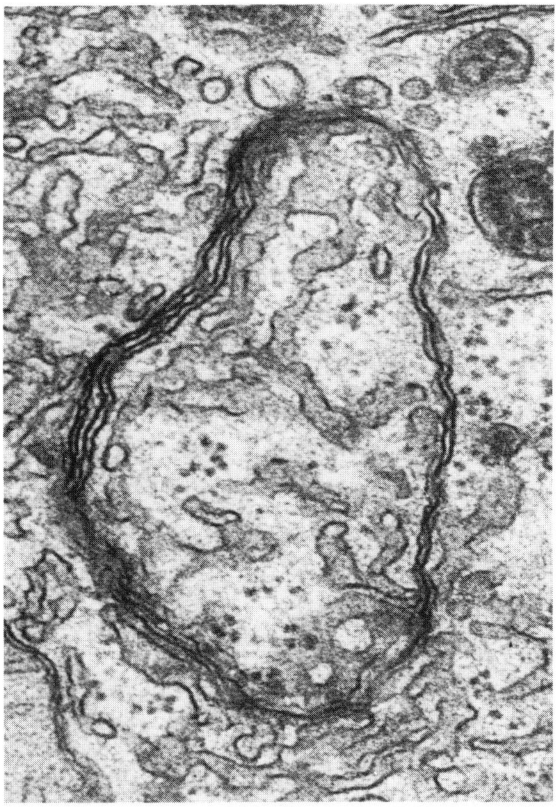

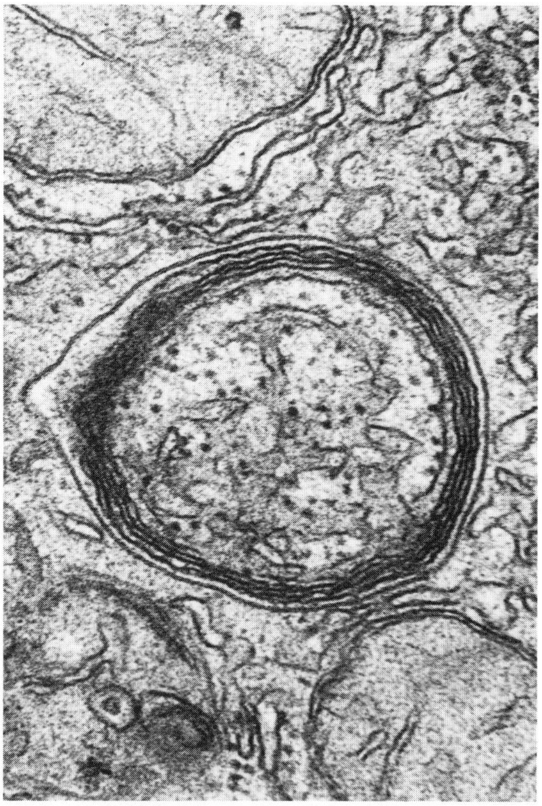

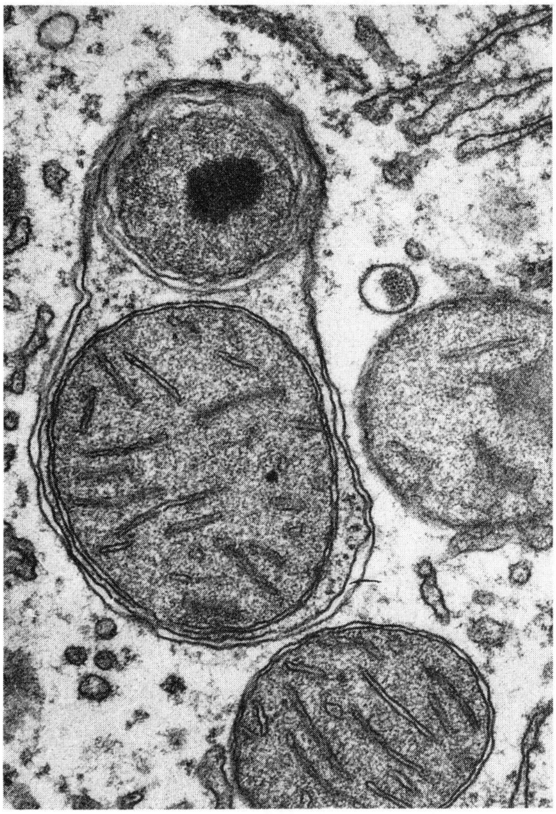

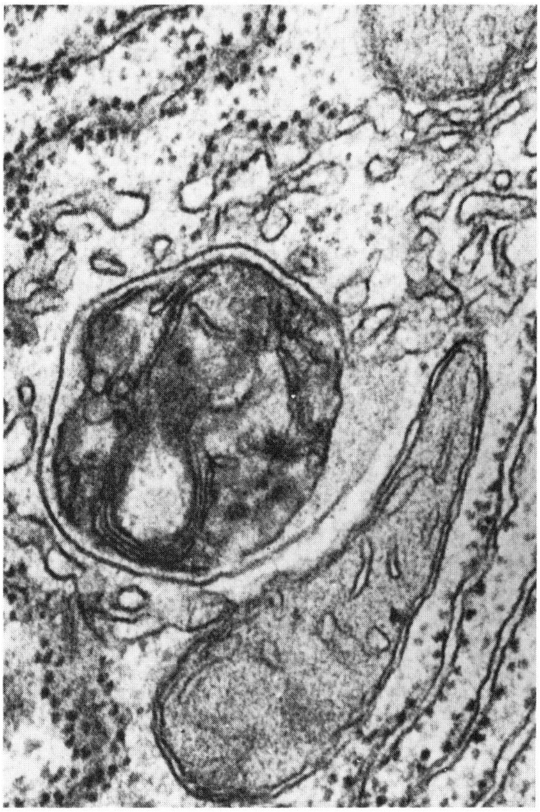

Figure 266, *upper left* **Figure 267,** *upper right*

Figure 268, *lower left* **Figure 269,** *lower right*

Some epithelia contain membrane-limited spherical granules that appear to be large primary lysosomes, but their moderately dense homogeneous content makes them morphologically indistinguishable from the secretory granules of glandular cells. They occur, however, in cells not known to have a secretory function; they show no tendency to accumulate at the luminal surface and are never observed in the process of exocytosis. They are therefore presumed to be either lysosomes or peroxisomes. Cytochemical demonstration of acid phosphatase or peroxidase activity is required for their unambiguous identification.

Figure 270. Cytoplasm of a cell from a ductus efferens of the ground squirrel *Citellus lateralis*. (Micrograph courtesy of Jeffrey Pudney.)

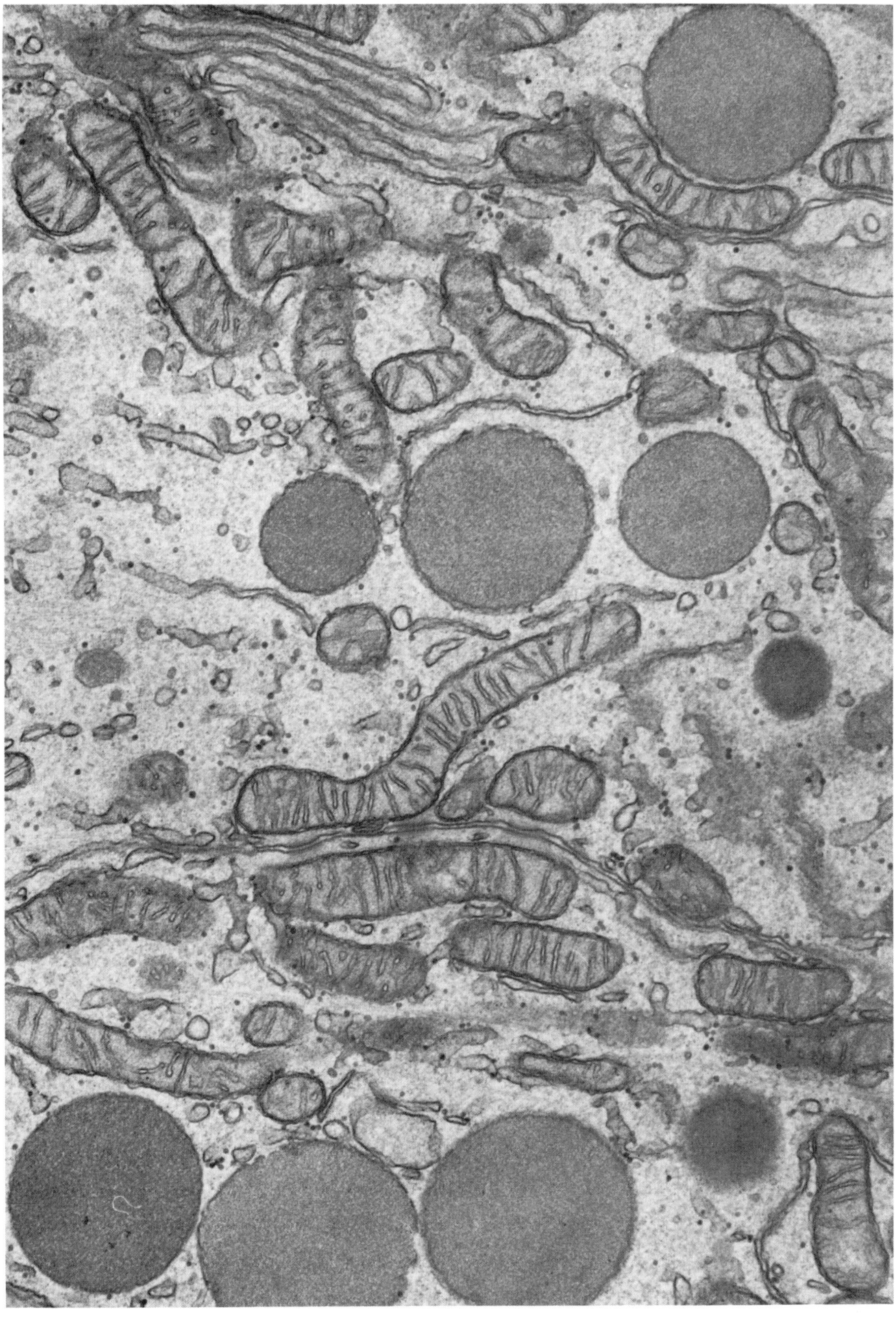

Figure 270

Certain segments of the epididymal epithelium normally have large numbers of dense, spherical lysosomes concentrated in the supranuclear region of the columnar cells, as illustrated in the accompanying micrograph. Other segments are characterized by numerous multivesicular bodies in the apical cytoplasm. The physiological significance of the presence of these two categories of lysosome in the same epithelium is not known.

Figure 271. Micrograph of epithelium from the proximal portion of the cauda epididymis of rabbit. (Micrograph courtesy of Roy Jones.)

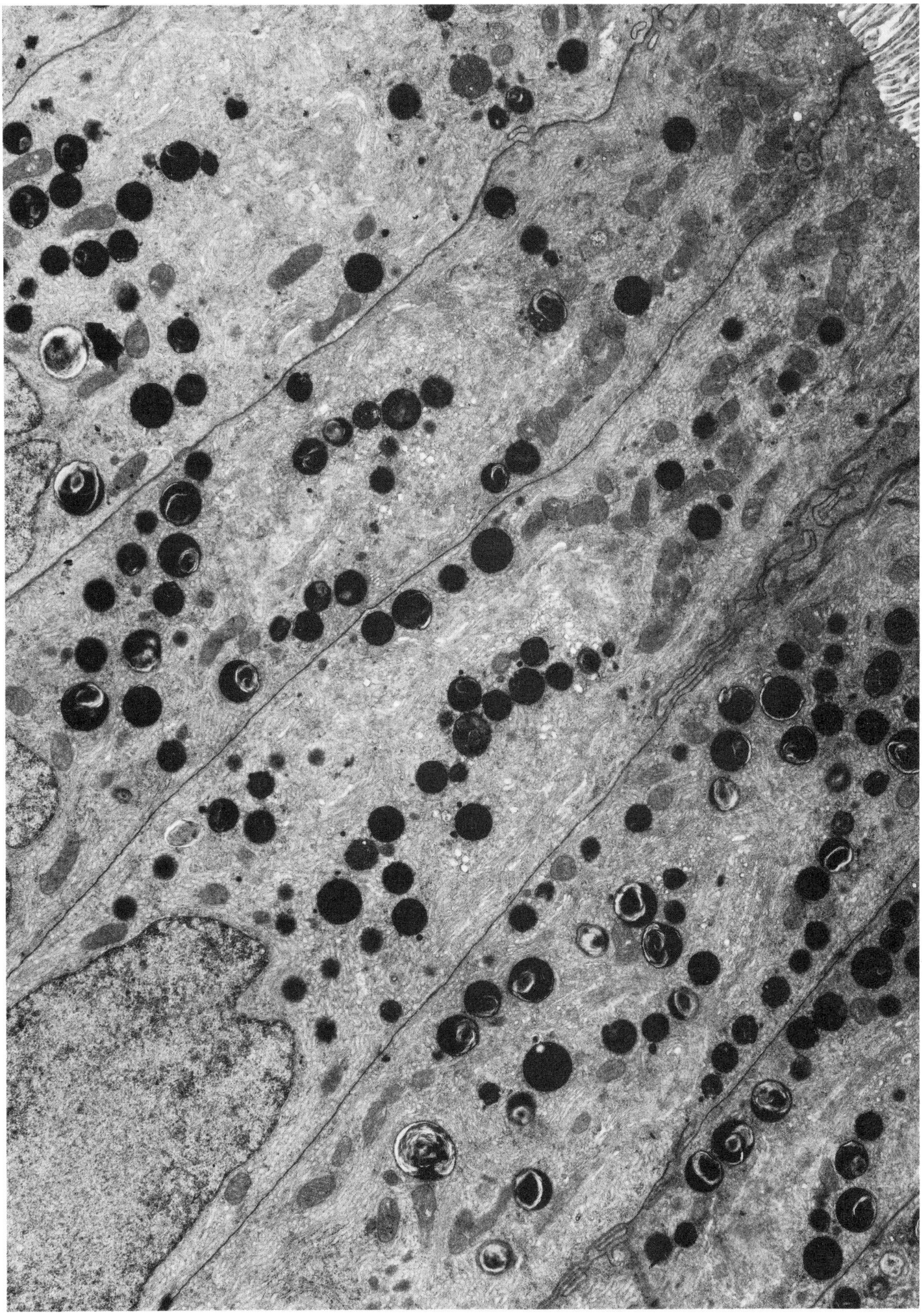

Figure 271

Among the commonest inclusions in secondary lysosomes of all cell types are concentric systems of thin dense lamellae that are interpreted as myelin forms of hydrated phospholipid. The accompanying micrographs include extreme examples routinely observed in epithelial cells in the distal portion of the cauda epididymis in rabbits. Similar inclusions are found in the smaller lysosomes on the previous page, but in this more distal region of the duct they may attain diameters of several microns.

Figures 272 and 273. Lysosomes in the basal cytoplasm of the principal cells in the cauda epididymis of the rabbit. (Micrographs courtesy of Roy Jones.)

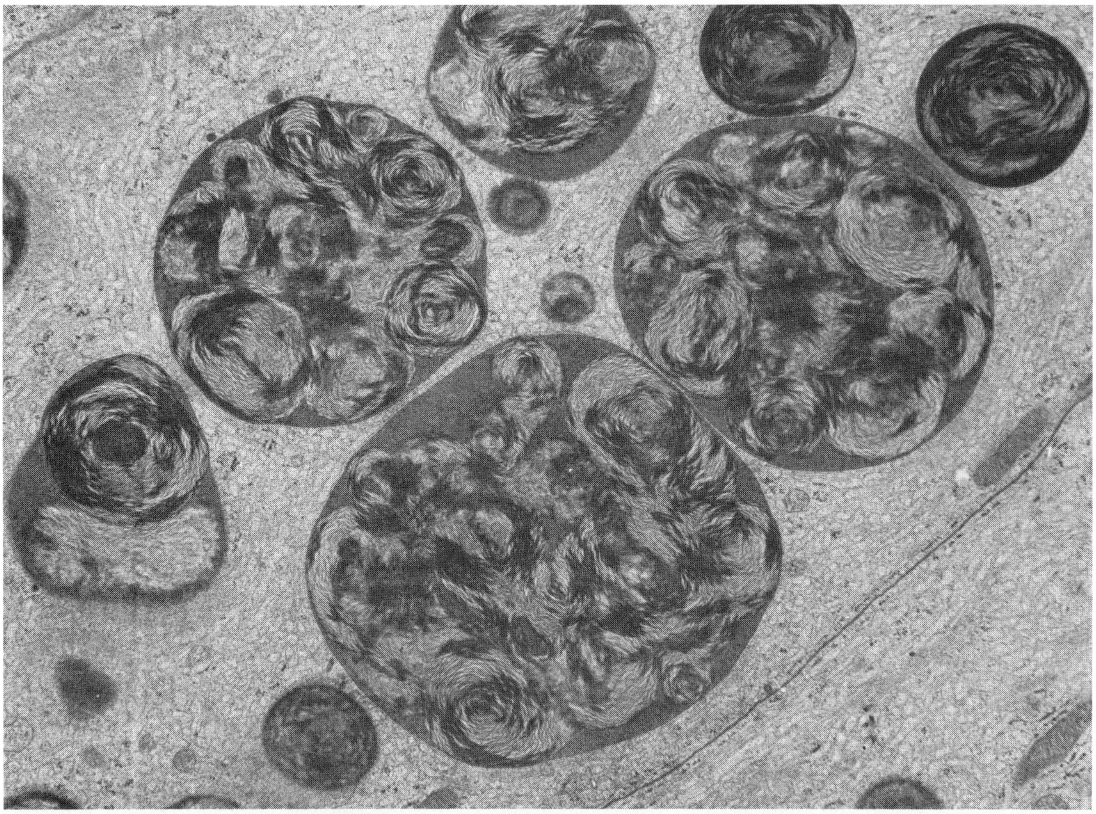

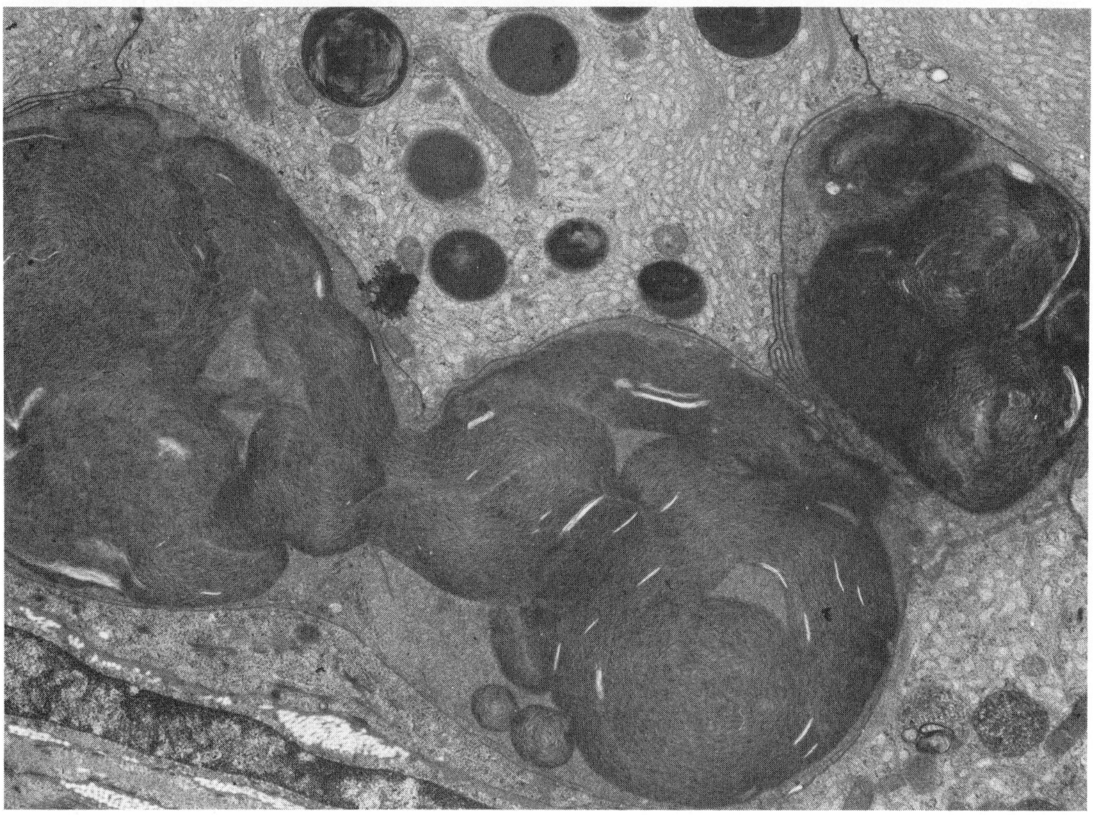

Figure 272, *upper* **Figure 273,** *lower*

Additional examples of laminated inclusions in lysosomes are illustrated here in micrographs from Leydig cells which, like other steroid-secreting cells, are rich in lysosomes. The clear spaces interposed among the lamellae are negative images of thin tabular crystals of unknown nature that have been dissolved in the course of specimen preparation.

Figure 274. Lysosomes in the interstitial cells of Leydig in the testis of the domestic boar, *Sus scrofa*.

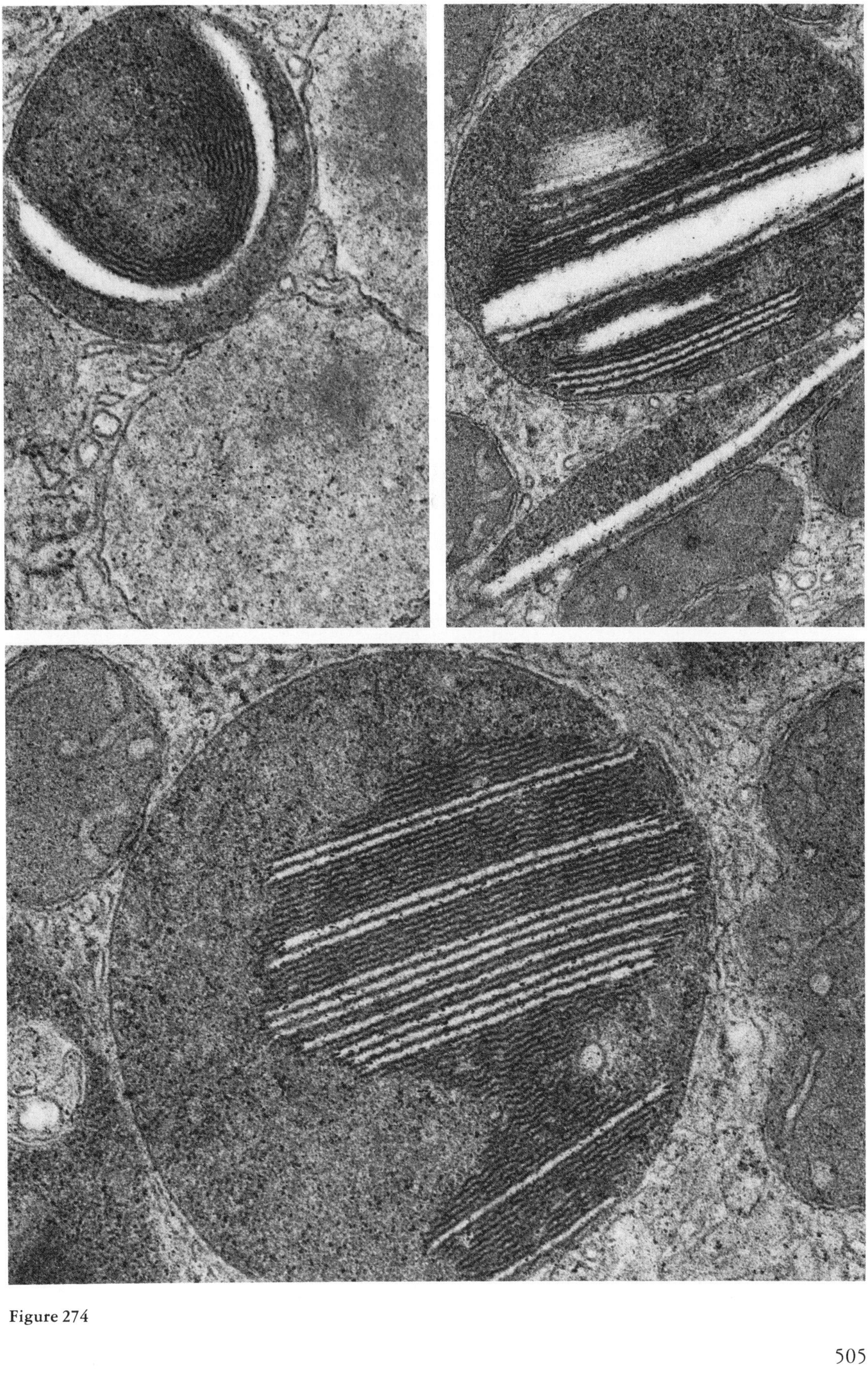

Figure 274

In cells that are actively engaged in heterophagy, the lysosomes are very heterogeneous, exhibiting a broad spectrum of primary lysosomes, heterophagic vacuoles, residual bodies, and lipofuscin pigment. The Sertoli cell of the seminiferous epithelium degrades the residual cytoplasm of spermatids left behind in the release of spermatozoa. The accompanying micrographs from the basal region of Sertoli cells show the typical appearance of primary lysosomes and residual bodies.

Figure 275. Sertoli cells from the testis of chinchilla.

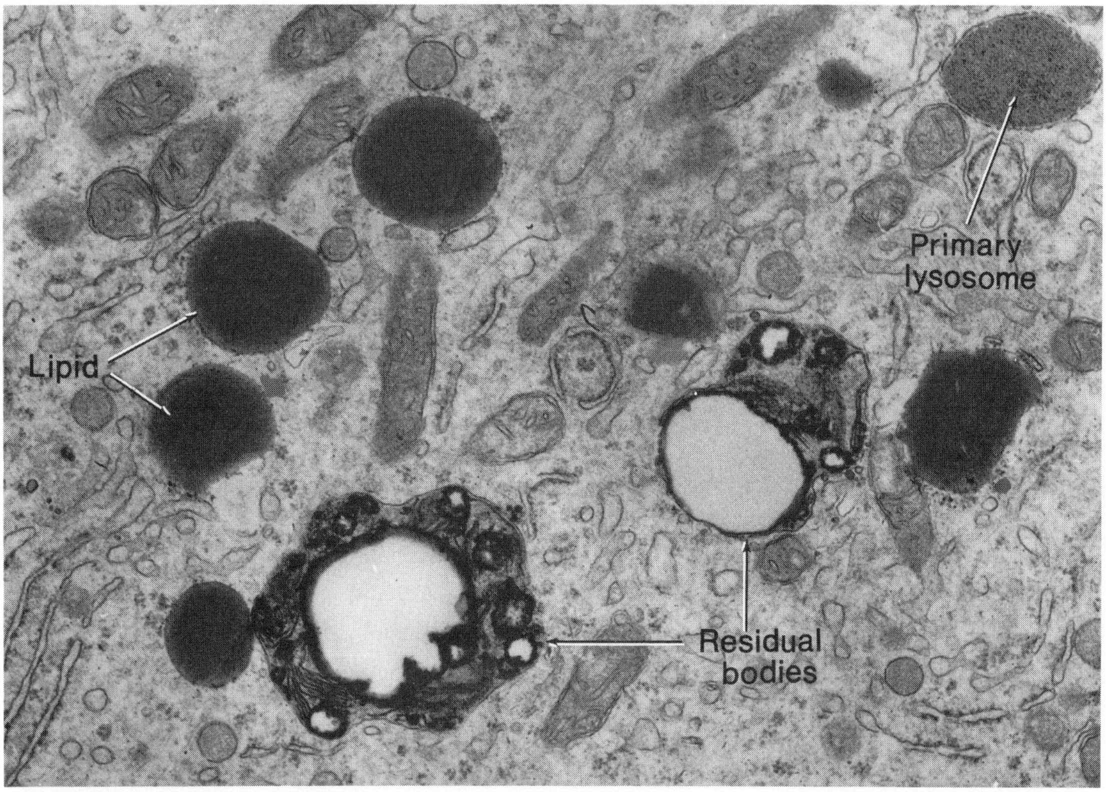

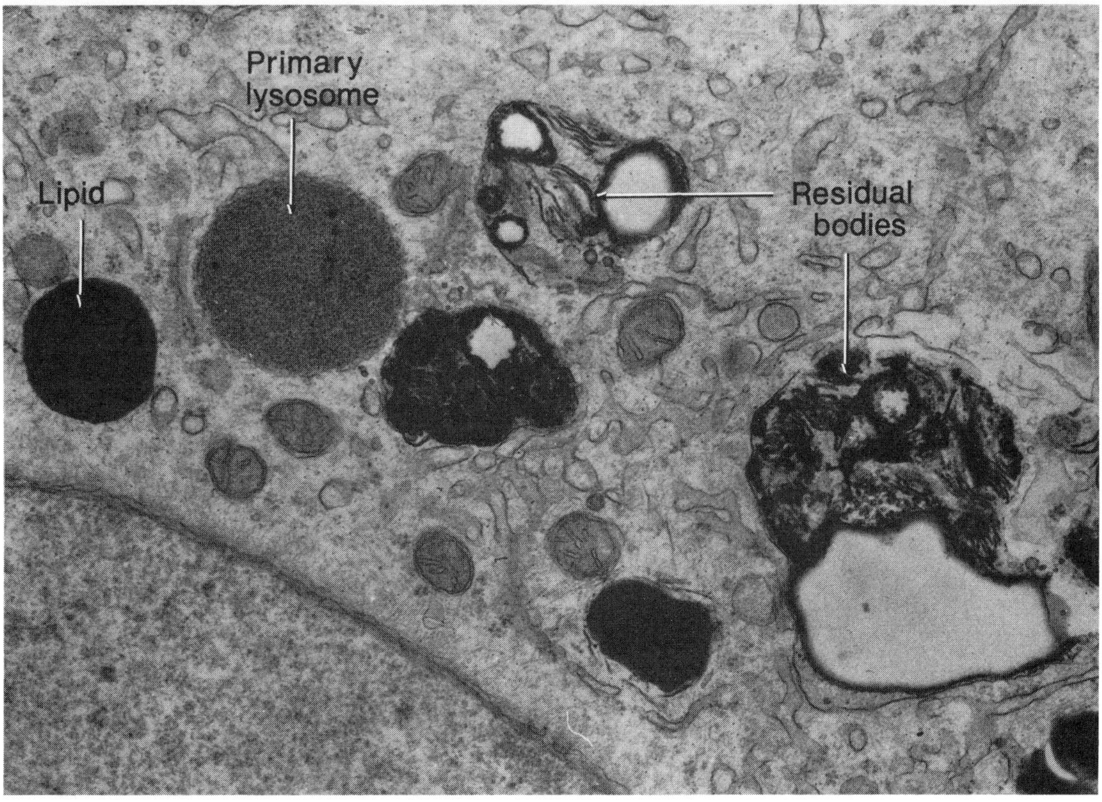

Figure 275, *upper and lower*

The specific granules of neutrophilic leucocytes were formerly considered to be lysosomes. Further cytochemical study demonstrated that two types of granules are formed sequentially in the differentiation of these cells. In promyelocytes and early myelocytes, the so-called azurophil granules contain peroxidase, acid phosphatase, β-glucuronidase, and esterase and therefore are considered to be lysosomal in nature. Later in differentiation, only specific granules are formed and these come to outnumber the azurophil granules. The specific granules of neutrophils contain antibacterial proteins and alkaline phosphatase and therefore are not lysosomes.

The accompanying micrograph of an early myelocyte from human bone marrow contains numerous azurophil granules which have the cytochemical properties of lysosomes.

Figure 276. Myelocyte from human bone marrow.

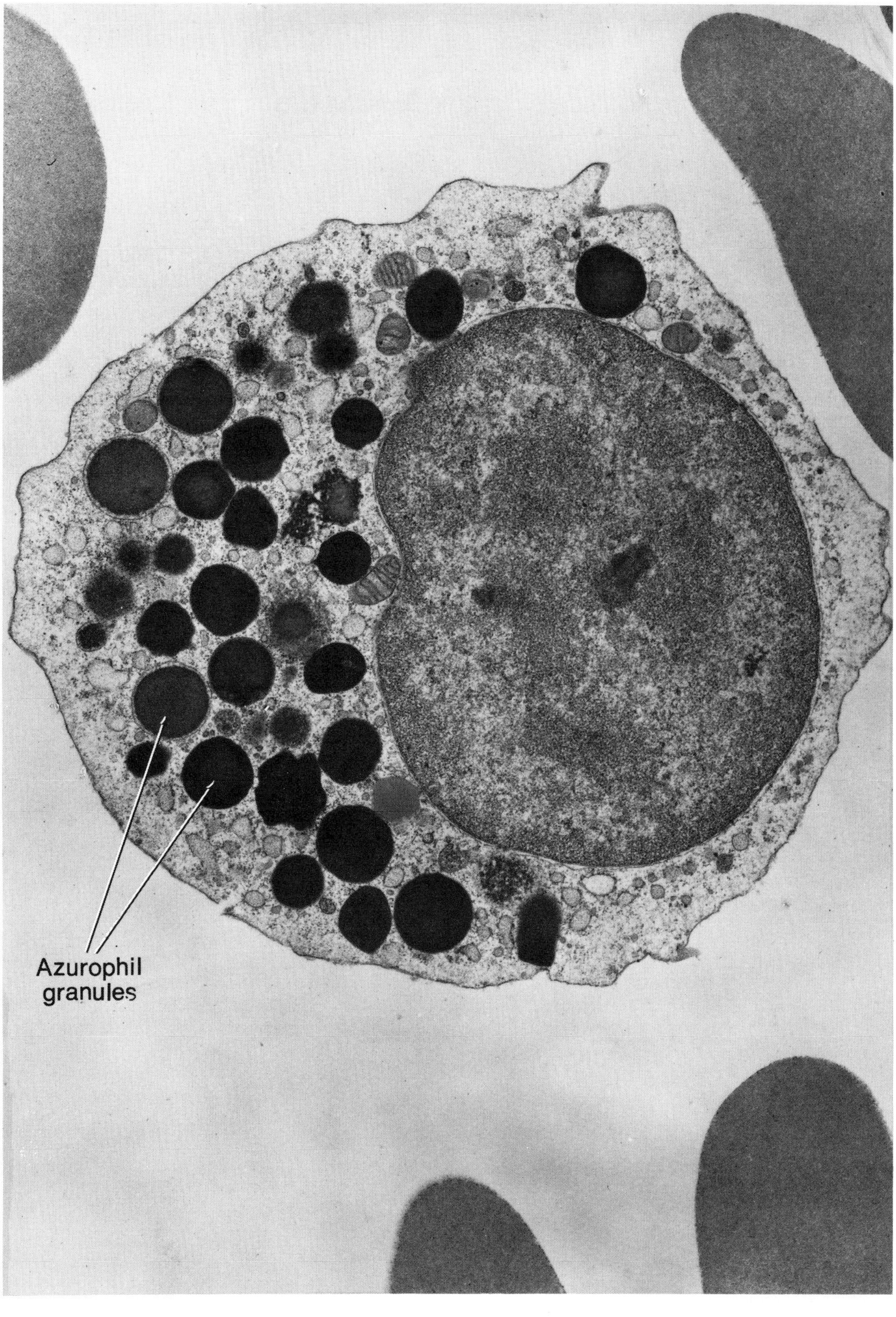

Figure 276

MULTIVESICULAR BODIES

Membrane-bounded vacuoles containing small vesicles are observed in many cell types. These are called *multivesicular bodies,* and because they stain with histochemical reactions for acid phosphatase, they are regarded as a morphologically distinct category of lysosomes. Short segments of their limiting membrane may be coated, suggesting that fusion of coated vesicles may contribute to their enlargement.

Multivesicular bodies are invariably present in considerable numbers in the epithelium lining the epididymal duct. This epithelium is active in endocytosis; coated vesicles and multivesicular bodies are a conspicuous feature of their apical cytoplasm (at arrows). The more common dense lysosomes are also abundant.

Figure 277. Principal cells of the lining epithelium of the human epididymal duct containing multivesicular bodies and other lysosomes. (Micrograph courtesy of Anita Hoffer.)

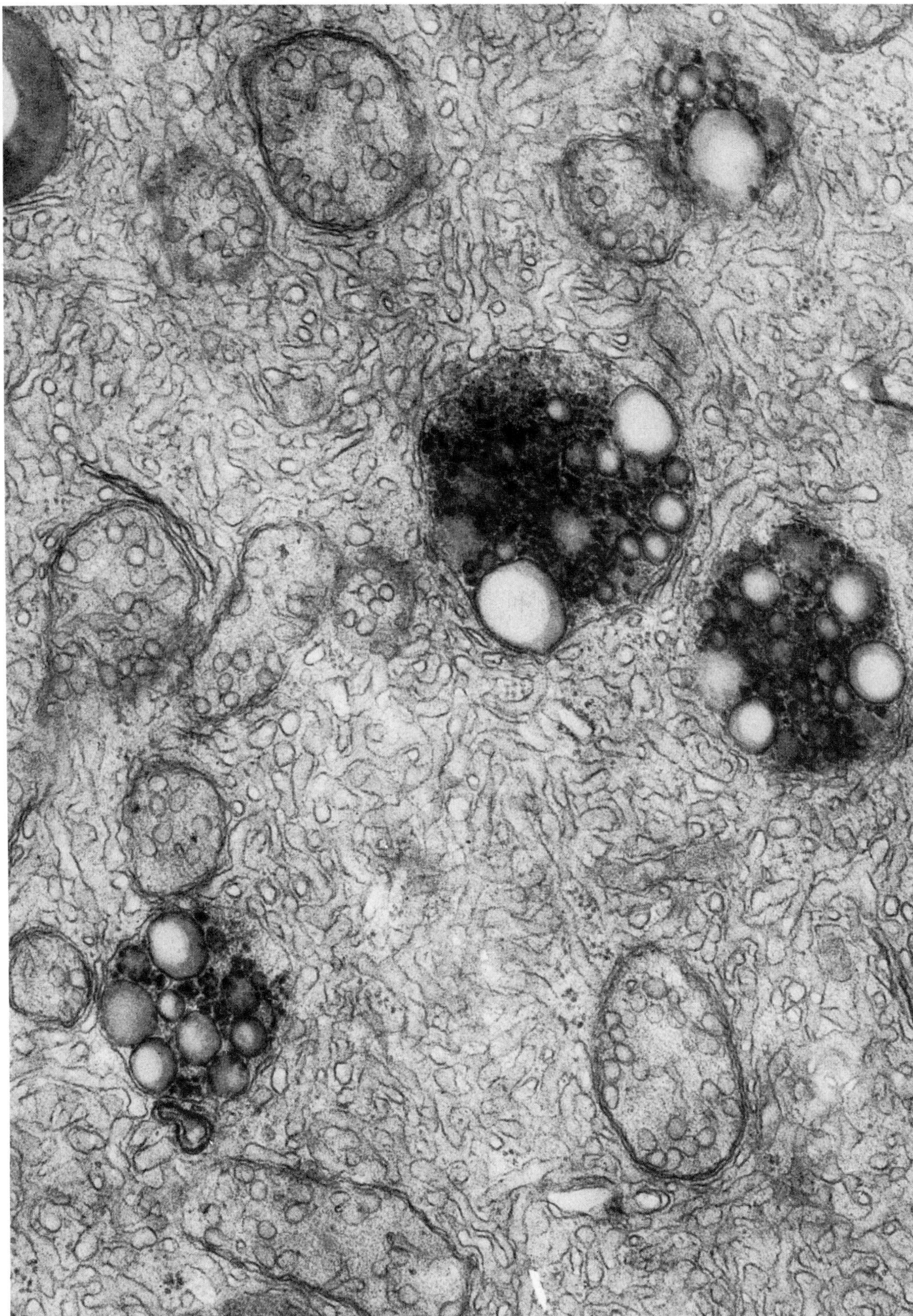

Figure 277

The upper figure on the facing page permits a comparison of the familiar dense lysosomes with a typical multivesicular body. The contents of the latter organelle may consist of a few vesicles or many. Occasional pits (at arrows) in the otherwise smooth contour on the surface suggest that the vesicles in the interior may form by invagination and budding off from the limiting membrane. The vesicles tend to have a fuzzy-coated membrane but smooth vesicles also occur.

The origin of multivesicular bodies is unclear, but it is apparent from their broad range of sizes that they grow by accretion, probably as a result of fusion and incorporation of vesicles into their limiting membrane. The variation in multivesicular body size and a possible developmental sequence is illustrated in the examples numbered 1 to 5 in the lower figure.

Figures 278 and 279. Multivesicular bodies in human epididymal epithelium. (Micrograph courtesy of Anita Hoffer.)

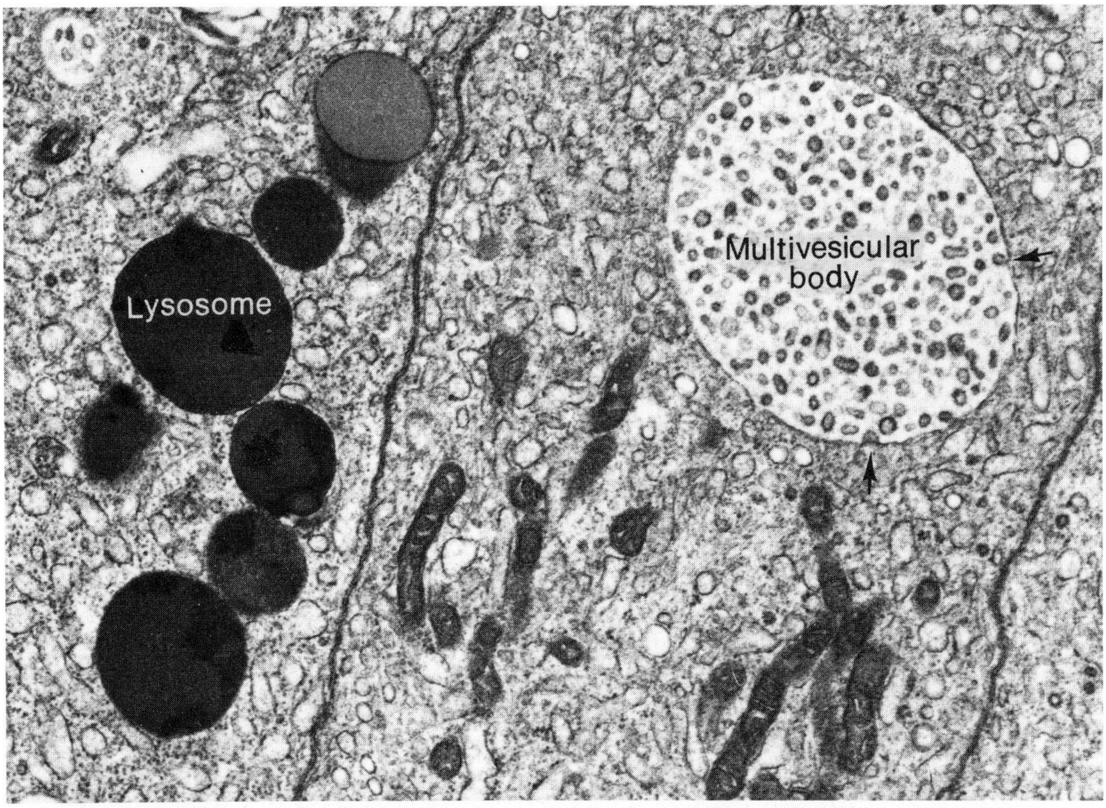

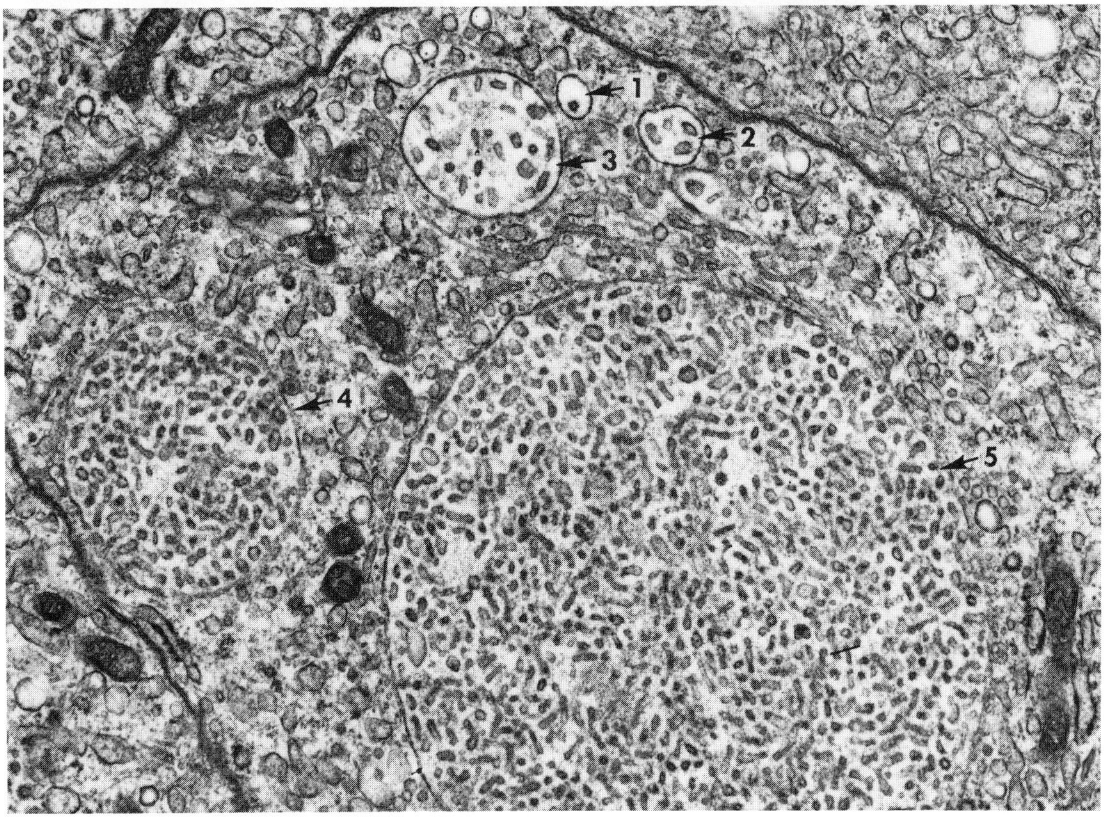

Figure 278, *upper* **Figure 279,** *lower*

REFERENCES

Lysosomes

Allison, A. C. Lysosomes. Carolina Biology Readers, Burlington, North Carolina, 1977. (Review)

Ashford, T. P. and K. R. Porter. Cytoplasmic components in hepatic cell lysosomes. J. Cell Biol. 12:198–202, 1962.

Barka, T. On the acid phosphatases of liver and reticuloendothelial cells. J. Histochem. Cytochem. 8:320, 1960.

Cohn, Z. A. and J. G. Hirsch. The influence of phagocytosis on the intracellular distribution of granule-associated components of polymorphonuclear leucocytes. J. Exp. Med. 112:1015–1022, 1960.

Davies, P. and A. C. Allison. Secretion of macrophage enzymes in relation to the pathogenesis of chronic inflammation. In Immunobiology of the Macrophage (D. S. Nelson, ed.), pp. 427–461, Academic Press, New York, 1976.

DeDuve, C. The lysosome concept. In Lysosomes. Ciba Foundation Symposium, Little, Brown and Company, Boston, 1963. (Review)

DeDuve, C., J. Berthet, F. Appelmans and R. Wattiaux. Enzymic content of the mitochondria fraction. Nature 172:1143–1144, 1953.

DeDuve, C., B. C. Pressman, R. Gianetto, R. Wattiaux and F. Appelmans. Intracellular distribution patterns of enzymes in rat-liver tissue. Biochem. J. 60:604–617, 1955.

DeDuve, C. and R. Wattiaux. Function of lysosomes. Ann. Rev. Physiol. 28:435–492, 1966. (Review)

Dingle, J. T., H. B. Fell and R. T. Dean, editors. Lysosomes in Biology and Pathology, Vol. I–V, (1969–1976), North Holland Publishing Company, Amsterdam.

Essner, E. and A. B. Novikoff. Localization of acid phosphatase activity in hepatic lysosomes by means of electron microscopy. J. Biophys. Biochem. Cytol. 9:773–784, 1961.

Friend, D. S. and M. G. Farquhar. Function of coated vesicles during protein absorption in the rat vas deferens. J. Cell Biol. 35:357–376, 1967.

Gomori, G. Microscopic Histochemistry. University of Chicago Press, Chicago, 1952.

Hirsch, J. G. Cinematographic observations on granule lysis in polymorphonuclear leucocytes during phagocytosis. J. Exp. Med. 116:827–834, 1962.

Holt, S. J. Factors governing the validity of staining methods for enzymes, and their bearing upon the Gomori acid phosphatase technique. Exp. Cell Res. 7:Suppl. 1–27, 1959.

Holtzman, E. Lysosomes, a Survey. Springer-Verlag, New York, 1976. (Review)

Neufeld, E. F., T. W. Lim and L. J. Shapiro. Inherited disorders of lysosomal metabolism. Ann. Rev. Biochem. 44:357–363, 1975.

Nichols, B. A., D. F. Bainton and M. G. Farquhar. Differentiation of monocytes. Origin, nature and fate of azurophil granules. J. Cell Biol. 50:498–515, 1971.

Novikoff, A. B. Lysosomes and related particles. In The Cell, Vol. II (J. Bracket and A. E. Mirsky, eds.), pp. 423–488. Academic Press, New York, 1961. (Review)

Novikoff, A. B., E. Podber, J. Ryan and E. Noe. Biochemical heterogeneity of the cytoplasmic particles isolated from rat liver homogenate. J. Histochem. Cytochem. 1:27–46, 1953.

Novikoff, A. B., H. Beaufay and C. DeDuve. Electron microscopy of lysosome-rich fractions from rat liver. J. Biophys. Biochem. Cytol. 2:Suppl. 179–184, 1956.

Novikoff, A. B. and E. Essner. Cytolysosomes and mitochondrial degeneration. J. Cell Biol. 15:140–146, 1962.

Palade, G. E. Intracellular distribution of acid phosphatase in rat liver cells. Arch. Biochem. 30:144–158, 1951.

Schneider, W. C. and G. H. Hogeboom. Cytochemical studies of mammalian tissues. The isolation of cell components by differential centrifugation. A review. Cancer Res. 11:1–22, 1951.

Schneider, W. C. and G. H. Hogeboom. Intracellular distribution of enzymes. IX. Certain purine-metabolizing enzymes. J. Biol. Chem. 195:161–166, 1952.

Weissmann, G., R. B. Zurier, P. J. Spieler and I. M. Goldstein. Mechanisms of lysosomal enzyme release from leucocytes exposed to immune complexes and other particles. J. Exp. Med. 134:149s–165s, 1971.

PEROXISOMES

In an early electron microscopic study of the kidney, Rhodin (1954) described membrane-limited, spherical cytoplasmic particles 0.2 to 0.4 μm in diameter that did not correspond to any of the traditional cell organelles. For lack of a better term, they were called *microbodies*. Similar structures were soon observed in hepatic cells, but in addition to a homogeneous, finely granular matrix these also contained an inclusion termed the *nucleoid,* which in some planes of section exhibited a lamellar substructure (Gansler and Rouiller, 1956). Noting that microbodies were more abundant in regenerating liver, Rouiller and Bernhard (1956) considered it likely that they represented formative stages of mitochondria and that the parallel striations in the nucleoid might be precursors of the cristae, but this interpretation gained little support.

In their studies on lysosomes, DeDuve and his coworkers developed centrifugation procedures with improved resolving power and were able to separate from the lysosome fraction particles that lacked the typical acid hydrolases but were rich in d-aminoacid oxidase and catalase. It was recognized that this new class of cytoplasmic particles probably corresponded to the "microbodies" previously described by microscopists. It was suggested that the term *peroxisome* was more appropriate for an organelle in which two hydrogen peroxide–generating enzymes were associated with catalase. This rapidly became the preferred term.

Correspondence of DeDuve's peroxisomes to microbodies was soon validated by the demonstration that the ultrastructure of the nucleoid in the microbody was identical to that of crystalline uricase (Hruban and Swift, 1964) and the finding that birds, reptiles, and man — all of which are known to lack uricase — also lack nucleoids in their hepatic microbodies (Afzelius, 1965). The nucleoids of rat hepatic microbodies were then isolated in a high degree of purity and urate oxidase was the only enzymatic activity detected. The isolated nucleoids consisted of parallel bundles of thick-walled, hollow tubules (15 nm outside diameter, 5 nm inside diameter). These were precisely arranged in a repeating pattern throughout the crystalline lattice with 10 such tubules around longitudinal channels 20 nm in diameter. Thus in transverse sections the nucleoids had a honeycomb pattern, and in longitudinal sections they presented an array of alternating dark and light striations (Tsukada et al., 1966). This structure applies only to the rat. The nucleoids of other species have a different organization.

Investigation of the origin of peroxisomes has been considerably facilitated by the availability of experimental methods for inducing their proliferation and by a cytochemical reaction for detecting their enzymatic activity. Administration of various hypocholesterolemic drugs results in a rapid increase in hepatic microbodies (Hruban et al., 1970;

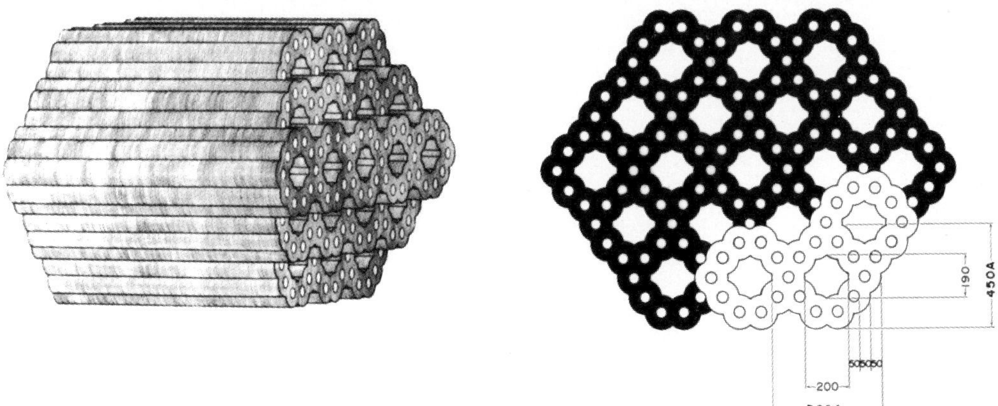

Diagrammatic reconstruction of the three-dimensional form of the nucleoid in rat hepatic peroxisomes. (From Tsukada et al., J. Cell Biol. 28:449–460, 1966.)

Reddy and Krishnahantha, 1975), and they can be intensely and selectively stained with the diaminobenzidene reaction for peroxidase (Fahimi, 1968). Despite the specificity of this method and the clarity of the resulting electron micrographs, the origin of peroxisomes remains a subject of controversy. Some investigators insist that their membrane is often continuous at some point with that of the smooth endoplasmic reticulum and that peroxisomes arise as evaginations from that organelle (Novikoff and Shin, 1964; Essner, 1967). Others find no convincing evidence of continuity and suggest instead that new peroxisomes arise from preexisting peroxisomes by a process of budding. Catalase is believed to be synthesized on ribosomes and transferred directly to the peroxisomes without involvement of the reticulum and Golgi complex (Legg and Wood, 1970). Others observe catalase reaction product on ribosomes in the vicinity of weakly stained, small peroxisome-like bodies that lack a distinct limiting membrane, and suggest that catalase which is synthesized on free and bound ribosomes accumulates locally and is then segregated by development of a limiting membrane (Fahimi, 1971).

Additional support for the concept of synthesis of peroxisomal enzymes in the cytoplasmic matrix also comes from biochemical studies which indicate that a catalase precursor is discharged into the cytosol and subsequently transferred to the peroxisomes (Lazarow and DeDuve, 1973). Catalase has been localized on free ribosomes but not on membrane-bound ribosomes. Thus it is the prevailing view that uricase and catalase are transferred to the interior of the peroxisome by a posttranslational mechanism (Goldman and Blobel, 1978). The origin of the limiting membrane is not clear, but budding from the rough reticulum or from GERL remains a likely possibility.

Application of the cytochemical staining reaction for peroxidase has shown that, in addition to the peroxisomes of kidney and liver, there are very large numbers of small membrane-bounded particles 0.15 to 0.25 μm in the absorptive cells of the intestine and in many other epithelia. Their membranes were originally reported by Novikoff and coworkers to be continuous with those of the smooth endoplasmic reticulum and they were considered to be local diverticula or varicosities of the reticulum. Owing to their small size and lack of nucleoids, it was suggested that these diminutive peroxidase-positive bodies should be designated *microperoxisomes* (Novikoff and Novikoff, 1972). Other workers have confirmed their widespread occurrence, but in the absence of reaction product in the lumen of the reticulum, their interpretation as local evaginations of the reticulum has not gained general acceptance.

It is clear that peroxisomes occur in a broad spectrum of different sizes, with nucleoids present in some cell types and lacking in others. As in the case of lysosomes, their positive identification cannot be made on morphological criteria alone but requires histochemical demonstration of one or more of their constitutent enzymes. Their functional significance in the metabolism of cells remains obscure.

Peroxisomes are abundant in hepatic cells and were first isolated from the crude lysosome fraction of liver homogenates. The micrographs on the facing page show their characteristic appearance *in situ* (at arrows). They are generally spherical or ovoid and have a homogeneous content of relatively low density. In the rat they usually have a single crystalloid inclusion called the *nucleoid*. A nucleoid is also present in the hamster but is a thin sheet, with a linear profile in section. Not infrequently it is folded and presents a C or V shape profile in sections. The nucleoid consists of urate oxidase. When present, the nucleoid is helpful in distinguishing peroxisomes from primary lysosomes, but in some species that have no urate oxidase, the nucleoid is absent. In those species in which nucleoids are present in the hepatic peroxisomes, they may be absent in the peroxisomes of other organs.

Figure 280. An area of cytoplasm from a rat liver cell containing peroxisomes. (Micrograph courtesy of Robert Bolender.)

Figure 281. A comparable micrograph from hamster liver.

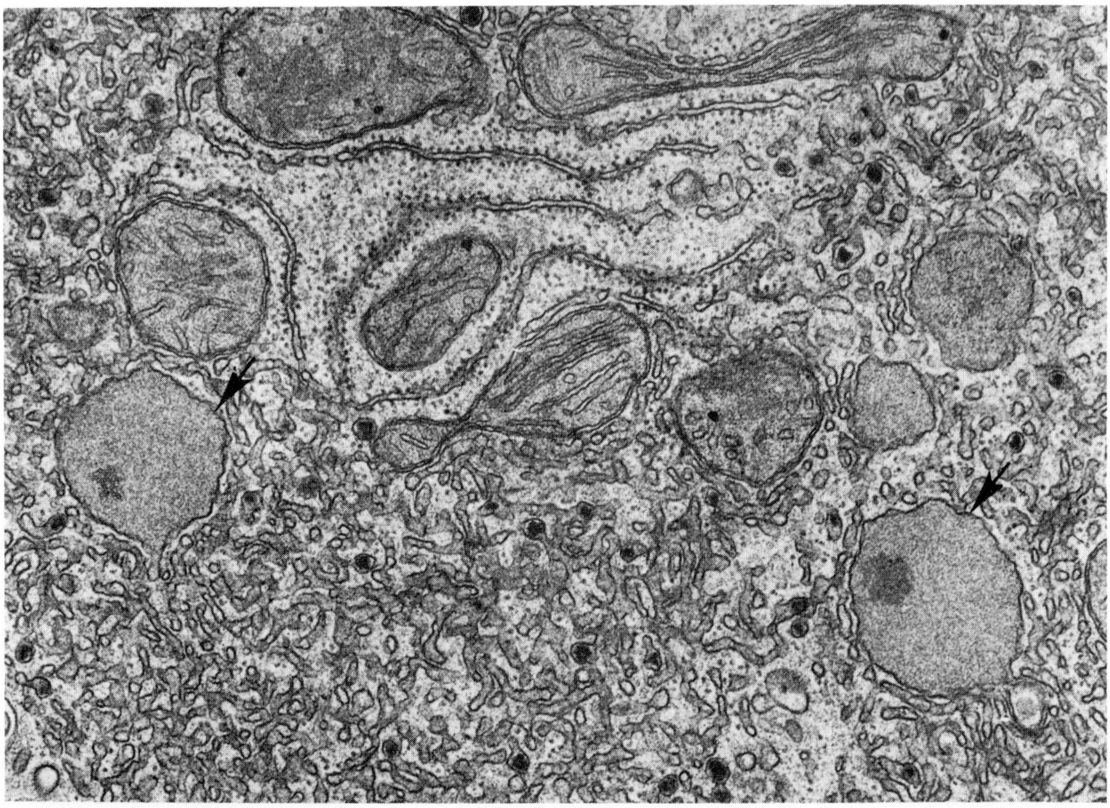

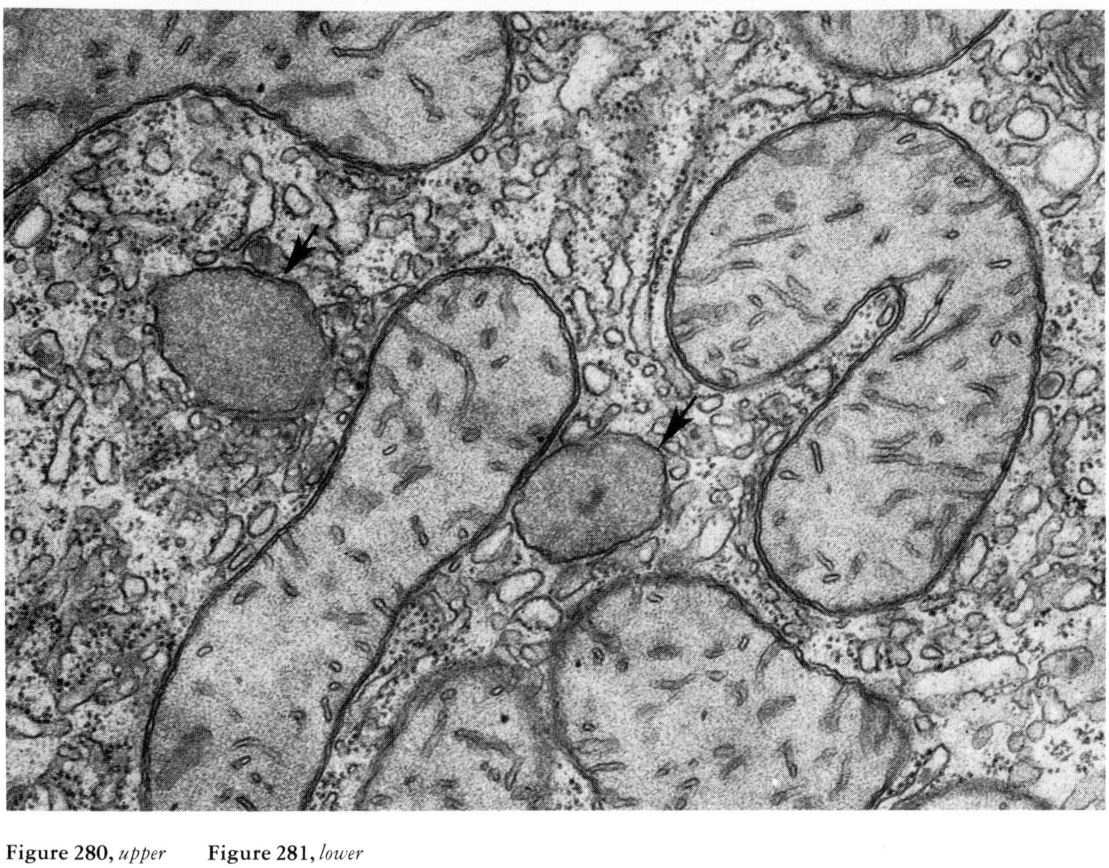

Figure 280, *upper* **Figure 281,** *lower*

Representative examples of microbodies or peroxisomes are shown here at higher magnification. The plane of section may not include the nucleoid, as in the one at the top of the figure. The great majority of nucleoids are sectioned obliquely and show only a faint longitudinal striation. With patient examination of scores of micrographs, rare longitudinal and transverse sections can be found. In this way the lattice of the urate oxidase crystal in the rat peroxisome has been worked out in considerable detail.

Figure 282. Rat liver peroxisomes. (Micrograph courtesy of Daniel Friend.)

Figure 283. Rat liver peroxisome. (Micrograph courtesy of Richard Wood.)

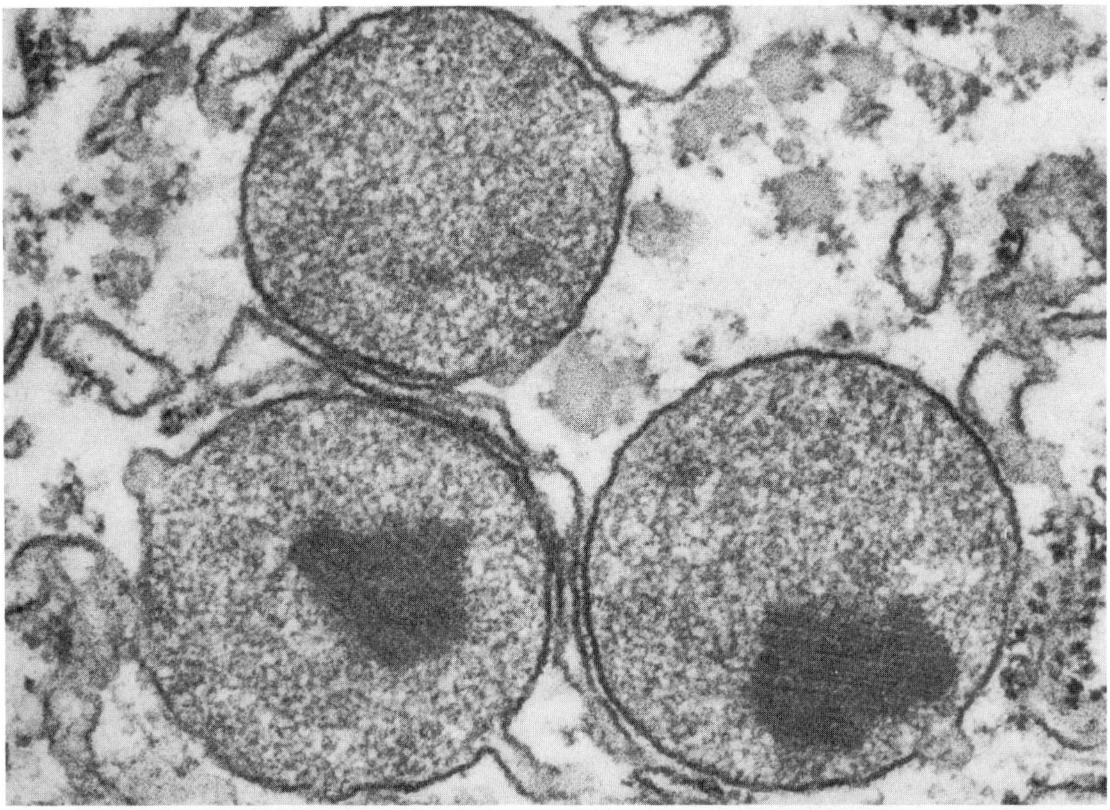

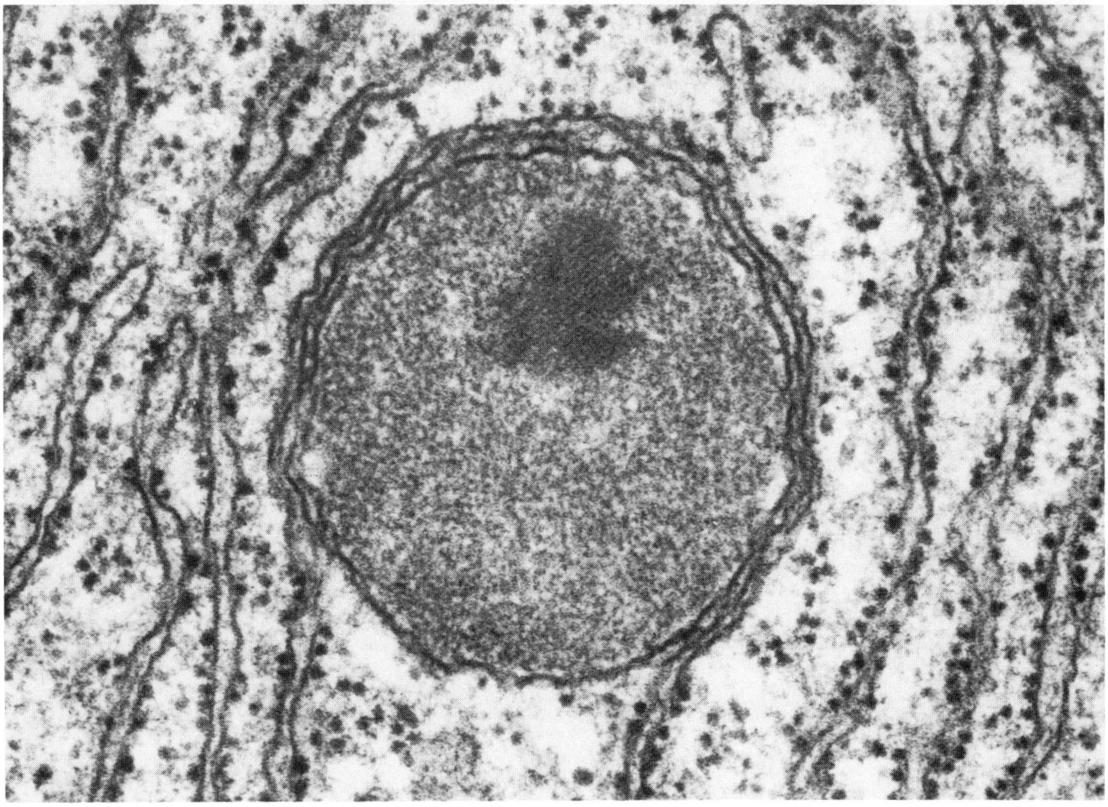

Figure 282, *upper* **Figure 283,** *lower*

Some cell types may contain lysosomes and peroxisomes. In routine preparations, these may be similar in size and density. But when stained with the diaminobenzidene reaction for peroxidase activity, the dense reaction product clearly distinguishes the peroxisomes from the lysosomes.

Figures 284 and 285. Leydig cells from boar testis stained with the cytochemical reaction for peroxidase.

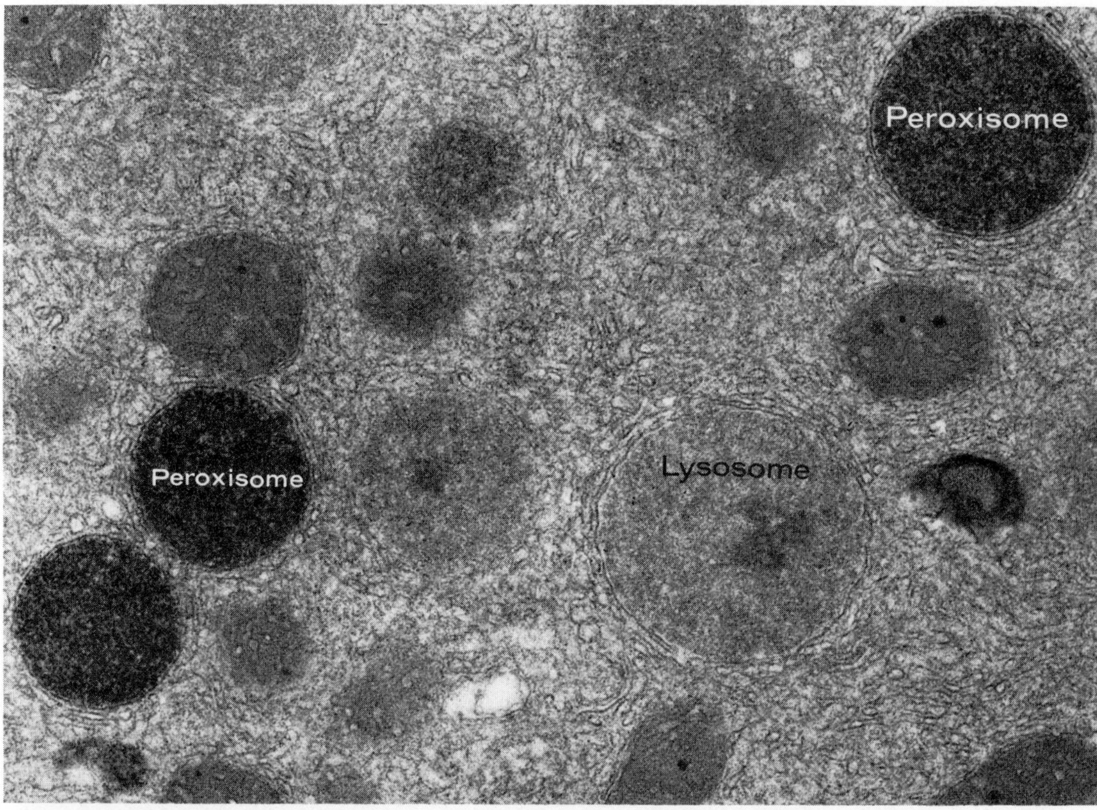

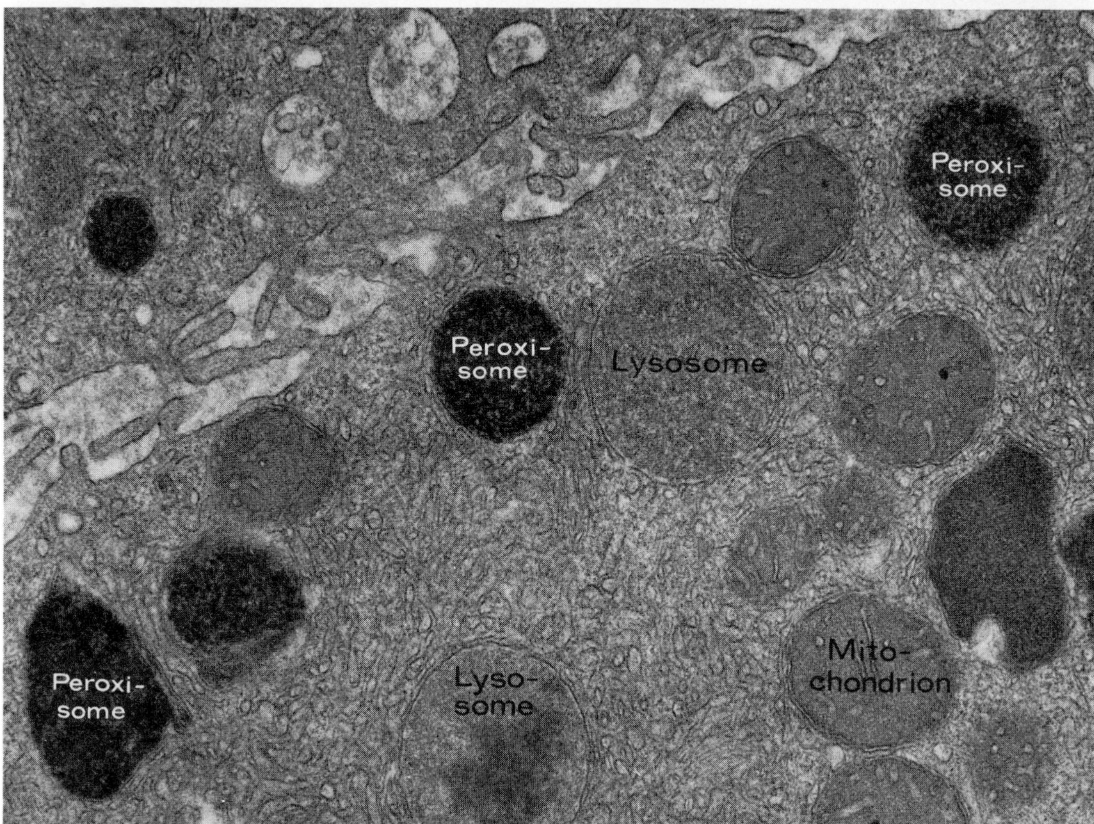

Figure 284, *upper* **Figure 285,** *lower*

Although the function of peroxisomes is poorly understood, they appear to be especially abundant in cells involved in cholesterol metabolism and synthesis of steroids — liver, adrenal, ovary, and interstitium of the testis. A relationship to cholesterol metabolism is suggested by the observation that the number of hepatic peroxisomes is dramatically increased by administration of several unrelated hypocholesterolemic drugs.

The upper micrograph on the facing page from normal rat liver stained with the diaminobenzidene reaction illustrates the normal size and number of peroxisomes. The lower micrograph shows the remarkable increase after the administration of [4-chloro-6-(2.3-xylidine) 2-pyrimidinylthio] acetic acid (Wy–14643).

Figure 286. Peroxidase-stained liver cytoplasm. (Micrograph courtesy of Darius Fahimi.)

Figure 287. Peroxidase stain of liver from an animal treated with Wy–14643. (Micrograph courtesy of Janardan Reddy, from Reddy and Krishnahantha, Science 190:787–789, 1975.)

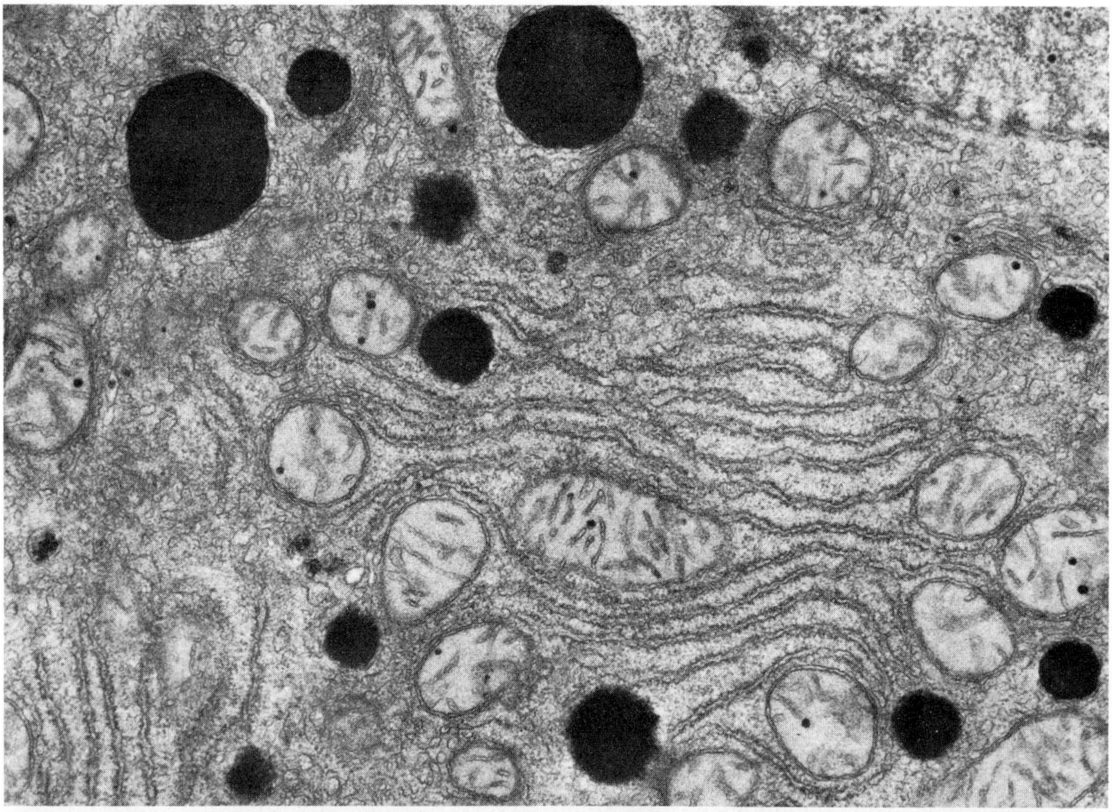

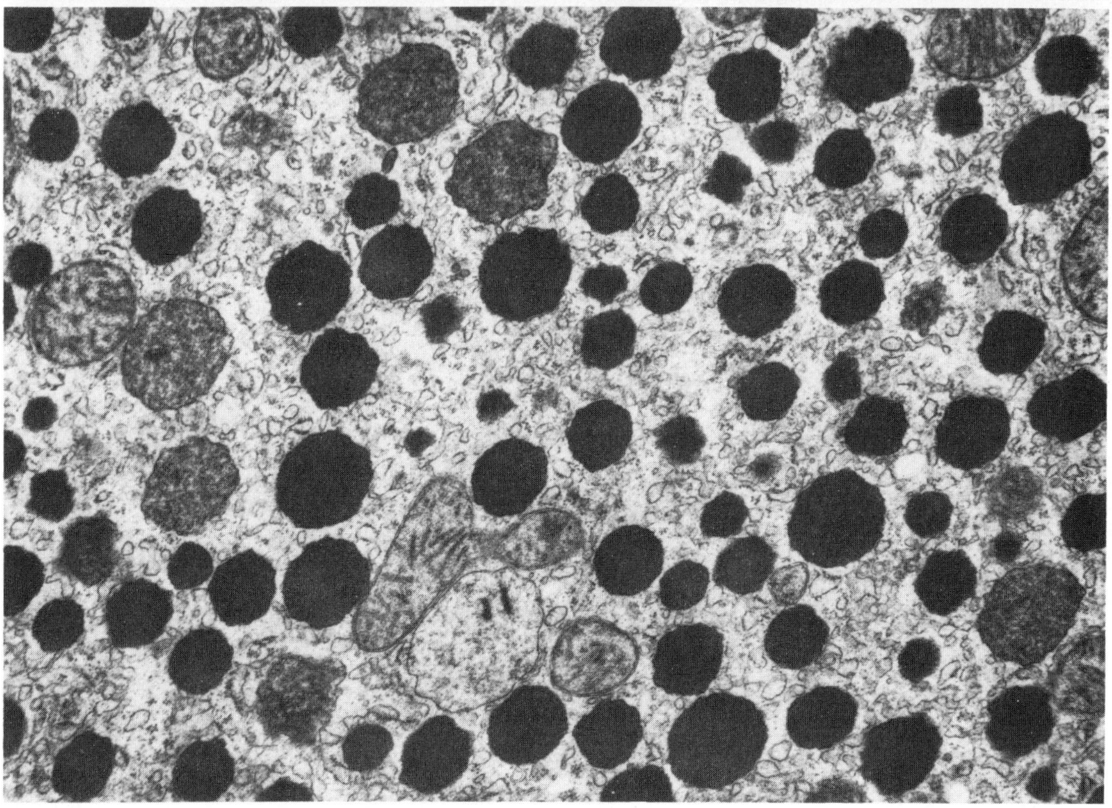

Figure 286, *upper* **Figure 287,** *lower*

Peroxisomes (microbodies), first observed in cells of the mouse proximal convoluted tubule, were described as spherical bodies with a homogeneous, finely granular matrix. Nucleoids were not a characteristic feature of those peroxisomes. More recent investigations of comparable organelles in the rat distinguish two types of formed structures in their matrix — cylindrical inclusions 85 to 140 nm in diameter and large tabular crystals varying in thickness and ranging up to 3 μm in length. Some contain mainly crystalline inclusions, some mainly cylinders, and others contain both. The inclusions tend to be at the periphery, and where they are crystals, they tend to deform the organelle into geometric shapes.

The accompanying micrograph of a cell from the proximal convoluted tubule shows a peroxisome near the cell base containing both crystals and a few circular profiles of cylinders in cross section. Despite their unusual appearance, these organelles can be identified as peroxisomes by positive staining for catalase and a negative reaction for acid phosphatase.

Figure 288. Proximal convoluted tubule cell from rat. (Micrograph courtesy of Michael Barrett and Paul Heidger, from Cell and Tissue 157:283–305, 1975.)

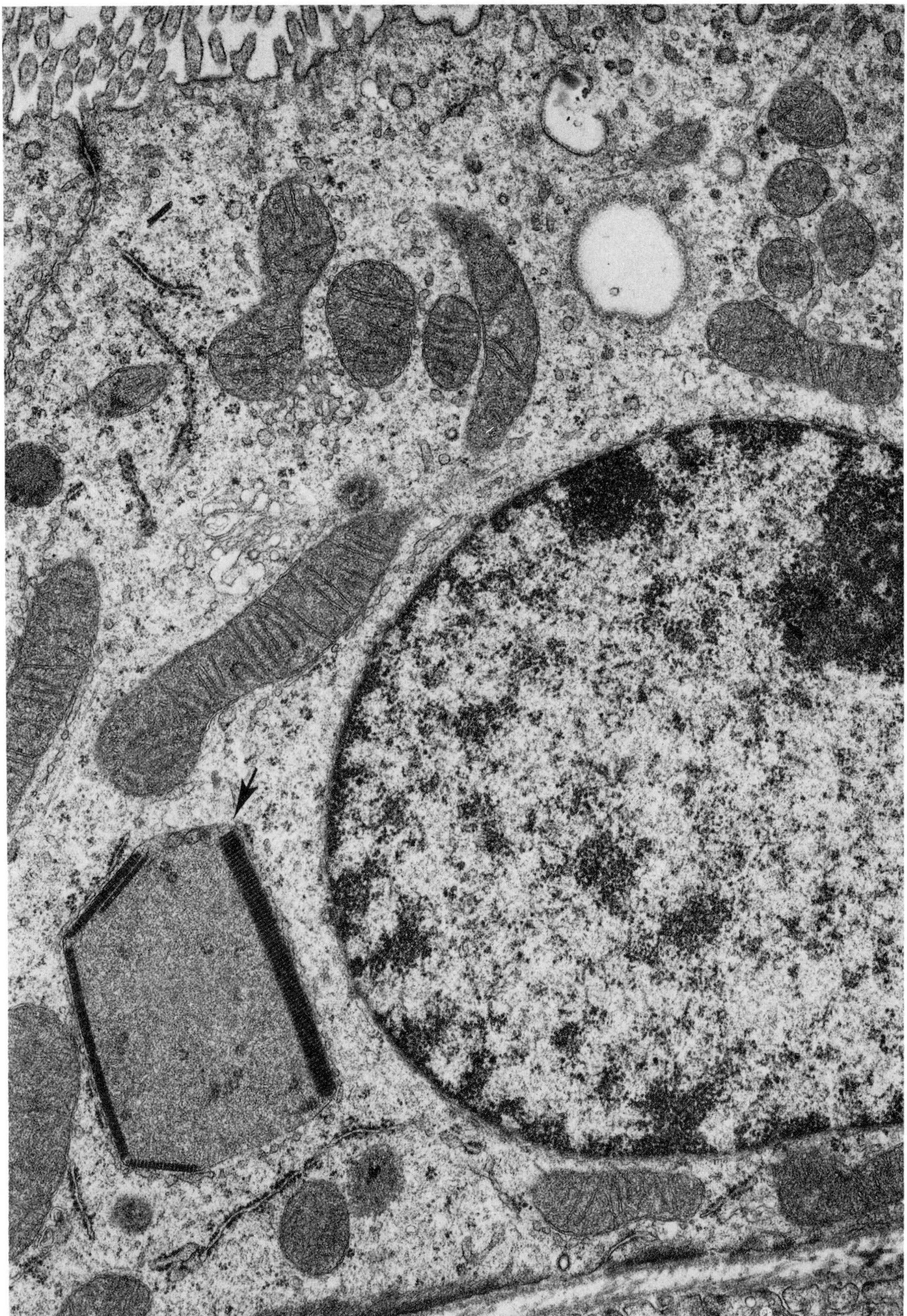

Figure 288

Two additional examples of peroxisomes from the proximal convoluted tubule of rat kidney, one containing only crystals and the other containing cylindrical inclusions seen here in transverse section.

Figures 289 and 290. Peroxisomes from the proximal convoluted tubule of rat kidney. (Micrograph courtesy of Michael Barrett and Paul Heidger, from Cell and Tissue 157:283–305, 1975.)

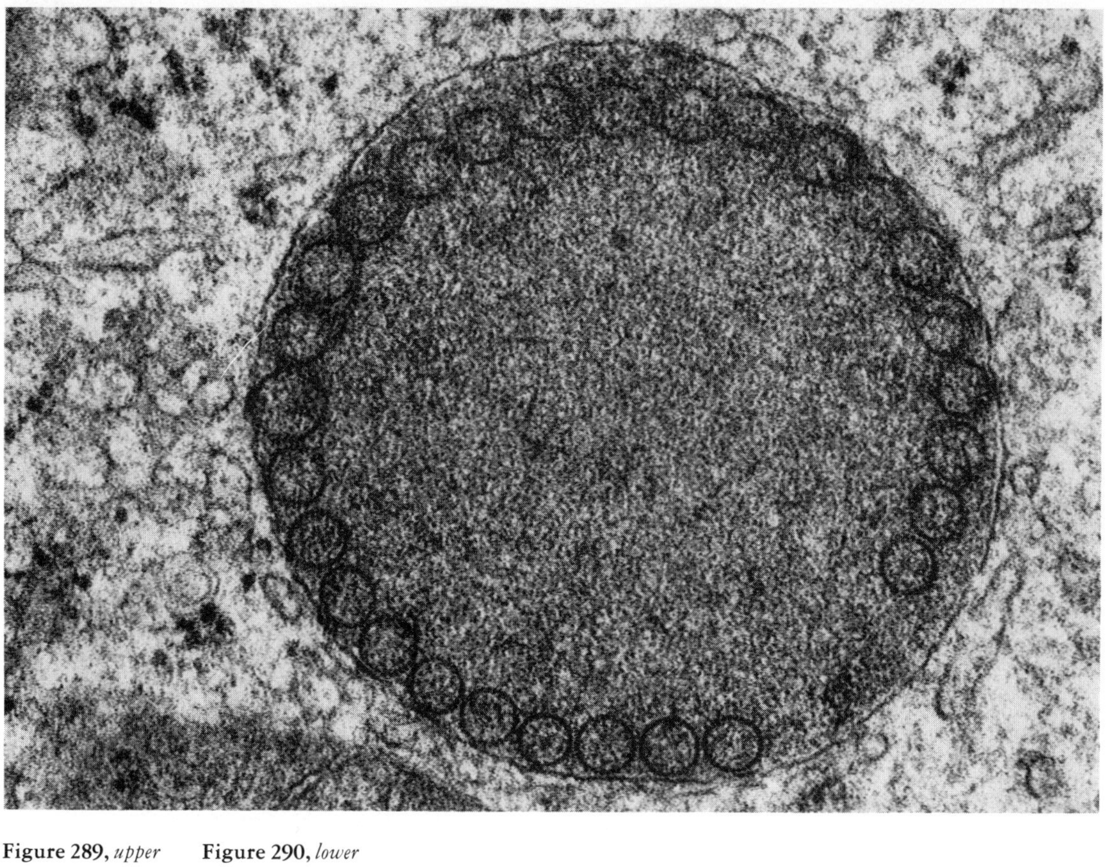

Figure 289, *upper* **Figure 290,** *lower*

REFERENCES

Peroxisomes

Afzelius, B. The occurrence and structure of microbodies. A comparative study. J. Cell Biol. 26:835–843, 1965.

Barrett, J. M. and P. M. Heidger. Microbodies of the rat renal proximal convoluted tubule: Ultrastructural and cytochemical investigations. Cell Tissue Res. 157:283–305, 1975.

DeDuve, C. The peroxisome: A new cytoplasmic organelle. Proc. Roy. Soc. Lond. 173:71, 1969.

DeDuve, C. and P. Baudhuin. Peroxisomes (microbodies and related particles). Physiol. Rev. 46:323–357, 1966. (Review)

DeDuve, C., H. Beaufay, P. Jacques, Y. Rahman-Li, O. Z. Sellinger, R. Wattiaux and S. DeConnick. Intracellular localization of catalase and of some oxidases in rat liver. Biochim. Biophys. Acta 40:186–187, 1960.

Essner, E. Endoplasmic reticulum and the origin of microbodies in fetal mouse liver. Lab. Invest. 17:71, 1967.

Fahimi, H. D. Cytochemical localization of peroxidase activity in rat hepatic microbodies. J. Histochem. Cytochem. 16:547–550, 1968.

Fahimi, H. D. Morphogenesis of peroxisomes in rat liver. J. Cell Biol. 87A, 1971.

Gansler, H. and C. Rouiller. Modifications physiologiques et pathologiques du chondrome. Etude an microscope électronique. Schweitz Z. Pathol. Bakt. 19:217–243, 1956.

Goldman, B. M. and G. Blobel. Biogenesis of peroxisomes: Intracellular site of synthesis of catalase and uricase. Proc. Nat. Acad. Sci. 75:5066–5070, 1978.

Hruban, Z. and M. Rechigl. Microbodies and related particles. Int. Rev. Cytol. Suppl. 1, 1969. (Review)

Hruban, Z., E. L. Vigil, A. Slesers and E. Hopkins. Microbodies. Constituent organelles of animal cells. Lab. Invest. 27:184–191, 1972. (Review)

Hruban, Z. and H. Swift. Uricase: Localization in hepatic microbodies. Science 146:1316–1317, 1964.

Hruban, Z., Y. Mochizuki, J. R. Esterly and T. W. Wong. Some effects of hypocholesterolemic and related agents. Fed. Proc. 29:385, 1970.

Lazarow, P. P. and C. DeDuve. The synthesis and turnover of rat liver peroxisomes. V. Intracellular pathway of catalase synthesis. J. Cell Biol. 59:507–524, 1973.

Legg, P. G. and R. L. Wood. New observations on microbodies. A cytochemical study on CP1B-treated rat liver. J. Cell Biol. 45:118–129, 1970.

Leighton, F., B. Poole, H. Beaufay et al. Large scale separation of peroxisomes, mitochondria, and lysosomes from the liver of rats injected with Triton WR–1339. J. Cell Biol. 37:483–513, 1968.

Novikoff, P. M. and A. B. Novikoff. Peroxisomes in absorptive cells of mammalian small intestine. J. Cell Biol. 53:532–560, 1972.

Reddy, J. K. and T. P. Krishnahantha. Hepatic peroxisome proliferation: Induction by two novel compounds structurally unrelated to chlofibrate. Science 190:787–789, 1975.

Rhodin, J. Correlation of ultrastructural organization and function in normal and experimentally changed proximal convoluted tubule cells of mouse kidney. A. B. Godvil, Stockholm, 1954.

Rouiller, C. and W. Bernhard. "Microbodies" and the problem of mitochondrial regeneration in liver cells. J. Biophys. Biochem. Cytol. 2:Suppl. 355–360, 1956.

Shmitka, T. K. Comparative ultrastructure of hepatic microbodies in some mammals and birds in relation to species differences in uricase activity. J. Ultrastr. Res. 16:598, 1966.

Svaboda, D. J. and D. L. Azarnoff. Response of hepatic microbodies to a hypolipidemic agent ethyl chlorophenoxyisobutyrate (CPIB). J. Cell Biol. 30:442–450, 1966.

Tsukada, T., Y. Mochizuki and S. Fujiwara. The nucleoids of rat liver cell microbodies. J. Cell Biol. 28:449–460, 1966.

LIPOCHROME PIGMENT

The cells of many organs contain brown cytoplasmic inclusions that classical histologists described as *lipochrome pigment,* or *lipofuscin.* These coarse granules exhibit a golden brown fluorescence in ultraviolet light; they stain lightly with lipid-soluble dyes; they give a positive staining reaction for carbohydrates; they are insoluble in acid and alkali and in most lipid solvents. Lipofuscin deposits are rare or absent in the cells of young animals but increase in number with advancing age. They were formerly referred to as "wear-and-tear pigment" in the belief that they represented breakdown products of cell components. Weak acid phosphatase and esterase activities can be demonstrated in some of these pigment granules.

Since the discovery of lysosomes and their characterization as an intracellular digestive system capable of degrading damaged or excess organelles, lipofuscin pigment granules have come to be regarded as end-stages of lysosomal autophagic activity. Accordingly, they are now believed to be accumulations of the metabolically inert residues of cellular material that could not be completely degraded by the lysosomal hydrolases.

Lipofuscin deposits are frequently seen in the brains of older animals and humans and are one of the few available cytological manifestations of the aging process. The granules vary in size and shape. They are limited by a membrane and their content is, for the most part, very dense and coarsely granular, but homogeneous spherical bodies with little natural density or osmiophilia are often present (see at the arrows).

Figure 291. Lipofuscin pigment in a neuron of rat brain. (Micrograph courtesy of Enrico Mugnaini.)

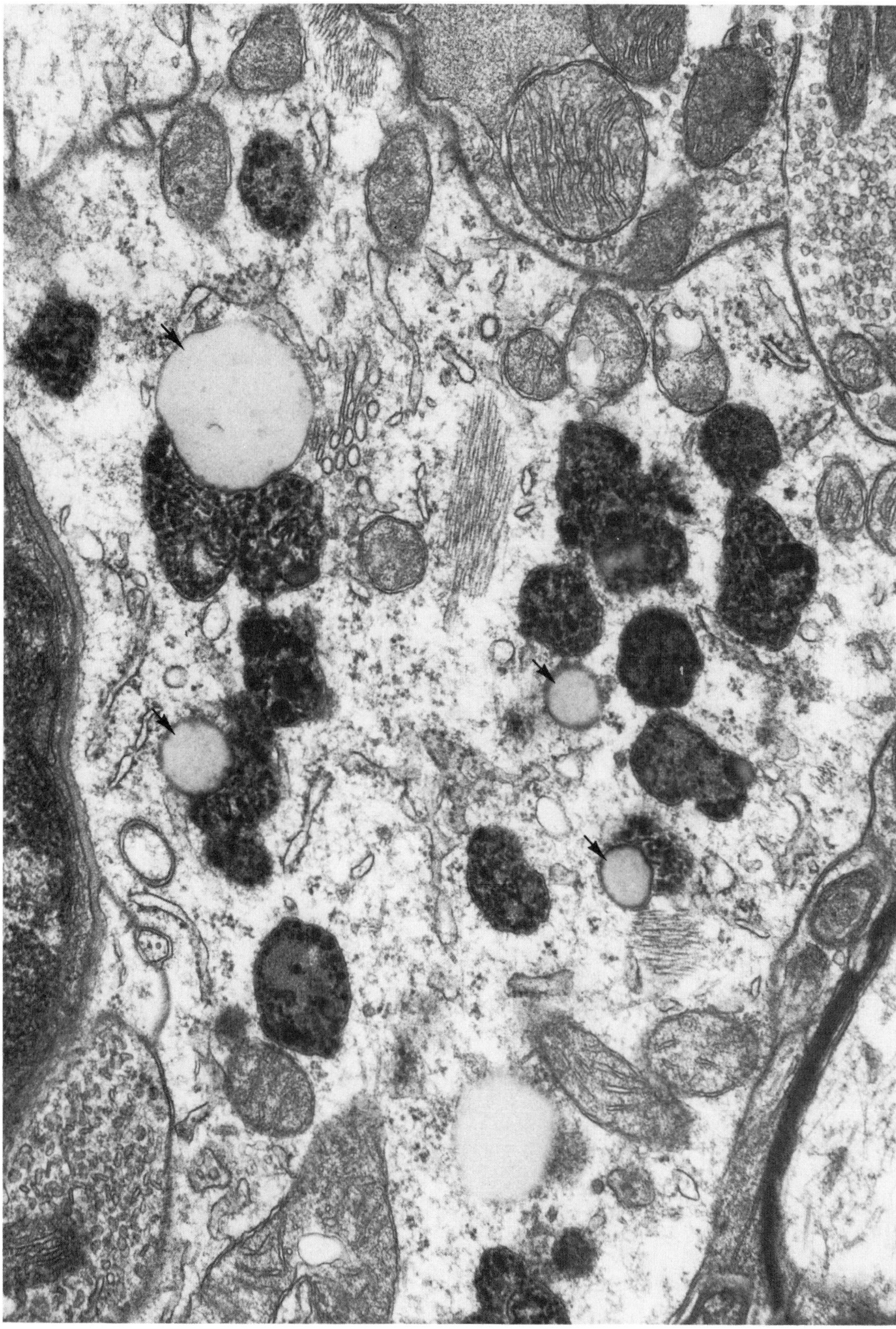

Figure 291

Long-lived cells, such as the Sertoli cells of the testes, which are normally engaged in phagocytosis and degradation of the residual cytoplasm of spermatids, often accumulate conspicuous deposits of lipochrome pigment as illustrated in the accompanying micrograph.

Steroid secreting endocrine glands are also especially prone to develop large amounts of pigment. Massive accumulation of such pigment in the inner part of the adrenal cortex and in the interstitial cells of the ovary of old mice was formerly described by light microscopists as "brown degeneration." Evidently, cells can accommodate a very large burden of lipochrome pigment without seriously jeopardizing their function but ultimately reach a point at which they are no longer viable.

Figure 292. Lipofuscin pigment in a cultured Sertoli cell. (Micrograph courtesy of Wayne Vogl.)

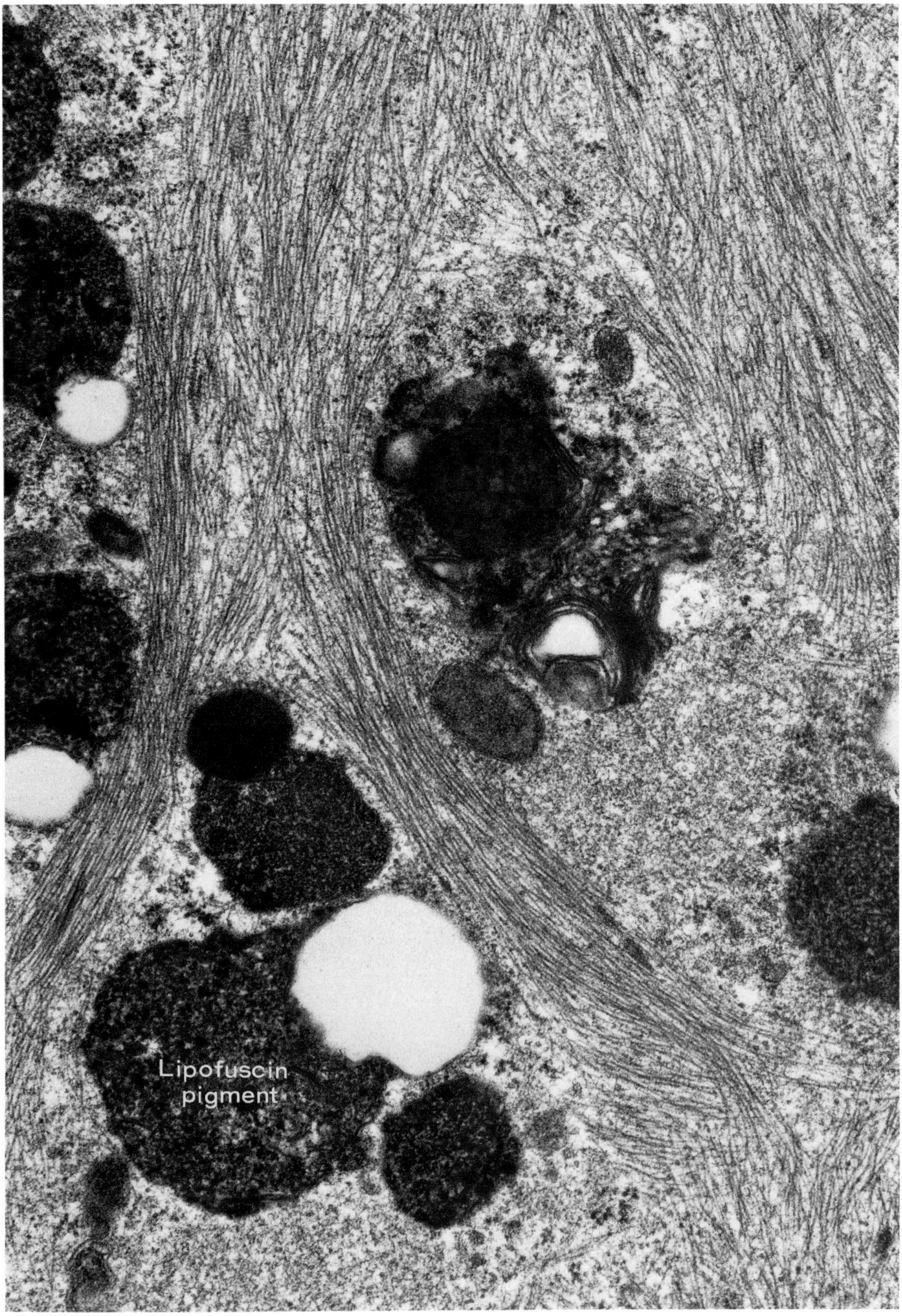

Figure 292

A common site of accumulation of lipofuscin pigment is in the cells of the myocardium. Pigment granules are typically located in the conical zones of sarcoplasm at the poles of the nucleus. With advancing age, they become increasingly abundant and may come to represent as much as 30 per cent of the total solids of human heart muscle fibers.

The accompanying micrographs of cardiac muscle cells from an old cat show lipofuscin pigment bodies and numerous mitochondria in a sarcoplasm that is rich in small dense particles of glycogen.

Figures 293 and 294. Lipofuscin pigment in right ventricular papillary muscle of an old cat.

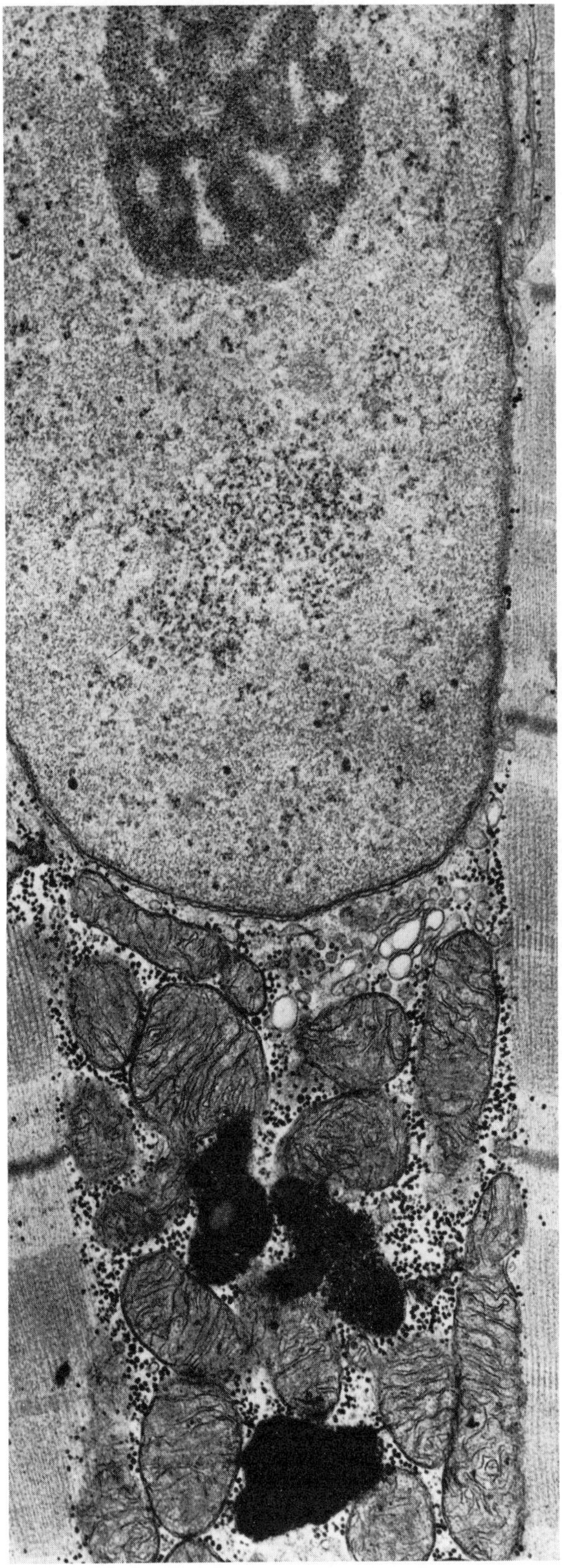

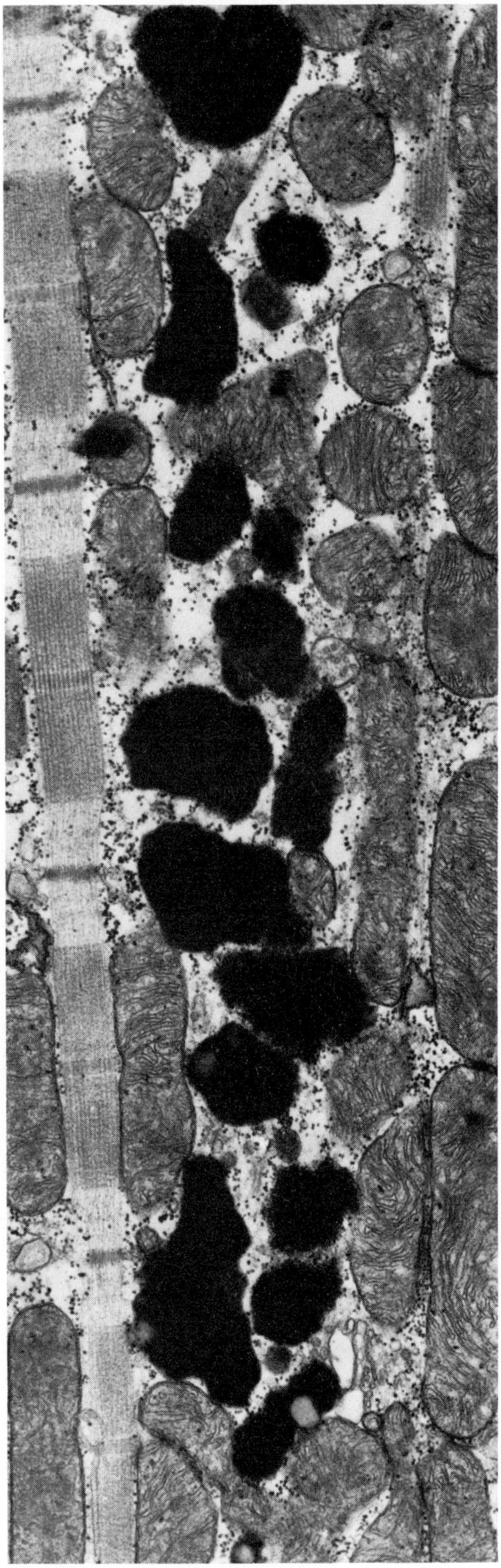

Figure 293, *left* Figure 294, *right*

REFERENCES

Lipochrome Pigment

Björkerud, S. The isolation of lipofuscin granules from bovine cardiac muscle. J. Ultrastr. Res. Suppl. 5:5–49, 1963.

Connor, C. L. Studies on lipochromes. Am. J. Path. 4:293–308, 1928.

Deane, H. W., and D. W. Fawcett. Pigmented interstitial cells showing "brown degeneration" in the ovaries of old mice. Anat. Rec. 113:239–245, 1952.

Hasan, M., and P. Glees. Genesis and possible dissolution of neuronal lipofuscin. Gerontologia 18:217–236, 1972. (Review)

Jones, I. C. Variation in the mouse adrenal cortex with special reference to the zona reticularis and to brown degeneration, together with a discussion of the "cell migration" theory. Quart. J. Micros. Sci. 89:53–74, 1948.

Malkoff, D., and B. Strehler. The ultrastructure of isolated and in situ human cardiac age pigment. J. Cell Biol. 16:611–616, 1963.

Mann, D. M. A., and P. O. Yates. Lipoprotein pigments — their relationship to aging in the human nervous system. I. Lipofuscin content of nerve cells. Brain 97:481–488, 1974.

Nanda, B. S., and K. Getty. Occurrence of aging pigment (lipofuscin) in the nuclei and cortices of the canine brain. Exp. Gerontol. 8:1–17, 1973.

Nandy, K. Properties of neuronal lipofuscin pigment in mice. Acta Neuropathol. 19:25–32, 1971.

Sekhon, S. S., and D. S. Maxwell. Ultrastructural changes in neurons of the spinal anterior horn of aging mice with particular reference to the accumulation of lipofuscin pigment. J. Neurocytol. 3:59–72, 1974.

Strehler, B. L., D. Mark, A. S. Mildvan and M. V. Gee. Rate and magnitude of age pigment accumulation in human myocardium. J. Gerontol. 14:430–439, 1959.

Tobin, C. E., and J. P. Birnbaum. Some factors influencing brown degeneration of the adrenal cortex in the Swiss albino mouse. Arch. Pathol. 44:269–281, 1947.

MELANIN PIGMENT

Melanin is the brown pigment of skin and hair, the choroid coat and ciliary body of the eye, and the leptomeninges and substantia nigra of mammalian brain. In amphibia and reptiles it is much more widely distributed, occurring also in the lining of serous cavities and in the stroma of liver and many other organs. Melanin is formed in specialized cells, the *melanocytes,* which arise in the neural crest of the embryo and subsequently migrate to many sites in the body. Melanocytes are especially abundant in the basal layer of the epidermis and underlying dermis and are responsible for the pigmentation of the skin. Melanin is synthesized in these cells by the action of the enzyme tyrosinase, which catalyzes the oxidation of the amino acid tyrosine to dihydroxyphenylalanine (DOPA) and to dopa quinone. Further elaboration of melanin occurs from these intermediates. It is a polymer which does not occur free but is invariably complexed with structural protein.

The pigment occurs in the form of extremely dense ellipsoidal granules 0.3μ by 0.7μ called *melanosomes.* The first recognizable precursors of these organelles are vesicles associated with the Golgi complex of the melanocyte. The contents of these vesicles acquire tyrosinase activity, and as pigment synthesis begins they take on an elongated form and develop a complex lamellar internal structure with a 10 nm periodicity. Progressive accumulation of melanin obscures this internal structural framework and the mature melanin granules generally appear homogeneous and extremely dense.

In melanocytes of invertebrates, amphibia, and some of those in mammals, the granules accumulate in great numbers throughout the cytoplasm but are retained within the cell. However, those of mammalian skin exhibit a unique form of secretory activity in which membrane-limited packets of melanin granules are budded off from dendritic processes of the melanocyte and transferred to the cytoplasm of the neighboring keratinocytes, which lack the capacity to synthesize melanin. This bizarre behavior is sometimes referred to as *cytocrine secretion.*

In its formative stages the melanosome is limited by a membrane and contains a highly organized internal structure of longitudinally oriented strands or concentric lamellae that have a regular 10 nm periodic structure along their length. This structural framework is believed to be composed of a protein endowed with tyrosinase activity. As melanization proceeds, increasing amounts of the melanin polymer are deposited upon this enzymatically active protein framework until the latter is embedded in its own dense reaction products and is no longer visible in thin sections.

Figure 295. A melanosome from a human fetus. (Micrograph courtesy of Aiden Breathnach.)

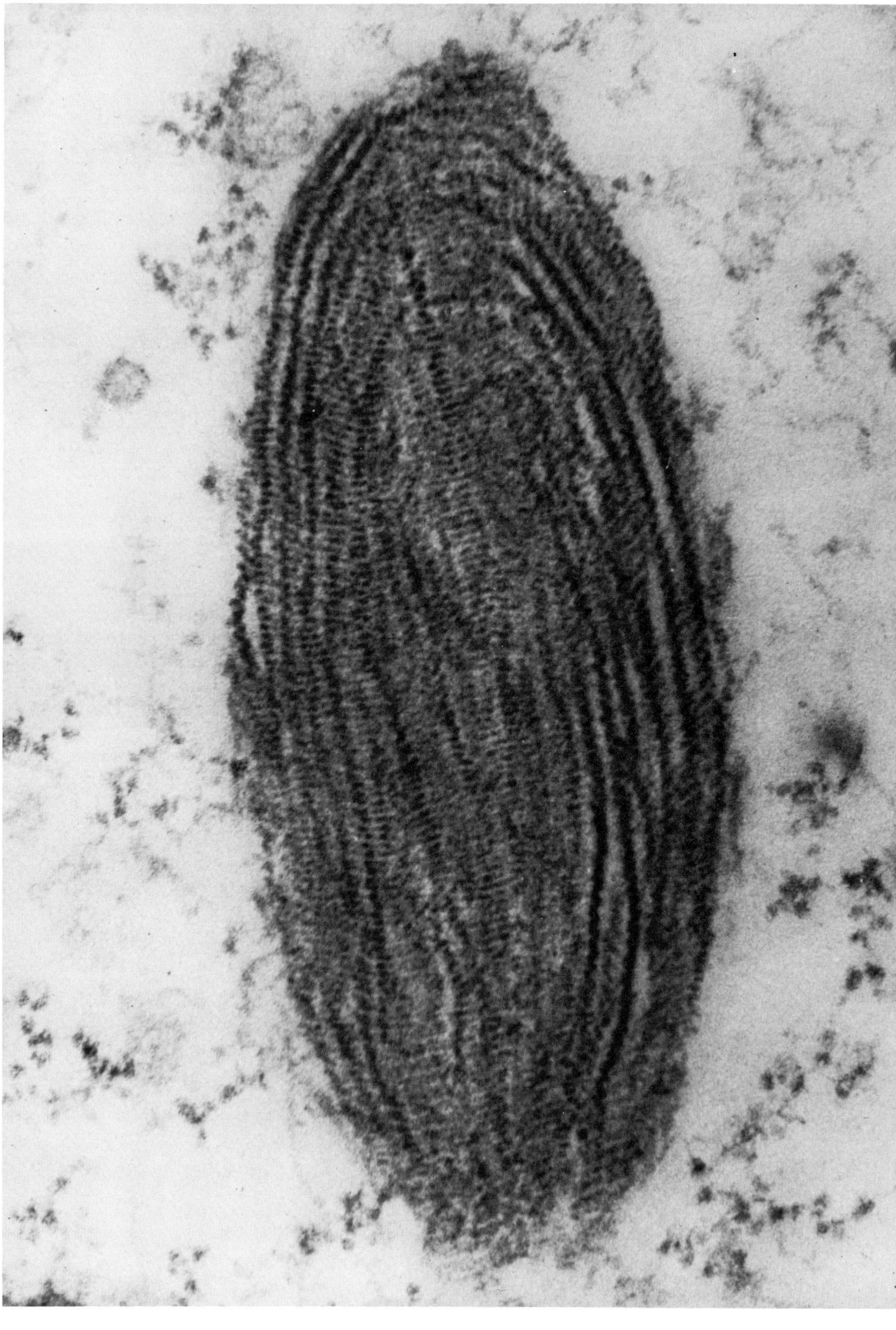

Figure 295

In mammalian skin, melanocytes are situated in the basal layers of the epidermis, and their dendritic processes extend into the interstices among the surrounding epithelial cells which are commonly called *keratinocytes* because of their capacity to synthesize the fibrous protein keratin. These latter cells contain melanosomes but are unable to synthesize melanin. Their melanosomes are produced by melanocytes and transferred from the tips of their dendritic processes to the surrounding epidermal cells. The mechanism of transfer is not completely understood, but membrane-limited packets of melanosomes appear to be budded off from the melanocytes and taken into the epithelial cells.

The micrograph reproduced here shows a melanocyte at the dermoepidermal junction and portions of two adjacent epithelial cells. The melanocytes are invaders of the epithelium and are not attached to neighboring cells by desmosomes. Their melanosomes usually do not accumulate in large numbers but are transferred at about the same rate as they are produced. Those present tend to be located at the periphery of the cell or in its dendritic processes.

Figure 296. Melanocyte from black human skin. (Micrograph courtesy of George Szabo.)

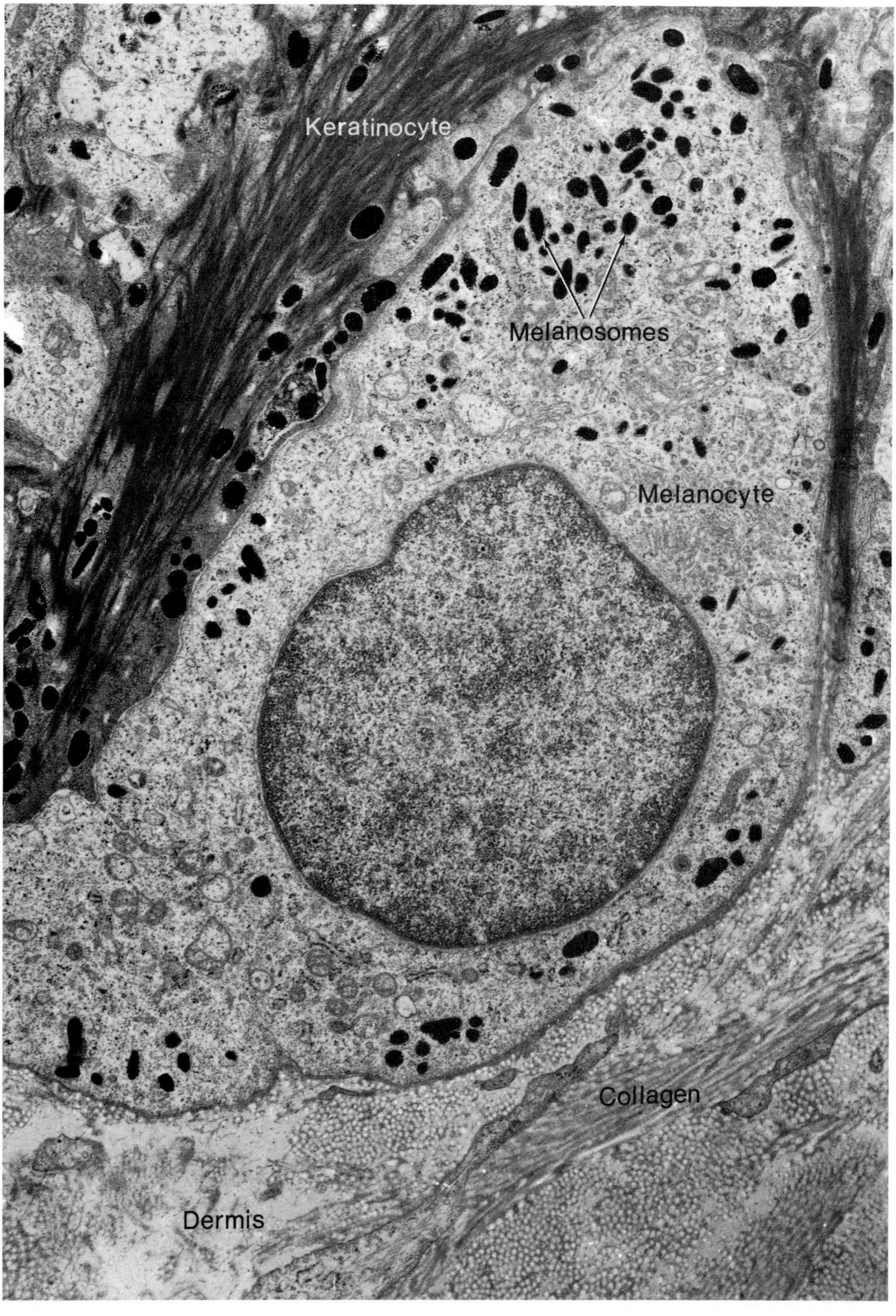

Figure 296

The accompanying micrograph illustrates again the interstitial location of a melanocyte in the basal portion of the epidermis. Its outline is indicated by the arrows. The concentration of melanosomes in the surrounding keratinocytes is greater than in the cell body of the melanocyte. While desmosomes are abundant between epithelial cells (at upper left), none are found around the periphery of the melanocyte.

Figure 297. Epidermal melanocyte from facial skin of an orang-utan. (Micrograph courtesy of George Szabo.)

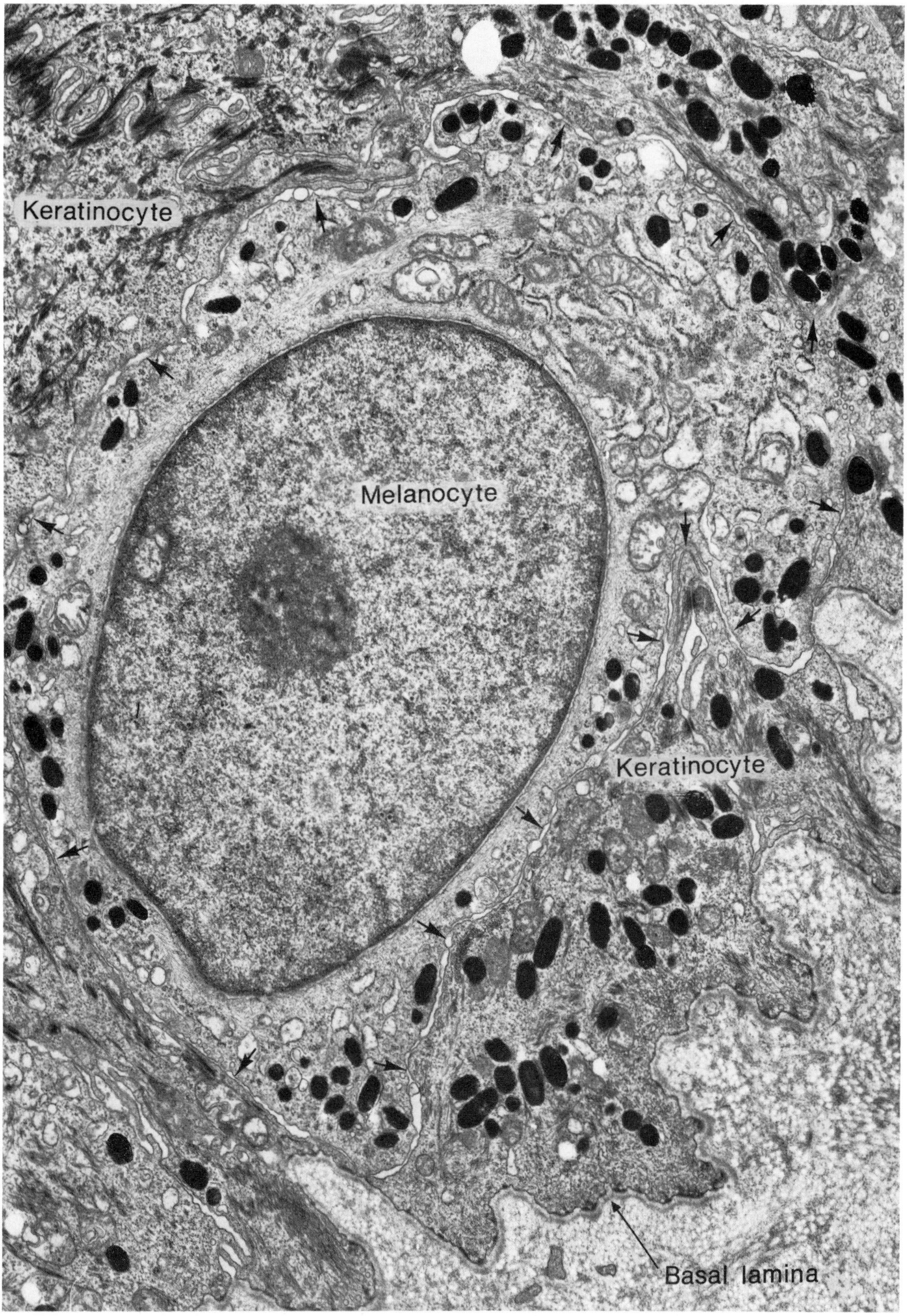

Figure 297

Melanin pigment is responsible for eye color. The anterior portion of the iris is a highly vascular connective tissue consisting of fibroblasts and melanocytes in a loose stroma. The color of the iris depends in large measure upon the amount of melanin pigment in the melanocytes of this layer. Its posterior surface is covered with a double layer of heavily pigmented epithelium.

The accompanying micrograph is of a melanocyte in the connective tissue at the root of the iris in the eye of a macaque. Its cytoplasm is crowded with melanosomes.

Figure 298. Pigment cell from the iris of *Macaca mulatta.* (Micrograph courtesy of Giuseppina Raviola.)

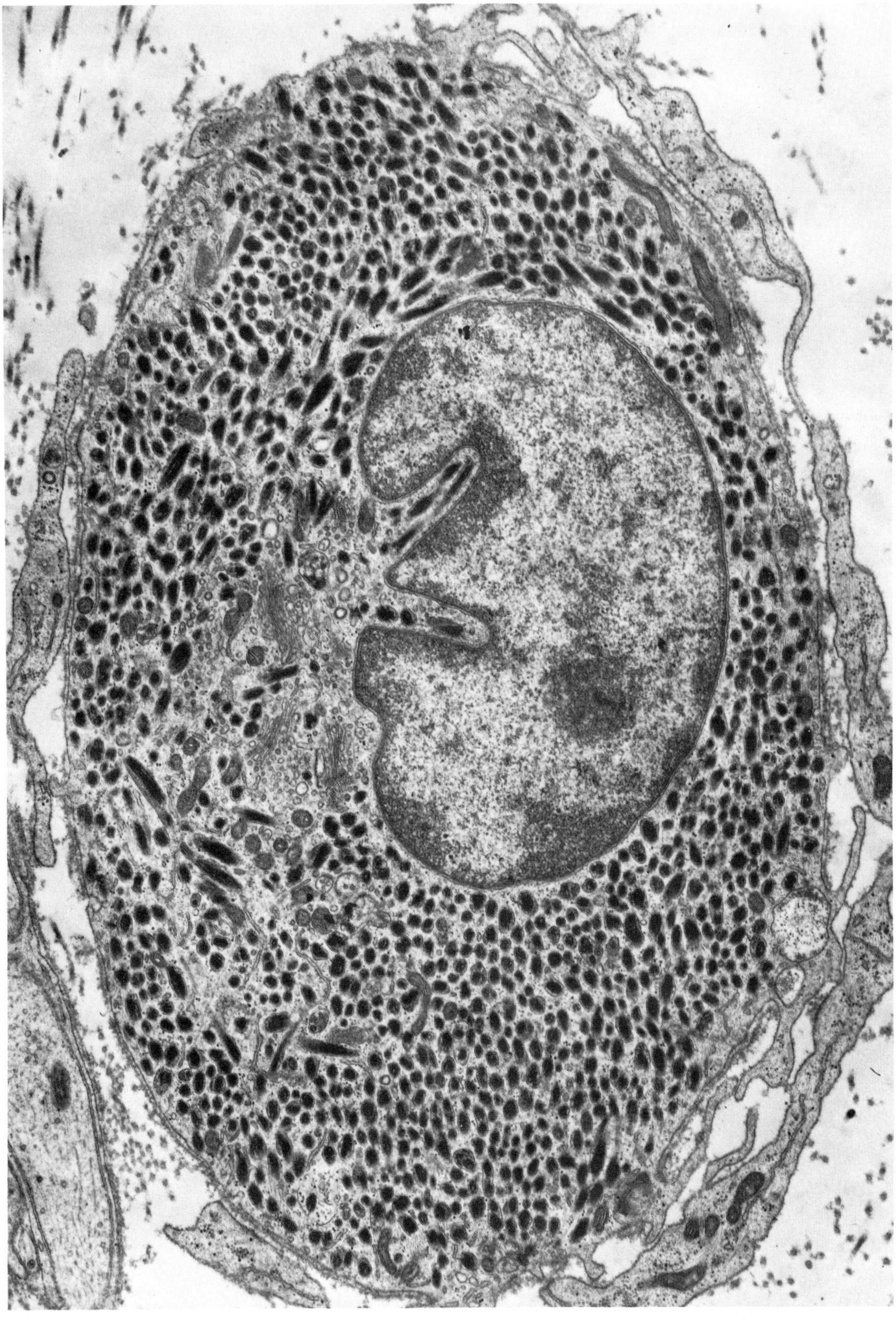

Figure 298

A melanocyte at higher magnification shows the melanosomes, all limited by a membrane, and containing varying concentrations of melanin. Where some have been sectioned longitudinally, they are seen to be quite long. The shape of melanosomes is subject to considerable variation from species to species and even in different strains of the same species.

The enzyme tyrosinase and the structural proteins of melanosomes are believed to be synthesized on ribosomes associated with rough endoplasmic reticulum. In the Golgi complex, these are packaged in small vesicles which elongate as the lamellar framework of the melanosome is assembled. Melanin then progressively accumulates, ultimately obscuring the ordered internal structure of the melanosome. The vesicles indicated by arrows near the juxtanuclear Golgi complex in this micrograph probably represent early stages in the formation of melanosomes.

Figure 299. Melanocyte from the iris of *Macaca mulatta.* (Micrograph courtesy of Giuseppina Raviola.)

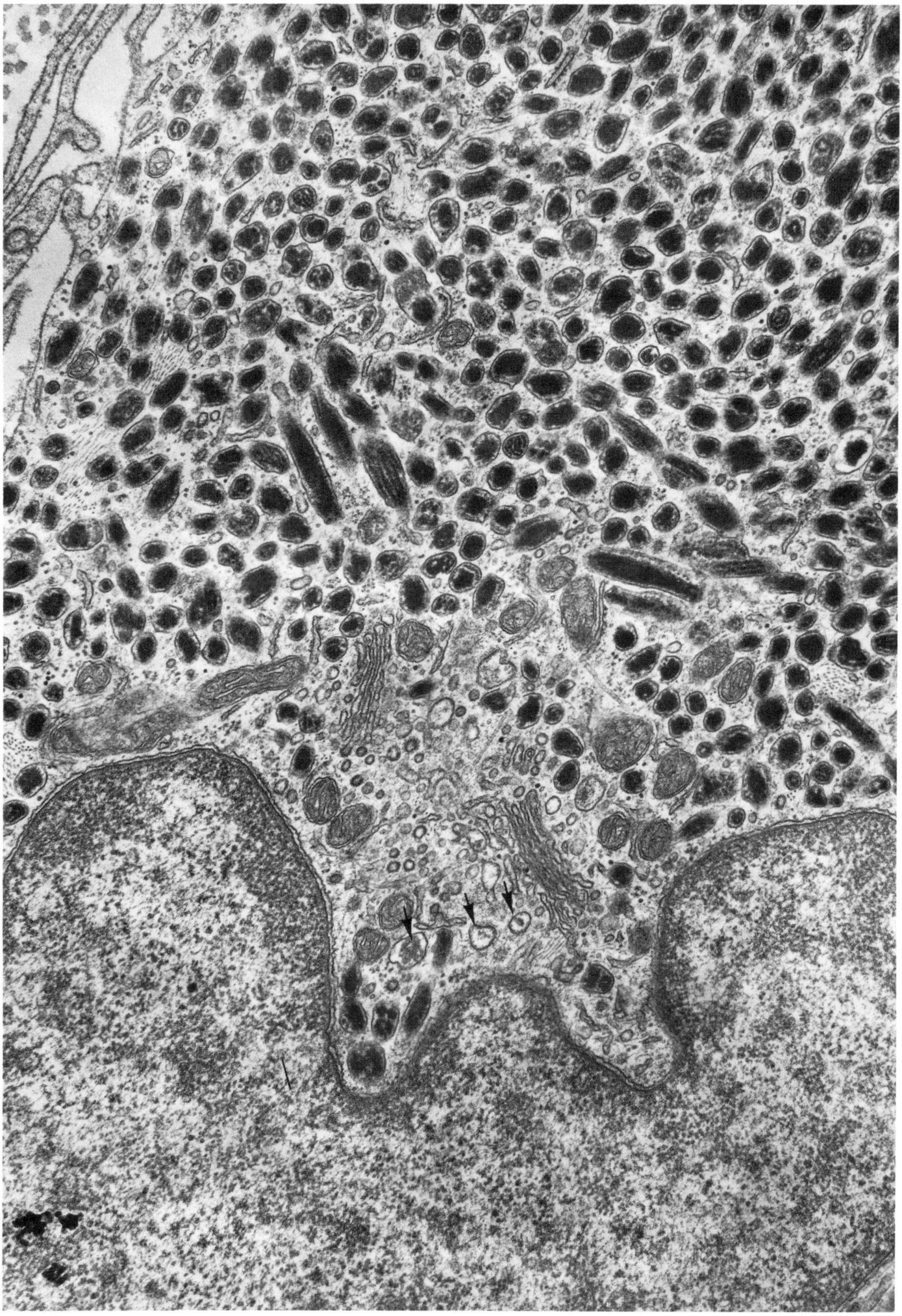

Figure 299

The melanosomes donated to other cells by melanocytes may be distributed individually in the cytoplasm or they may occur in groups enclosed in a limiting membrane.

In the accompanying micrograph of conjunctival epithelium from the limbal region of a macaque eye, the intercellular spaces are occupied by horseradish peroxidase that was injected intravenously and escaped from vessels in the episclera, permeating the epithelium and outlining the cells. Within the cells are a number of membrane-bounded clusters of melanosomes.

Figure 300. Melanin in cells of the conjunctival epithelium of *Macaca mulatta.* (Micrograph courtesy of Giuseppina Raviola.)

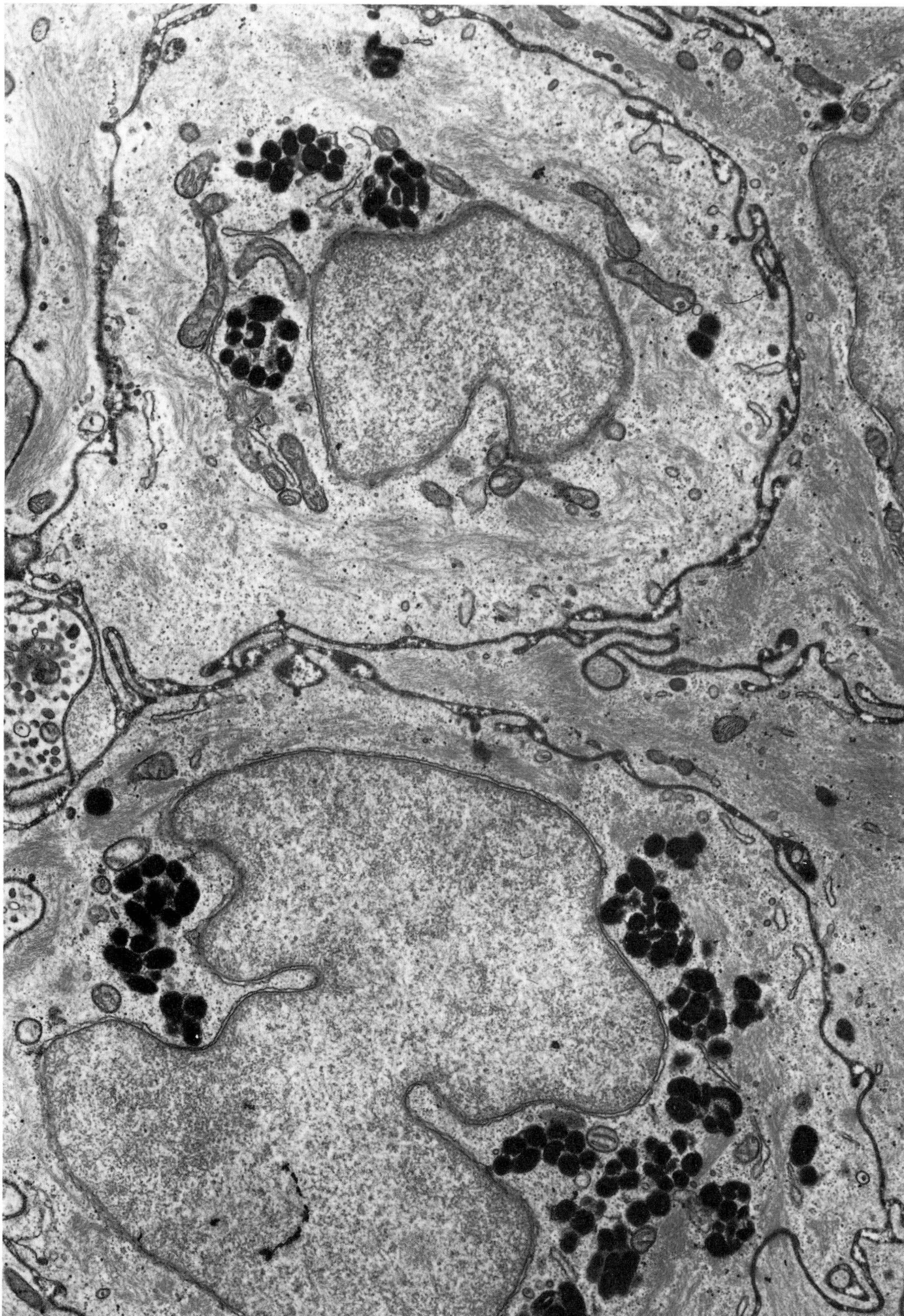

Figure 300

REFERENCES

Melanin

Barnicot, N. A. and M. S. C. Birbeck. The electron microscopy of human melanocytes and melanin granules. In The Biology of Hair Growth (W. Montagna and R. A. Ellis, eds.), p. 259, Academic Press, New York, 1958.

Cohen, J. and G. Szabo. Study of pigment donation in vitro. Exp. Cell Res. 50:418, 1968.

Fitzpatrick, T. B. and G. Szabo. The melanocyte: Cytology and cytochemistry. J. Invest. Dermatol. 32:197–209, 1959.

Seiji, M., K. Shimao, M. S. C. Birbeck and T. B. Fitzpatrick. Subcellular localization of melanin biosynthesis. Ann. N.Y. Acad. Sci. 100:497–533, 1963.

Seiji, M., T. B. Fitzpatrick and M. S. C. Birbeck. The melanosome: A distinctive subcellular particle of mammalian melanocytes and the site of melanogenesis. J. Invest. Dermatol. 36:243–252, 1961.

Szabo, G. The biology of the pigment cell. In The Biological Basis of Medicine (E. B. Bittar and N. F. Bittar, eds.), Vol. 6, Academic Press, New York, 1969.

Zelickson, A. S. Melanocyte, melanin granule and Langerhans cells. In Ultrastructure of Normal and Abnormal Skin (A. S. Zelickson, ed.), Lea and Febiger, Philadelphia, 1968.

CENTRIOLES

In suitably stained histological sections examined with the light microscope a pair of short rods, the *centrioles*, are seen in a specially differentiated region of the juxtanuclear cytoplasm called the *centrosome*, or *cell center*. The position of the centrosome was long used as an indication of the polarity and symmetry of the cell, and a line passing through the center of the nucleus and through the centrosome was defined as the *cell axis*. In secretory epithelial cells the pair of centrioles (*diplosome*) is often located in the supranuclear cytoplasm partially surrounded by the Golgi apparatus. The position of these organelles was believed to determine the polarity of the cell and the direction of its secretion.

The centrioles were observed to double in number immediately before cell division, one pair migrating to the opposite pole of the nucleus. After the breakdown of the nuclear envelope the mitotic spindle developed with a diplosome at each pole. The centrioles were thus considered to play an important role in cell division, serving as centers for organization of the spindle apparatus. It was logical to attribute the doubling of the centrioles immediately before cell division to a process of division comparable to binary fission of bacteria. They were therefore regarded as self-duplicating organelles exhibiting continuity from one cell generation to the next. A diploid cell generally had a single pair of centrioles, but in polyploid cells there was a pair for each chromosome set so that megakaryocytes of bone marrow or multinucleate giant cells of bone may have 30 or more. An exception to this correspondence between ploidy and number of centrioles was recognized in the case of ciliated epithelial cells. A remarkable proliferation of centrioles was observed as an initial event in the differentiation of cilia. The resulting centrioles took up a position immediately beneath the plasmalemma and served as the *basal bodies* of the developing cilia.

Electron microscopy demonstrated that centrioles were hollow cylindrical structures with nine evenly spaced triplet microtubules in their walls (de Harven and Bernhard, 1956; Bessis et al., 1968). Their duplication in the mitotic cycle and their continuity in the daughter cells were confirmed but the traditional views on the mechanism of centriole duplication proved to be incorrect.

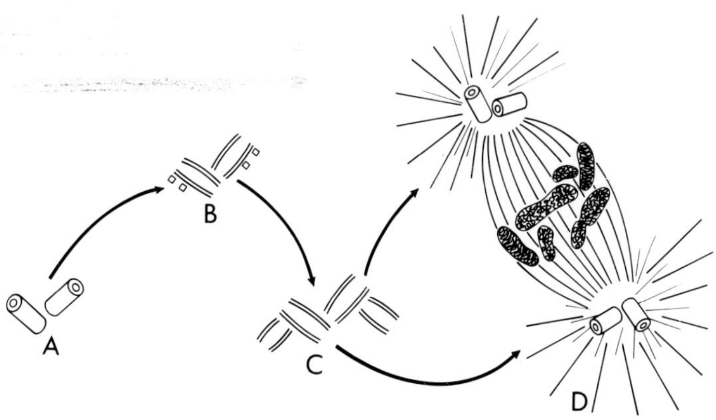

Diagram of centriole replication prior to cell division. Each member of the original pair (A) induces formation of a new centriole perpendicular to a specific region of its wall (B and C). The two diplosomes so formed take up positions at opposite poles of the mitotic spindle (D). In B and C the centrioles are depicted in longitudinal section.

Centrioles do not undergo transverse division nor do they split longitudinally. Instead, the new centriole develops in end-to-side relationship to a specific region of a preexisting centriole but separated from it by a narrow electron lucent space. The anlage from which it develops, called the *procentriole*, is an annular condensation of dense material approximately the same diameter as a mature centriole but initially devoid of microtubules. The procentriole elongates by accretion of material to its free end, and the pinwheel arrangement of triplet microtubules gradually appears within the previously homogeneous dense ring. The forming centriole maintains its orientation perpendicular to the parent centriole. After reduplication the original members of the diplosome separate and each, with its newly formed centriole, moves to one pole of the division figure. Thus in the cell cycle, centrioles are duplicated but not, as previously believed, by division. A template mechanism cannot be invoked, since the new centriole arises perpendicular to its precursor and not in intimate contact with it. The preexisting centriole seems to act merely as a site of induction of nucleation for self-assembly of the new centriole from precursors synthesized elsewhere in the cell.

Studies of ciliogenesis have shown that centrioles can arise apart from preexisting centrioles (Stockinger and Cirelli, 1965; Steinman, 1968; Sorokin, 1968). The great majority of the basal bodies develop around dense spherical bodies variously called *deuterosomes*, or *procentriole organizers*. These in turn arise by condensation of smaller aggregations of filamentous material (*filosomes*). Several annular procentrioles may arise radially around the same procentriole organizer. They elongate rapidly and acquire the microtubular internal structure typical of centrioles. When they have attained their definitive length they dissociate from the organizer and move to the cell surface, where each initiates polymerization of the nine doublet microtubules forming the axoneme of a cilium.

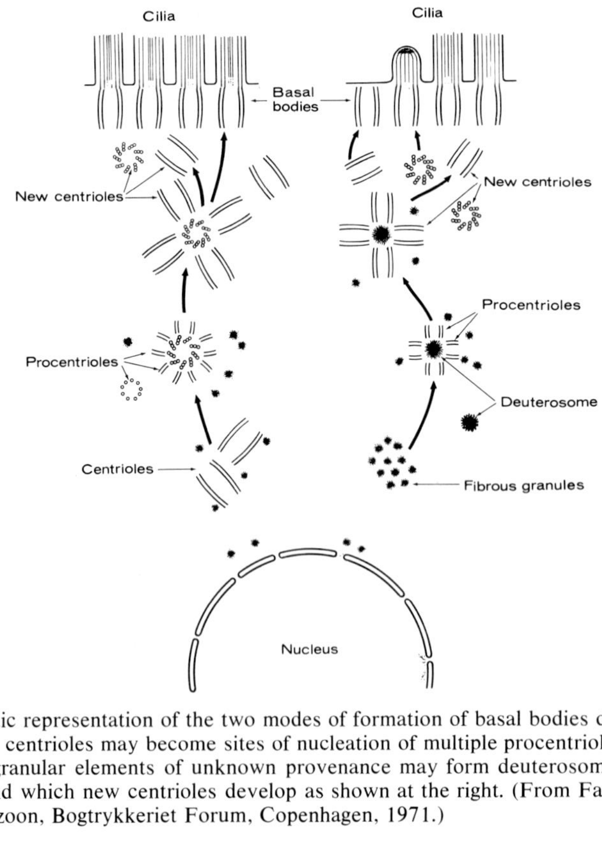

Diagrammatic representation of the two modes of formation of basal bodies during ciliogenesis. The two original centrioles may become sites of nucleation of multiple procentrioles as shown at the left. Also, fibrogranular elements of unknown provenance may form deuterosomes or procentriole organizers around which new centrioles develop as shown at the right. (From Fawcett, in Genetics of the Spermatozoon, Bogtrykkeriet Forum, Copenhagen, 1971.)

In its role as basal body of a cilium or flagellum, the centriole can be considered to serve a template function, since the polymerization of tubulin takes place directly on the distal end of the triplet microtubules in its wall and the ninefold symmetry of the centrioles is expressed in the nine doublets of the axoneme. For reasons that remain obscure only two members of each centriolar triplet nucleate the assembly of tubulin under these conditions, since doublet and not triplet microtubules are formed as peripheral elements of the axoneme.

Assembly of single microtubules evidently does not require a preexisting site of nucleation, for the axial pair of single microtubules in the axoneme arises without a corresponding structure in the axis of the centriole. The nucleating function of centrioles for microtubule assembly has been demonstrated *in vitro* using centrioles isolated from human cells and tubulin extracted from bovine brain (McGill and Brinkley, 1975).

The important role traditionally assigned to the centrioles in formation of the mitotic spindle seems not to have been valid. Mitosis occurs with formation of a typical spindle in both lower and higher plants in the absence of centrioles (Pickett-Heaps, 1969, 1971). In electron micrographs of animal cells, the spindle microtubules do converge upon the centriolar region but they rarely contact the centrioles themselves. They terminate in the pericentriolar cytoplasm in dense spherical granules called *centriolar satellites*. Selective damage to the pericentriolar material by laser microbeam irradiation in prophase results in failure of the chromosomes to separate at anaphase. This occurs with little or no morphological evidence of structural damage to the centrioles themselves (Berns et al., 1977). Investigative attention has therefore shifted to the surrounding specialized zone of cytoplasm which constitutes the other major component of the centrosome. It has been possible to isolate centrosomes from certain strains of cultured cells that have been treated with colchicine to block spindle formation. When these are incubated *in vitro* with tubulin extracted from brain, large numbers of microtubules polymerize around each centrosome. These emanate from the pericentriolar material and not from the centrioles (Gould and Borisy, 1977). It is concluded therefore that centriolar satellites and possibly other components of the pericentriolar cytoplasm can initiate assembly of microtubules both *in vivo* and *in vitro*. This does not necessarily mean that the centrioles are inactive during cell division. They may control aggregation of the pericentriolar material or may be involved in its activation at the appropriate time in the cell cycle. The interrelations of the several components of the centrosome have yet to be worked out.

The ubiquitous occurrence of centrioles, their replication, and their continuity from generation to generation of cells prompted the speculation that they might possess their own DNA, as do mitochondria and chloroplasts. In several early cytochemical and biochemical analyses of basal bodies isolated in bulk from protozoa, the presence of DNA was reported (Seaman, 1960; Randall and Disbrey, 1965). This could not be confirmed in later investigations with purer fractions (Flavell and Jones, 1970). Further cytochemical studies using acridine orange staining suggested, however, that RNA was associated with basal bodies of cilia (Hartman et al., 1974). RNA in centrioles and in their satellites was also reported on the basis of the sensitivity of certain of their components to digestion with ribonuclease (Stubblefield and Brinkley, 1967; Brinkley and Stubblefield, 1970; Dippel, 1976).

Evidence of a functional role for RNA in centrioles has been presented more recently. When basal bodies isolated from protozoa are injected into Xenopus eggs, they induce the formation of conspicuous radial arrays of microtubules (asters). The aster-inducing activity of basal bodies was eliminated by prior treatment with ribonuclease, but this treatment did not interfere with their capacity to serve as templates for polymerization of microtubules at their ends (Heidemann et al., 1977). It was concluded that centrioles contain RNA and that this component is necessary for initiation of aster formation. Other interpretations of the observations are possible and arguments based upon analysis of partially purified preparations have the inherent weakness of possible contamination with components of the neighboring cytoplasm. Although the evidence is highly suggestive of the presence of RNA, a resolution of this problem will require further study.

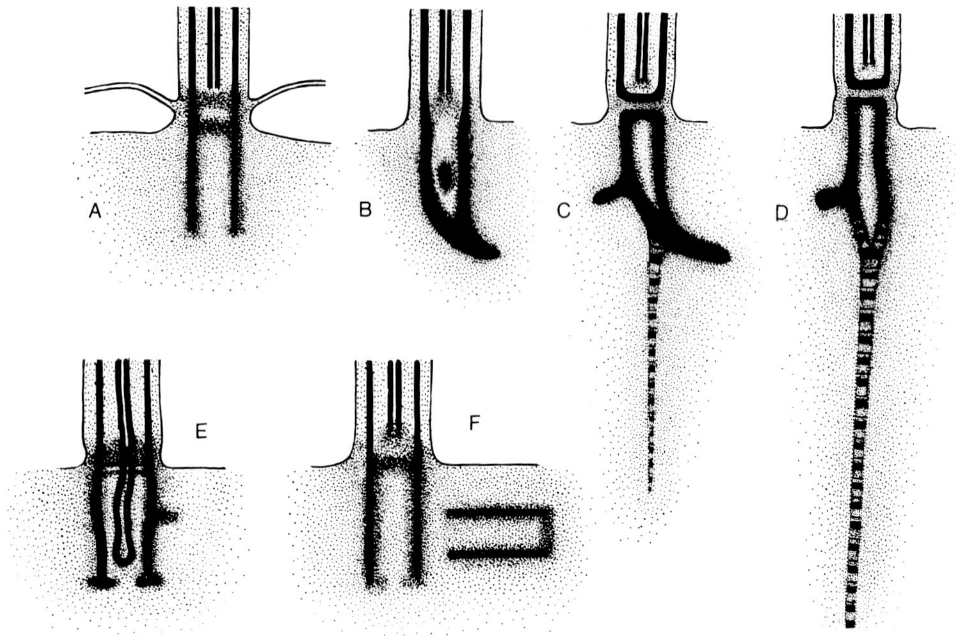

Centrioles serving as basal bodies for cilia or flagella usually retain their simple cylindrical form (A), but their shape may be modified by the development of lateral appendages and one or both ends may be closed (B, C, D). In many invertebrates and some vertebrates, cross-striated ciliary rootlets extend from the lower end of the basal body for variable distances into the apical cytoplasm (C, D). In rare instances in protozoa (viz, *Euplotes*), the central pair of microtubules of the axoneme may form a loop extending into the central cavity of the basal body (E). In cells with a single flagellum, a second centriole is often at right angles to the one serving as basal body (F). (From Fawcett, in The Cell, Vol 2. J. Brachet and A. Mirsky, eds., Academic Press 1961.)

Centrioles are usually positioned so that their long axes form a right angle. Their perpendicular orientation is maintained even though they may be half a micrometer or more apart and have no visible structural elements connecting them. The nature of the long-range forces or organization of the centrosomal cytoplasm that are responsible for this relationship are unknown. Departures from the usual orthogonal arrangement are occasionally encountered in normal cells and are reported to be common in malignant tumors.

Centrioles vary somewhat in length from one cell type to another but are usually about 0.5 μm long and 0.2 μm in diameter. In the accompanying micrograph a pair of centrioles near the lumenal surface of an epithelial cell shows the usual perpendicular orientation. The discrepancy in length of the two members of this pair is unusual.

The two ends of a centriole can often be distinguished in that one is slightly narrower and appears to be closed and the other appears open.

Figure 301. Intestinal epithelium of a chicken embryo. (Micrograph courtesy of Sergei Sorokin.)

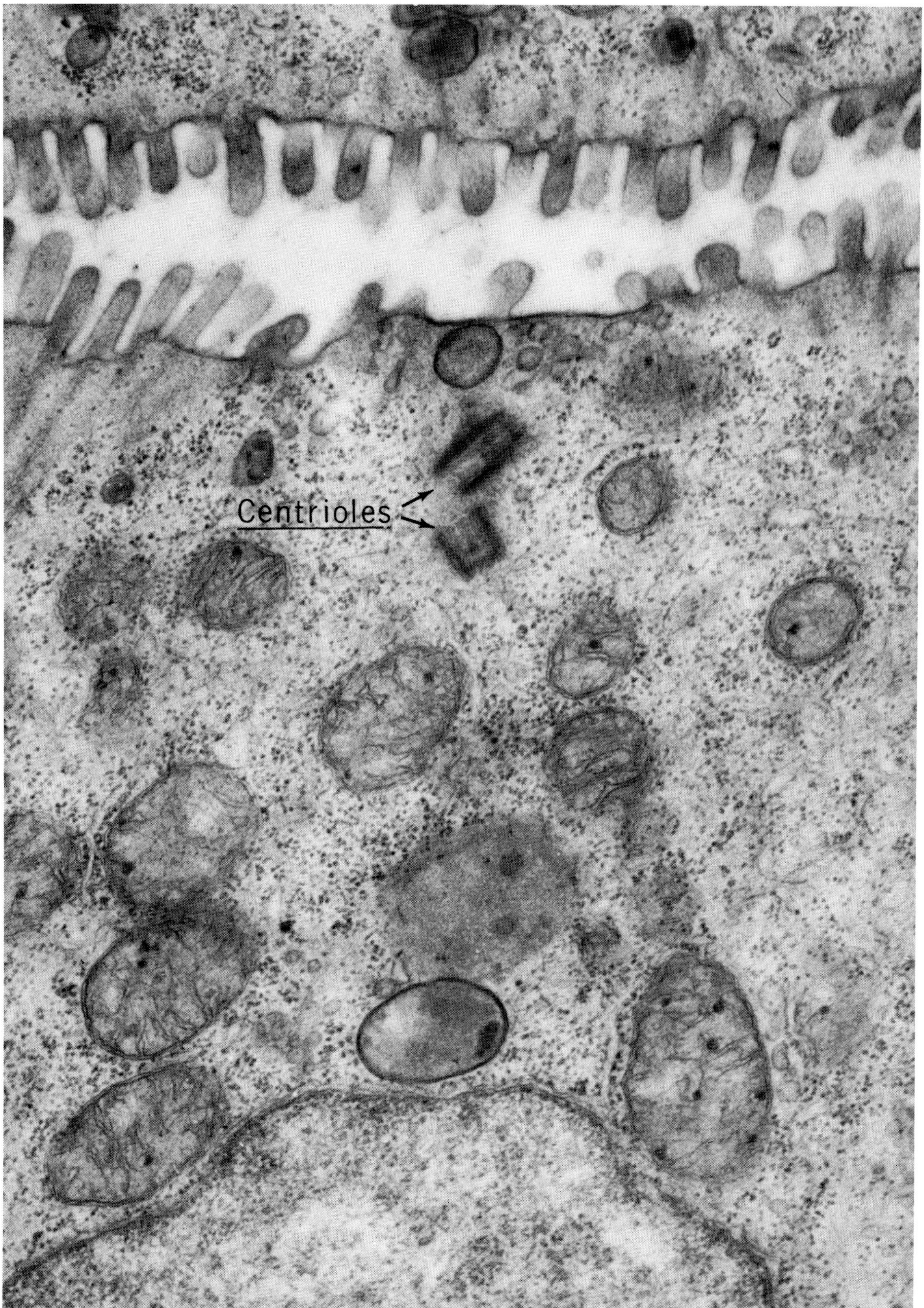

Figure 301

The centrioles replicate early in cell division and take up positions at either pole of the division figure. Concurrently with the condensation of the chromosomes and breakdown of the nuclear envelope, microtubules polymerize, extending from the chromosomes to the poles and from pole to pole, to form the mitotic spindle.

In the accompanying micrograph made before introduction of aldehyde fixatives, the microtubules have not been well preserved.The plane of section is a fortunate one, however, in that three of the centrioles are included. The second centriole at the lower pole is in a plane perpendicular to this section and is not seen. Because the two diplosomes usually differ in their orientation, all four centrioles are rarely, if ever, included in the same thin section.

Figure 302. Centrioles of dividing spermatocyte from cock testis. (Micrograph courtesy of Toshio Nagano.)

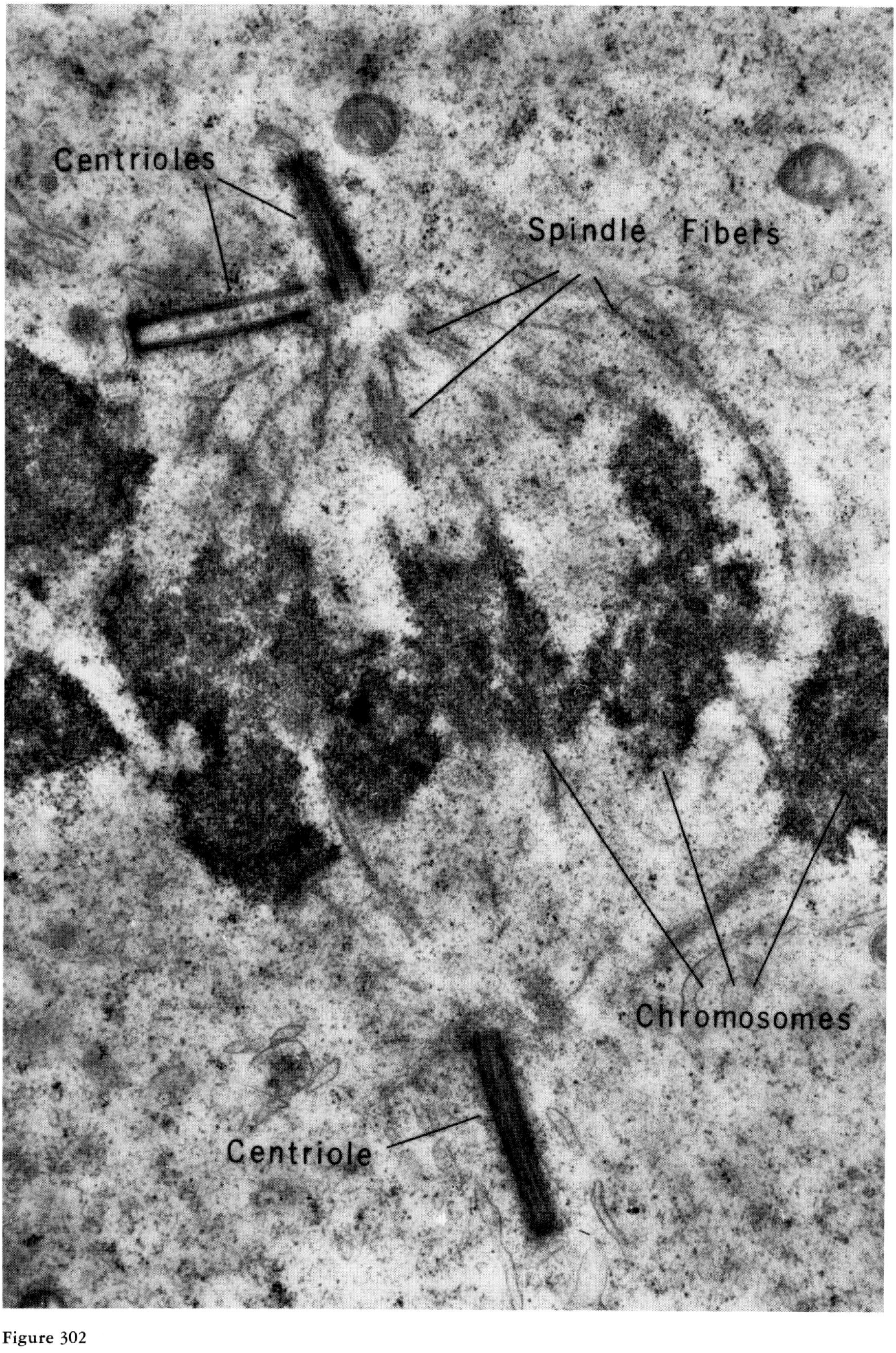

Figure 302

In transverse section the cylindrical wall of the centriole is made up of nine longitudinally oriented triplet microtubules. A dense material of unknown nature occupies the interstices between the triplets. Each triplet is at a constant angle of about 40 degrees to its respective tangent. A pattern is thus formed that resembles the vanes of a turbine or the charges of a pyrotechnic "pinwheel." The innermost microtubule of each triplet has a circular cross section and is designated subunit *a* and the other two *b* and *c*. The latter two each share a segment of the wall of the adjacent microtubule and therefore have a C-shaped profile instead of a circular cross section. Subunit *a* has two short diverging projections resembling the arms on the doublets of the ciliary axoneme. One of these is directed inward along a radius and appears to have a free end pointing toward the center of the centriole. The other projection connects with subunit *c* of the next triplet. The successive triplets are thus linked together *a* to *c* around the circumference of the centriole by a series of linear densities. It is not known whether these latter correspond to the dynein arms or nexin links of flagellar axonemes.

Figure 303. Centriole. (Micrograph courtesy of Susumu Ito.)

Figure 304. Centriole from embryonic chick pancreas. (Micrograph courtesy of Jean André.)

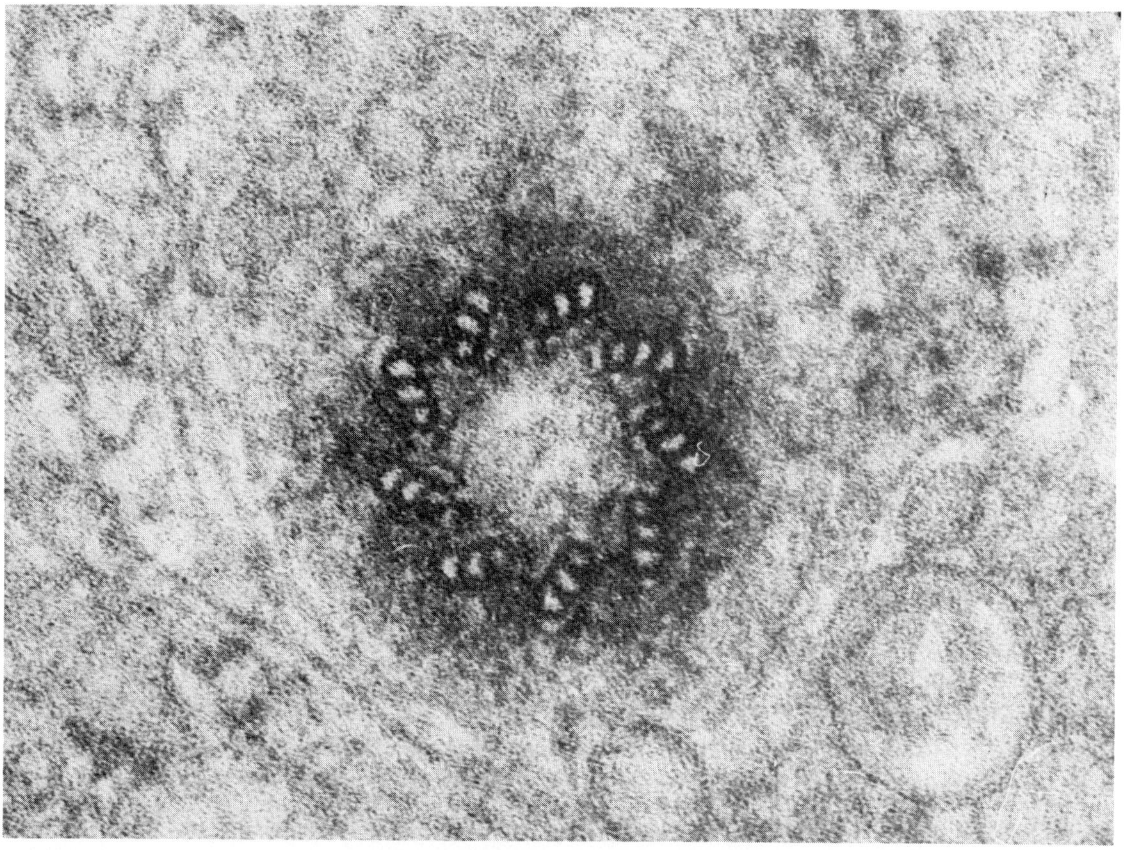

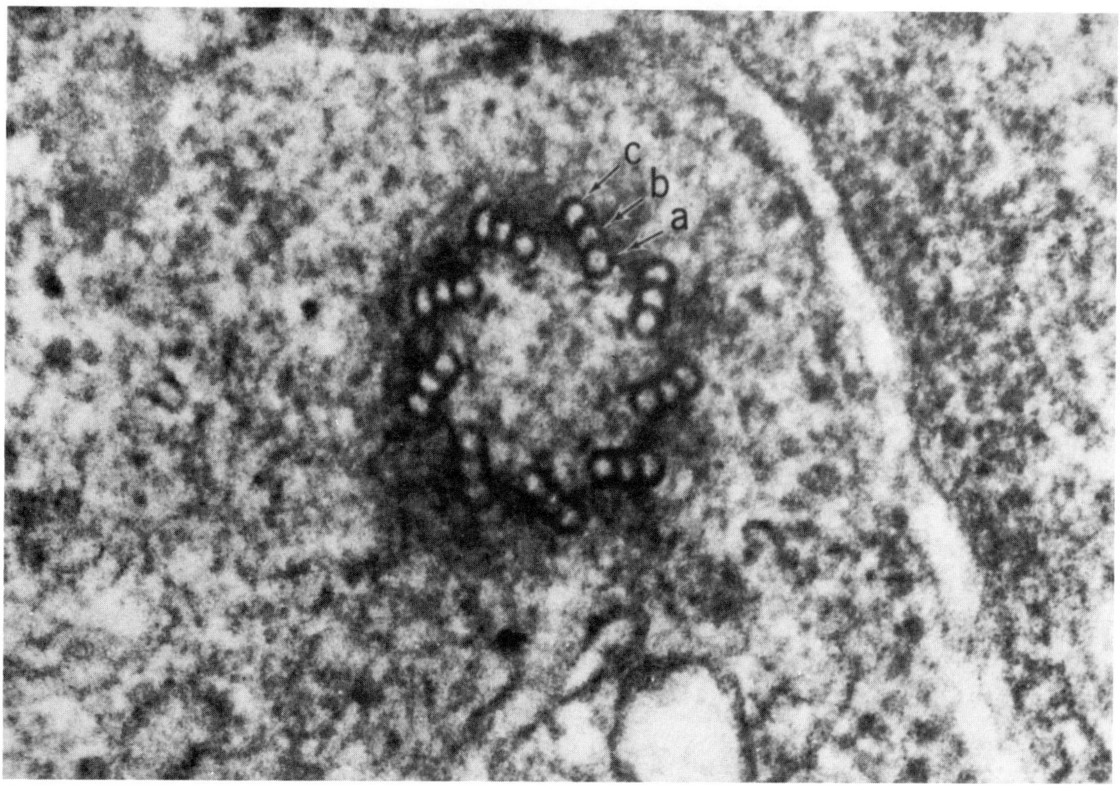

Figure 303, *upper* **Figure 304,** *lower*

Images of radially symmetrical periodic structures can be intensified and made more coherent by a technique involving rotation between each of several successive exposures in the enlarger. The cross section of a centriole at the upper left (*A*) has been intensified by summation of fractional exposures through nine steps of rotation to produce the image at the upper right (*B*). This has succeeded in accentuating the hub-and-spokes pattern of densities in the interior of the centriole. This pattern is clearly visible in some cross sections of centrioles and absent in others. It is not clear whether there are real differences among centrioles in this respect or whether this variation in their appearance in electron micrographs reflects differences in preservation or different planes of section along the length of the centriole.

The cross section of a centriole stained with tannic acid in the lower figure shows the protofilaments of the triplets in negative image. Like other microtubules, the wall of subunit *a* has 13 tubulin protofilaments, while *b* and *c* each have 10. The core structure and radiating spokes are illustrated with unusual clarity. Although the dimensions of the hub are similar to those of a microtubule, no protofilaments are detectable in its wall.

Figure 305. Centriole from H_3 polyoma tumor. *A*, Original image. *B*, Image reinforced by rotation. (Micrographs courtesy of Etienne de Harven.)

Figure 306. Centriole fixed in glutaraldehyde and tannic acid. (Micrograph courtesy of Vitauts Kalnins.)

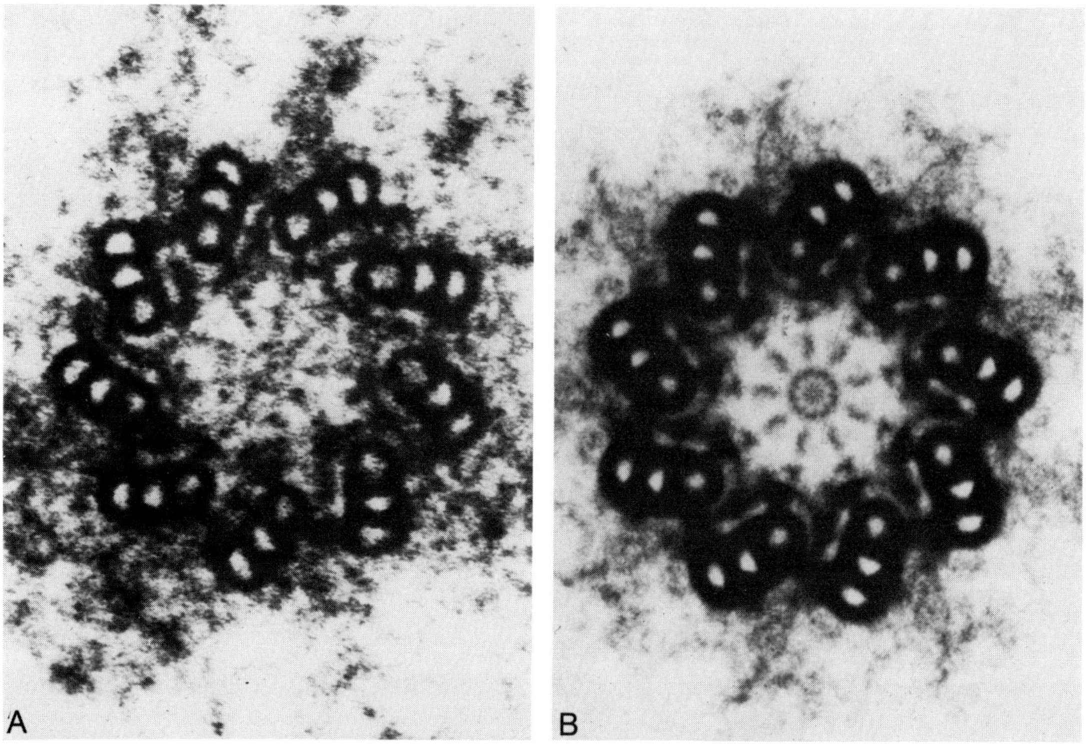

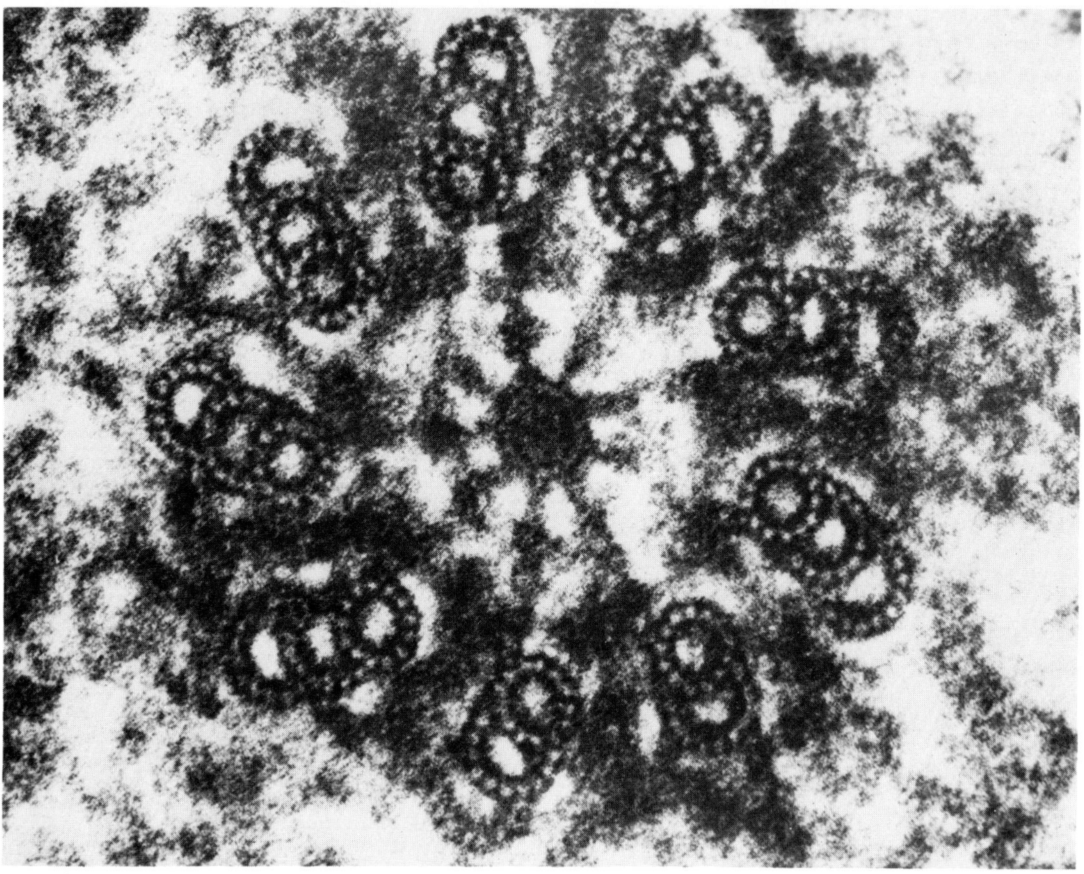

Figure 305, *upper* **Figure 306,** *lower*

In interphase cells, microtubules commonly radiate from the centrosome, and in dividing cells the microtubules of the mitotic spindle converge toward a diplosome at either pole of the division figure. The accompanying micrographs of dividing cells show numerous microtubules radiating from the immediate vicinity of a centriole. Such images have led to the assumption that the centrioles serve as nucleation sites for microtubule assembly. However, detailed studies of this region suggest that the microtubules do not actually contact the centrioles but end in a specialized pericentriolar zone of cytoplasmic matrix and are often associated with small dense bodies called *centriolar satellites* (indicated by arrows).

Laser microbeam irradiation of acridine orange–treated living cells results in a disorganization of the pericentriolar material but causes no detectable alteration of the centrioles. This treatment disturbs formation of interpolar microtubules and interferes with anaphase movement of chromosomes. It is concluded that although centrioles may not be directly involved as sites of nucleation of microtubules, they may nevertheless contribute indirectly to spindle mechanics as agents of synthesis or organization of the pericentriolar material.

Figures 307 and 308. Centriole regions irradiated with a laser microbeam without acridine orange sensitization. (Micrographs from Berns et al., J. Cell Biol. 72:351–367, 1977.)

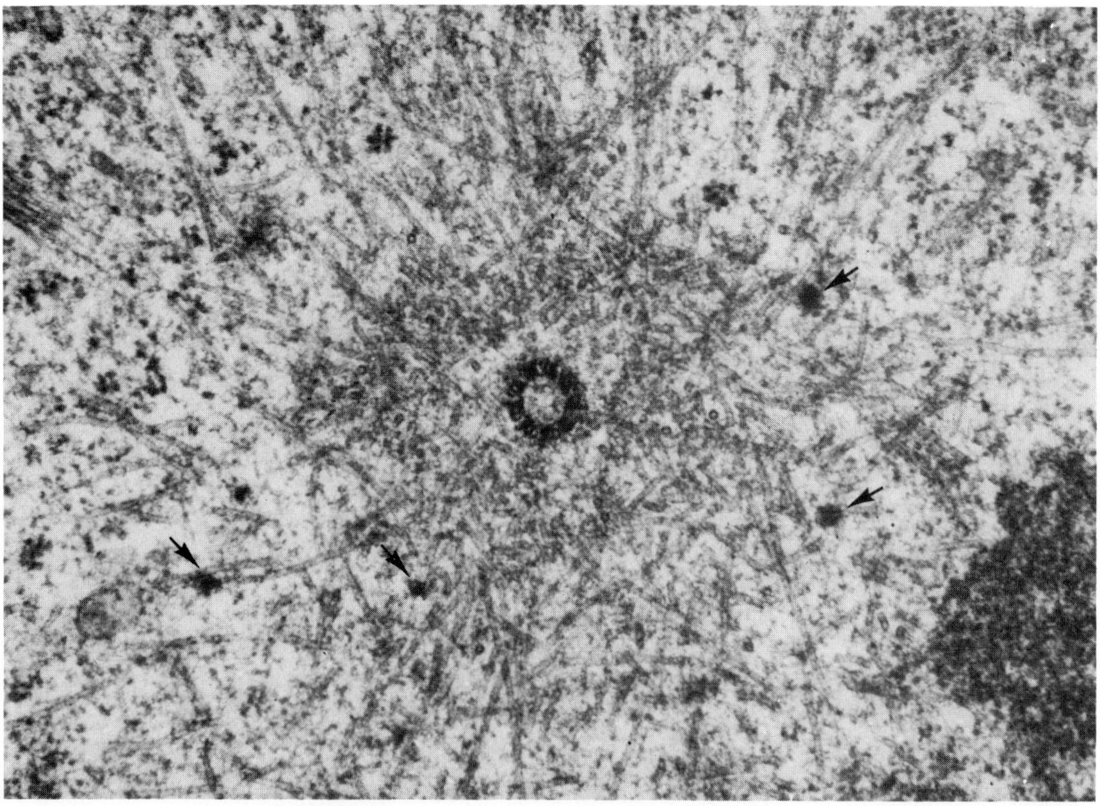

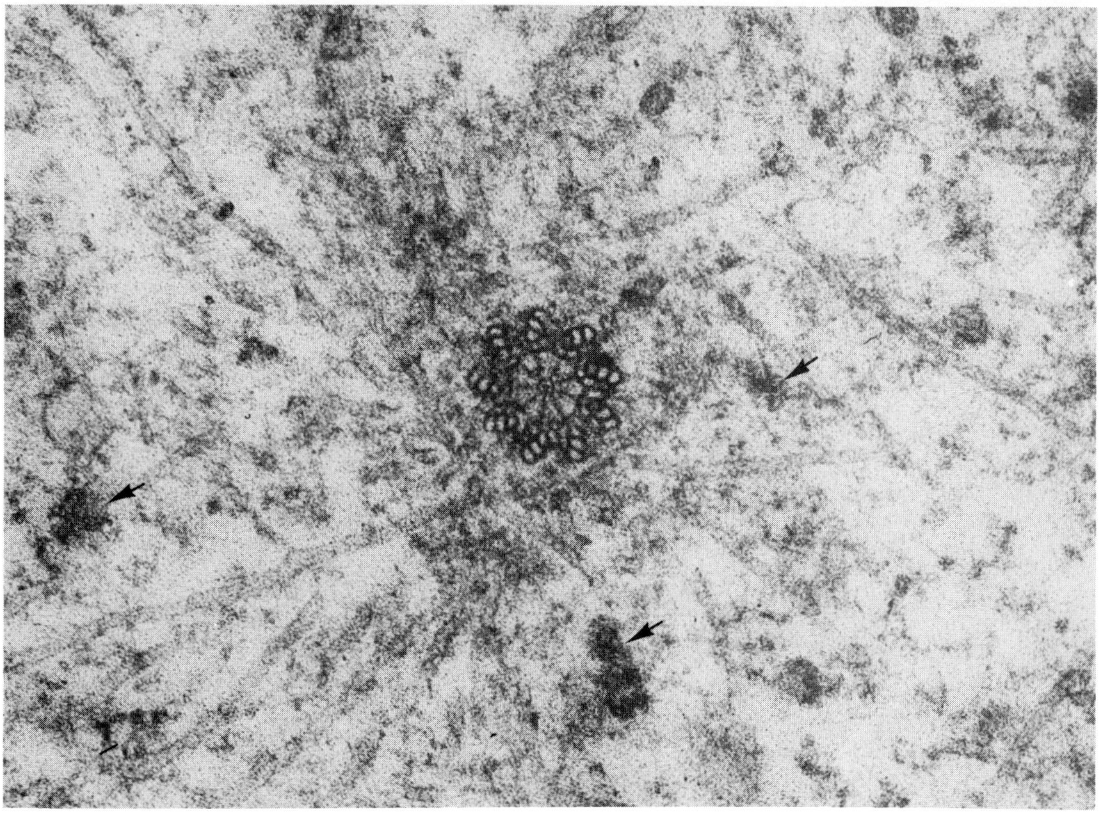

Figure 307, *upper* **Figure 308**, *lower*

The plane of a thin section only rarely happens to coincide with the long axis of both members of a pair of centrioles. It is more common for one to be cut longitudinally and the other transversely, as in the accompanying micrographs. In both of these examples, the centrioles are oriented at right angles even though they are some distance apart.

The pair of centrioles in the upper micrograph on the facing page is closely associated with the Golgi apparatus and occupies a concavity in the nucleus. This is a common relationship in the hemopoietic cell line and in other cell types as well.

The lower figure illustrates the special character of the pericentriolar cytoplasm which often contains numerous satellites. The labeled microtubules end in one of the satellites.

Figure 309. A pair of centrioles and the associated Golgi complex in a myelocyte from guinea pig bone marrow.

Figure 310. Centrosomal region of an ascites tumor cell. (Micrograph from Guy de Thé, J. Cell Biol. 23:265–275, 1964.)

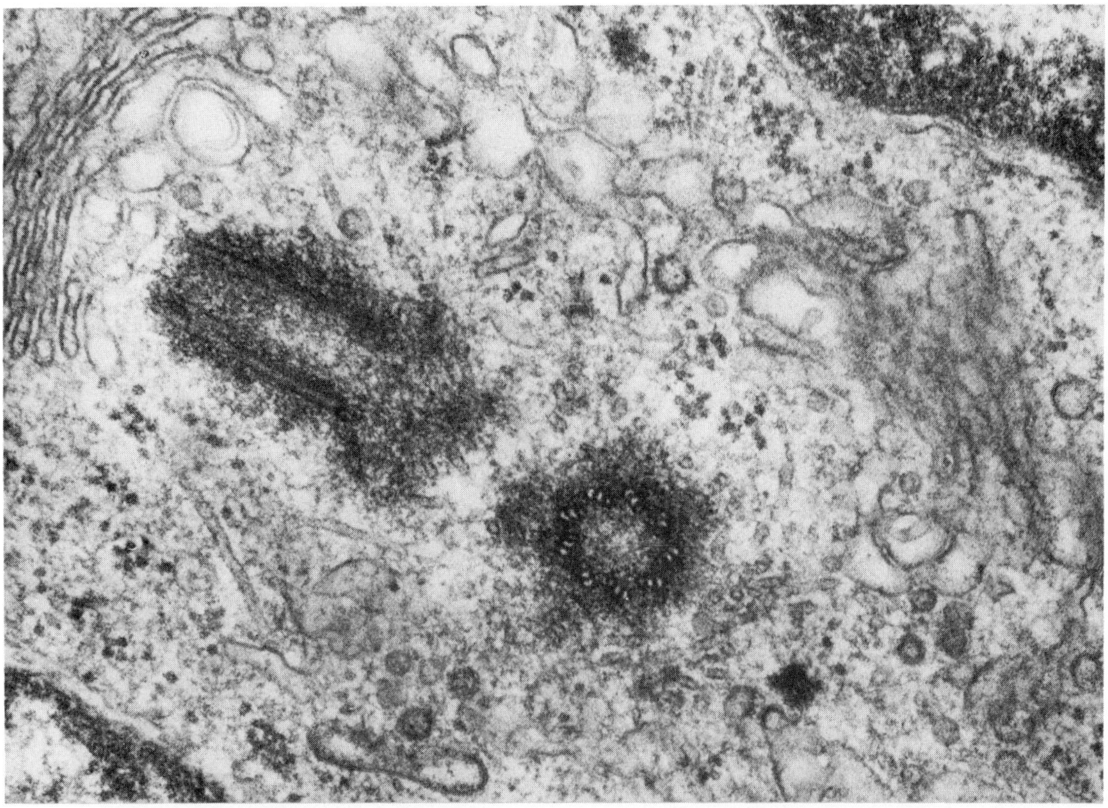

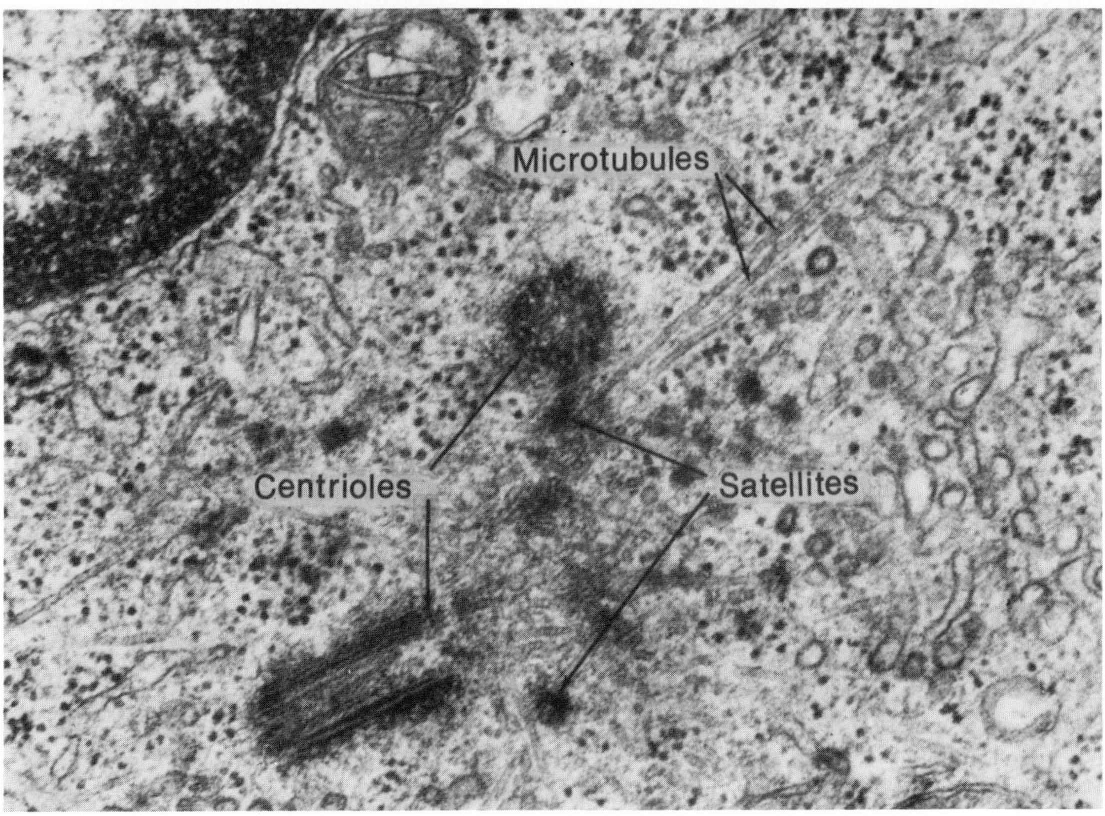

Figure 309, *upper* **Figure 310**, *lower*

In ciliogenesis, newly formed single centrioles, serving as basal bodies, are arranged in rows and oriented perpendicular to the cell surface. The end that is adjacent to the membrane functions as a site of nucleation for microtubule protein which polymerizes on subunits *a* and *b* of the triplets to form the nine doublet microtubules of the axoneme. The centrioles arrive at the cell apex at different times and the polymerization of tubulin evidently begins as soon as the end of the organelle is juxtaposed to the membrane. Thus in the accompanying micrograph four developing cilia are seen in successive stages of growth.

Flagella usually arise from one member of a pair of centrioles positioned at right angles. The axoneme grows out from the one perpendicular to the membrane. The lower figure on the facing page shows early stages in formation of a sperm flagellum. The differing lengths of the distal centrioles illustrated is attributable to the fact that the micrographs are not all from the same species. In one example (*B*) the second centriole is out of the plane of section. In sperm tail development the centrioles are involved in organizing other components in addition to the axoneme. In the most advanced stage shown (*D*), both centrioles are sectioned longitudinally. The thin line above the proximal centriole (*) is the anlage of the capitulum which will later articulate with the implantation fossa of the sperm nucleus. The periodic densities in the bracket (**) represent an early stage in assembly of the cross-striated columns of the connecting piece. The adjunct has begun to develop at the other end of the proximal centriole (at arrow). The centrioles thus appear to be involved in the organization of four distinct structures — axoneme, capitulum, centriolar adjunct, and connecting piece.

Figure 311. Ciliated cell of the mammalian oviduct (Micrograph courtesy of Everett Anderson.)

Figure 312. Successive stages of flagellum formation in spermatids.

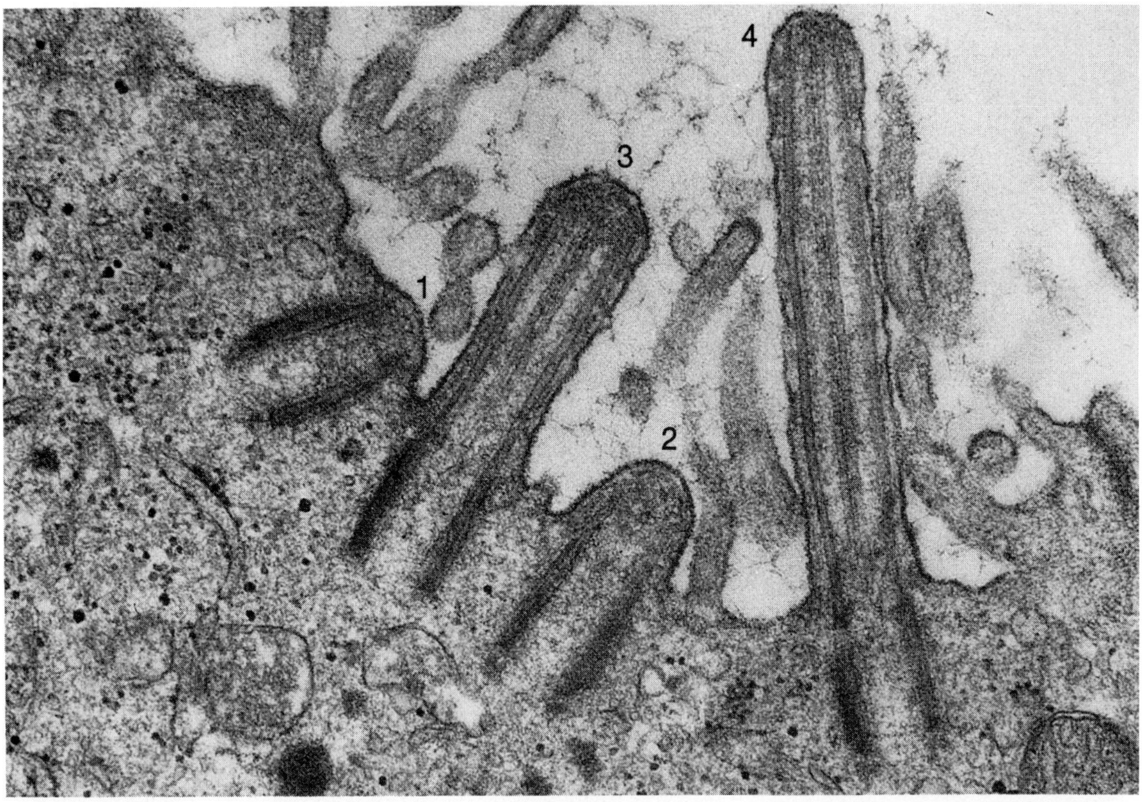

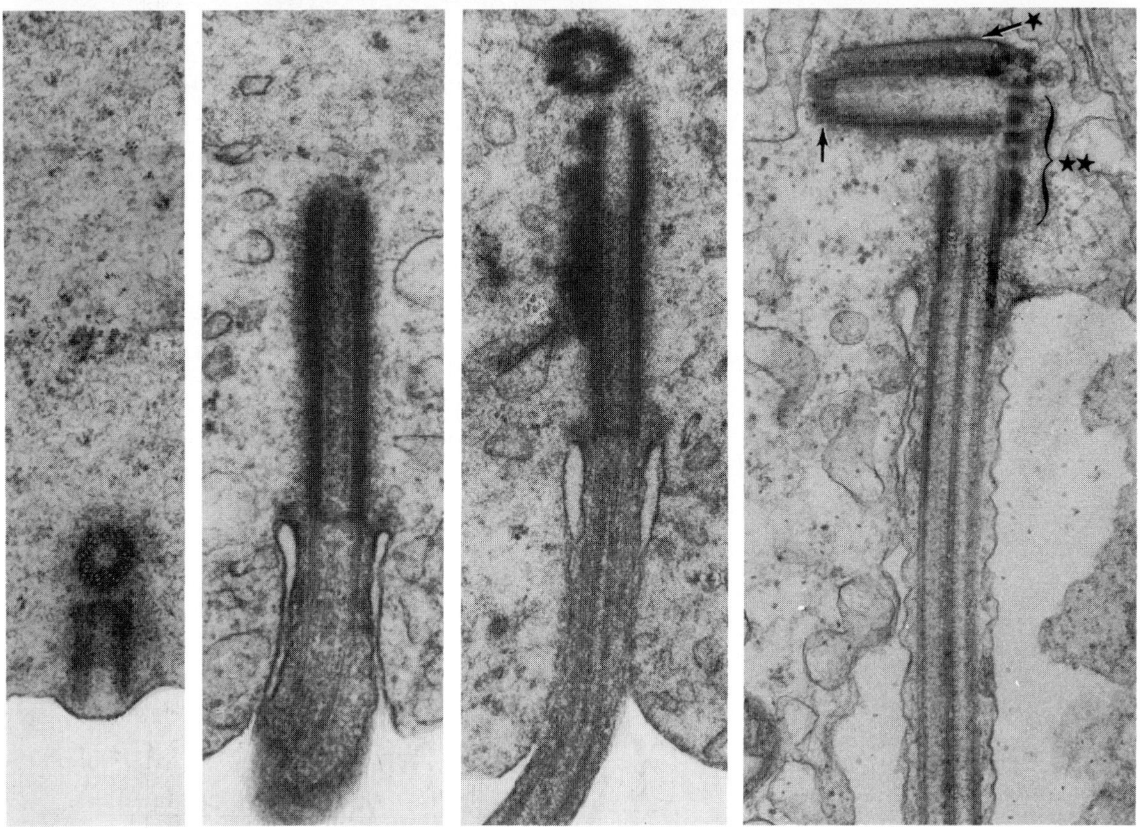

A B C D

Figure 311, *upper* **Figure 312,** *lower*

CENTRIOLAR ADJUNCT

During formation of a cilium or flagellum, only subunits *a* and *b* of the centriolar triplets serve as nucleation sites for formation of the doublets of the axoneme. A unique behavior of the proximal centriole during spermatogenesis demonstrates, however, that subunit *c* can serve as a site for nucleation of tubulin during elongation of the triplets to form the transient organelle known as the *centriolar adjunct*.

Early in mammalian spermiogenesis, the pair of centrioles takes up a position in the peripheral cytoplasm with the end of one member of the pair adjacent to the cell membrane. This relationship initiates the development of the flagellum at the end of the *distal centriole (A)*. Soon thereafter, fine filamentous material gathers around the distal end of the *proximal centriole (B)*. The centriole then increases in length by accretion and polymerization of this material at its end. This process continues for some time after the centrioles and base of the flagellum have moved inward and become lodged in the implantation fossa in the caudal pole of the nucleus (*C*). As will be illustrated in subsequent micrographs, this newly formed appendage is not identical to the centriole from which it arises.

The centriolar adjunct may attain a length two or three times that of the centriole. It then disappears, leaving no residue in the mature spermatozoon *(D)*. It has been found in all mammalian species thus far examined but its function remains unknown.

Figure 313. Diagram of formation of the centriolar adjunct. (From Fawcett, in Genetics of the Spermatozoon, Bogtrykkeriet Forum, Copenhagen, 1971.)

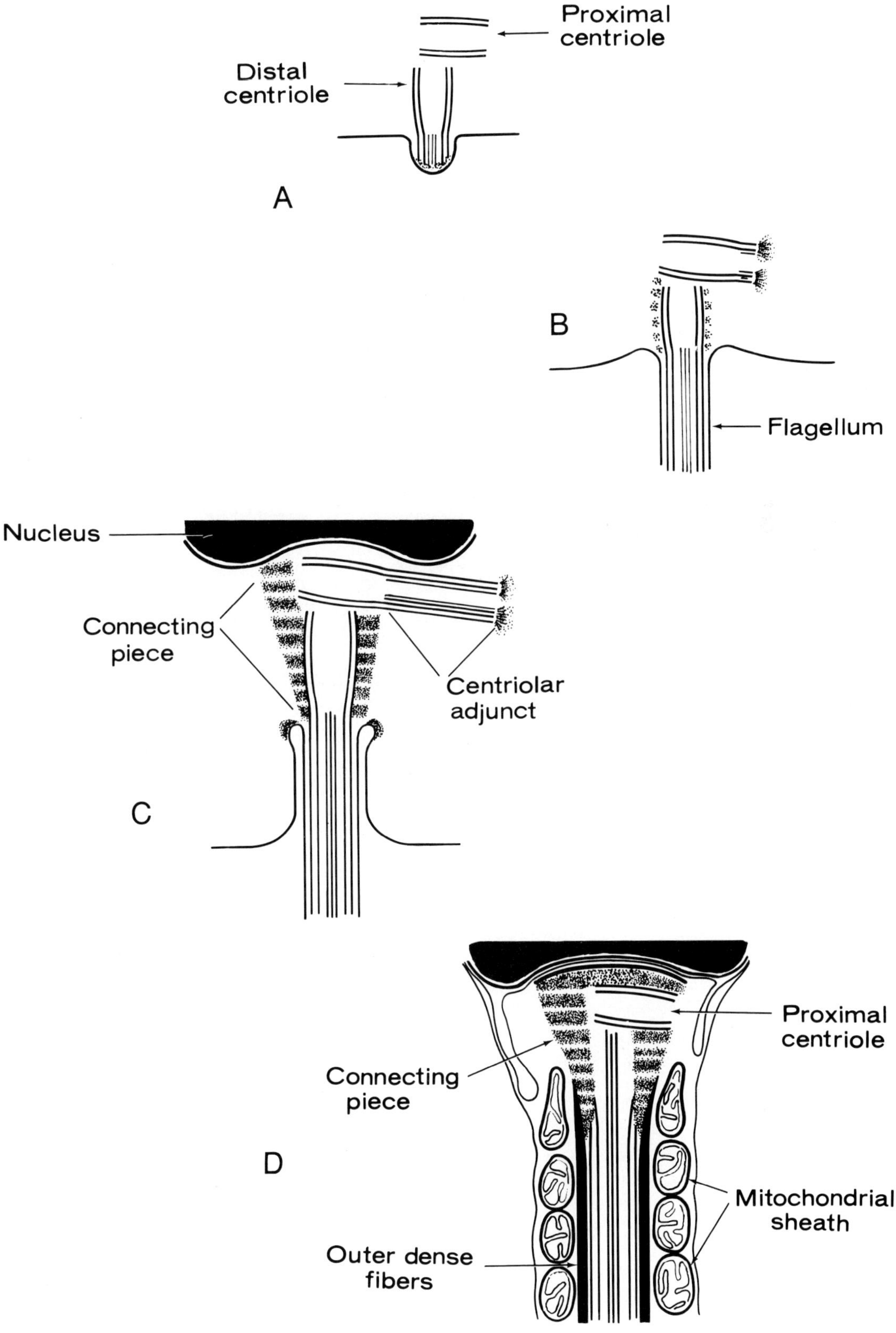
Proximal centriole
Distal centriole
A
B
Flagellum
Nucleus
Connecting piece
Centriolar adjunct
C
Proximal centriole
Connecting piece
D
Mitochondrial sheath
Outer dense fibers

Figure 313

At the base of the flagellum of a mammalian spermatid, the juxtanuclear proximal centriole is continuous (at the arrow) with a structure of similar configuration, the centriolar adjunct. At the end of the latter, there is an accumulation of loose textured material that appears to be condensing to contribute to elongation of this organelle. The cavity in the centriolar adjunct is narrower than that of the centriole proper.

At the lower left a transverse section through the centriole (at the level of the white line in the upper figure) shows the familiar pinwheel arrangement of closed triplet microtubules. At the lower right a comparable section through the adjunct (at the level of the black line in the upper figure) reveals some distinctive structural differences. Subunit *a* of the triplets is a typical closed microtubule, but *b* and *c* are usually open, presenting a free edge that has failed to fuse with the wall of the adjacent microtubule. In addition, the centriolar adjunct has a lining layer of complex ultrastructure which is lacking in the centriole. This accounts for the differences in diameter of their central cavities.

The formation of atypical triplet microtubules in the centriolar adjunct of spermatids attests to the capacity of subunit *c* of the centriolar triplets to serve as a site for nucleation of microtubule protein. Why it is inactive during generation of an axoneme so that only doublets are formed remains unexplained.

Figure 314. Implantation fossa, centriolar adjunct, and base of the flagellum in a chinchilla spermatid. (From D. W. Fawcett and D. M. Phillips, Anat. Rec. 165:153–184, 1969.)

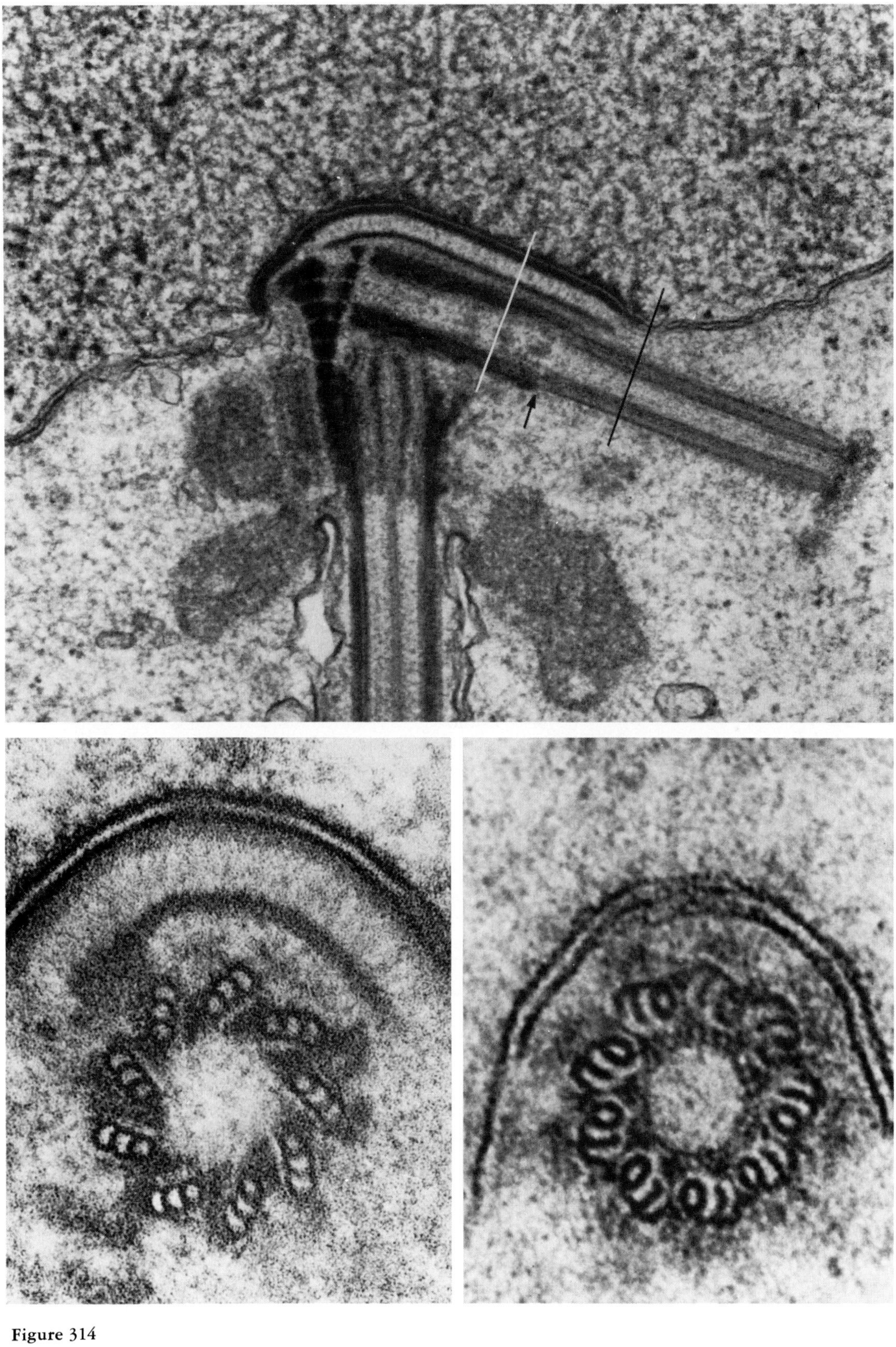

Figure 314

The dense material that forms the capitulum and the cross-striated columns of the connecting piece in spermatozoa is formed in intimate association with the proximal and distal centrioles, respectively. These versatile organelles are believed to play a dominant role in the organization of these components of the spermatozoon. The cross-striated columns of the connecting piece are probably homologous to the cross-striated rootlets that often extend downward from the basal bodies into the cytoplasm of ciliated epithelial cells.

Traditionally the basal bodies were regarded as kinetic centers responsible for initiating the beat of cilia and flagella. This interpretation is no longer tenable. It has been shown that sperm tail movements continue after laser microbeam destruction of the centriolar region (Lindemann and Rikmenspoel, 1972). Moreover, electron microscopic studies of spermatogenesis have shown that the distal centriole which initiates development of the flagellum later disintegrates and is no longer present in the mature spermatozoon. The juxtanuclear centriole usually persists, occupying a niche just beneath the capitulum that attaches the tail to the sperm head. In some species, however, this centriole also disintegrates during sperm maturation (Woolley and Fawcett, 1973). Thus centrioles are essential for development of cilia and flagella but are not required to initiate and maintain their beat. Observe in the accompanying micrograph that no distal centriole can be identified at the base of the axoneme.

Figure 315. Nearly mature spermatozoon from the testis of a Chinese hamster. (From D. W. Fawcett and D. M. Phillips, Anat. Rec. 165:153–184, 1969.)

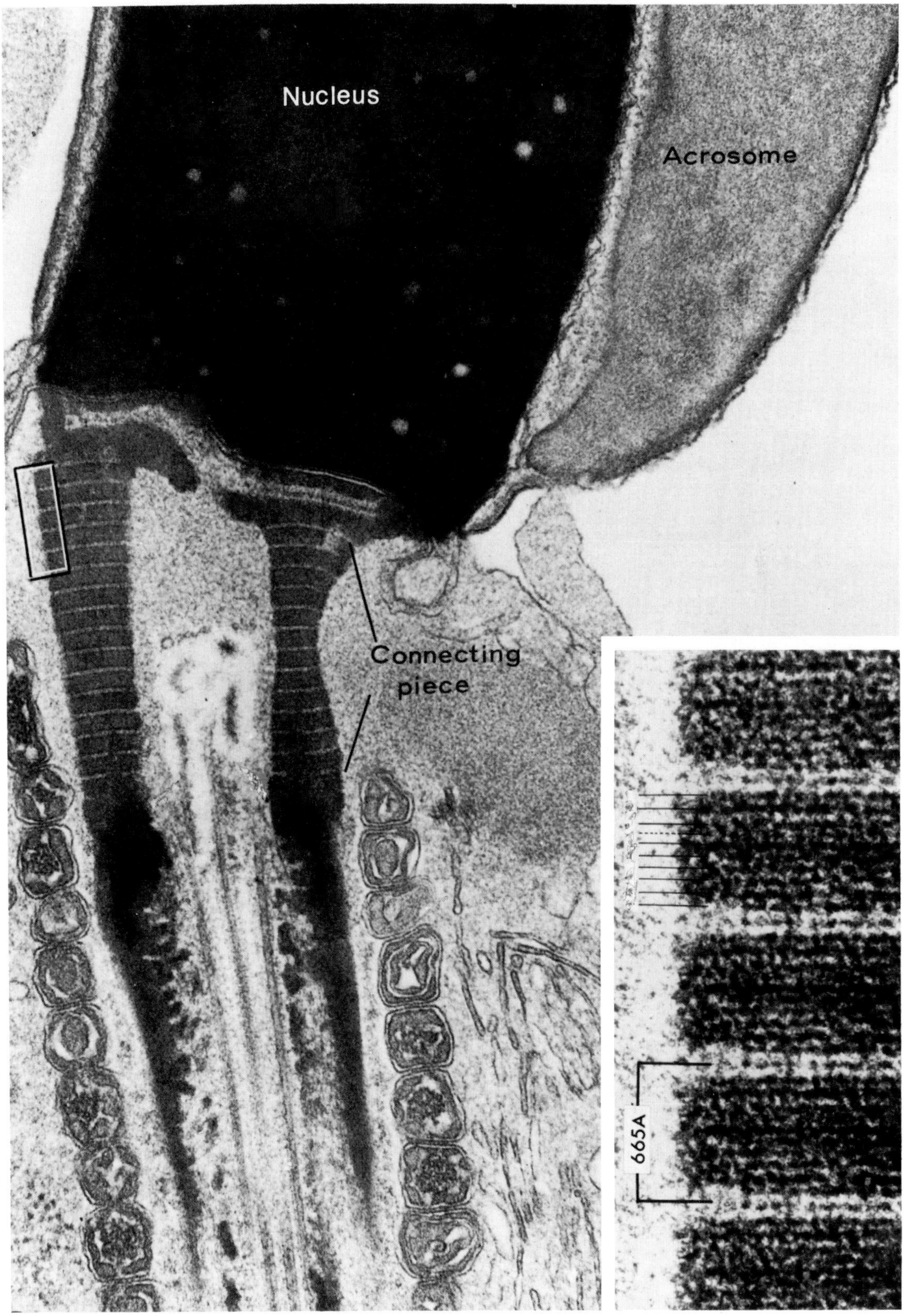

Figure 315

REFERENCES

Centrioles

Anderson, R. G. W. and R. M. Brenner. The formation of basal bodies in the Rhesus monkey oviduct. J. Cell Biol. 50:10–34, 1971.

Berns, M. W., J. B. Ratner, S. Brenner and S. Meredith. The role of the centriolar region in animal cell mitosis. A laser microbeam study. J. Cell Biol. 72:351–367, 1977.

Bessis, M., J. Briton-Gorins and J. P. Thiery. Centriole, corps de Golgi, et aster des leucocytes. Rev. Hemat. 13:363–386, 1958.

Brinkley, B. R. and E. Stubblefield. Ultrastructure and interaction of the kinetochore and centriole in mitosis and meiosis. *In* Advances in Cell Biology (D. M. Prescott, ed.), Vol. 1, Appleton-Century-Crofts, New York, 1970.

Cande, W. Z., J. Snyder, D. Smith, K. Summers and J. R. McIntosh. A functional mitotic spindle prepared from mammalian cells in culture. Proc. Nat. Acad. Sci. 71:1559–1563, 1974.

Dippell, R. V. Effects of nuclease and protease on the ultrastructure of Paramecium basal bodies. J. Cell Biol. 69:622–637, 1976.

Fawcett, D. W. and D. M. Phillips. The fine structure and development of the neck region of the mammalian spermatozoon. Anat. Rec. 165:153–184, 1969.

Fawcett, D. W. Observations on cell differentiation and organelle continuity in spermatogenesis. *In* The Genetics of the Spermatozoon (R. A. Beatty and S. Gluecksohn-Waelsch, eds.), pp. 37–68, Bogtrykkeriet Forum, Copenhagen, 1971.

Flavell, R. A. and I. G. Jones. DNA from isolated pellicles of Tetrahymena. J. Cell Biol. 9:719–726, 1971.

Fulton C. Centrioles. *In* Origin and Continuity of Cell Organelles (J. Reinert and H. Ursprung, eds.), Springer-Verlag, New York, 1971.

Gould, R. R. and G. G. Borisy. The pericentriolar material in Chinese hamster ovary cells nucleates microtubule formation. J. Cell Biol. 73:601–615, 1977.

Hartman, H., J. D. Puma and T. Gurney. Evidence for the association of RNA with the ciliary basal bodies of Tetrahymena. J. Cell Sci. 16:241–259, 1974.

de Harven, E. and W. Bernhard. Etude au microscope electronique de l'ultrastructure du centriole chez les vertebres. Z. Zellforsch. Mikr. Anat. 45:378–398, 1956.

Heidemann, S. R., G. Sander and M. W. Kirschner. Evidence for a functional role of RNA in centrioles. Cell 10:337–350, 1977.

Lindemann, C. B. and R. Rikmenspoel. Sperm flagella: Autonomous oscillations of the contractile system. Science 175:337–338, 1972.

McGill, M. and B. R. Brinkley. Human chromosomes and centrioles as nucleating sites for the in vitro assembly of microtubules from bovine brain tubulin. J. Cell Biol. 67:189–199, 1975.

Pickett-Heaps, J. D. The evolution of the spindle apparatus: An attempt at comparative ultrastructural cytology in dividing plant cells. Cytobios 1:257–280, 1969.

Pickett-Heaps, J. D. The autonomy of the centriole: Fact or fancy? Cytobios 3:205–214, 1971.

Pickett-Heaps, J. D. Aspects of spindle evolution. Ann. N. Y. Acad. Sci. 253:352–361, 1975.

Randall, J. T. and C. Disbrey. Evidence for the presence of DNA in basal body sites in *Tetrahymena pyriformis*. Proc. Roy. Soc. Ser. B. 162:473–491, 1965.

Seaman, G. R. Large-scale isolation of kinetosomes from the ciliated protozoon *Tetrahymena pyriformis*. Exp. Cell Res. 31:292–302, 1960.

Stubblefield, E. and B. R. Brinkley. Architecture and function of the mammalian centriole. Symp. Int. Soc. Cell Biol. 6:175–218, 1967.

Sorokin, S. Reconstruction of centriole formation and ciliogenesis in mammalian lungs. J. Cell Sci. 3:207–230, 1968.

Steinman, R. M. An electron microscopic study of ciliogenesis in developing epidermis and trachea in embryos of Xenopus. Am. J. Anat. 122:19–55, 1968.

Stockinger, L., and E. Cirelli. Ein bisker unbekannte Art der Zentriolenvermehrung. Z. Zellforsch. Milr. Anat. 68:733–740, 1965.

Woolley, D. M. and D. W. Fawcett. The degeneration and disappearance of the centrioles during the development of the rat spermatozoon. Anat. Rec. 177:289–302, 1973.

CILIA AND FLAGELLA

The minute motile processes that occur on some cell types are designated by different names depending upon their length, the character of their movement, and the number per cell. When there are only one or two and they are long in relation to the size of the cell, they are called *flagella*. When they are shorter and numerous, they are called *cilia*. Cilia are 6 to 10 μm long, whereas protozoan flagella range up to 150 μm in length and some sperm tails are several times this length. Flagella beat independently and exhibit an *undulating* motion that results from the propagation of successive waves of bending from base to tip. Cilia have an oscillating to-and-fro motion in which they are relatively straight and stiff during the forward or effective stroke and more curved and whip-like during the return stroke.

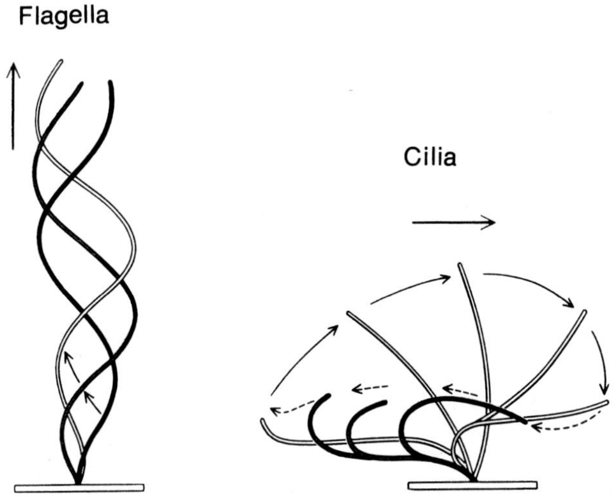

Diagram illustrating the difference in pattern of movement of cilia and flagella (From "How Cilia Move," P. Satir. Copyright October 1974 by Scientific American, Inc. All rights reserved).

The beating of cilia is highly coordinated and the rate varies on different biological material from 10 to 30 per second. All of the cilia on the same cell may beat in phase, *isochronal rhythm,* or in an out-of-phase *metachronal rhythm*. In the latter, which is characteristic of ciliated epithelia, successive cilia along each row in the direction of beat are slightly more advanced in their cycle than the preceding cilium. This results in appearance of waves that sweep over the epithelial surface like the waves that move before the wind across a field of wheat, except that for cilia, the metachronal waves are perfectly regular in their occurrence and they move at a uniform rate. It is not clear whether this is due to sequential activation of the cilia or is imposed upon them by external hydrodynamic and viscous forces of the fluid medium.

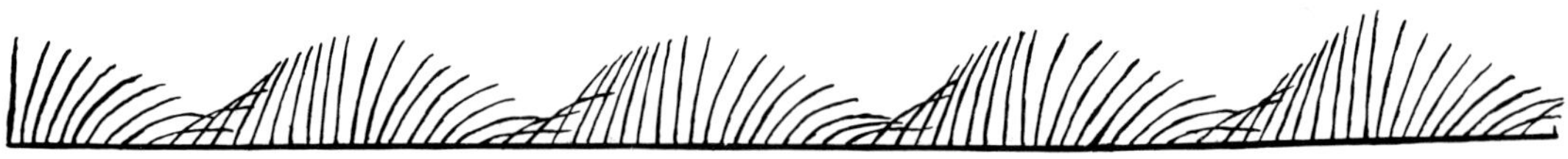

Illustration of the relative positions of successive cilia along the rows resulting in the metachronal wave observed on some ciliated epithelia. (From Fawcett, D. W. Cilia and Flagella. in The Cell, vol. 2., J. Brachet and A. E. Mirsky, eds., Academic Press, New York, 1961.)

Cilia are essential to locomotion in a large class of free living protozoa and to feeding in sessile protozoa. In multicellular invertebrates, they are involved in such diverse physiological processes as alimentation, circulation, and respiration. Flagella propel a large class of protozoa, and flagellate male gametes are involved in reproduction throughout most of the animal kingdom and in some primitive plants.

Few cellular activities have proved more fascinating to cytologists than ciliary and flagellar motion. The occurrence of a continuous flickering movement on the surface of epithelia of the respiratory tract produced by cilia was first described by Purkinje and Valentin in 1834. Soon thereafter Sharpey (1836) established that the direction of beat of tracheal cilia is always toward the pharynx.

Although with the light microscope cilia appeared devoid of internal structure, Engelmann (1868) postulated that their contractile protoplasm was probably composed of longitudinally oriented fibrillar elements that shortened into globular form during the active phase of the beat. The first actual observation of fibrillar components in motile cell processes is attributed to Jensen (1887) and Ballowitz (1888), who described fraying of the tips of sperm tails into a number of fibrils when the flagella were disrupted by pressure on the coverglass. Dellinger (1909) confirmed the presence of several longitudinal internal fibrils by microdissection of protozoan cilia, and fibrils were reported in stained intact epithelial cilia by Korschekov (1923) and Grave and Schmitt (1925).

Motile microorganisms were among the first biological materials examined when the electron microscope was developed. Disrupted protozoa and spermatozoa were simply dried onto a Formvar substrate. Crude though this method of preparation was, it permitted Harvey and Anderson in 1943 to reaffirm the presence of internal fibrils in cilia and flagella. Studying disrupted sperm flagella with slightly improved technique, Hodge (1949) found that the number of fibrils was always 11 and that two of these differed from the other nine in appearance and stability. In a series of comparative studies on plant cilia, Manton and Clarke (1950–1952) also found 11 fibrils and concluded that this number was probably of universal occurrence in plant cilia. They published in 1952 a three-dimensional diagram of the probable arrangement of components based upon their extensive experience with images of disrupted cilia. In this reconstruction, nine fibrils were evenly spaced around a central pair. The nine peripheral fibrils were depicted as though tubular and each divided by a radially oriented septum. Two similar fibrils in the axis of the cilium were enclosed in a membrane and surrounded by a slender helically wound fiber, forming a sheath around the central pair.

Technical progress in fixation, embedding, and microtomy of tissues for electron microscopy in the early 1950s made it possible for Fawcett and Porter (1954) to study epithelia from a wide range of animal species in ultrathin sections. Cross sections of cilia substantiated the general features of Manton's hypothetical reconstruction of their internal organization. From invertebrates to humans, the number of fibrils was 11, with

2 in the center and nine around the periphery. The outer fibrils appeared hollow and had a double-barreled, figure-of-eight cross section. The two central fibrils were single and had a circular cross section. As methods of fixation and staining improved, additional components were resolved. Afzelius (1959) and Gibbons and Grimstone (1960) drew attention to the fact that the subunits of the doublets were unequal in size. The smaller designated subfiber A was slightly closer to the axis of the cilium than the larger, called subfiber B. Subunit A had a circular cross section, subunit B a C-shaped cross section closed by a sector of the wall of subunit A. Two small diverging projections, or *arms,* extended from the wall of subfiber A toward the next doublet in the row (Afzelius, 1959), and these were found to project clockwise from the point of view of an observer looking along the cilium from base toward tip. A radial pattern of slender *spokes* extends from the sheath around the central pair to subfiber A of each doublet.

A return to the study of dissociated cilia and flagella using negative staining with phosphotungstate (Pease, 1963; Andre and Thiery, 1963) showed that the contrast medium penetrated into the interior of the axonemal elements, thus establishing that the "fibrils" are in fact microtubules. Their cylindrical wall was seen to be made up of about 13 beaded *protofilaments* 3.5 nm in diameter with a repeating axial periodicity of about 8 nm. In the meantime, microtubules with similar substructure had been observed in the cytoplasm of many cell types (Ledbetter and Porter, 1963) and it was realized that the elements comprising the axoneme of cilia and flagella were an unusually stable form of a tubular structural component of nearly universal occurrence in cells of plants and animals.

Investigation of the mechanism of ciliary motion required a model system that could be subjected to experimental manipulation *in vitro.* This was provided by Hoffman-Berling (1950) who showed that ciliated cells extracted with glycerin were killed but that addition of ATP resulted in reactivation of their cilia. It was subsequently shown by Child and Satir that glycerination removes the ciliary membrane but leaves the axonemes intact, suggesting that the principal function of the membrane is to maintain around the axoneme the appropriate concentration of ions and ATP to sustain its function. In more recent refinements of this approach, the membrane is removed with detergent, leaving the axonemes accessible to experimental alterations in their environment. On addition of ATP to the bare axonemes of sea urchin spermatozoa, they are induced to swim like living sperm (Gibbons, 1963). Thus, as in other cellular processes, the energy required for ciliary motion appeared to depend upon the hydrolysis of ATP to ADP and phosphoric acid. In further studies on the biochemical dissection of flagella, Gibbons showed that the arms of the doublets could be selectively removed by sequestering magnesium ion with a chelating agent. The protein extracted, called *dynein,* was shown to have ATPase activity (Gibbons and Rowe, 1965) and the axonemes deprived of arms were no longer capable of movement on addition of ATP. When bare axonemes were exposed to dynein in the presence of Mg^{+} ions, the arms were restored to the doublets and ATPase activity was again demonstrable in the axonemes.

Skeletal muscle was long believed to shorten as a result of a conformational change in the molecules of its contractile proteins, but it was shown by Huxley and Hanson (1954) that muscle contraction takes place by sliding of interdigitating sets of actin and myosin filaments of constant length. Similarly, it was long assumed that the fibrillar components of the ciliary axoneme were capable of shortening. But if the ciliary fibrils were contractile, those on the concave side during the effective stroke would be expected to shorten. In electron microscopic studies on the tips of cilia in different phases of the beat, Satir (1966) found that the fibrils on this side extended farther into the tips. It was therefore evident that bending occurred without shortening of the fibrils. It was concluded that ciliary and flagellar motion involves a sliding-microtubule mechanism comparable to the sliding-filament mechanism of muscular contraction.

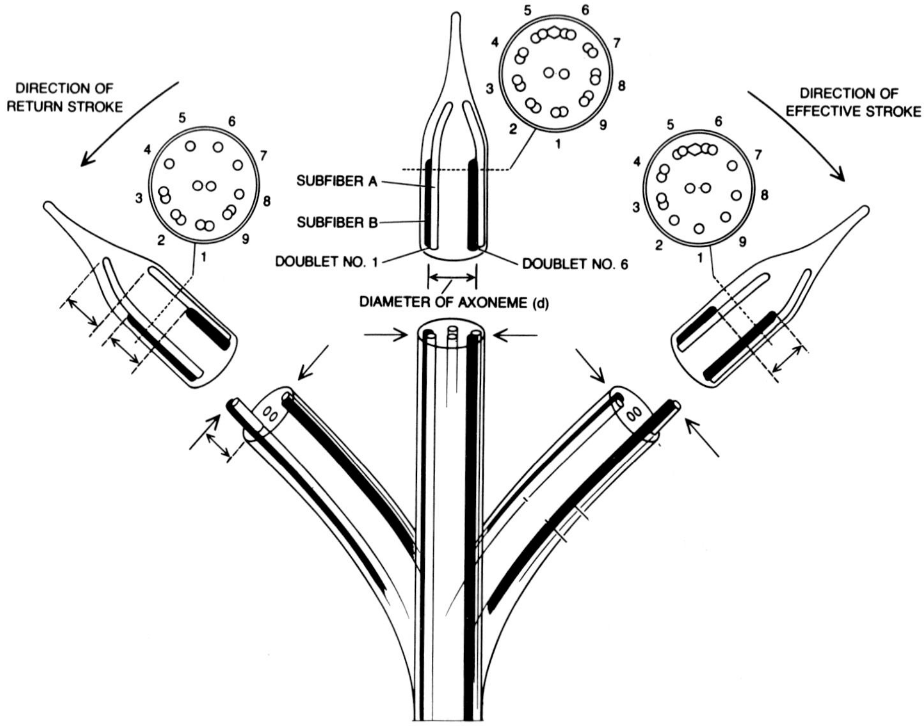

Diagram illustrating the relative positions of the ends of the axonemal doublets at the tip of a cilium in different phases of its beat. If movement were produced by microtubule shortening those on the concave side of the bend should be shorter. Instead there is no change in length of the doublets in different phases of the cycle – a finding consistent only with a sliding microtubule mechanism. (Slightly modified from P. Satir, J. Cell Biol. 26:805–834, 1965.)

More direct evidence was provided by ingenious experiments of Summers and Gibbons (1971). Segments of flagellar axonemes were briefly exposed to trypsin to release the doublets from their attachment to the central sheath. Upon the addition of ATP to this preparation, tubules rapidly emerged from opposite ends of the axonemal segments as a result of active sliding movements between the doublets. The trypsin treatment evidently digested the radial spokes and interdoublet nexin links, uncoupling the sliding of the doublets from axonemal bending and permitting ATP-energized sliding to propel microtubules out of the ends of the segments of axoneme.

Clinical observations on a congenital form of human infertility have now validated the assumption that dynein is essential for motility of cilia and flagella. A disease syndrome described by Kartagener (1933) is characterized by chronic sinusitis, bronchiectasis, situs inversus viscerum, and male infertility. The ejaculate of such a patient, analyzed by Pedersen and Rebbe (1975), showed normal numbers of spermatozoa but no motility. When the spermatozoa were examined with the electron microscope, they exhibited an absence of arms on the doublets of the axoneme. Biopsy of ciliated cells of the respiratory tract of several such patients has shown a similar defect in the cilia. An absence of ciliary and flagellar motion accounts for the chronic sinusitis, bronchiectasis, and male infertility in these patients, who appear to have an inherited defect in synthesis of dynein (Afzelius et al., 1975; Eliasson et al., 1977).

Current research is concentrated on the structure, composition, and function of the radial spokes and nexin links. It seems likely that these components provide the shear resistance within the axoneme that converts sliding of microtubules into bending. Consistent with this interpretation is the observation that in longitudinal sections of straight regions, the spokes are perpendicular to the doublet microtubules and in bent regions they are oblique as might be expected if they were attached to the doublet at one end and to the central sheath at the other when the doublet was displaced by sliding (Warner and Satir, 1974).

The differences in size and shape of cilia and microvilli are well illustrated by scanning micrographs of the lumenal surface of the epithelium lining the mammalian oviduct. The tufts of cilia associated with individual ciliated cells project several microns above the convex apices of nonciliated cells covered with short microvilli. The number of ciliated cells in this epithelium is under hormonal control by estrogens.

Figure 316. Scanning micrograph of the epithelium of the human oviduct, at mid cycle. (Micrograph from Gaddum-Rosse, Blandau and Tiersch, Am. J. Anat. 138:269–275, 1973.)

Figure 316

The inconsistent orientation of the cilia in scanning micrographs does not accurately reflect their state in the living cell. All of the cilia on a cell would normally be in the same phase of their cycle of beating. Routine methods of fixation are incapable of instantly immobilizing them in the attitude typical of their effective stroke or recovery phase. They therefore tend to come to rest in a more or less straight neutral position.

Figure 317. Scanning micrograph of human oviductal epithelium. (Micrograph courtesy of Penelope Gaddum-Rosse.)

Figure 317

On cells of protozoa and other invertebrates, the cilia are often closely spaced in precisely ordered rows with no intercalated microvilli and little intervening space. The doublet microtubules of the axoneme have two short diverging *arms* that project from subunit A, toward subunit B of the adjacent doublet. Radial *spokes* extend inward from subunit A toward the central pair of microtubules. The central pair are joined at intervals by short *bridges*. The orientation of the central pair is the same in all cilia on the same cell and the direction of beat is perpendicular to a line joining their centers.

Figure 318. Lateral cilia of a gill from the mussel, *Anodonta cataracta.* (Micrograph from Ian Gibbons, Biophys. Biochem. Cytol. 11:179–205, 1961.)

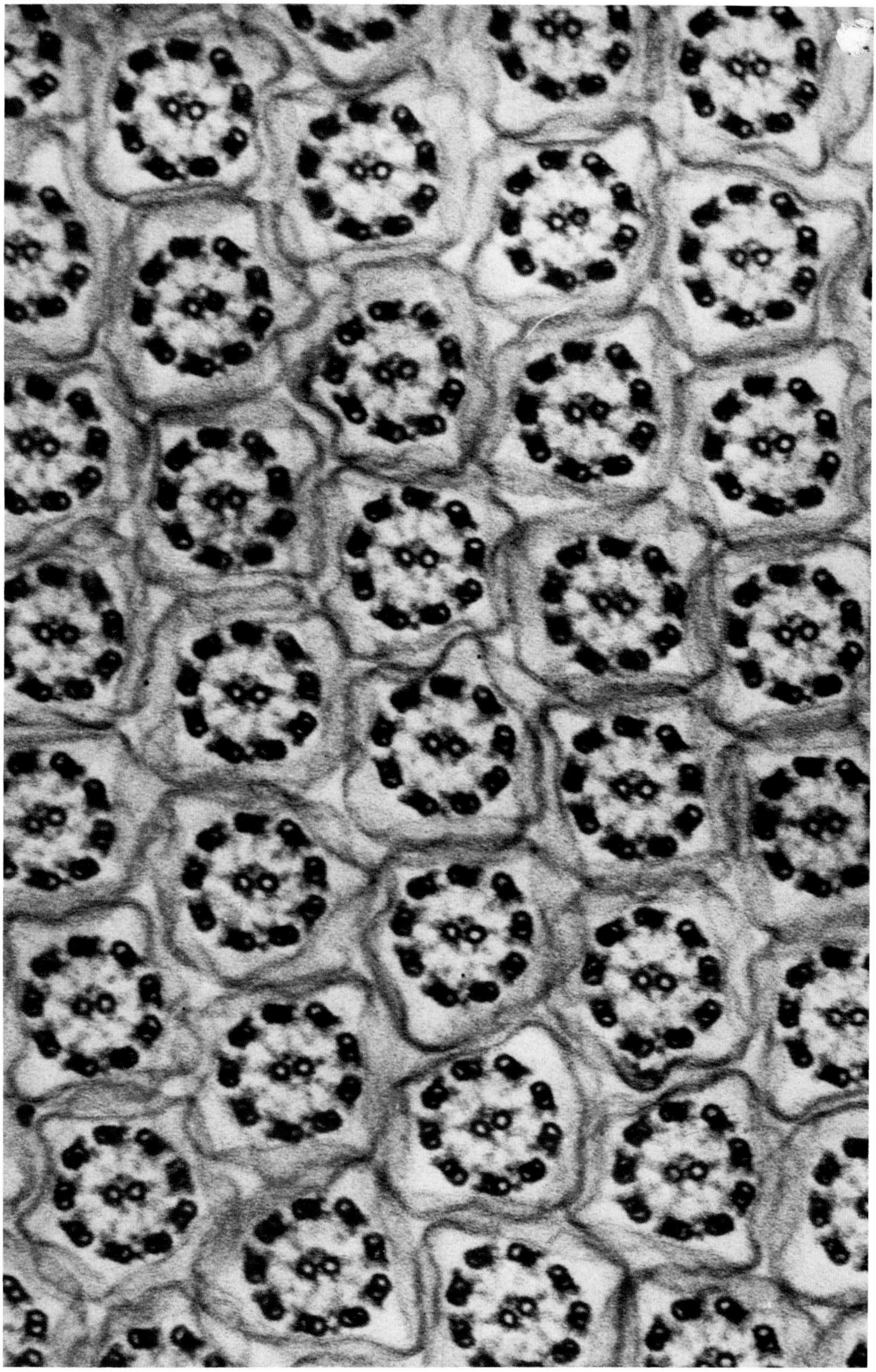

Figure 318

The ciliated cells of mammals often have microvilli interspersed among the cilia. These transverse sections near the cell surface permit a comparison of the size and internal structure of the two types of cell process. Both are limited by a trilaminar unit membrane and have a cytoplasmic matrix of similar density. The nonmotile microvilli have no resolvable internal structure, whereas the motile cilia have a consistent pattern of nine doublet microtubules uniformly spaced around a central pair of singlet microtubules. This nine plus two complex was visualized by the early light microscopists as a single axial fiber and was therefore called the *axoneme*. The term has been retained even though it does not accurately describe the microtubule complex revealed by the electron microscope.

Figure 319. Epithelial cilia and microvilli from rat trachea, fixed in aldehyde and osmium tetroxide and exposed to low molecular weight galloyl-glucoses (tannic acid). (Micrograph from Simionescu and Simionescu, J. Cell Biol. 70:608–621, 1976.)

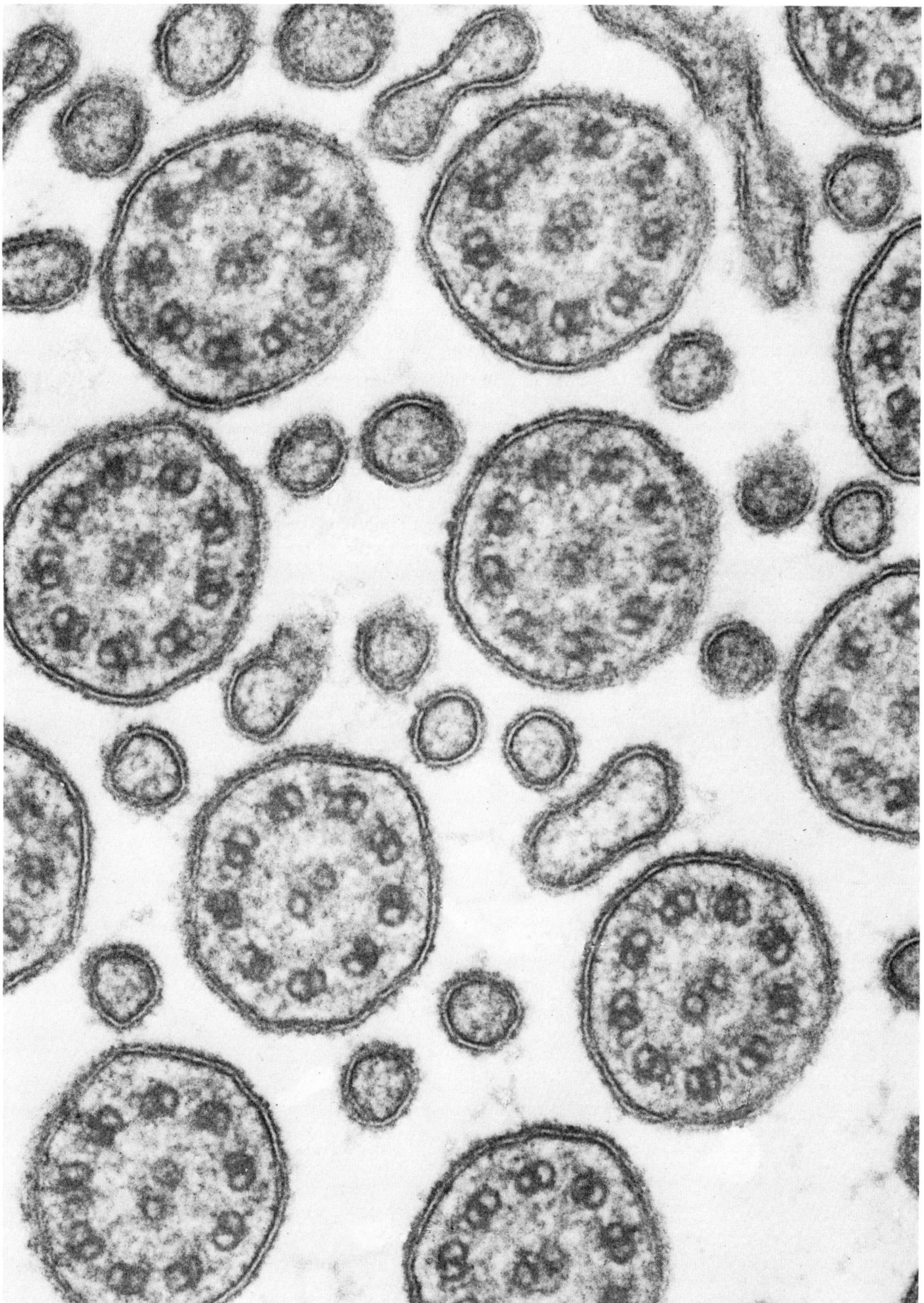

Figure 319

MATRIX COMPONENTS OF CILIA

The microtubules of the axoneme are adequately preserved by routine fixation for electron microscopy of thin sections. These methods are less satisfactory for their subunits and appendages and have contributed little to our understanding of the organization of the ciliary matrix. Addition of tannic acid to the fixative results in an apparent thickening and a more intense staining of these components. The dynein arms are made more conspicuous, the radial spokes more robust, and the unstained protofilaments in the wall of the microtubules are clearly revealed in negative image.

The upper figure on the facing page is a diagrammatic representation of components of the axoneme and matrix identifiable in high magnification micrographs of tannic acid–treated cilia such as illustrated in the lower figure. The nexin links depicted are seldom clearly visualized. The foreshortening of the radial spokes in the diagram reflects the prevailing uncertainty as to the exact site of linkage of the spoke heads to the central microtubule complex. The central sheath is interpreted by some investigators as a thin filament wound helically around the central pair, but it usually appears in cross-sections as two symmetrical arciform elements joining the walls of the microtubules instead of a circumferential sheath. The distance between the central microtubules is somewhat exaggerated in the diagram and the connecting bridge is therefore longer than it appears in micrographs. Negatively stained preparations of disrupted cilia have proved more informative than sections for visualizing the bridges between the central pair and the appendages of the doublets (see Fig. 320), but the linkage of the spokes to the central microtubule complex is broken in dissociation of the axonemes.

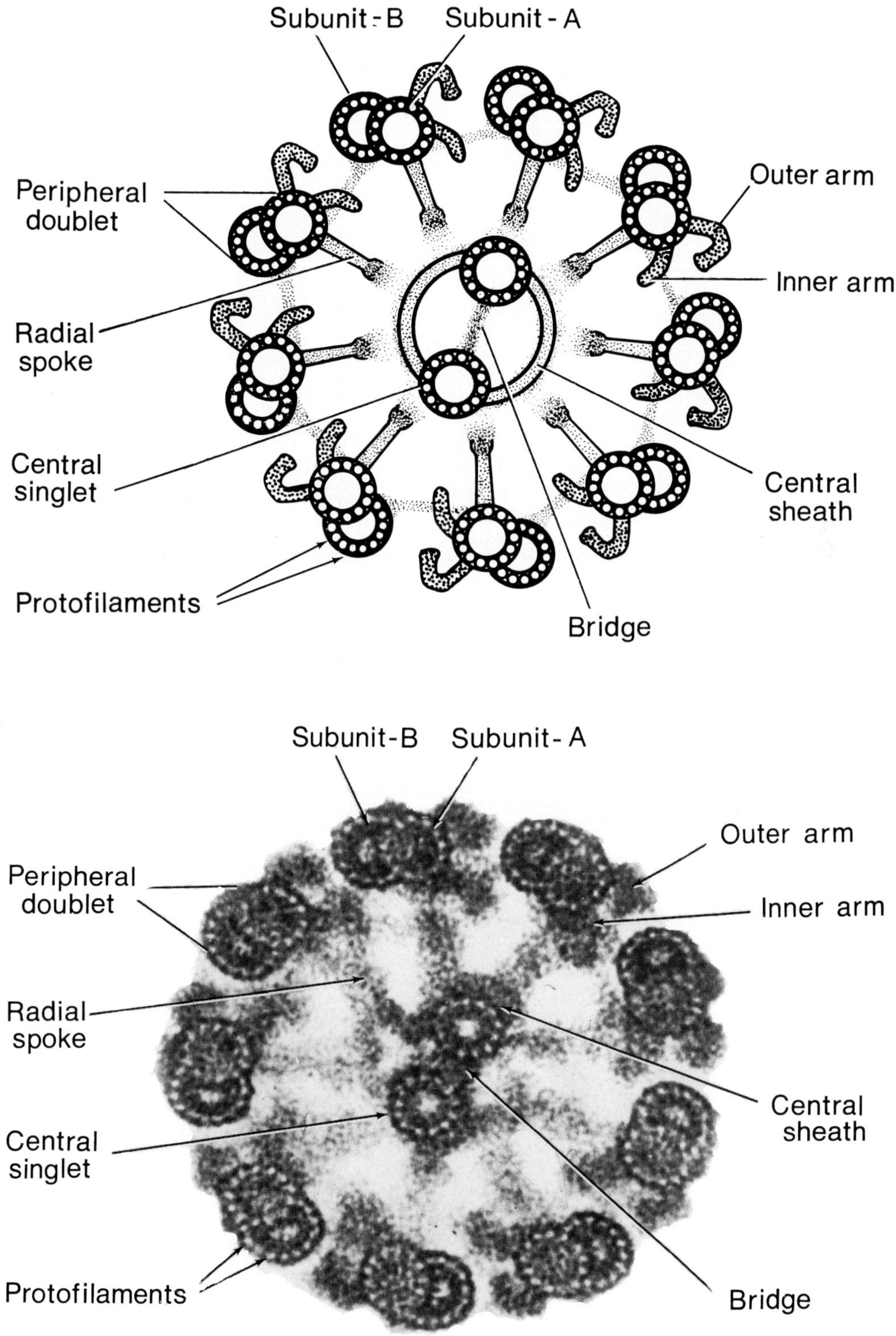

Diagram of components of the axoneme recognizable in cross sections, such as that shown below, after aldehyde and tannic acid fixation.

The disposition of the dynein arms along the length of the doublets is studied to best advantage in dissociated cilia examined with negative staining. The upper figure (*A*) shows parts of three overlapping doublets from a demembranated cilium disintegrated in the presence of ATP. Figure 320*B* is a reinforced image of the same micrograph produced by a two-step linear translation. Free dynein arms can be seen projecting from the lowermost doublet at intervals of 24 nm along its length. Somewhat less clearly shown are arms bridging the interspaces between the three doublets. During active sliding, the space between doublets is considerably less than that usually seen in electron micrographs of thin cross sections. The microtubule sliding responsible for ciliary bending is believed to be due to attachment of arms on subunit A of one doublet to subunit B of the adjacent doublet followed by a conformational change resulting in arm deflection.

Figure 321*A* is a thin longitudinal section of a cilium showing the radial spokes joining subunit A of a doublet to a row of projections along one of the central pair of microtubules. The spokes occur in groups of three with a repeat of 86 nm. Figure 321*B* is a reinforced image of the same micrograph resulting from linear translation in steps corresponding to 86 nm between successive exposures.

Figure 320. Negatively stained preparation of dissociated axonemal doublets. (From F. D. Warner and D. R. Mitchell, J. Cell Biol. 76:261–277, 1978.)

Figure 321. Longitudinal section of a cilium. (From F. D. Warner and P. Satir, J. Cell Biol. 63:35–63, 1974.)

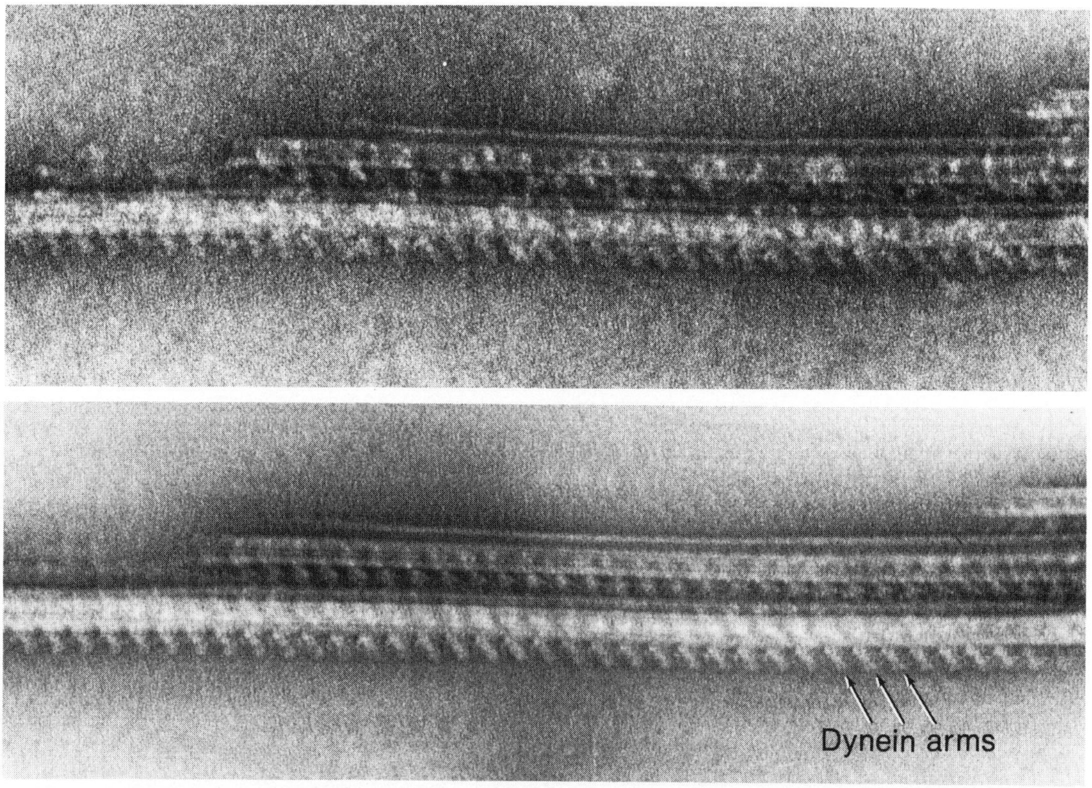

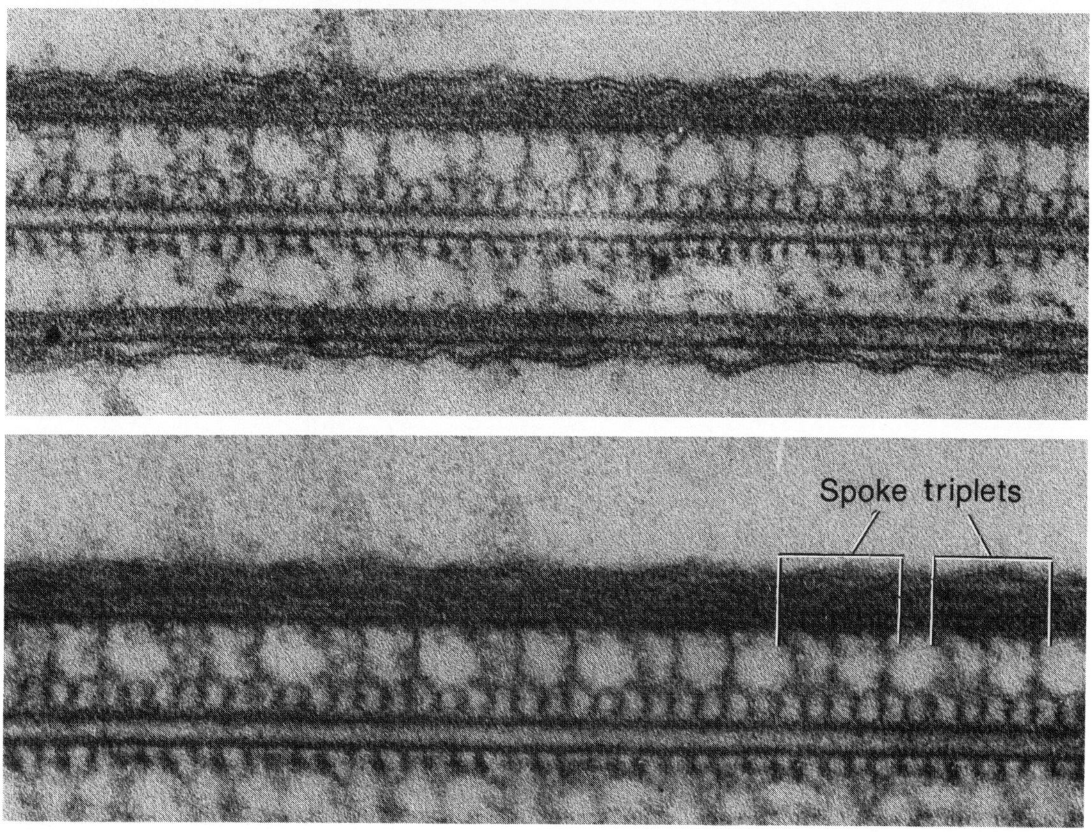

Figure 320, *upper two* **Figure 321**, *lower two*

Our current concept of the complex organization of a doublet microtubule and its appendages is presented in the accompanying diagram based on data from optical diffraction analysis and computer reconstruction of ciliary and flagellar axonemes. The wall of the microtubule is made up of heterodimers of tubulin arranged in tandem to form protofilaments. Two rows of dynein arms project from the A-tubule with an axial repeat of 24 nm. The arms are staggered such that the outer arms are 9 nm nearer the flagellar base than the corresponding inner arm. Three subunits are discernible in each arm. The members of the radial spoke triplets are unevenly spaced 32 nm and 24 nm apart with the longer spacing nearer the flagellar base. The exact sites of attachment of the arms and spokes on the surface lattice of the A tubule have yet to be established.

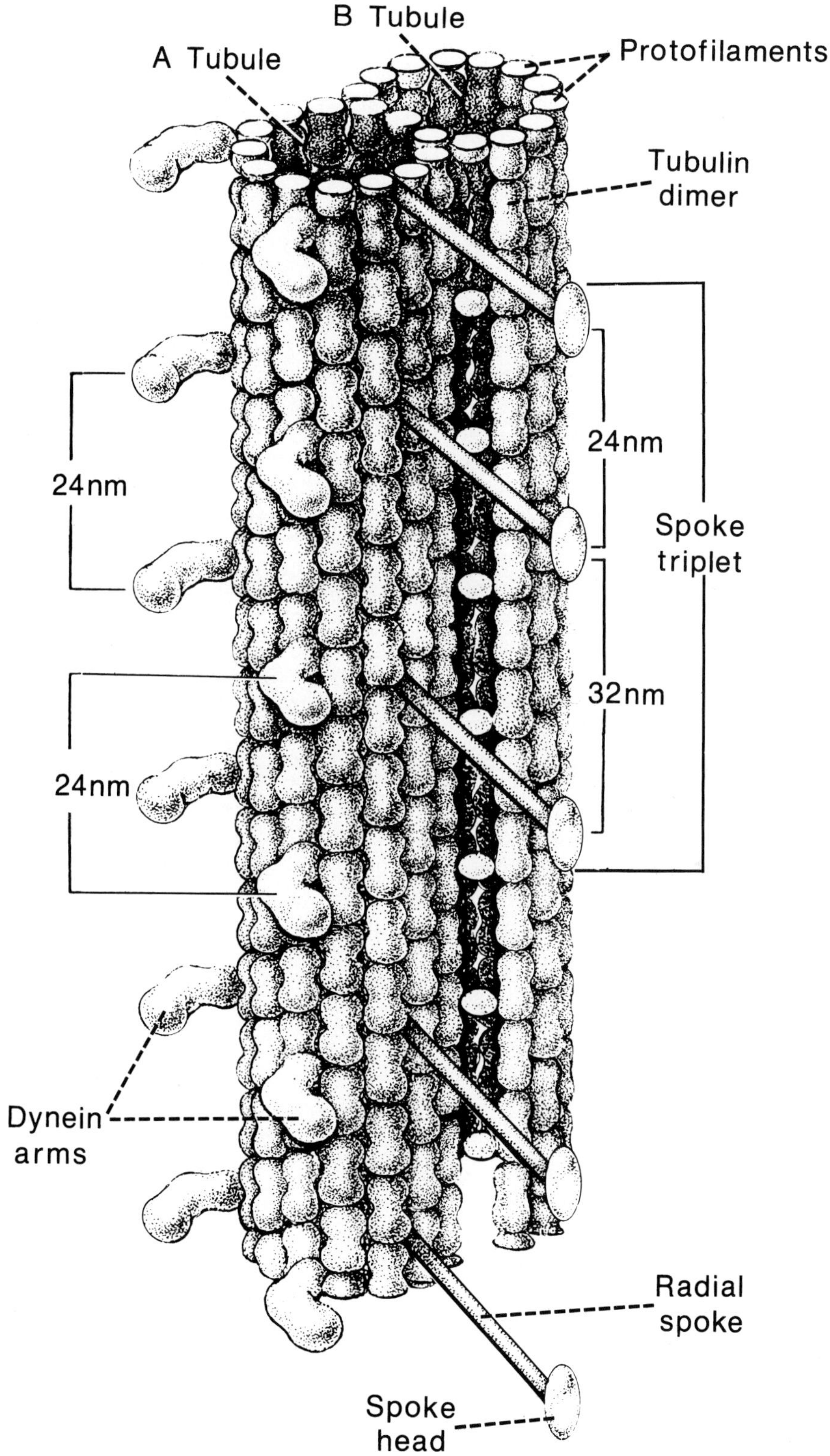

Drawing of an axonemal doublet and its appendages. (From R. Linck, based upon data of Amos, Linck, and Klug, 1971, and Olson and Linck, 1977. Reproduced from *The Spermatozoon,* D. W. Fawcett and M. Bedford, eds., Urban and Schwarzenberg, Inc., Baltimore, Md. 1979.)

ABERRANT SOLITARY CILIA

Cilia and flagella usually occur on motile cells or on the free surfaces of epithelia. However, a solitary cilium is a normal occurrence on the cells lining the intrahepatic bile ducts, the intercalated ducts of the pancreas, the rete testis, and certain segments of the nephron. The functional significance of these processes is not evident, since it seems unlikely that their motility could do more than create local turbulence in the fluid contents of these excretory ducts.

Rudimentary cilia other than those that project from the free surface of epithelia were first observed in electron micrographs of cells in the anterior pituitary (Barnes, 1961) and on fibroblasts and smooth muscle cells of chicks (Sorokin, 1962). They have since been reported in some thirty-six avian and mammalian tissues (Scherft and Daems, 1967) and in several cell lines in tissue culture (Brinkley and Stubblefield, 1967; Rieder et al. 1979; Albrecht-Buehler and Bushnell, 1980).

Especially puzzling is the occurrence of rudimentary cilia on mesenchymal derivatives such as smooth muscle cells and fibroblasts. As illustrated in the accompanying micrograph, these grow out from one of the centrioles in the juxtanuclear Golgi region of the cell. The doublet microtubules of the axoneme terminate a short distance beyond the basal body. The tapering cilium shown here passes out of the plane of section, but its slender terminal portion, devoid of an axoneme, is seen again at the arrow.

Such abortive cilia usually lack the central pair of microtubules and are probably immotile. Because modified cilia are found in photoreceptors and in olfactory epithelia, it has been suggested that these solitary cilia might have a sensory function but there is no evidence to support this speculation. It seems probable that they are merely anomalous, rudimentary structures with no function.

Figure 322. Aberrant cilium on a mesenchymal cell in the interstitium of the testis.

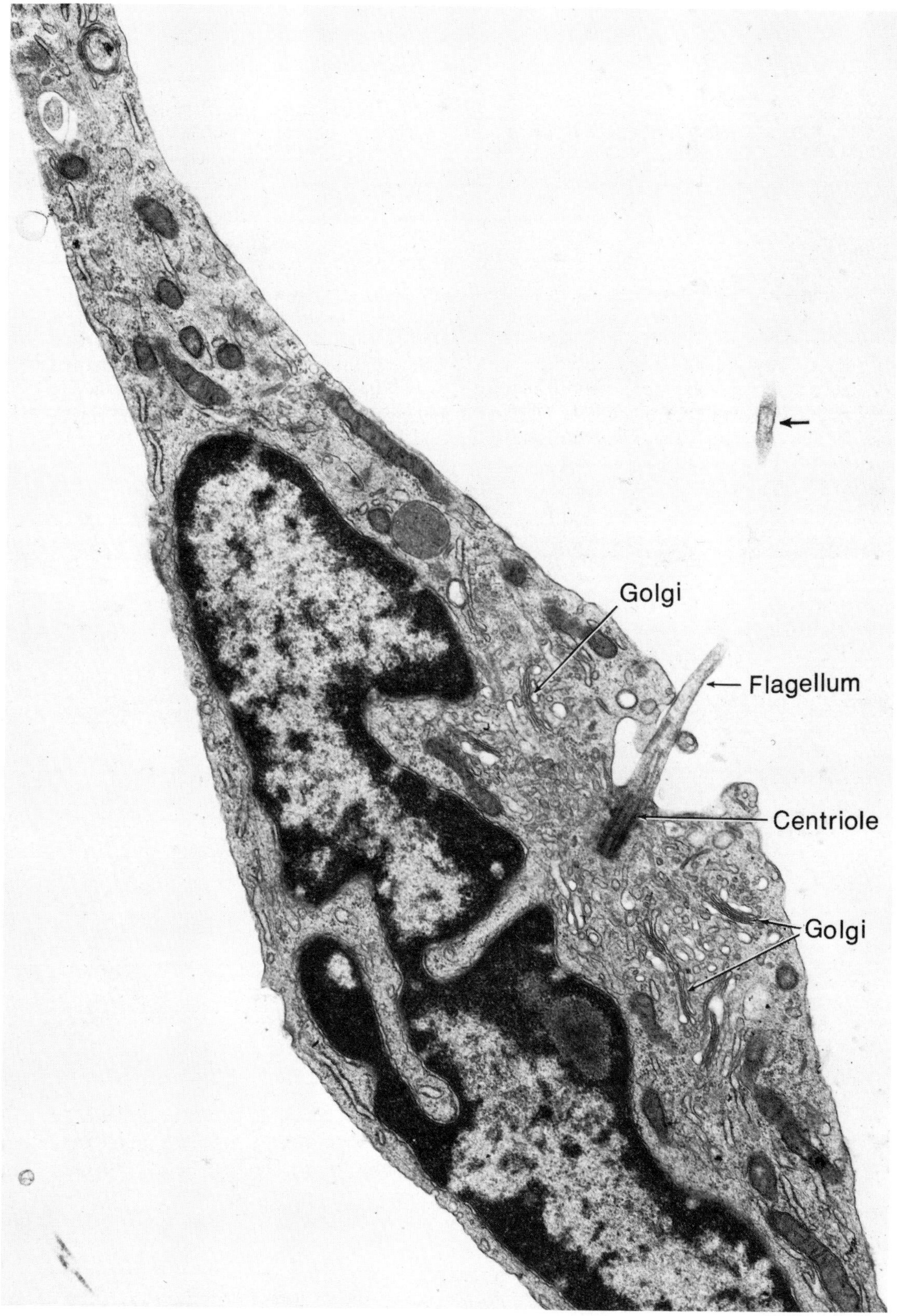

Figure 322

MODIFIED CILIA

The outer segments of the rods and cones of the vertebrate retina and many photoreceptors of invertebrates begin their differentiation as single cilia with a typical basal body and doublet microtubules forming a more or less complete axoneme. In their subsequent development, membranous discs containing the visual pigment accumulate within the confines of the ciliary membrane and become closely packed in parallel array to constitute the outer segment of the photoreceptor cell. Vestiges of the axonemal microtubules (at arrows) persist along one side of the outer segment which is connected to the inner segment by the proximal portion of the shaft of the original cilium.

Figure 323. Longitudinal section of a rod outer segment.

Figure 324. Longitudinal section of a cone outer segment. (Micrographs courtesy of T. Kuwabara.)

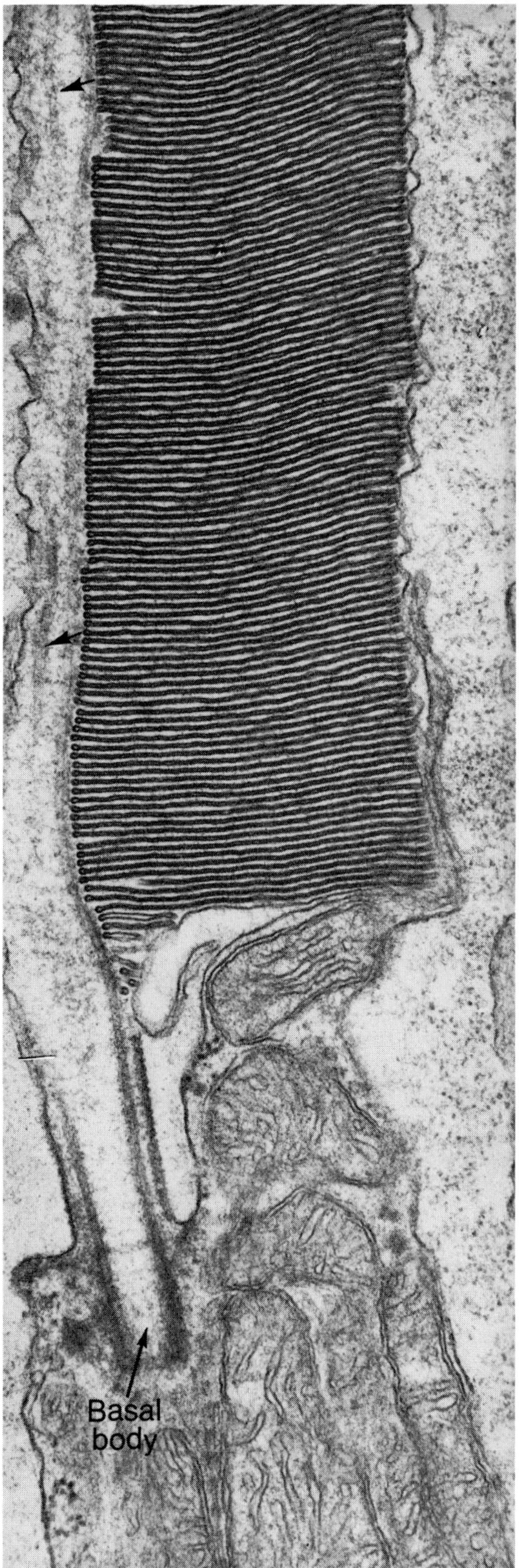

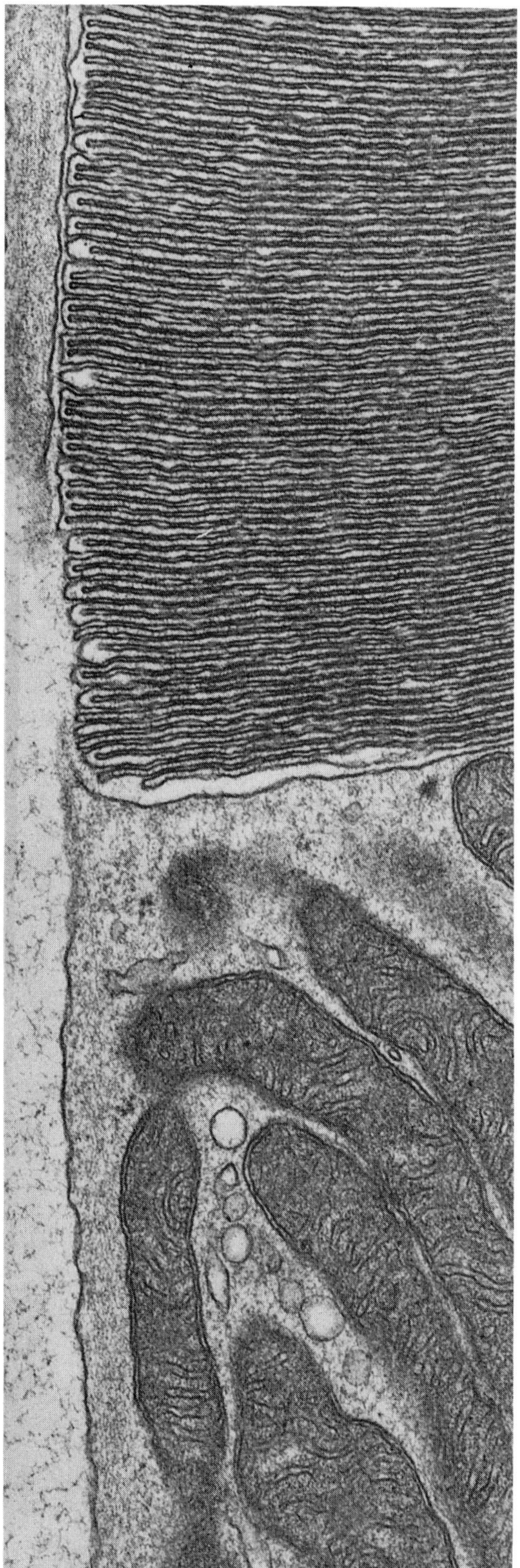

Figure 323, *left* **Figure 324**, *right*

STEREOCILIA

Histologists using the light microscope recognized a category of epithelial cell processes which had the appearance of cilia but showed no motility. These were called *stereocilia* to distinguish them from the motile form, or *kinocilia*. The stereocilia have a rather limited distribution, being found on the epididymal epithelium of mammals, and in certain sense organs, notably the organ of Corti, the crista ampullaris, macula utriculi, and macula sacculi of the mammalian inner ear; the lateral line organs of fish; and the sensilla of insects.

With the electron microscope, the stereocilia in these several organs were found to differ in their fine structure. Those of the epididymis were found to be as long or longer than kinocilia but more slender and quite flexible. They have no obvious internal structure other than an axial bundle of actin filaments which often extends some distance downward into the apical cytoplasm. They differ from microvilli only in their length and greater flexibility. Those of sensory organs are relatively thick, stiff bristle-like appendages, with a shaft that is often narrower at the base than in its more distal portion. The interior is filled with closely packed, longitudinally oriented filaments.

Stereocilia in the mammalian epididymis are slender, flexible processes resembling unusually long microvilli, but they are more variable in their orientation and less closely packed than the microvilli of a brush border. They are assumed to be a device for amplifying the surface of this absorptive epithelium.

Figure 325. Stereocilia of the epithelium of rabbit epididymis. (Micrograph courtesy of Roy Jones.)

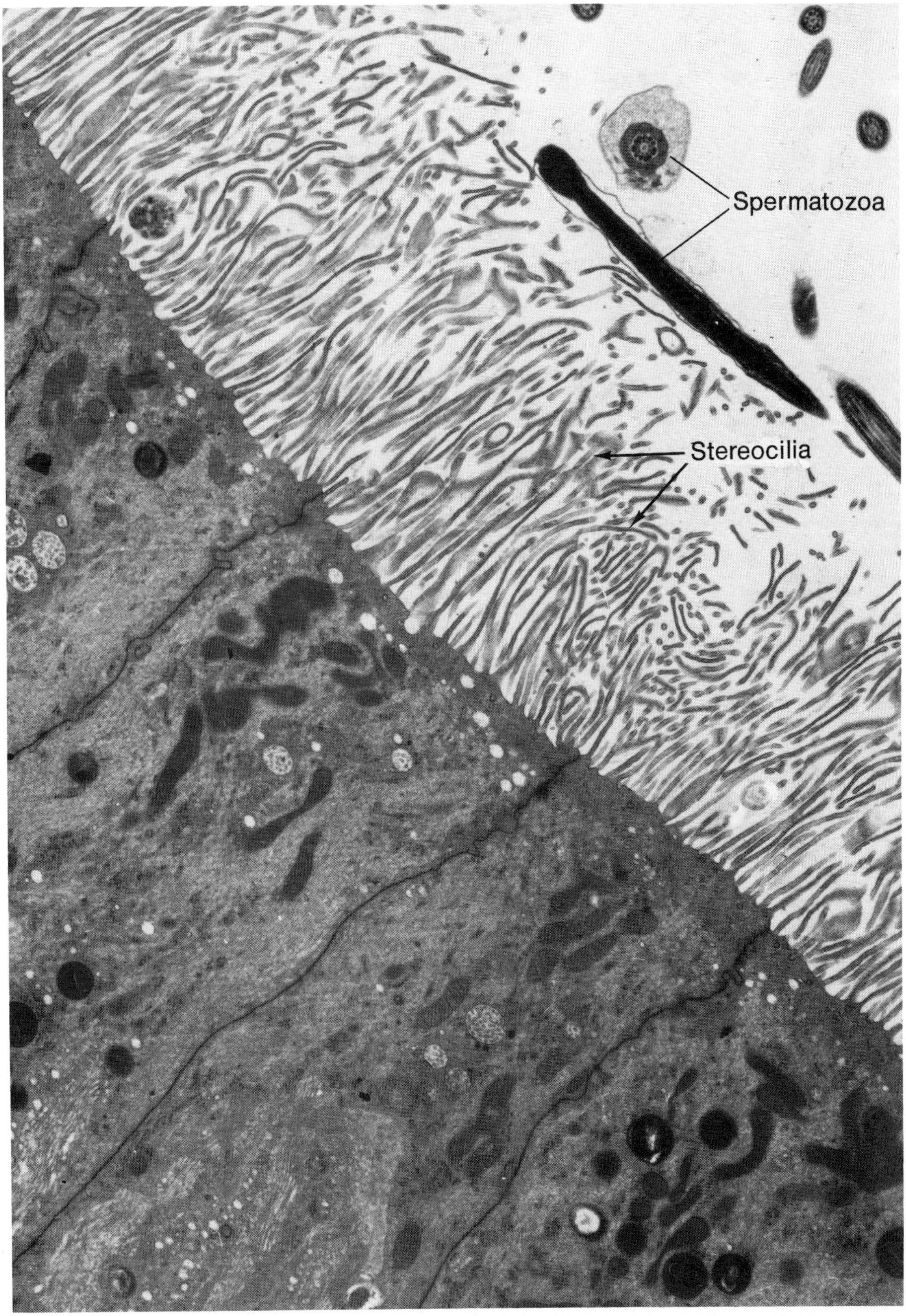

Figure 325

Cells of sense organs detecting vibration are provided with stiff surface projections, the size of cilia but lacking their internal structure. The accompanying low-power scanning micrograph presents a surface view of the organ of Corti in the third turn of the guinea-pig cochlea. Three rows of outer hair cells are shown, each bearing a V-shaped array of sensory hairs. The inset shows at high magnification the hairs on two cells from the first row in the second turn of the organ of Corti. Their relative size can be judged by comparison with the microvilli on neighboring cells. These specializations do not arise from centrioles and never contain microtubules. Therefore, they cannot be regarded as modified cilia, but appear to be a unique specialization for mechanoelectrical transduction.

Figure 326. Scanning micrographs of guinea pig organ of Corti. (Micrographs courtesy of Kogaku Saito and Kiyoshi Hama.)

Figure 326

In longitudinal section, the sensory hairs are narrower at their base and contain a large number of parallel filaments that converge into a dense bundle in the narrow basal region of the hair and then diverge again as they continue into the cuticular plate in the apical cytoplasm.

Figure 327. Longitudinal section of the apex of a hair cell from the saccular macula of the goldfish. (Micrograph courtesy of Kogaku Saito and Kiyoshi Hama.)

References for Cilia and Flagella and Sperm Flagellum are found at the end of the next part, pp. 638–639.

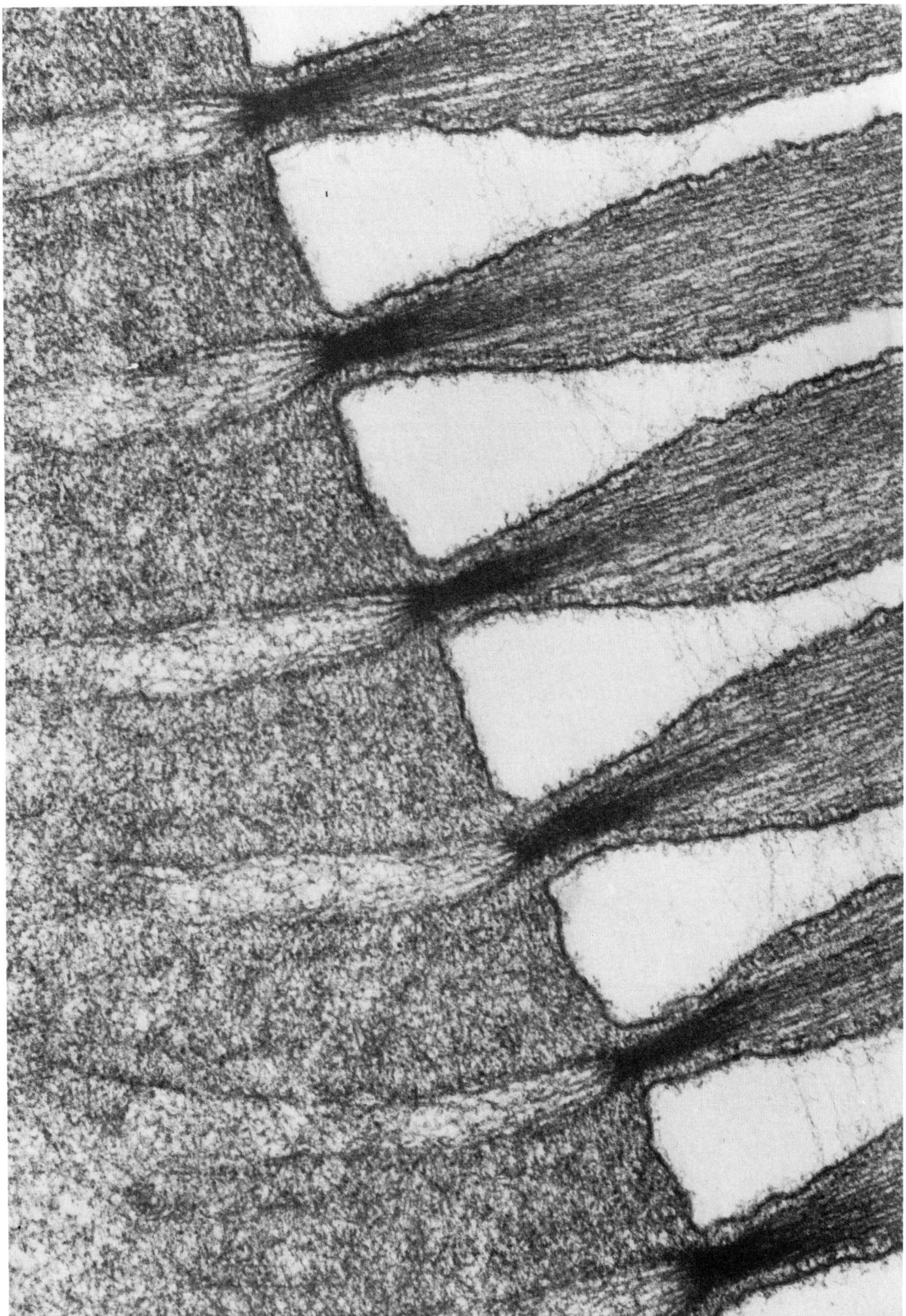

Figure 327

SPERM FLAGELLUM

MAMMALIAN SPERM FLAGELLUM

The spermatozoon was first observed in 1677 by Leeuwenhoek, who described it as progressing by undulating movements of its long tail, like eels swimming in water. Flagella were seen by him and other early microscopists on protozoa. However, no internal structure was resolved in these slender processes until the latter part of the 19th century, when Jensen (1887) and Ballowitz (1890) observed that the tip of sperm flagella occasionally frayed during preparation into a tuft of exceedingly fine fibrils. Little significance was attached to this observation until it was verified in the earliest electron micrographs of dissociated and desiccated mammalian sperm tails (Seymour and Benmosche, 1941; Baylor et al., 1943). Estimates of the number of fibrils ranged from 8 to 24 due to the different degrees of fragmentation and artifacts of drying.

When satisfactory methods of fixation and thin sectioning were developed, there was general agreement on the nine peripheral and two central fibrils of the axoneme (Watson, 1952; Fawcett, 1954) but the mammalian sperm flagellum was found to have an additional set of nine *outer dense fibers* around the axoneme (Challice, 1953; Bradfield, 1953). In transverse section these vary in size and shape, with three of the nine usually being larger than the other six. With the conventional system of numbering, the larger fibrils are numbers 1, 5, and 6. The presence of nine outer fibers in addition to the axoneme (9+9+2) was found to be characteristic of the majority of animal species with internal fertilization, while lower forms that release their sperm directly into sea water retain the more primitive 9+2 pattern of flagellar microtubules. It was speculated that the outer fibers were accessory motor elements that evolved to overcome the greater resistance encountered in the secretions of the female reproductive tract. This interpretation persisted until the outer fibers were isolated and analyzed (Price, 1973; Baccetti et al., 1973). They were found to consist of four polypeptides ranging from 11,000 to 55,000 daltons. Their amino acid profile was unlike that of any known contractile protein, and the high content of cysteine was more suggestive of a scleroprotein like keratin. Therefore the outer dense fibers are now interpreted as passive stiffening elements, but it is not obvious what advantage is conferred by adding resilient elastic components around the axoneme of the sperm flagellum.

A coiled strand wrapped around the proximal part of the sperm tail was described by Jensen (1887), who was among the first to recognize that this was composed of mitochondria. Subsequent investigators were divided among those who regarded it as a continuous sheath formed by end-to-end fusion of mitochondria (Retzius, 1909; Schnall, 1952) and those who insisted that the mitochondria are coiled helically around the axoneme but retain their individuality (Gresson and Zlotnik, 1945; Challice, 1953; Yasuzumi, 1956). Electron microscopy established the latter interpretation as the correct one for mammals, but fusion of the mitochondria is common in reptiles, birds, and invertebrates (Fawcett, 1958). The segment of the tail enclosed by the *mitochondrial sheath* is called the *middle piece,* or *midpiece.* Its length varies greatly from species to species, ranging from about a dozen mitochondrial turns in humans to over 200 in some rodents.

At the distal end of the midpiece is the *annulus (Jensen's ring, terminal ring).* This dense circular structure is firmly attached to the flagellar membrane, where the latter is reflected from the midpiece onto the next segment, called the *principal piece.* The

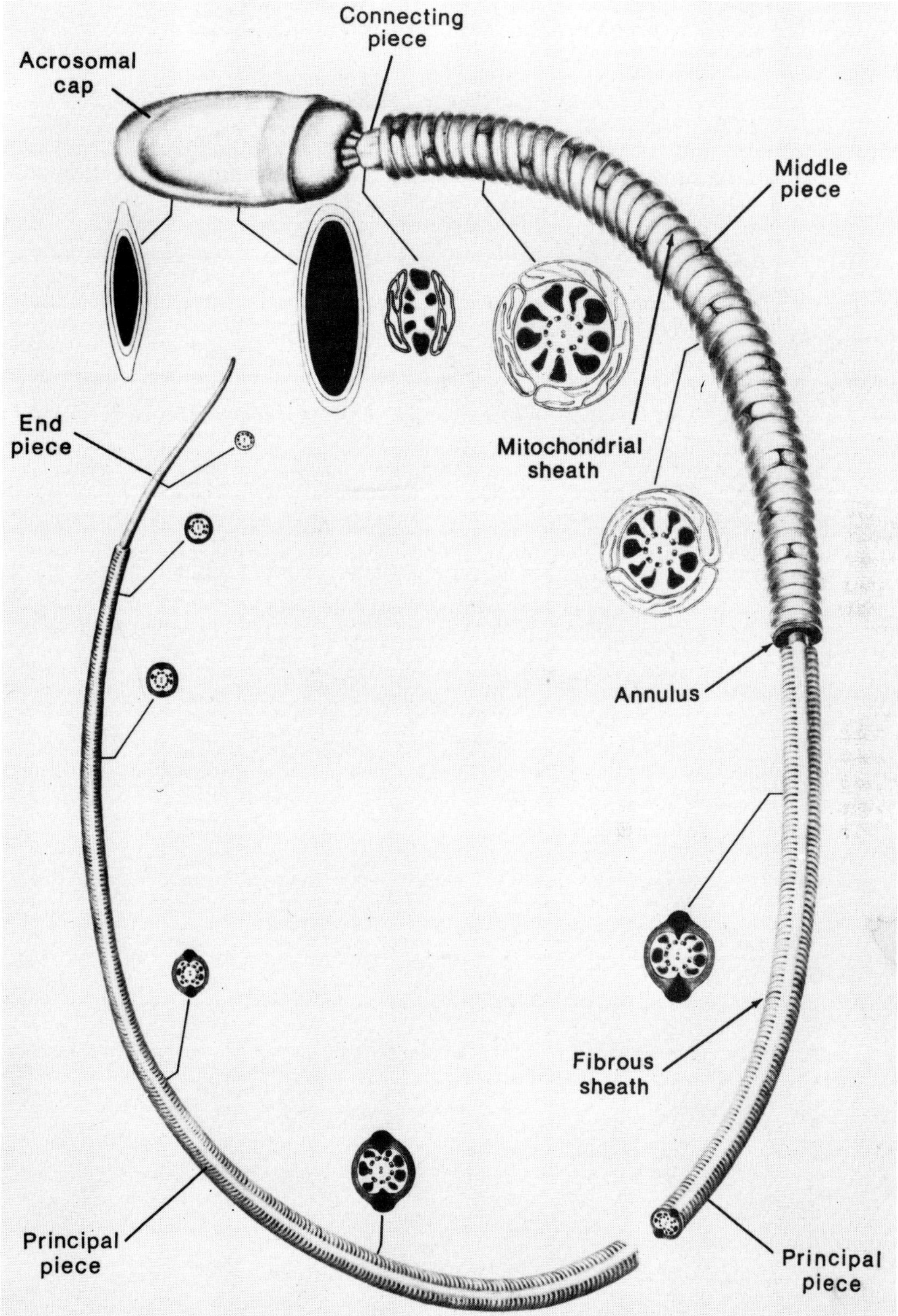

Schematic representation of a generalized mammalian spermatozoon as it would appear with cell membrane removed, and in cross sections at various levels. (From D. W. Fawcett, The Mammalian Spermatozoon, Dev. Biol. 44:394–436, 1975.)

shape of the annulus and its firm attachment to the cell membrane suggest that it serves to stabilize the mitochondrial sheath and prevent caudad displacement of mitochondria during vigorous movements of the tail.

In the long principal piece of the flagellum, the axoneme is invested by a *fibrous sheath (tail helix, cortical helix)*. Light microscopists described it as consisting of one or more continuous fibers wound around the axoneme in a tight helix. With the electron microscope, it proved to consist of a series of circumferentially oriented *ribs* that extend half way around the tail and terminate in two *longitudinal columns* which run along the dorsal and ventral aspects of the sheath for its entire length. The closely spaced ribs are usually of uniform thickness but may occasionally branch and anastomose with neighboring ribs (Bradfield, 1953; Fawcett, 1954, 1958). The outer dense fibers 3 and 8 terminate near the end of the midpiece. In the principal piece, their place is occupied by inward extensions from the longitudinal columns of the fibrous sheath that attach to appendages on doublets 3 and 8. The position of the longitudinal columns would seem to impose some restraint to flexion in a dorsoventral plane, but the principal plane of tail bending is perpendicular to the line joining the central pair of microtubules, and movement in this plane can be accommodated by alternate widening and narrowing of the interspaces between ribs on the two sides of the sheath. The function of the fibrous sheath is unclear, but it has been suggested (Challice, 1953) that if lateral flexion of the tail is to be produced by a propagated wave of contraction passing along the axoneme, there should be parallel resistances to provide the necessary couple. If the fibrous sheath is endowed with elastic properties, it may provide the needed resistance. This explanation is less than convincing, however, since flagella of primitive sperm function effectively without a fibrous sheath. Although the longitudinal columns and ribs of the fibrous sheath appear to have a different substructure, fibrous sheaths have now been isolated and found to consist of a single polypeptide (Olsen, 1977).

The fibrous sheath ends about 10 μm from the tip of the tail and the terminal segment of the axoneme enclosed only by the flagellar membrane constitutes the *end-piece* of the sperm tail.

In a scanning electron micrograph of spermatozoa on the surface of the uterine endometrium, the smooth contour of the long tapering tails gives no hint of their regional differentiation or of their complex internal organization.

Figure 328. Scanning electron micrograph of rabbit spermatozoa on the endometrium of the uterus. (Micrograph courtesy of David Phillips.)

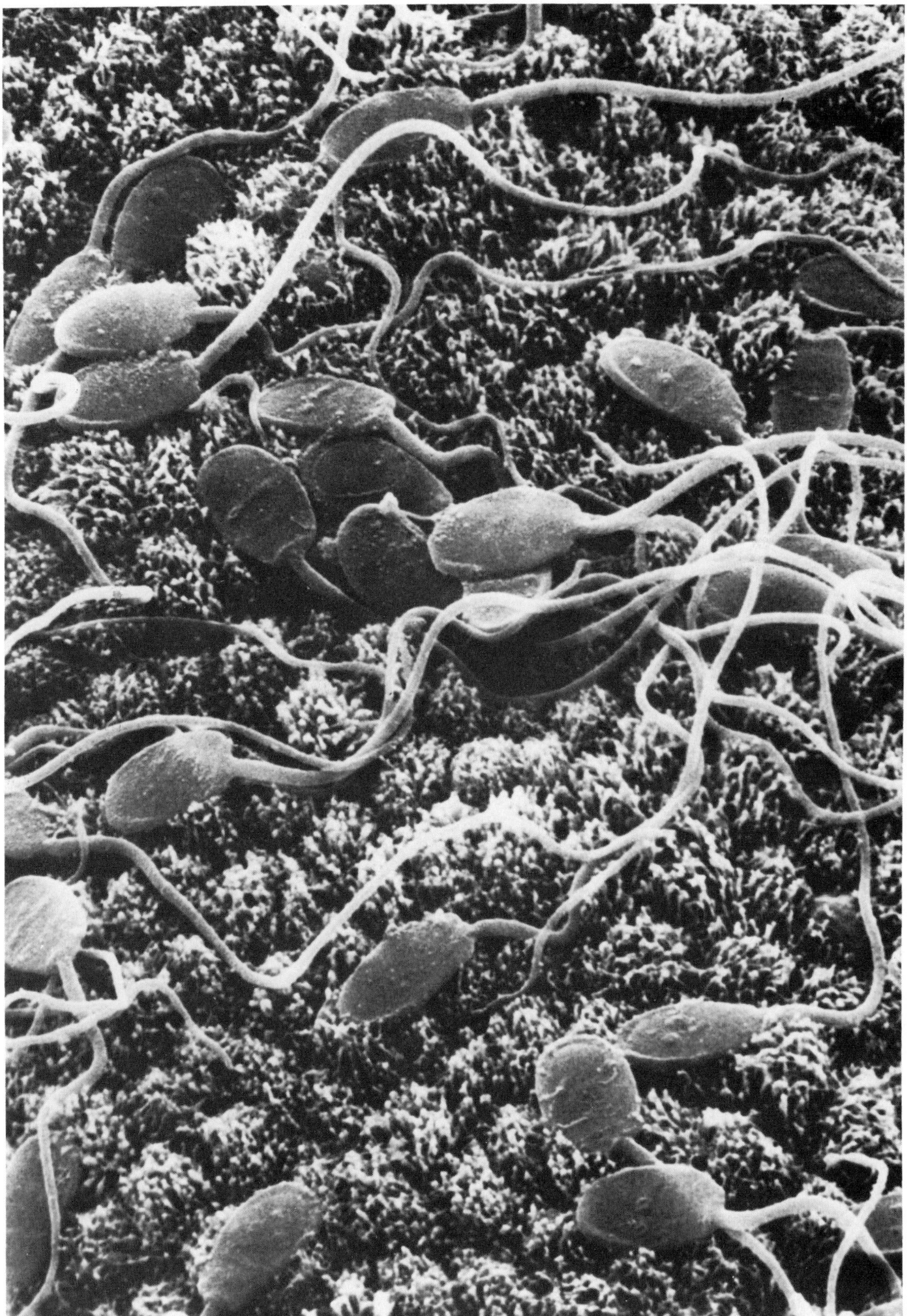

Figure 328

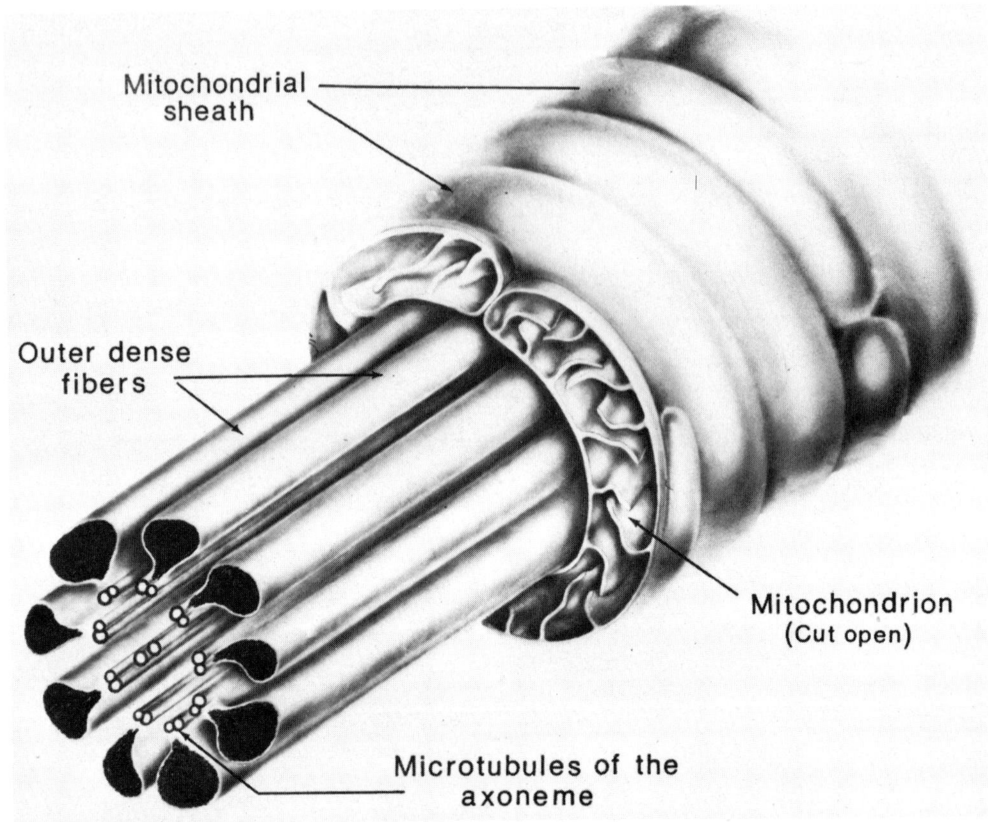

An extended diagram of the components in the midpiece of a mammalian spermatozoon tail. (From D. W. Fawcett, Dev. Biol. 44:394–436, 1975.)

In the extended diagram (above) showing the arrangement of components in the midpiece of the mammalian sperm tail, the helical mitochondrial sheath is seen to be closely applied to the longitudinal dense fibers. This relationship is illustrated in the transverse sections on the facing page. The dense fibers display distinctive differences in size and shape. In most species, numbers 1, 5, and 6 are somewhat larger than the others. A variable number of small *satellite fibrils* of unknown provenance and function are found in the interstices between the outer dense fibers.

Figure 329. Transverse sections through the midpiece of Chinese hamster spermatozoa. (Micrograph courtesy of David Phillips.)

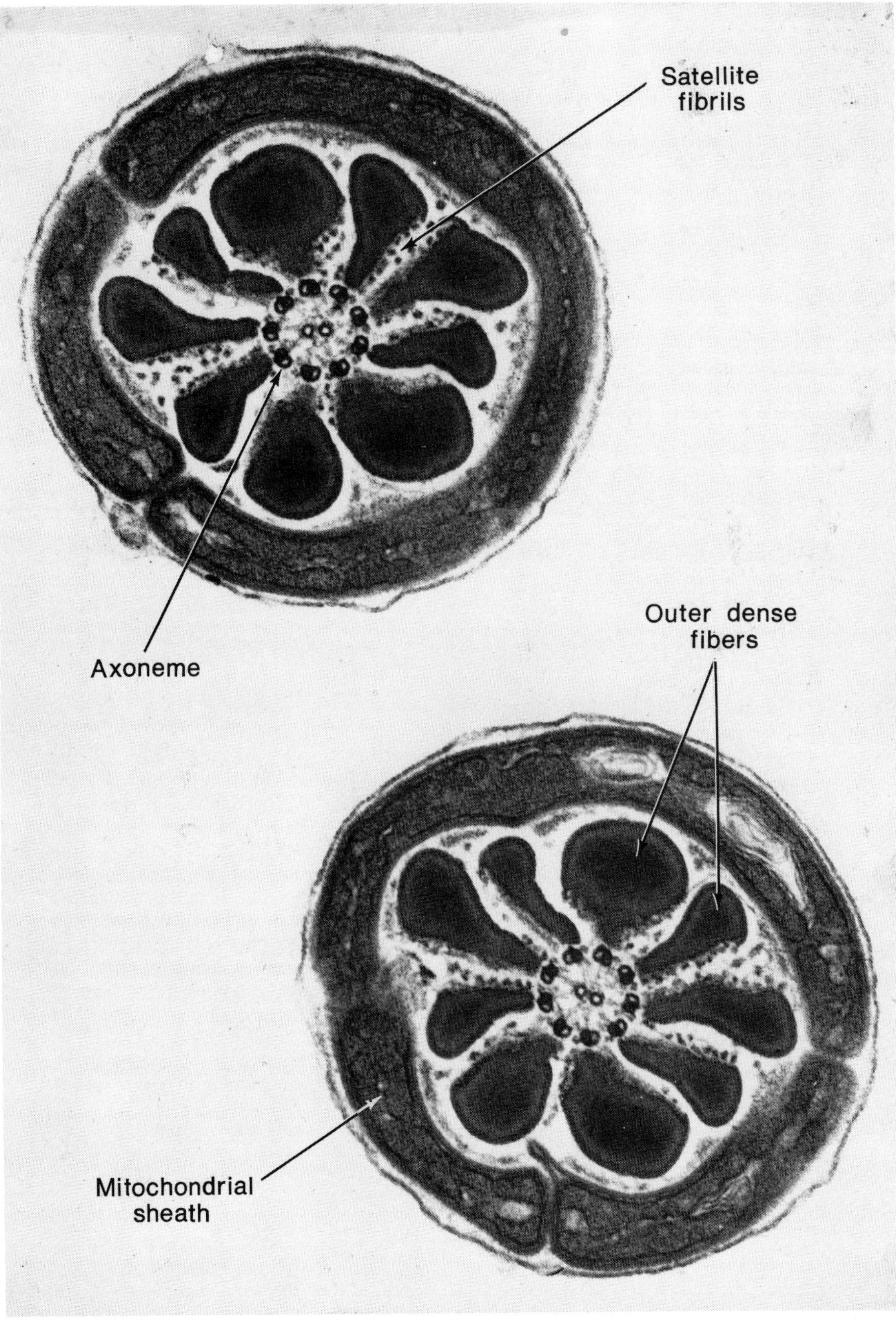
Satellite
fibrils
Axoneme
Outer dense
fibers
Mitochondrial
sheath

Figure 329

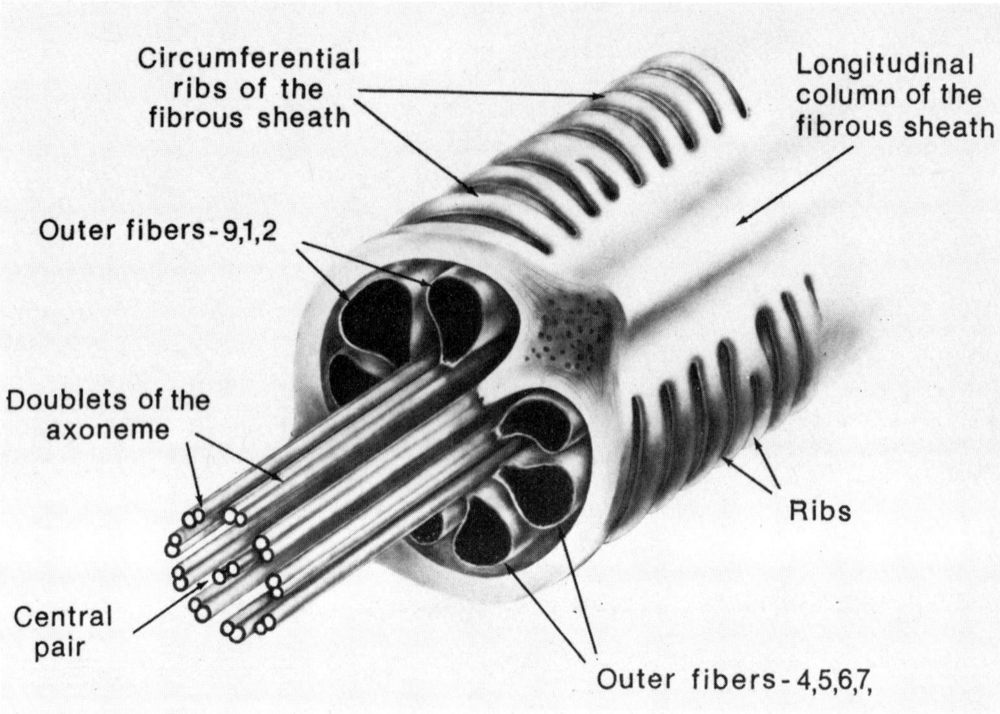

An extended diagram of components of the principal piece of a mammalian spermatozoon. (From D. W. Fawcett, Dev. Biol. 44:394–396, 1975.)

The topographical relationship of components in the principal piece of the mammalian spermatozoan is shown above. The fibrous sheath consists of continuous dorsal and ventral longitudinal columns joined by circumferential ribs. The columns are coplanar with the central pair of microtubules and hence offer little resistance to bending in the plane at right angles to the central pair. Bending in this plane is accommodated by widening or narrowing of the spaces between the ribs of the fibrous sheath.

The micrograph on the opposite page presents cross sections at various levels in the tapering principal piece of several spermatozoa. Note that outer dense fibers 3 and 8 have terminated and their positions are occupied by inward projections from the longitudinal columns, which attach to the corresponding doublets of the axoneme. This attachment probably prevents these doublets from participating in the microtubule sliding responsible for wave generation.

Figure 330. Transverse sections through the principal piece of testicular sperm in Chinese hamster. (Micrograph courtesy of David Phillips.)

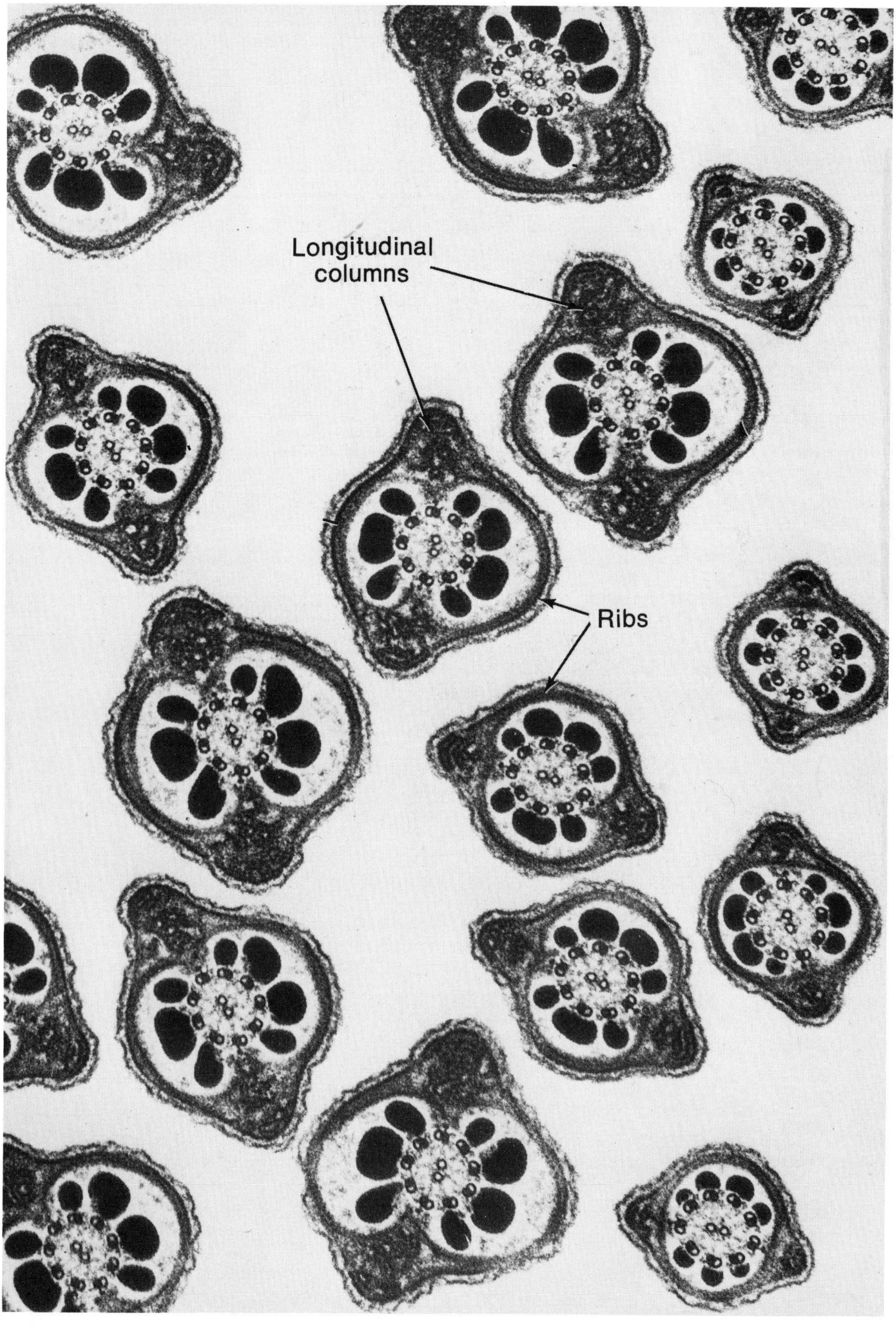

Figure 330

The accompanying micrographs permit a comparison of longitudinal sections of the midpiece (left) and principal piece (right). The principal piece shown here is not entirely typical, for in this species there is considerable fusion of neighboring ribs of the fibrous sheath near their insertion into the longitudinal columns. The individual ribs are seen more clearly in the micrographs that follow.

Figure 331. Longitudinal section of midpiece from sperm of the suni *Nesotragus moschatus.*

Figure 332. Longitudinal section, principal piece, mouse sperm.

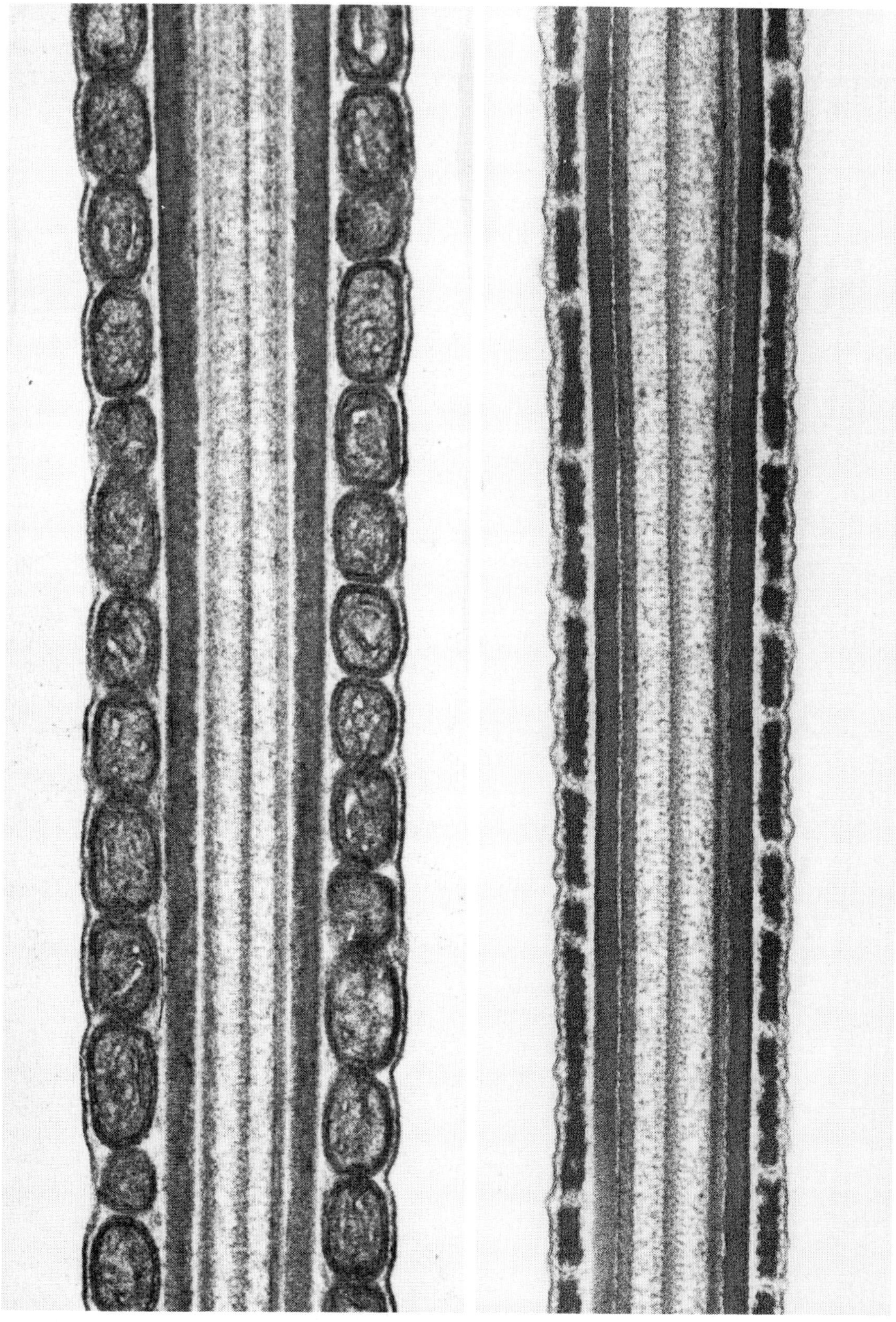

Figure 331, *left* **Figure 332**, *right*

At the junction of the midpiece and principal piece of mammalian sperm is the annulus (Jensen's ring), a dense structure firmly attached to the overlying flagellar membrane. The configuration of this region shows considerable species variation. In some there is a groove or recess between the annulus and the motor apparatus. In others, the surface contour is flat and the annulus is triangular in section, with the base at the plasma membrane and the apex projecting beneath the last gyre of the mitochondrial sheath. It is speculated that the annulus is a stabilizing structure preventing caudal displacement of mitochondria in actively motile sperm tails.

Figure 333. Longitudinal section of the junction of midpiece and principal piece in spermatozoon of chinchilla, *(Chinchilla laniger)*.

Figure 334. Comparable region of the spermatozoon of the suni *(Nesotragus moschatus)*.

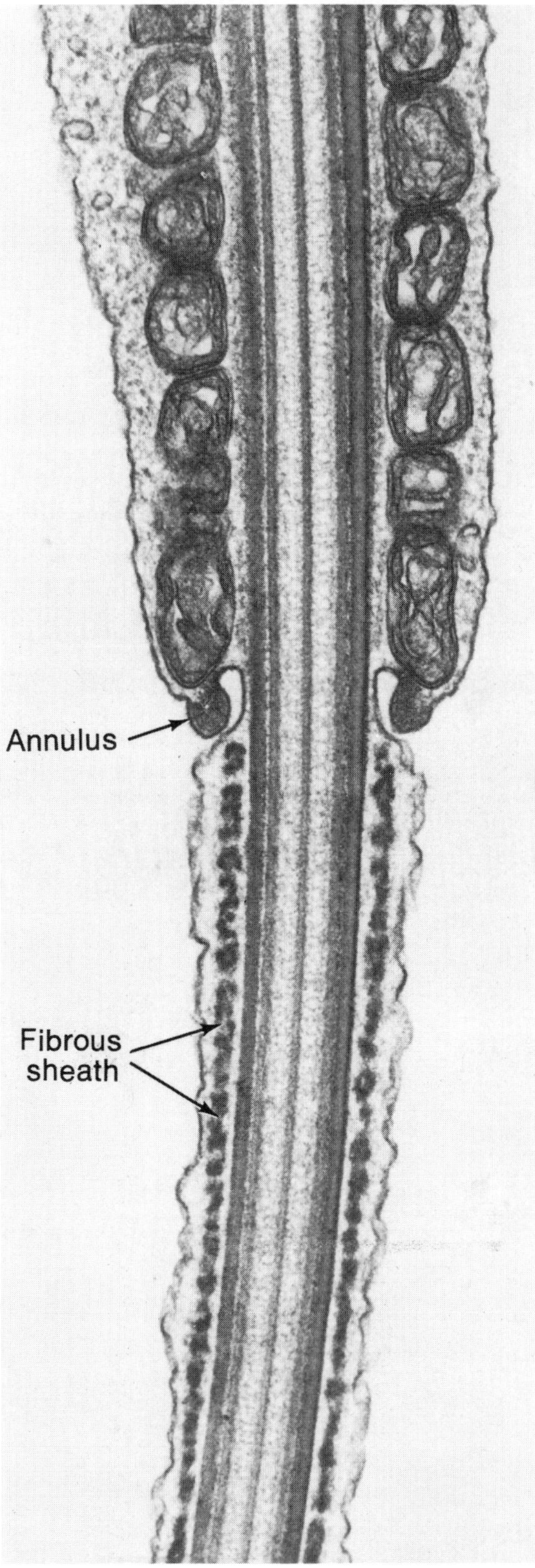

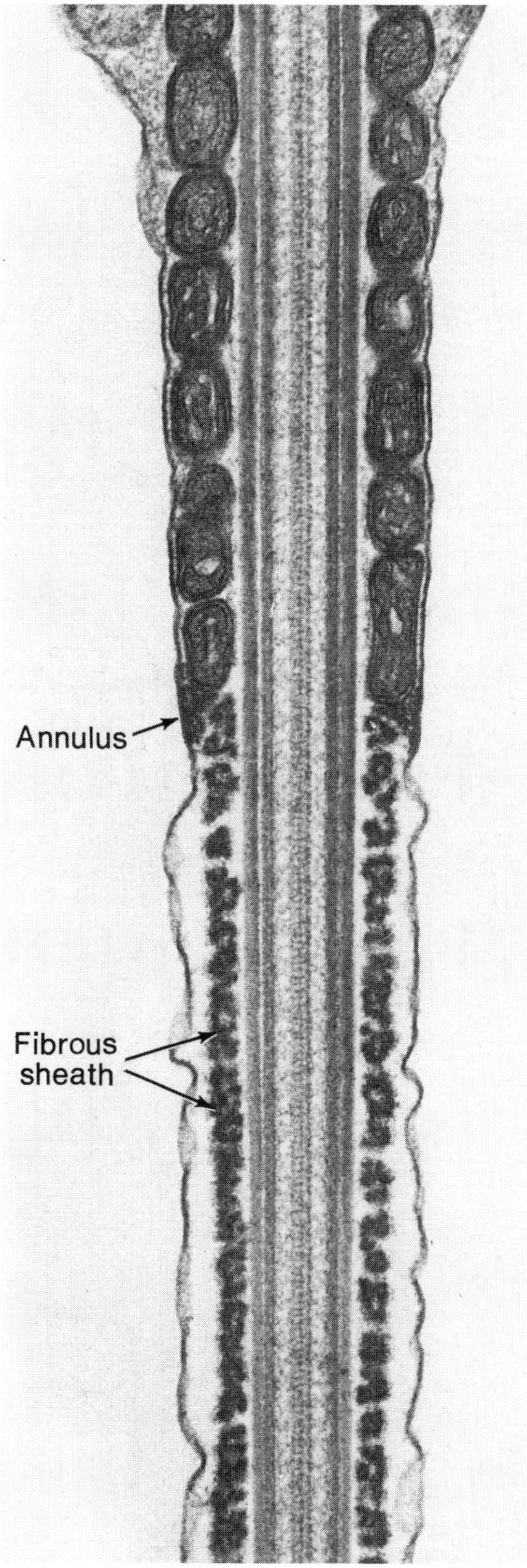

Figure 333, *left* **Figure 334,** *right*

The organization of the axoneme of sperm flagella appears to be identical to that of cilia. Visualization of its components is facilitated by aldehyde fixation in the presence of tannic acid, which serves as a mordant for subsequent staining with osmium and uranyl acetate (Mizuhira, 1971). This preparative procedure improves the contrast of the image but appears to result in some thickening of the doublets, arms, and spokes. The protofilaments in the walls of the microtubules are seen in negative image. Subunits of similar size are seen in negative image in the satellite fibrils (at arrows). This has led to the suggestion that these may also be composed of tubulin, but this requires verification.

Figure 335. Cross-section of guinea pig sperm tail at the level of the midpiece, treated with tannic acid. (Micrograph courtesy of Daniel Friend.)

Figure 336. Cross-section of sperm tail at the level of the principal piece. Tannic acid preparation. (Micrograph courtesy of David Phillips.)

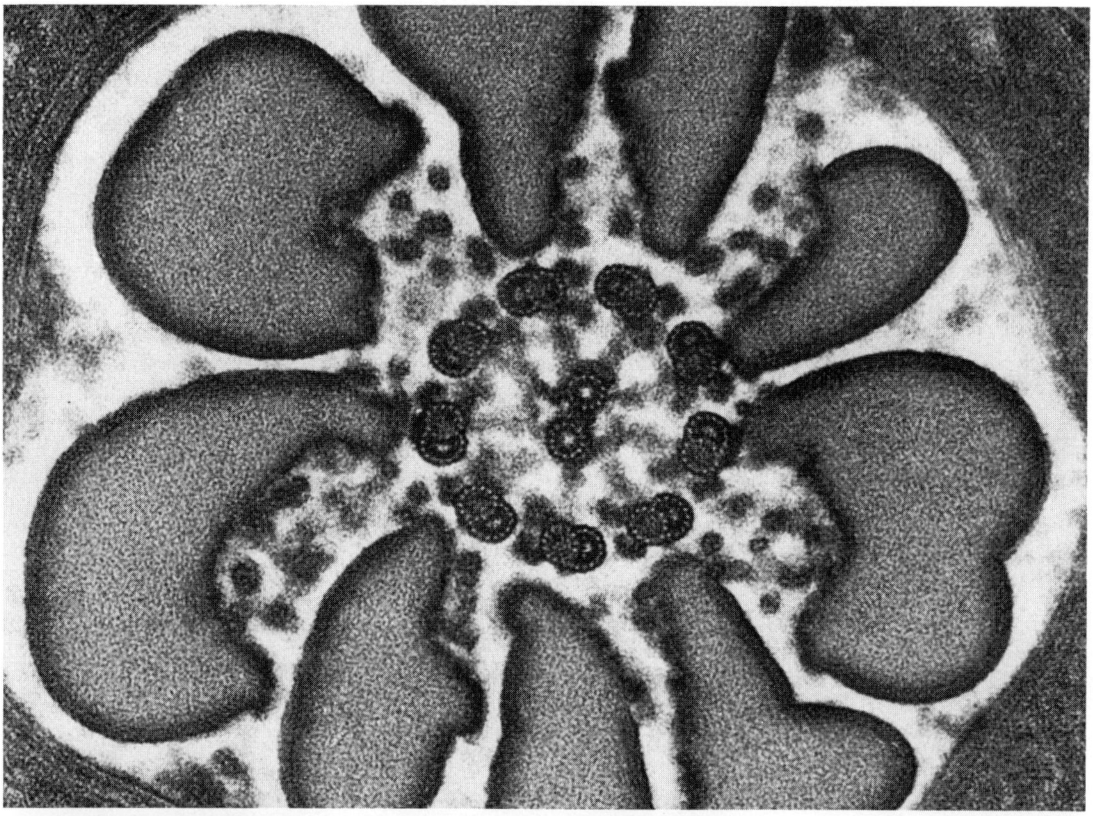

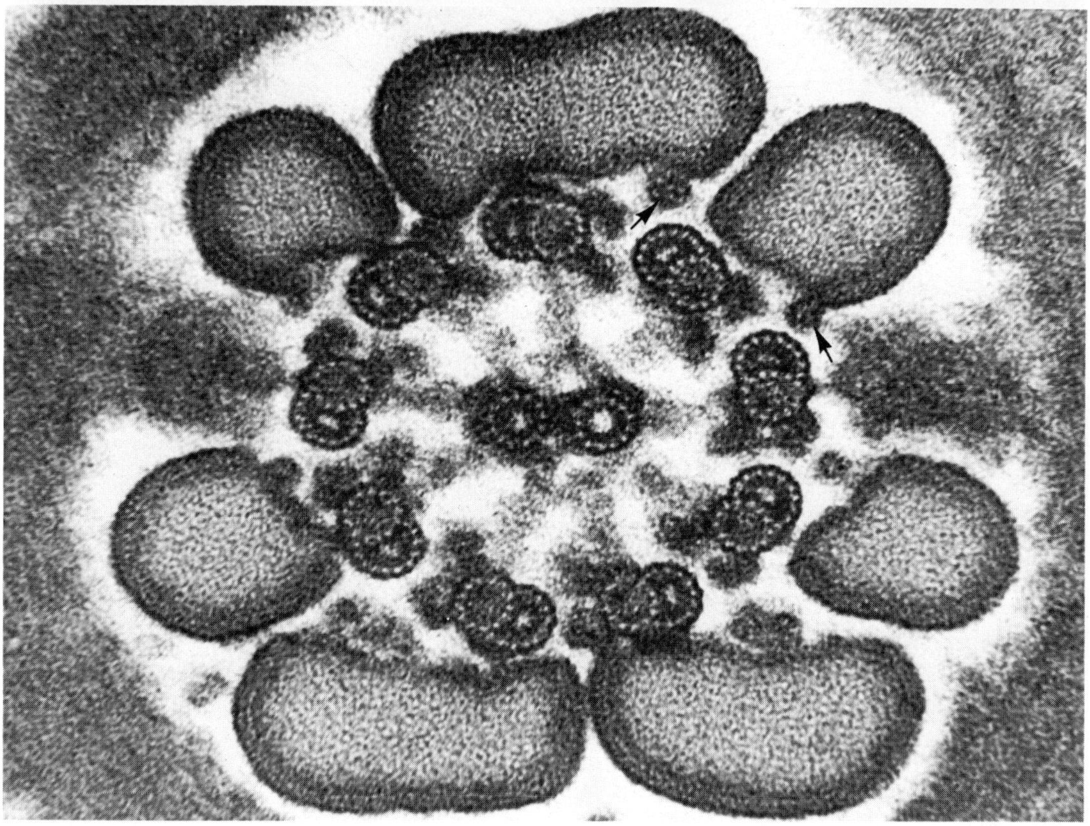

Figure 335, *upper* Figure 336, *lower*

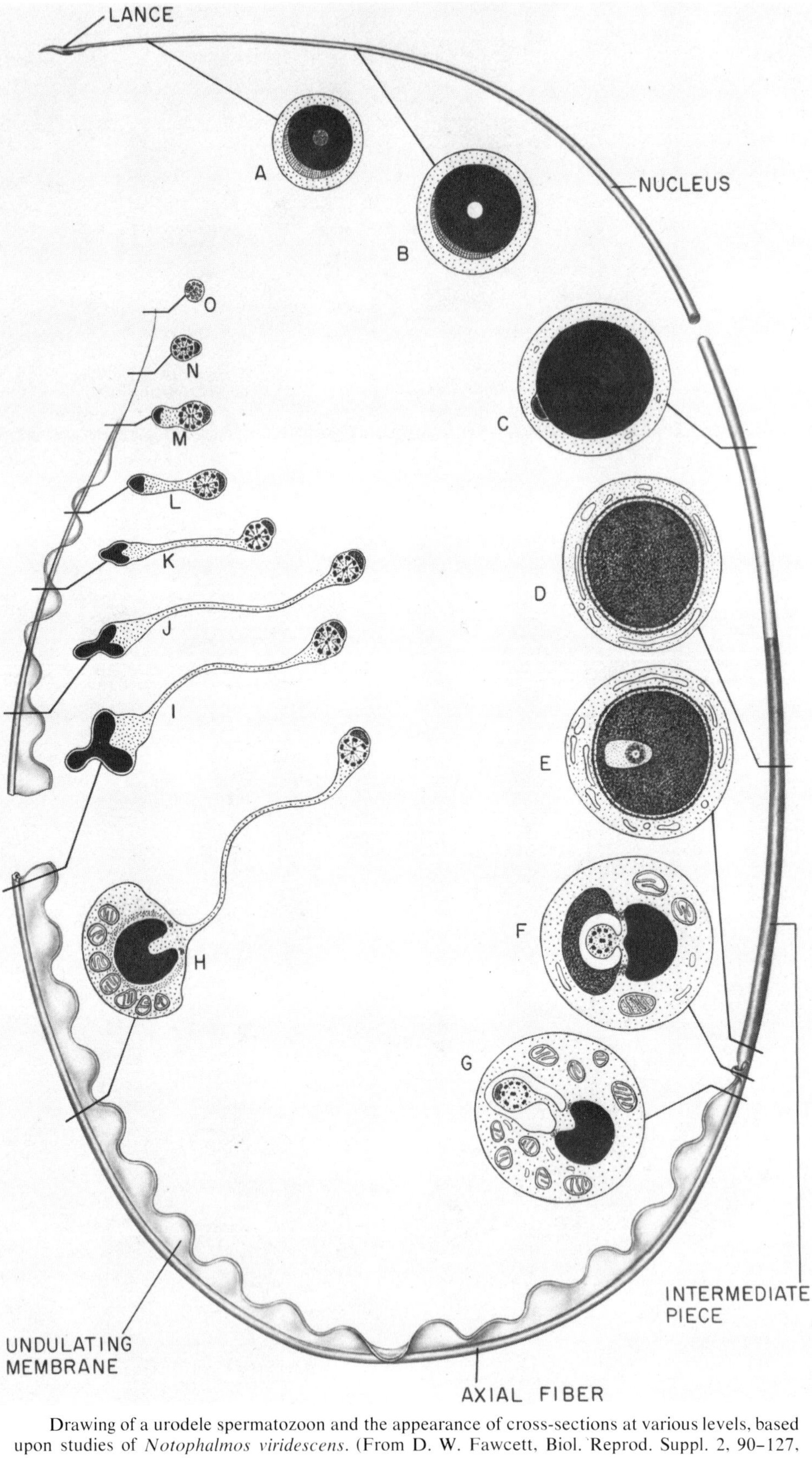

Drawing of a urodele spermatozoon and the appearance of cross-sections at various levels, based upon studies of *Notophalmos viridescens*. (From D. W. Fawcett, Biol. Reprod. Suppl. 2, 90–127, 1970.)

URODELE SPERM FLAGELLUM

In the sperm tail of urodele amphibians, a large dense fiber or rod continues caudally in the axis of the nucleus and intermediate piece. This *axial fiber* is homologous with one of the nine outer dense fibers (number 8) of mammalian sperm tails. It tapers gradually toward the tip of the tail and its cross-sectional outline changes from horseshoe shaped (H) to a trefoil (J, K) and finally a thin crescent (L, M). Throughout the long middle piece of the spermatozoon, mitochondria are closely applied to the convex surface of the axial fiber (H).

The plasmalemma is attached to the axial fiber along the two ridges that form the heels of the horseshoe-shaped cross section, but between these lines of attachment it converges to form the *undulating membrane,* a thin fold that encloses the axoneme in its thickened free margin. A slender *marginal fiber* with a crescentic cross section is closely associated with doublet 3 of the axoneme. In the living spermatozoon, the thick axial fiber is immotile, but waves of bending rapidly propagated along the axoneme result in rapid undulations of the membrane that propel the sperm slowly forward.

Thus in urodeles, the 9+9+2 formula has been modified to 2+9+2. In toads, there is no marginal fiber and the axial fiber is thinner and more flexible than in urodeles. The activity of the axoneme not only produces rapid undulations of the membrane but results in low amplitude secondary bends in the tail as a whole. Motility resides entirely in the axoneme.

Urodele sperm occur in the testis in bundles, and a section through one of these at the level of the midpiece provides micrographs of the kind shown here. The cytoplasm surrounding the axial fiber contains numerous mitochondria. In both mammalian and urodele spermatozoa the mitochondria that provide the energy for locomotion are more closely associated with the dense fibers than they are with the motile axoneme. This relationship encouraged the speculation that the thick fibers or rods were accessory contractile elements. It is now known that they are non-contractile components of the sperm tail endowed with limited flexibility.

Figure 337. Transverse section through a portion of a sperm bundle of the newt, *Notophthalmos viridescens.*

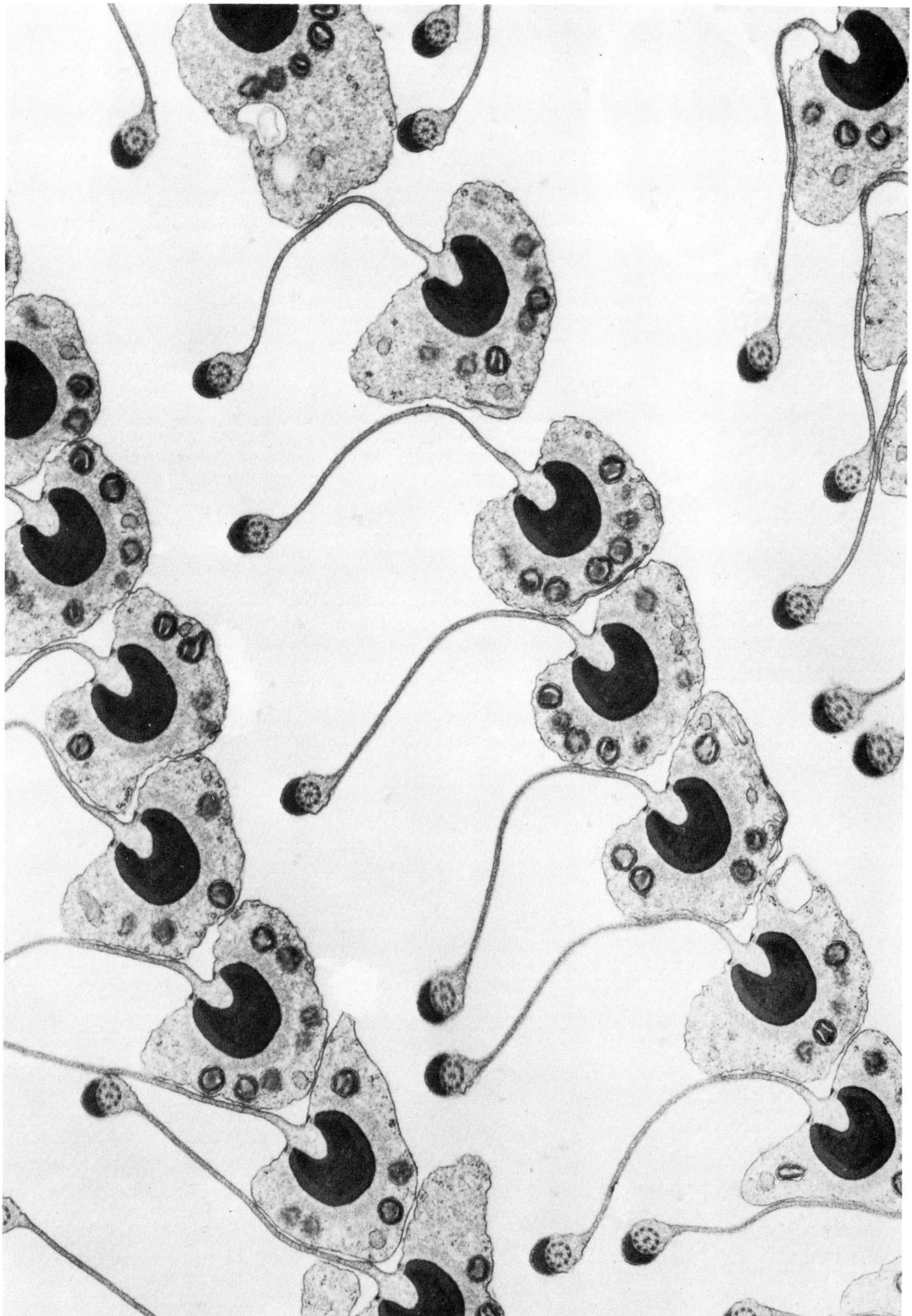

Figure 337

The accompanying micrographs present transverse sections of the midpiece and principal piece of salamander sperm. Even though the flagellar membrane appears to be attached to parallel ridges of the axial fiber or rod, this evidently does not prevent ATP generated by the mitochondria from diffusing between the leaves of the undulating membrane to the axoneme.

In the relatively short principal piece, the axial rod assumes the form of a trefoil and the membrane is closely applied to its surface without intervening cytoplasm, except at the base of the undulating membrane. The axial fiber and marginal fiber evidently correspond to outer fibers 3 and 8 of the mammalian sperm tail.

Figure 338 and **339.** Electron micrographs of transverse sections of the midpiece and principal piece of several sperm from *Notophthalmos viridescens.* (From Fawcett, Biol. Reprod. [Supp. 2, 90–127, 1970.])

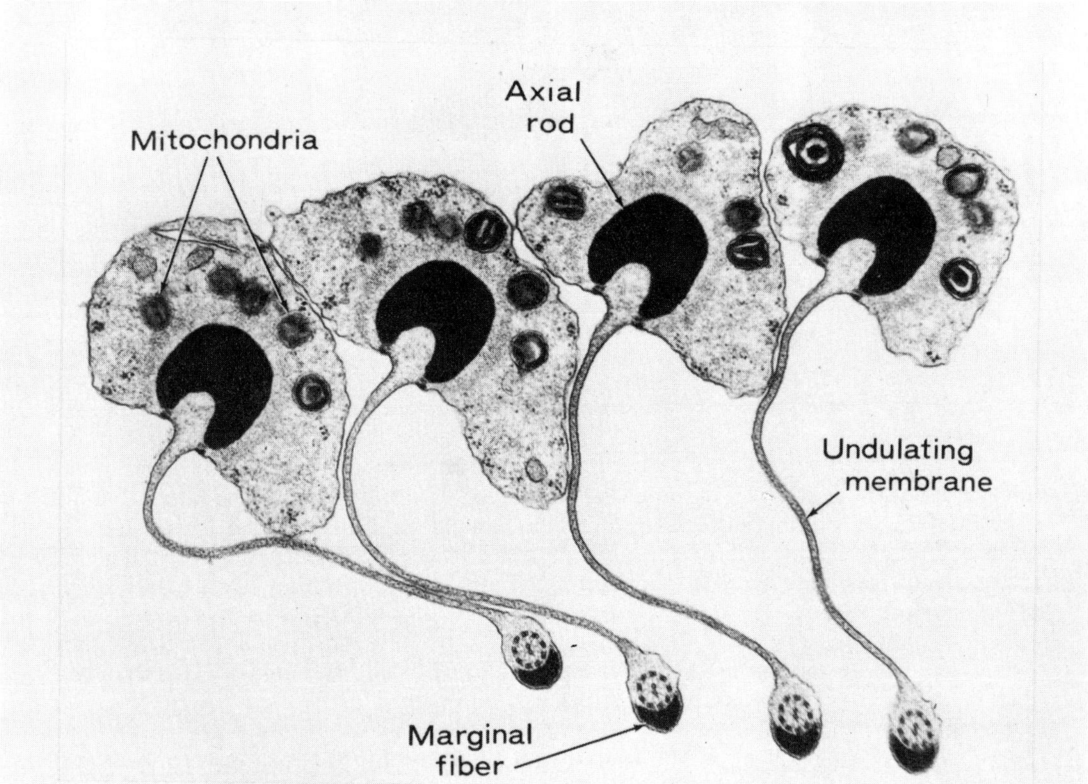

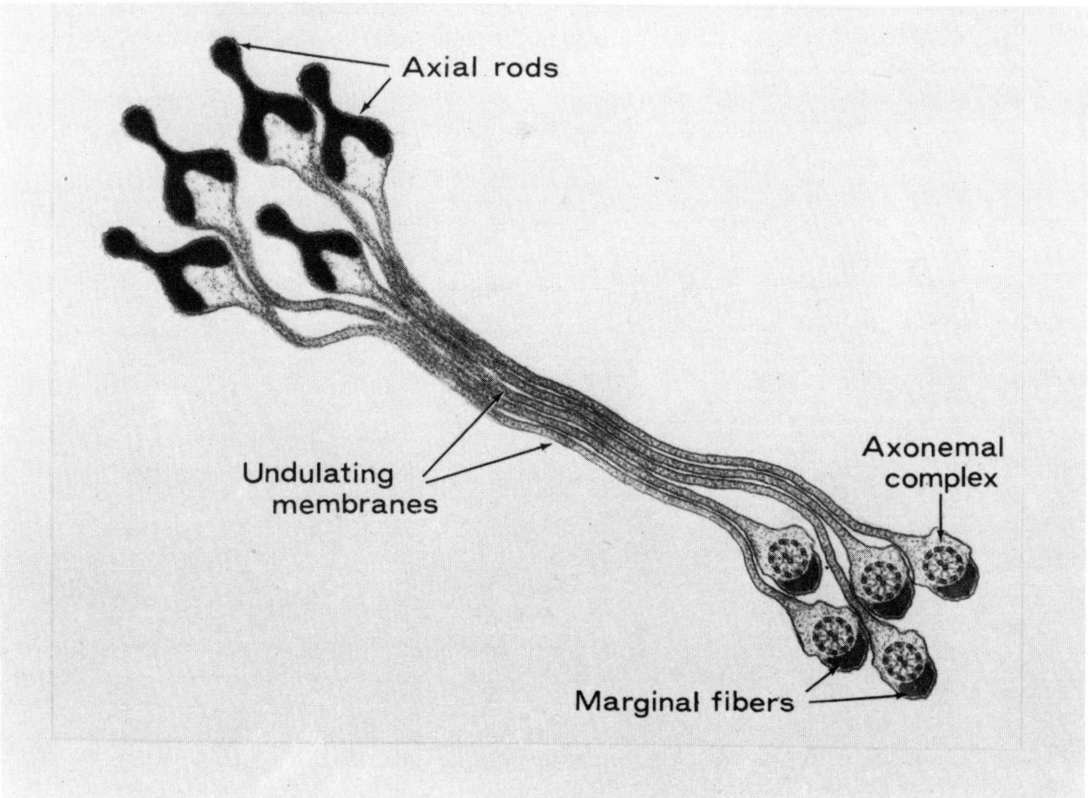

Figure 338, *upper* **Figure 339,** *lower*

INSECT SPERM FLAGELLUM

The flagella of insect spermatozoa vary greatly in their length and in the details of their cross-sectional organization, but the great majority have certain common features that are illustrated in the accompanying drawing based upon a caddis fly. The motor apparatus is a typical 9+2 axoneme, but in place of the nine outer dense fibers that are characteristic of mammalian sperm tails, the insects have nine *accessory tubules*. In cross section, these resemble other microtubules in having discernable protofilaments in their wall, but they vary in diameter from species to species with the number of protofilaments ranging from 13 to 16. Although composed of tubulin, these do not appear to be accessory motor elements, for they possess no dynein arms, and they would seem to be too far apart to interact in a sliding tubule mechanism. It is assumed that they serve as elastic stiffening structures analogous to the outer dense fibers of mammalian sperm flagella.

Instead of a helical mitochondrial sheath, insect sperm flagella have one or two modified mitochondria that run longitudinally along one side of the axoneme for nearly its entire length, which may be several hundred microns in some species. In the course of spermiogenesis, electron dense material accumulates in the mitochondrial matrix and forms one or more extensive paracrystalline structures that replace the usual pattern of membranous cristae. These *mitochondrial derivatives* are subject to considerable interspecific variation, and usually bear little resemblance to conventional mitochondria.

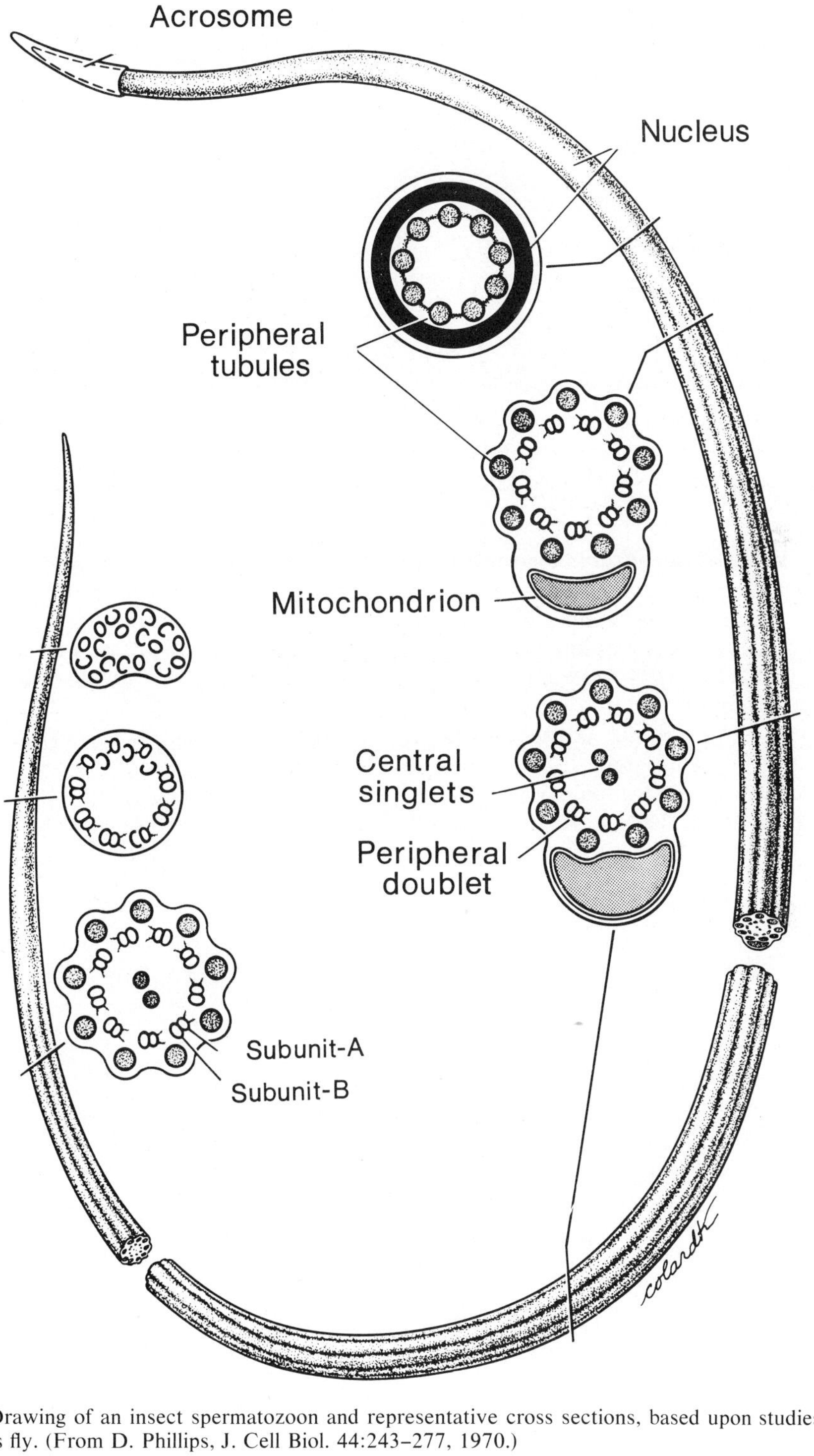

Drawing of an insect spermatozoon and representative cross sections, based upon studies of a caddis fly. (From D. Phillips, J. Cell Biol. 44:243–277, 1970.)

Sperm flagella of insects also differ from those of mammals in the position of the nine outer elements in relation to the doublets of the axoneme. In the mammal, each of the outer dense fibers remains directly radial to the corresponding doublet. In insect sperm, the accessory tubules arise as extensions from the wall of subunit B of each doublet, but when they detach, they take positions opposite the spaces between the axonemal doublets.

The position of the mitochondrial derivative is constant with respect to the components of the axoneme. It is centered on doublet number 3. A line passing through the centers of the central pair of microtubules is about 15 degrees oblique to the axis of symmetry passing through the mitochondrial derivative and between doublets 7 and 8.

Figure 340. Cross sections of sperm tails of a caddis fly. (Micrograph courtesy of David Phillips.)

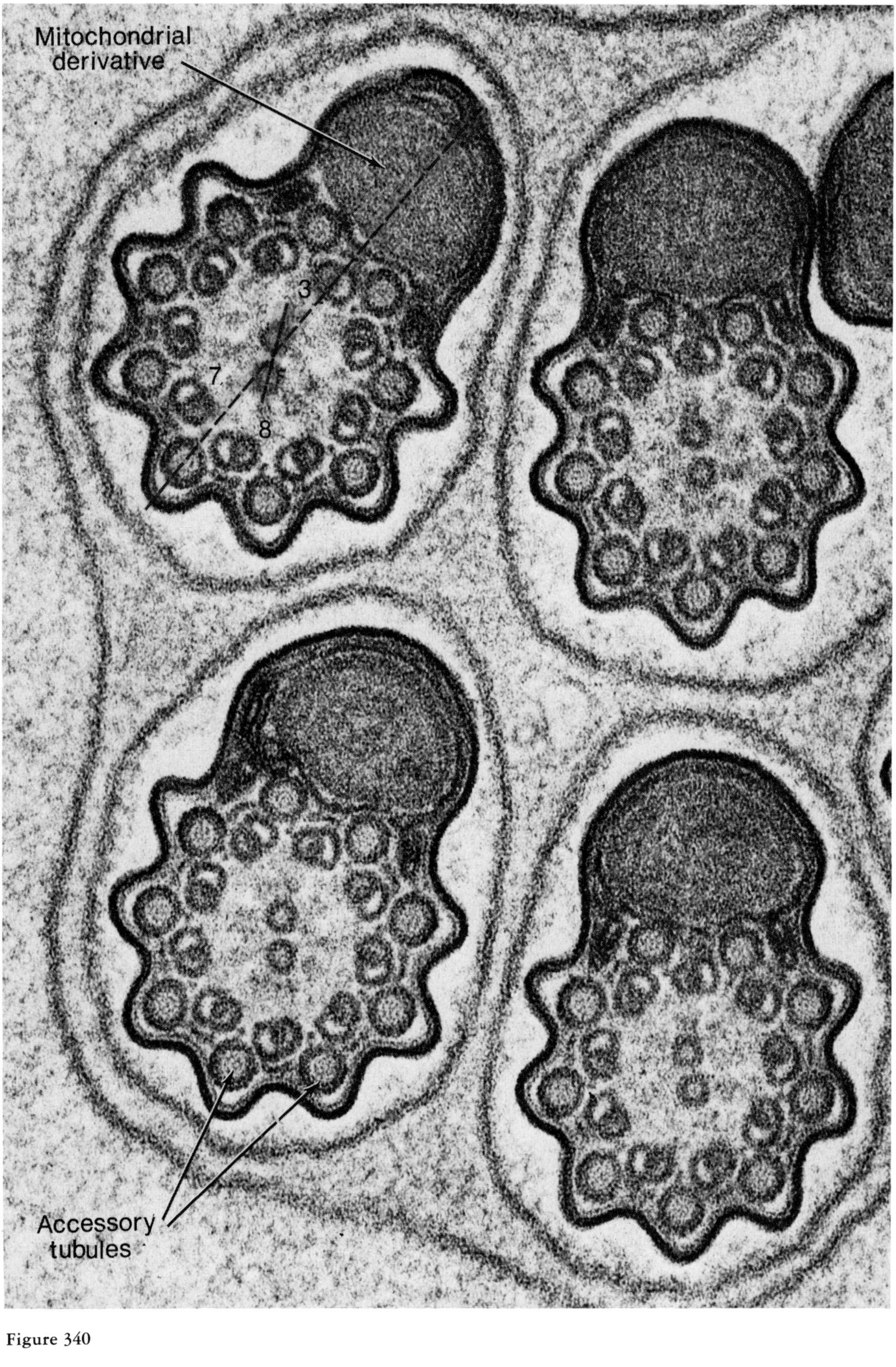

Figure 340

In some insect species, the lumen of the accessory tubules is filled with a dense material which gives them the appearance of dense fibers, but at higher magnification the protofilaments of the tubule wall can be resolved. The central pair of microtubules may also have a dense core.

The flagellar membrane of many Lepidopteran spermatozoa is decorated by radiating projections with a highly regular periodic structure. The chemical nature and significance of these appendages are obscure.

Figure 341. Cross sections of spermatozoa in the testis of a moth. (Micrograph courtesy of David Phillips.)

Figure 341

Sliding of the doublets of the axoneme has been conclusively shown to provide the motive force for flagellar motility. There is suggestive evidence that the regulatory function needed to convert doublet sliding to propagated waves of bending may involve some form of interaction between the radial spokes and the central pair of microtubules.

The nearly universal occurrence in motile flagella of the 9+2 or 9+9+2 patterns of axonemal components, and the observation that flagella with genetic absence of the central pair are immotile, has fostered the belief that the presence of two central microtubules is essential for motility. However, comparative studies on insect spermatozoa indicate that the number 2 is certainly not a *sine qua non*. The spermatozoa of a caddis fly in the upper figure on the facing page have a 9+7 formula while those of the mosquito, illustrated below, exhibit a 9+9+1 pattern. The spermatozoa of both species are motile, but detailed analysis of the tail movements has not been carried out to detect possible qualitative differences attributable to these unusual patterns.

Figure 342. Cross sections of spermatozoa from the caddis fly *Polycentropus*. (Micrograph courtesy of David Phillips.)

Figure 343. Cross sections of spermatozoa of a mosquito of the genus *Culex*. (Micrograph from Phillips, J. Cell Biol. 40:28–43, 1969.)

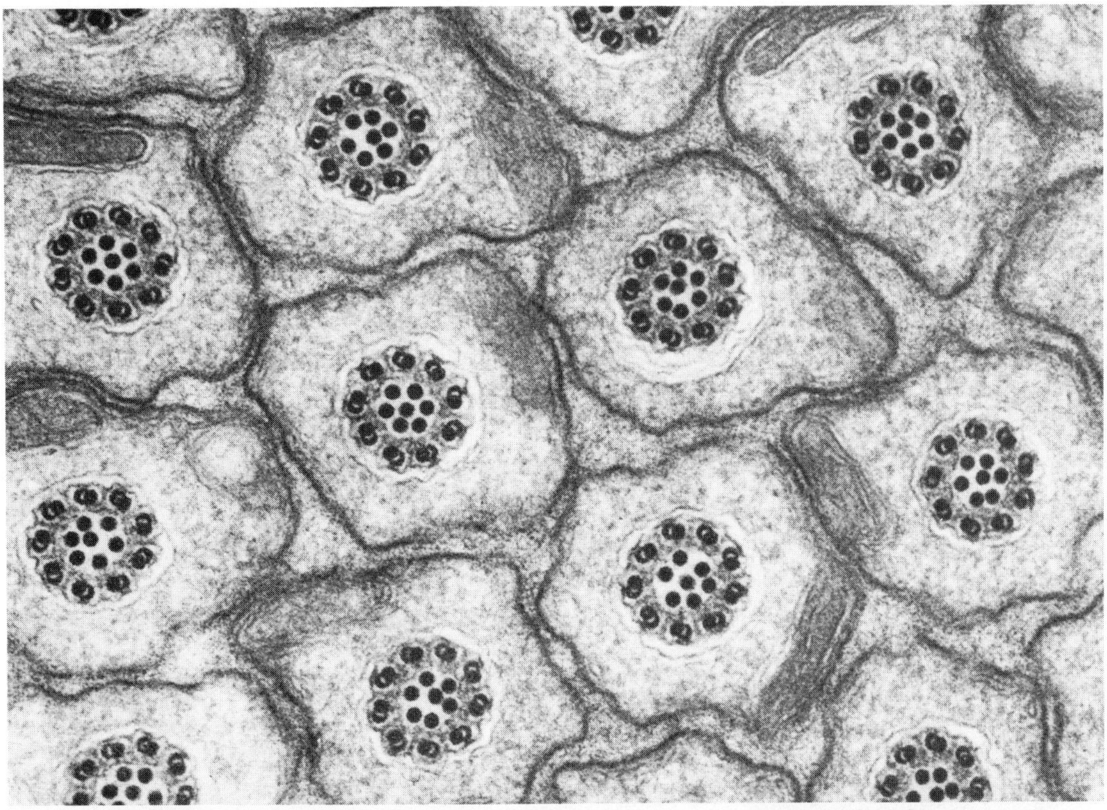

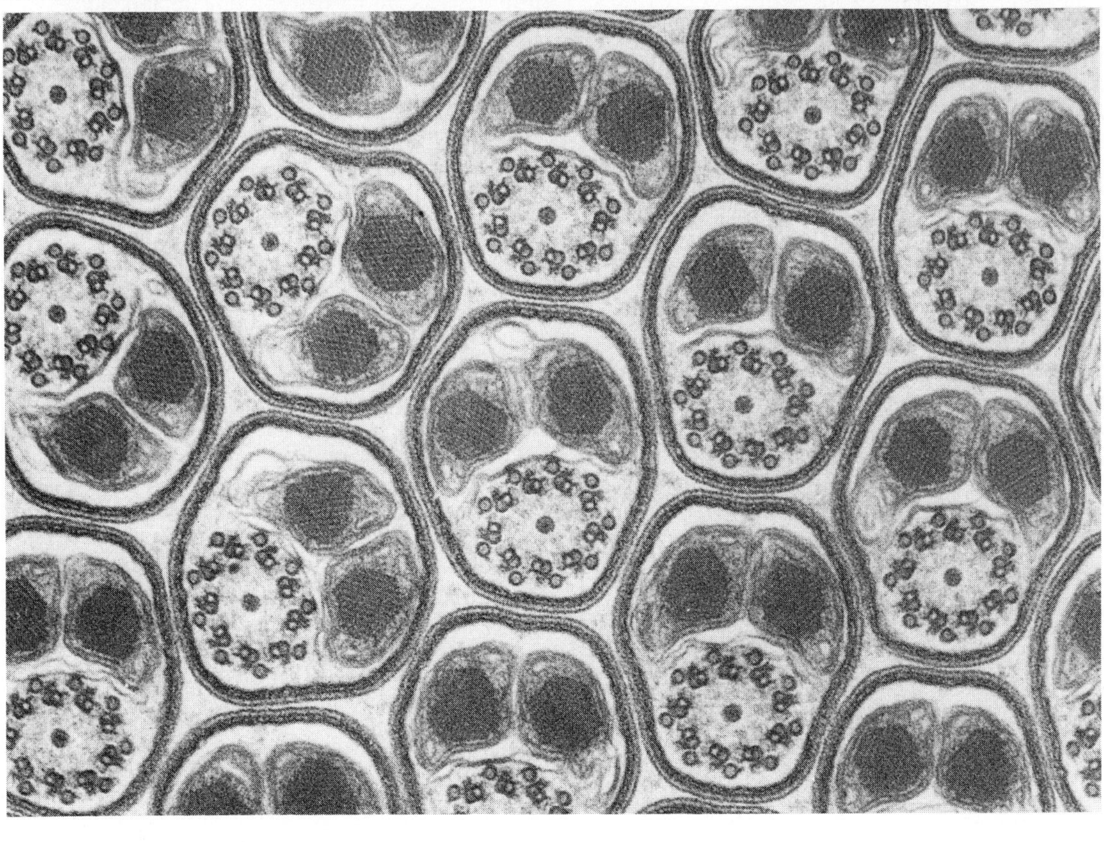

Figure 342, *upper* **Figure 343,** *lower*

The most invariable components of motile flagella are the nine doublets. Rare exceptions are found among gall midges in which the sperm flagellum in some species has an extraordinary and varying pattern consisting of rows of doublets without any structure corresponding to the central pair in conventional axonemes. These spermatozoa exhibit a short period of vibratile motility in the female tract but do not show propagated waves of bending — an observation consistent with the view that the spokes and central pair in conventional 9+2 flagella are involved in translation of sliding into waves of bending.

It is evident in the lower figure that the 170 doublets are spaced at the normal intervals along the rows, but they possess only the outer dynein arm and no obvious spokes or other appendages.

Figures 344 and 345. Cross sections of the sperm tail of the Cecidomyid fly, *Monarthropalpus buxi.* (Micrographs courtesy of R. Dallai, from Baccetti et al., Tissue Cell 6:269–278, 1974.)

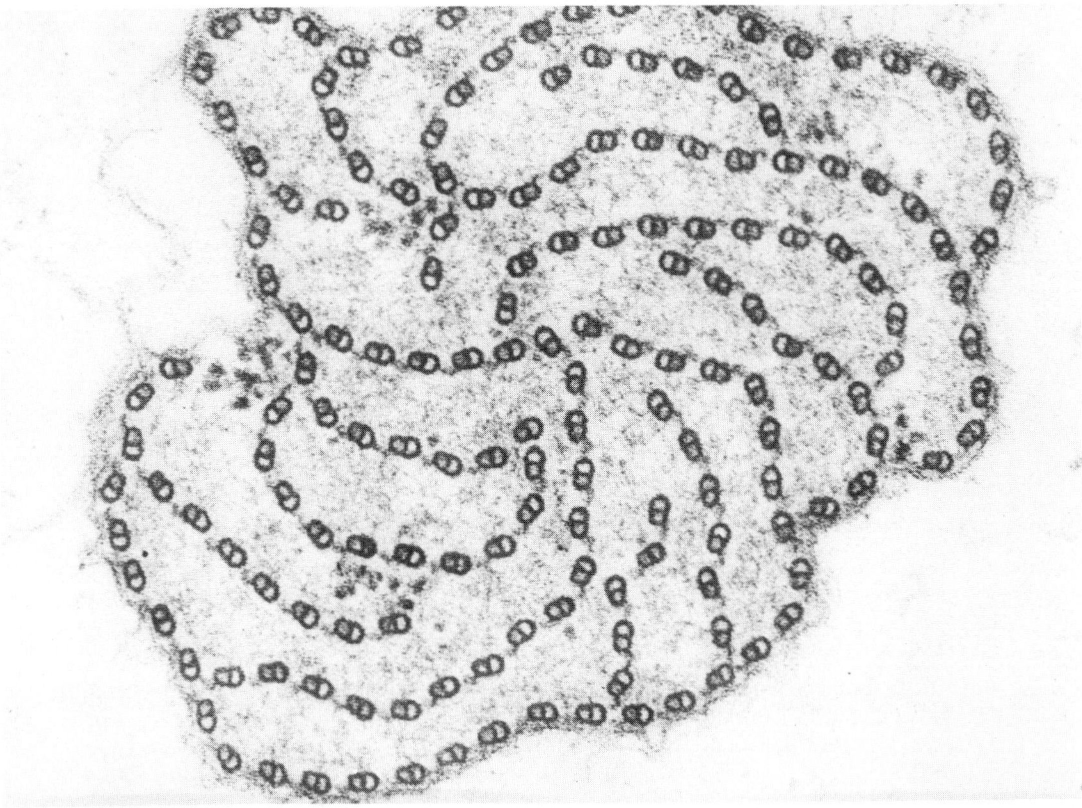

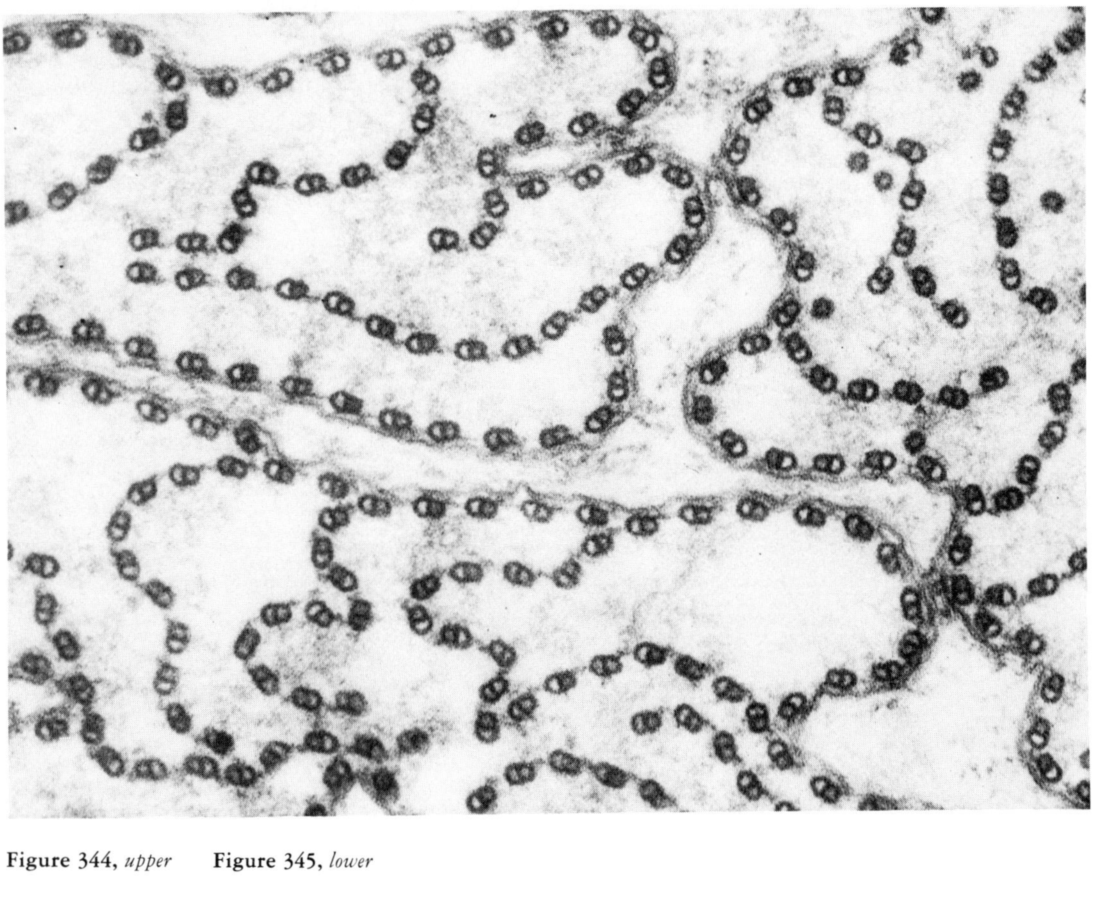

Figure 344, *upper* **Figure 345,** *lower*

One of the most extraordinary of the exceptions to the prevailing 9+9+2 pattern of insect sperm flagella is that of another cecidomyid fly illustrated in the upper micrograph on the facing page. The lobulated, elliptical cross section of the sperm tail contains mitochondria scattered among about 1000 closely packed doublet microtubules arranged in poorly defined rows.

In a related genus shown in the lower figure, approximately 100 doublets are arranged in parallel concentric rows or a spiral. These spermatozoa are believed to have a vibratile motion like that of other cecidomyid flies, but their motility has not been observed.

Figure 346. Cross section of sperm tail of the cecidomyid fly, *Diplolaboncus tumorificus*. (Micrograph from Baccetti and Dallai, J. Ultrastr. Res. 55:50–69, 1976.)

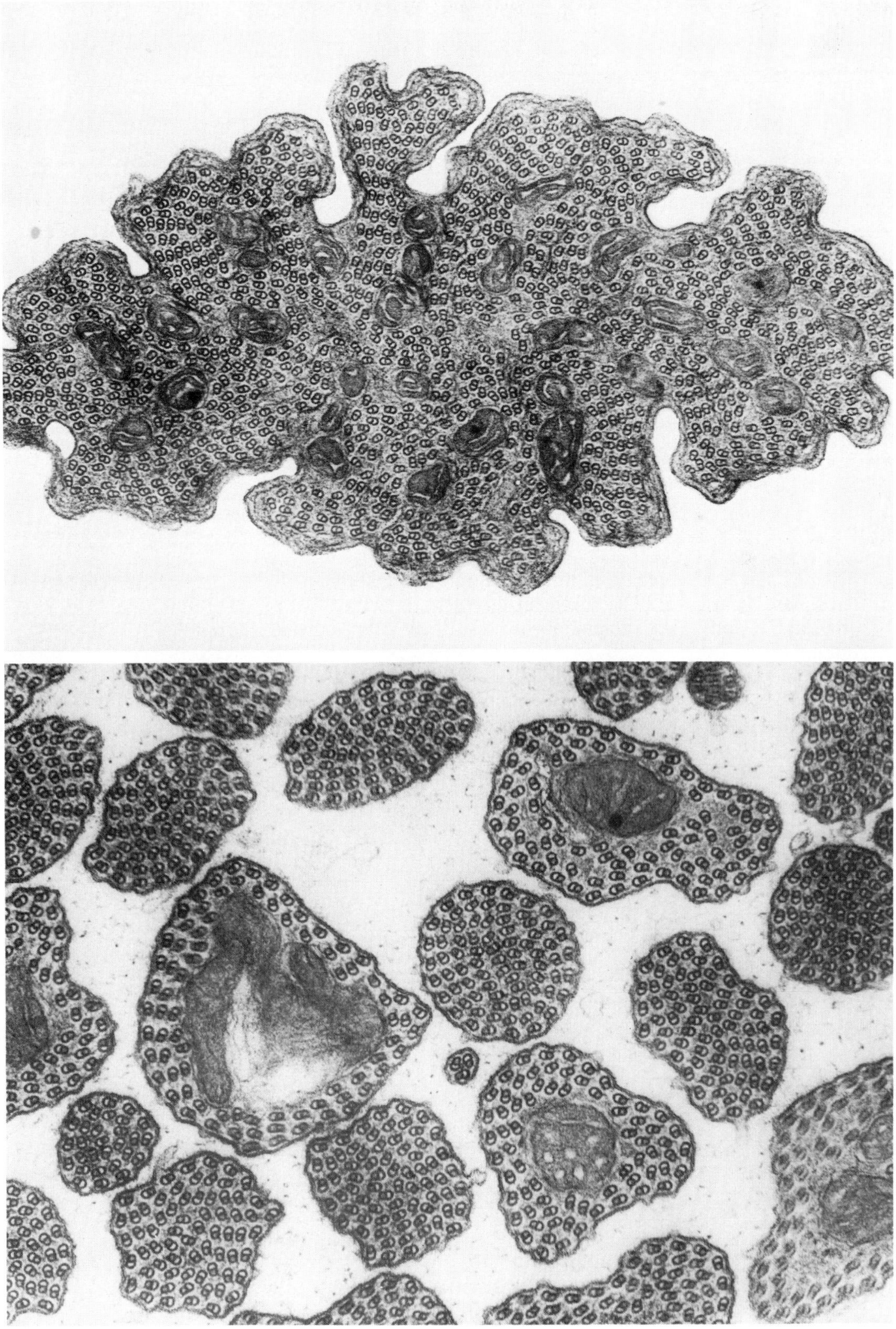

Figure 346

At higher magnification, it is apparent that the doublets have only the outer dynein arm (see at arrows). The arms on doublets in the same row are consistent in their orientation, but those in neighboring rows may either point in the same or in opposite directions.

The motor apparatus is still more aberrant in scale insects (coccids). It consists of singlet microtubules arranged in concentric rings around the nucleus or an amorphous central core.

Figure 347. An area from a cross section of sperm tail from the cecidomyid fly, *Diplolaboncus tumorificus.* (Micrograph courtesy of Romano Dallai.)

Figure 348. Transverse sections of spermatozoa of the oyster shell scale insect, *Lipidosaphes.* (From Phillips, Spermiogenesis, Academic Press, New York, 1974.)

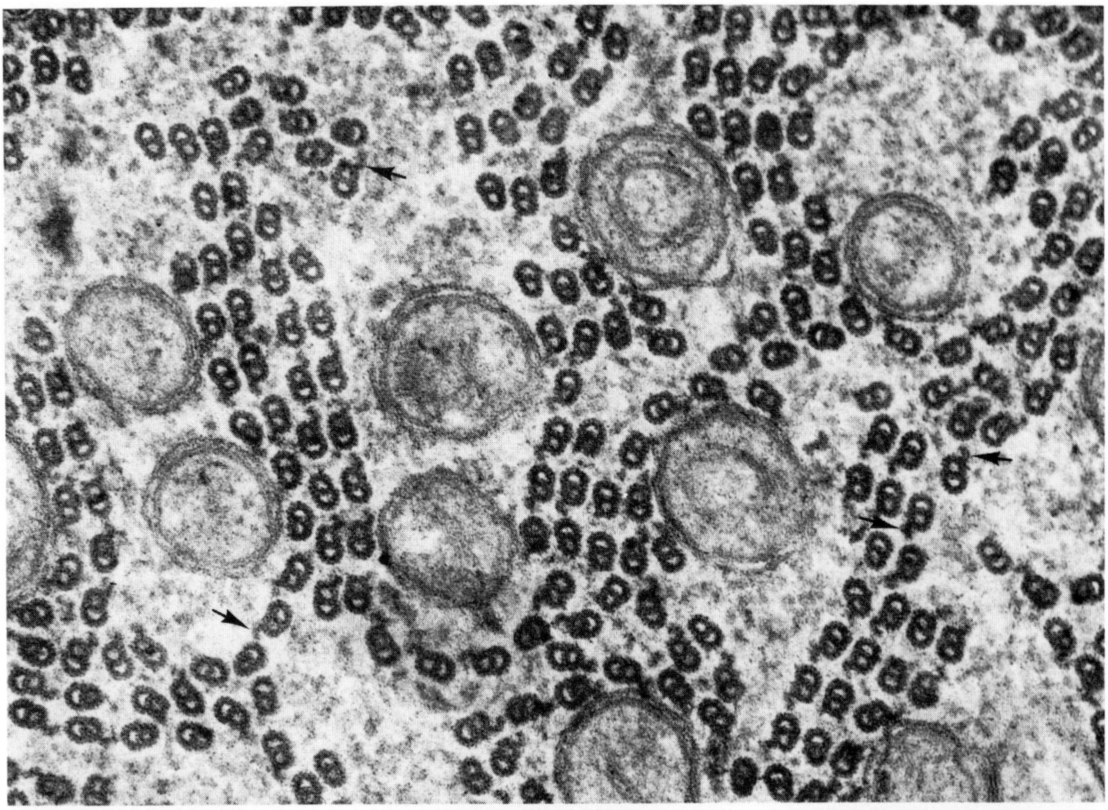

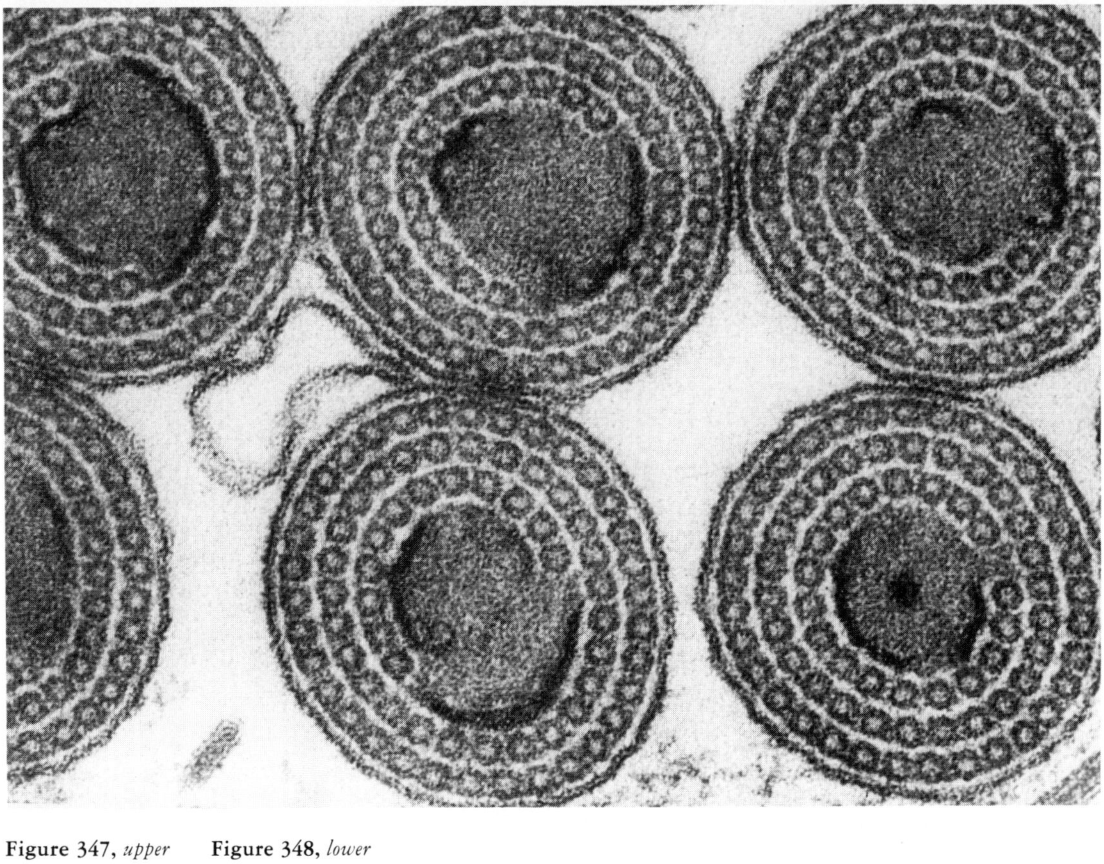

Figure 347, *upper* **Figure 348**, *lower*

REFERENCES

Cilia and Flagella and Sperm Flagellum

Afzelius, B. Electron microscopy of the sperm tail. Results obtained with a new fixative. J. Biophys. Biochem. Cytol. 5:269–278, 1959.

Afzelius, B., R. Eliasson, O. Johnson et al. Lack of dynein arms in immotile human spermatozoa. J. Cell Biol. 66:225–232, 1975.

Albrecht-Buehler, G. and A. Bushnell. The ultrastructure of primary cilia in quiescent 3T3 cells. Exp. Cell Res. 126:427–437, 1980.

Allen, R. D. A reinvestigation of cross sections of cilia. J. Cell Biol. 37:825–831, 1968.

Amos, L. A. and A. Klug. Arrangements of subunits in flagellar doublet microtubules. J. Cell Sci. 14:523–549, 1974.

Amos, L. A., R. S. Linck and A. Klug. Molecular structure of flagellar microtubules. *In* Cell Motility, Book C, (R. D. Goldman, T. D. Pollard, and J. L. Rosenbaum, eds.), pp. 847–867, Cold Spring Harbor Laboratory, New York, 1976.

Andre, J. and J. P. Thiery. Mis en evidence d'un sous-structure fibrillaire dans les filaments axonematiques des flagelles. J. Miscrosc. 2:71, 1963.

Bacetti, B., R. Dallai, F. Giusti and F. Bernine. The spermatozoon of Arthropoda XXV. A new model of tail having up to 170 doublets: *Monarthropalpus buxi*. Tissue Cell 6:269–278, 1974.

Baccetti, B., V. Pallini and A. G. Burrini. The accessory fibers of the sperm tail. III. High sulfur and low sulfur components in mammals and cephalopods. J. Ultrastr. Res. 57:289–308, 1976.

Baccetti, B. and R. Dallai. The spermatozoon of Arthropoda XXVII. Uncommon axoneme patterns in different species of cecidomyid flies. J. Ultrastr. Res. 55:50–69, 1976.

Baccetti, B. and B. A. Afzelius. The Biology of the Sperm Cell. S. Karger, Basel, 1976.

Ballowitz, E. Untersuchunger über die Struktur der Spermatozoön zugleich ein Beitrag sur Lehre von feinerem Base der contraktilen Elements. Arch. Mikroskop. Anat. Entwicklungsmech. 32:401–473, 1888.

Barnes, B. G. Ciliated secretory cells in the pars distalis of the mouse hypophysis. J. Ultrastr. Res. 5:453–467, 1961.

Bishop, D. W. and H. Hoffmann-Berling. Extracted mammalian sperm models. I. Preparation and reactivation with ATP. J. Cell. Comp. Physiol. 53:445–466, 1959.

Bishop, D. W. Sperm motility. Physiol. Rev. 42:1–59, 1962.

Brinkley, B. R., E. J. Stubblefield and T. C. Hsu. The effects of colcemid inhibition and reversal on the fine structure of the mitotic apparatus of chinese hamster cells in vitro. J. Ultrastr. Res. 19:1–18, 1967.

Brokaw, C. J. Movement of flagella of Polytoma uvella. J. Exp. Biol. 40:149–156, 1963.

Brokaw, C. J. Non-sinusoidal bending waves of sperm flagella. J. Exp. Biol. 43:155–169, 1965.

Brokaw, C. J. Bend propagation by a sliding filament model for flagella. J. Exp. Biol. 55:289–304, 1971.

Brokaw, C. J. Computer simulation of flagellar movement. Biophys. J. 12:564–586, 1972.

Chasey, D. Observations on the central pair of microtubules from the cilia of *Tetrahymena pyriformis*. J. Cell Sci. 5:453–458, 1969.

Chasey, D. Further observations on the ultrastructure of cilia from *Tetrahymena pyriformis*. Exp. Cell. Res. 74:471–479, 1972.

Dallai, R. An overview of atypical spermatozoa in insects. *In* The Spermatozoon (D. W. Fawcett and J. M. Bedford, eds.), pp. 253–265, Urban and Schwarzenberg, Baltimore, 1979.

Fawcett, D. W. Cilia and flagella. *In* The Cell, vol. 2 (J. Brachet and A. E. Mirsky, eds.), pp. 217–298, Academic Press, New York, 1961.

Fawcett, D. W. The anatomy of the mammalian spermatozoon with particular reference to the guinea pig. Z. Zellforsch. 67:279, 1965.

Fawcett, D. W. A comparative view of sperm ultrastructure. Biol. Reprod. 2, Suppl. 90–127, 1970.

Fawcett, D. W. The mammalian spermatozoon. A review. Dev. Biol. 44:394–436, 1975.

Fawcett, D. W. and S. Ito. The fine structure of bat spermatozoa. Am. J. Anat. 116:567, 1965.

Fawcett, D. W. and D. M. Phillips. Recent observations on the ultrastructure and development of the mammalian spermatozoon. *In* Comparative Spermatology (B. Baccetti, ed.), Academic Press, New York, 1969.

Fawcett, D. W. and K. R. Porter. A study of the fine structure of ciliated epithelia. J. Morphol. 94:221–281, 1954.

Gibbons, I. R. The relationship between fine structure and beat in the gill cilia of a lamellibranch mollusc. J. Biophys. Biochem. Cytol. 11:179–205, 1961.

Gibbons, I. R. Studies on the protein components of cilia from *Tetrahymena pyriformis*. Proc. Nat. Acad. Sci. 50:1002–1010, 1963.

Gibbons, I. R. Chemical dissection of cilia. Arch. Biol. (Liège) 76:317–352, 1965.

Gibbons, I. R. Studies on the adenosine triphosphatase activity of 14S and 30S dynein from cilia of *Tetrahymena*. J. Biol. Chem. 241:5590–5596, 1966.

Gibbons, I. R. The biochemistry of motility. Ann. Rev. Biochem. 37:521–546, 1968.

Gibbons, B. H. and I. R. Gibbons. Flagellar movement and adenosine triphosphatase activity in sea urchin sperm extracted with Triton X-100. J. Cell Biol. 54:75–97, 1972.

Gibbons, I. R. and A. V. Grimstone. On flagellar structure in certain flagellates. J. Biophys. Biochem. Cytol. 7:697–716, 1960.
Gibbons, I. R. and A. J. Rowe. Dynein: A protein with adenosine triphosphatase activity from cilia. Science 149:424–426, 1965.
Greenwood, A. D., I. Manton and B. Clarke. Observations on the structure of the zoospores of Vaucheria. J. Exp. Bot. 8:71–86, 1957.
Grimstone, A. V. and A. Klug. Observations on the substructure of flagellar fibres. J. Cell Sci. 1:351–362, 1966.
Jensen, O. S. Untersuchunger über die Samenkörper der Saugethiero, Vögel, und Amphibian. I. Saugetheire. Arch. mikroskop. Anat. Entwicklungsmech. 301:379–425, 1887.
Linck, R. W. Flagellar doublet microtubules: Fractionation of minor components and α-tubulin from specific regions of the A-tubule. J. Cell Sci. 20:405–439, 1976.
Linck, R. W. Advances in the ultrastructural analysis of the sperm flagellar axoneme. *In* The Spermatozoon (D. W. Fawcett and J. M. Bedford), Urban and Schwarzenberg, Baltimore, 1979.
Manton, I. The fine structure of plant cilia. Symp. Soc. Exp. Biol. 6:306–319, 1952.
Manton, I. and B. Clarke. Observations with the electron microscope in the internal structure of the spermatozoid of Fucus. J. Exp. Bot. 7:416–432, 1956.
Ogawa, K., T. Mohri and H. Mohri. Identification of dynein as the outer arms of sea urchin sperm axonemes. Proc. Nat. Acad. Sci. 74:5006–5010, 1977.
Olson, G. E., D. W. Hamilton and D. W. Fawcett. Isolation and characterization of the fibrous sheath of rat epididymal spermatozoa. Biol. Reprod. 14:517–530, 1976.
Olson, G. E., and R. W. Linck. Observations of the structural components of flagellar axonemes and central pair microtubules from rat sperm. J. Ultrastr. Res. 61:21–43, 1977.
Pease, D. C. The ultrastructure of flagellar fibrils. J. Cell Biol. 18:313–326, 1963.
Pedersen, H. and H. Rebbe. Absence of arms on the axoneme of immobile human spermatozoa. Biol. Reprod. 12:541–544, 1975.
Phillips, D. M. Exceptions to the prevailing pattern of tubules (9+9+2) in the sperm flagella of certain insect species. J. Cell Biol. 40:28–43, 1969.
Phillips, D. M. Fine structure of *Sciara coprophila* sperm. J. Cell Biol. 30:499–517, 1966.
Phillips, D. M. Insect sperm: Their structure and morphogenesis. J. Cell Biol. 44:243–277, 1970.
Phillips, D. M. Spermiogenesis. Academic Press, New York, 1974.
Picheral, B. Les éléments cytoplasmique au cours de la spermiogenèse du Triton Pleurodeles waltlii. III. L'evolution des formation caudales. Zeitsch. Zellforsch. 131:399–416, 1972.
Picheral, B. Structural, comparative and functional aspects of spermatozoa in urodeles. *In* The Spermatozoon (D. W. Fawcett and J. M. Bedford, eds.), pp. 267–287. Urban and Schwarzenberg, Baltimore, 1979.
Purkinje, J. F. and G. Valentin. Entdeckung kontinuierlicher durch Wimperhaare erzengter Flimmerbewegungen als eines allgemeinen Phänomens in der Klassen der Amphibien, Vögel, und Sangethieren. Arch. Anat. Physiol. 1:391–400, 1834.
Rieder, C. L., C. G. Jensen and L. Jensen. The resorption of primary cilia during mitosis in a vertebrate (Ptk_1) cell line. J. Ultrastr. Res. 68:173–185, 1979.
Rumery, R. E. and E. M. Eddy. Scanning electron microscopy of the fimbriae and ampullae of rabbit oviducts. Anat. Res. 178:83–102, 1974.
Sale, W. S. and P. Satir. The direction of active sliding in *Tetrahymena* cilia. Proc. Nat. Acad. Sci. 74:2045–2049, 1977.
Satir, P. Studies on cilia. The fixation of the metachronal wave. J. Cell Biol. 18:345–365, 1963.
Satir, P. Structure and function in cilia and flagella, pp. 1–52. Protoplasmatologia III E, 1965.
Satir, P. Studies on cilia. II. Examination of the distal region of the ciliary shaft and the role of the filaments in motility. J. Cell Biol. 26:805–834, 1965.
Satir, P. Studies on cilia. III. Further studies on the cilium tip and a "sliding filament model" of ciliary motility. J. Cell Biol. 39:77–94, 1968.
Scherft, J. P. and W. T. Daems. Single cilia in chondrocytes. J. Ultrastr. Res. 19:546–555, 1967.
Sleigh, M. A. Cilia and Flagella. Academic Press, London, 1974.
Sorokin, S. Centrioles and the formation of rudimentary cilia by fibroblasts and smooth muscle cells. J. Cell Biol. 15:363–377, 1962.
Stephens, R. E. Thermal fractionation of outer fiber doublet microtubules into A- and B-subfiber components: A- and B-tubulin. J. Molec. Biol. 47:353–363, 1970.
Stephens, R. E. Isolation of nexin — the linkage protein responsible for the maintenance of the ninefold configuration of flagellar axonemes. Bio. Bull. 139:438, 1970.
Summers, K. E. and I. R. Gibbons. Adenosine-triphosphate-induced sliding of tubules in trypsin-treated flagella of sea urchin sperm. Proc. Nat. Acad. Sci. 68:3092–3096, 1971.
Warner, F. D. New observations on flagellar fine structure. The relationship between matrix structure and the microtubule component of the axoneme. J. Cell Biol. 47:159–182, 1970.
Warner, F. D. and D. R. Mitchell. Structural conformation of ciliary dynein arms and the generation of sliding forces in *Tetrahymena* cilia. J. Cell Biol. 76:261–277, 1978.
Warner, F. D. and P. Satir. The structural basis of ciliary bend formation. Radial spoke positional changes accompanying microtubule sliding. J. Cell Biol. 63:35–63, 1974.
Witman, G. B., J. Plummer and G. Sander. *Chlamydomonas* flagellar mutants lacking radial spokes and central tubules. Structure, composition, and function of specific axonemal components. J. Cell Biol. 76:729–747, 1978.

CYTOPLASMIC INCLUSIONS

GLYCOGEN

Glycogen is the polysaccharide that constitutes the storage form of carbohydrate in animal cells. It is broken down to yield glucose, and the enzymatic degradation of glucose in turn provides energy and also supplies short-chain carbon skeletons that are reused in the synthesis of various basic components of protoplasm. Although the major carbohydrate stores of the body are in the liver and in skeletal muscle, glycogen is found in smaller amounts in a great many cell types.

Particles 20 to 35 nm in diameter and of low density were described in electron micrographs of liver by Fawcett in 1956 and tentatively interpreted as the *in situ* counterpart of the particulate glycogen that had been isolated by Lazarow in 1942 in early experiments on differential centrifugation of liver homogenates. Their identity was subsequently established in correlated morphological and biochemical studies (Revel, Napolitano and Fawcett, 1956). With improved procedures for staining ultrathin sections with lead, glycogen was soon routinely demonstrable as a particulate component of the cytoplasm distinguishable from the ribonucleoprotein particles later called ribosomes. Its appearance in electron micrographs is considerably influenced by the method of specimen preparation, but after treatment of tissue with most of the fixatives in current use, glycogen is preserved in the form of roughly isodiametric 20 to 30 nm particles, often irregular in outline. These so-called *beta particles* occur singly in some cell types, while in others they form aggregates of larger size commonly referred to as *alpha particles*, or rosettes. These vary greatly in size, depending upon the number of constituent unit particles, and may reach diameters up to 0.1 μm. Nothing is known about the local factors that determine whether a given cell type will have glycogen in the form of individual particles or aggregates. Nor is it known whether the chemical and metabolic properties of these two morphological forms differ significantly.

Glycogen is usually confined to the cytoplasm, but in the liver of amphibian larvae, in animals and humans with severe diabetes, and in the heritable glycogen storage diseases, intranuclear glycogen is observed. In such cases the particles are usually smaller than those in the cytoplasm of the same cells and do not form aggregates. The physical properties or ionic environment of the nucleoplasm may result in polymerization of glycogen that is shorter chained or less highly branched. If it is synthesized in the cytoplasm and enters the nucleus secondarily, the size of the nuclear pore complexes may conceivably be responsible for some selectivity in the size of the molecules that enter the nucleoplasm.

In mammalian liver, glycogen appears in the form of coarse rosettes or alpha particles. These are not uniformly distributed in the cytoplasm but tend to be concentrated in areas that are rich in smooth endoplasmic reticulum. As seen in the accompanying micrograph, the glycogen occupies the interstices among the membrane-limited circular and elongated profiles of the reticulum. This association is discussed further in the section on the endoplasmic reticulum.

Figure 349. Electron micrograph of a hepatic cell from a fasted hamster after an injection of cortisone. (Micrograph courtesy of Albert Jones.)

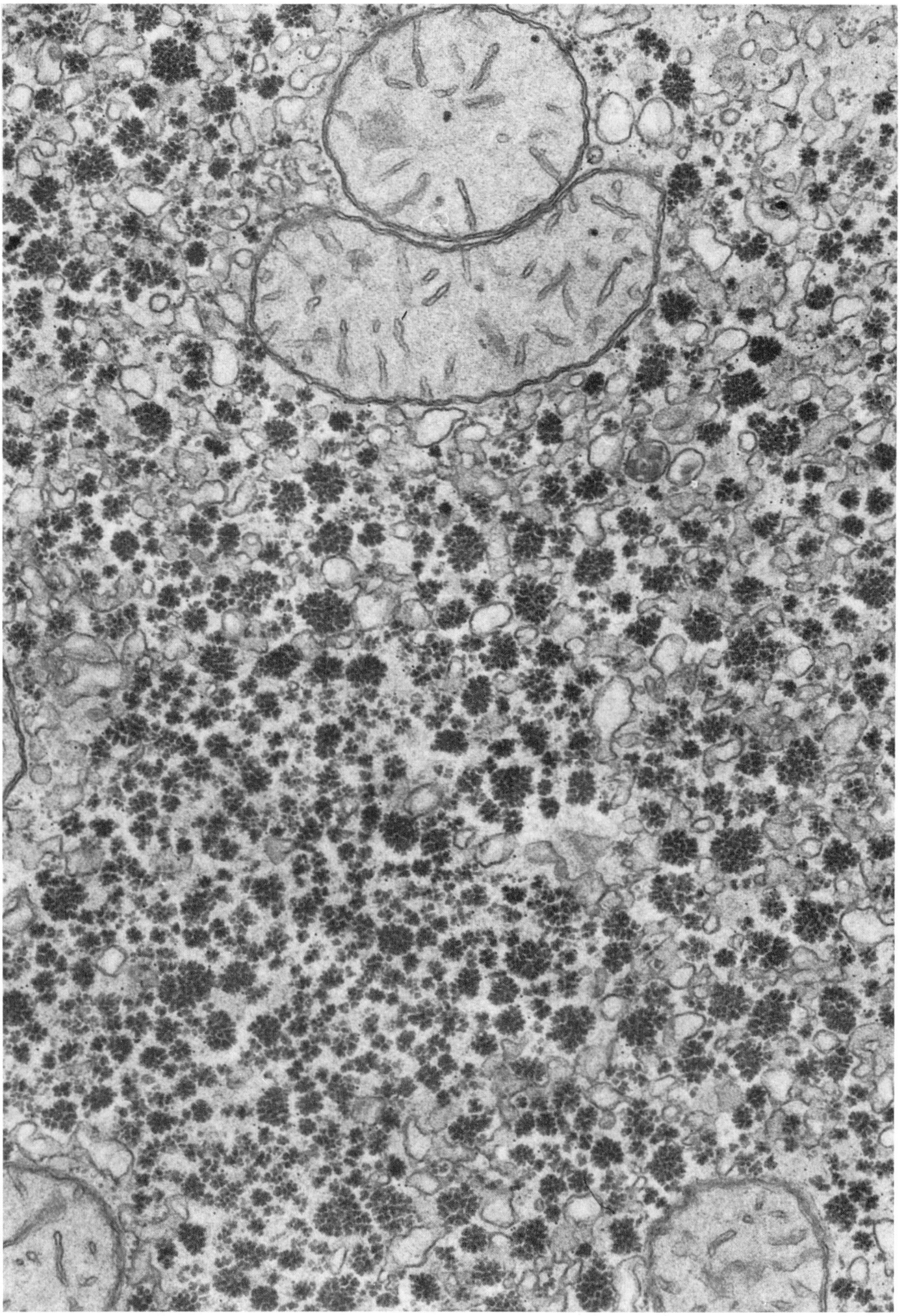

Figure 349

There are significant species differences in the size and appearance of glycogen. In amphibian liver, illustrated here, the particles are so small and uniform in size that they might easily be mistaken for unit beta particles. In micrographs of higher magnification, however, they are found to be aggregates like the alpha particles of mammalian liver, but made up of much smaller subunits.

Figure 350. Glycogen in the liver of the salamander, *Batrachoseps attenuatus.*

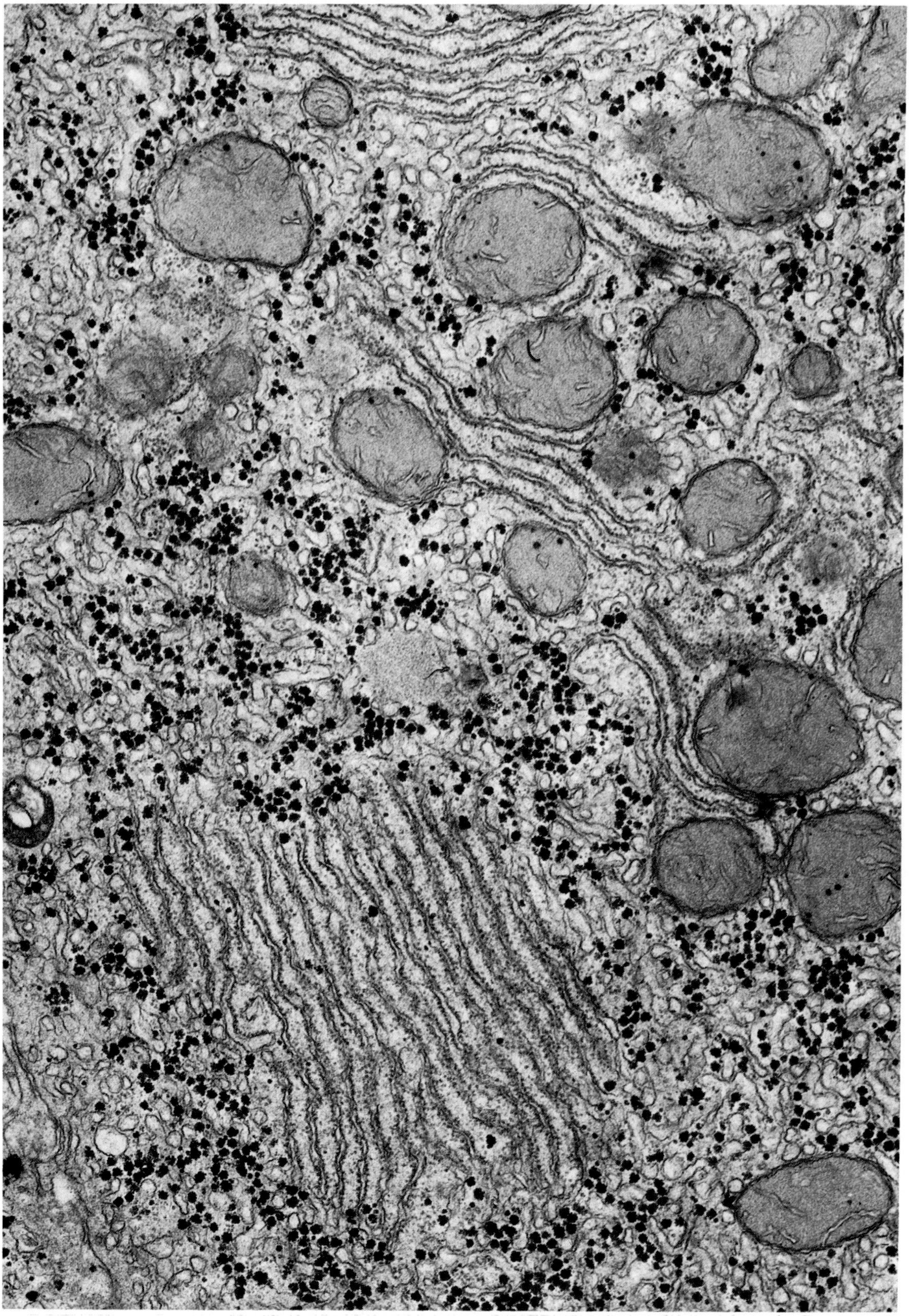

Figure 350

Glycogen is an important energy source for cardiac muscle. In this tissue, it does not aggregate to form the alpha particles but occurs as individual particles distributed throughout the sarcoplasm in close association with the mitochondria and the sarcoplasmic reticulum. A few particles also occur within the contractile material, most frequently among the actin filaments in the I-band. But they may also be aligned in short rows in the interstices between myosin filaments in the H-band. They are not found in the region of interdigitation of actin and myosin filaments.

Figure 351. Glycogen in ventricular papillary muscle of cat heart.

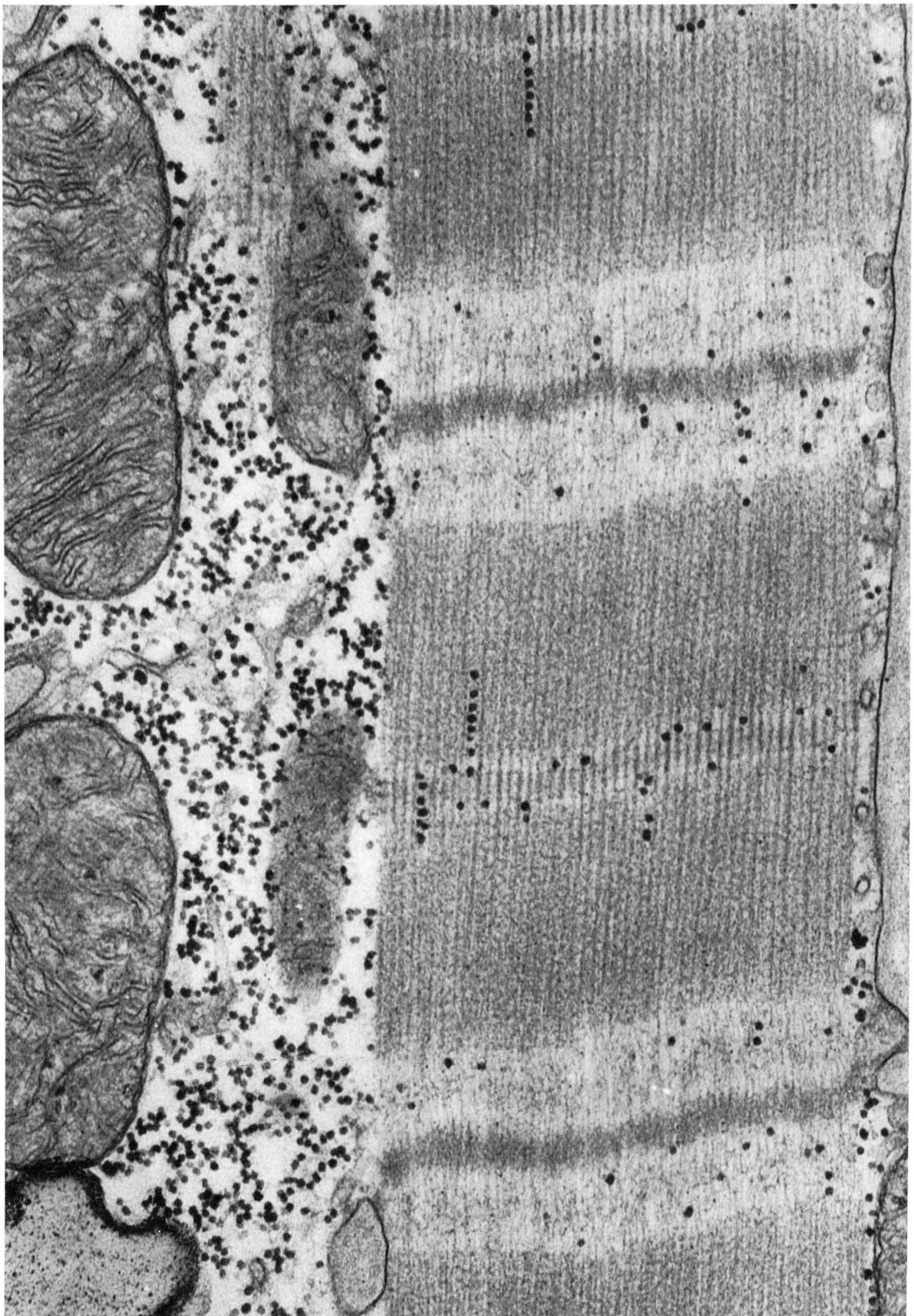

Figure 351

Glycogen occurs in small amounts in the mammalian central nervous system in protoplasmic and fibrous astrocytes and in presynaptic axon terminals, but conspicuous accumulations are rare. It is considerably more abundant in cold-blooded vertebrates, especially in winter dormancy. The accompanying micrograph of a mossy fiber from the turtle cerebellum in winter shows abundant glycogen in the presynaptic cytoplasm and small amounts in the postsynaptic elements.

Figure 352. Glycogen in a mossy fiber from cerebellum of a winter turtle *Chrysemys picta*. (Micrograph courtesy of Enrico Mugnaini.)

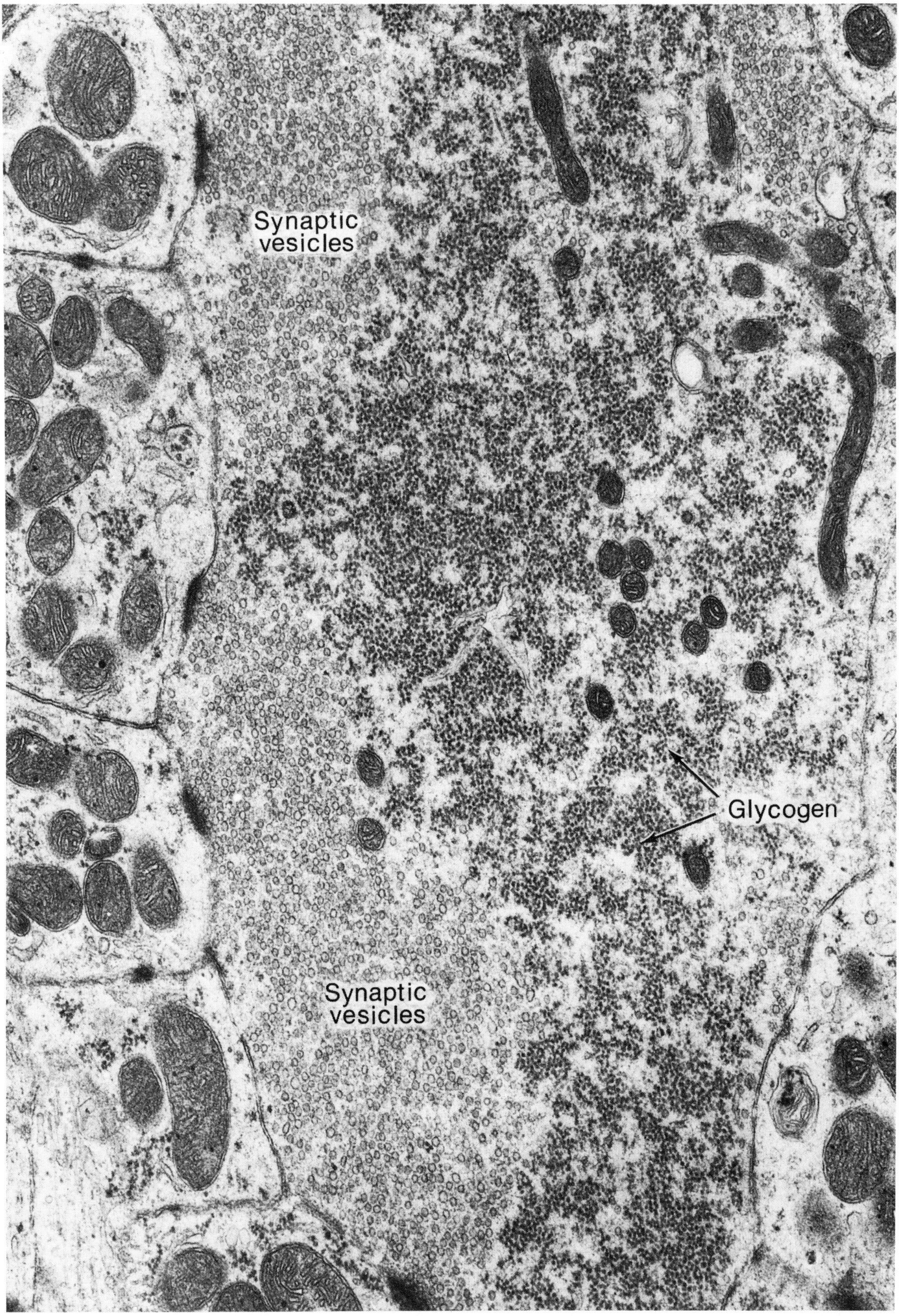

Figure 352

Glycogen is seldom uniformly distributed in the cytoplasm but tends to accumulate in dense regional deposits. It is generally excluded from the cell center and Golgi region, a narrow zone around the nucleus and the ectoplasmic layer immediately beneath the plasma membrane.

Figures 353 and 354. Glycogen deposits in cells of the guinea pig rete testis. (From Fawcett and Dym, J. Reprod. Fertil., 38:401–409, 1974.)

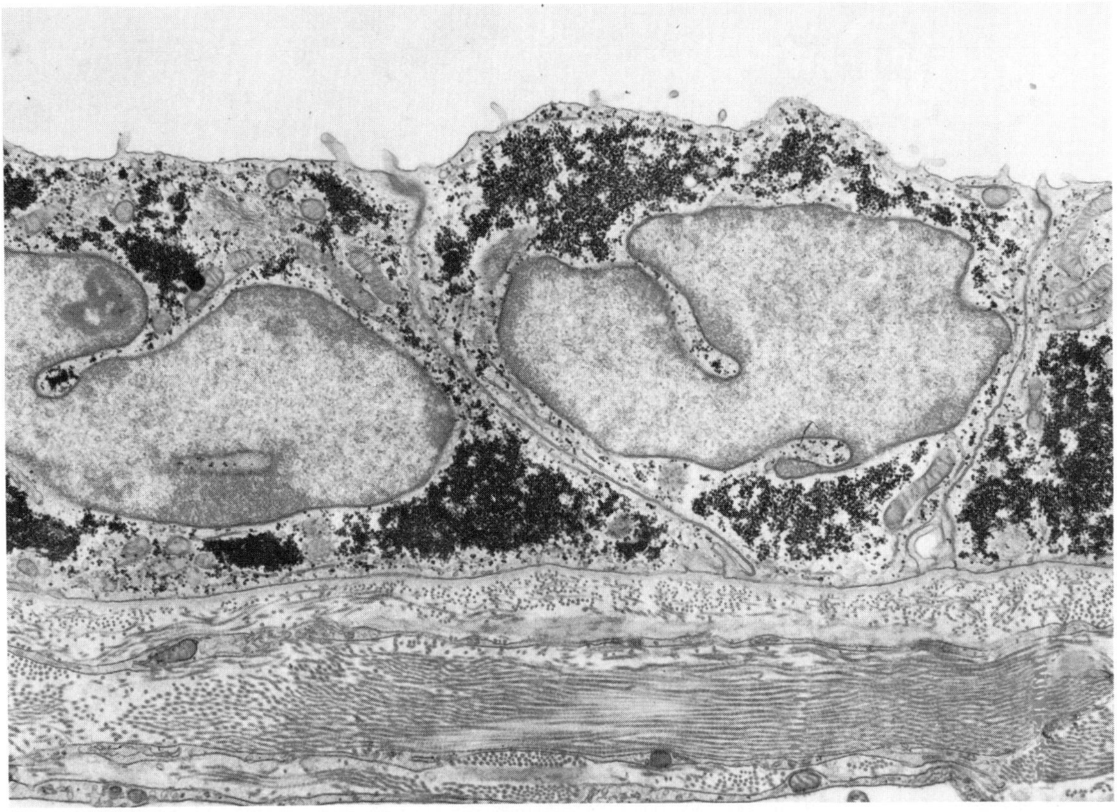

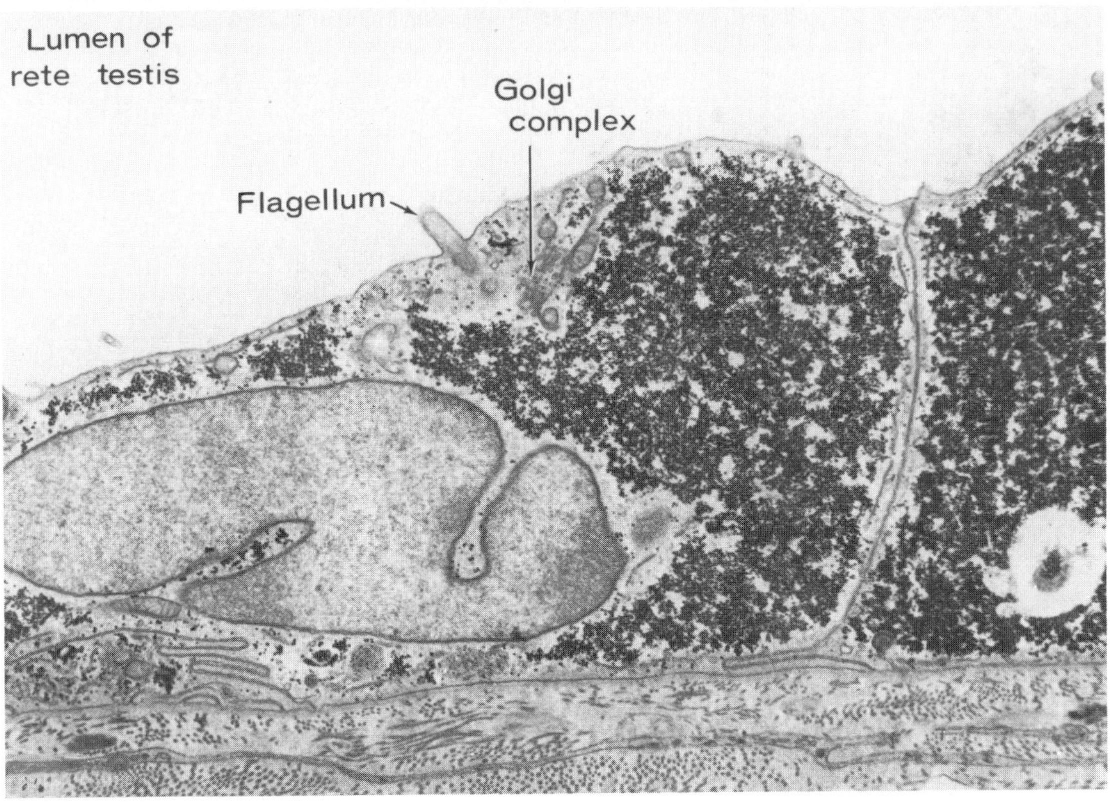

Figure 353, *upper* **Figure 354,** *lower*

The most extreme example of accumulation of glycogen is in the glycogen body of the nerve cord of the developing chick (Doyle and Watterson, 1949). In this transient organ of unknown function on the dorsal aspect of the spinal cord, glycogen may amount to 60 to 80 per cent of its fat-free dry weight. Glycogen fills the cell body to much the same extent as lipid does in an adipose cell, displacing the nucleus and organelles to the cell periphery.

A comparable example in an adult mammal is illustrated in the accompanying micrograph of the epithelial cells that line the tubuli recti in guinea pig testis. An extraordinary accumulation of glycogen beta particles fills the cell body, deforming and displacing the nucleus to the cell apex and leaving only a very thin shell of cytoplasmic matrix immediately beneath the plasma membrane. The centrioles and a small Golgi apparatus are found near the nucleus and occasional mitochondria occur in the thin peripheral layer of cytoplasm. This glycogen-rich segment of the excurrent duct system of the testis has, thus far, been reported only in the guinea pig. The functional significance of its remarkable degree of glycogen storage is unknown.

Figure 355. Glycogen-storing cells of a tubulus rectus of the guinea pig testis. (From Fawcett and Dym, J. Reprod. Fertil. 88:401–409, 1974.)

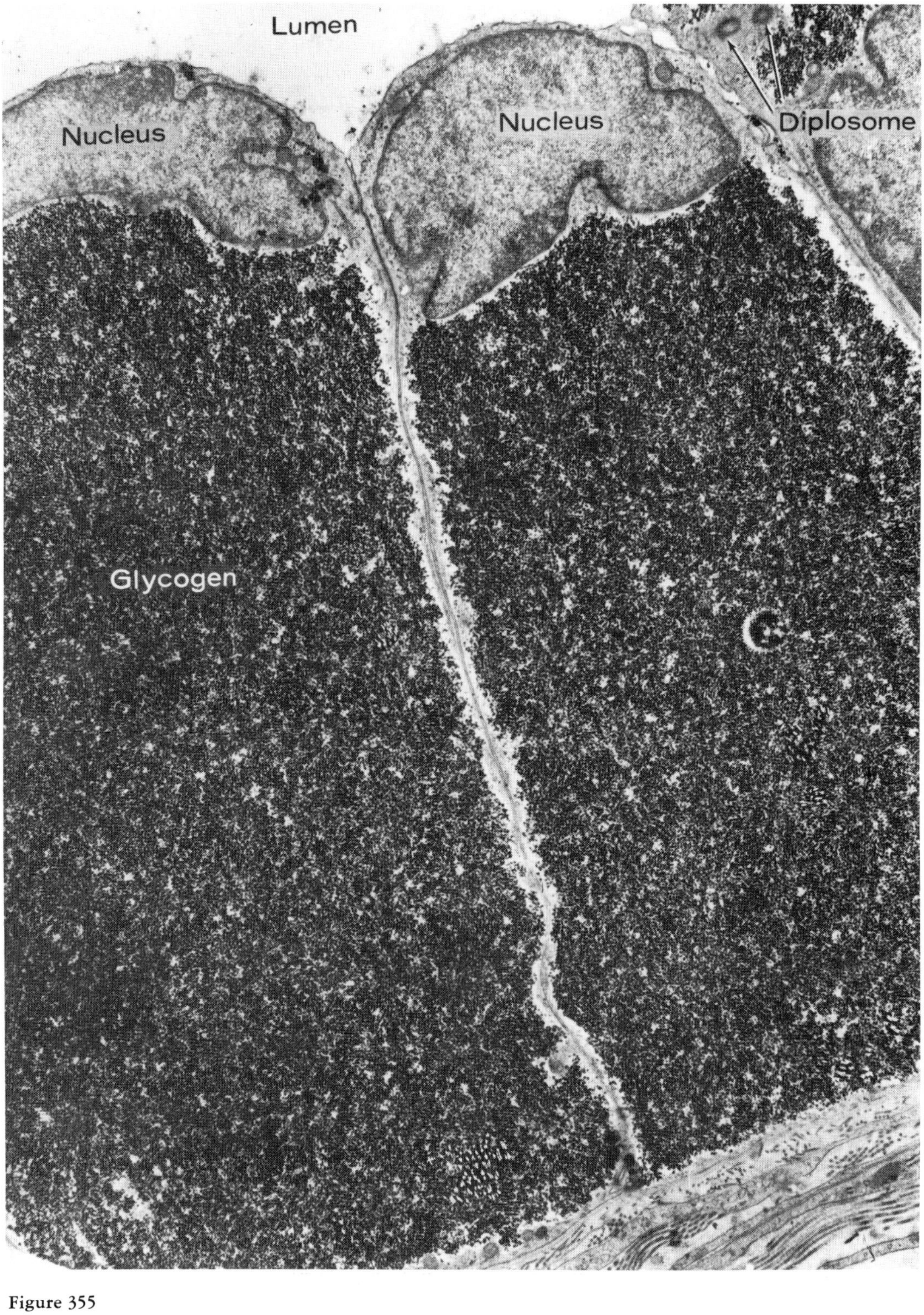

Figure 355

LIPID

The commonest lipid inclusions found in cells are triglycerides. These are fluid at body temperature and appear in living cells as refractile spherical droplets. They reach their largest size in adipose cells which specialize in synthesis and storage of fat as a reservoir of potential chemical energy. Lipid yields more than twice as many calories per gram as does carbohydrate. Although the bulk of the body's store of lipid is in adipose tissue, nearly all cell types contain a few lipid droplets. These serve as a local store of energy and a potential source of short carbon chains that are used by the cell in its synthesis of membranes and other lipid-containing structural components or secretory products.

When cells are exposed to primary osmium fixation, lipid droplets that are rich in unsaturated fatty acids are blackened by reduction of osmium and rendered insoluble to alcohol and other reagents used in the subsequent processing for electron microscopy. Under favorable conditions, the droplets retain their spherical form, but occasionally they take on distorted and irregular outlines that are not representative of their form in the living cell. The degree of blackening of the lipid with osmium varies with its composition. Failure of the fixative to penetrate to the center of large droplets may result in extraction of lipid from their interior in later steps of preparation, leaving a clear center surrounded by a dense black rim of insolubilized lipid.

With aldehyde fixatives, lipid is well preserved but when subsequently exposed to osmium, it may reduce relatively little. In micrographs of such preparations, the lipid usually has a homogeneous grey appearance. Because of variations in the degree of saturation of triglycerides and the several sources of artifacts in fixation and staining, one must be prepared to encounter a variety of appearances of lipid in electron micrographs.

The lipid that accumulates within cells for internal use is not limited by a membrane but takes the form of spherical oil droplets suspended in the aqueous cytoplasmic matrix.

In the accompanying micrograph, the lipid was intensely blackened by osmium even though primary fixation was with glutaraldehyde. The droplets have maintained their normal spherical form.

Figure 356. Lipid in a Sertoli cell from opossum testis, *Didelphis virginiana*.

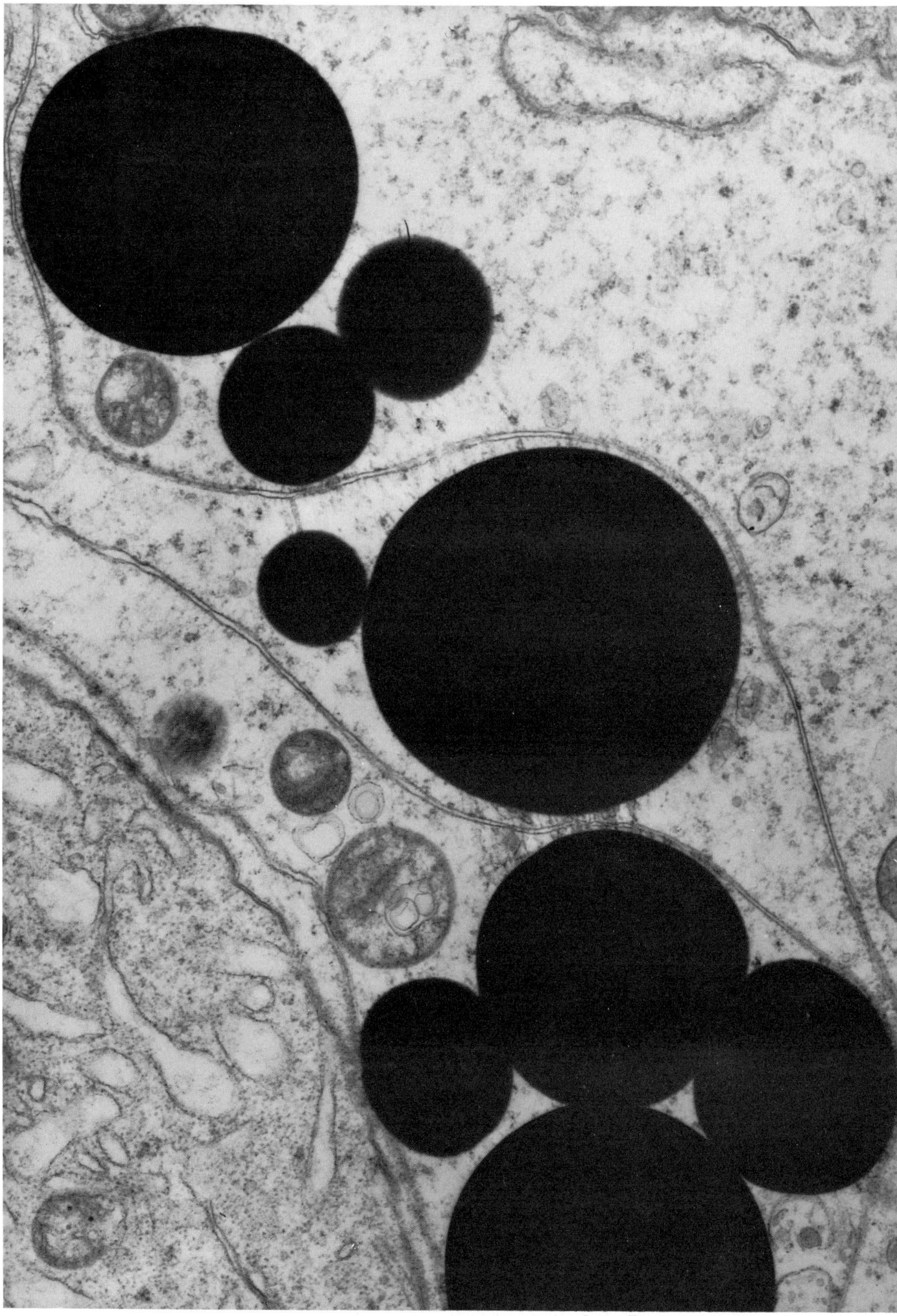

Figure 356

Lipid droplets in intestinal epithelial cells are an exception to the general rule that intracellular lipid is not membrane bounded. In this case, however, the lipid is not destined for local use by the cells in which it resides. In the course of intestinal absorption of dietary fat, intralumenal fatty acids and monoglycerides diffuse through the plasma membrane of the brush border and are reesterified to triglyceride by enzymes in the membrane of the smooth endoplasmic reticulum. The small lipid droplets so formed are lodged within the lumen of the reticulum and not in the cytoplasmic matrix. After further processing in the Golgi complex, they are released into the intercellular clefts of the epithelium as chylomicrons, which continue their passage through the epithelium via a paracellular route.

Figure 357. Intestinal epithelium of rat in an advanced stage of fat absorption. (Micrograph from S. Palay and J. P. Revel in *Lipid Transport* [H. C. Meng, ed.], Charles C Thomas, Springfield, Illinois, 1964.)

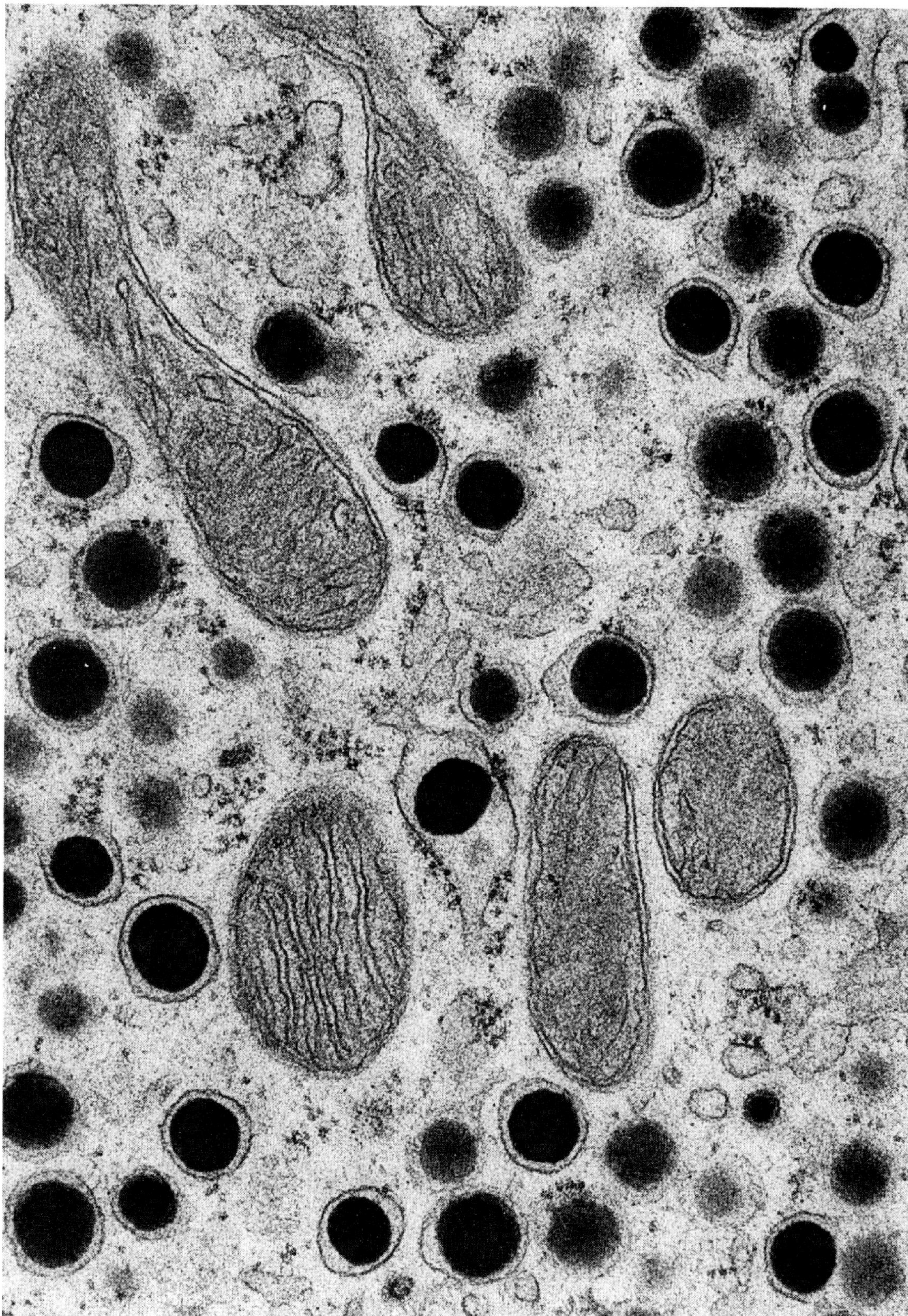

Figure 357

A given cell type may contain abundant lipid droplets in one species and relatively few in another. Such is the case with the Sertoli cells of the testis illustrated here. In some species such as Rhesus monkey or the gerenuk shown here, they have large numbers of small lipid droplets in the basal cytoplasm, while in the domestic boar, one or two droplets several microns in diameter are not uncommon, and in laboratory rodents a few droplets of intermediate size are present but only at certain stages of the spermatogenic cycle.

The lipid droplets are often surrounded by concentric cisternae of smooth endoplasmic reticulum as shown in the lower figure. Since agranular reticulum is known to be involved in lipid metabolism, this intimate relationship to the droplets is no doubt functionally significant.

Figure 358. Multiple small and medium-sized lipid droplets at the base of Sertoli cells from the gerenuk *(Litocranius walleri).*

Figure 359. Common relationship of the smooth reticulum to a lipid droplet in a Seroli cell from the ram *(Ovis aries).* (Figures from Fawcett in Handbook of Physiology, Sect. 7, Vol. 5, Am. Physiol. Soc., Washington, D. C., 1975.)

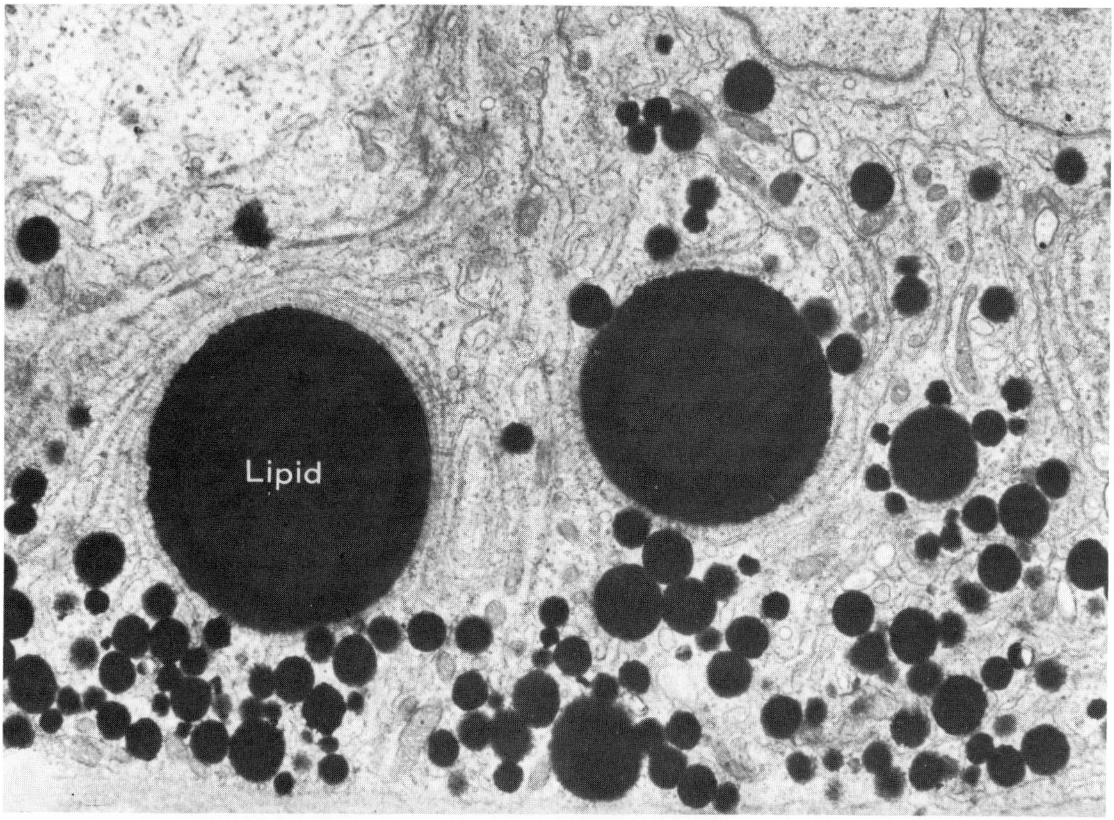

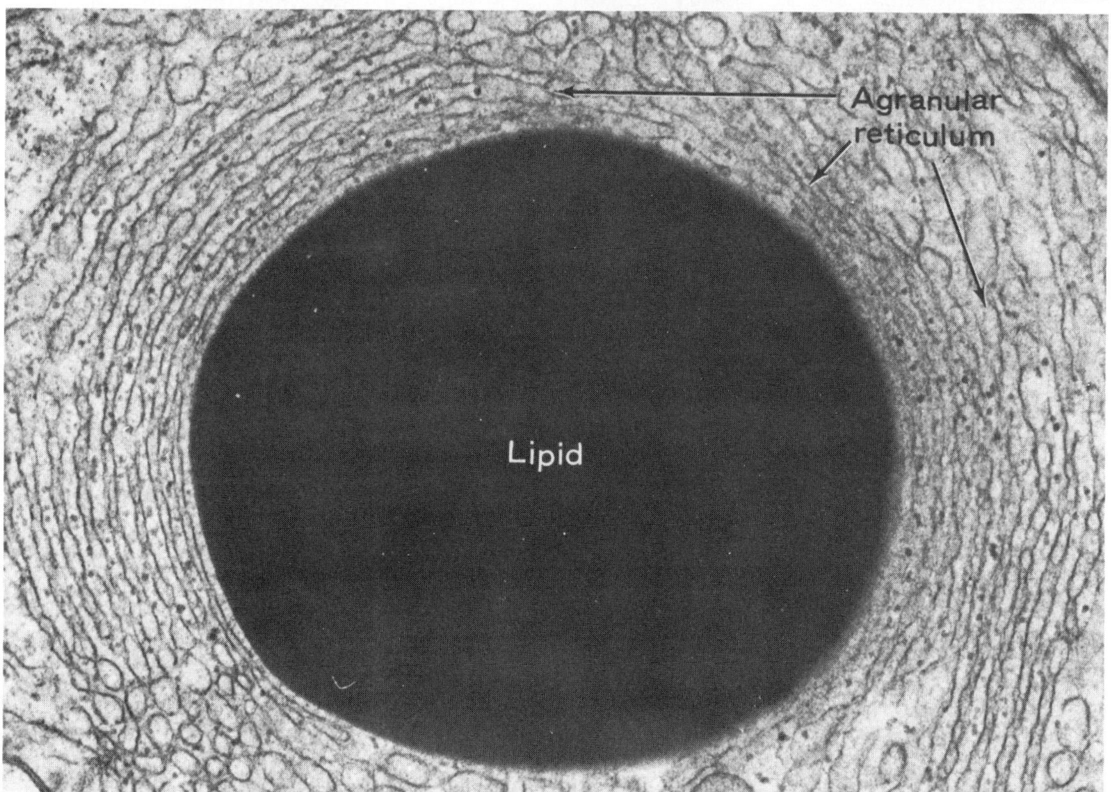

Figure 358, *upper* **Figure 359,** *lower*

The same cell type may contain differing amounts of lipid in different phases of physiological activity. This is especially true of steroid-secreting endocrine glands in which cholesterol esters that are precursors in the biosynthesis of steroid hormones are stored in droplets of triglyceride.

The accompanying micrograph shows a portion of a Leydig cell from a reproductively active seasonal breeder. The cell contains numerous lipid droplets of modest size. On the other hand, in the quiescent phase of the annual breeding cycle, it would contain one or two enormous lipid droplets occupying the greater part of the cell body.

Similarly, the cells of the unstimulated adrenal cortex are rich in lipid, whereas a stimulated actively secreting gland contains relatively little. These variations are consistent with the function of the lipid droplets in these cells as sites of accumulation of precursors that are depleted during active synthesis and release of steroid hormone.

Figure 360. Leydig cell from a reproductively active ground squirrel, *Citellus lateralis.*

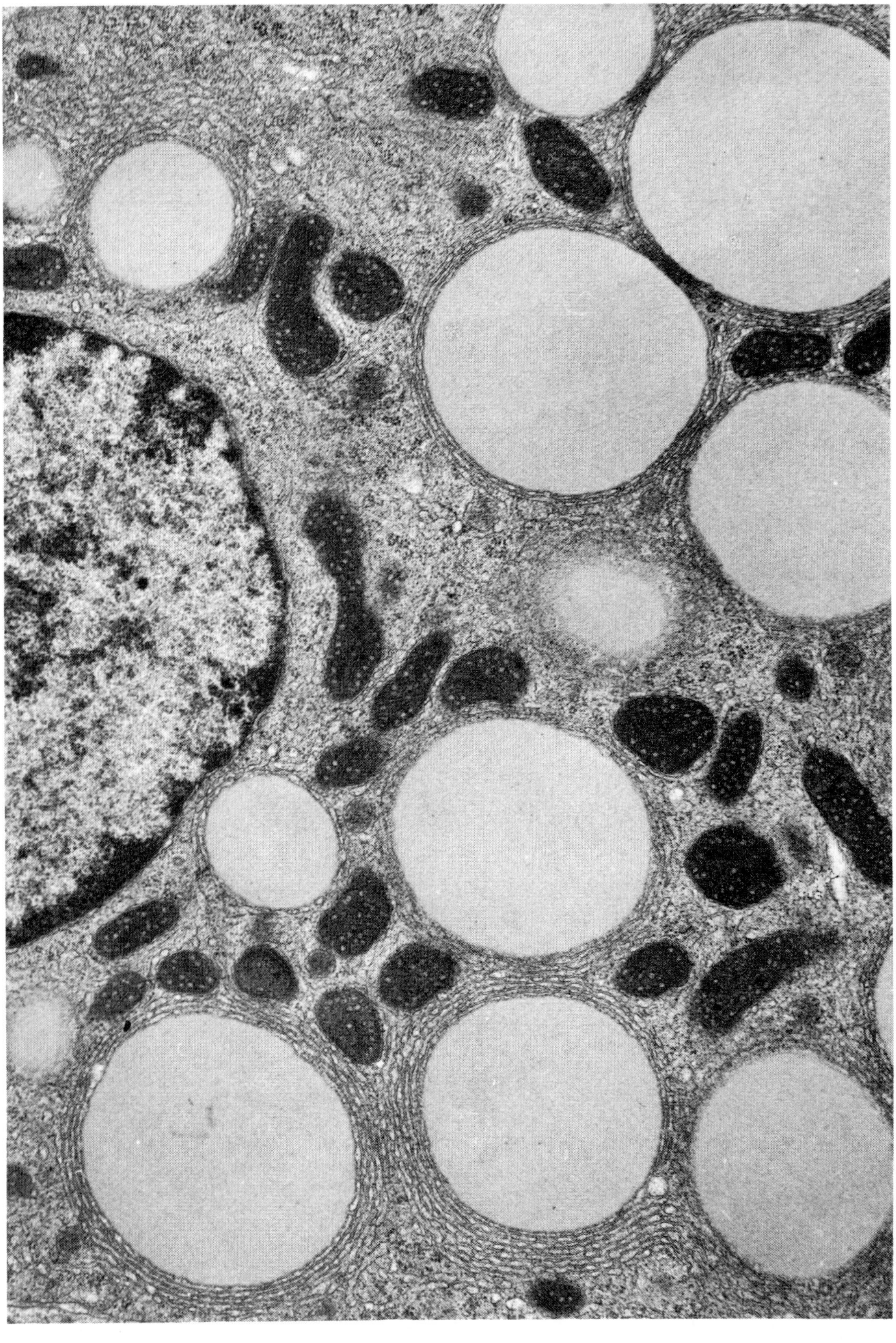

Figure 360

In contrast to jet-black appearance of lipid in several of the foregoing illustrations, the lipid droplets in the testicular interstitial cell shown here exhibit little or no osmiophilia and are instead a homogeneous pale grey. Since the fixation and preparative procedures were much the same, it seems likely that the differences in osmiophilia of the lipid is a reflection of differences in the degree of saturation of their constituent fatty acids.

Notice that here again one or more cisternae of smooth endoplasmic reticulum are in close apposition to the lipid, creating an erroneous impression that the droplets themselves have a limiting membrane.

Figure 361. Lipid in Leydig cell of the ground squirrel, *Citellus lateralis*.

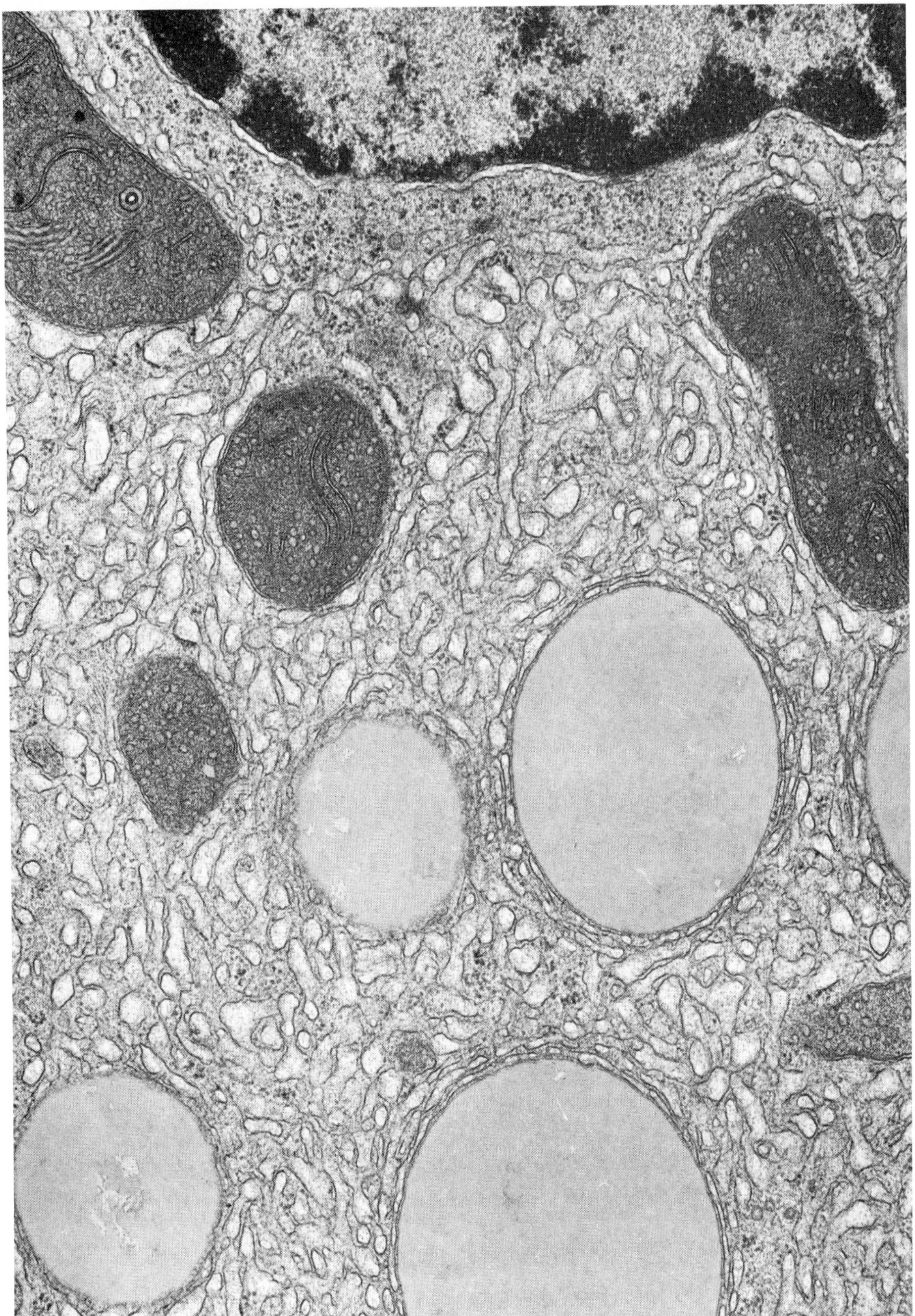

Figure 361

Adipose cells are among the largest cells in the body, but less than one fortieth of their total volume is metabolically active. The nucleus is displaced to the periphery and the cytoplasm is reduced to a thin layer surrounding a large globule of accumulated lipid. The very large relative size of adipose cells can be appreciated in the accompanying micrograph by comparison of three leucocytes in the field with portions of two fat cells included in the lower half of the figure.

Newly synthesized lipid first forms small droplets in the peripheral layer of cytoplasm (at arrows) and these subsequently coalesce with the large central lipid drop.

Figure 362. Portions of two adipose cells and associated connective tissue from the epididymal fat pad of a rat.

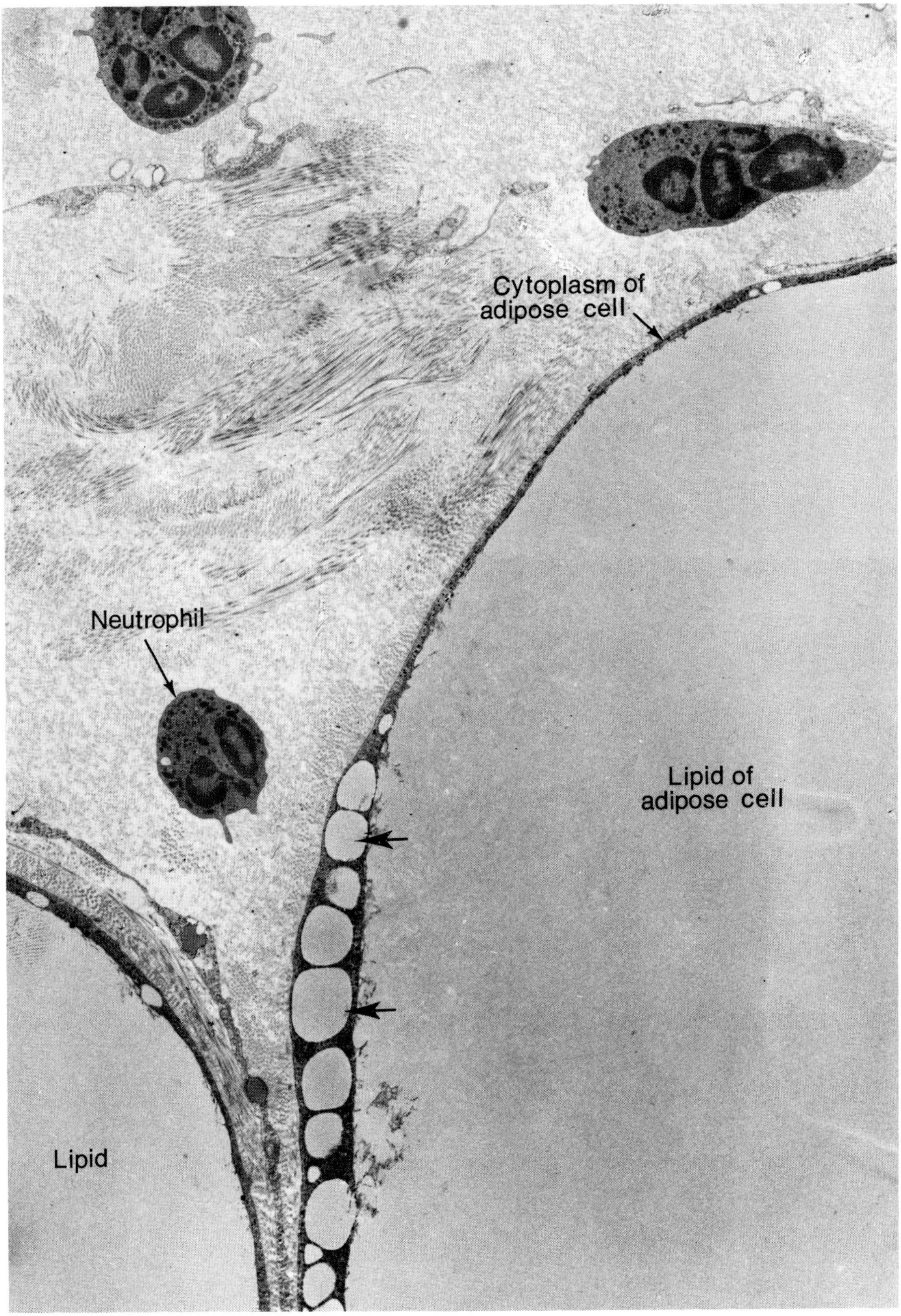
Cytoplasm of
adipose cell
Neutrophil
Lipid of
adipose cell
Lipid

Figure 362

CRYSTALLINE INCLUSIONS

Crystalline inclusions have long been recognized by light microscopists as normal constituents of certain cell types, such as the Sertoli cells and Leydig cells of the human testis, rabbit oocytes, and the nuclei of liver cells of dogs, foxes, and jackals.

The electron microscope has now revealed crystalline inclusions in many additional types of cells. Crystals of one kind or another have been reported to occur in nearly all compartments of the cell — in the nucleus, in mitochondria, the endoplasmic reticulum, the Golgi complex, in secretion granules, and free in the cytoplasmic matrix. The crystals have usually been assumed to be protein, but in only a few cases has the specific protein been identified. In mammalian tissues, iron salvaged from the degradation of hemoglobin may be sequestered by phagocytes in crystalline deposits of hemosiderin. Hemoglobin itself under certain conditions may form crystals in the erythrocytes. Apoferritin and other proteins that take up iron occur in crystalline form in the tissues of several invertebrates. However, since the chemical nature and physiological significance of the majority of the crystals that have been described are unknown, it seems pointless to devote space here to more than a few representative examples.

A striking example of intracellular crystals is found in the liver of the slender salamander, *Batrachoseps attenuatus*. The prevalence of these inclusions varies from individual to individual and with the seasonal cycle of estivation and activity, but it is not uncommon to find large numbers of crystals in every hepatic cell. They are never free in the cytoplasmic matrix but are invariably enclosed by a closely conforming membrane, whose outer surface is often studded with ribosomes. The crystals thus appear to be a product of protein synthesis that accumulates in high concentration in the endoplasmic reticulum and crystallizes there. They are believed to be a storage form of protein, but the nature and ultimate fate of the protein are unknown.

Figure 363. Liver cell from the slender salamander, *Batrachoseps attenuatus*.

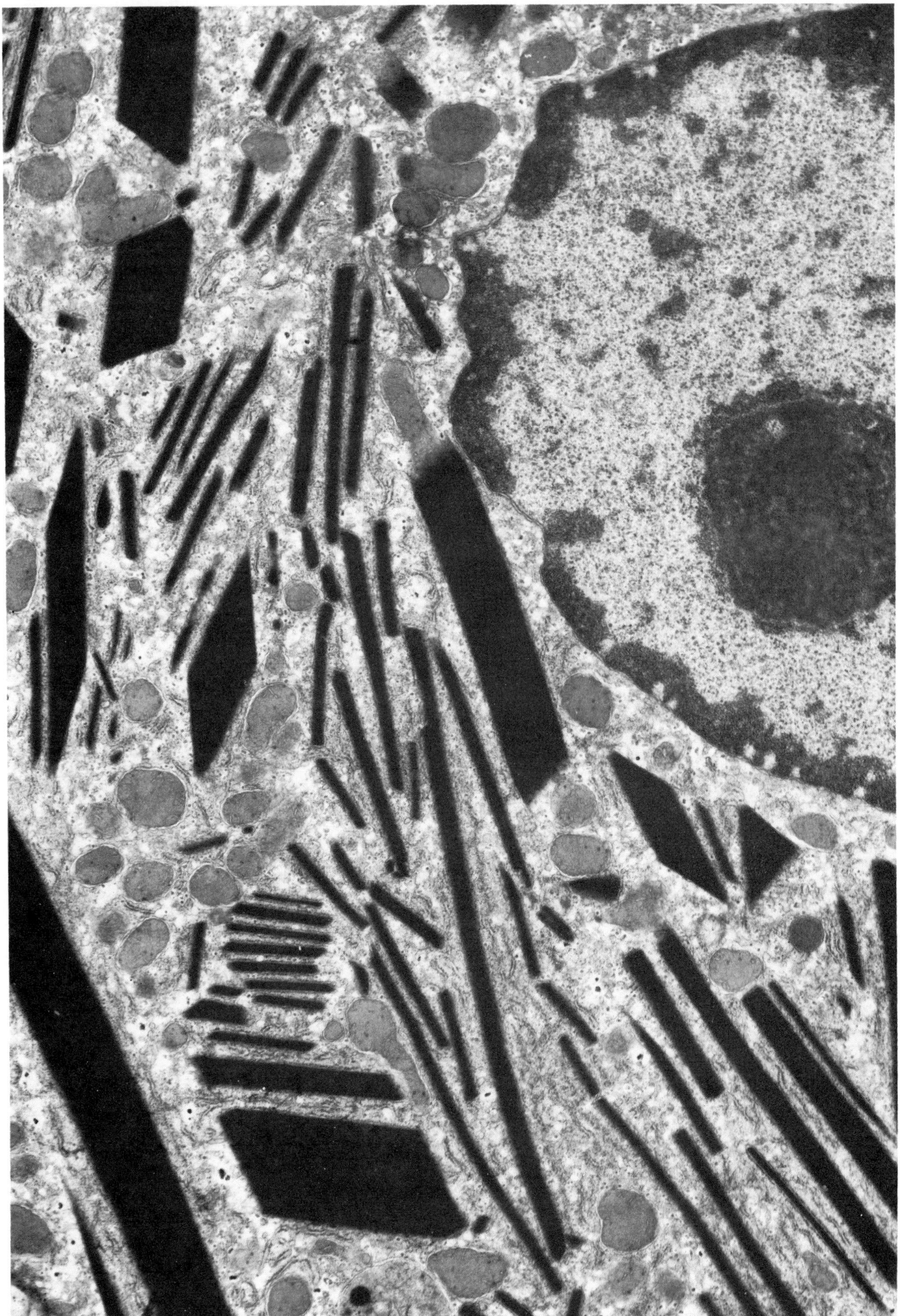

Figure 363

An additional example of protein crystals in amphibian liver: in this micrograph, the occurrence of thin tabular crystals in cisternae of granular endoplasmic reticulum is evident at the lower right of the figure.

Figure 364. Liver cell of *Batrachoseps attenuatus.*

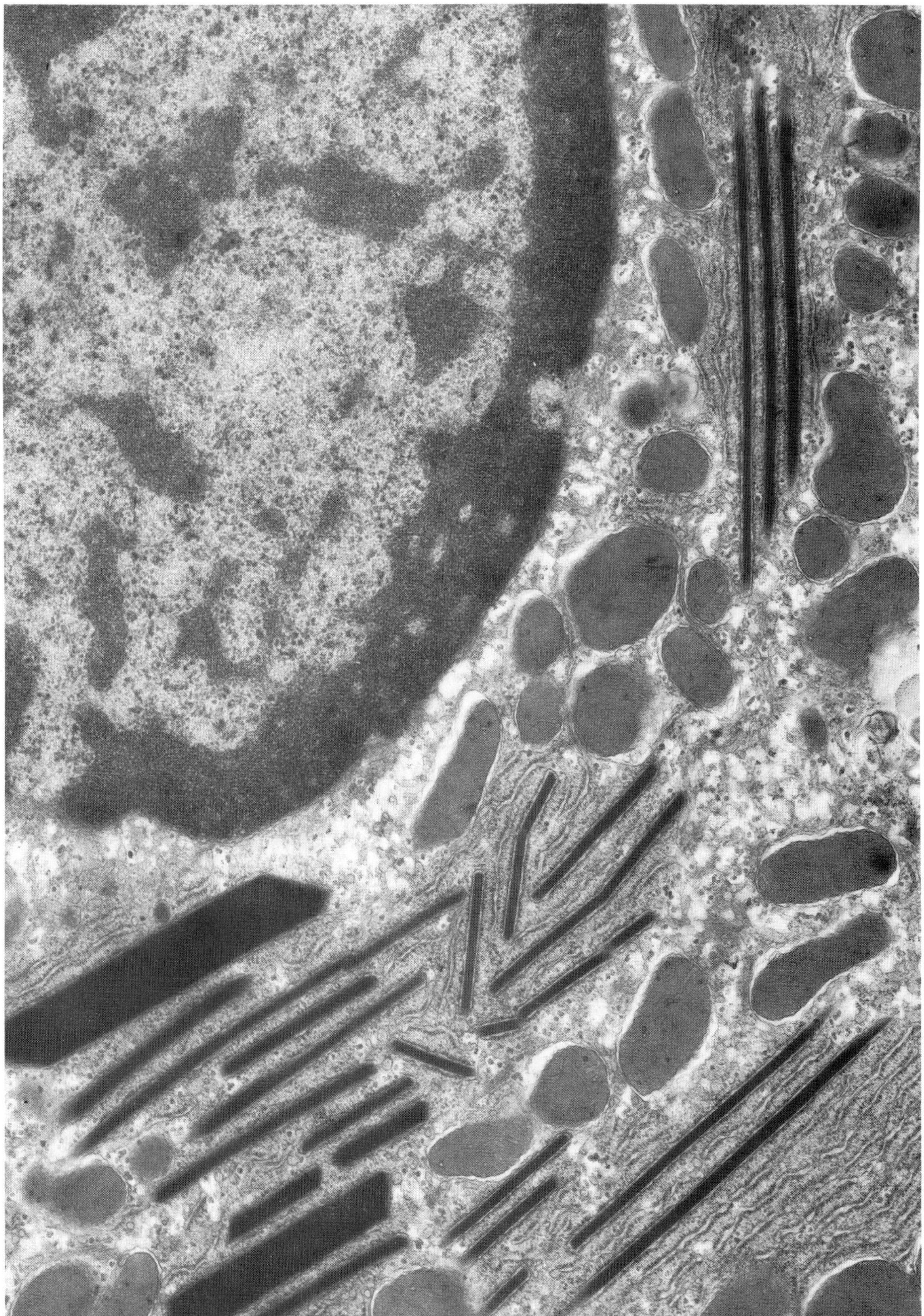

Figure 364

Reinke in 1895 described the crystalline inclusions in human testicular interstitial cells. He concluded from their staining and solubility properties that they were protein in nature and we know no more of their chemistry today. They occur in variable numbers in all men of reproductive age but are not found in other species. They were first observed in electron micrographs in 1956 and with the limited resolution then attainable appeared to be composed of 15 nm globular macromolecules (Fawcett and Burgos, 1956). They were later reported to consist of 20 nm tubular hexagons (Yamada, 1962), and 5 nm filaments (Sisson and Fahrenbach, 1967). The diversity of interpretations was due in part to the variation in patterns observed when the crystal lattice is sectioned in different planes. Improved techniques for extending the analytical capabilities of the electron microscope led to a consensus that the crystals of Reinke consist of 5 nm filaments and are trigonal in their lattice (a = 30 nm, c = 45 nm). The accompanying micrograph includes several crystals sectioned parallel to the (001) plane showing a fine honeycomb pattern composed of 5 nm filaments in a hexagonal arrangement. The arrows indicate a dislocation in one of the crystals.

Figure 365. Crystals of Reinke from human Sertoli cell. (From T. Nagano and I. Ohtsuki, J. Cell Biol. 51:148–161, 1971.)

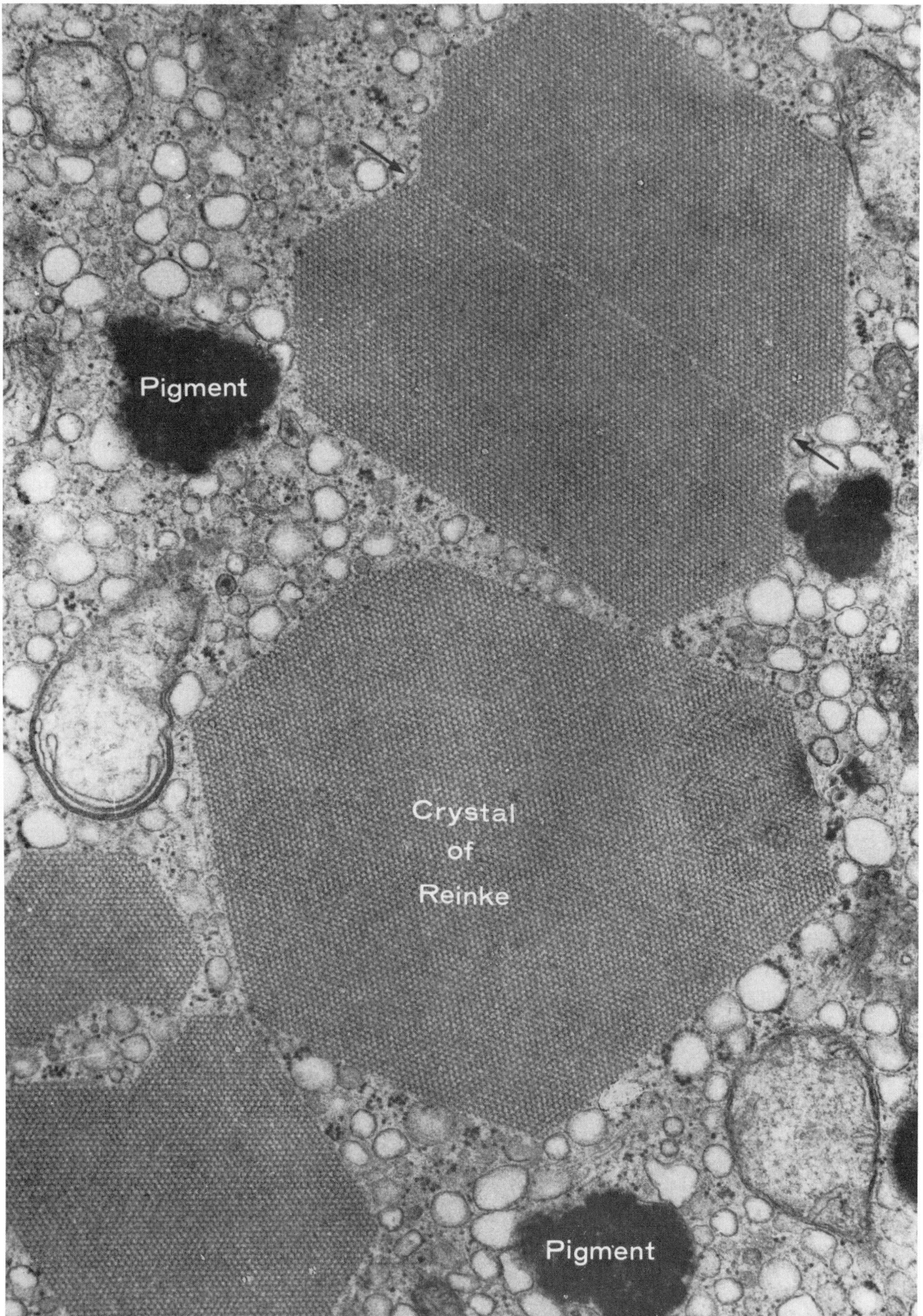

Figure 365

Crystals are rarely observed in the nucleus but the accompanying micrograph shows a cluster of small crystals of Reinke in a Leydig cell nucleus. This is circumstantial evidence that macromolecules can pass into as well as out of the nucleoplasm through the pore complexes in the nuclear envelope.

Intranuclear uric acid crystals occur normally in the liver of dogs and other canids and are occasionally observed in the cornea in humans suffering from gout. Uric acid, however, is a relatively small diffusable molecule.

Figure 366. Intranuclear crystals of Reinke in a human Leydig cell. (Micrograph courtesy of Wolf Fahrenbach.)

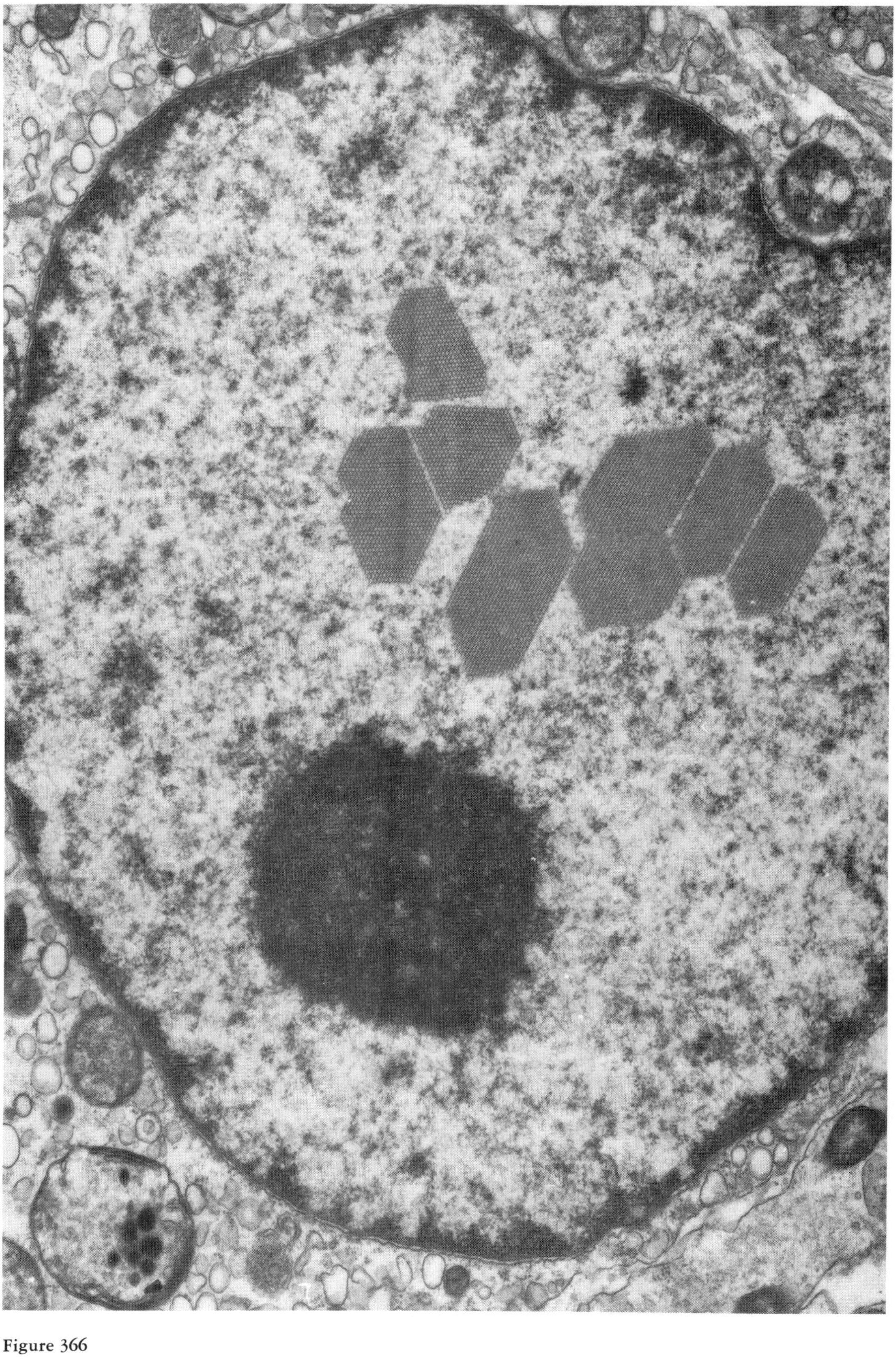

Figure 366

The presence of crystalloids in the germ cells of the human male was first noted by Lubarsch in 1896. Rather inappropriately, he referred to them as Charcot-Böttcher crystalloids, honoring Charcot, who had earlier described crystals in various other tissues, and Böttcher, who had first observed crystals of different nature in seminal fluid. Although of little descriptive value, the eponym has persisted in the literature, being applied indiscriminately to crystals in both spermatogonia and Sertoli cells, even though the two may not be identical.

The crystalloids in Sertoli cells were first described at the electron microscope level by Nagano (1966). They consist of closely packed fusiform bundles of filaments 8 to 15 nm in diameter arranged more or less parallel. The packing of the subunits is less precisely ordered than in many other crystals, and their continuity is often interrupted by areas of lower density containing loosely associated filaments intermingling with granular and vesicular components of the cytoplasm. In cross sections viewed at high magnification, the "filaments" appear tubular. These crystalloids are seen mainly in the Sertoli cells of men, but superficially similar structures have been reported in boar testis. Their chemical nature is unknown.

Figure 367. Crystalloid of Charcot-Böttcher in a human Sertoli cell. (From Nagano, Z. Zellforsch, 73:89–106, 1966.)

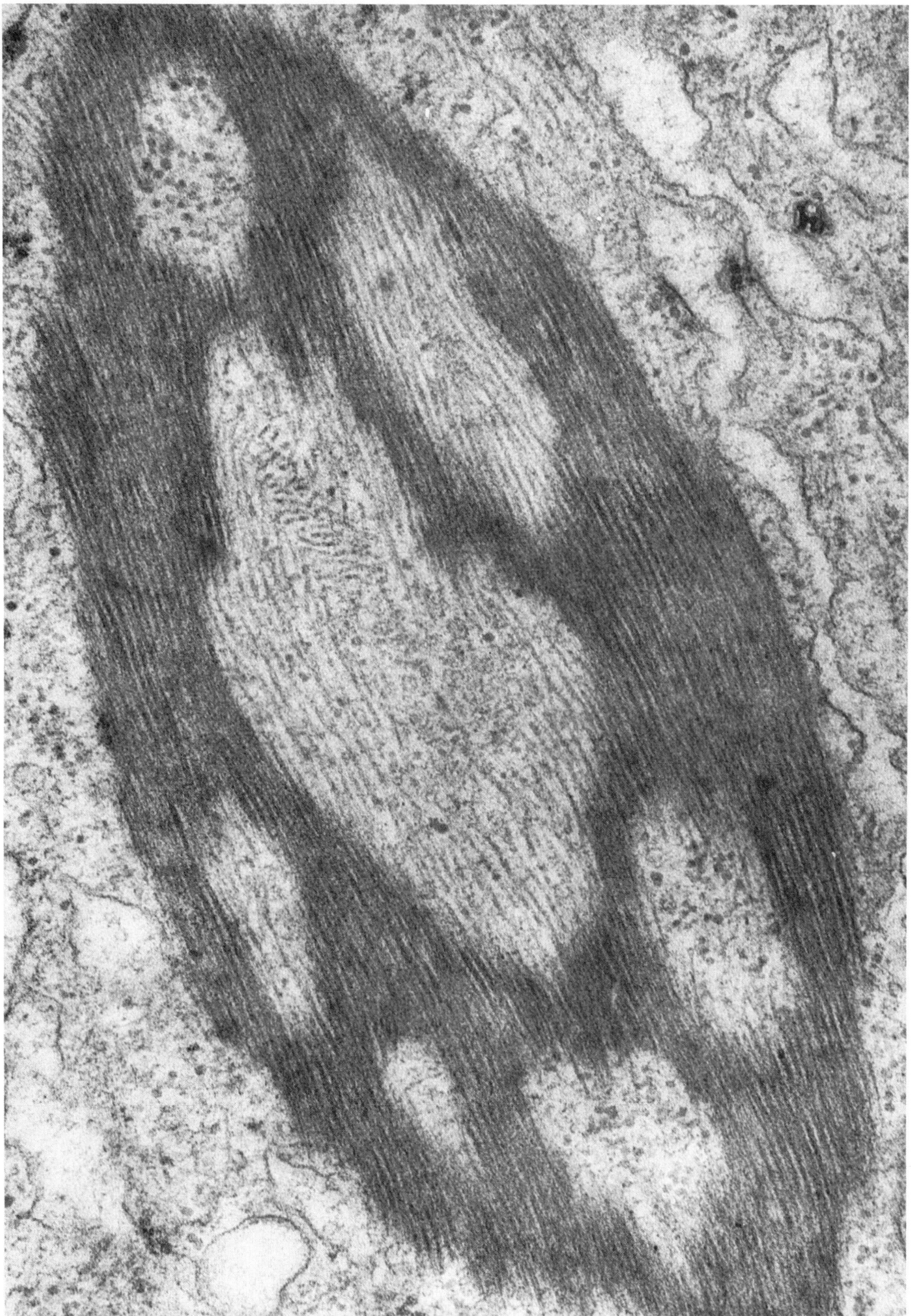

Figure 367

Crystals are normal constituents of at least one cell organelle. The microbody, or peroxisome, of animal cells described in an earlier section of the book (p. 515) may contain a crystalline nucleoid that has been identified in the rat as urate oxidase. The accompanying micrograph is of a corresponding organelle in a plant cell. The organelle is 0.5 to 1.5 μm in diameter and often contains crystalline inclusions. In the example shown here the favorable plane of section shows the crystal lattice with unusual clarity. This organelle has been described in the leaves of several dicotyledonous and monocotyledonous plants and is interpreted as a form of plant microbody containing enzymes probably involved in the metabolism of glycolate, a chloroplast product. They also give an intense histochemical reaction for catalase.

Figure 368. A microbody or peroxisome from a cell in a leaf of tobacco (*Nicotiana tabacum*). (From S. E. Frederick and E. H. Newcomb, Science 163:1353, 1969. Copyright 1969 by American Association for Advancement of Science.)

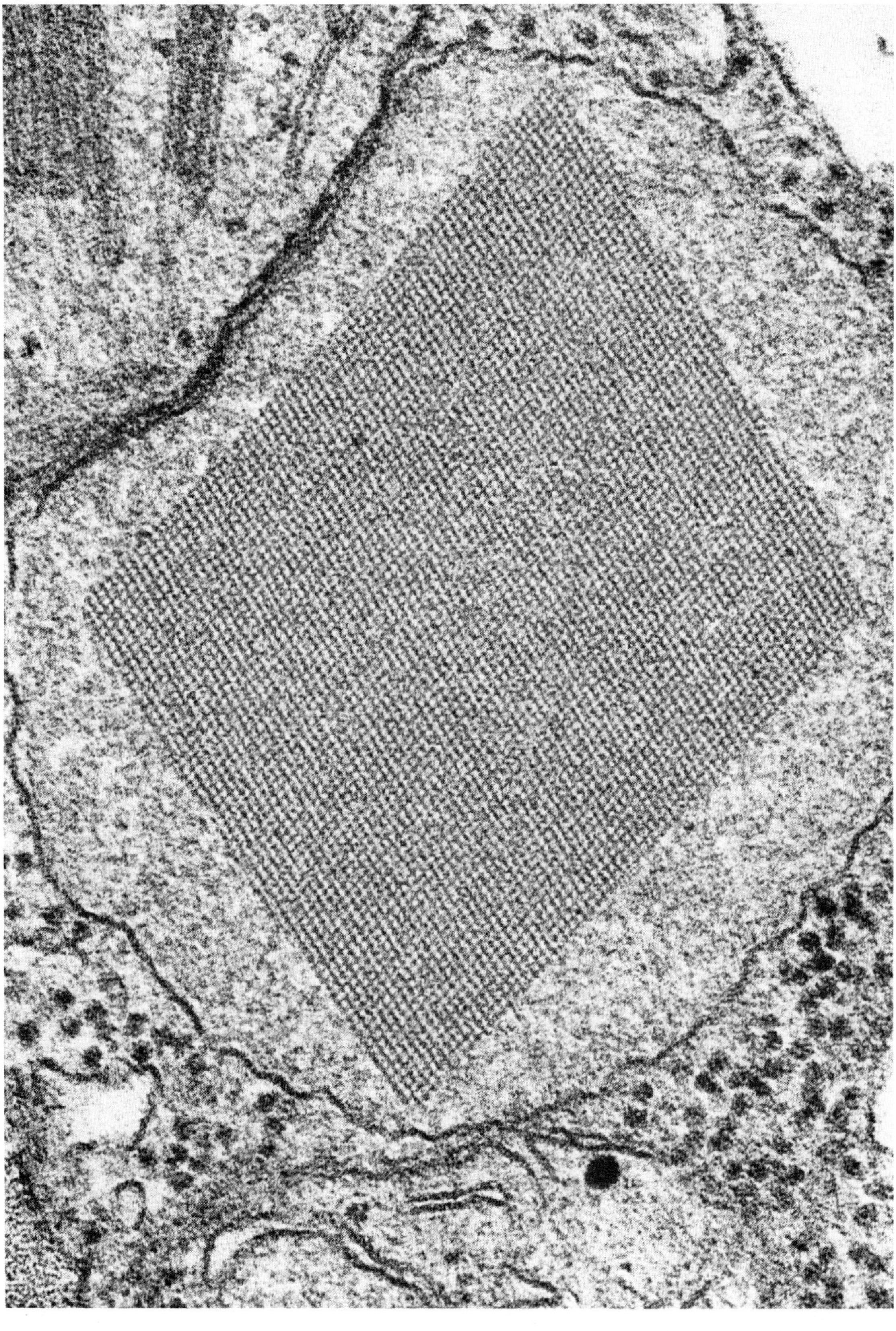

Figure 368

Hemoglobin is uniformly dispersed in normal erythrocytes and is present in high concentration. In electron micrographs it appears as a moderately dense, amorphous material filling the interior of the erythrocyte. The factors normally responsible for keeping the hemoglobin in solution are poorly understood. Under conditions of partial hemolysis, even without significant change in its concentration, hemoglobin may come out of solution and form large polygonal crystals within the erythrocytes. One such is illustrated in the accompanying micrograph.

In the longitudinal section of a capillary shown here, portions of two erythrocytes are seen. The lower one is normal and its hemoglobin is uniformly distributed throughout the cell. In the upper erythrocyte, the hemoglobin has formed a large single crystal. With the hemoglobin crystallized out, the surrounding matrix is quite pale. The erythrocyte membrane here appears to be morphologically intact, but one cannot exclude the possibility that there may have been some change in its permeability that initiated the crystallization of the hemoglobin.

Figure 369. Erythrocyte in a capillary from the myocardium of a cat.

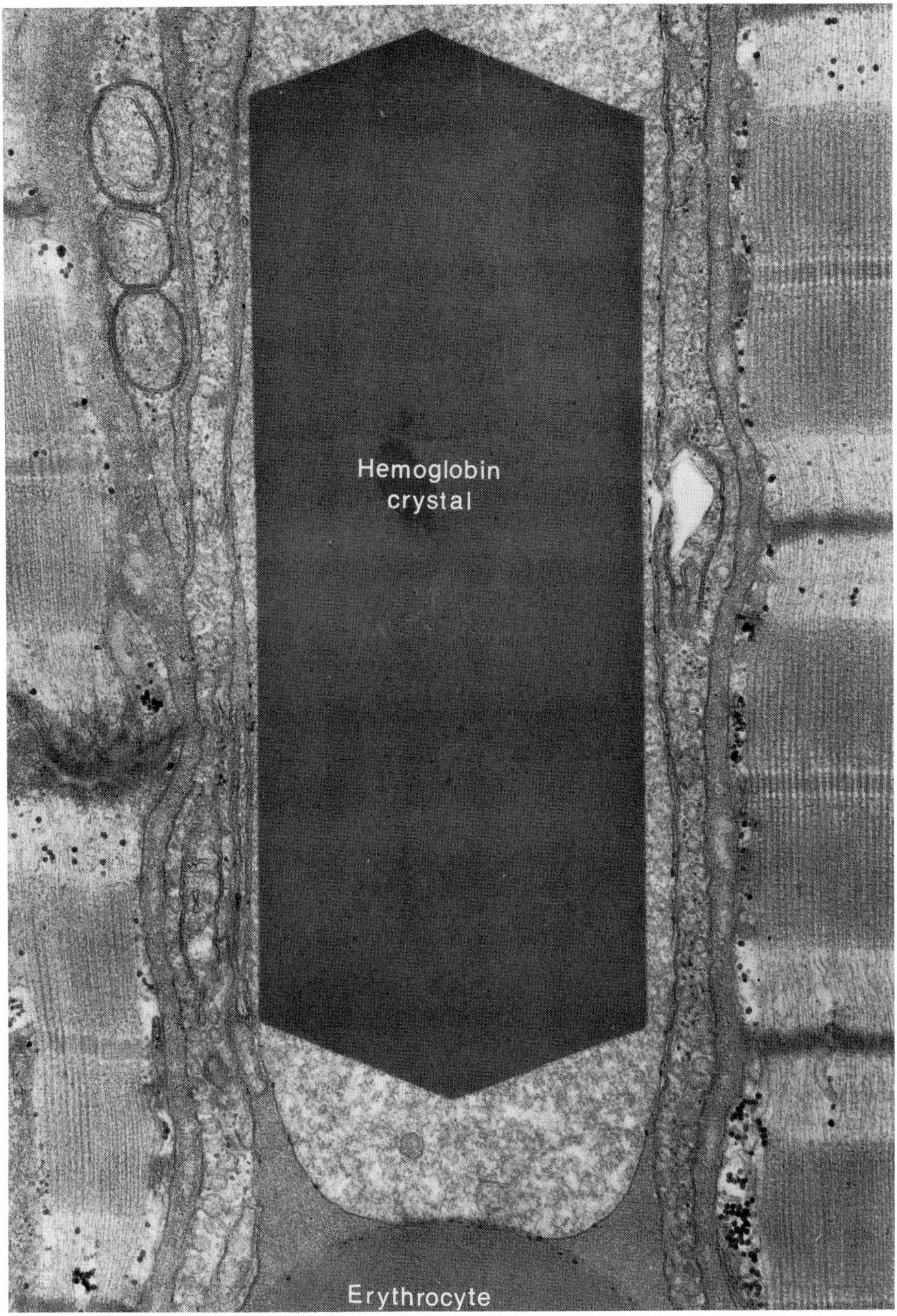

Figure 369

The yolk platelets of all amphibians studied to date contain crystals. The accompanying micrograph of a portion of a cell from the presumptive ectoderm of a frog shows a number of yolk platelets, several lipid droplets, and many small pigment granules. Each yolk platelet is enclosed by a membrane and contains one or more dense crystals embedded in a filamentous matrix. At higher magnification, in the inset, the lattice of one of the yolk crystals is evident.

Figure 370. Ectoderm cell from a blastula of *Rana pipiens*.(From S. J. Karaseki, J. Cell Biol. 18:135, 1963.)

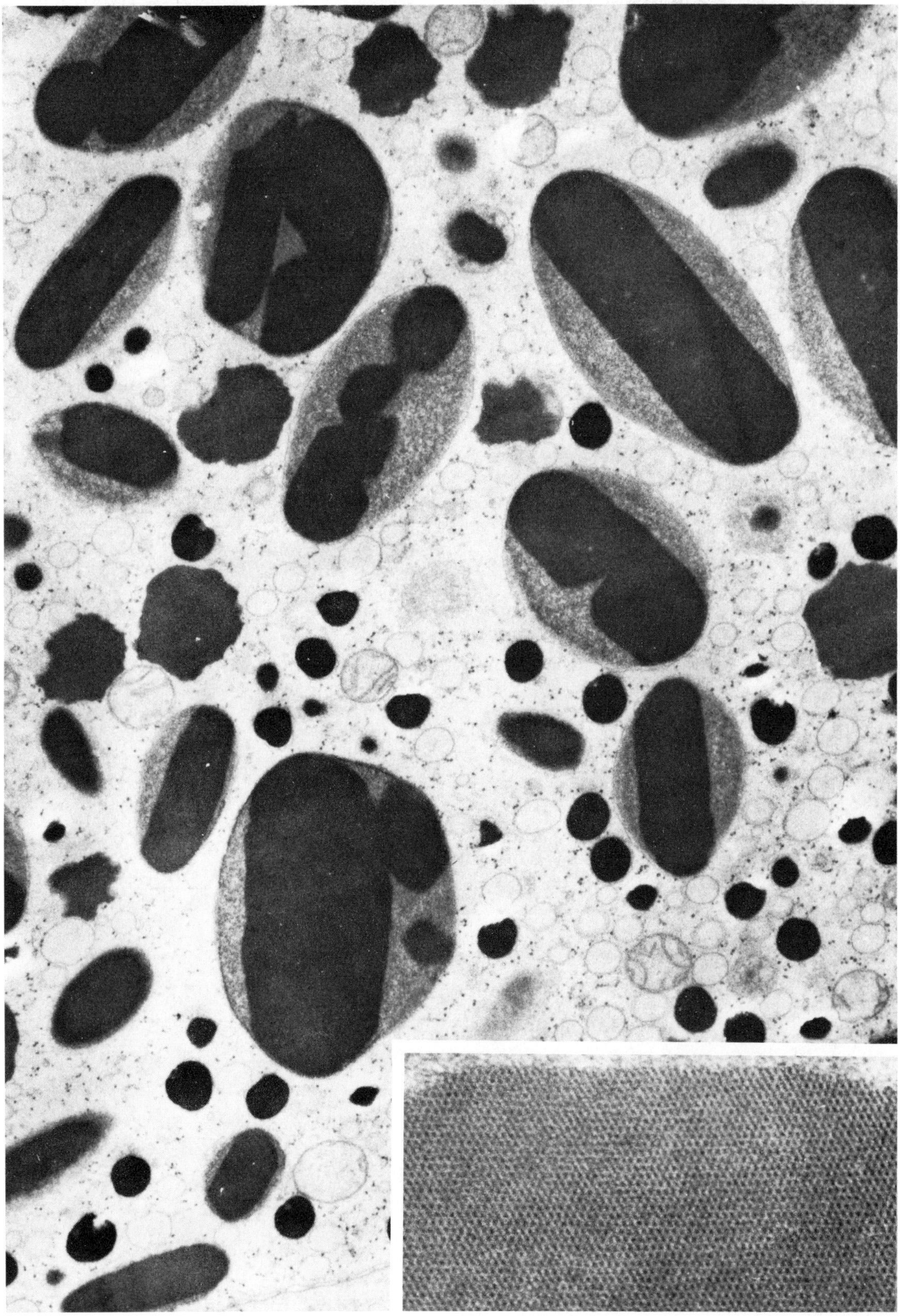

Figure 370

Iron absorbed in the intestinal mucosa of mammals combines with a protein apoferritin to form the substance ferritin, the form in which iron is stored in cells. *Ferritin* is a large molecule visible in unstained electron micrographs because of its high iron content (23 per cent). Because there appears to be no excretory pathway for excess iron, ferritin may accumulate in cells in considerable quantity. In mammals, iron accumulating in excess of the capacity of the ferritin storage mechanism forms granular aggregates of *hemosiderin*.

Crystallization of ferritin is uncommon in vertebrate tissues, but in some insects, large crystals are formed in cells of the intestinal tract. In the accompanying micrograph of a cell from the midgut of a Homopteran insect, large numbers of ferritin particles are visible in the cytoplasmic matrix and a large crystal of ferritin occupies the center of the figure.

Figure 371. Ferritin crystal in midgut cell of *Campylenchia latipes*. (Micrograph courtesy of J. Gouranton and R. Folliot.)

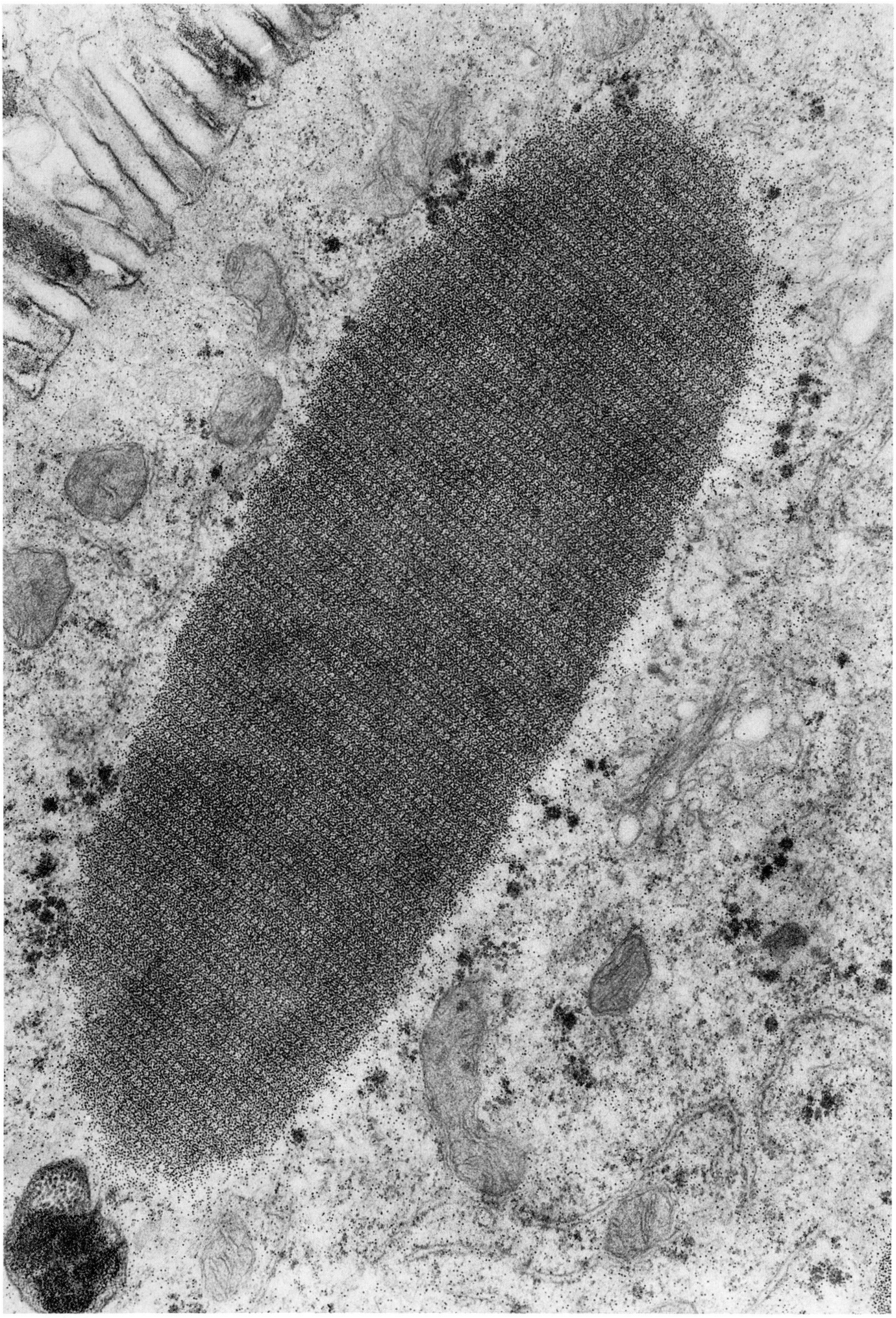

Figure 371

A number of crepuscular and nocturnal animal species have a mirror-like layer behind the retina called the *tapetum lucidum*. In dim light, the light which is not absorbed as it passes through the retina is reflected back to the photosensitive elements. Vision in dim light is thereby improved. The eyes of animals having a tapetum lucidum shine brightly in the dark when caught in the beam of a flashlight or automobile headlights.

There are two types of tapetum lucidum; in some species it consists of extracellular fibers, whereas in others it is made up of several layers of cells whose cytoplasm is filled with highly ordered rod-like crystals. These contain a peculiar compound rich in zinc and cysteine.

In the electron micrograph presented here, the majority of these hexagonally packed inclusions are seen in cross section and present a circular profile, but at the upper left some have been cut obliquely, revealing their rodlike form. The mitochondria and other organelles are displaced to the periphery of the cell.

At the lower right are portions of two melanin-containing pigment cells.

Figure 372. Tapetum lucidum of a cat's eye. (Micrograph courtesy of Giuseppina Raviola.)

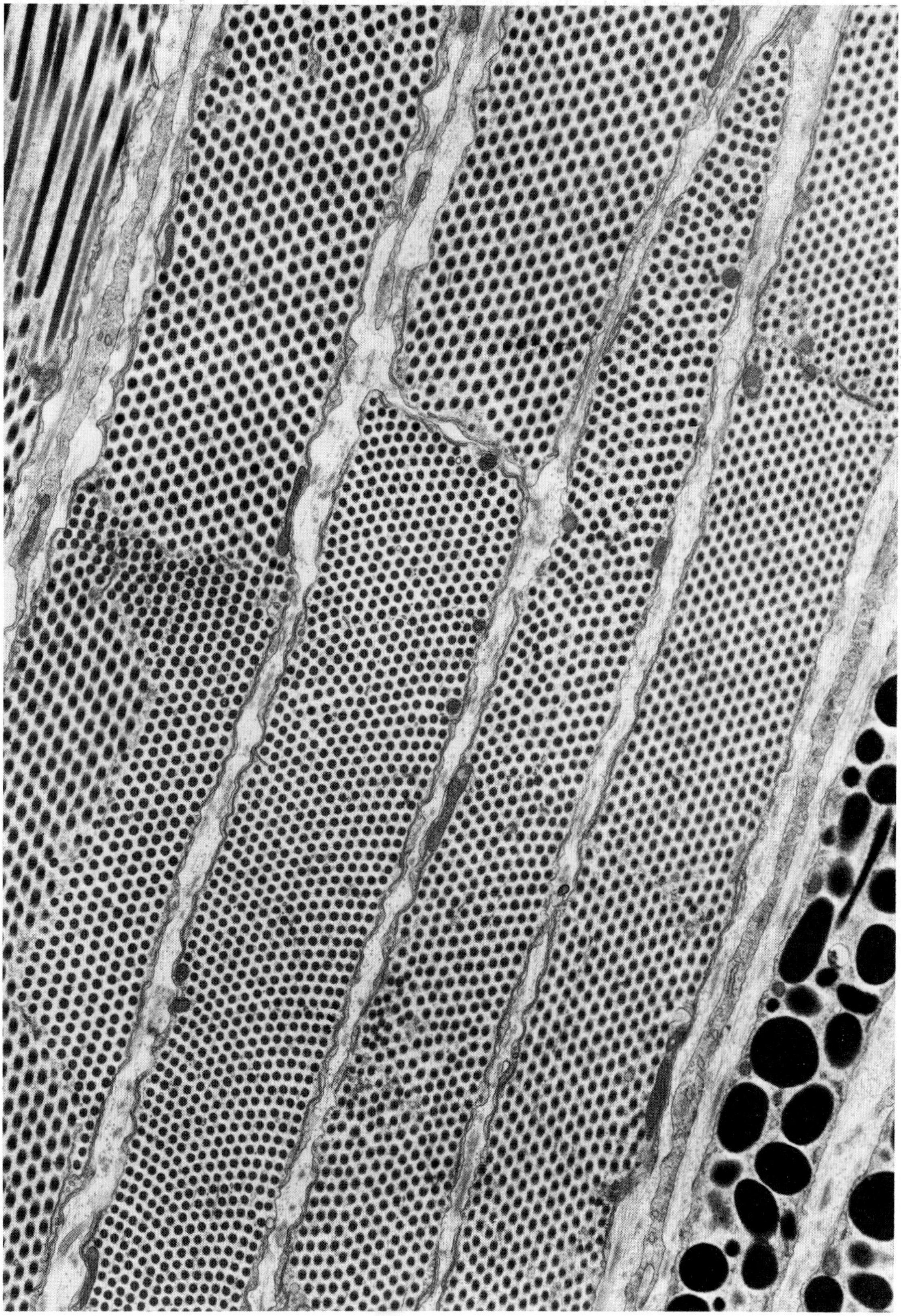

Figure 372

At higher magnification, the crystals in the cells of the tapetum lucidum are seen to have a surrounding membrane. They apparently reside in tubular elements of the endoplasmic reticulum. It is not clear how their uniform spacing and highly regular arrangement is established and maintained.

Figure 373. Cat tapetum lucidum. (Micrograph courtesy of Giuseppina Raviola.)

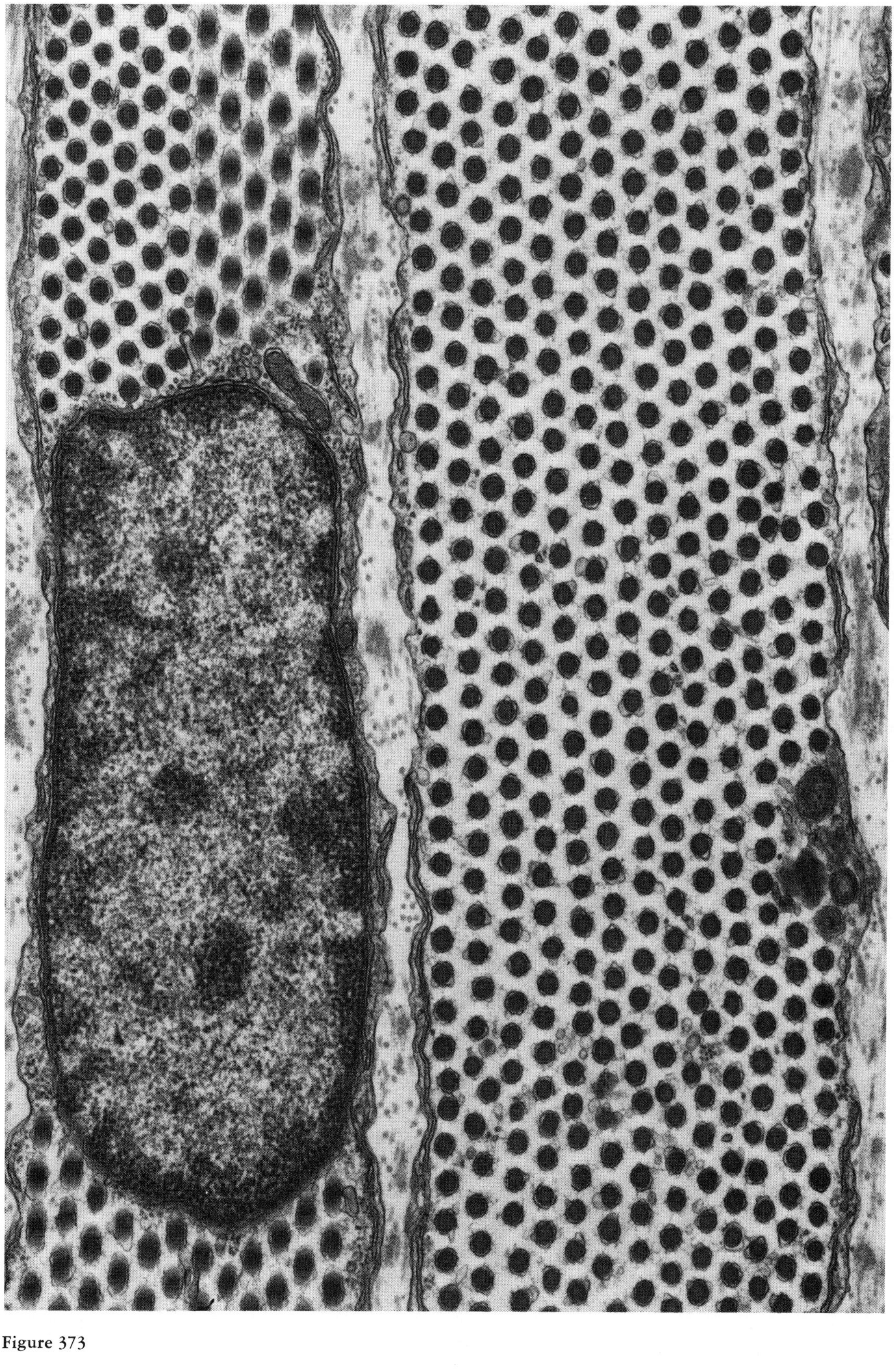

Figure 373

SECRETORY PRODUCTS

The secretions of cells are extremely varied, including products as diverse as tears, milk, hydrochloric acid, digestive enzymes, hormones, neurotransmitters, mucus, and fibrous proteins. Classical descriptions of secretion were confined to those glandular epithelial cells that contained stainable secretory droplets or granules. Not all secretory activities of cells are accompanied by microscopically detectable accumulation of the product in the cytoplasm. For example, steroid-secreting endocrine glands do not store their product, but produce and release it on demand. Cells engaged in elaboration of copious watery secretions may exhibit no cytological evidence of secretory activity and little change in appearance from the resting to the active state. In addition, we now know from electron microscopic and autoradiographic studies that many cell types (viz., fibroblasts, chondroblasts, osteoblasts, smooth muscle cells) do synthesize and release into their environment collagen, elastin, and various glycosaminoglycan constituents of the extracellular matrix. Precursors of these products are usually not detectable in the cytoplasm microscopy.

In this chapter, we will be concerned with those secretory cells in which the product or its precursor is concentrated in obvious secretory granules or vesicles and stored for a time in this form. To this category belong exocrine glands such as pancreatic acinar cells, salivary glands, mucous and chief cells of the gastric glands, and goblet cells of the intestinal mucosa; also endocrine cells such as those of the anterior pituitary and pancreatic islets and various neurosecretory cells. Limitations of space permit inclusion of only a few representative examples.

Portions of the intracellular secretory pathway have already been discussed in the separate sections on the organelles. It will suffice here to reiterate that protein-rich secretions are synthesized on membrane-associated polyribosomes and segregated in the lumen of the endoplasmic reticulum. The material is then channeled to the Golgi region, where small vesicles bud off from transitional elements of the reticulum and carry quanta of the product to the Golgi complex. In the Golgi, the product is concentrated and often modified by addition of carbohydrate moieties synthesized in that organelle. In condensing vacuoles associated with the secretory face of the Golgi, the product is further compacted to form secretory granules, limited by a membrane of Golgi origin capable of fusing with the apical plasmalemma during exocytosis.

Exocrine Glands

In glandular acini the cells are typically pyramidal in form, with secretory granules concentrated in the apex and the basal cytoplasm occupied by extensive parallel arrays of granular endoplasmic reticulum. The Golgi apparatus is usually supranuclear. Cytologists have traditionally relied upon the position of the Golgi relative to the nucleus to identify the secretory pole of the cell. But in some circumstances this criterion may be misleading.

The type-f cell of the tick salivary gland shown here becomes very active in the synthesis and secretion of its glycoprotein product for a short period at the beginning of the blood-meal. Early in its active phase secretory granules and endoplasmic reticulum are distributed throughout the cytoplasm, and the Golgi (not shown here) is displaced to a paranuclear or basal position. At a later stage, the organelles and granules become localized in a pattern more typical of continuously secreting exocrine glands.

The secretion granules are enclosed in a close-fitting membrane which is often obscured by the matching density of the mature granule but can be seen on condensing granules in the Golgi region. A prominent nucleolus shown here is typical of cells engaged in synthesis of large amounts of protein for export.

Figure 374. Cell type-f from Acinus III in the salivary gland of the tick *Rhipicephalus appendiculatus.*

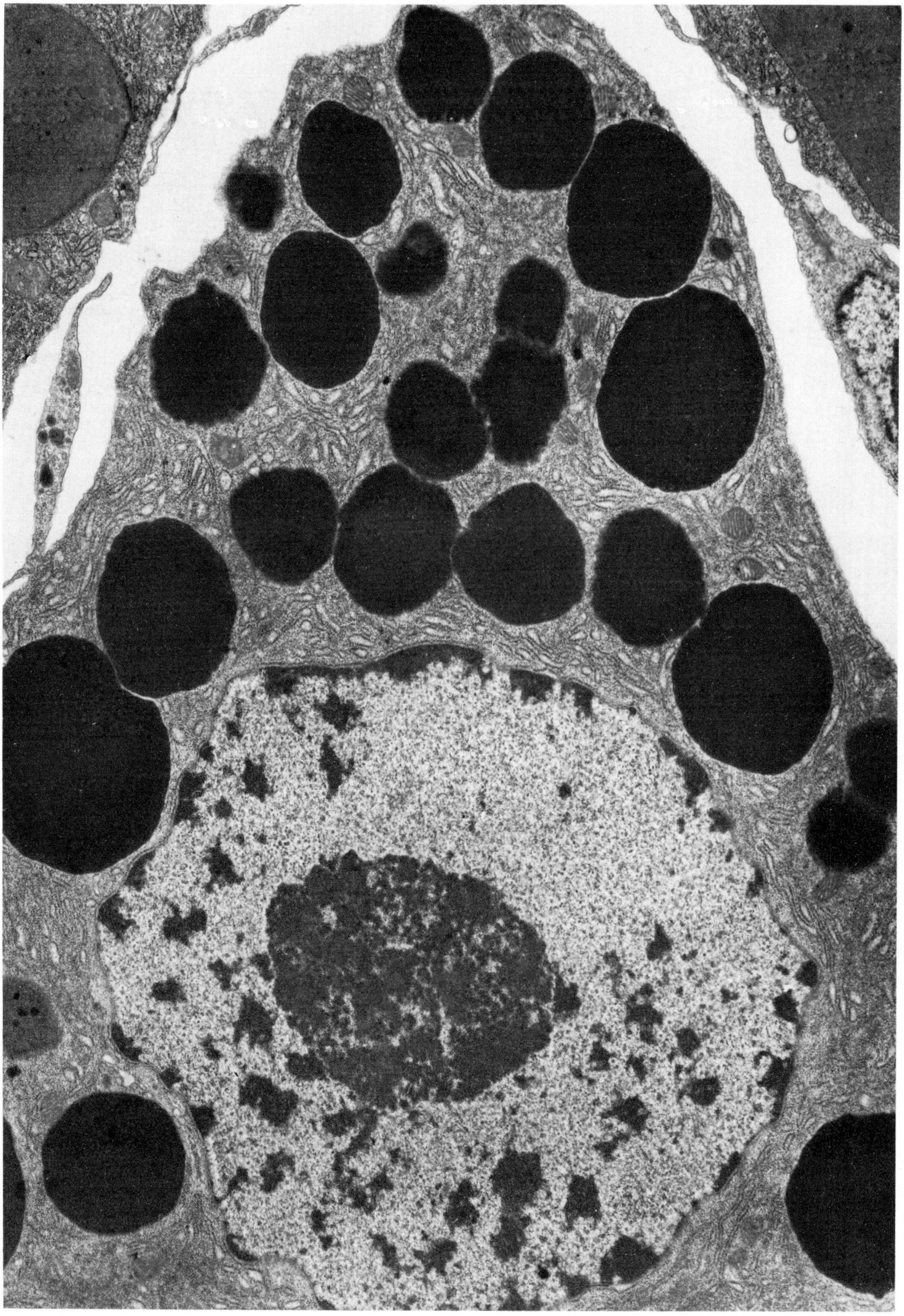

Figure 374

Secretory granules in tissue section may appear oval or irregular due to deformation in sectioning or shrinkage during dehydration. In a freeze-fracture preparation illustrated on the facing page these sources of distortion are eliminated and the secretion granules are smooth-contoured spheres. Their limiting membrane contains intramembrane particles in numbers comparable to those of Golgi cisternae. The nucleus at the lower right shows numerous pore complexes in its fractured inner membrane.

Figure 375. Freeze-fracture replica of a cell from rat adrenal medulla. (Micrograph courtesy of Daniel Friend.)

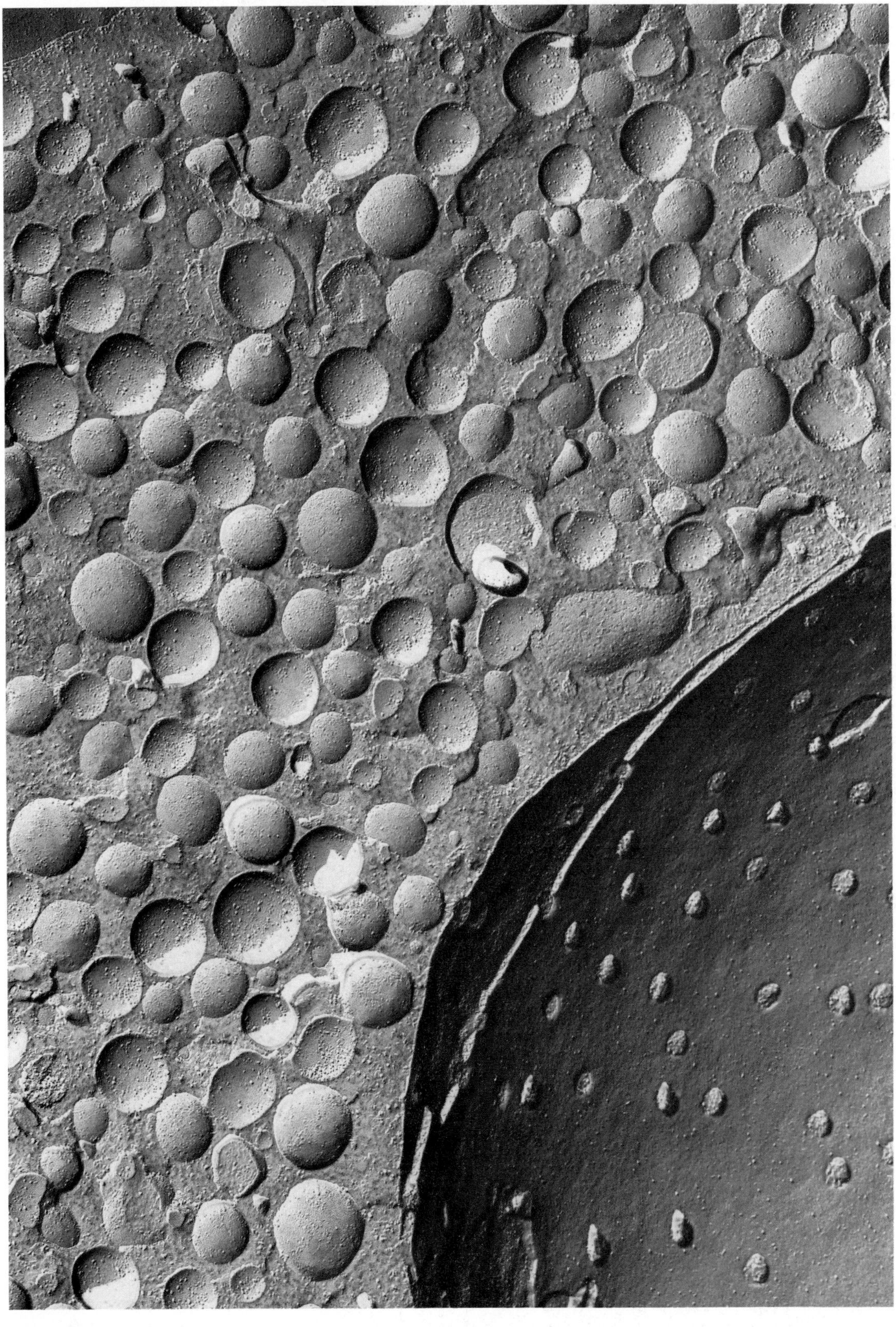

Figure 375

The glandular cell that has been most extensively studied by both biochemical and morphological approaches is the pancreatic acinar cell shown here. The cell apex is crowded with spherical zymogen granules of uniform size. Cisternae of the endoplasmic reticulum that follow a meandering course among the secretory granules are more variable in width and less highly ordered than those at the cell base.

Figure 376. Human pancreatic acinar cell. (Micrograph courtesy of Susumu Ito and Arthur Like.)

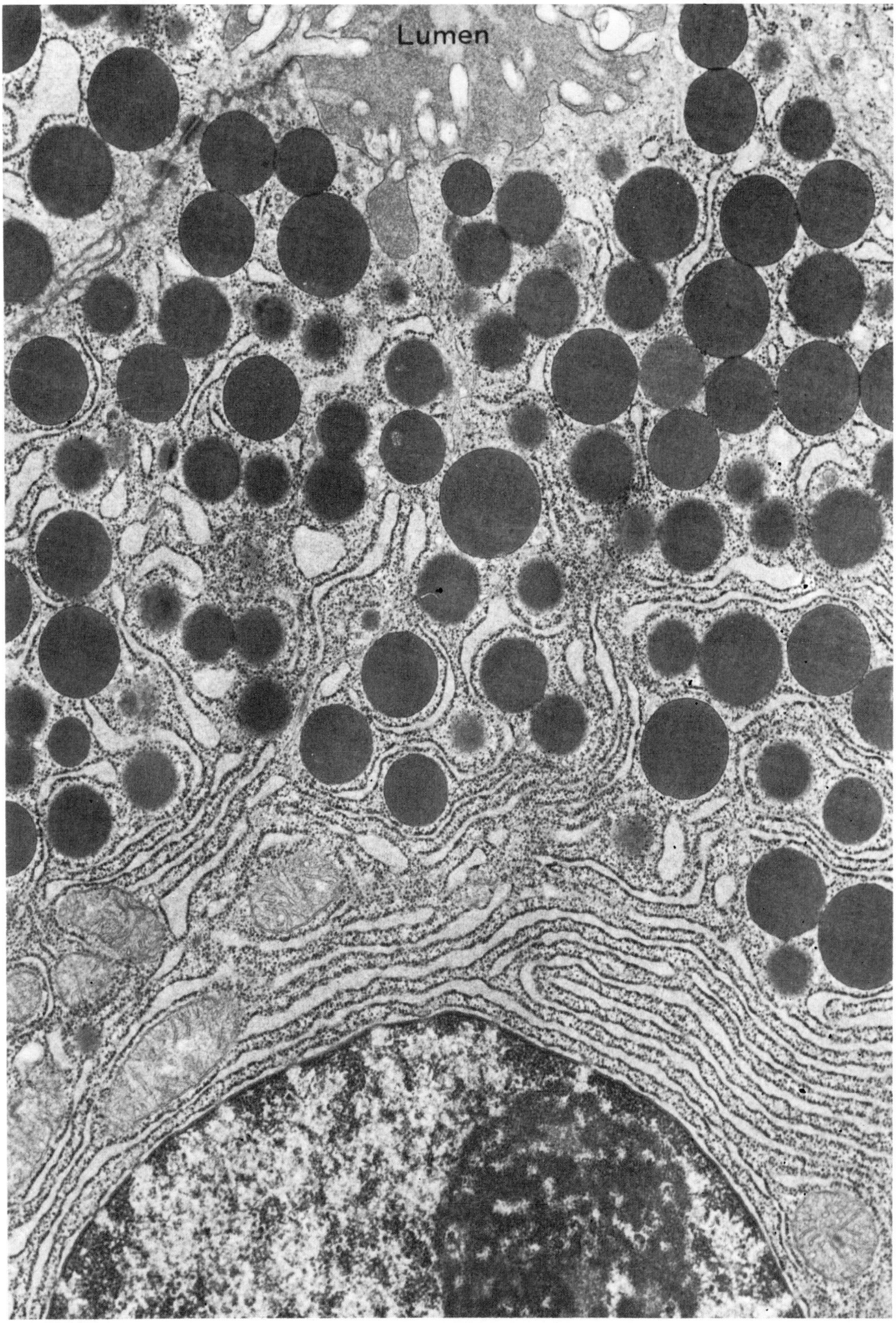

Figure 376

The secretory granules of the same cell type in different species may vary considerably in size and density. The accompanying micrographs are not of the same magnification and do not permit a valid comparison of size, but it is evident that the zymogen of the human pancreatic acinar cell above is much more dense or osmiophilic than that of the guinea pig pancreas below. Where the zymogen is of relatively low density the limiting membrane of the secretory granules is clearly resolved.

Figure 377. Zymogen granules from human pancreatic acinar cell. (Micrograph courtesy of Susumu Ito and Arthur Like.)

Figure 378. Zymogen granules from guinea-pig pancreas. (Micrograph courtesy of Robert Bolender. J. Cell. Biol. 61:269–287, 1974.)

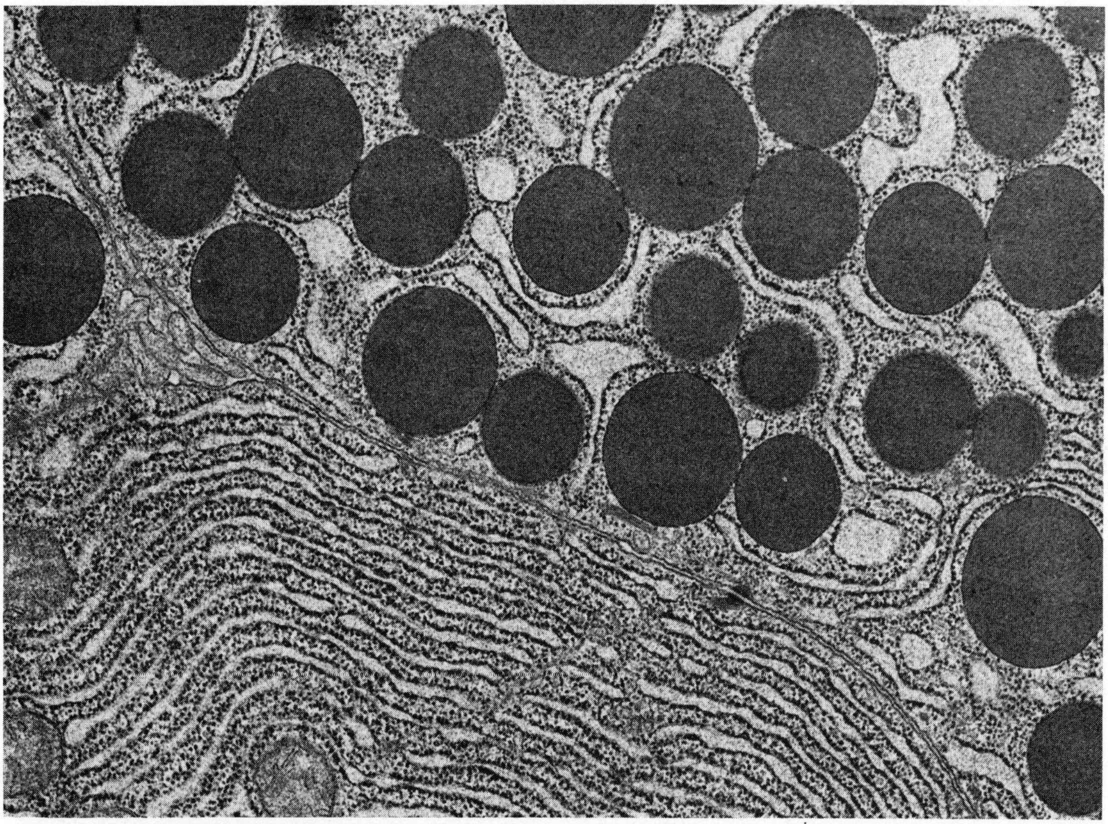

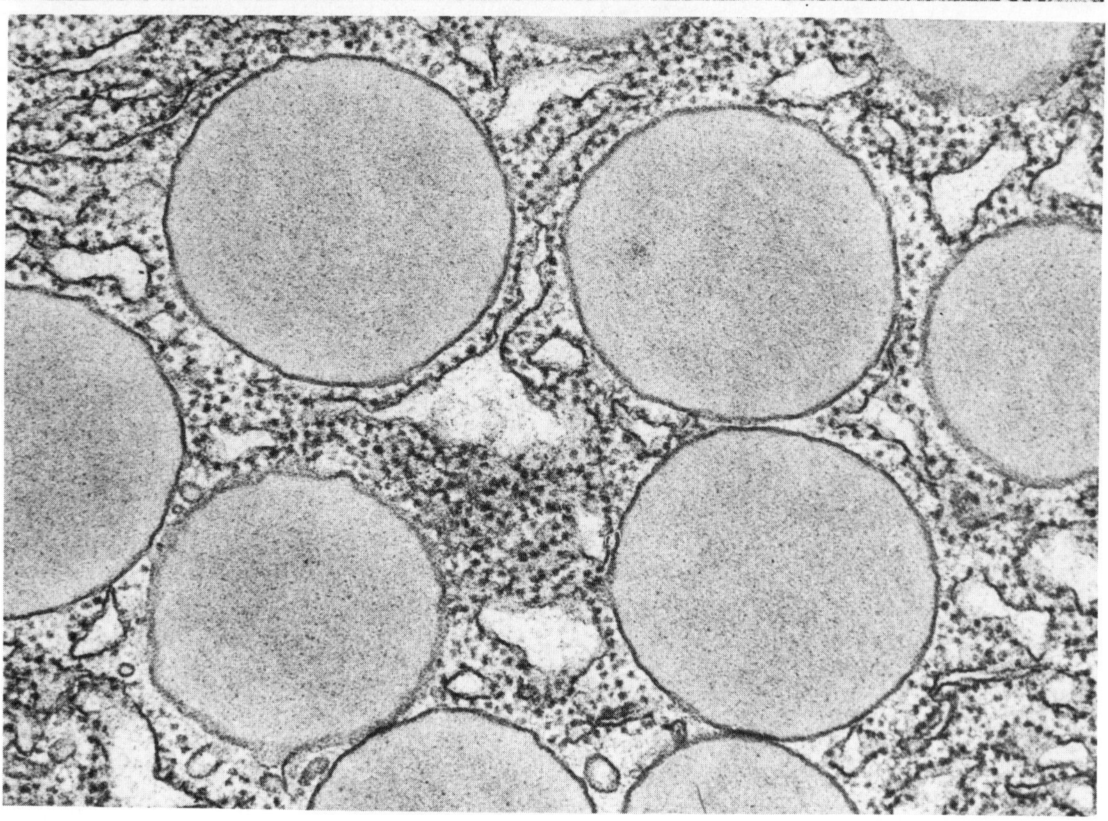

Figure 377, *upper* **Figure 378,** *lower*

Secretory granules of some glands have a dense medulla and a paler cortical zone. The interpretation of this appearance has been in dispute, some investigators regarding it as evidence of chemical heterogeneity while others have considered it an artifact of specimen preparation attributable to differential penetration or partial extraction by the fixative. In the preparations illustrated here the possibility of artifact has been minimized by rapid freezing with liquid nitrogen, followed by freeze substitution. The fact that the zonation persists under these conditions is strong evidence for a true heterogeneity in composition of the secretory granule. This interpretation is further supported by histochemical evidence for a higher concentration of carbohydrate in the cortical zone, a finding which argues against the density difference being due to different degrees of concentration of the same product.

These observations raise interesting questions as to whether the zonation of the granules is the result of temporal discontinuity in the nature of the product synthesized or segregation within the condensing vacuole of two components secreted simultaneously.

In the upper figure, continuity (at white arrow) between the innermost Golgi cistern and the smaller of the two condensing vacuoles favors the interpretation that these structures arise from the Golgi and not from GERL. The association of two transport vesicles with the limiting membrane of the larger condensing vacuole (at black arrows) suggests that secretory product from the endoplasmic reticulum may be added without transit through the Golgi cisternae. The zonation characteristic of the mature secretory granules is foreshadowed by a slight difference in density between the outer zone and the interior of the condensing vacuole.

Figure 379 and 380. Freeze substitution preparations of gerbil parotid gland. (Micrographs courtesy of Atsuchi Ichikawa.)

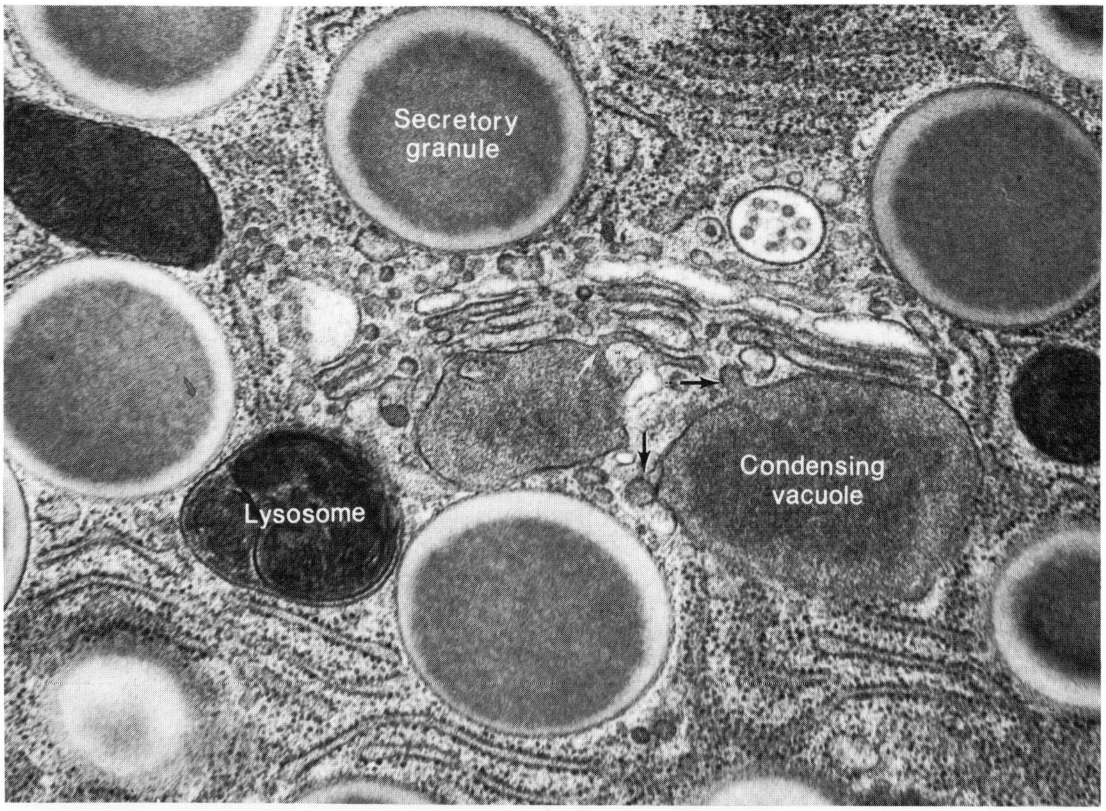

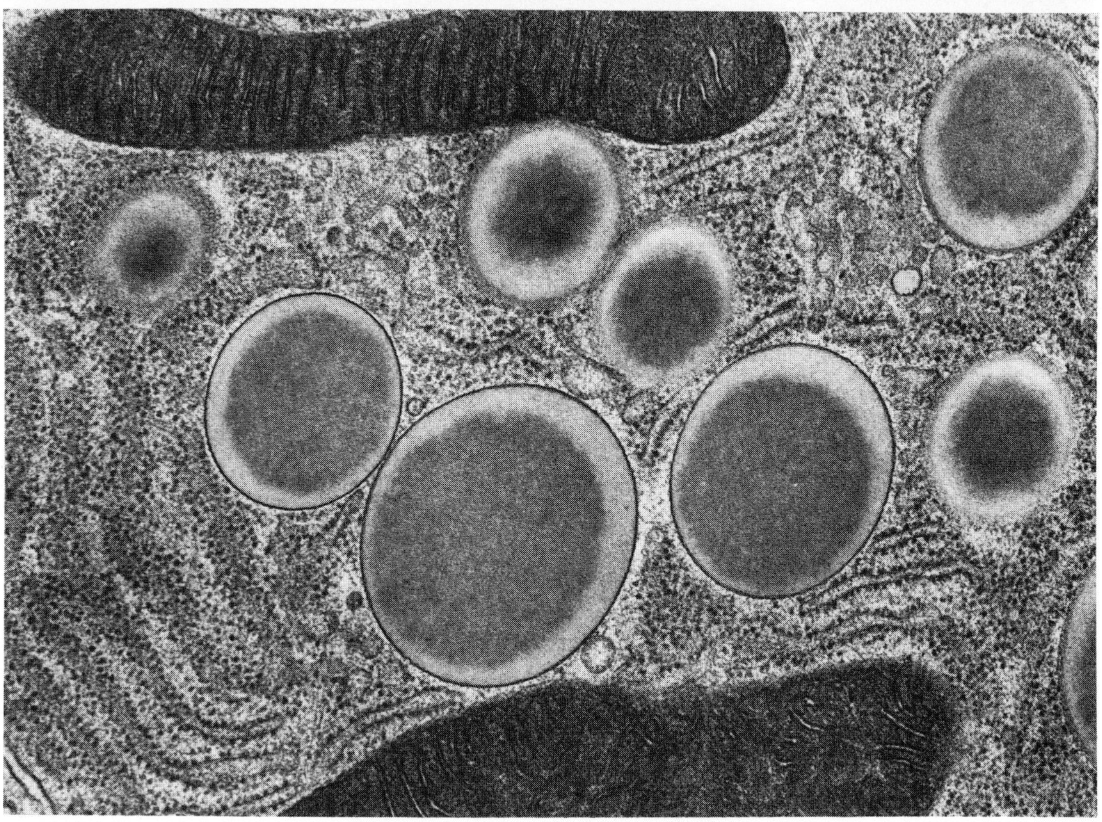

Figure 379, *upper* **Figure 380,** *lower*

Another pattern of inhomogeneity is occasionally encountered in secretory granules. The accompanying micrograph illustrates the compound secretory granules in a cell of the tick salivary gland that is believed to secrete a cement substance involved in attachment to the mammalian host. The membrane bounded granules are composed of multiple dense subunits in a less dense matrix. The secretion granules of most glands are formed in condensing vacuoles somewhat larger than the definitive granules and once formed these never coalesce with one another. The compound granules illustrated here arise from minute membrane limited granules formed in the Golgi apparatus (see at arrows). Their membranes and contents coalesce to form larger membrane-bounded granules. These in turn coalesce to give rise to compound structures in which the granular subunits are not individually membrane-bounded but are enclosed in a common membrane. This process thus bears some resemblance to formation of the acrosome in spermatids but there the contents of the multiple proacrosomal granules coalesce into a single large acrosomal granule.

Figure 381. Type-d cell from acinus type III of the salivary gland of *Rhipicephalus appendiculatus.*

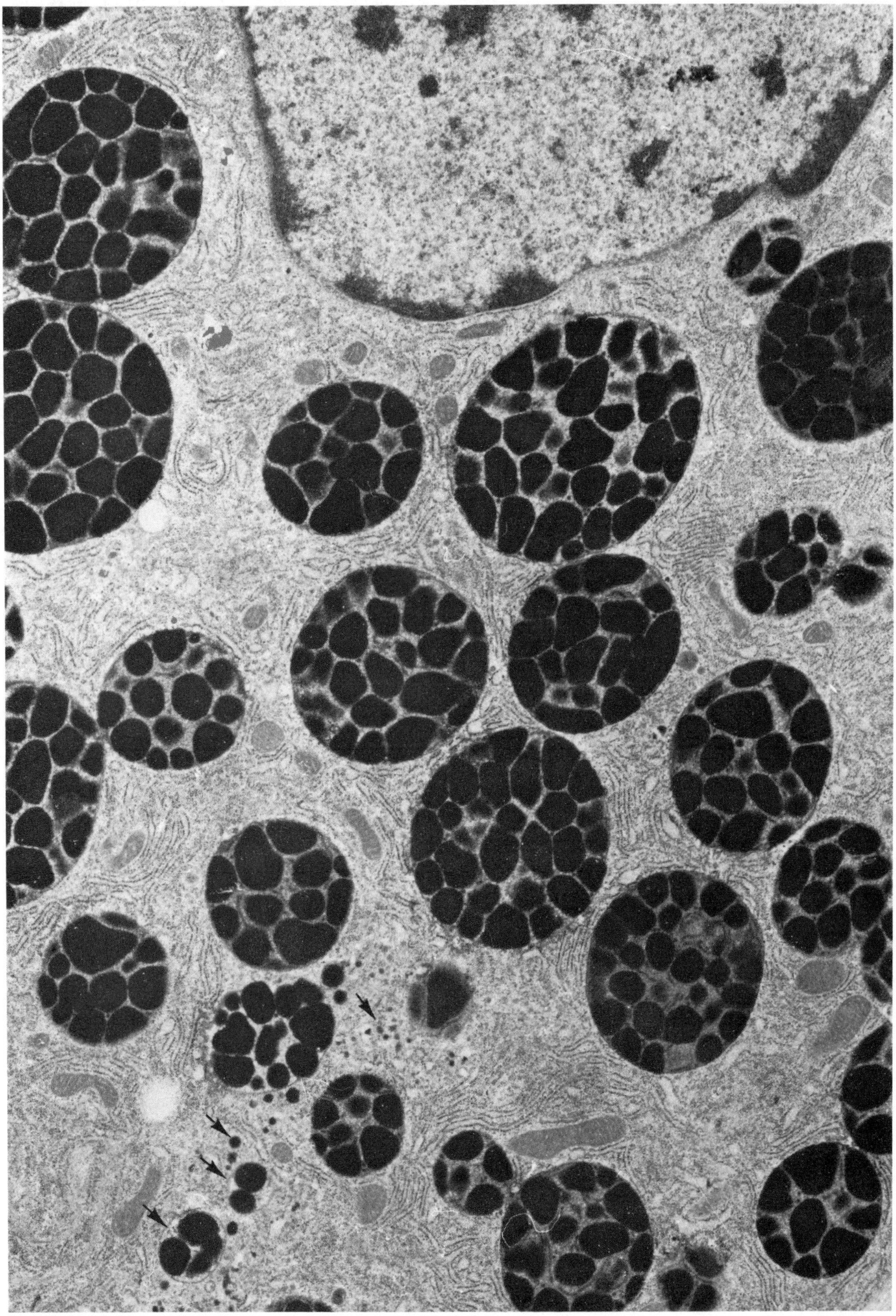

Figure 381

The content of some secretion granules that appears amorphous at low magnification is found on higher magnification to be composed of minute granular subunits as shown in the inset in the upper figure. It is not clear whether these represent the quanta of product transported from the reticulum to the Golgi cisternae and condensing vacuoles or whether the product precipitates in granular aggregates during condensation.

A unique feature of these secretory granules is the presence of a second dense granular component that appears to be added to the fully formed granules during storage. These are indicated in the lower figure by arrows in one of the granules. The arrow at the left of the figure points to one free in the cytoplasm and the star indicates one in process of fusion with the secretory granule.

Figures 382 and 383. Secretory granules of cell type-e of acinus type III in the salivary gland of *Rhipicephalus appendiculatus.*

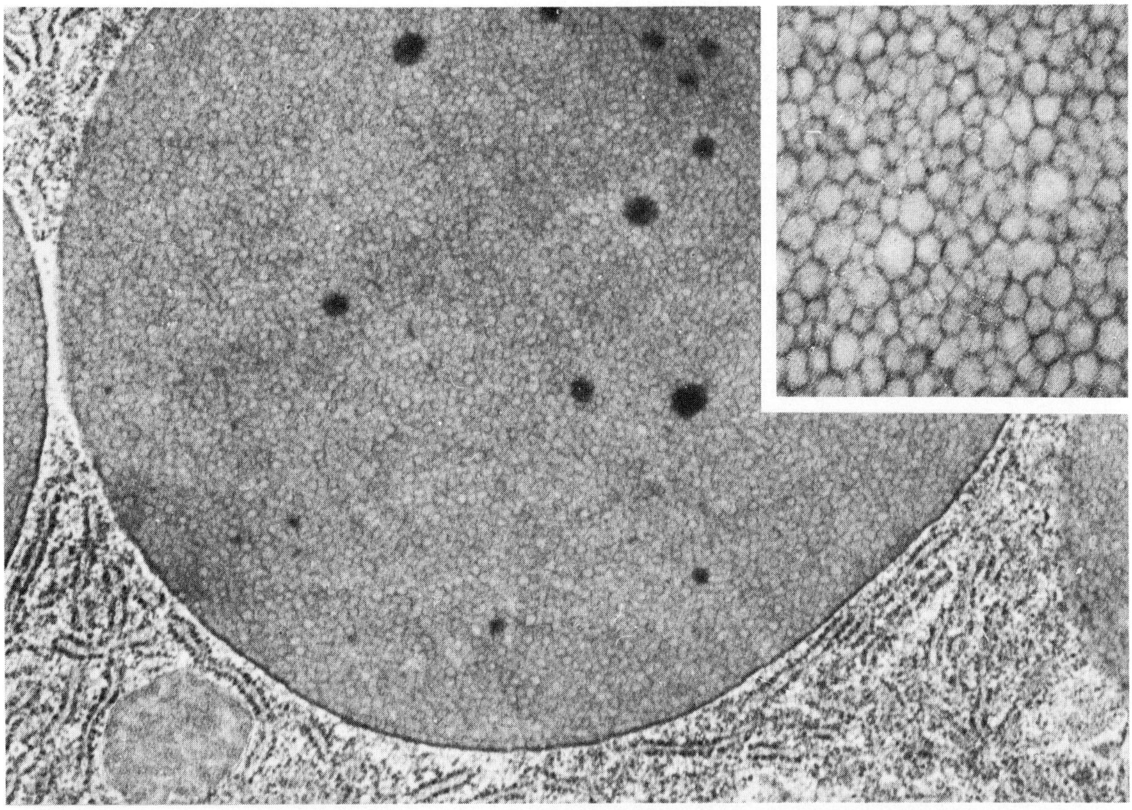

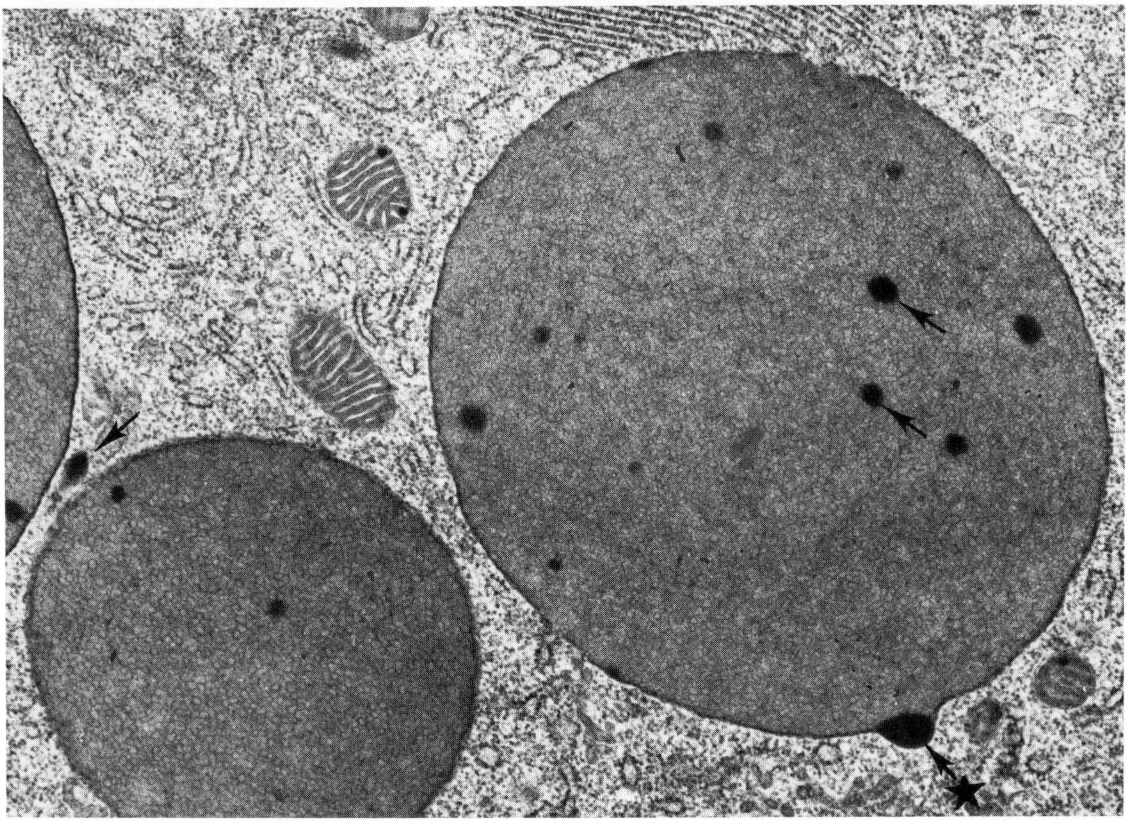

Figure 382, *upper* **Figure 383**, *lower*

There are many ultrastructurally distinct cell types in the respiratory, gastrointestinal, and reproductive tracts that produce a slimy secretion called *mucus*. The principal components of mucus responsible for its physicochemical properties are glycoproteins of very high molecular weight. These consist of extended polypeptide chains with oligosaccharide units projecting from them, and the carbohydrate accounts for more than half the molecular weight. The glycosyl transferases responsible for conjugating the sugar residues to the protein core reside mainly in the Golgi complex. Therefore this organelle has an important synthetic role in the secretory pathway. Various categories of glycoprotein are identified histochemically — neutral, acid, sialated, and sulfated glycoproteins. These may be present in varying proportions in the different mucus-secreting cells of the body. It is not surprising therefore that the ultrastructural appearance of the secretory granules of mucous cells varies considerably. Some are as dense as those of zymogenic cells, others are electron lucent. Some are homogeneous, others display a variety of patterns of internal structure.

The surface mucous cells of the gastric mucosa in some species have secretory granules with a dense core surrounded by a wide zone of lower density. It is not known whether the granules are biochemically heterogeneous or whether the disparity in density reflects a difference in the physical state of a single component. The degree of mutual deformation of the closely packed granules and the variable position of the dense core suggest that they may consist of a solid granule suspended in a semisolid or fluid matrix.

Figure 384. Surface mucous cells from gastric mucosa of rat. (Micrograph courtesy of Susumu Ito.)

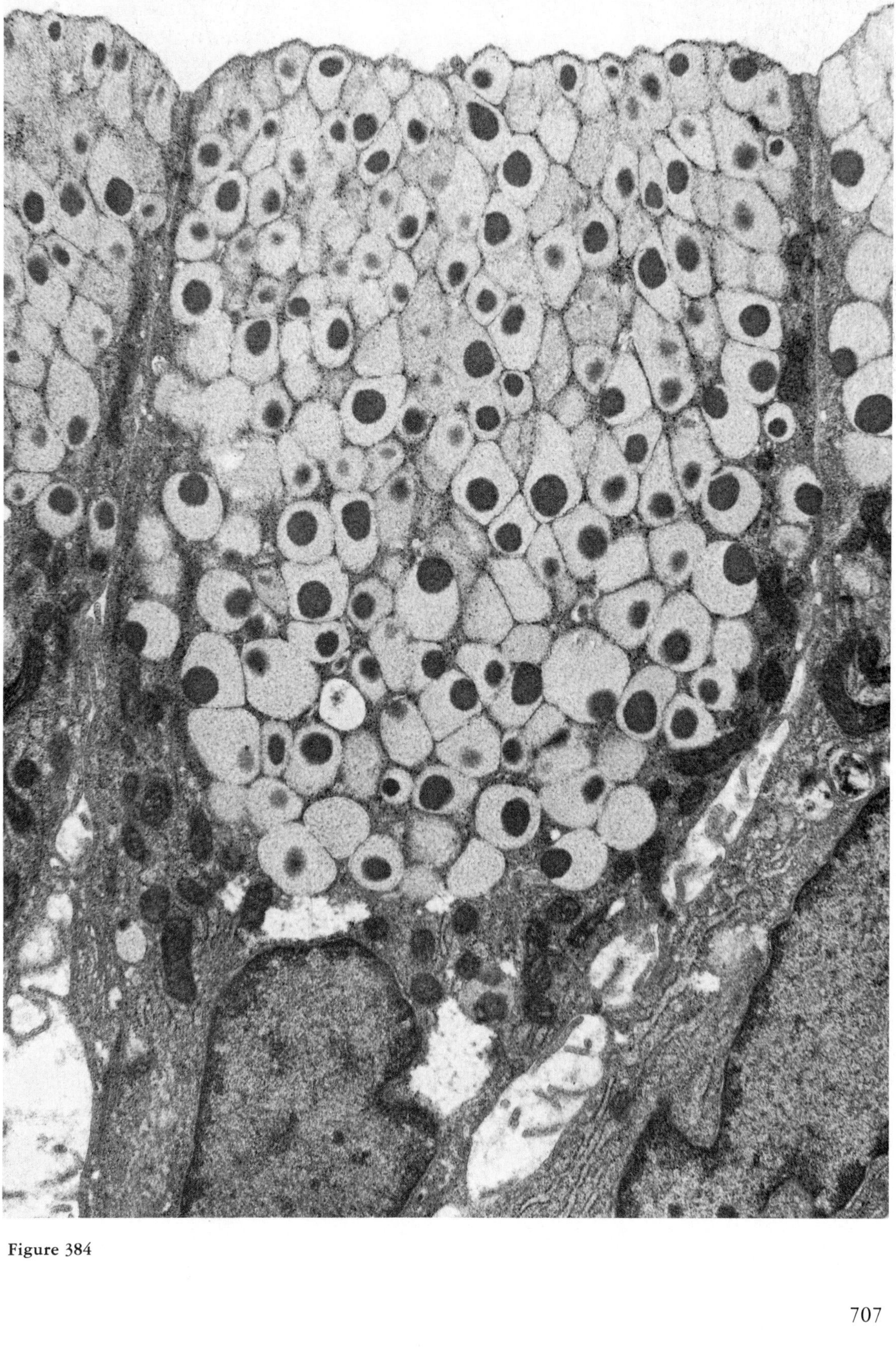

Figure 384

Surface mucous cells of guinea pig stomach have large granules of varying shape. Their pleomorphism suggests that they have a malleable semisolid content, for fluid-filled vesicles would be spherical or, if closely packed, polygonal. This degree of osmiophilia is unusual in a product rich in carbohydrate.

Figure 385. Surface mucous cells from guinea pig stomach. (Micrograph courtesy of Dixon Hingson.)

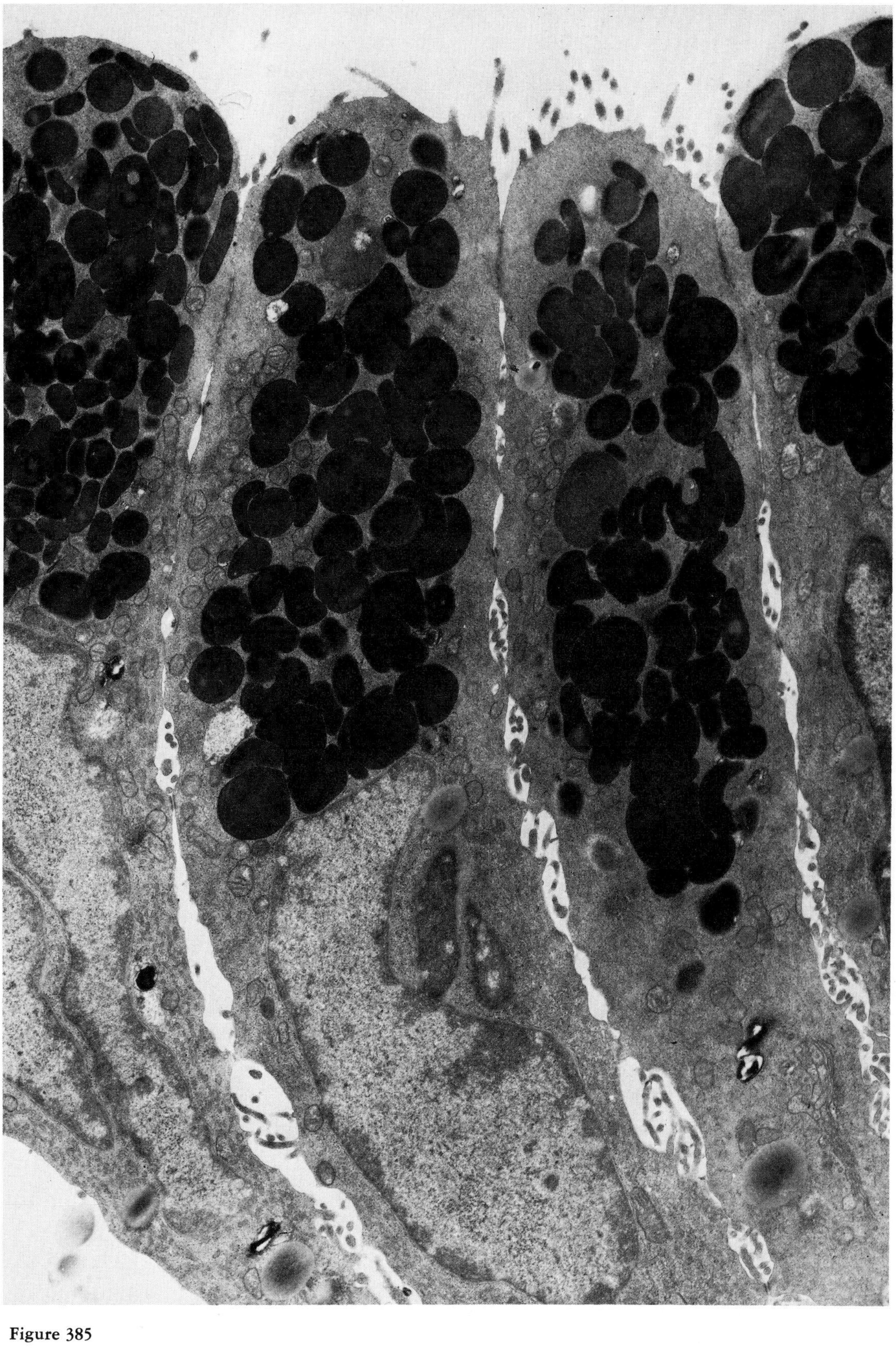

Figure 385

Exocytosis of Secretory Products

The terms *endocytosis* and *exocytosis* were introduced by DeDuve (1963) to refer respectively to the uptake of material by interiorization in vesicles formed at the cell surface, and to extrusion of material transported to the surface in vesicles formed within the cell. The majority of secretory products are released by this latter process. The secretory protein or glycoprotein is concentrated in the Golgi region in membrane-bounded secretory vesicles or granules. During secretion, the limiting membrane of the secretory granule fuses with the plasma membrane, permitting outflow of the product. The accompanying micrograph shows mucus being discharged into the lumen from one of the secretory granules in the apical cytoplasm of a mucus-secreting cell from the colon. The limiting membrane of the granule is subsequently incorporated into the plasmalemma. The excess surface membrane accumulated during exocytosis is believed to be recycled to the Golgi complex in small vesicles.

Figure 386. Exocytosis in a mucus-secreting cell from the colon. (Micrograph courtesy of Marian Neutra and Robert Specian.)

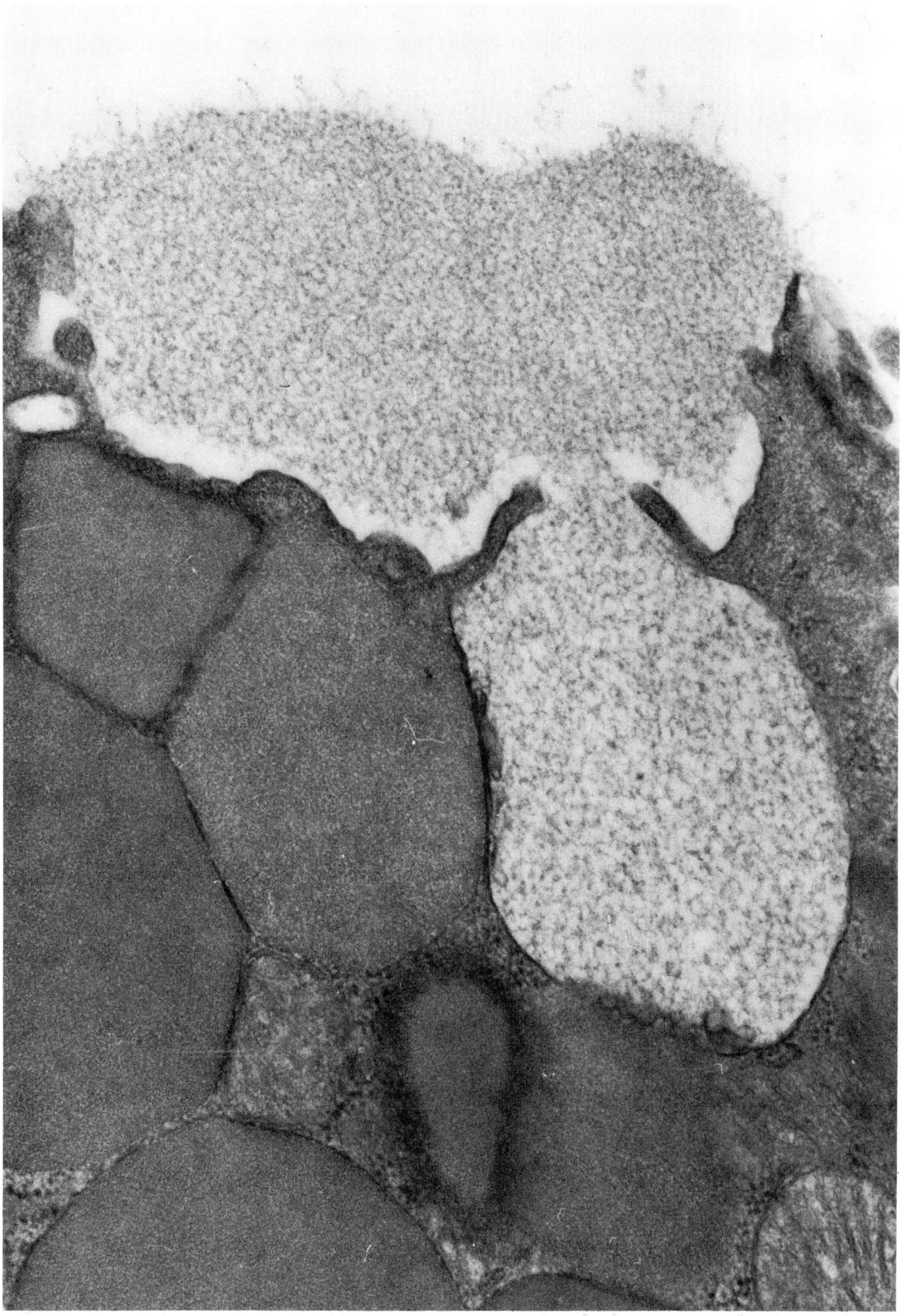

Figure 386

A necessary precondition for secretion by exocytosis is packaging of the product in a membrane capable of fusing with the plasmalemma. This is believed to be a function of the Golgi apparatus. Several related problems still require explanation. Among these is the basis for recognition that insures selective discharge of the secretion at the apical plasma membrane and nowhere else on the cell surface. Equally puzzling is the fact that even though secretory granules are closely packed in the apical cytoplasm, they do not normally fuse with one another. However, as illustrated in the accompanying micrograph of a highly stimulated pancreatic acinar cell, when a granule has opened onto the cell surface, the properties of its membrane are apparently changed in such a manner that other secretory granules can fuse with it. The membrane of the second granule is, in turn, altered so that a third and fourth can fuse with it, and so on. In this way a secretory pathway to the apical surface from deep in the cell may be established through a chain of communicating secretory granules. A similar phenomenon has been observed in experimentally stimulated mast cells.

Figure 387. Apical region of an acinar cell from perfused canine pancreas stimulated with pancreozymin. (From Ichikawa, J. Cell Biol. 24:369, 1965.)

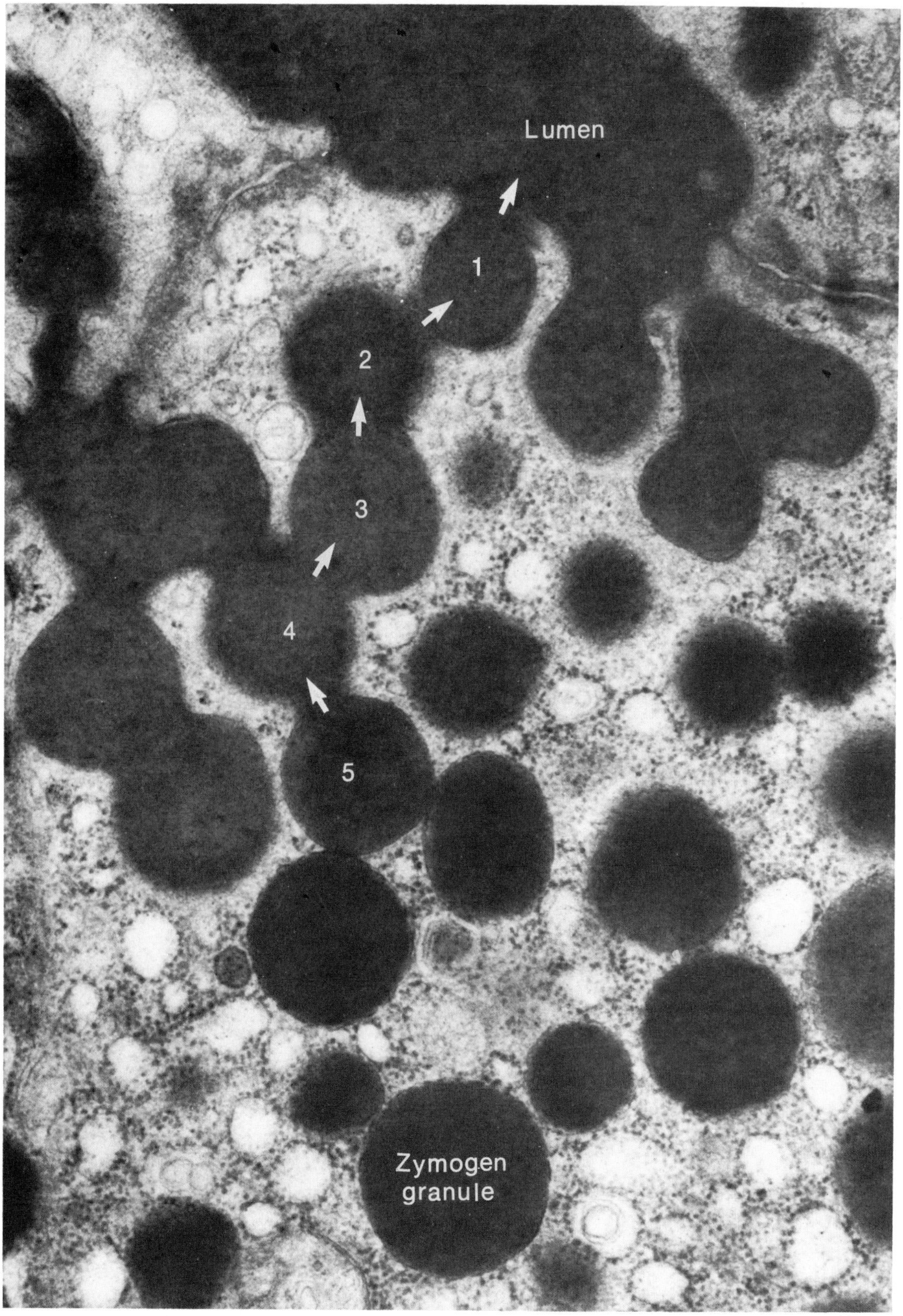

Figure 387

The continuous basal rate of mucus secretion involves intermittent fusion of single secretion granules with the plasma membrane, as illustrated in the previous micrograph. Experimental stimulation of secretion in organ cultures of colonic mucosa induces a progressive and massive depletion of mucus from crypt goblet cells. In this process, exocytosis of a few centrally situated granules is followed by an orderly sequential fusion of successive granules toward the cell center, often sparing the more peripheral granules. This results in a deep concavity in the apex of the goblet cell, such as is illustrated in the accompanying micrograph.

Study of mucus-secreting cells by freeze-fracturing has shown the presence of particle-poor potential fusion sites in the membranes between adjacent mucous granules closely packed in the cell apex. It is suggested therefore that the cell is prepared for multiple membrane fission-fusion events to produce granule confluence during stimulated massive secretion (Neutra and Schaeffer, 1977).

Figure 388. Mucus-secreting cell from a crypt of rabbit colonic mucosa after stimulation *in vitro* with acetylcholine. (Micrograph courtesy of Marian Neutra and Robert Specian.)

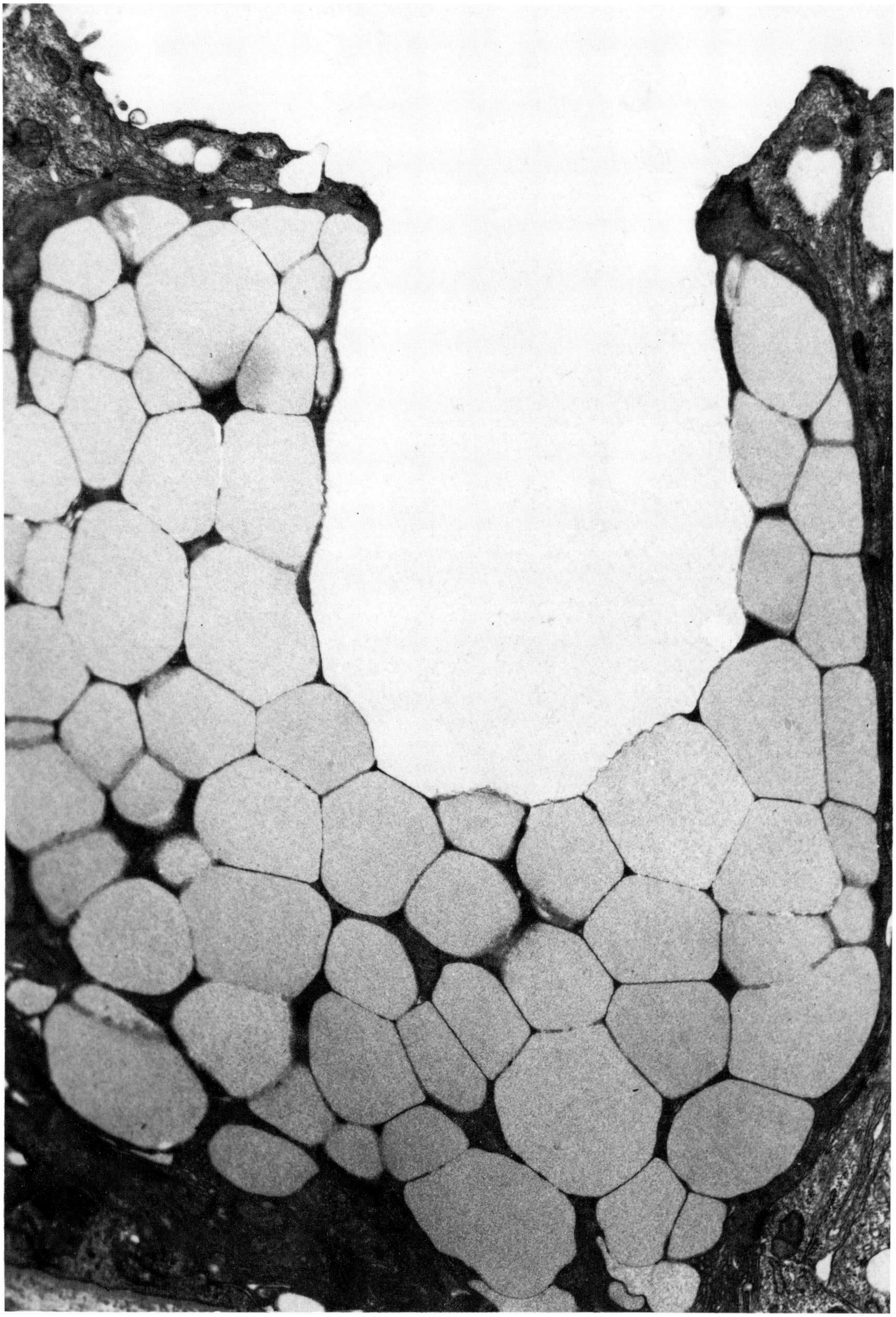

Figure 388

Endocrine Glands

In a number of endocrine glands, the small dense granules are separated from their limiting membrane by a clear zone. It seems likely that this appearance is due to loss of fluid content or extraction of less condensed soluble protein during specimen preparation. This interpretation is made the more plausible by the observation that better fixation of the same cell type sometimes succeeds in preserving material of appreciable density between the granule and the limiting membrane, as illustrated in the following figure.

Figure 389. Secretory granules in an alpha cell of an islet of Langerhans in human pancreas. (Micrograph courtesy of Arthur Like.)

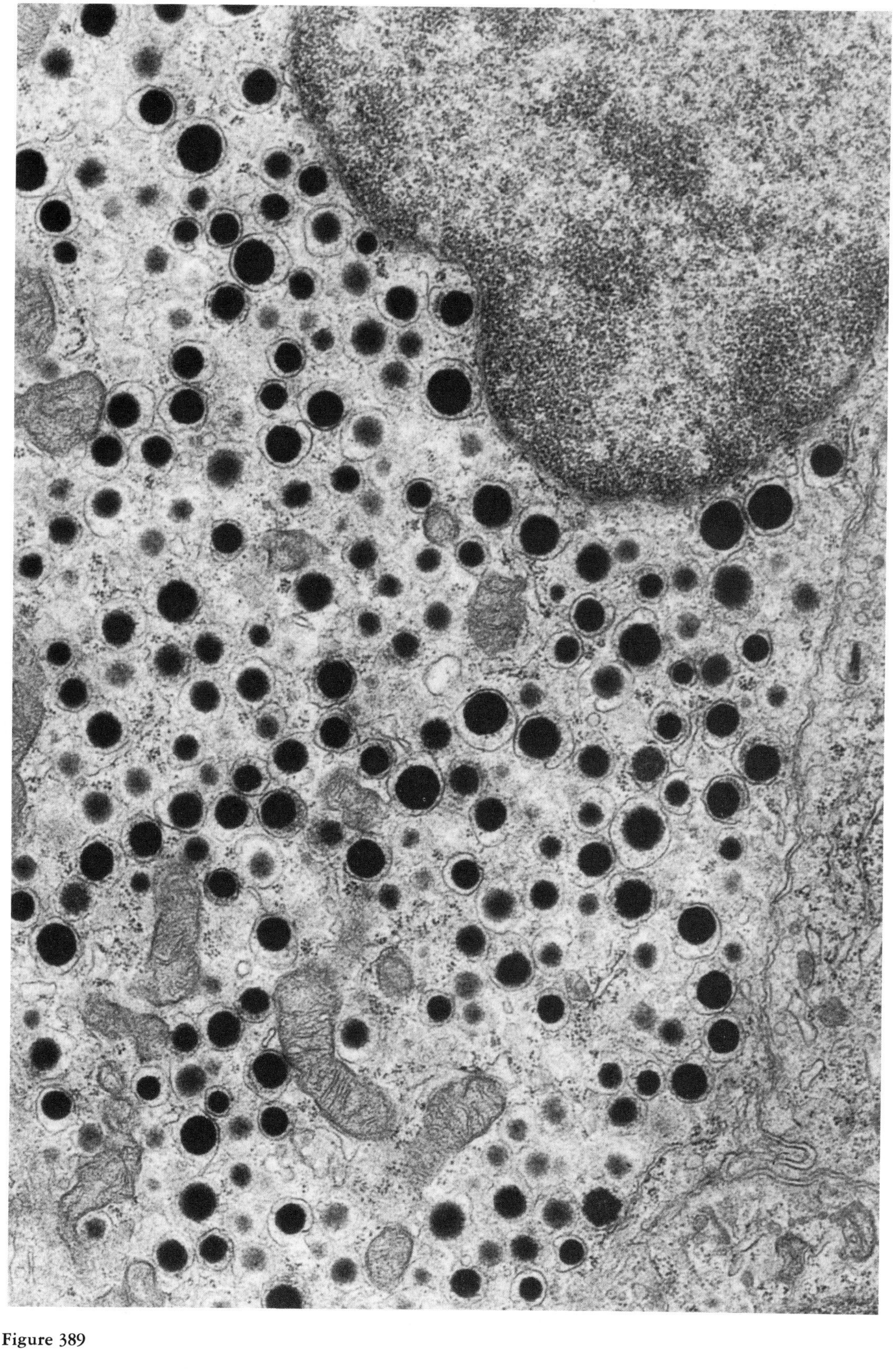

Figure 389

The accompanying micrograph includes portions of an alpha cell (above) and two beta cells (below). In this preparation, the dense alpha cell granules are surrounded by material of somewhat lower density. This component was apparently extracted in the preparation illustrated in the previous figure.

In the beta cells, on the other hand, the condensation of the secretory product, proinsulin, results in its crystallization. The human beta cell granules therefore consist of single or multiple tabular crystals in a membrane-bounded spherical vesicle, the fluid content of which has been extracted in the course of dehydration. In other species, the beta cells may contain homogeneous spherical granules similar in appearance to those of the alpha cells but differing from them in size.

Figure 390. Secretory granules of human alpha and beta cells of the endocrine pancreas. (Micrograph courtesy of Arthur Like.)

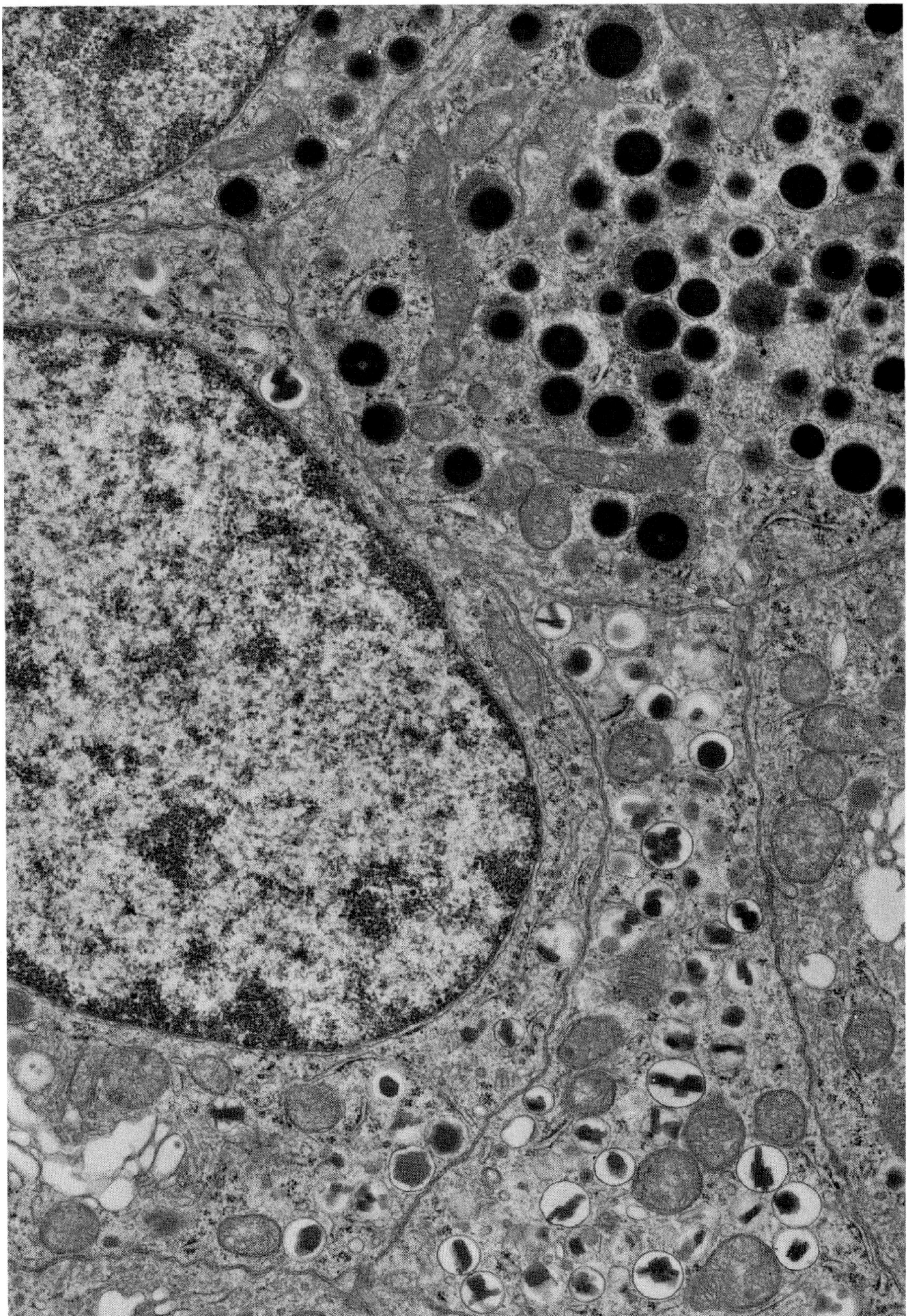

Figure 390

Other Secretory Cells

Secretion has been most thoroughly studied in cells of glandular epithelia. In such cells, secretion is polarized toward the lumenal surface, and release of the product occurs only on this limited region of the cell surface. A number of wandering cells of the blood and connective tissue are also secretory. Their secretory granules are not concentrated toward one pole of the cell but are distributed throughout the cytoplasm and can be discharged anywhere on the cell surface.

The mast cell shown here is an example. A variety of agents induce release of their histamine- and heparin-containing granules. The membrane of the granules can interact with any part of the cell membrane to establish an opening through which the granule contents escape.

Figure 391. Secretory granules of a mast cell from rat connective tissue.

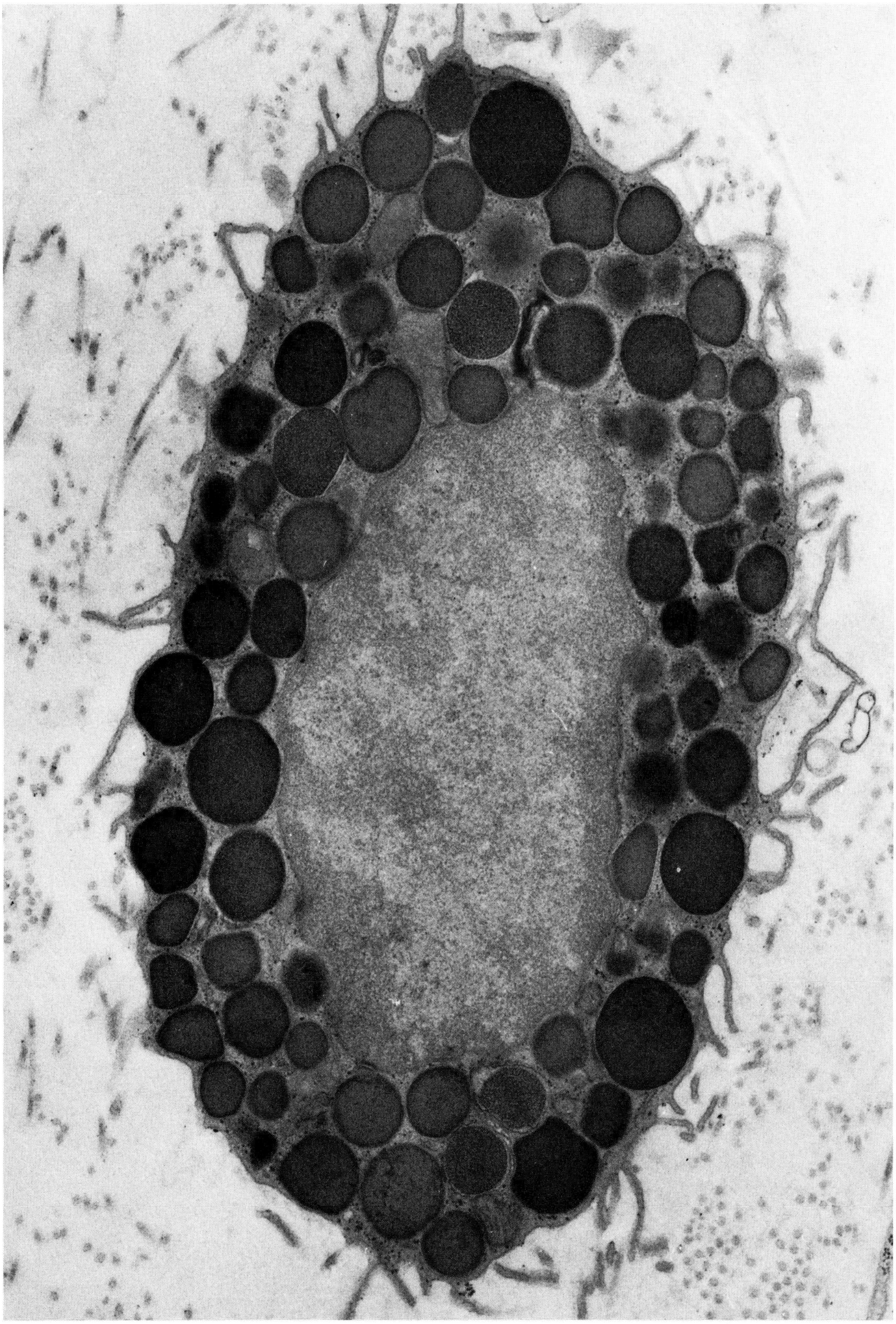

Figure 391

SYNAPSES

The term *synapse* was introduced by Sir Charles Sherrington (1897). It was derived from the Greek word meaning "to clasp or fasten together" and used to describe the site where the axon terminal of one neuron establishes functional contact with the cell body or a dendrite of another neuron. The term was subsequently extended to include the functional contact between the neuron and its effector cells, such as the contact of an axon terminal with a muscle fiber. At its ending, the axon is often expanded into an *end bulb* or *bouton terminal*. Early in this century, it was shown by Sir Henry Dale and others (1936) that when the nerve impulse reaches the synapse, a chemical transmitter substance is released which diffuses across the intercellular cleft to activate the postsynaptic element. Although the release of a chemical transmitter is not strictly analogous to secretion by a glandular cell, the mechanisms involved have enough in common to justify inclusion of this process here rather than in the section on cell junctions.

The development of recording techniques using microelectrodes (Young, 1939; Fatt and Katz, 1950) made it possible to study the electrical activity and the time course of events on the presynaptic and postsynaptic elements of the giant synapses of invertebrates and at myoneural junctions. Measurement of synaptic delay, and pharmacological interruption of the process, led to the conclusion that the mechanism of transmission at vertebrate synapses was chemical and not electrical. The transmitter was liberated in unit amounts (quantal release), suggesting that it might be in small packets of uniform size that could be released one or more at a time (Fatt and Katz, 1950; del Castillo and Katz, 1954).

With the advent of electron microscopy, the structural basis for quantal release became apparent. The axon terminals at interneuronal and neuromuscular synapses were found to contain large numbers of 40 to 50 nm spherical membrane-bounded vesicles with clear centers (Palade and Palay, 1954; Palay, 1956). These were called *synaptic vesicles* and it was inferred that they contained the transmitter. Although a direct cytochemical demonstration of transmitter in them was not possible, strong circumstantial evidence was soon obtained by isolation of nerve endings and of synaptic vesicles from brain homogenates and the identification of transmitter in these preparations (DeRobertis et al., 1963; Whittaker, 1965). It was therefore proposed that the vesicles move toward the synaptic membrane releasing their content into the synaptic cleft (Eccles, 1964). The release of the contents of individual synaptic vesicles was assumed to correspond to the quanta of transmitter previously postulated to account for the miniature end-plate potentials. The transmitter diffusing across the cleft reacts with specific receptors in the postsynaptic membrane, or sarcolemma, resulting in a transient increase in its permeability to ions and a consequent depolarization and initiation of an action potential (Katz and Miledi, 1967).

Electron micrographs also showed characteristic specializations of the synaptic membranes. In interneuronal synapses, there are local aggregations of dense material and fine filaments beneath the membrane of both the presynaptic and postsynaptic elements. In section these bear a superficial resemblance to the zonula adherens of epithelia, but they are asymmetrical. The layer of dense filamentous material associated with the postsynaptic membrane is often, but not invariably, thicker than that on the presynaptic side. DeRobertis and coworkers refer to it as the *subsynaptic web*. The synaptic vesicles are often very closely associated with the thinner presynaptic cytoplasmic density. The assemblage of vesicles and the junctional density are collectively referred to as a *synaptic complex* (Palay, 1958). Because these appear to be sites of preferential accumulation of vesicles and release of transmitter, they are also called *active zones* of the synaptic junction (Couteaux, 1961). There is considerable variation in the width of the synaptic cleft and in the linear extent and degree of asymmetry of the densities of the synaptic junctions from one region of the brain to another. When specially stained and viewed in sections tangential to the synaptic cleft, the presynaptic density is not homogeneous but displays a hexagonal or triagonal pattern in which the interstices are occupied by synaptic vesicles (Gray, 1963; Akert and Pfenninger, 1969). This ordered structure is called the *presynaptic grid*.

In addition to the densities of the synaptic complexes, there are smaller attachment devices called *puncta adherentia*. These densities are thicker than those of the synaptic complexes but are of very limited linear extent and they are symmetrical. Synaptic vesicles do not associate with them and they seem to be accessory specializations for maintaining coherence of the synaptic membranes.

In routine preparations the synaptic cleft usually appears empty in interneuronal synapses, but with special staining methods a thin intermediate layer of intercellular material can sometimes be seen. It appears to be glycoprotein in nature. In the neuromuscular synapse, the synaptic cleft is considerably wider and contains a distinct lamina externa that extends down into the conspicuous clefts of the corrugated subneural region of the muscle fiber.

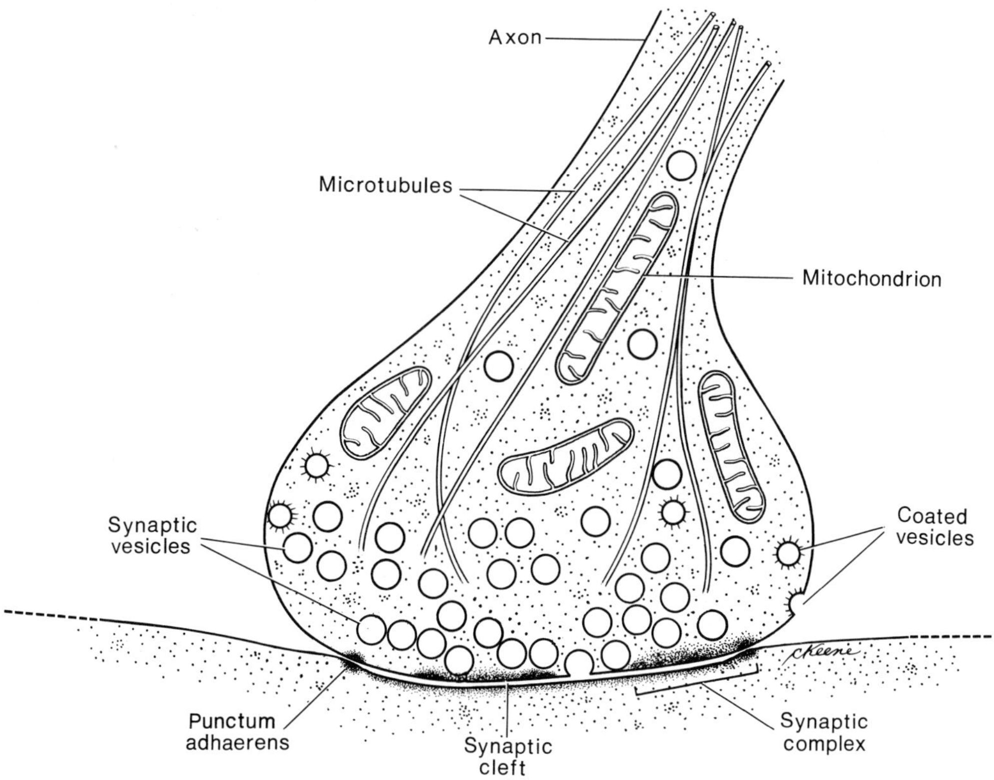

Schematic representation of the structural components of a synapse.

Neuromuscular Synapse

The accompanying micrograph illustrates the characteristic features of a neuromuscular junction, in a specimen prepared by rapid freezing with liquid helium followed by freeze substitution. The nerve terminal is covered (above) by a portion of a Schwann cell and is separated from the muscle fiber (below) by a synaptic cleft containing a conspicuous external lamina. The cytoplasm of the nerve ending contains several mitochondria, a few microtubules, a number of neurofilaments, and a large aggregation of synaptic vesicles on the side toward the synaptic cleft.

Figure 392. Frog myoneural junction prepared by freeze substitution. (Micrograph courtesy of John Heuser and Tom Reese.)

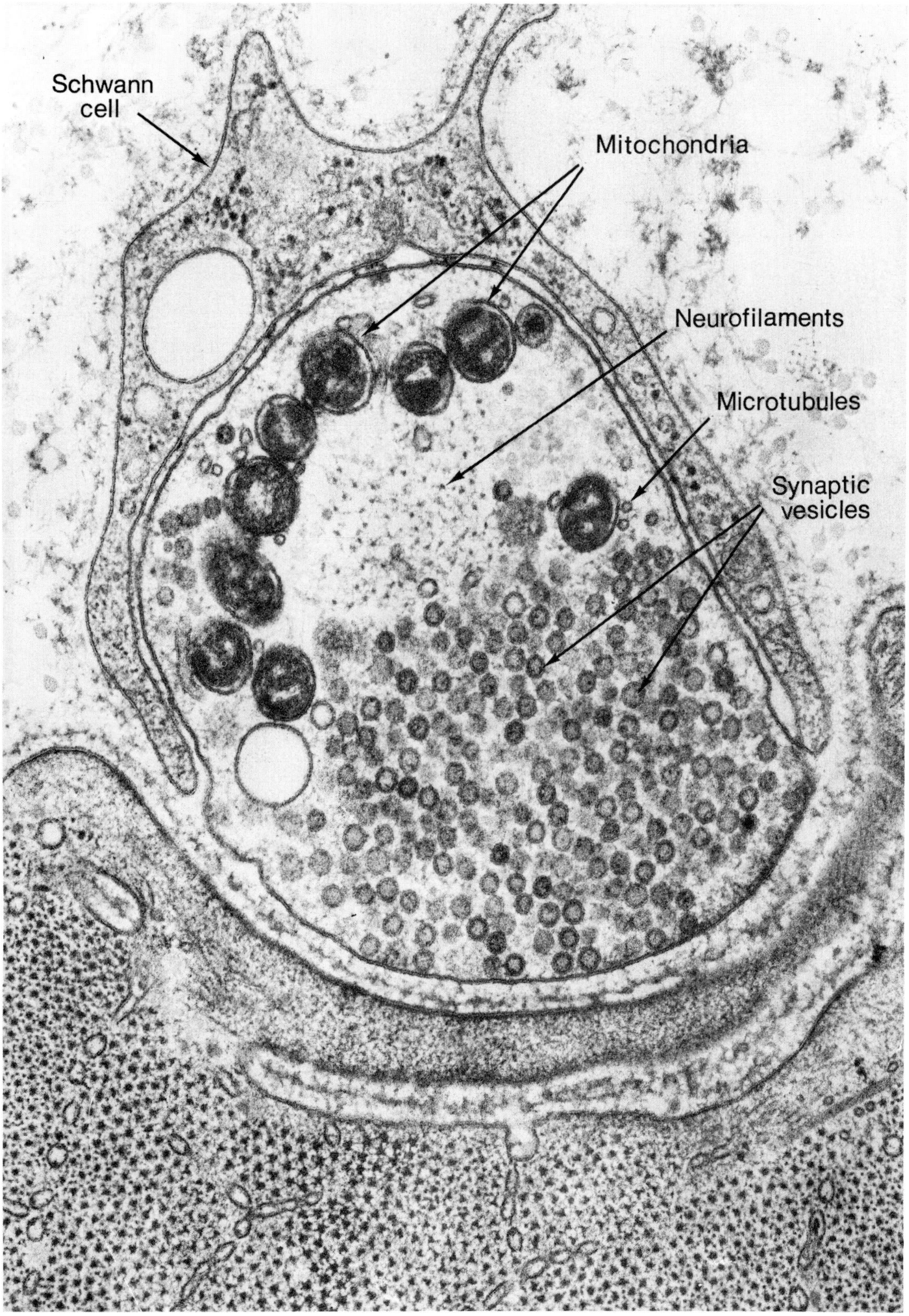
Schwann cell
Mitochondria
Neurofilaments
Microtubules
Synaptic vesicles

Figure 392

A longitudinal section of the nerve ending at a frog myoneural junction prepared by rapid freezing and freeze substitution, showing a profusion of small spherical synaptic vesicles. In such unstimulated nerve terminals, synaptic vesicles are rarely seen discharging their content of transmitter into the synaptic cleft.

Figure 393. Longitudinal section of frog myoneural junction prepared by freeze substitution. (Micrograph courtesy of John Heuser and Tom Reese.)

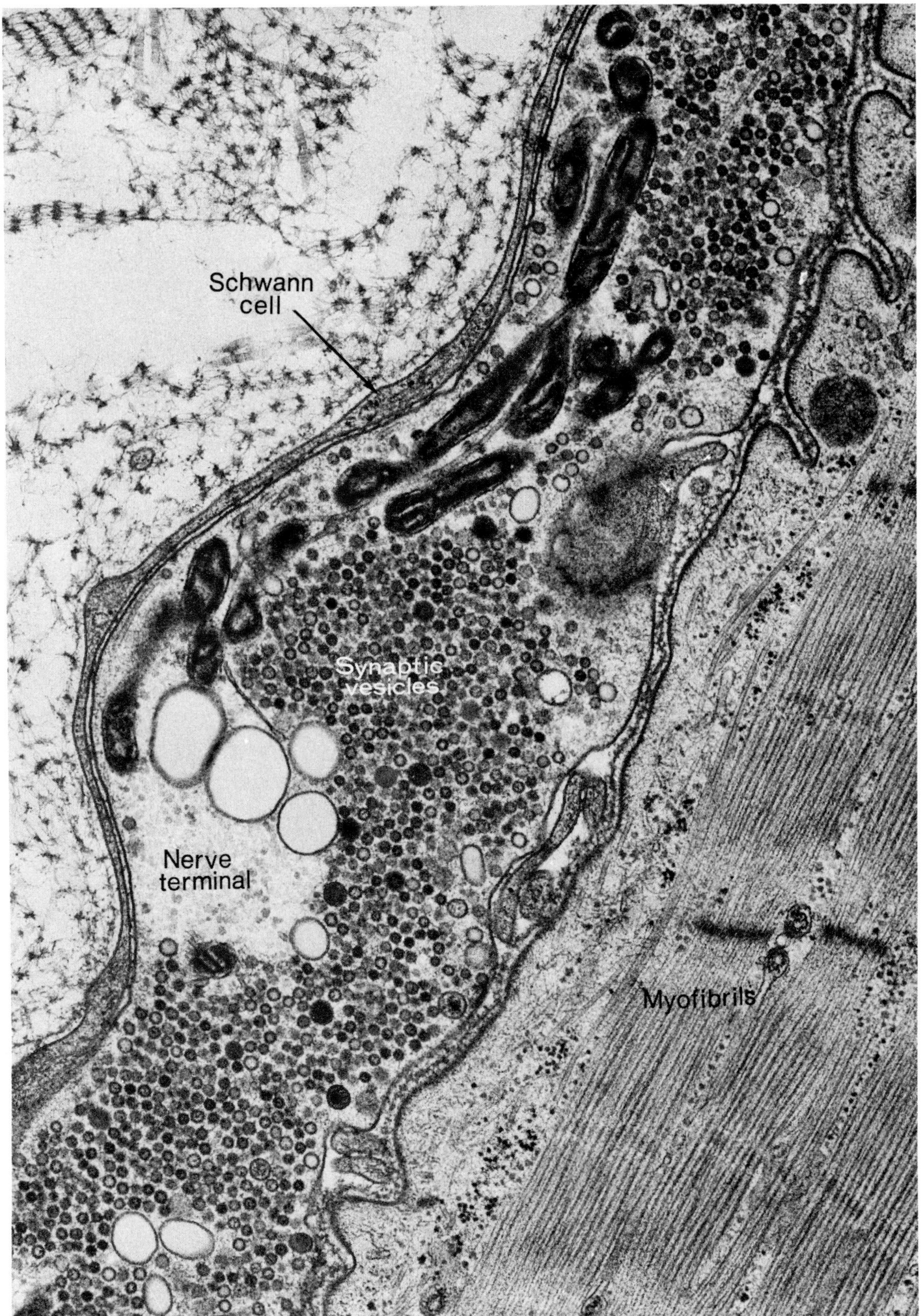

Schwann cell
Synaptic vesicles
Nerve terminal
Myofibrils

Figure 393

Convincing morphological evidence for the hypothesis that nerve terminals discharge transmitter from synaptic vesicles by exocytosis could not be obtained with routine methods of fixation for electron microscopy. After such treatment, vesicles opening into the synaptic cleft were rare and their number could not be correlated with the number of quanta of transmitter released. It is now evident that this absence of correlation is due to the fact that chemical fixation is a relatively slow process incapable of capturing events in the presynaptic membrane having a time course measured in milliseconds.

By augmenting transmitter release with stimulation in the presence of 4-aminopyridine and rapid freezing of the myoneural junction, it has been possible to capture synaptic vesicles in the process of exocytosis. Under these conditions there is an excellent correlation between their number and the level of stimulus-evoked transmitter release (Heuser and Reese, 1979).

The accompanying high magnification micrograph shows the presynaptic (right) and postsynaptic (left) elements of frog neuromuscular junction quick frozen during a large burst of acetylcholine release 5 milliseconds after a single nerve stimulus. Several synaptic vesicles are seen discharging their contents into the synaptic cleft.

Figure 394. High magnification micrograph of the presynaptic and postsynaptic membranes of freeze-substituted frog myoneural junction after nerve stimulation in the presence of 4-aminopyridine. (Micrograph courtesy of John Heuser and Tom Reese.)

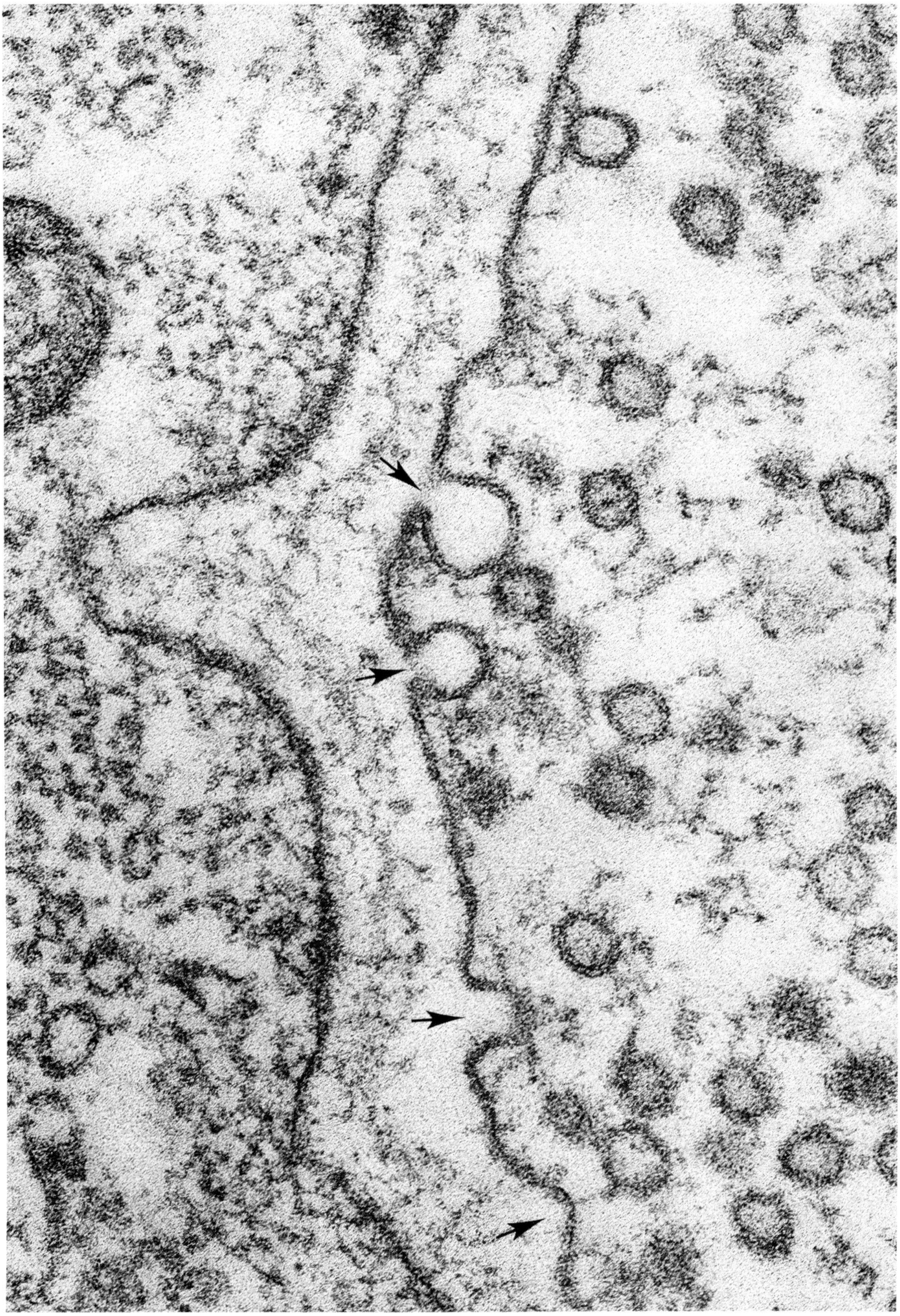

Figure 394

Central Nervous System Synapses

A wide variety of synaptic relationships between neurons are found in the central nervous system. The most common are *axodendritic synapses.* These may occur anywhere along the dendrites and the same axon may synapse with dendrites of many neurons. The dendrites of some types of neurons are provided with numerous lateral projections with bulbous tips and narrow stalks, called *spines,* or *thorns.* These are commonly the sites of axodendritic synapses. The terminal boutons of axons frequently end on the cell bodies of other neurons to form *axosomatic synapses.* Although relatively uncommon, *axoaxonal synapses* are occasionally observed. In these, synaptic vesicles usually occur in only one of the elements and this is described as the presynaptic component. In some instances, however, synaptic vesicles are present in both. Although it was traditionally assumed that dendrites could not be presynaptic elements, rare instances of *dendodendritic* and *somatoaxonic synapses* have been reported. A consideration of all the various interneuronal relationships is beyond the scope of this book. Only a few examples of the more common types of synapse will be presented. For a more detailed treatment of this subject the interested reader is referred to texts and atlases of neurocytology (e.g., Peters, Palay and Webster's Fine Structure of the Nervous System, Philadelphia, W. B. Saunders Company, 1976).

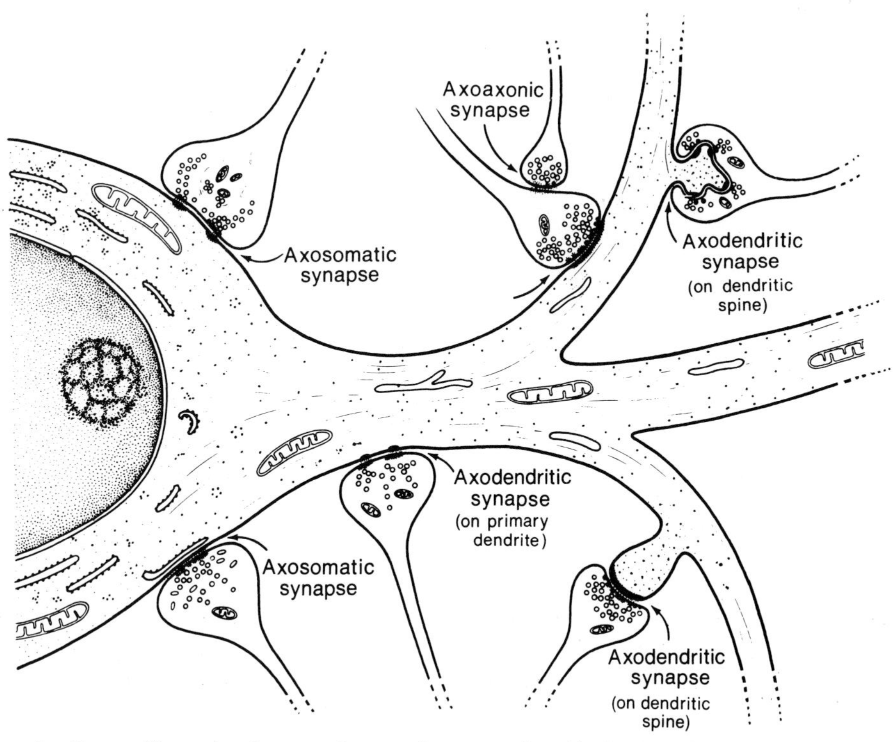

Composite diagram illustrating the several types of synapses found in the central nervous system.

Striking examples of axodendritic synapses are seen in the accompanying micrograph of the giant axon of a Müller cell in the central nervous system of a lamprey ammocoete and a portion of the adjoining neuropil. Along the periphery of the giant axon are four synapses with neighboring dendrites. The two synapses in the lower half of the figure consist of sharply circumscribed aggregations of synaptic vesicles closely associated with specialized regions of the axolemma and of the postsynaptic membrane (at arrows). The two clusters of synaptic vesicles in the upper half of the figure are farther from the membrane and are evidently associated with synaptic complexes that are out of the plane of the section.

Another typical synapse between a small axon and a smaller dendrite is seen at the lower right (arrow).

Figure 395. Section of a portion of the axon of a Müller cell and adjacent neuropil from the central nervous system of an ammocoete of *Petromyzon marinus*. (From Smith, Järlfors and Beranék, J. Cell Biol. 46:199, 1970.)

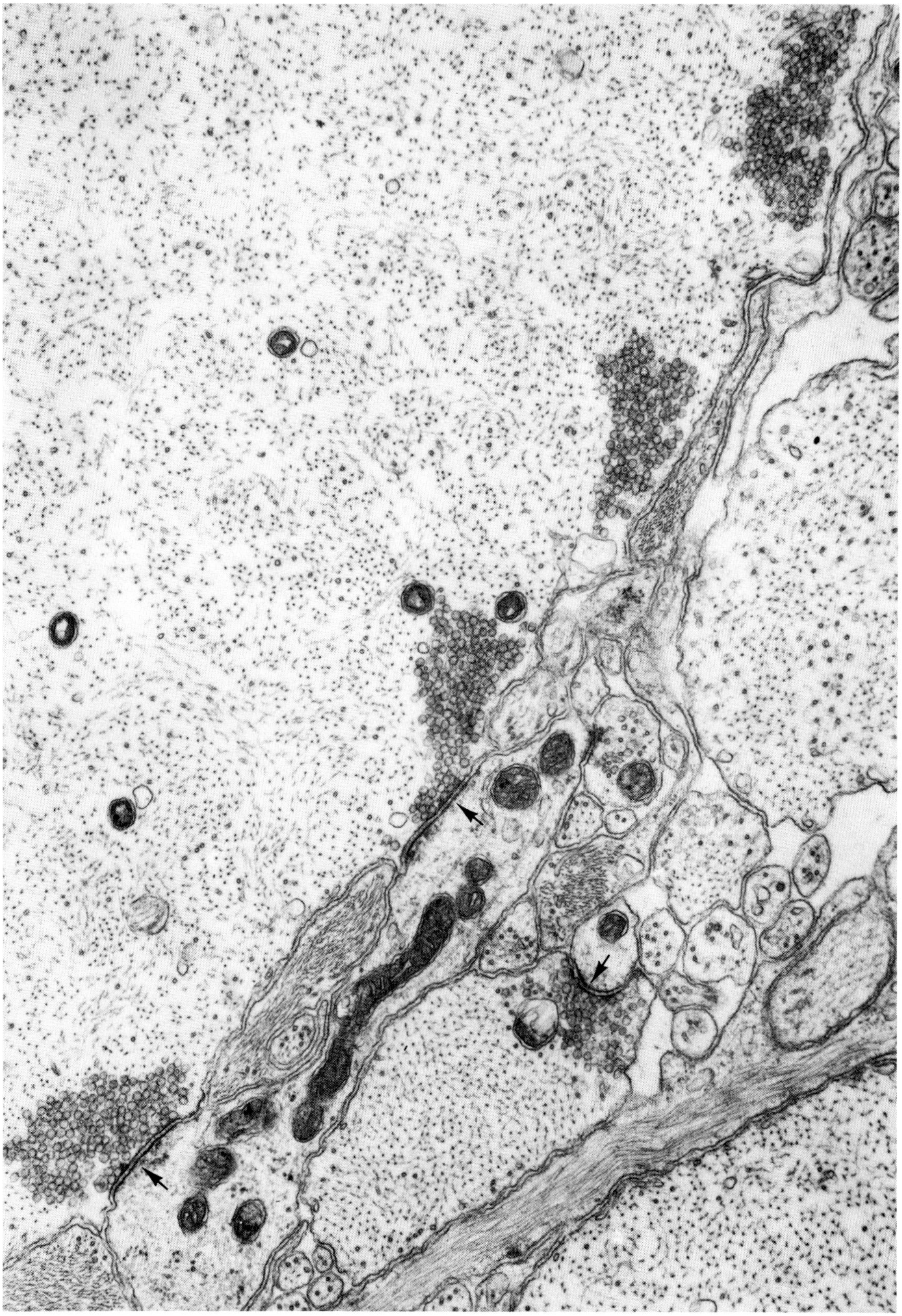

Figure 395

Shown here is an axodendritic synapse from mammalian central nervous system. A nerve enters the field at the upper right. Its myelin sheath terminates and its axon expands into a terminal bouton closely applied to the surface of a large dendrite at the left. The nerve terminal is filled with synaptic vesicles. The synaptic complex (at the arrow) is of relatively limited extent, but similar local membrane specializations are probably present in other planes of section.

Figure 396. Axodendritic synapse from rat central nervous system. (Micrograph courtesy of Enrico Mugnaini.)

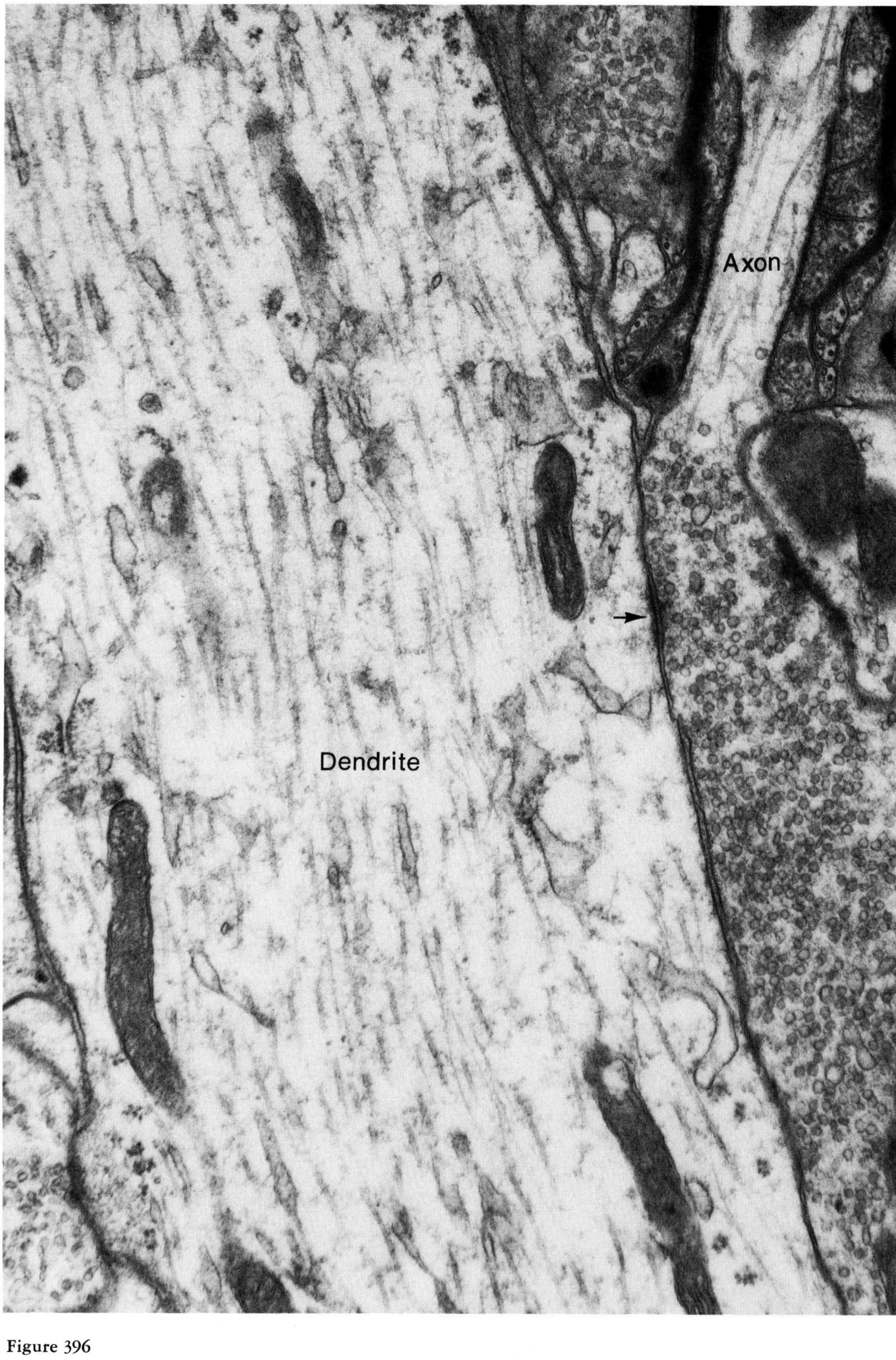

Figure 396

Axosomatic synapses are illustrated on the facing page. The nerve terminal in the center of the figure synapses with the cell body of a neuron at the lower right. The nerve ending is filled with mitochondria and synaptic vesicles. Two synaptic complexes on the interface with the cell body are indicated by arrows. At the left, the same axon synapses with a dendrite. Another axodendritic synapse is seen at the top of the figure.

Figure 397. Axosomatic and axodendritic synapses in the central nervous system of rat. (Micrograph courtesy of Enrico Mugnaini.)

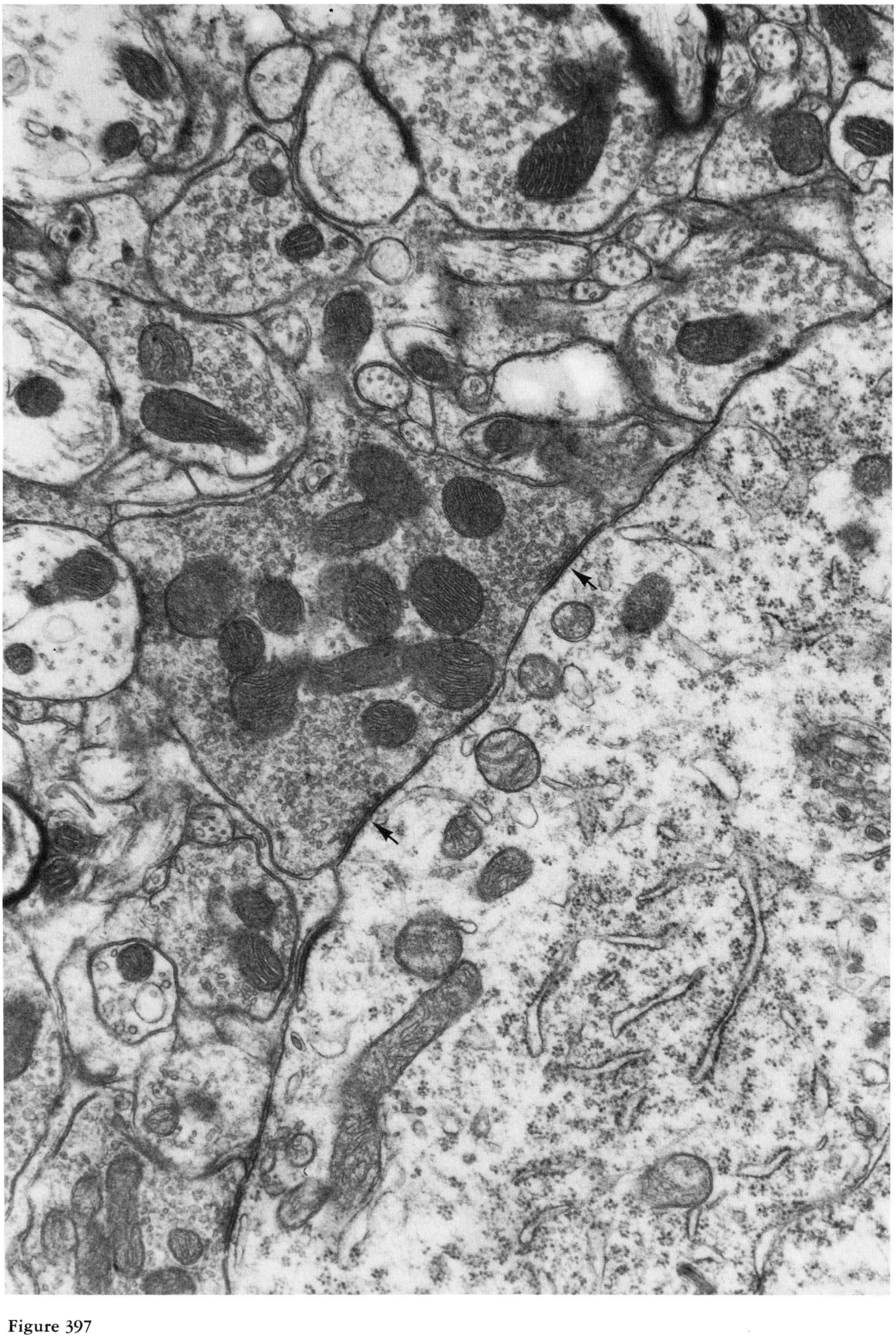

Figure 397

Extending down the middle of this field is a small branch of a Purkinje cell dendrite. Two thorns or spines can be seen extending laterally from the dendrite. The axon terminals of parallel fibers form synapses with the thorns. Isolated profiles of numerous other thorns are identifiable in the surrounding neuropil.

Figure 398. Synapses with the thorns on a Purkinje cell dendrite from rat cerebellum. (From Peters, Palay and Webster, Fine Structure of the Nervous System, Philadelphia, W. B. Saunders Company, 1976.)

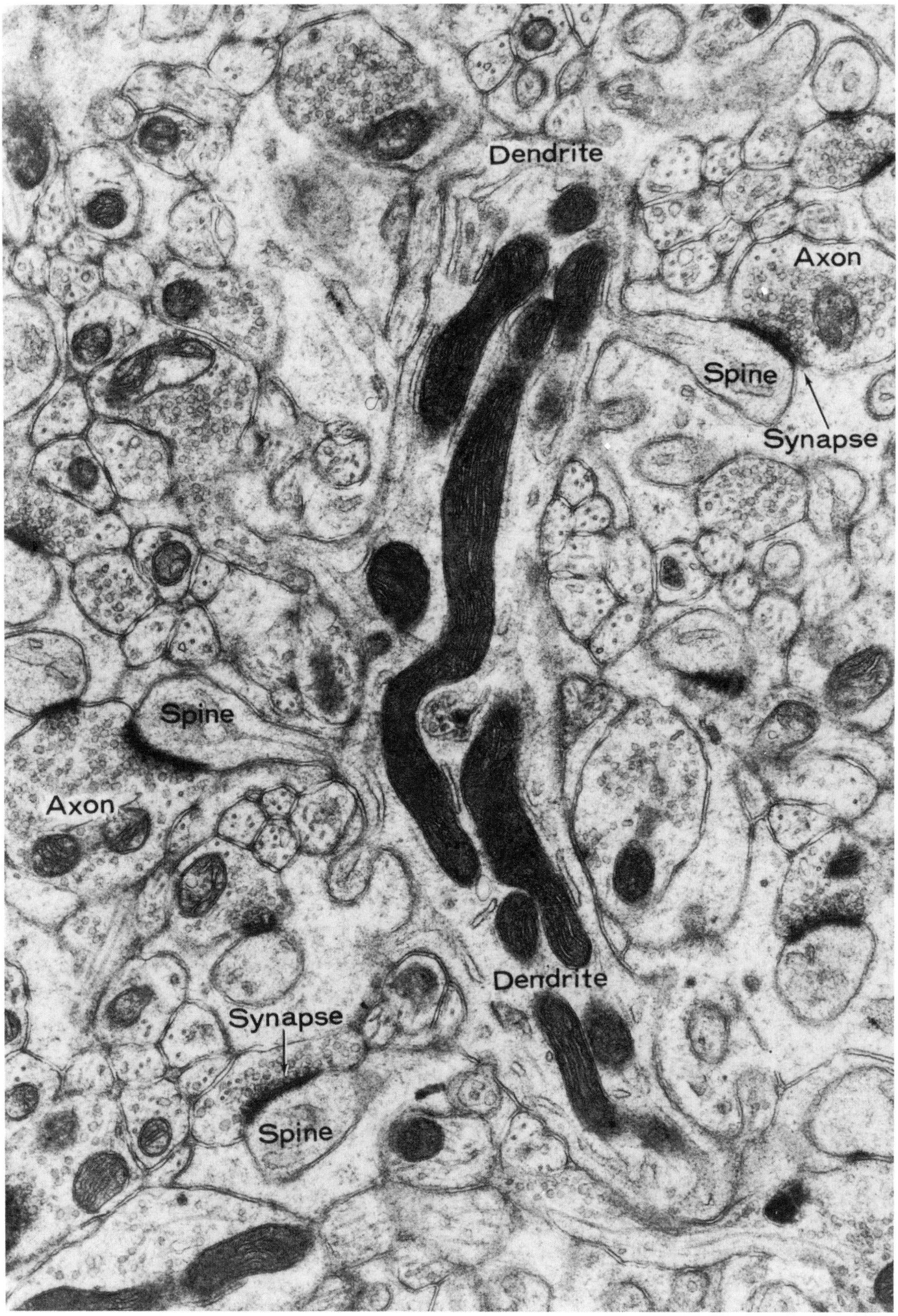
Dendrite
Axon
Spine
Synapse
Spine
Axon
Dendrite
Synapse
Spine

Figure 398

REFERENCES

Glycogen

Fawcett, D. W. Observations on the cytology and electron microscopy of hepatic cells. J. Nat. Cancer Inst. 15(suppl):1475, 1955.
Fawcett, D. W. Identification of particulate glycogen and ribonucleoprotein in electron micrographs. J. Histochem. Cytochem. 6:95, 1958.
Lazarow, A. Particulate glycogen, a submicroscopic component of the guinea pig liver cell; its significance in glycogen storage and the regulation of blood sugar. Anat. Rec. 84:31, 1942.
Porter, K. R. and C. Bruni. An electron microscope study of the early effects of 3'Me DAB on rat liver cells. Cancer Res. 19:997, 1959.
Revel, J. P., L. Napolitano and D. W. Fawcett. Identification of glycogen in electron micrographs of thin tissue sections. J. Biophys. Biochem. Cytol. 8:575–589, 1960.
Richter, W. R., R. J. Stein, E. J. Rdzok, S. M. Moize and M. E. Bischoff. Ultrastructural studies of intranuclear crystalline inclusions in the liver of the dog. Am. J. Pathol. 47:587, 1965.

Lipid

Cushman, S. W. Structure-function relationships in the adipose cell. I. Ultrastructure of the isolated adipose cell. J. Cell Biol. 46:326–346, 1970.
Cushman, S. W. Structure-function relationships in the adipose cell. II. Pinocytosis and factors influencing its activity in the isolated adipose cell. J. Cell Biol. 46:342–353, 1970.
Imaizumi, M. On the fine structure of the surface of lipid droplets in adipose cells. Arch. Histol. Jpn. 30:353–365, 1969.
Jeanrenaud, B. Dynamic aspects of adipose tissue metabolism. A review. Metab. Clin. Exp. 10:535, 1961.
Luckenbill, L. M. and A. S. Cohen. The association of lipid droplets with cytoplasmic filaments in avian subsynovial adipose cells. J. Cell Biol. 31:195, 1966.
Napolitano, L. The differentiation of white adipose cells. J. Cell Biol. 18:663–679, 1963.
Napolitano, L. and D. W. Fawcett. The fine structure of brown adipose tissue in the newborn mouse and rat. J. Biophys. Biochem. Cytol. 4:685–692, 1958.
Napolitano, L. and H. T. Gagne. Lipid-depleted white adipose cells. Anat. Rec. 147:273, 1963.
Slavin, B. G. The cytophysiology of mammalian adipose tissue. Int. Rev. Cytol. 33:297–334, 1972.
Suter, E. The fine structure of brown adipose tissue. Lab. Invest. 21:246–258, 1969.
Suter, E. R. and W. Staubli. An ultrastructural histochemical study of brown adipose tissue from prenatal rats. J. Histochem. Cytochem. 18:100–106, 1970.
Williamson, J. R. Adipose tissue: Morphological changes associated with lipid mobilization. J. Cell Biol. 20:57–74, 1964.

Crystalline Inclusions

Fawcett, D. W. and M. H. Burgos. Observations on the cytomorphosis of the germinal and interstitial cells of the human testis. In Ciba Foundation Symposium on Aging, Vol. 2 (G. E. W. Wolstenholme and E. C. P. Millar, eds.), p. 36, Little Brown & Company, Boston, 1956.
Fawcett, D. W. and M. H. Burgos. Studies on the fine structure of the mammalian testis. II. The human interstitial tissue. Am. J. Anat. 107:270, 1960.
Honjin, R. and T. Nakamura. A refinement of the values of the lattice parameters in the crystal structure of amphibian yolk platelets by x-ray crystallography. J. Ultrastr. Res. 20:400, 1967.
Lubarsch, O. Uber das Vorkonimen krystallinischer und krystalloider Bildungen in den Zellen des menschlichen Hodens. Virchow's Arch. Path. Anat. 145:316–338, 1896.
Nagano, T. Some observations on the fine structure of the Sertoli cell of human testis. Z. Zellforsch. 73:89–106, 1966.
Nagano, T. and I. Ohtsuki. Reinvestigation of the fine structure of Reinke's crystal in the human testicular interstitial cell. J. Cell Biol. 51:148–161, 1971.
Reinke, F. Beiträge zur Histologie des Menschen. Arch. Mikr. Anat. 47:34–44, 1896.
Sisson, J. K. and W. H. Fahrenbach. Fine structure of steroidogenic cells of a primate cutaneous organ. Am. J. Anat. 121:337, 1967.
Sohval, A. R., Y. Suzuki, J. L. Gabrilove and J. Churg. Ultrastructure of crystalloids in spermatogonia and Sertoli cells of normal human testis. Ultrastr. Res. 34:83–102, 1971.
Toyama, Y. Ultrastructural study of crystalloids in Sertoli cells of normal, intersex and experimental cryptorchid swine. Cell Tissue Res. 158:205–213, 1975.
Yamada, E. Some observations on the fine structure of the interstitial cell in the human testis as revealed by electron microscopy. Gunma Symp. Endocrinol. 2:1, 1965.

Secretory Products

Bainton, D. F. and M. G. Farquhar. Segregation and packaging of granule enzymes in eosinophilic leucocytes. J. Cell Biol. 45:54–73, 1970.

Farquhar, M. G. Processing of secretory products by cells of the anterior pituitary gland. Mem. Soc. Endocrinol. 19:79–124, 1970.

Holtzman, E., S. Schachter, J. Evans et al. Origin and fate of membranes of secretion granules and synaptic vesicles — membrane circulation in neurons, gland cells and retinal photoreceptors. In Synthesis, Assembly and Turnover of Cell Surface Components (G. Poste and G. L. Nicolson, eds.) North Holland Publishing Company, Amsterdam, New York, 1977.

Jamieson, J. D. Membranes and secretion. Hosp. Pract., pp. 71–80, Dec. 1973.

Jamieson, J. D. and G. E. Palade. Condensing vacuole conversion and zymogen granule discharge in pancreatic exocrine cells: Metabolic studies. J. Cell Biol. 48:503–522, 1971.

Jamieson, J. D. and G. E. Palade. Synthesis intracellular transport and discharge of secretory proteins in stimulated pancreatic exocrine cells. J. Cell Biol. 50:135–158, 1971.

Jamieson, J. D. Transport and discharge of exportable proteins in pancreatic exocrine cells: *In vitro* studies. Current Topics in Membranes and Transport (F. Bonner and A. Kleinzeller, eds.), Academic Press, New York, 1972.

Lagunoff, D. Membrane fusion during mast cell degranulation. J. Cell Biol. 57:352–372, 1970.

Meldolesi, J. Secretory mechanisms in pancreatic acinar cells. Role of the cytoplasmic membranes. Adv. Cytopharm. 2:71, 1974.

Palade, G. E. Intracellular aspects of the process of protein secretion. Science 189:347–358, 1975.

Röhlich, P., P. Anderson and B. Uvnäs. Electron microscopic observations on compound 48/80 induced degranulation in rat mast cells. Evidence for sequential exocytosis of storage granules. J. Cell Biol. 51:465–483, 1971.

Synapses

Akert, K. and K. Pfenninger. Synaptic fine structure and neuronal dynamics. In Symposium of International Society for Cell Biol. Vol 8, Cellular Dynamics of the Neuron (S. H. Barondes, ed.), pp. 245–260, Academic Press, New York, 1969.

Couteaux, R. Principaux critères morphologiques et cytochimiques utilisables aujourd 'hui pour définir les divers types de synapses. Actual. Neurophysiol. 3e serie, 145–173, 1961.

Dale, H. H., W. Feldberg and M. Vogt. Release of acetylcholine at voluntary nerve endings. J. Physiol. (Lond.) 86:353–380, 1936.

del Castillo, J. and B. Katz. On localization of acetylcholine receptors. J. Physiol. (Lond.) 128:157–181, 1955.

DeRobertis, E., G. Rodriguez de Lores Arnaiz, L. Salganicoff and A. Pellegrino de Iraldi. Isolation of synaptic vesicles and structural organization of the acetylcholine system within brain nerve endings. J. Neurochem. 10:225–235, 1963.

Eccles, J. C. The Physiology of Synapses. Springer-Verlag, Berlin, 1964.

Fatt, P. and B. Katz. Some observations on biological noise. Nature 166:597–598, 1950.

Gray, E. G. Electron microscopy of presynaptic organelles of the spinal cord. J. Anat. 97:101–106, 1963.

Gray, E. G. Electron microscopy of excitatory and inhibitory synapses: A brief review. Prog. Brain Res. 31:141–155, 1969.

Heuser, J. Morphology of synaptic vesicle discharge and reformation at the frog neuromuscular junction. In Motor Innervation of Muscle (S. Thesleff, ed.) pp. 51–115, Academic Press, New York, 1976.

Heuser, J. E. Synaptic vesicle exocytosis revealed in quick-frozen frog neuromuscular junctions treated with 4-aminopyridine and given a single electrical shock. Neuroscience Symposia. Vol. III (W. M. Cowan and M. A. Ferrendeli, eds.) Soc. For Neuroscience, Bethesda, Md.

Heuser, J. E. and T. S. Reese. Evidence for recycling of synaptic vesicle membrane during transmitter release at the frog neuromuscular junction. J. Cell Biol. 57:315–344, 1973.

Heuser, J. E., T. S. Reese and D. M. D. Landis. Functional changes in frog neuromuscular junctions studied by freeze-fracture. J. Neurocytol. 3:109–131, 1974.

Katz, B. and R. Miledi. The release of acetylcholine from nerve ending by graded electrical pulses. Proc. Roy. Soc. Ser. B. 167:23–38, 1967.

Palade, G. E. and S. L. Palay. Electron microscope observations of interneuronal and neuromuscular synapses. Anat. Rec. 118:335–336, 1954.

Palay, S. L. Synapses in the central nervous system. J. Biophys. Biochem. Cytol. 2: (Suppl.) 193–202, 1956.

Palay, S. L. The morphology of synapses in the central nervous system. Exp. Cell Res. Suppl. 5:275–293, 1958.

Pappas, G. D. and D. P. Purpura. Structure and Function of Synapses. Raven Press, New York, 1972.

Peters, A., S. L. Palay and H. deF. Webster. The Fine Structure of the Nervous System: The Neurons and Supporting Cells. W. B. Saunders Company, Philadelphia, 1976.

Whittaker, V. P., I. A. Michaelson and R. J. Kirkland. The separation of synaptic vesicles from disrupted nerve-ending particles ('synaptosomes'). Biochem. J. 90:293–303, 1964.

Young, J. Z. Fused neurons and synaptic contacts in the giant nerve fibers of cephalopods. Phil. Trans. Roy. Soc. Lond. B. 229:465–503, 1939.

CYTOPLASMIC MATRIX AND CYTOSKELETON

MICROTUBULES

Microtubules, as now defined, have a relatively short history in cytology, but, as members of a broader category of filamentous components of the cytoplasm, they were certainly observed by many of the early cytologists. The spindle fibers and astral rays of dividing cells were extensively studied; longitudinal filamentous structures were observed in nerve axons (Freud, 1882) and in the axonemes of sperm flagella (Ballowitz, 1890); circumferential bands of filaments were described in amphibian erythrocytes (Meves, 1910). Although the components of these structures are now known to be microtubules, they could be identified only as "filaments" with the light microscope. Even after the introduction of the electron microscope, the tubular nature of the core structures of cilia and sperm flagella escaped detection in the dissociated, air-dried preparations first examined (Harvey and Anderson, 1948; Manton, 1952). It was not until it became possible to cut ultrathin transverse sections that the hollow nature of the axonemal fibers was suggested (Fawcett and Porter, 1954). Their cross sectional images showed a dense rim or wall and a light center, but there was a curious reluctance to describe them as tubules. It was suspected that incomplete penetration of a fibril by osmium might result in a dense periphery and a lighter center. Moreover, the axonemal "fibrils" were considered to be the active agents of flagellar motion and it was difficult to conceive of a contractile tubule. For several years they continued to be described as fibrils.

With the development of satisfactory procedures for embedding tissue and cutting thin sections this became the principal method of biological electron microscopy. An ancillary technique called *negative staining* was soon introduced (Huxley, 1956; Horne and Brenner, 1958), which involved drying down of disrupted cells on the specimen grid together with a solution of sodium phosphotungstate. The electron density of the metal then served as a contrast medium for the less dense cell components and revealed, in negative image, details that were otherwise invisible. This stimulated a revival of interest in the examination of fragmented and dispersed cell components. Observation of the frayed-out tips of the axonemes of sperm flagella by this method showed that the phosphotungstate penetrated into the interior of the axonemal "fibrils" and this eliminated all lingering doubt about their hollow nature (Pease, 1963; Andre and Thiery, 1963). They appeared to be cylinders whose walls consisted of about 11 longitudinally oriented beaded filaments 3.5 nm in diameter with a repeating axial periodicity of about 8 nm.

Microtubules in the cytoplasm of cells were only rarely preserved by the osmium-containing fixatives then in general use. Among the first to be described were those comprising the manchette of mammalian spermatids (Burgos and Fawcett, 1955), the marginal band of teleost and amphibian erythrocytes (Fawcett and Witebsky, 1959), those forming the centrioles and mitotic spindles (de Harven and Bernhard, 1956; Roth and Daniels, 1962; Harris, 1962), and those in the axons of neurons (Palay, 1960). The term *microtubule* was proposed in a paper describing those associated with the developing nematocysts of hydra (Slautterback, 1963). It was not until the introduction of glutaraldehyde as a fixative (Sabatini et al., 1963) that microtubules were routinely preserved and came to be recognized as a nearly universal constitutent of the cell cytoplasm. They were then observed in the cortex of plant cells, and subunits were seen in their walls in thin transverse sections (Ledbetter and Porter, 1963) at about the same time that their protofilaments were observed by negative staining.

It soon became apparent that microtubules are highly versatile organelles involved in maintenance and change of cell shape, chromosomal movements in mitosis, and in directional guidance or active translocation of particulate components of the cytoplasm (Porter, 1966). They rapidly became a subject of intense investigative interest.

Selective solubilization of ciliary and flagellar components provided the first methodological approach to isolation of microtubules for investigation of their protein (Gibbons, 1963, 1965). A subunit with a molecular weight of 60,000 daltons was described which appeared to correspond to the 4 to 5 nm particles responsible for the beaded appearance of protofilaments in negatively stained preparations (Grimstone and Klug, 1966). Characterization of the more labile cytoplasmic microtubules was made possible by taking advantage of the discovery that the antimitotic drug colchicine binds to the subunits and prevents their polymerization to form spindle microtubules (Taylor, 1965). A procedure was then worked out for purification of microtubule subunits from brain on the basis of their colchicine-binding activity (Weisenberg, Borisy and Taylor, 1968). This became the basis for many of the subsequent chemical studies. The microtubule protein was named *tubulin* (Mohri, 1968), and whether it was extracted from flagella, brain, or spindles, the isolated native protein had a molecular weight of about 110,000 daltons. It proved to be a dimer, and the 55,000 dalton molecular weight monomer was found to occur in two forms, *α and β tubulin.* The dimers probably consist of one molecule each of the α and β form (Bryan and Wilson, 1971) and these are arranged in tandem along the length of the protofilaments. The possible occurrence of homodimers and of other arrangements within the wall of the tubule has not been ruled out.

Low temperature, increased pressure, and various antimitotic alkaloids were found to result in disappearance of microtubules or to prevent their formation. From these observations has emerged the concept that polymerized microtubules in the cytoplasm are in a state of equilibrium with their monomeric subunits (Inoué, 1952; Inoué and Sato, 1967). The factors that regulate the equilibrium to cause microtubule polymerization or depolymerization in different physiological states of the cell remain largely unknown. The polymerization of the microtubules of cilia and flagella can be prevented, but unlike microtubules of the general cytoplasm, once formed, they are considered to be stabilized by some mechanism yet to be elucidated.

Morphological studies have suggested that ordered arrays of microtubules often arise in close relationship to centrioles or associated small dense bodies, and it has been speculated that these serve as sites of nucleation or initiation of microtubule assembly. However, it remains unclear whether such initiation sites are necessary for formation of the microtubules that are randomly oriented in the ground cytoplasm. To shed further light on this question numerous studies on *in vitro* assembly of microtubules have been undertaken. Cytoplasmic microtubules from brain have been solubilized and the resulting preparations of tubulin have been extensively used for study of *in vitro* assembly of microtubules (Borisy and Olmsted, 1972; Weisenberg, 1972).

Partially purified subunits derived from rat brain polymerize *in vitro* into microtubules nearly indistinguishable from those seen in cells *in vivo*. Polymerization proceeded optimally at 35 to 37° C, required the presence of guanosine triphosphate and magnesium ions, and seemingly did not require nucleation (Weisenberg, 1972). In other work, on assembly of microtubules from the supernatant of porcine brain homogenates, it was reported that small disc-like structures of the same diameter as microtubules were needed as nucleation sites. When these were removed by high-speed centrifugation, little polymerization took place (Borisy and Olmsted, 1972). The question remains unsettled, but it is likely that a nucleation site or microtubule-organizing center is

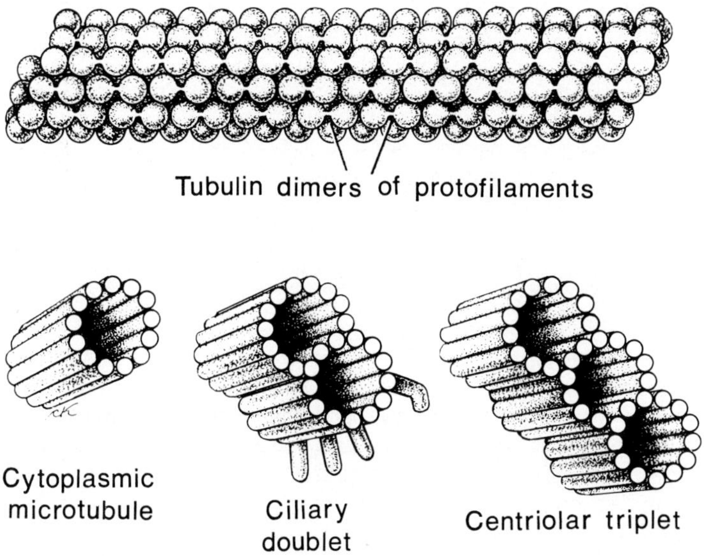

Schematic depiction of the linear polymers of tubulin that make up the protofilaments in the wall of a microtubule. Shown below are three configurations in which the protofilaments may be assembled: individual cytoplasmic microtubules, doublets in ciliary and flagellar axonemes, and triplets in the wall of centrioles.

necessary *in vivo* for formation of organized arrays of microtubules such as the mitotic spindle, but it is apparent that the organizing center need not be a centriole.

When tubulin is purified from brain, the preparation contains small amounts (15 per cent) of at least two other proteins of high molecular weight (300,000 to 350,000 daltons). These *microtubule-associated proteins* (MAPs) persist through repeated cycles of *in vitro* microtubule disassembly and reassembly (Murphy and Borisy, 1975; Vallee and Borisy, 1978). Addition of these proteins to solutions of tubulin devoid of MAPs promotes microtubule assembly (Sloboda et al., 1976). Fine filamentous projections from microtubules that were observed earlier in electron micrographs of peripheral nerve axons and other cell types are now believed to correspond to the MAPs that copurify with tubulin. The presence of these proteins uniformly distributed along the length of microtubules *in situ* can be demonstrated immunocytochemically (Connolly et al., 1978). Microtubules assembled *in vitro* in the absence of MAPs appear smooth surfaced, but when these proteins are present, the microtubules are decorated with very slender lateral projections (Binder and Rosenbaum, 1978). A 300,000 dalton microtubule-associated protein (MAP_2) isolated from neurotubules promotes *in vitro* assembly, binding in a constant ratio of 1 mole of MAP to 9 moles of tubulin dimer, and forms on the resulting microtubules very slender lateral projections exhibiting a 32 nm axial periodicity (Kim, Binder and Rosenbaum, 1979).

The functional significance of these filamentous proteins is not well understood, but it is speculated that where microtubules are laterally associated in organized arrays, as in the spindle, they may form stabilizing bridges between them (Hepler, McIntosh and Cleland, 1970). They may also be involved in attachment of organelles or other cytoplasmic particles that are translocated along bundles of microtubules (Smith et al., 1975; Heggeness et al., 1978).

In the upper figure on the opposite page is a cytoplasmic microtubule in cross section showing the usual 13 protofilaments. The lower figure is a preparation of microtubules assembled *in vitro* from flagellar tubulin. The inset presents one of these at high magnification showing 15 protofilaments. No explanation is available for the increase of one or two protofilaments upon assembly *in vitro*. A similar phenomenon has been reported for *in vitro* assembly of brain tubulin (Pierson et al., 1978). Flagellar microtubules differ from cytoplasmic microtubules in their greater resistance to solubilization by high pressure, low temperature, or drugs such as colchicine, which bind to tubulin. It is possible, however, to obtain soluble tubulin subunits from flagellar doublets by extensive sonication (Kuriyama, 1976). Under suitable conditions, this flagellar tubulin will assemble *in vitro* into single microtubules but not into doublets. Their walls contain 14 or 15 protofilaments instead of the 13 characteristic of microtubules *in vivo*.

Figure 399. Cytoplasmic microtubule with the usual 13 protofilaments. (Micrograph courtesy of Vitauts Kalnins.)

Figure 400. Microtubules assembled *in vitro* from flagellar tubulin. (From Binder and Rosenbaum, J. Cell Biol. 79:500–515, 1978.)

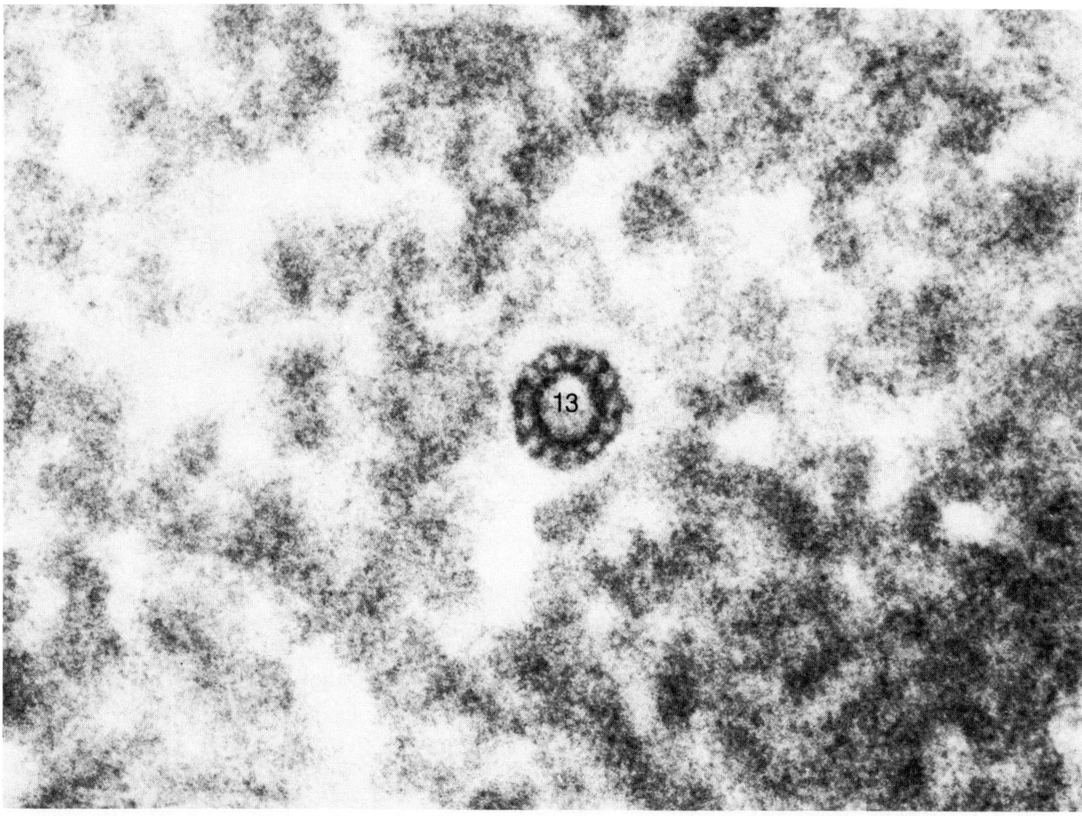

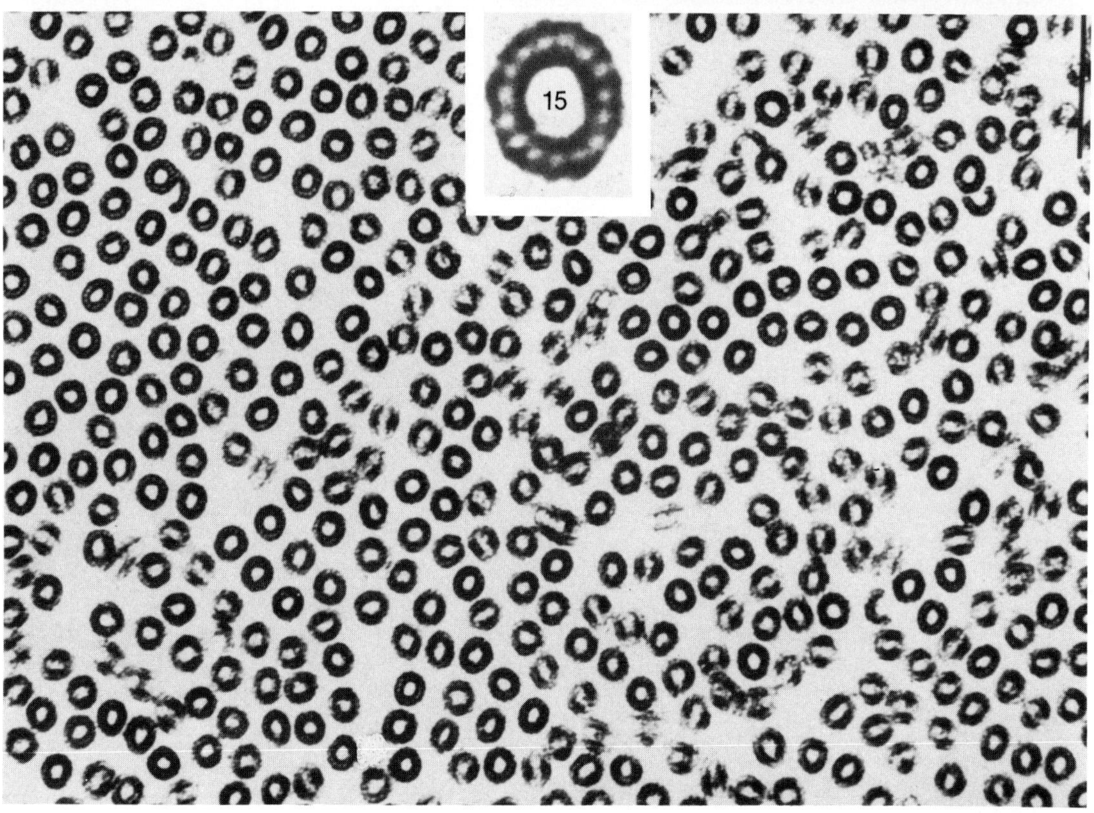

Figure 399, *upper* **Figure 400,** *lower*

The accompanying micrographs permit a comparison of the smooth-walled appearance of microtubules assembled in the absence of MAPs (top figure) with the fuzzy appearance of microtubules assembled in their presence (middle figure). In the lower figures, the radiating filaments visible in the figure on the right are evidently responsible for a more uniform spacing of the microtubules than that seen on the left with microtubules lacking MAPs.

Figure 401. Microtubules assembled from flagellar tubulin in the absence of MAPs.

Figure 402. Microtubules assembled in the presence of MAPs.

Figure 403. *A.* Cross sections of MAP-free microtubules. *B.* Cross sections of MAP-decorated microtubules. (Figures from Binder and Rosenbaum, J. Cell Biol. 79:500–515, 1978.)

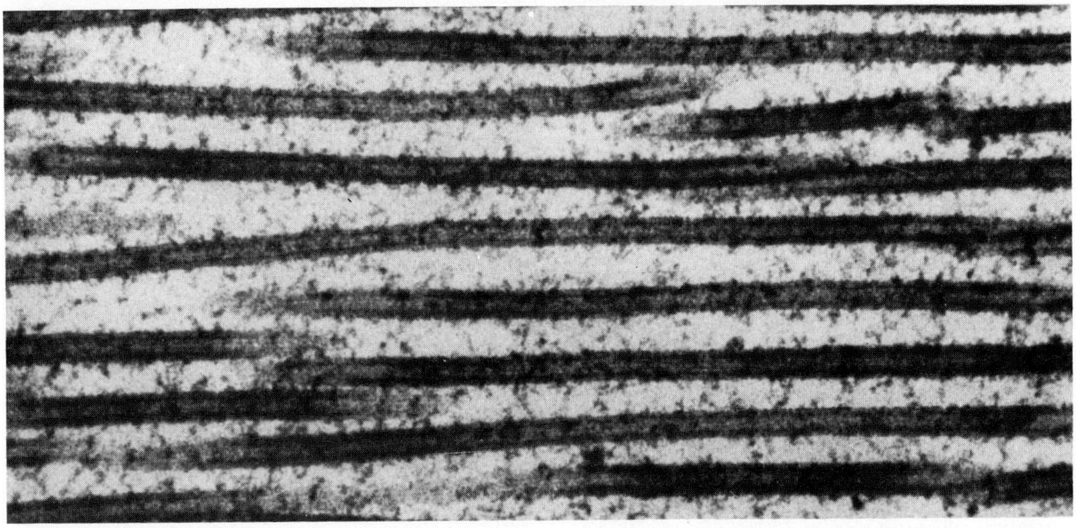

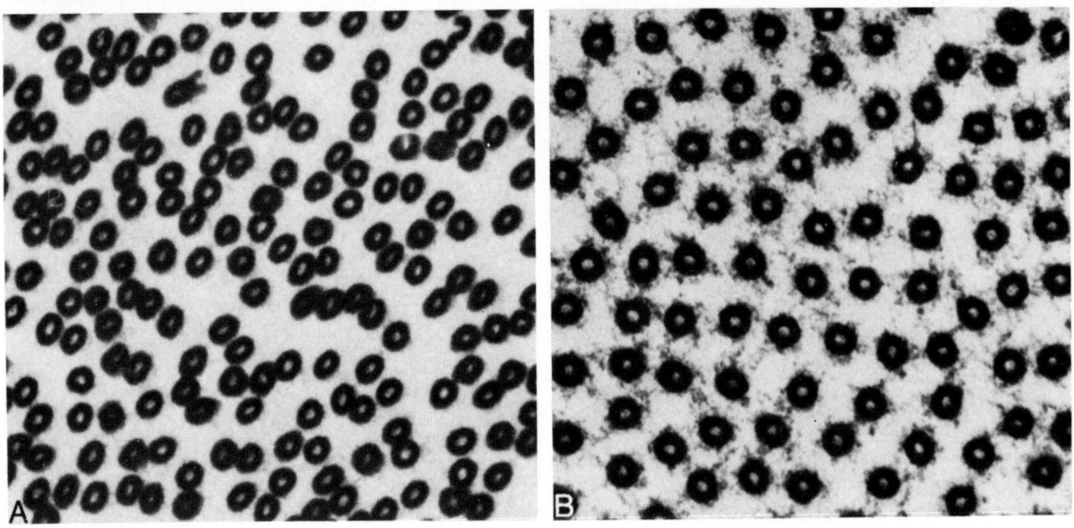

Figure 401, *upper* **Figure 402**, *center* **Figure 403**, *lower*

Microtubules in Particle Movement

Cytoplasmic streaming in animal cells is less apparent than in plants in which it has been extensively studied, particularly in the green algae. Nevertheless time-lapse cinematography of mammalian cells in culture has clearly established that a variety of particulate structures in the cytoplasm exhibit linear displacements along paths arranged radially with respect to the cell center. Examination of the same cells in electron micrographs reveals microtubules deployed in a radiating pattern centered on the centrosome. Directed saltatory movements of particles cease after treatment of cells with drugs that depolymerize microtubules (Freed and Lebowitz, 1970). Perhaps the most dramatic of the numerous examples of the association of particle migration with microtubules is seen in the melanophores and erythrophores of fish. These cells have an extraordinary array of microtubules radiating from the cell center. When stimulated to pale, the pigment granules rapidly move to the cell center and on disaggregation move slowly outward in a series of saltatory movements. Microtubules persist during aggregation but in reduced number and are restored during dispersion (Bikle, Tilney and Porter, 1966; Junquiera, Raker and Porter, 1974; Schliwa, Osborn and Weber, 1978). In these and other examples, there is no doubt that microtubules define pathways along which particles move, but whether they provide a motive force remains uncertain. There is a possibility that actin filaments associated with the microtubule are involved, but a force-generating interaction between the two has yet to be demonstrated.

The form of transport that has been the subject of most investigation is that which occurs in the processes of nerve cells. Two classes of transport are generally accepted, a slow axoplasmic flow at a rate of 1 to 2 mm/day and faster saltatory movement of particles at up to 400 mm/day. Of these only the faster component is affected by microtubule disassembly. Close association between synaptic vesicles and microtubules has been reported in axons of some vertebrate species (Smith, Järlfors and Beránek, 1970; Gray, 1975) and short cross-bridges linking the two structures have been described (Smith, 1971). It has been established that the close topographical relationship between mitochondria and microtubules in cross sections of insect axons is statistically significant, and linkages were also found between the two comparable to those previously described for the synaptic vesicles (Smith, Järlfors and Cayer, 1977). Very close association of mitochondria with microtubules has also been demonstrated in several unrelated types of cultured vertebrate cells by indirect immunofluorescence techniques, which permit their simultaneous visualization. No such association with actin microfilaments was found. It was suggested therefore that microtubules may be involved in movement and intracellular distribution of mitochondria (Heggeness, Simon and Singer, 1978). While there is general agreement that microtubules determine the direction along which cytoplasmic streaming preferentially occurs, no satisfactory mechanism has been put forward to explain how they might contribute to the generation of force for translocation of membrane-limited particles.

The interference contrast images at the top of the facing page illustrate the dramatic transformation in a living fish chromatophore from the dispersed (*A*) to the aggregated state (*B*) of its pigment granules. This occurs with no significant change in the outline of the cell (indicated by the arrows). Immunofluorescence microscopy with antitubulin antibody, in the middle figure, shows the striking array of radially disposed microtubules in a chromatophore with dispersed pigment. In the cell with aggregated pigment in the lower figure, it is evident that the radial dispositon of microtubules remains unchanged.

The rapid movement of pigment granules along oriented tracts of microtubules has made chromatophores a favored system for studying the role of microtubules in the translocation of particulate components of the cytoplasm. There is no doubt that they are involved in pigment migration, because their depolymerization with colchicine or low temperature reduces or abolishes pigment movement, but the exact role microtubules play remains uncertain. Although radially oriented microtubules persist in both states, pigment aggregation is associated with a reduction of about 50 per cent in their numbers, suggesting that microtubule assembly and disassembly may be involved.

Figure 404. *A*. Chromatophore with dispersed pigment. *B*. The same cell with pigment aggregated.

Figure 405. Indirect immunofluorescence of a melanophore with dispersed pigment.

Figure 406. Immunofluorescence of chromatophore with aggregated pigment. (Figures from Schliwa, Osborn and Weber, J. Cell Biol. 76:229–236, 1978.)

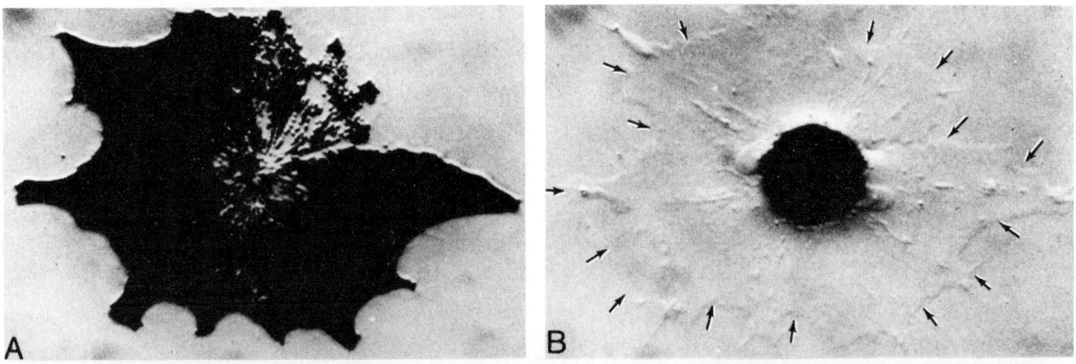

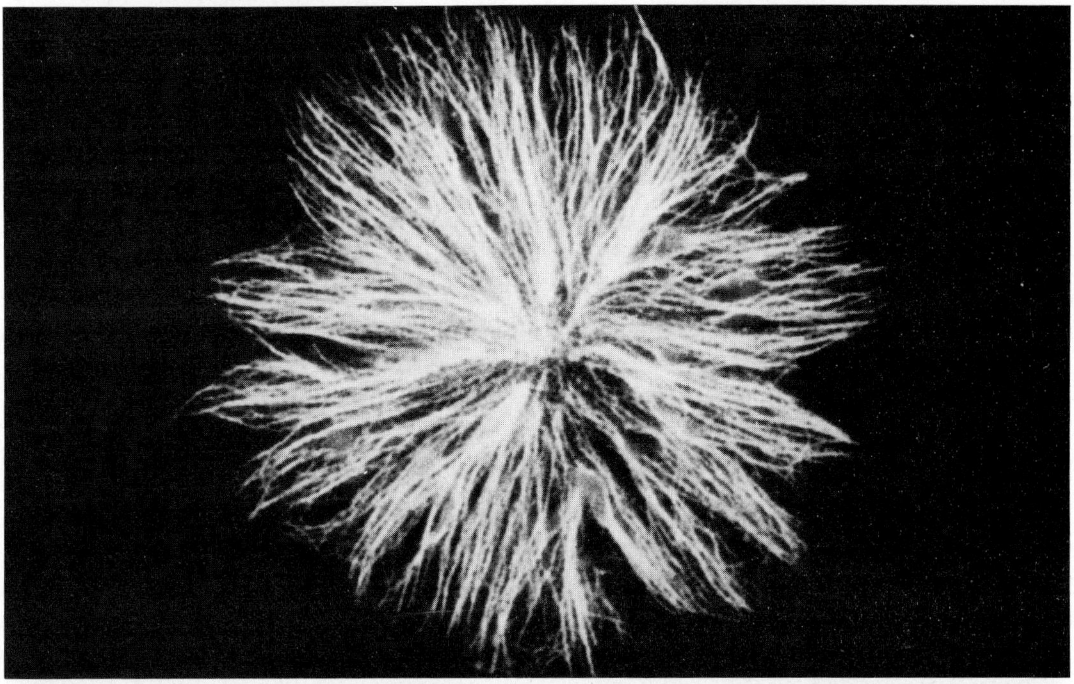

Figure 404, *upper* **Figure 405**, *center* **Figure 406**, *lower*

Neurons must maintain synaptic terminals which are often located at great distances from the cell body. Streaming of the cytoplasm, which is particularly conspicuous both in the perikaryon and along the length of the axon, is probably essential to solve the logistic problems inherent in an extensive system of axonal branches. A slow bulk flow of axoplasm and a more rapid transport of specific substances are well documented.

In the large dendrite in the accompanying micrograph, numerous cross-sectional profiles of longitudinally oriented microtubules are uniformly distributed throughout. These are believed to play an essential role in transport both in dendrites and axons — a conclusion supported by the observation that colchicine and other agents that depolymerize microtubules also inhibit axoplasmic translocation. As in other forms of cytoplasmic movement, it is not known whether the microtubules merely serve as tracks determining the direction of movement or whether they interact with some other filamentous component to supply motive force.

Also present in the cytoplasm are loose aggregations of 10 nm neurofilaments. These are assigned no role in transport but are thought to have a supportive cytoskeletal function.

Figure 407. Proximal dendrite of an anterior horn cell. (Micrograph courtesy of Raymond Wuerker.)

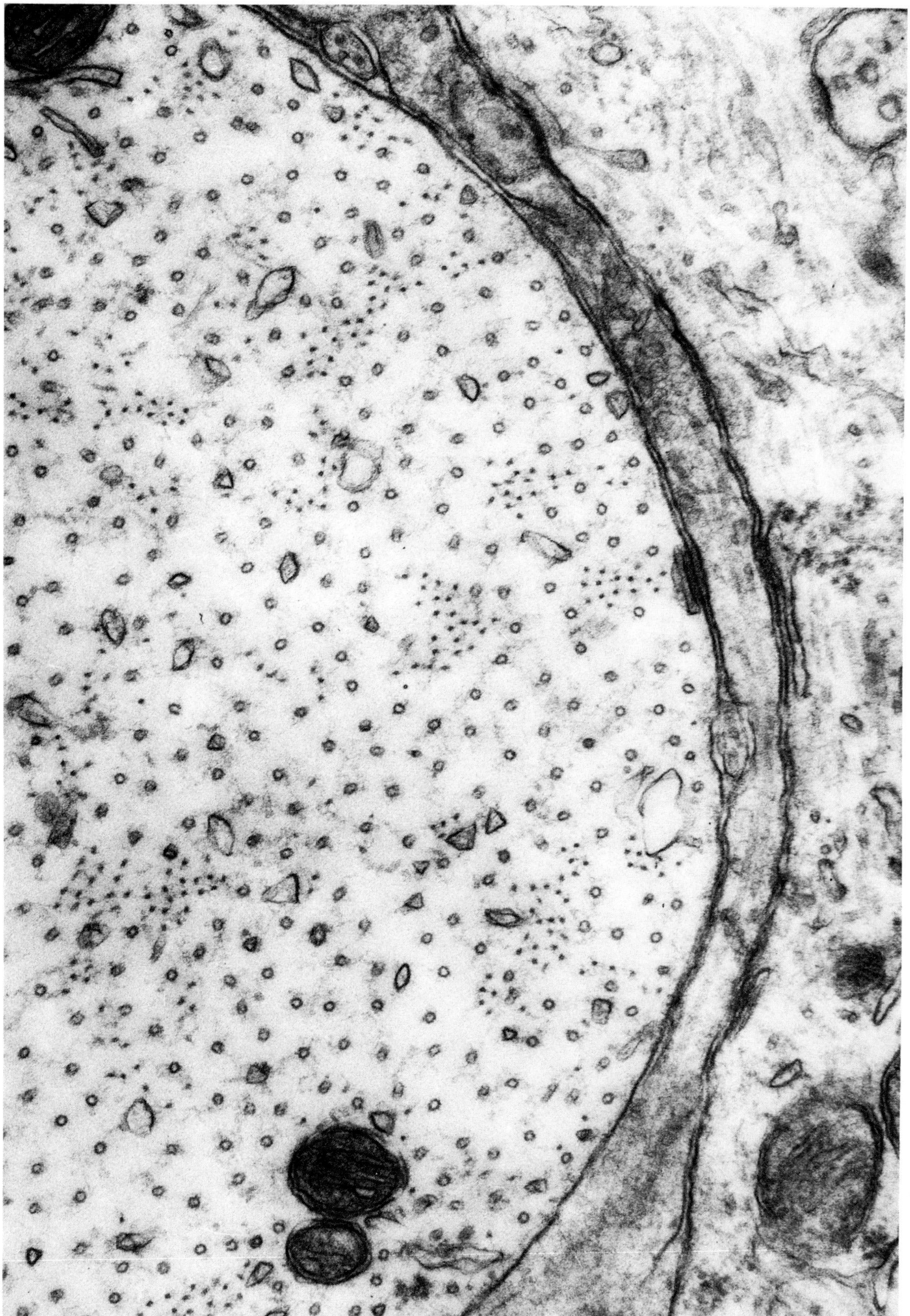

Figure 407

It is generally assumed that synaptic vesicles are restricted to nerve terminals and that synthesis and sequestration of transmitter in their interior occur locally. However, there are observations suggesting that vesicles and their contents may arise centrally and be transported along microtubules to the nerve terminals. Isolated or sparsely distributed vesicles are also observed at nonsynaptic sites, and the striking alignment of vesicles along microtubules illustrated in the upper figure on the facing page strongly suggests that they are in transit. Schmitt (1968) has speculated that a motive force may be generated at the site of contact between a microtubule and a particle which might be responsible for the observed movement of particles along the axon. No experimental confirmation of this has been forthcoming, but in the material illustrated here, small bridges, or links, have been demonstrated between the vesicles and the microtubules.

Microtubules are not especially abundant at synapses, but the lower figure shows vesicles closely clustered around a few microtubules at the periphery of the synaptic area.

Figures 408 and **409.** Giant axons in the central nervous system of the lamprey (*Petromyzon marinus*). (From Smith, Järlfors and Beránek. J. Cell Biol. 46:199–219, 1970.)

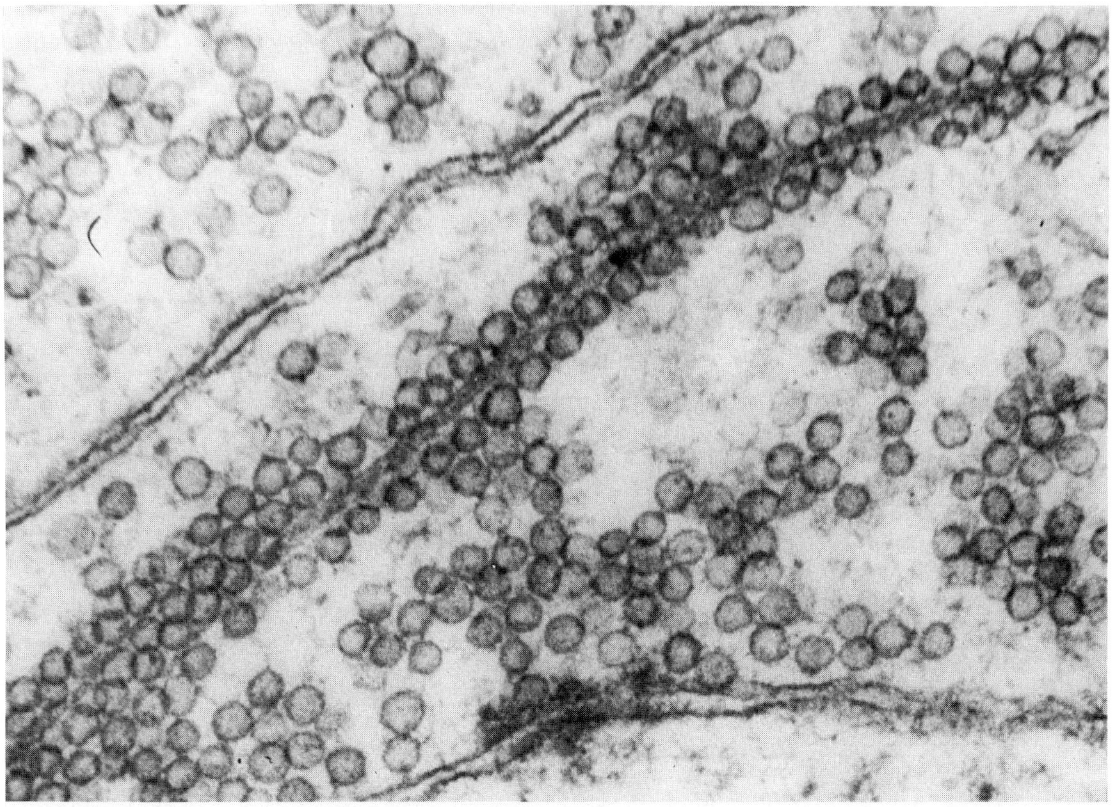

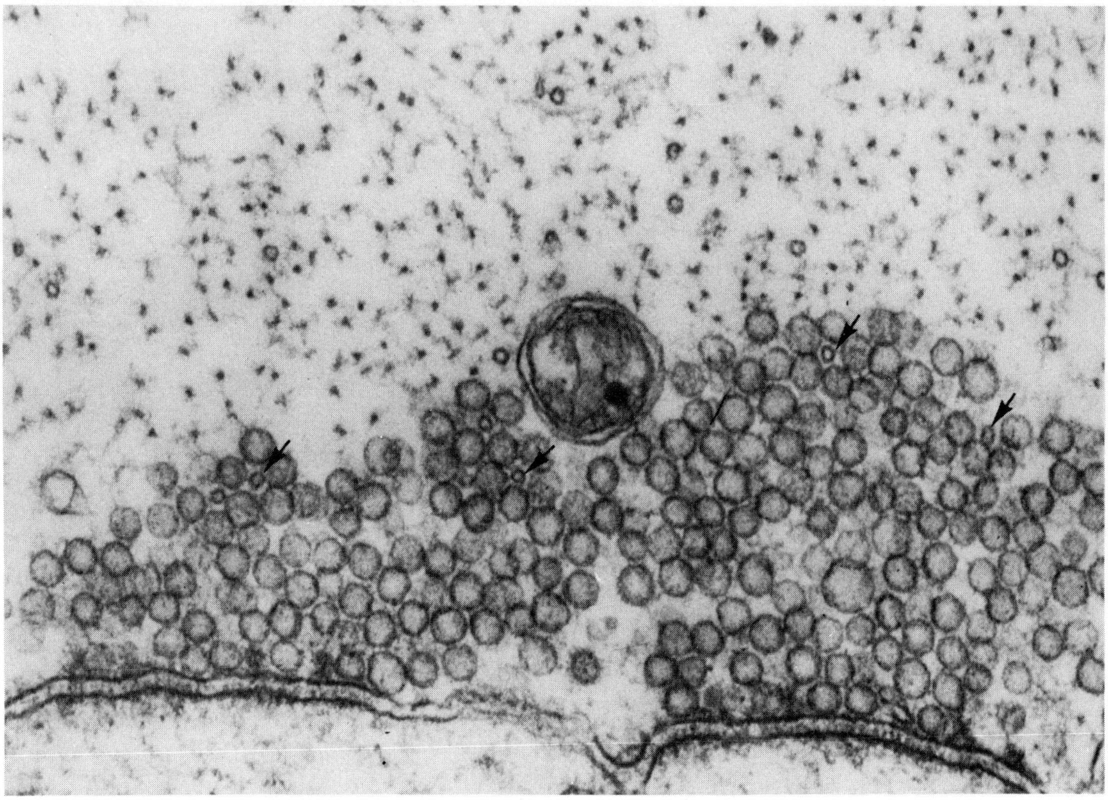

Figure 408, *upper* **Figure 409,** *lower*

Microtubules in Mitosis

As traditionally defined by light microscopists, the *mitotic apparatus* included a *spindle* composed of fibers converging upon a pair of centrioles at both poles, and *asters* made up of fibers radiating from the centrioles. The spindle included *chromosomal fibers* from the chromosomes to the poles and *continuous fibers,* or *interzonal fibers,* extending from pole to pole.

Electron micrographs reveal that the spindle "fibers" and astral "rays" of classical cytology are bundles of microtubules. The microtubules are fixed to local differentiations of the chromosomes called *kinetochores*. The polar ends of the chromosomal and continuous microtubules terminate in a specialized zone of cytoplasmic matrix around the centrioles. At anaphase of mitosis, the motion of the chromosomes has two components: a chromosome-to-pole movement accompanied by a shortening of the chromosomal microtubules and a movement of the poles away from one another. The generation of these movements remains a subject of controversy, with some investigators favoring a sliding microtubule mechanism (McIntosh et al., 1969) and others favoring a polarized polymerization and depolymerization resulting in shortening of the chromosomal microtubules and lengthening of the pole-to-pole microtubules (Inoué, 1976).

The accompanying micrograph shows groups of microtubules extending from the pole (out of the field at the upper left) to chromosomes (at the lower right). The kinetochores are not visible.

Figure 410. Chromosomal microtubules in midanaphase of mitotic division in *Pelomyxa.* (Micrograph courtesy of Evans Roth.)

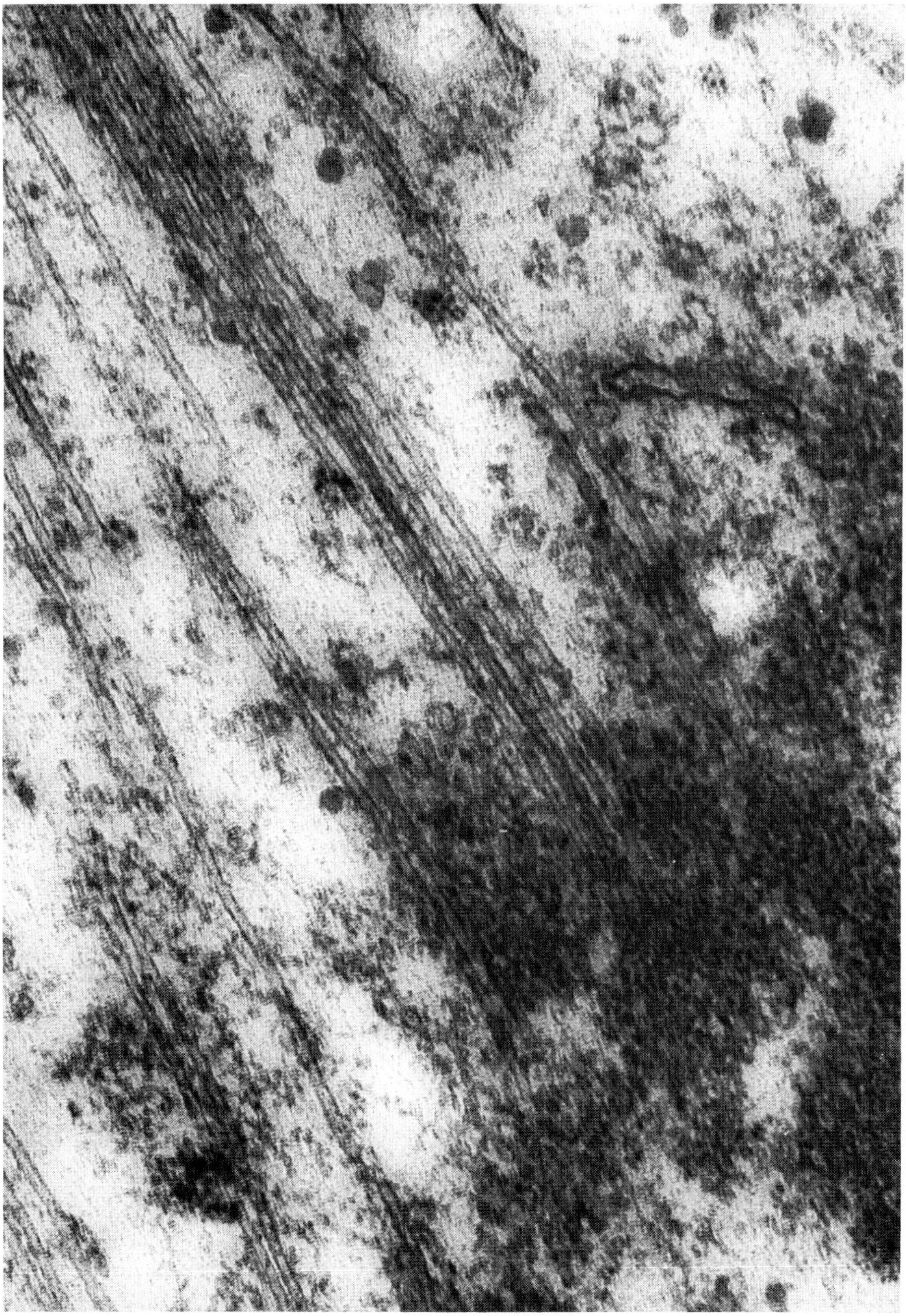

Figure 410

When the chromosomes have moved toward the poles in late anaphase, the cleavage furrow constricts around the midpoint of a large bundle of interzonal spindle microtubules that are surrounded by a dense amorphous material. This shaft of microtubules, enclosed at its midregion by the plasma membrane, persists for some time and connects the daughter cells well into interphase. The dense zone in the intercellular connection corresponds to the *midbody* (Fleming's body, *Zwischenkörper*) described by classical cytologists.

Figure 411. Microtubules associated with the midbody in telophase of mammalian cell division. (Micrograph courtesy of David Phillips.)

Figure 411

The accompanying micrographs present additional examples of persisting interzonal spindle microtubules in telophase of dividing cells. When the connection between daughter cells is finally severed, the midbody is incorporated into one daughter cell. The dense matrix material disperses and the microtubules depolymerize.

Figure 412. Segment of the bridge connecting daughter cells in HeLa cell mitosis. (Photomicrograph courtesy of Breck Byers.)

Figure 413. Persisting interzonal microtubules in a dividing guinea pig erythroblast.

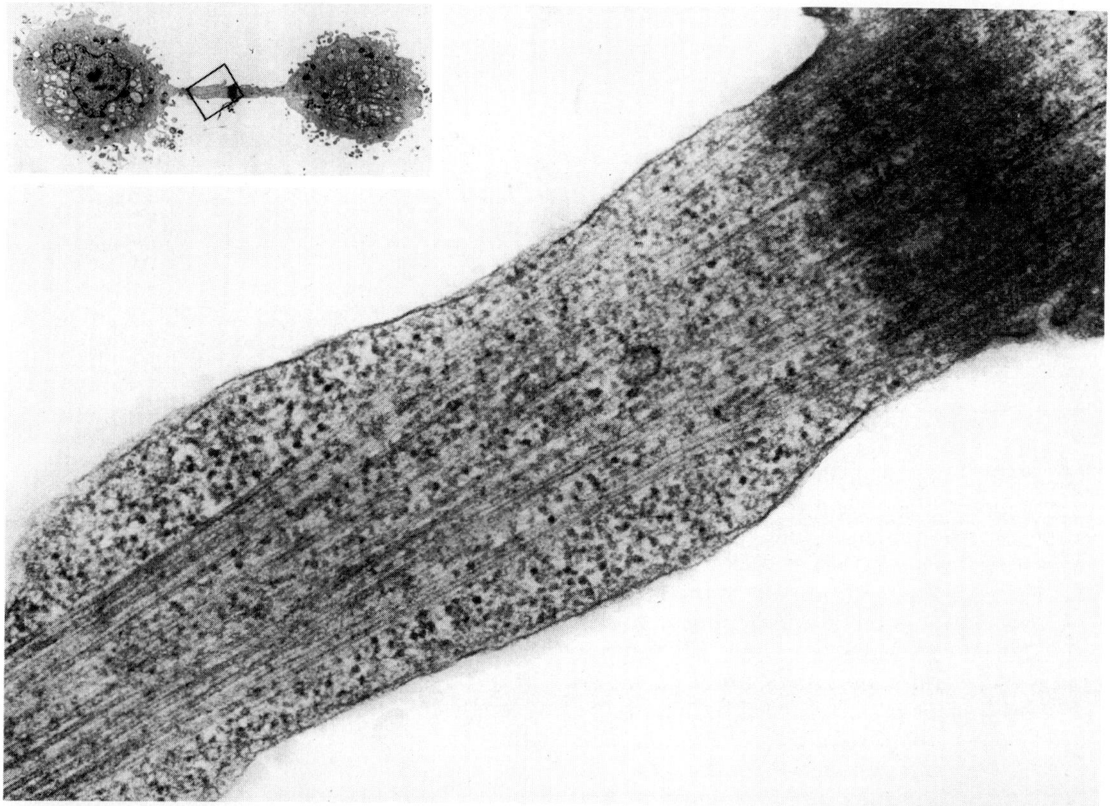

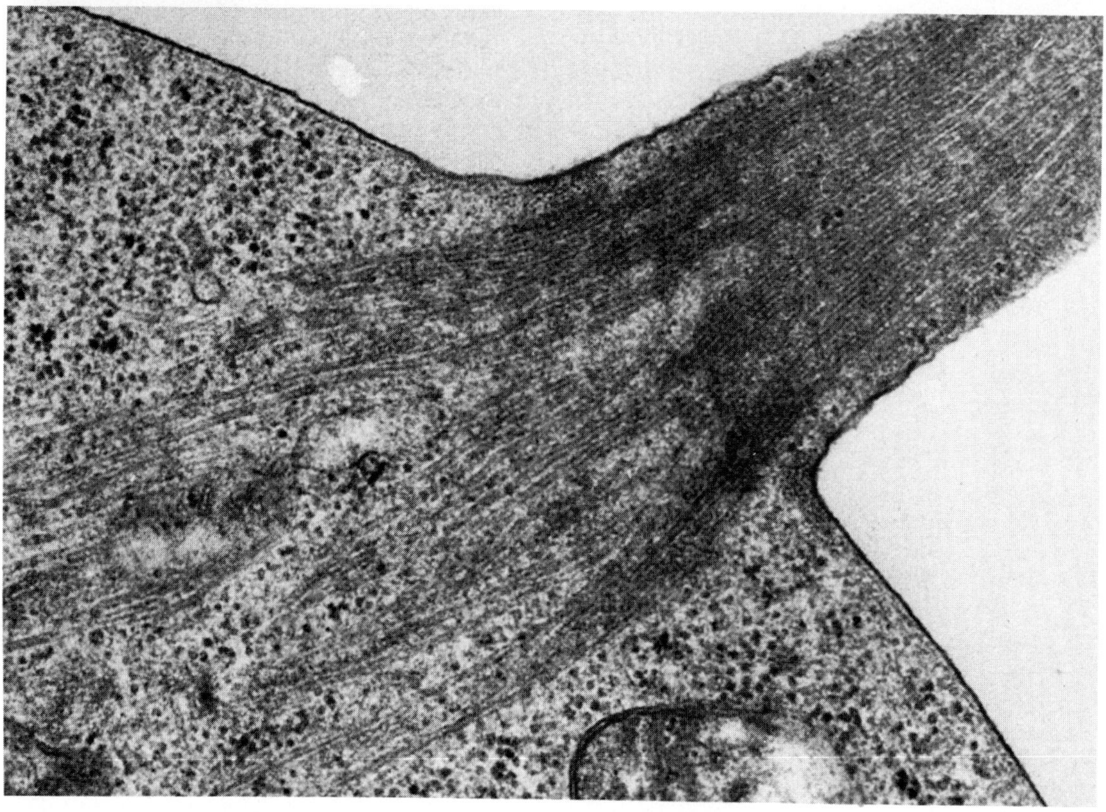

Figure 412, *upper* **Figure 413,** *lower*

Microtubules in Cell Motility

Shortening of skeletal muscle depends upon sliding of interdigitating sets of actin and myosin filaments. Their displacement relative to one another is caused by chemomechanical changes in orientation of minute cross bridges between the filaments (Huxley, 1969). The generation of bending movements in cilia and flagella is attributed to a sliding microtubule mechanism comparable to the sliding filament mechanism of striated muscle (Satir, 1968). Sliding results from interaction of arms projecting from each doublet microtubule of the axoneme with the adjacent doublet (p. 577).

A comparable interaction of microtubules is in part responsible for large-scale changes of shape in certain ciliate protozoa, the one most thoroughly studied being *Stentor coerulus* (Huang and Pitelka, 1973). From its trumpet-shaped extended state, this organism can contract to a quarter of its length in a few milliseconds.

Parallel rows of cilia extend the full length of the organisms. Immediately beneath the cell surface between the rows of cilia, light microscopists described linear structures which they designated *Km fibers*. When examined with the electron microscope, these proved to be highly ordered arrays of microtubules. The accompanying micrograph shows a cilium flanked by two such microtubule arrays seen in cross section. The microtubules are interconnected by short bridges to form discrete layers, or ribbons, which in turn are attached to basal bodies of neighboring cilia. Adjacent ribbons are connected by longer bridges between a few of the microtubules at the ends of the ribbons nearest the cell surface. During extension the overlapping ribbons of stiff microtubules slide with respect to one another, decreasing their degree of overlap and causing the organism to elongate.

More deeply situated in the cell cortex are longitudinal bundles of filaments comprising the *myonemes,* one of which is shown here in cross section. Shortening of the organism results from contraction of the myonemes.

Figure 414. Transverse section of a portion of the cortex of a partially extended *Stentor coerulus.* (From Huang and Pitelka, J. Cell Biol. 57:704–728, 1973.)

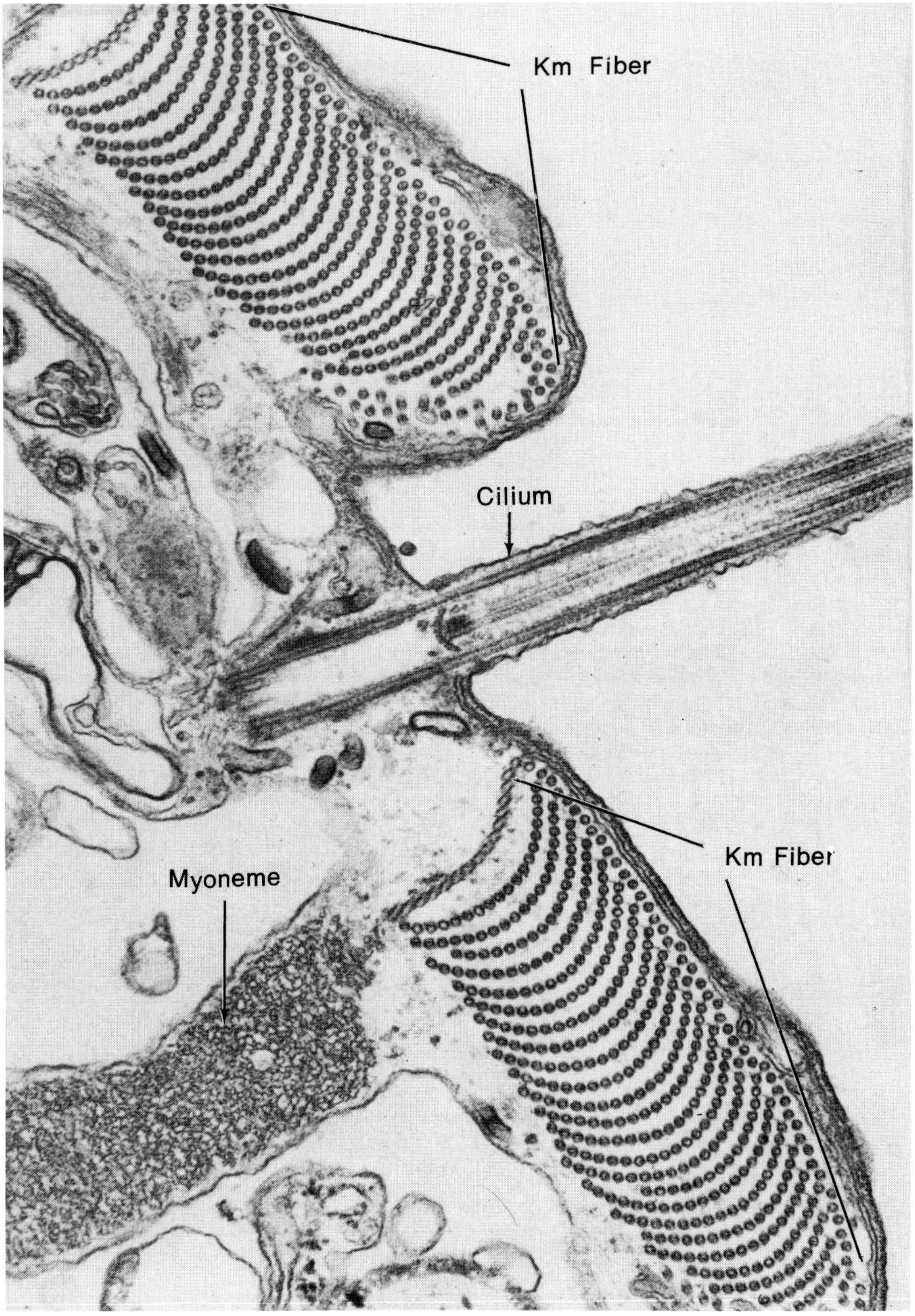

Figure 414

The micrograph on the facing page shows a longitudinal section of a Km fiber in a contracted *Stentor*. The ribbons diverge laterally to insert into the basal bodies of closely spaced cilia. Forty or more overlapping microtubule ribbons are seen here, but in comparable views of a fully extended cell, the basal bodies would be spaced at longer intervals and the number of overlapping ribbons would be reduced to about six. These unusually straight and rigid microtubules, like those of cilia and flagella, are not depolymerized by colchicine.

The microtubule and filament systems in this organism function as antagonistic elements in contraction and extension. The motive force for extension depends upon microtubule-to-microtubule sliding generated by specific intertubule bridges, whereas the active shortening appears to result from a conformational change in the filaments comprising the myonemes. This is of interest because it is possible that the movements involved in shape change in less specialized cells of higher organisms may prove to be due to similar antagonistic effects of microtubules and contractile filamentous components of the cytoplasm.

Figure 415. Longitudinal section of a Km fiber in the cortex of a contracted Stentor. (From Huang and Pitelka, J. Cell Biol. 57:704–728, 1973.)

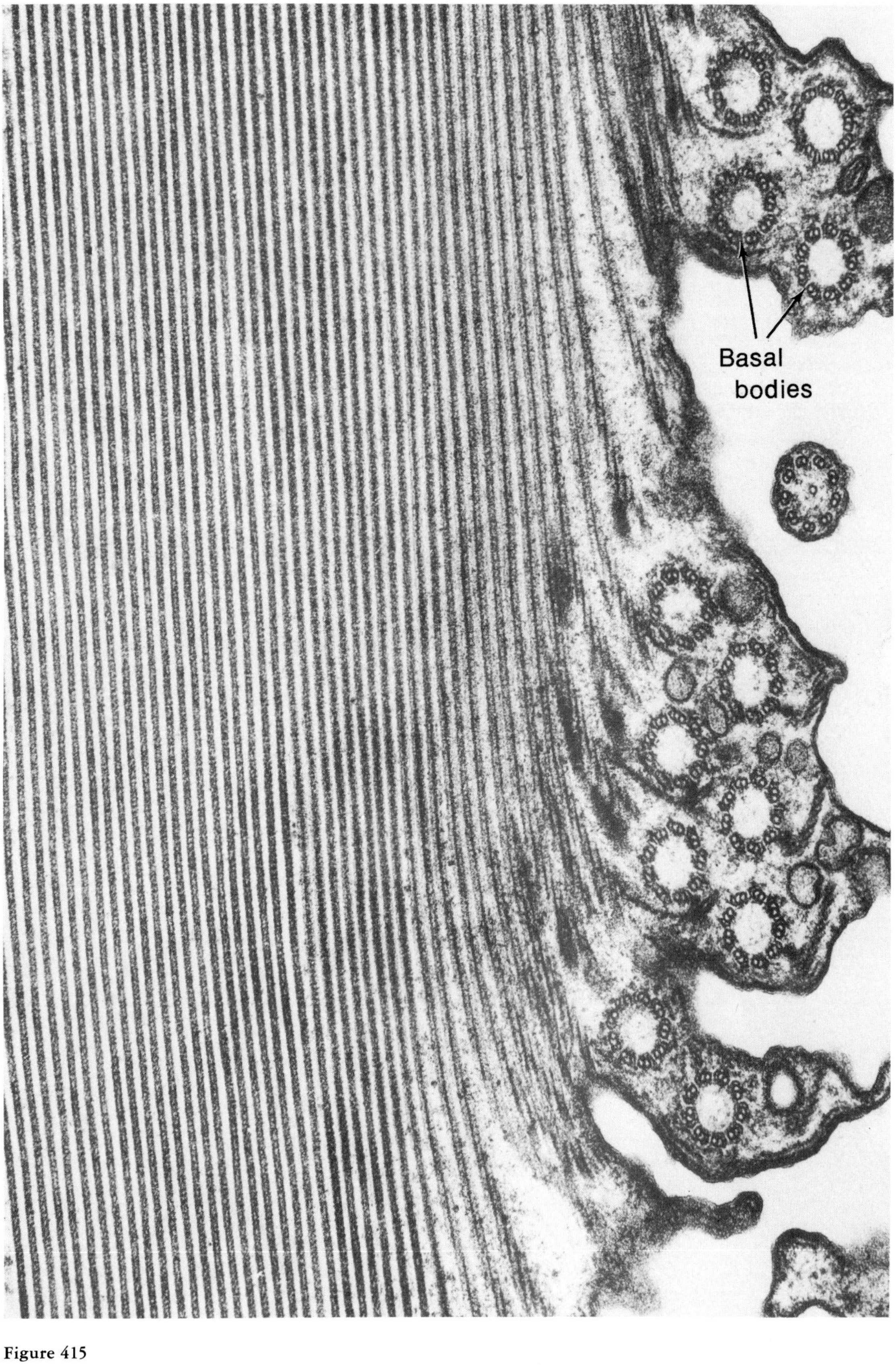

Figure 415

Cytoskeletal Function of Microtubules

One of the first functions ascribed to microtubules was that of providing an internal framework or cytoskeleton serving to determine and maintain cell form. Later, as cell elongation and the extension of cytoplasmic processes were found to be associated with the appearance of axial microtubules, greater emphasis was placed upon their role in shape change. Nevertheless, there remain a number of examples wherein microtubules are responsible for maintaining a distinctive and stable cell shape.

A dark line called the *marginal band (Randreifen; striae bordante)* was observed by Ranvier (1875), Dethier (1875), and others around the periphery of nucleated erythrocytes of birds and amphibians. Meves (1904) attributed elastic properties to this structure and suggested that it served to maintain the flattened elliptical form of these cells. He concluded from studies on supravitally stained preparations that the marginal band was made up of fibrillar subunits. These remarkable observations were discounted by influential contemporary cytologists who considered the marginal band to be a staining artifact (Weidenreich, 1905). Interest in this structure therefore rapidly declined.

Fifty years later when fish erythrocytes were examined with the electron microscope, Meves' findings were substantiated by the finding of a circumferential bundle of microtubules where he had described a marginal band (Fawcett and Witebsky, 1964). In sections perpendicular to the plane of cell flattening, a cluster of cross-sectional profiles of microtubules was found at both ends of the cell.

In the anucleate biconcave discoid erythrocytes of mammals, a marginal band is ordinarily not present. An interesting exception is found in Camelidae, which have anucleate erythrocytes that are elliptical rather than circular. These are said to possess a marginal bundle of microtubules (Barclay, 1966).

Figure 416. Erythrocyte from the toadfish *Opsanus Tau.* (From Fawcett and Witebsky, Z. Zellforsch. 62:785–806, 1964.)

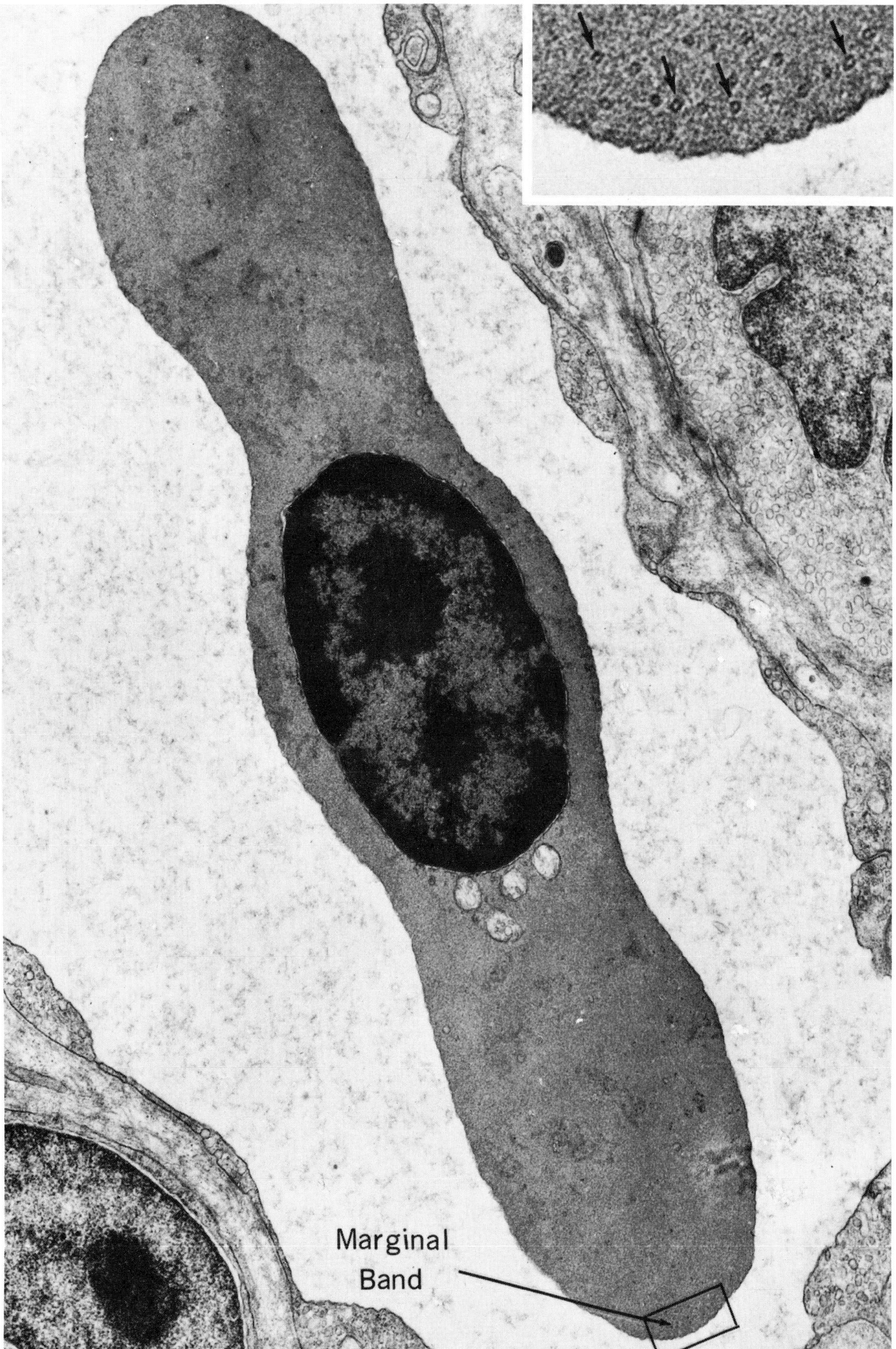

Figure 416

The assumption that a marginal bundle of microtubules in nucleated erythrocytes is related to maintenance of their discoid shape derives additional support from the observation that the flat blood platelets have a similar disposition of microtubules (Behnke, 1965). In the equatorial section on the facing page, the bundle forms a ring just inside of the membrane but not in contact with it. In sections perpendicular to the disc, the marginal bundle is seen as a cluster of microtubules at either end of the cross-sectional profile of the platelet (at the arrows and in the inset). Their number varies from 6 to 20 or more. It is not known whether the bundle consists of independent closed circles, open circles overlapping to varying extent, or a single long microtubule in a coil. The microtubules form a coherent structural unit which remains intact when the platelet membrane is disrupted, but if the bundle itself is broken, the microtubules straighten out, indicating that they have elastic properties and are under some tension when in the marginal bundle.

Cooling of platelets to 0 to 4° C results in disappearance of microtubules, but they reappear spontaneously upon rewarming and ultimately reconstitute a marginal bundle. Their reappearance is often associated with a resumption of the normal lenticular form of the platelet (Behnke, 1967).

Figure 417. Part of an equatorial section of a human blood platelet. (Micrograph courtesy of Olaf Behnke.)

Figure 418. Several platelets in a glomerular capillary of rat. (Micrograph courtesy of Greta Tyson and Ruth Bulger. Am. J. Anat. 135:319–343, 1972.)

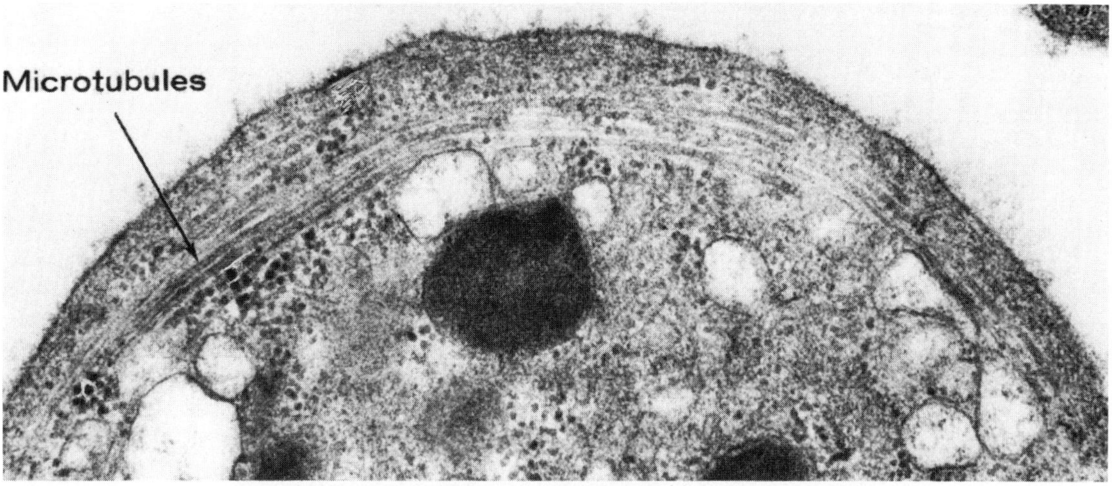

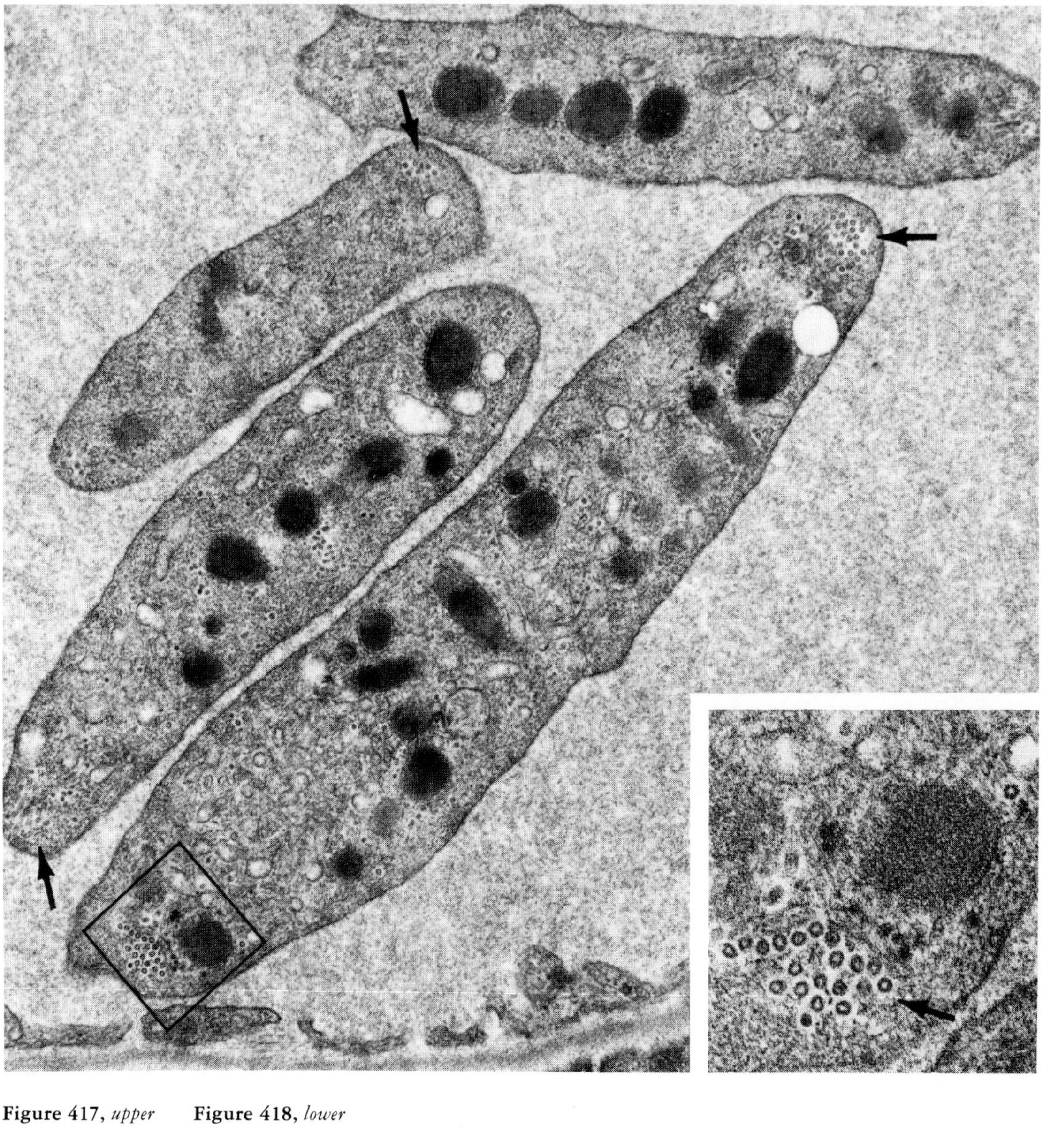

Figure 417, *upper* **Figure 418**, *lower*

In the organ of Corti in the inner ear, the hair cells responsible for transduction of mechanical vibration to a nerve impulse are supported by rows of cells called *pillar cells*. The micrographs shown here illustrate the thick walled microtubules in a pillar cell. These highly ordered stable arrays of microtubules interrupted by occasional mitochondria and tubular elements of the endoplasmic reticulum are believed to exemplify a purely mechanical cytoskeletal function of microtubules.

Figures 419 and 420. Electron micrographs of pillar cells of the mammalian organ of Corti. (Micrographs courtesy of Hans Engström.)

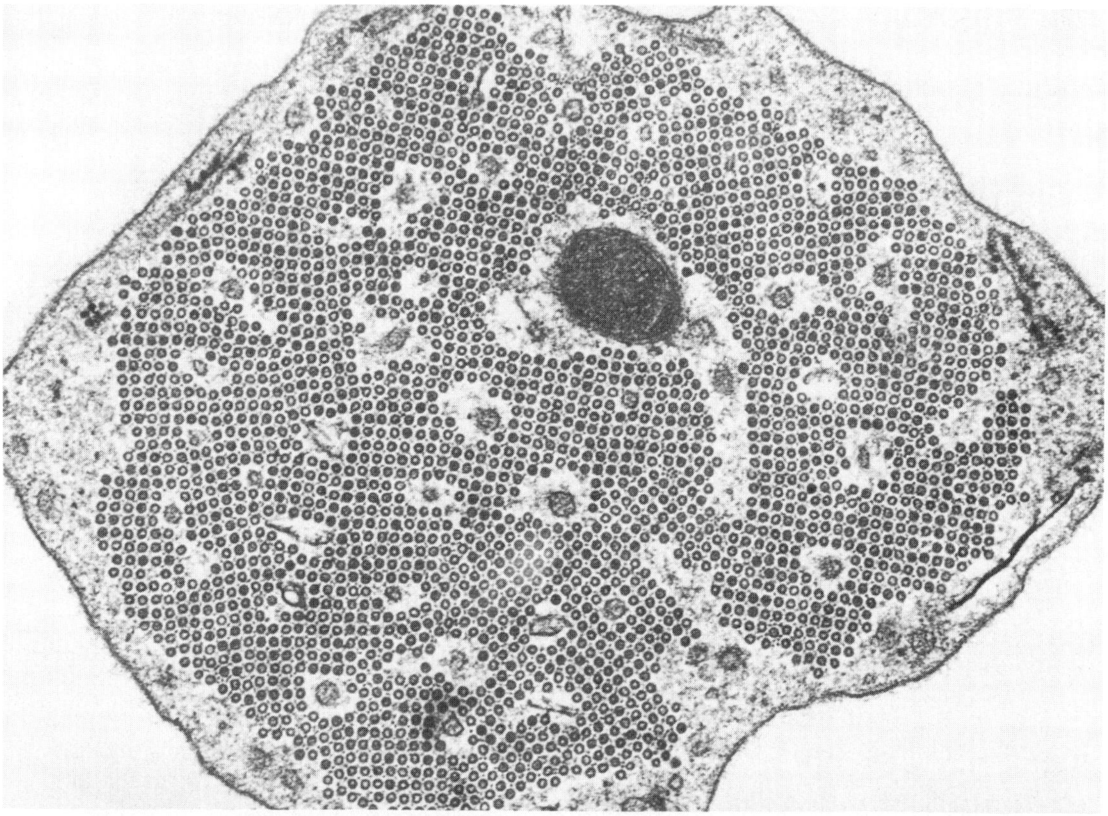

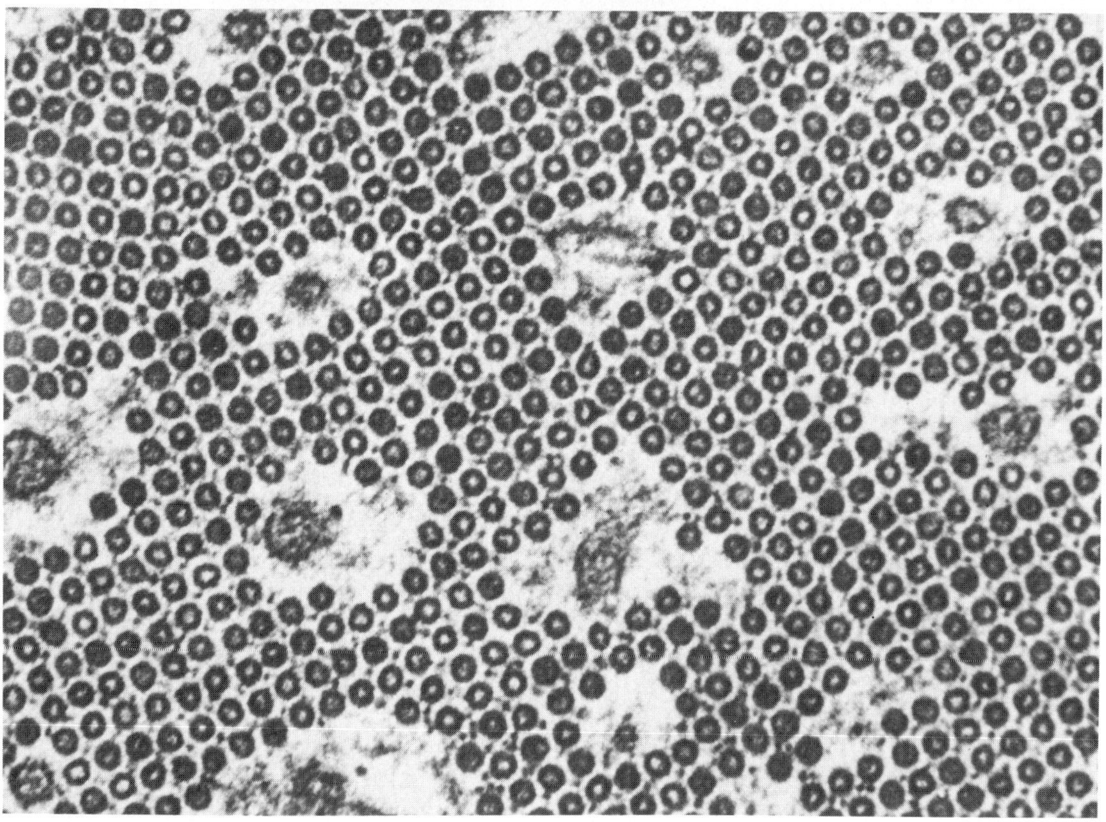

Figure 419, *upper* **Figure 420,** *lower*

The columnar Sertoli cells of the seminiferous epithelium function as nurse cells in metabolic and mechanical support of the developing male germ cell. Late in spermiogenesis, the heads of maturing spermatozoa occupy deep recesses in the apical surface of their supporting cells. The accompanying micrograph presents a section transverse to the axis of the Sertoli cells near the lumen. Several sperm heads are sectioned at different levels, and the surrounding Sertoli cell cytoplasm contains large numbers of cross-sectional profiles of microtubules that are oriented parallel to the vertical axis of the cell. Microtubules vary in abundance in different stages of the spermatogenic cycle, but when the Sertoli cells are hosts to advanced spermatids, cytoplasmic microtubules are always abundant and consistent in their orientation. It is assumed that the microtubules in such supporting cells have an important cytoskeletal function.

Figure 421. Horizontal section through the apical cytoplasm of a Sertoli cell in the testis of the toad, *Bufo marinus.*

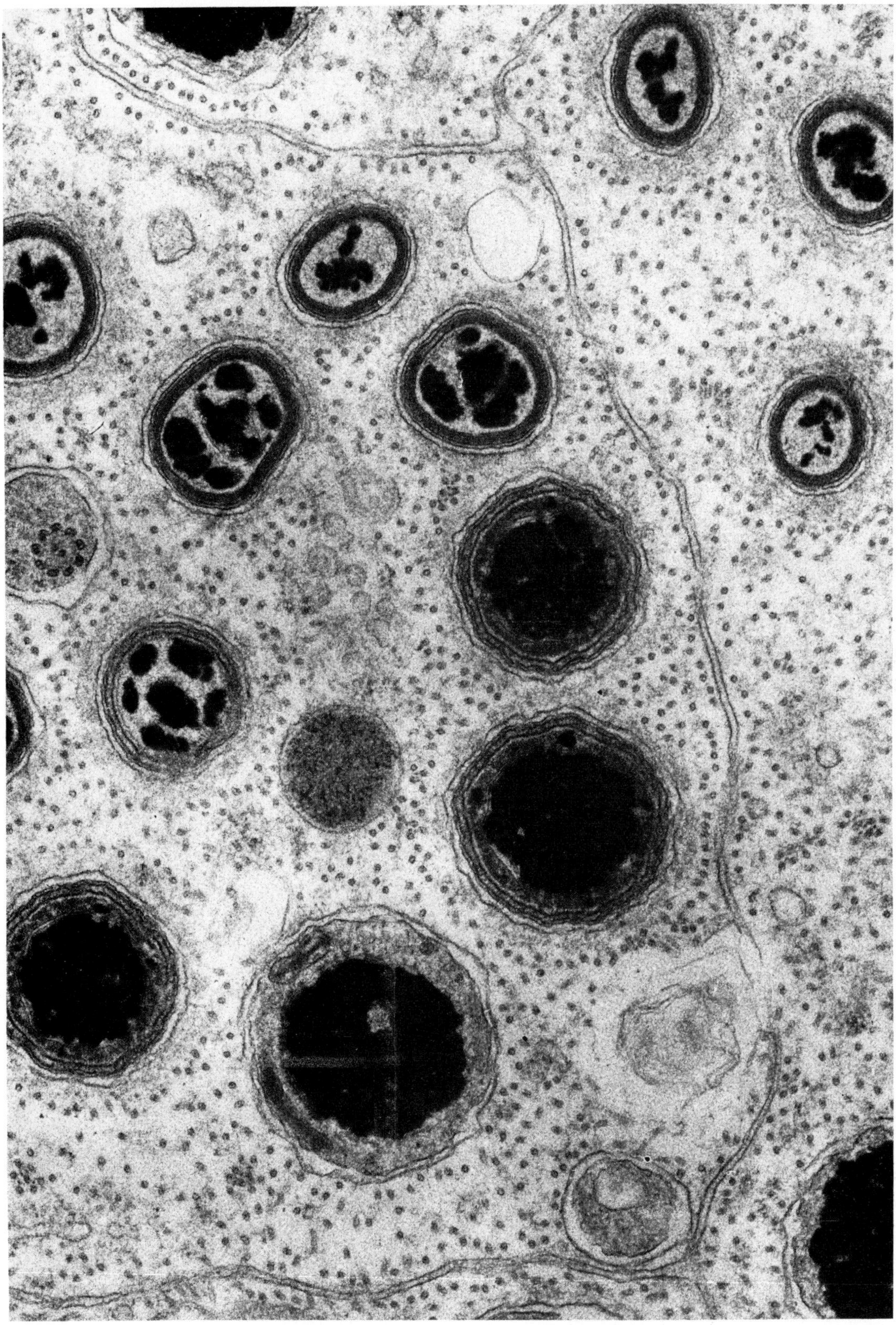

Figure 421

Microtubules in Morphogenesis

Studies of mammalian spermatogenesis provided one of the first observations implicating microtubules in a change of cell shape. The germ cells are approximately spherical during meiosis and the early postmeiotic stages of development. The spermatids then undergo a rapid elongation. This change is associated with the appearance of a new organelle which classical cytologists called the *caudal sheath,* or *manchette.* It was described as a sleeve-like cylindrical structure arising from an annular thickening of the cell membrane around the equator of the nucleus and extending back into the cytoplasm of the spermatid around the base of the flagellum.

In early electron micrographs of this region, the manchette was found to be a cylindrical array of parallel microtubules (Burgos and Fawcett, 1954). It is a transient structure arising at the beginning of spermatid elongation and disappearing completely when this stage of morphogenesis is completed. The elongation of the microtubules from a fixed origin proximally evidently contributes to lengthening of the cell. The close association of vesicular elements with the outer and inner aspect of the manchette suggests the possibility that it may also participate in transport of particulates or at least serve as a framework guiding a caudal flow of cytoplasm.

The microtubules of the manchette are more stable than those of most other cell types. This may account for the fact that they were observed before the introduction of aldehyde fixatives.

Figure 422. Transverse section through the caudal cytoplasm of a Chinese hamster spermatid. (Micrograph courtesy of David Phillips.)

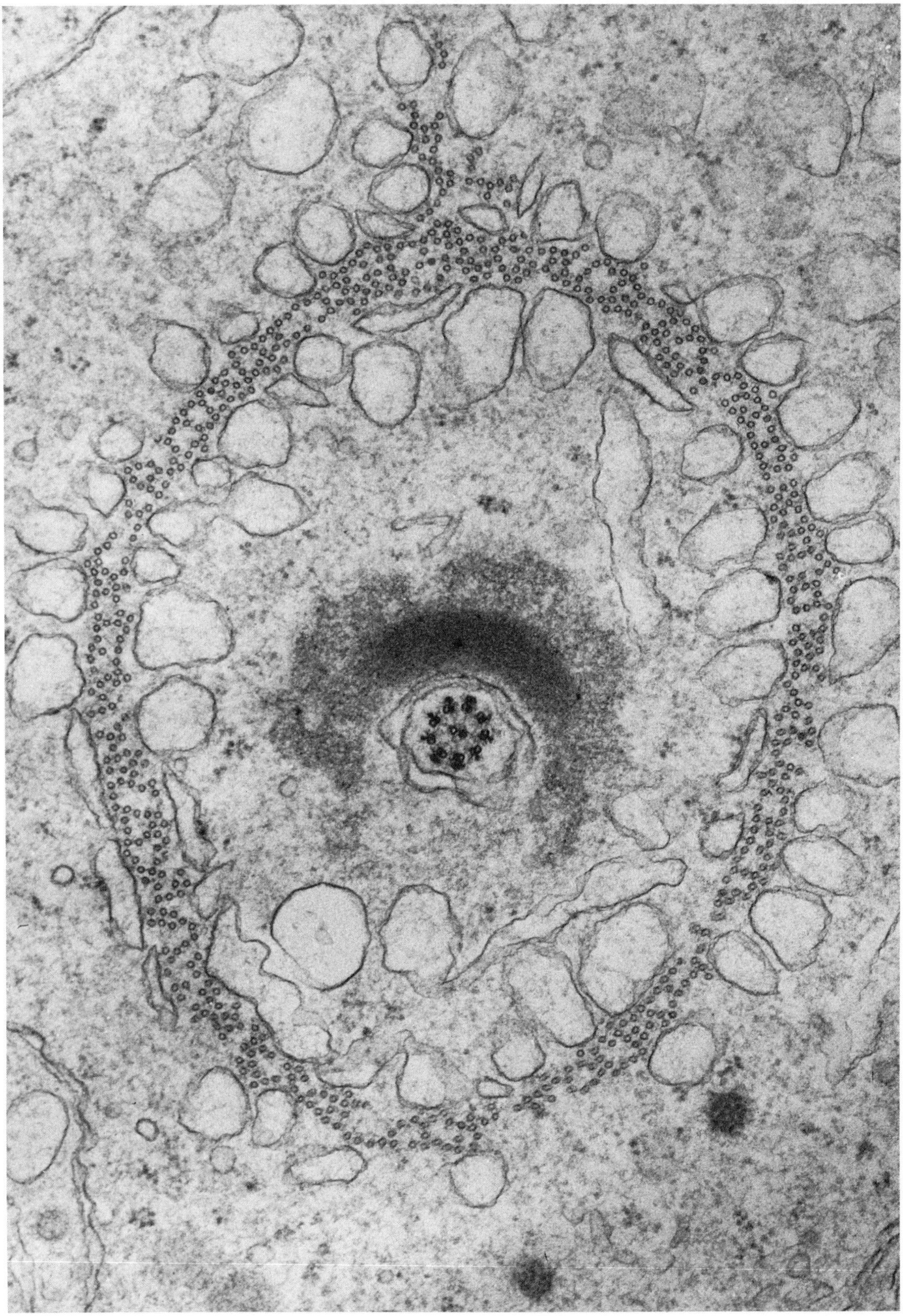

Figure 422

It has frequently been suggested that the mobility of certain classes of transmembrane proteins is limited by their attachment to microtubules and other cytoskeletal elements (Nicolson, 1976). Microtubules have also been assigned a role in guiding secretory granules to the surface for release (Lacy and Malaisse, 1973). However, convincing microscopic evidence for termination of microtubules at the inner aspect of the membrane has seldom been presented. The manchette is one of the few unambiguous examples of a microtubule organelle ending at the cell membrane. As illustrated in the accompanying micrograph, its constituent microtubules end in a dense subplasmalemmal matrix seen at the top of the figure. This material occupies the concavity in a prominent annular ridge on the cell surface around the caudal portion of the spermatid nucleus. In this case, attachment of microtubules does not seem to be an example of cytoskeletal control of transmembrane proteins but merely provides a fixed origin from which polymerizing microtubules lengthen to achieve the internal deformation necessary for spermatid elongation.

Figure 423. Longitudinal section of the manchette of a rodent spermatid.

Figure 423

The microtubules composing the manchette are joined together in an irregular pattern by slender cross bridges as seen at the arrows in the accompanying micrographs. Similar linkages are seen in various other organelles composed of microtubules, such as the axonemes of Heliozoa, the Km fibers of *Stentor*. There is evidence that the bridges connect to certain specified protofilaments and project at a fixed angle to the tubule. The limited variety of geometric patterns displayed by microtubule arrays is evidently an expression of these restraints (Fujiwara and Tilney, 1975). Bridges projecting from microtubules at the outer and inner aspects of the manchette frequently appear to attach to the membrane of the associated vesicles or tubular elements of the endoplasmic reticulum (at asterisks). It is not known whether these are enduring attachments or transient linkages involved in a transport mechanism (Fawcett, Anderson and Phillips, 1971).

Figures 424 and 425. Nucleus and adjacent manchette of rodent spermatids.

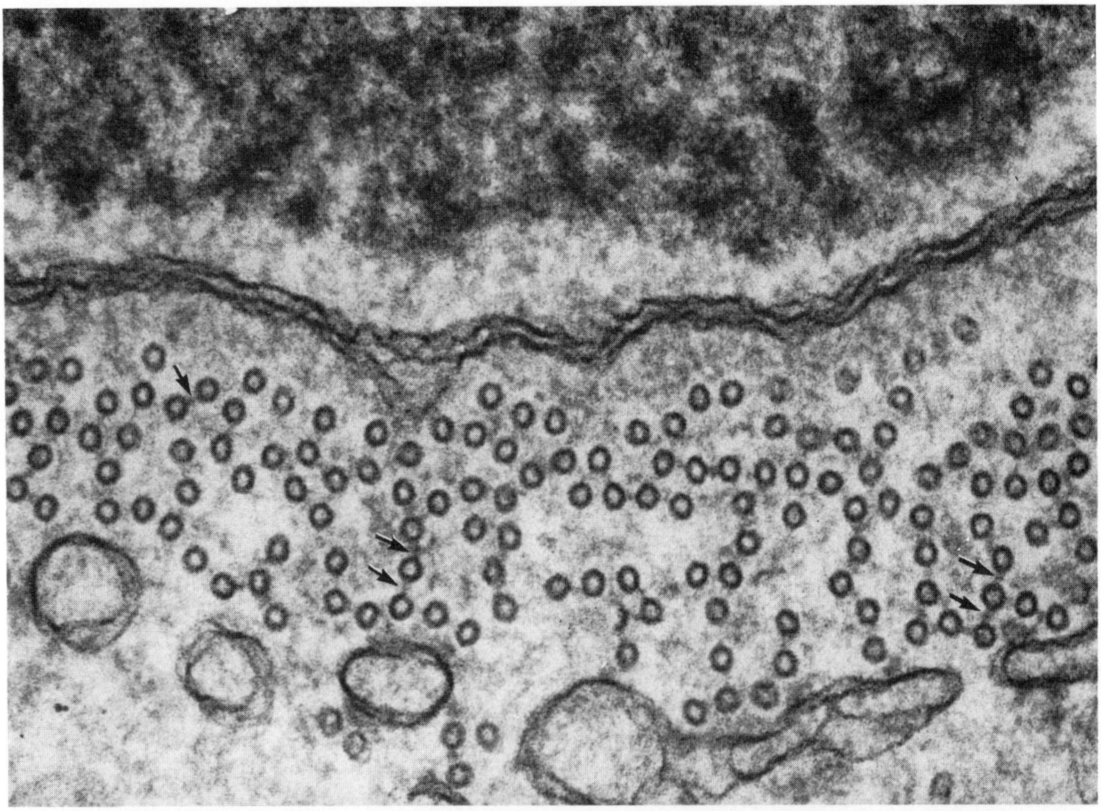

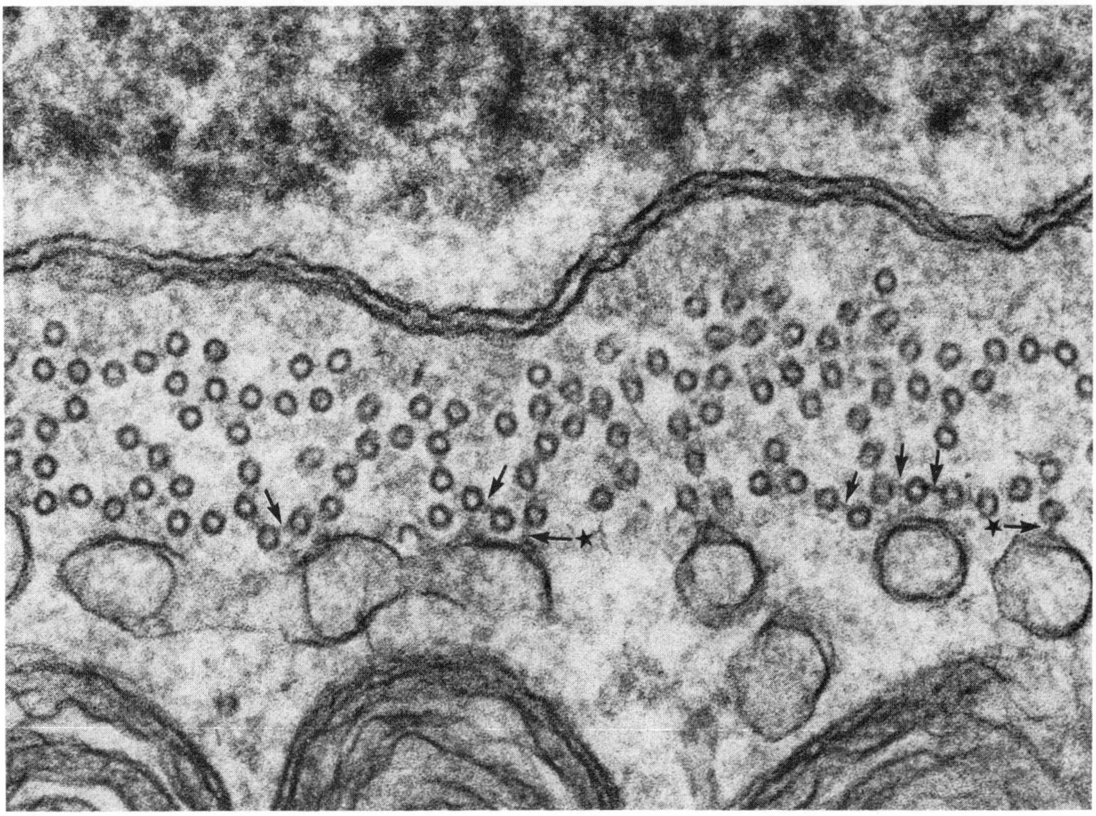

Figure 424, *upper* Figure 425, *lower*

In spermiogenesis of various species, there are numerous examples of the transient appearance of arrays of microtubules during the shaping of the nucleus and certain components of the sperm tail. The disposition of the microtubules suggests that if they were capable of generating movement, they might contribute to the observed shape changes. The accompanying micrographs from avian spermiogenesis illustrate this point.

The spermatozoa of finches have a corkscrew-shaped head and a very long-pitched mitochondrial helix extending throughout the greater part of the length of the tail. Early in spermiogenesis when the sperm head is beginning to take on its unusual form, a large bundle of microtubules pursues a helical course around the condensing nucleus as shown in A. The microtubules later disappear when the nuclear shaping is complete, but in the surface of the mature sperm head, shown in B, shallow depressions (at arrows) mark the former site of close apposition of the helical microtubule bundle. Although intrinsic morphogenetic factors associated with chromatin condensation may be largely responsible for the shape assumed by the nucleus, the relationship of the helical bundle of microtubules suggests that the latter may play an ancillary morphogenetic role.

The microtubule bundle continues posteriorly and winds around the midpiece, forming a double helix with the continuous strand of fused mitochondria (Panel C). The mitochondrial helix of the mature sperm (Panel D) retains the same pitch as the microtubule bundle after the latter has disappeared. It is difficult to escape the conclusion that this transient microtubule organelle is involved in determining the form of the mitochondrial sheath.

Figure 426. Stages in the differentiation of the finch spermatozoon. (From Fawcett, Anderson and Phillips, Dev. Biol., 26:220–251, 1971.)

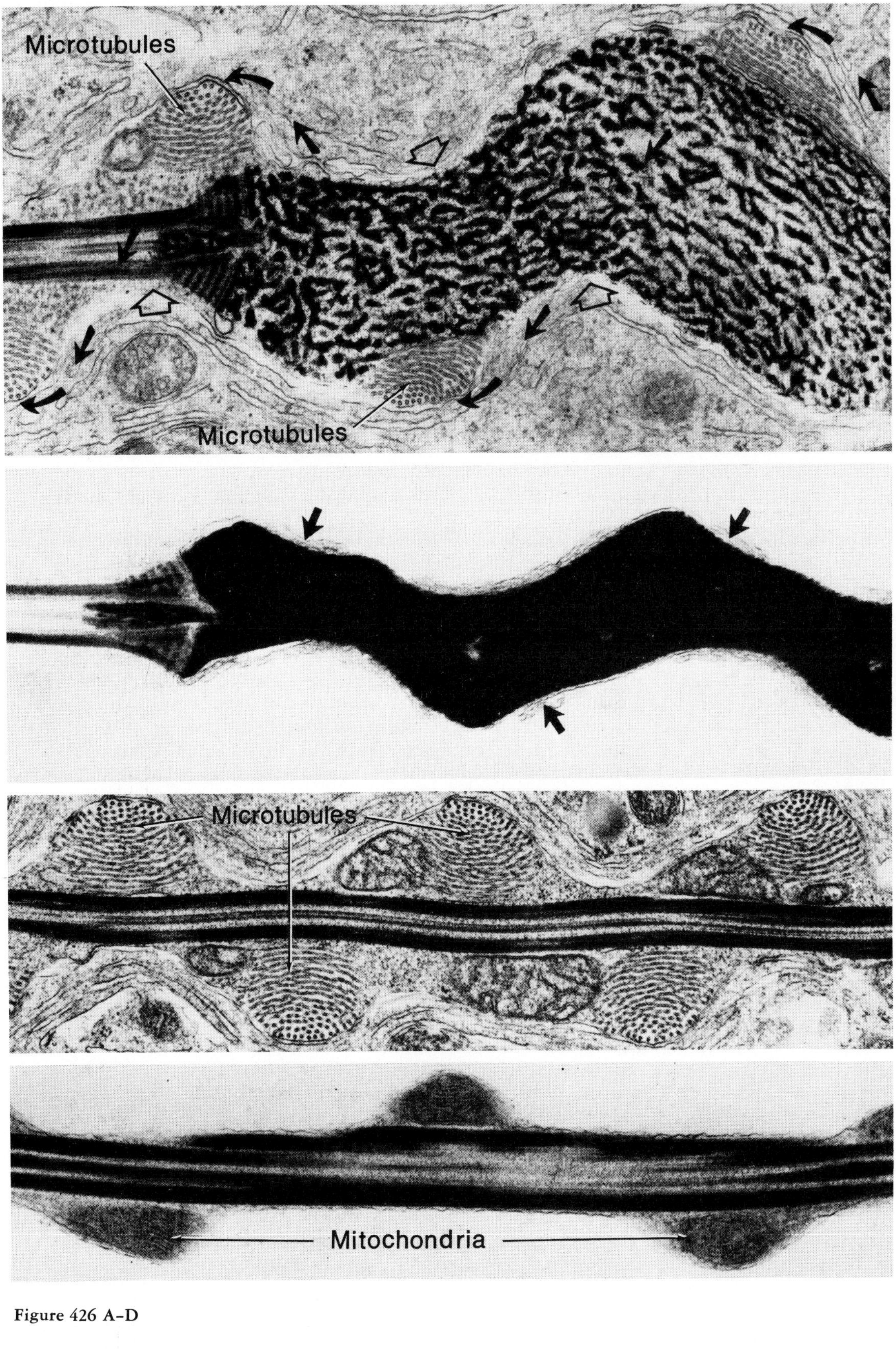

Figure 426 A–D

CYTOPLASMIC FILAMENTS

When the microscopic study of cells shifted in the later part of the 19th century from observation of living material to examination of chemically preserved and stained cells, many of the leading cytologists of that period held the view that the fundamental structure of protoplasm consisted of delicate fibrils traversing a homogeneous ground substance. Fleming, a leading proponent of the *fibrillar theory* of protoplasm, ascribed to these unbranched and discontinuous filar elements a major role in the activities of cells. Others, such as Carnoy and Van Beneden, argued just as strongly in favor of the *reticular theory*, which assumed that the fibrillae formed a continuous branching and anastomosing network. Fibrils of muscle were thought to be local modifications of a general fibrillar network, and spindles and asters were believed to arise around the centrioles by rearrangement of preexisting filaments. Considering the methods used and the limited resolution of their microscopes, it is likely that their interpretations were based in part upon a relatively coarse pattern of precipitation of proteins by the fixatives. But Remak had seen neurofibrils in nerve cells as early as 1843, and Froman demonstrated them by use of silver salts in 1864. There is no doubt that some of the coarser meshed networks observed in other cell types represented true filamentous components of the cytoplasm. Heidenhain (1899) concluded that these were supporting or cytoskeletal structures and therefore called them *tonofibrils*.

In the next half century, the list of separately named filamentous components lengthened to include neurofilaments, glial filaments, tonofilaments, myofilaments, and spindle fibers, but there was no evidence to indicate whether or not these were different fibrous proteins. Electron microscopy set apart microtubules and revealed two filament types in myofibrils. At least two categories of cytoplasmic filaments of general occurrence were distinguishable on the basis of size — 6 nm filaments and 10 nm filaments. It now appears from electrophoretic and immunocytochemical studies that there are multiple categories of 10 nm filaments. Research on the filamentous components of cytoplasm and their respective functions is now one of the most active areas in cell biology.

Filaments of Striated Muscle

The cross striations of muscle fibers were first observed by Leeuwenhoek (1677), who believed that the striae probably represented a spiral elastic coil capable of retraction to produce movement. Myofibrils teased from insect muscle were observed to contract under the microscope (Rollet, 1885; McDougall, 1897). The fact that they bulged between successive Z-lines under these conditions led to the speculation that contraction resulted from swelling (McDougall, 1898). What was known of the biochemistry of contraction seemed consistent with the view that the chemical changes necessary for contraction and relaxation were produced by accumulation of lactic acid within the muscle and its neutralization by buffer systems in the surrounding extracellular space (Meigs, 1912).

When examined at high magnification in stained preparations, the cross striations appeared as alternating dark and light bands. Under the polarizing microscope, bands that were dark in stained preparations appeared light, or anisotropic, and were therefore called *A-bands,* whereas the light bands of stained specimens appeared dark, or isotropic, and were designated *I-bands* (Brucke, 1858). The A-band in stained sections was frequently bisected by a lighter central region called the H-band and in favorable preparations the middle of the H-band was traversed by a slender M-line. A dark transverse line in the middle of the I-band was called the *Z-line*. A segment of myofibril between two successive Z-lines was termed a *sarcomere*.

Polarization optical studies of muscle were taken up again in the early 1900s, when its theory was better understood. The positive birefringence of the A-band in polarized light was interpreted to mean that it was composed of long molecules or micelles arranged parallel to the long axis of the myofibril. The I-band was also observed to be weakly birefringent after extraction of phospholipids and therefore probably also contained oriented asymmetrical molecules.

The first significant biochemical studies of muscle were undertaken in the 1940s when Szent-Györgyi extracted a protein, *actomyosin,* which displayed physical properties characteristic of long asymmetrical molecules, and was shown to be capable of contracting *in vitro* upon addition of ATP. By differential extraction, actomyosin could be separated into an asymmetrical protein, *myosin,* which possessed ATPase activity, and a globular protein, *actin*. If the ionic strength of the medium was adjusted to that of the sarcoplasm of muscle, actin polymerized to a higher molecular weight fibrous protein (F-actin).

The histochemical approach to the study of tissues was popular at that time among morphologists. It was shown that extraction of fresh muscle in solutions known to extract myosin resulted in loss of the birefringence of the A-bands. These observations localized myosin to this segment of the sarcomere and provided evidence that the anisotropism of the myofibrils was due in large part to orientation of long molecules of myosin. But the mechanism of contraction was still elusive. Some postulated a change in configuration of the myosin molecules involving a folding and cross-linking comparable to that of keratin molecules in the shrinkage of wool (Astbury, 1944). Others envisioned a process comparable to blood coagulation and clot retraction, but none of these theories could adequately explain relaxation.

The earliest studies of muscle by electron microscopy confirmed the presence of filamentous subunits in myofibrils (Hall et al., 1946; Draper and Hodge, 1949), but little more could be learned from dissociated, dried, and metal-shadowed preparations. The development of methods for cutting ultrathin sections made it possible to examine the structural components of muscle in their normal spatial relations and with high resolution. The *myofibril,* the smallest unit of the contractile material visible with the light microscope, was found by Huxley and Hanson (1954) to be composed of smaller units, the *myofilaments.* These were of two kinds, thick 11 nm *myosin filaments* and thin 5 nm *actin filaments.* The cross-banded pattern of striated muscle was found to be a reflection of the arrangement of these two categories of filaments. Arrays of parallel myosin filaments 1.5 μm long and 45 nm apart composed the A-band. They alternated along the length of the myofibril with arrays of actin filaments that extended about 1 μm in either direction from each Z-line, where they appeared to be attached laterally. Actin filaments not only made up the I-band, but they extended some distance into the adjacent A-bands, occupying the interstices between the hexagonally packed myosin filaments. The depth to which the actin filaments penetrated into the A-band was found to vary with the degree of contraction. The central region of the A-band not penetrated by the actin filaments corresponded to the H-band of light microscopy. In the region of their interdigitation at either end of the A-band, the thick and thin filaments were only 10 to 20 nm apart, and this narrow interval was spanned by evenly spaced cross bridges that radiated from each myosin filament toward the neighboring actin filaments.

When the ultrastructure of muscle was examined in different states of contraction and extension, it was found that the length of the A-band remains constant but the length of the I-band varies with the degree of contraction. In stretched myofibrils, the actin filaments are drawn out at either end of the array of myosin filaments and the H-band becomes very broad, whereas in the contracted state, the actin filaments penetrate more deeply into the A-band and the H-band is shortened. These observations led Huxley and Hanson to conclude that muscle shortening results from sliding of the actin filaments along the myosin filaments toward the center of the sarcomere. The cross bridges on the myosin filaments suggested further that these produced the sliding by a repetitive cycle of attachment and detachment to sites on the actin filament. With each cycle of operation of the cross-linkages, the actin filaments are forced to move a short distance farther into the A-band.

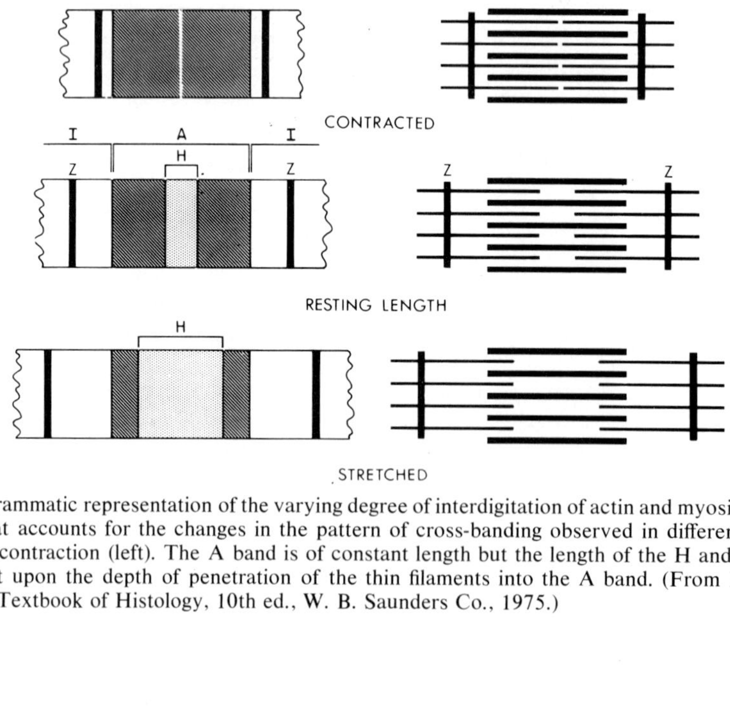

Diagrammatic representation of the varying degree of interdigitation of actin and myosin filaments (right) that accounts for the changes in the pattern of cross-banding observed in different states of myofibril contraction (left). The A band is of constant length but the length of the H and I bands is dependent upon the depth of penetration of the thin filaments into the A band. (From Bloom and Fawcett, Textbook of Histology, 10th ed., W. B. Saunders Co., 1975.)

The sliding-filament model of muscle contraction provided, for the first time, a logical and satisfying explanation of many aspects of the general physiology of muscle. Its evolution is an excellent example of the fundamental importance of morphological observations. In several decades of intensive physiological study and speculation on possible mechanisms of muscle contraction, a theory involving sliding of filaments of constant length had never been advanced, but when the topographical relations of the ultrastructural components of myofibrils were revealed by the electron microscope, such a mechanism at once became obvious and only needed experimental confirmation.

Dissociation of actin and myosin filaments into their molecular subunits and their examination with the electron microscope soon followed. Isolated myosin filaments are 1.5 μm long with a 0.2 μm smooth central segment, but toward each end there are numerous short lateral projections that correspond to the cross bridges between thick and thin filaments seen in thin sections of muscle. The myosin filaments can be further dissociated into myosin molecules. These are approximately 160 nm long and consist of slender rods 2 to 4 nm thick and a wider terminal segment.

Brief exposure to protease cleaves the molecule into two fragments called meromyosins. The rod-like tail, or backbone, of the molecule is *light meromyosin* (LMM) and a short segment including the laterally projecting head is *heavy meromyosin* (HMM). The heavy meromyosin fragment retains the adenosine triphosphatase activity of the myosin molecule and the capacity to bind to actin.

Isolated actin filaments are about 1 μm long and at high magnification appear to consist of globular subunits 5.5 nm in diameter arranged in two helically entwined strands. Reduction of the ionic strength of the medium results in further dissociation of filamentous actin (F-actin) into its globular subunits (G-actin). The thin filaments of the I-band of muscle are therefore polymers of G-actin molecules. Myosin and actin filaments similar to those of intact muscle can be reassembled *in vitro* from their molecular subunits.

Two physiologically important regulatory proteins are associated with the actin filaments, but these are not resolved in electron micrographs. Each of the helically entwined strands of actin comprising the thin filaments of muscle is believed to have an associated submicroscopic strand of *tropomyosin* molecules arranged end to end (Ebashi, Endo and Ohtsuki, 1969). Each molecule of tropomyosin is 40 nm long, spanning nine actin molecules. The second regulatory protein *troponin* is a globular molecule attached to each tropomyosin molecule near its end. Troponin binds calcium that is released from the sarcoplasmic reticulum in response to nerve stimulation. Calcium binding is believed to result in a change in configuration of tropomyosin molecules, exposing binding sites for the projecting heads of the myosin molecules of the adjacent thick filaments.

The cross-banded appearance of striated muscle illustrated in the accompanying micrograph of cardiac muscle results from the ordered arrangement of actin and myosin filaments. The dark A-bands are of constant length corresponding to the length of the myosin filaments. The lighter I-band consisting of actin filaments is bisected by the dense Z-line, where the actin filaments are joined laterally. During contraction, the I-band becomes narrower as the actin filaments penetrate more deeply into the interstices between the myosin filaments of the A-band. Higher magnification is required to resolve the relations of the two sets of filaments.

Figure 427. Papillary muscle of cat heart.

Figure 427

Here one sarcomere length of two adjacent myofibrils is shown at higher magnification. The A-band is seen to be composed of a parallel array of myosin filaments 10 to 11 nm thick and about 1.5 μm long. The light I-band consists of actin filaments about 6 nm thick extending in either direction from the central Z-line. Actin filaments are not confined to the I-band but penetrate some distance into the A-band, occupying the interspaces between the myosin filaments. The bracket marked X indicates the region of overlap of the two sets of interdigitating filaments. The bracket marked Y indicates the central region of the A-band, where there are only myosin filaments. This region corresponds to the H-band of light microscopists.

Figure 428. Papillary muscle of cat heart.

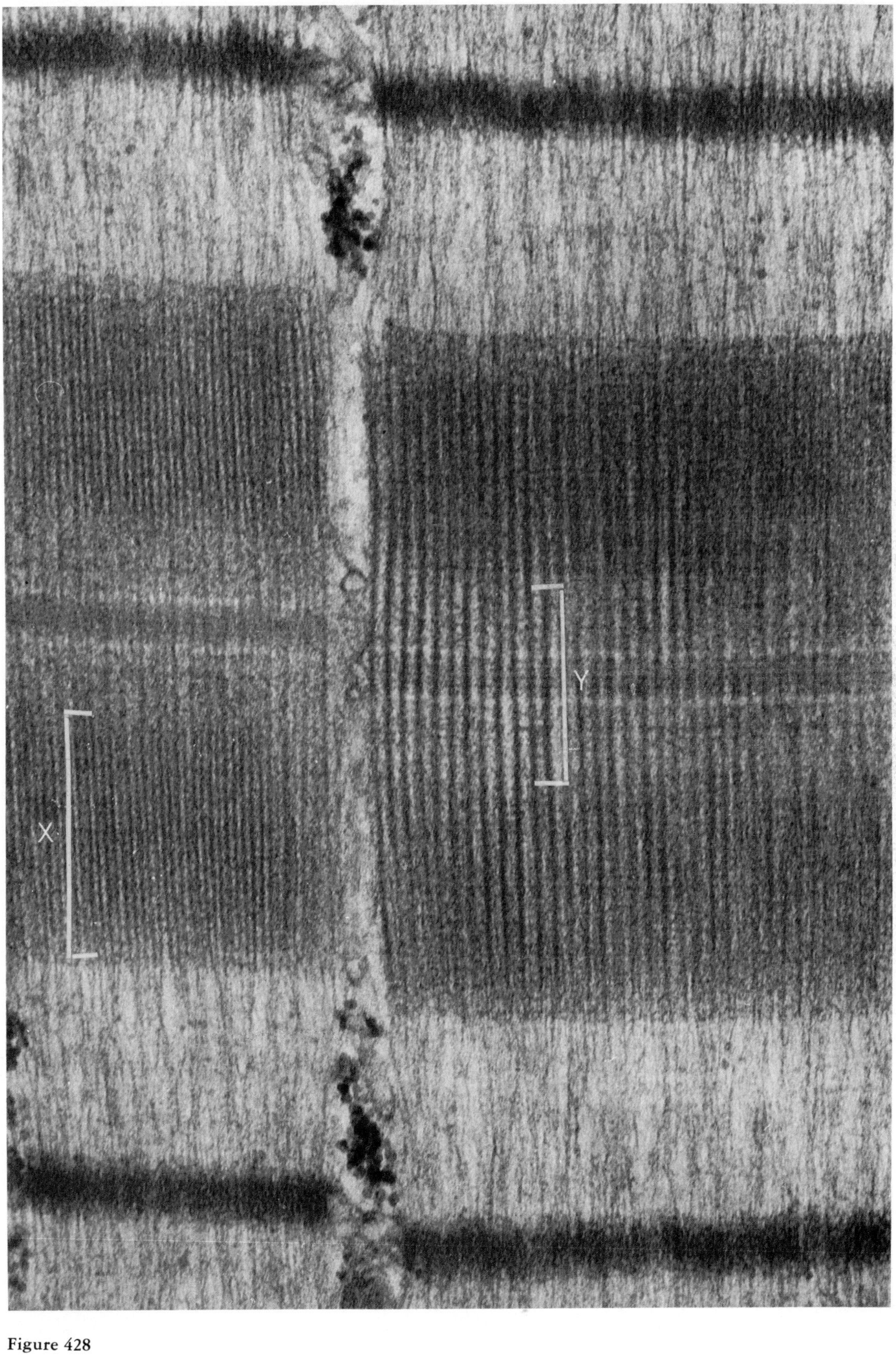

Figure 428

The accompanying micrograph shows at high magnification the region of interdigitation of actin and myosin filaments at the end of the A-band. The unfixed muscle was quickly frozen using liquid helium, then fractured and deep-etched. Cross bridges between the myosin and actin filaments are clearly visible. A fine axial periodicity can be seen in the actin filaments of the I-band at the left of the figure.

Figure 429. Skeletal muscle prepared by quick freezing and deep etching. (Micrograph courtesy of John Heuser.)

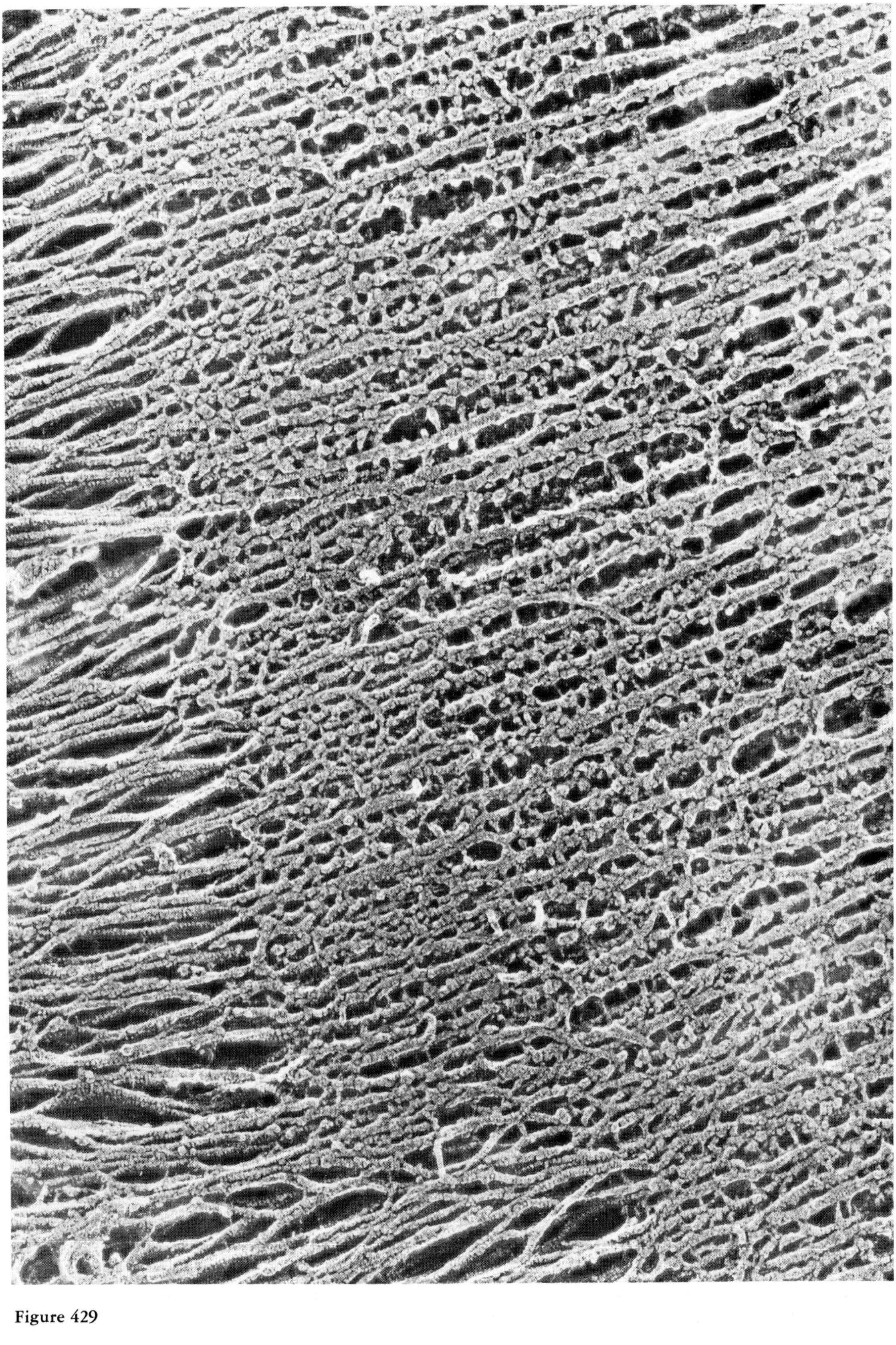

Figure 429

The highly ordered arrangement of thick and thin filaments is seen with unusual clarity in insect muscle. Indirect flight muscles shorten very little when they contract. The I-bands are shorter at resting length and the A-bands longer than in the striated muscle of vertebrates. The thin filaments penetrate far into the A-band, resulting in an unusually narrow H-band where only the thick filaments are present. In the accompanying micrograph, the cross bridges are barely discernible.

Figure 430. Flight muscle of *Drosophila melanogaster.* Section 0.25 μm thick, photographed at a magnification of 83,000 with a 1 million volt electron microscope. (Courtesy of Hans Ris, HVEM Facility, University of Wisconsin.)

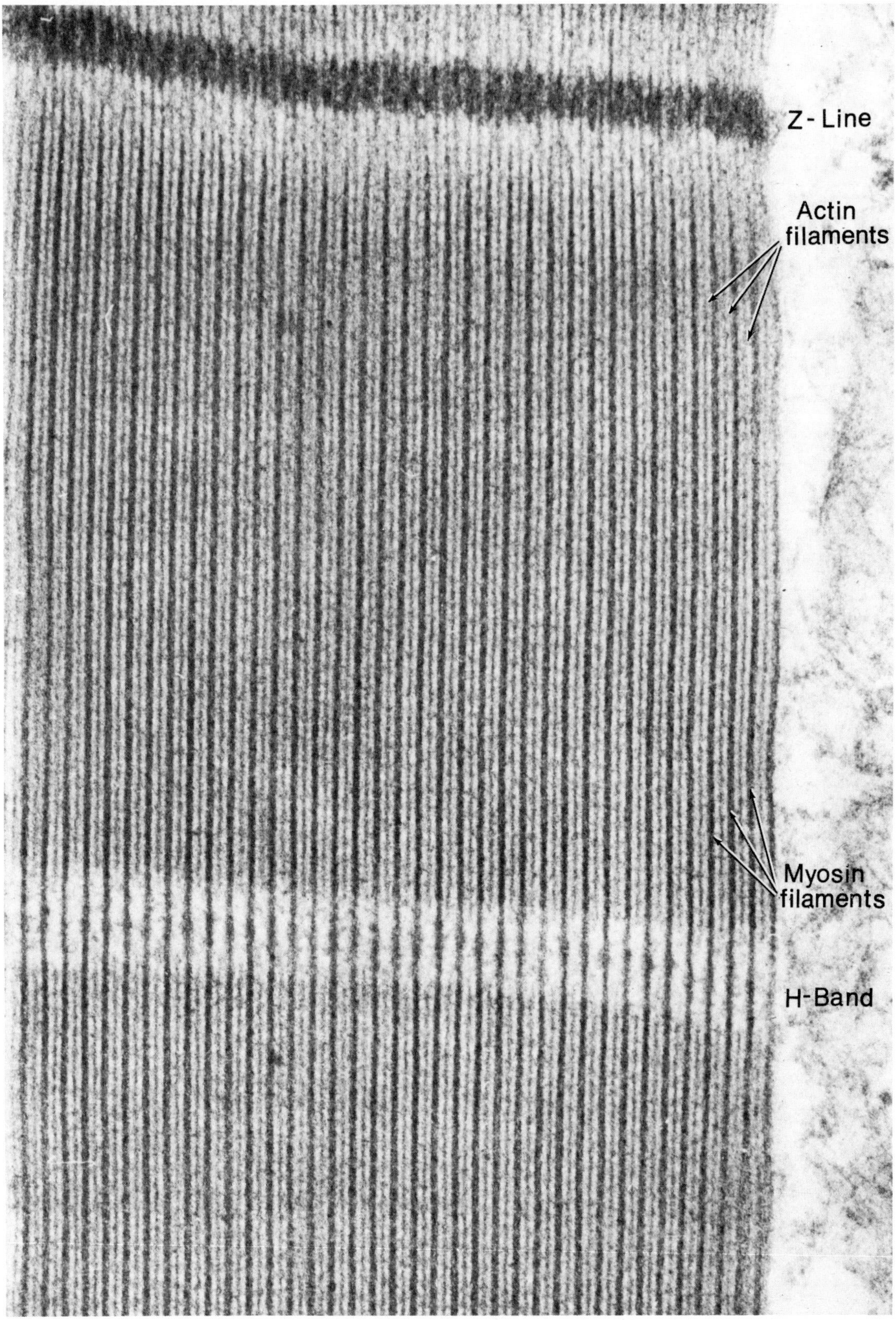

Figure 430

Transverse sections through the A-band of insect flight muscle display a remarkably regular lattice of filaments. The thick filaments are hexagonally arranged and each is surrounded by six thin filaments. In insect muscle, the actin filaments are in line with the rows of myosin filaments in all axes. In mammalian muscle, they are situated in the trigonal positions of the lattice so that each is shared by three equidistant thick filaments.

Figure 431. Asynchronous flight muscle of a fly, *Bombylius major.* (Micrograph courtesy of J. Auber, Am. Zool. 7:451–456, 1967.)

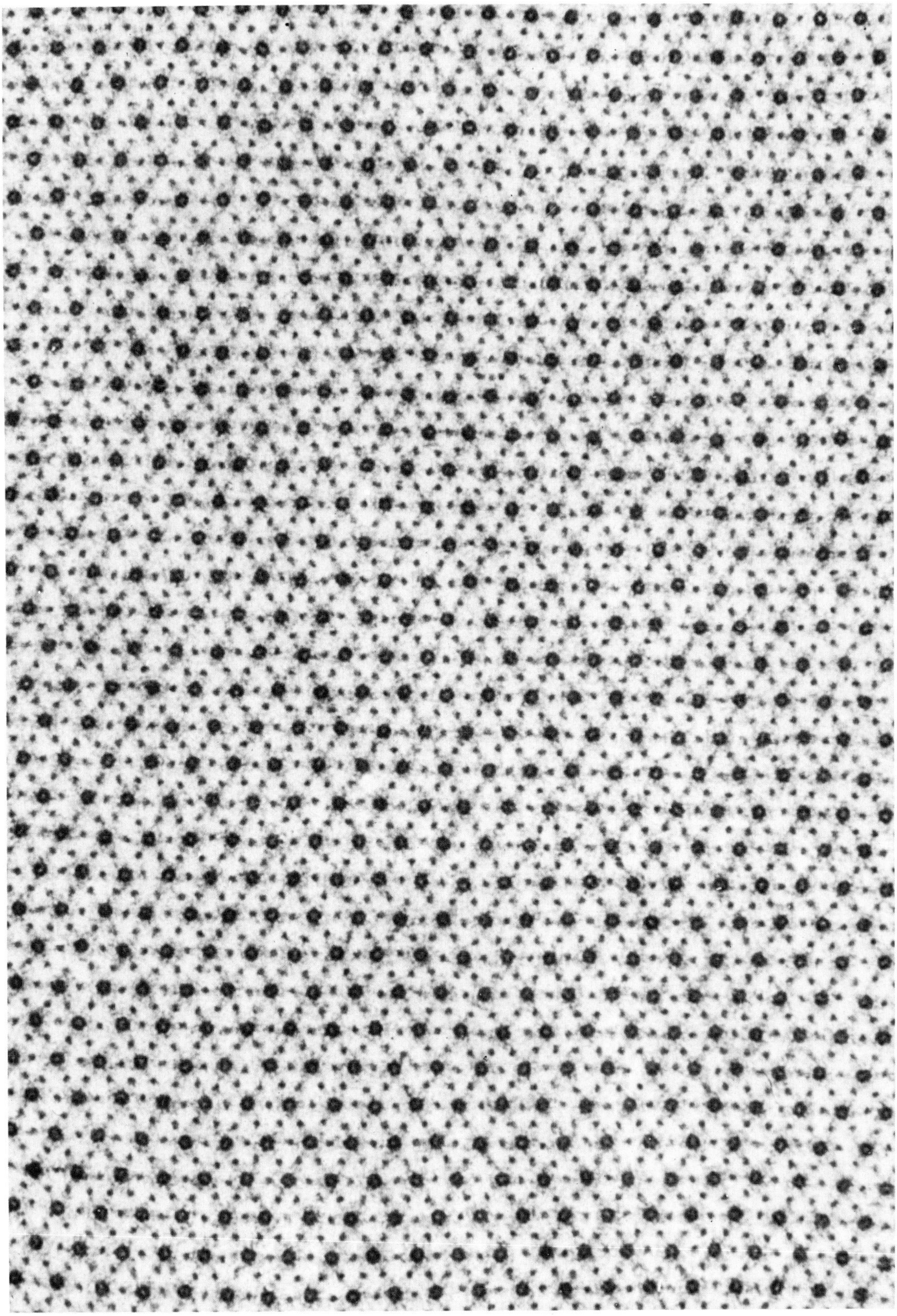

Figure 431

Filaments of Smooth Muscle

With routine methods of specimen preparation, the cytoplasm of smooth muscle cells often appears surprisingly homogeneous. There are clusters of ribosomes, a few mitochondria, and numerous focal densities scattered throughout the cytoplasm, but the 6 nm filaments normally present in great abundance are only resolved with some difficulty. Since it has been shown that osmium tetroxide destroys actin filaments *in vitro* (Maupin-Szamier and Pollard, 1978), it is likely that the full complement of filaments is seldom preserved in tissue processed for thin sectioning. However, in fortuitous preparations viewed at high magnification, the actin filaments can sometimes be seen. The area enclosed in the rectangle is shown at higher magnification in the next micrograph.

Figure 432. Smooth muscle of mouse epididymal duct.

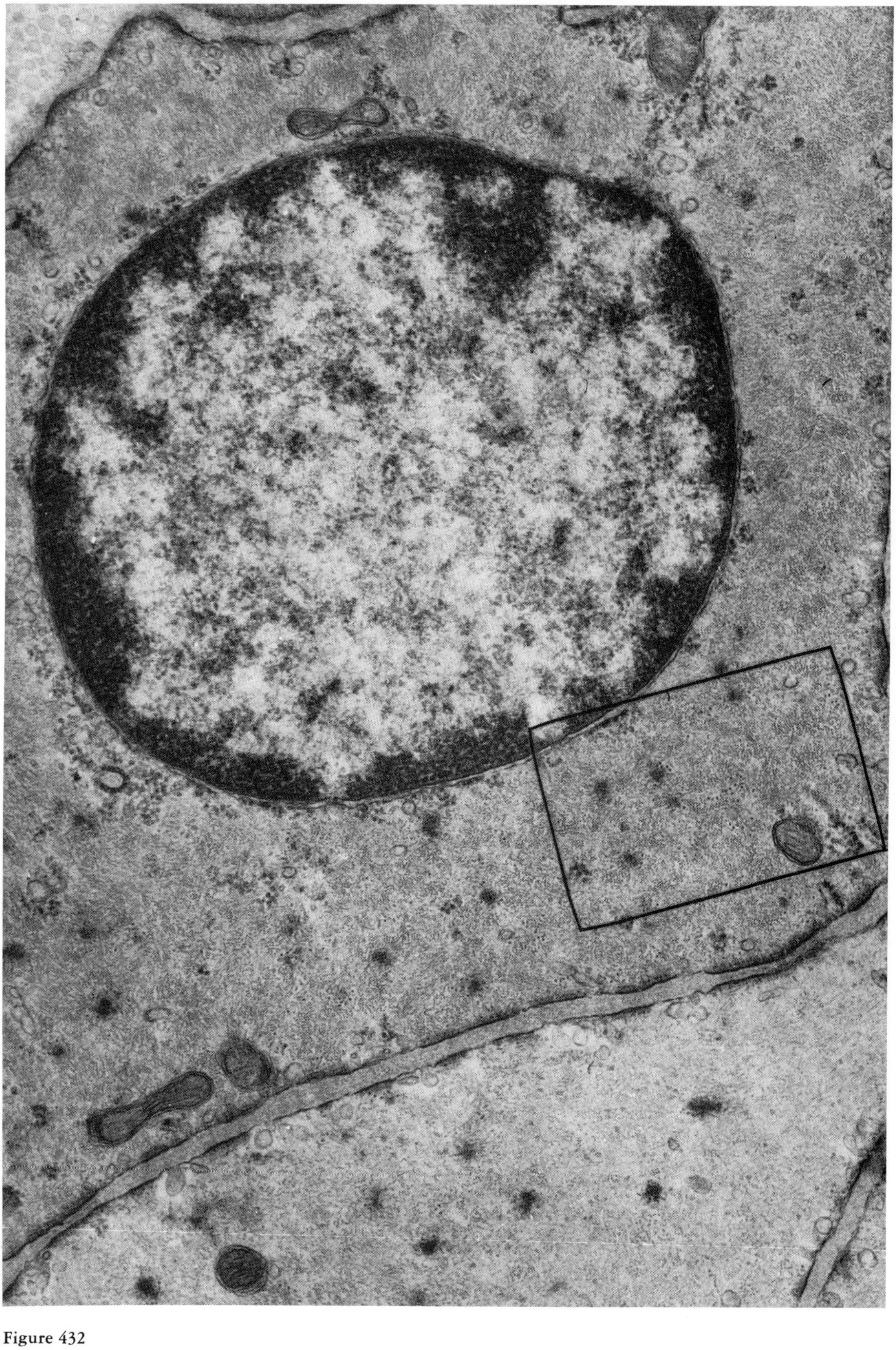

Figure 432

The cut ends of actin filaments appear as dark dots of uniform size. They are not distributed in a regular pattern as are those of striated muscle but occur in irregular groupings that represent bundles or tracts of parallel filaments. Their small size can be appreciated by comparison with the ribosomes (16 nm) at the upper right.

Thick filaments of myosin are known to be present from observations on smooth muscle glycerinated at pH 6.0 (Kelly and Rice, 1968) and from x-ray diffraction patterns (Small and Squire, 1972), but under conventional conditions of chemical fixation the myosin of smooth muscle is quite labile. The micrograph on the facing page was among the first (1966) purporting to show thick filaments (at arrows). It has since been learned that very carefully balanced conditions of divalent cation and ATP concentration, ionic strength, and pH must be maintained to preserve smooth muscle myosin in filamentous form (Small, 1977). Although the filaments shown here were assumed to be myosin, their size and sharp cross-sectional outline now make it likely that these are 10 nm filaments. The observation that they tend to be clustered around the dense nodes in the cytoplasm has suggested that these components may form a complex cytoskeleton distinct from the contractile apparatus (Small and Sobieszek, 1977).

Figure 433. Smooth muscle of mouse epididymis.

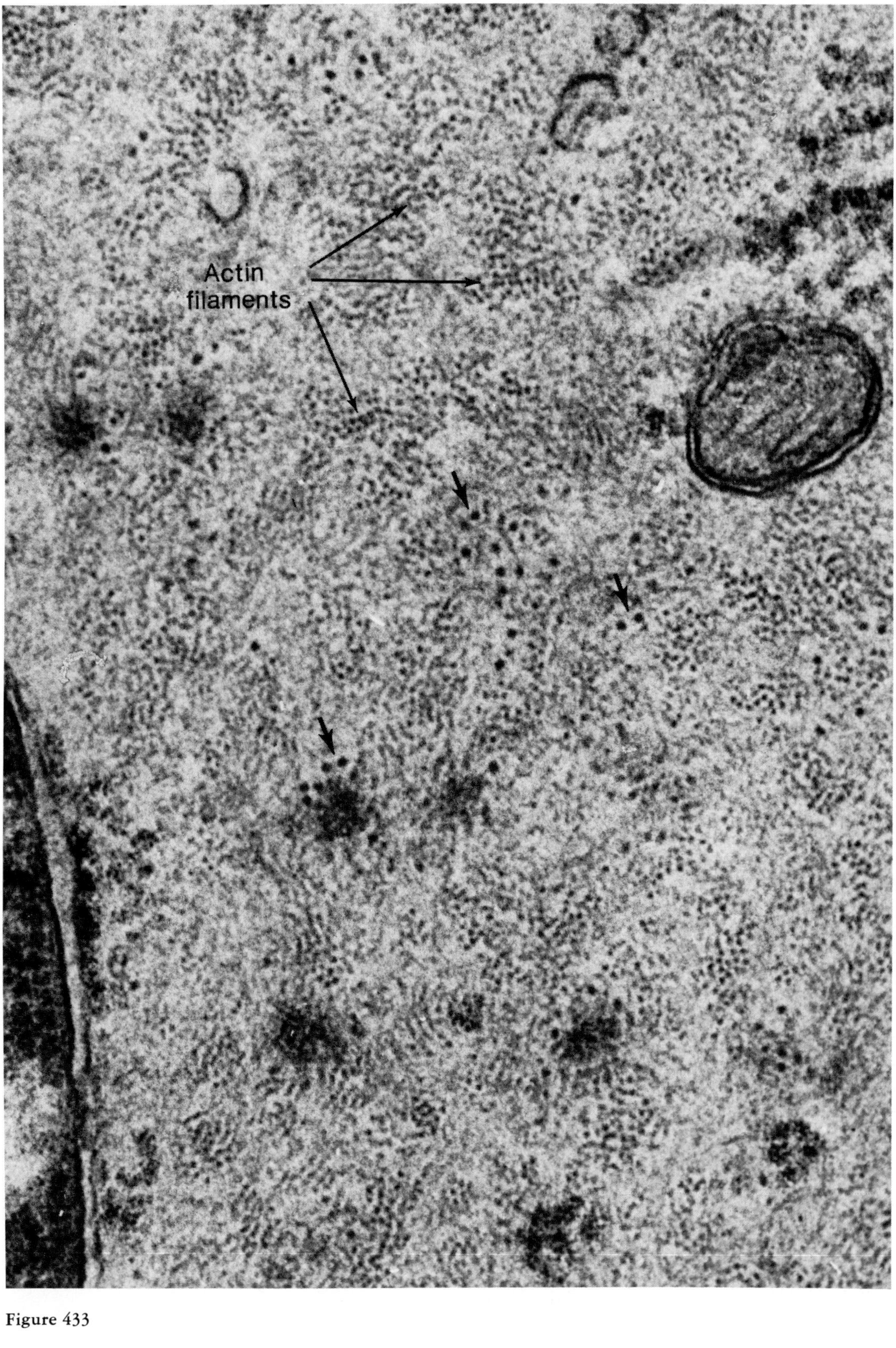

Figure 433

With appropriate preparative techniques, it is occasionally possible to preserve myosin filaments in smooth muscle. In the accompanying micrograph of vascular smooth muscle, numerous thick filaments are uniformly distributed among the actin filaments. As can be seen at higher magnification in the inset, the actin filaments outnumber the myosin filaments about 15:1 and the latter are rather irregular in outline.

Figure 434. Electron micrograph of a vascular smooth muscle cell in cross section. (Micrograph courtesy of Andrew Somlyo from *Vascular Smooth Muscle,* Springer-Verlag, Heidelberg, 1972.)

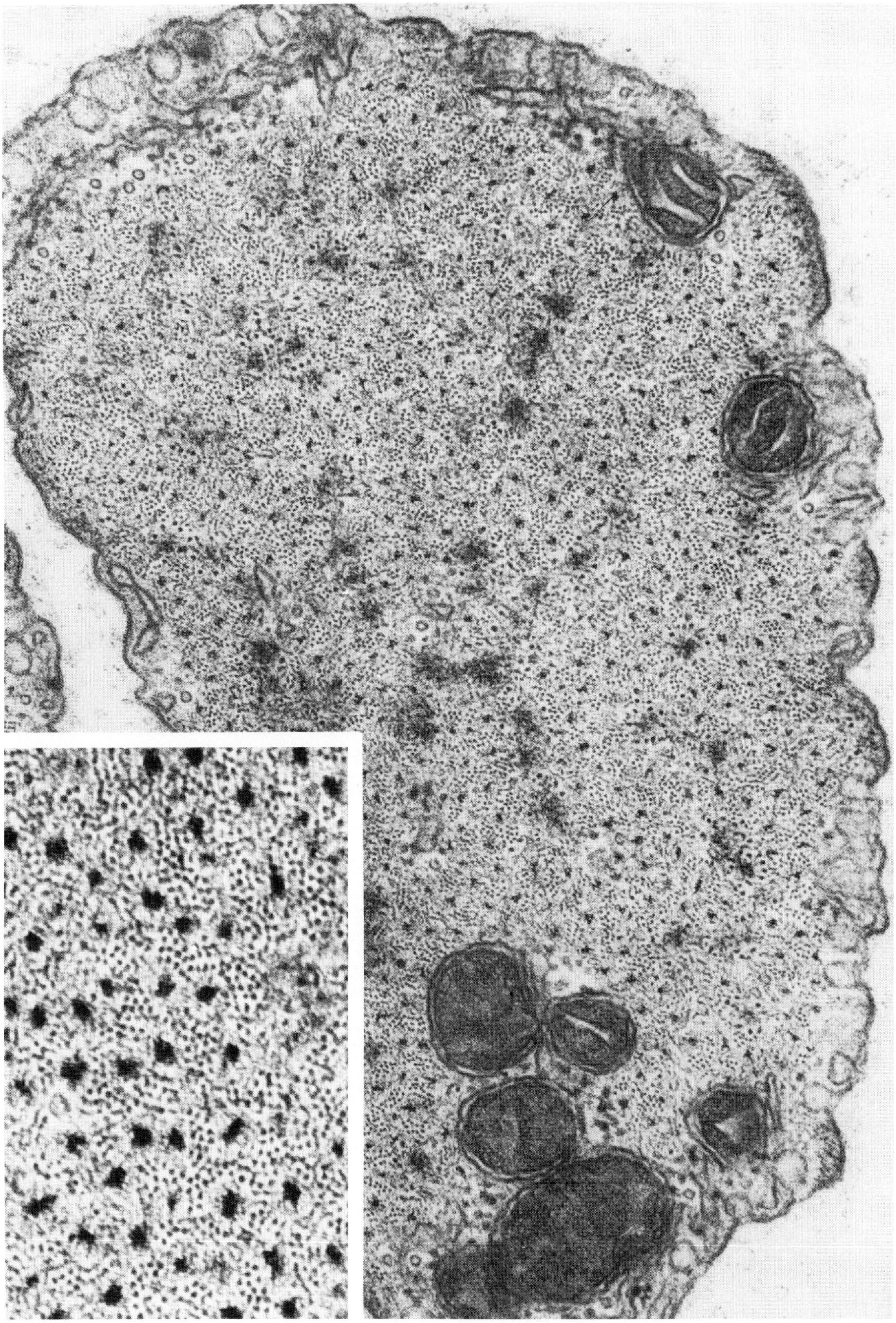

Figure 434

Actin and Myosin of Nonmuscle Cells

Contractility has long been recognized as one of the fundamental properties of protoplasm, expressed in cytokinesis, protoplasmic streaming, amoeboid locomotion, and morphogenetic movements of eukaryotic cells. But until recently nothing was known about the molecular mechanism involved.

In one of the early approaches to this problem, Loewy (1952) prepared an extract of the slime mold *Physarum,* which had some properties in common with actomyosin from muscle. The extract was able to hydrolyze ATP, and ATP had a demonstrable effect on its viscosity. These findings suggested a similarity in the basic mechanisms underlying muscle contraction and protoplasmic streaming of slime molds. Crude actomyosin preparations were soon extracted from human platelets (Bettex-Galland and Luscher, 1959).

The demonstration that nonliving glycerinated muscle and isolated myofibrils contracted when treated with ATP in the presence of Mg^{++} ions stimulated Hoffman-Berling to undertake similar experiments on cells other than muscle. In a variety of glycerin-extracted cell types, he observed feeble contractions upon addition of Mg^{++} and ATP. These studies attracted little attention because the highly heterogeneous extracts seemed to offer little possibility of isolation and analysis of the active components. But with improved biochemical methods for dealing with such complex mixtures, it was soon possible to purify actin (Hatano and Oosawa, 1966) and myosin (Hatano and Tazawa, 1966; Adelman and Taylor, 1969) from *Physarum* and to demonstrate their close relationship to the actin and myosin of muscle. The addition of gel electrophoresis and column chromatography to the armamentarium of the biochemist has enabled this field to move forward rapidly in the last decade.

Morphological identification of actin in cells was made possible by development of an ingenious technique which involves the use of the heavy meromyosin fragment of the myosin molecule as a histochemical reagent. Heavy meromyosin will enter cells after glycerin extraction and will bind to the cytoplasmic actin filaments as it does to muscle actin. It binds in a distinctive configuration suggestive of "arrowheads" repeating at regular intervals along the filament (Ishikawa, Bischoff and Holtzer, 1969). The "decoration" of cytoplasmic filaments with heavy meromyosin is considered a highly specific means of identification of filamentous actin.

Cytoplasmic actins from a number of cell types have been purified, and in all of their physicochemical parameters they are remarkably similar to muscle actin. In morphological studies by negative staining, actin occurs in filaments about 6 nm in thickness. These consist of 5 nm globular monomers arranged in a double helix. When decorated with heavy meromyosin, the arrowheads have a periodicity of about 38 nm corresponding to the half-pitch length of the double helix.

Neither actin nor myosin alone can generate force for movement. The presence of actin in cells therefore implies the coexistence of myosin. As expected, myosin has been found in several actin-containing cell types, including platelets, granulocytes, fibroblasts, and neurons, and it is able to bind to actin and possesses actin-activated ATPase activity. Unlike cytoplasmic actin, the myosins that have been isolated exhibit considerable variation in their physical, chemical, and enzymatic properties. Filaments of the size and shape of myosin filaments have been observed in the cytoplasm of a number of cell types, but in the absence of a specific method comparable to decoration of actin filaments, their identification is less certain. Antibodies have been prepared to myosin from striated muscle, platelets, and granulocytes, and to myosin from smooth muscle, and these have been used in the immunohistochemical localization of myosins in other cell types. Although useful, these lack the specificity of antibody prepared against pure myosin of the type characteristic of the cell type being studied.

In most cells, actin filaments are present in the cytoplasmic matrix, especially in a peripheral ectoplasmic zone immediately beneath the cell membrane. In some epithelial cells possessing a brush border, a bundle of actin filaments is found in the core of each microvillus. In these and other cells, the actin filaments appear to be attached to the cell membrane. This anchoring of the contractile apparatus to the cell surface may be essential for motility or for shape change. In cells in tissue culture, bundles of actin filaments are associated with the surface that is in contact with the substrate and extend out into the cell processes. These were designated "stress fibers" before their chemical nature was established, and the term continues to be used.

Until more information is available on the form and distribution of myosin in cells, little can be said about the mechanism by which actin and myosin interact to produce movement in cells other than muscle. Relatively little progress has been made to date in localizing myosin in electron micrographs of nonmuscle cells. Myosin filaments can sometimes be identified in the cytoplasm of amoebae, but routine methods of fixation apparently do not preserve myosin filaments in the cells of vertebrates and as a rule only actin filaments are seen. Myosin purified from such cells can be induced to form filaments *in vitro*, but it seems probable that in the cytoplasm it occurs instead in the form of very small aggregates. However, it is conceivable that if it is present in the form of randomly oriented filaments in low concentration, these would appear in sections as dots or short linear profiles indistinguishable from other cytoplasmic proteins precipitated by the fixative (Pollard and Weihing, 1974). The straight 6 nm actin filaments that are seen in electron micrographs probably represent only a fraction of the actin present in the living cell, for it is known that osmium-containing fixatives extract actin or transform it into a microfilament network that is difficult to identify in thin sections. Those straight 6 nm filaments that do resist extraction may have been stabilized by association with other proteins such as α-actinin or tropomyosin. Much of the current research on localization of actin and myosin therefore relies upon the use of fluorescent antibodies at the light microscope level of resolution.

When antibody against purified actin is reacted with flattened cells spread out upon the substrate in tissue cultures, the fluorescence is associated with long straight bundles of filaments traversing the perinuclear cytoplasm and extending out into cell processes. They correspond to the so-called "stress fibers" seen in living cells by phase contrast microscopy and to bundles of 6 nm filaments seen in electron micrographs (Goldman et al., 1975). Although the rigid appearance of these microfilament bundles might suggest that they are static cytoskeletal elements, their rapid reorganization during locomotor changes in cell-to-substrate contact indicates that they are part of a dynamic system involved in cell motility.

Figure 435. Thinly spread human skin fibroblast stained by indirect immunofluorescence with actin-specific antibody. (From Elias Lazarides, J. Cell Biol. 65:549–561, 1975.)

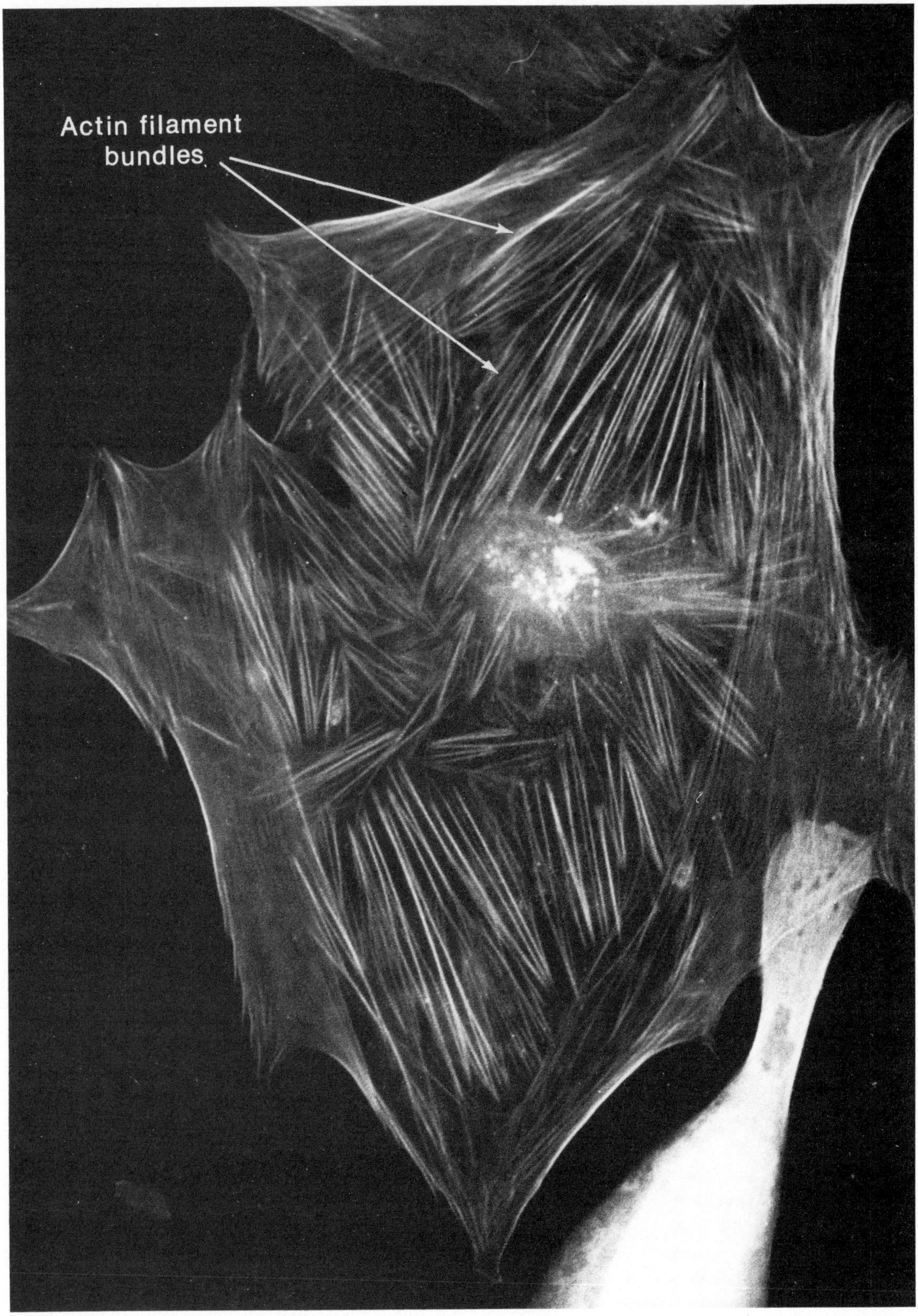

Figure 435

In cells in their normal environment *in vivo* 6 to 7 nm actin filaments occurring individually or in small groups interlace to form a complex meshwork in the cytoplasm. When the same cell types are explanted and maintained in culture, the filaments tend to aggregate into coarse bundles. In the accompanying micrograph of a Sertoli cell in vitro there are many individual filaments and several large aggregations of parallel filaments. The latter correspond to the "stress fibers" shown in the previous figure of a tissue culture cell stained with fluorescein conjugated antibody against actin.

Figure 436. Cytoplasm of a Sertoli cell from testis of *Citellus lateralis* in culture. (Micrograph courtesy of Wayne Vogl.)

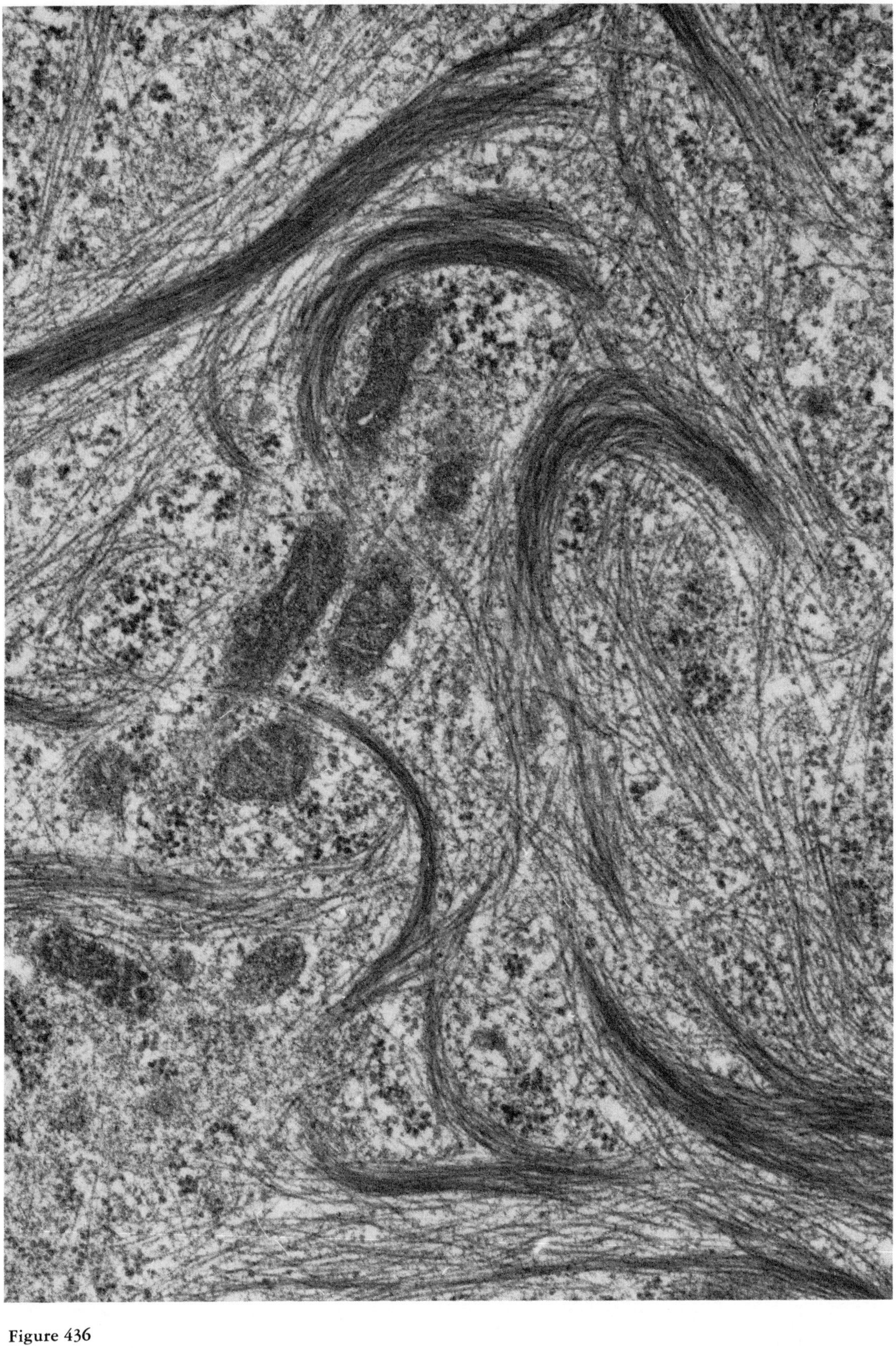

Figure 436

Filamentous Cytoskeleton

Now nearly a century after the controversy among light microscopists as to whether the fundamental organization of protoplasm was *fibrillar, reticular,* or *alveolar*, investigation of the organization of the cytoplasmic matrix has once again become one of the most active frontiers of cell biology. High-voltage electron microscopy for three-dimensional imaging of the cytoskeleton and immunocytochemical procedures are rapidly contributing to a better understanding of the structural basis for cell movement and for the translocation of organelles and other particulate components within the cytoplasm.

A three-dimensional network of fine filaments was described in transmission electron micrographs of the axoplasm of giant axons (Metuzals, 1969). From the study of stereo pairs taken with the high voltage electron microscope, Porter and his colleagues (1976) demonstrated in thinly spread tissue-culture cells slender, branching strands, or *microtrabeculae*, 4 to 10 nm in diameter forming a three-dimensional lattice throughout the cytoplasmic matrix. This microtrabecular lattice is attached to the plasmalemma and to membranous cytoplasmic organelles and its meshes are traversed by 10 nm filaments and microtubules.

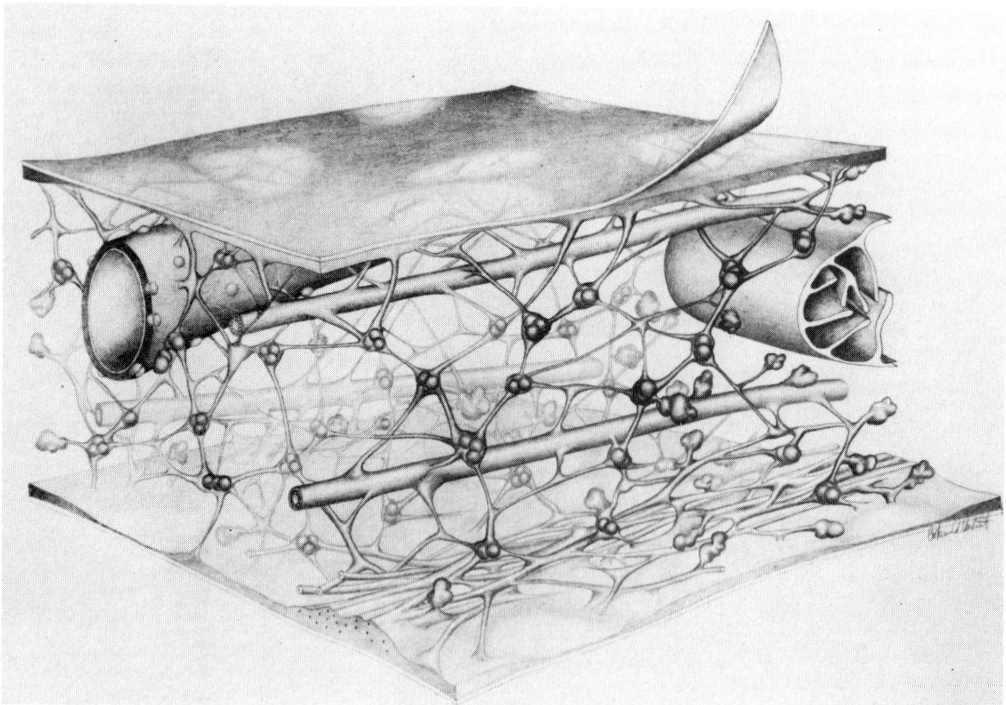

Schematic presentation of the microtrabecular lattice in the cytoplasm of a thinly spread cell in culture as seen with the high voltage electron microscope. The trabeculae are joined at thicker nodal points. The lattice is traversed by microtubules and the trabeculae appear to attach to these, to mitochondria, and to the cell membrane. (Drawing courtesy of Keith Porter.)

Concurrently with these electron microscopic observations, other cell biologists have exploited the potential of fluorescent antibodies for localization of specific structural proteins in tissue-culture cells at the level of the light microscope. With antiactin, long, straight, fluorescent strands were shown (Lazarides and Weber, 1974), corresponding in distribution to the "stress fibers" originally described in cultured cells by Warren Lewis and to the bundles of 6 nm microfilaments encountered in electron micrographs of such cells (Buckley and Porter, 1973). This pattern of conspicuous bundles of actin filaments was found most commonly in sessile cells firmly attached to the substrate, whereas in more motile cells or in regions with active undulating ruffles, few if any discrete microfilament bundles were visible and the actin fluorescence was more diffuse. The organization of the actin therefore appeared to be clearly related to the state of activity of the cell. The ultrastructural configuration of the actin corresponding to diffuse fluorescence at the light microscope level was not immediately apparent.

To observe the cytoskeleton with the resolution of the electron microscope and with the possibility of parallel immunocytochemical identification of its constituents, Weber and associates (1978) employed an ingenious technique involving removal of the membranes of cultured cells with detergent (Triton) and extraction of soluble protein components, followed by dehydration and critical-point drying. The resulting preparations showed a three-dimensional lattice of microtrabeculae comparable to that described by Porter. Ferritin-labeled antibody against actin not only decorated microfilament bundles but the finer lattice of microtrabeculae as well, suggesting the possibility that both consist of actin.

The application of the new technique of rapid freezing of unfixed cells followed by deep etching and rotary shadowing (Heuser, 1979) has provided further insight into the nature of the cytoskeleton. Such preparations reveal that both the "stress fibers" and the microtrabeculae are composed of filamentous subunits with an axial periodicity comparable to that of actin. Thus evidence is accumulating for a major cytoskeletal role for actin.

Evidence that the microtrabecular meshwork of cytoplasmic matrix consists mainly of actin has come from a novel approach, in which demembranated, extracted tissue-culture cells are quick frozen with liquid helium, deep etched, and rotary shadowed. The resulting micrographs viewed as negatives in stereo pairs have the three-dimensional quality of scanning electron micrographs and the high resolution of transmission electron microscopy (Heuser, 1979).

The accompanying low-power micrograph of the "cytoskeleton" of a portion of a cell in tissue culture does not do justice to the potential of the method for three-dimensional analysis, but it clearly shows a complex meshwork of microtrabeculae similar to that seen in more limited areas of fixed cells viewed by high-voltage electron microscopy. At higher magnification in succeeding figures, the strands representing microtrabeculae are seen to have a fine periodicity corresponding to that of actin.

Figure 437. Micrograph of the microtrabecular meshwork of a 3T3 cell, treated with Triton, washed, quick frozen with liquid helium, rotary shadowed. (Micrograph courtesy of John Heuser.)

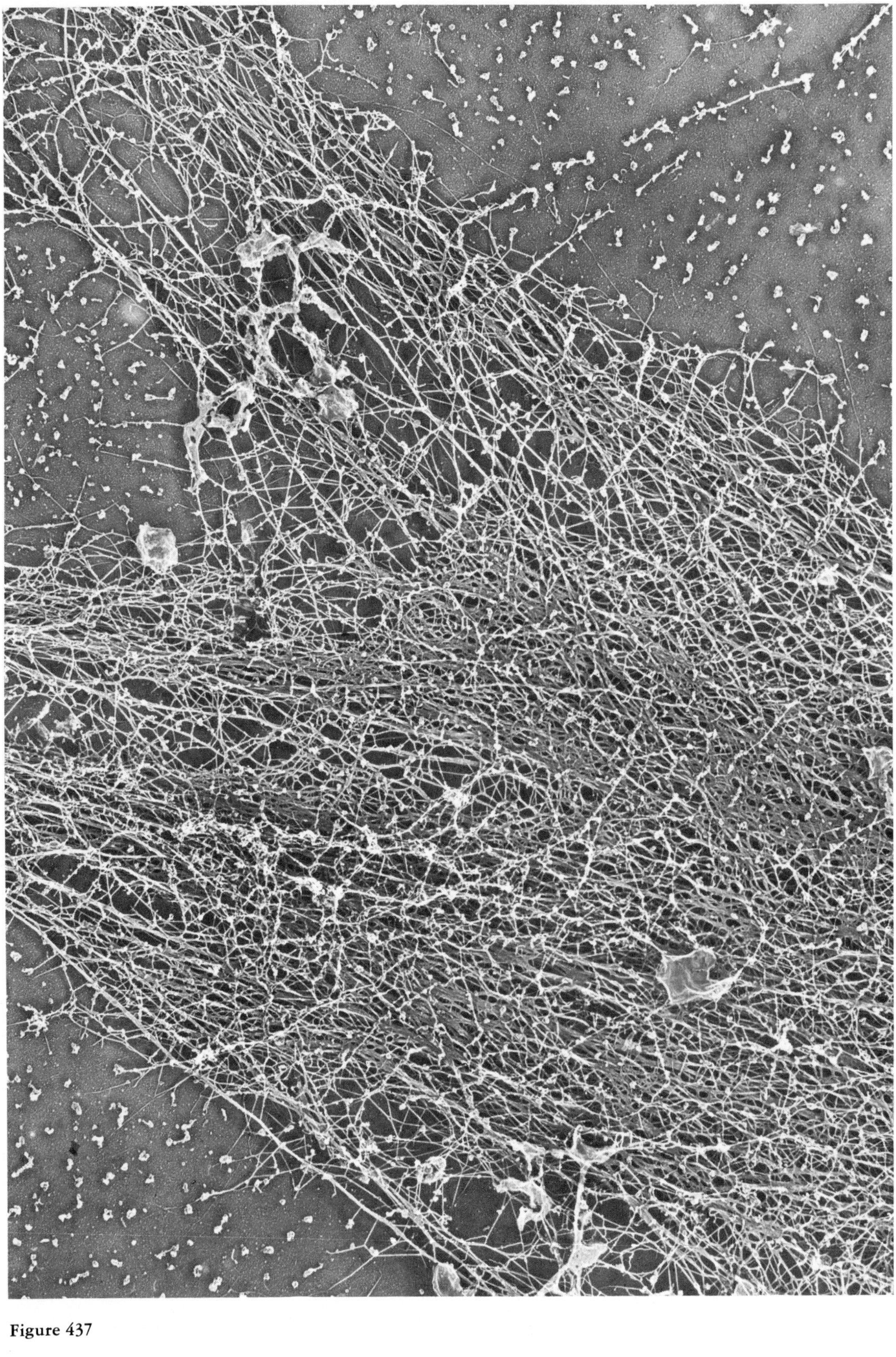

Figure 437

At higher magnification of the cytoskeleton in the accompanying micrograph, two "stress fibers" are seen to be composed of bundles of filaments that are continuous at their margins with elements of the general microtrabecular meshwork, thus strongly suggesting that the two are alternative configurations of the same structural protein.

The term *cytoskeleton* has been used to refer to passive stiffening, or stabilizing, filamentous components of the cytoplasm, involved in the maintenance of cell shape and providing a framework of fixed origins and insertions for the motor elements of the cytoplasm. If both the stress fibers and microtrabecular meshwork consist of actin, which under appropriate conditions can also serve as a component of the motor apparatus of the cell, then the appropriateness of referring to these elements as part of the "cytoskeleton" may have to be reconsidered.

Figure 438. Portion of the cytoskeleton of a 3T3 fibroblast, demembranated, depleted of soluble protein, quick frozen, deep etched, and rotary shadowed. (Micrograph courtesy of John Heuser.)

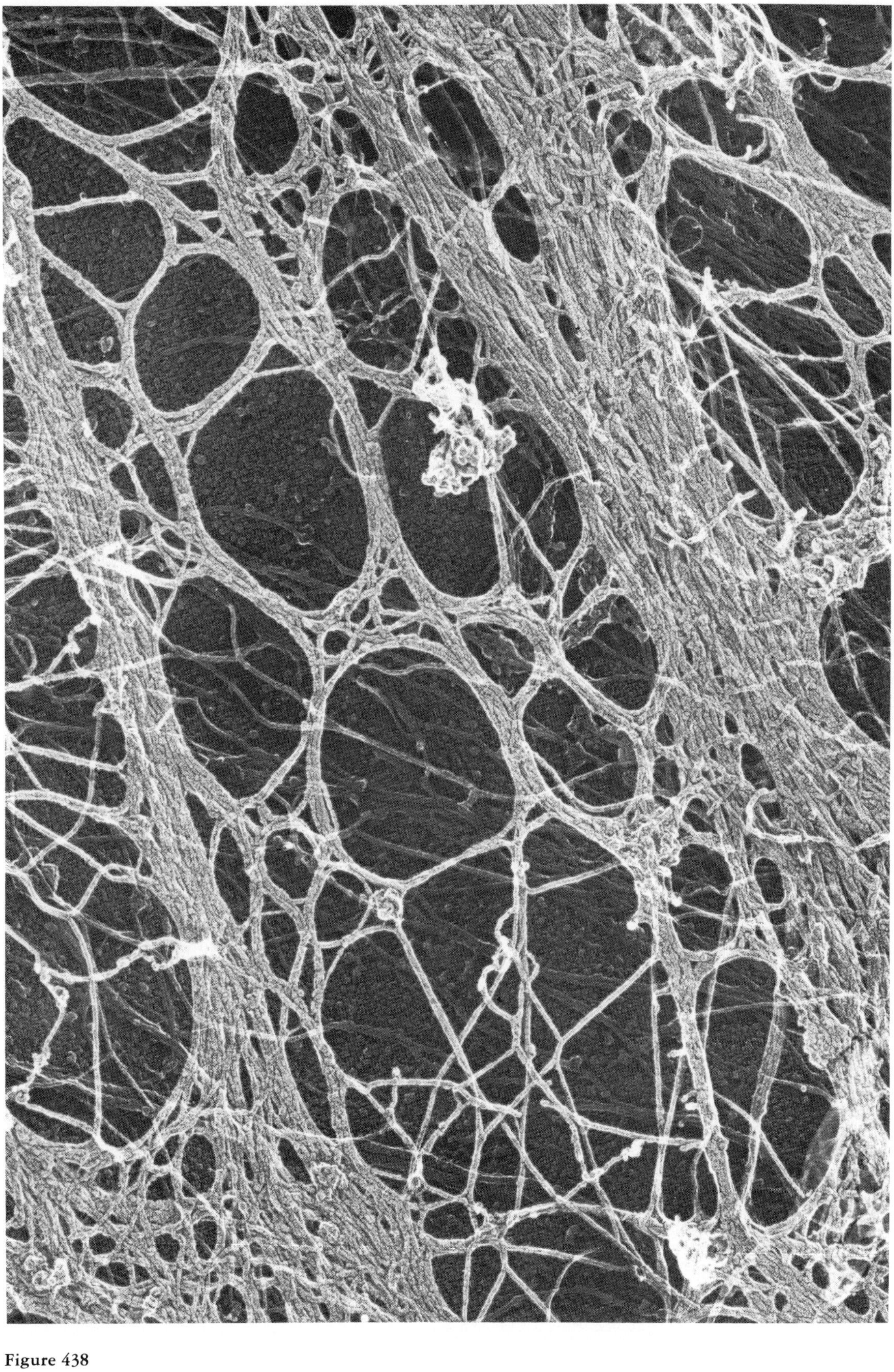

Figure 438

Shown here at high magnification are some large stress fibers. The filaments of which they are composed clearly show a 5 to 6 nm periodicity apparently corresponding to the globular subunits of fibrous actin.

Figure 439. Portion of the fibrous framework of the cytoplasm of a 3T3 fibroblast, demembranated, depleted of soluble protein, quick frozen, deep etched, and rotary shadowed. (Micrograph courtesy of John Heuser.)

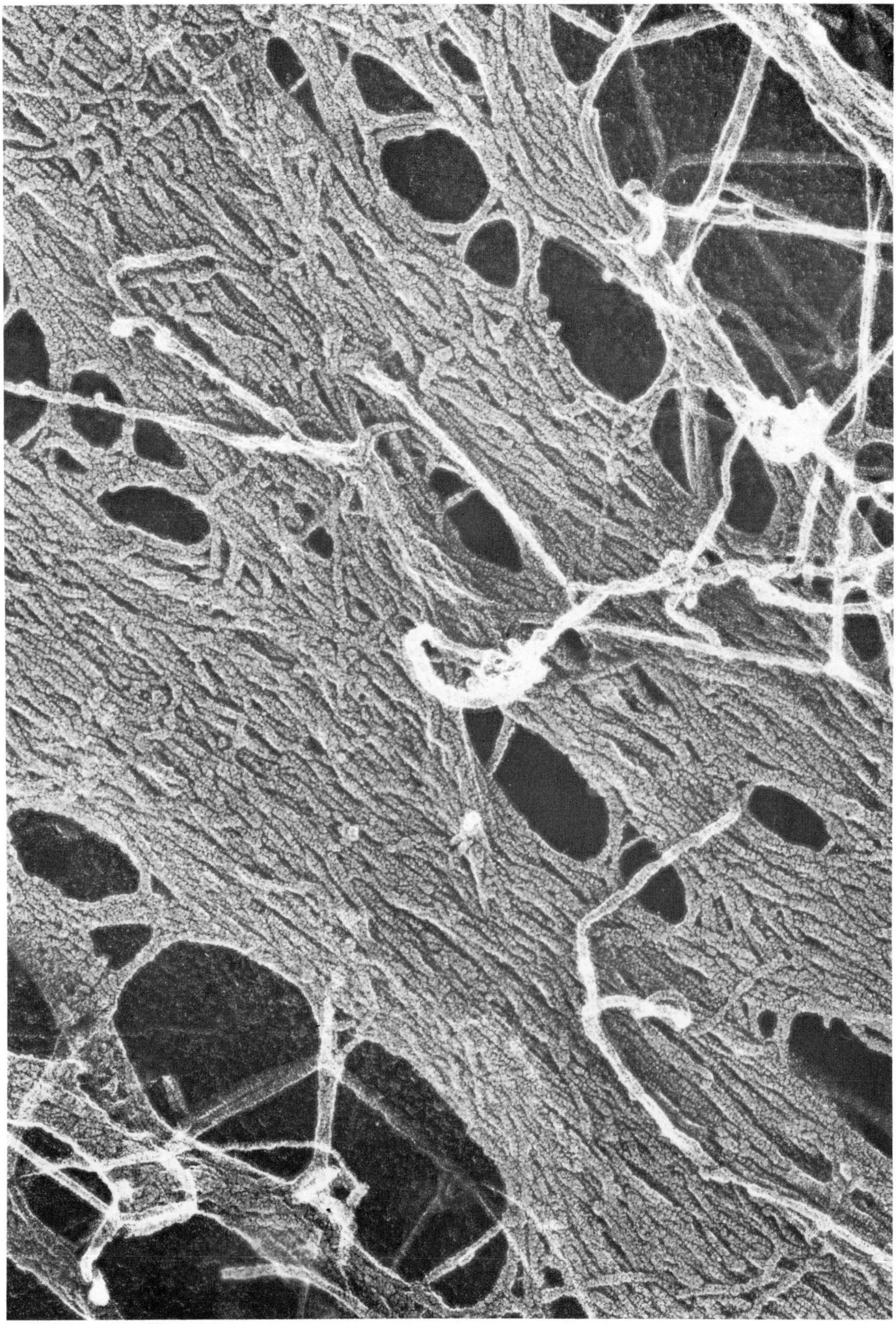

Figure 439

Presented here are two additional examples of the dense cytoskeleton in the peripheral cytoplasm of tissue culture cells. In the upper figure, two microtubules can be seen traversing the field and passing through the meshwork of filaments.

Figures 440 and 441. Cytoskeleton of tissue culture cells prepared by quick freezing, deep etching, and rotary shadowing. (Micrographs courtesy of John Heuser.)

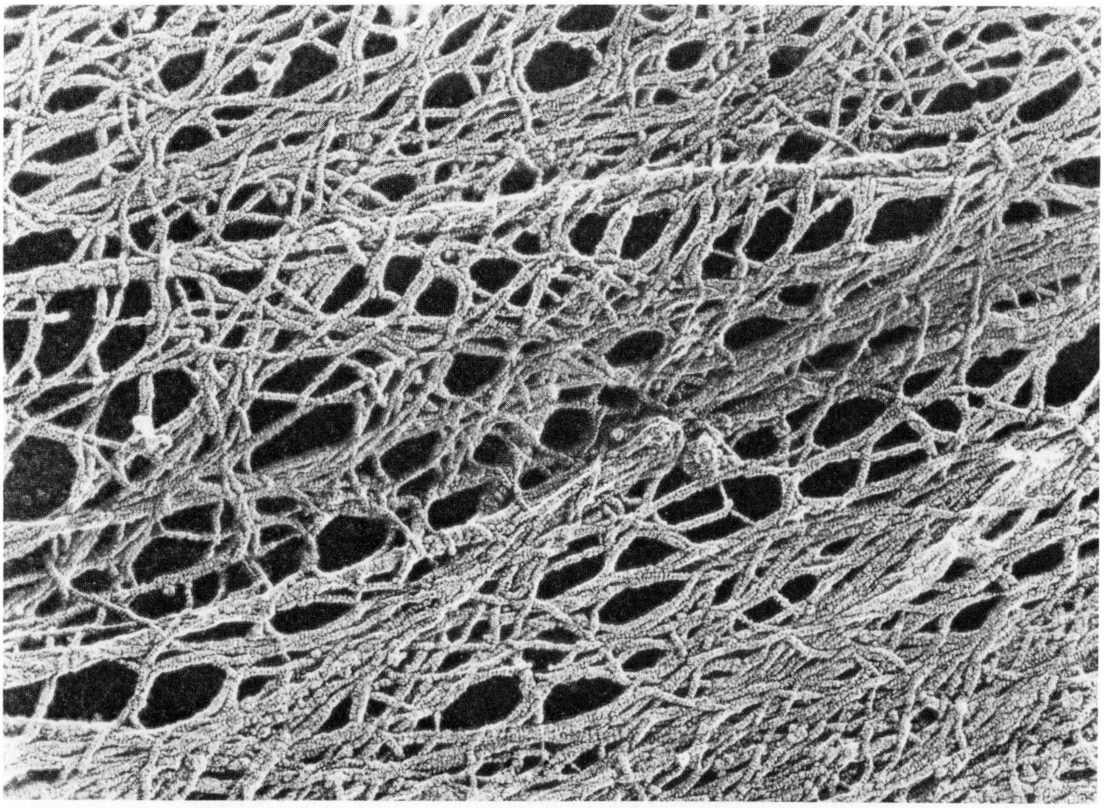

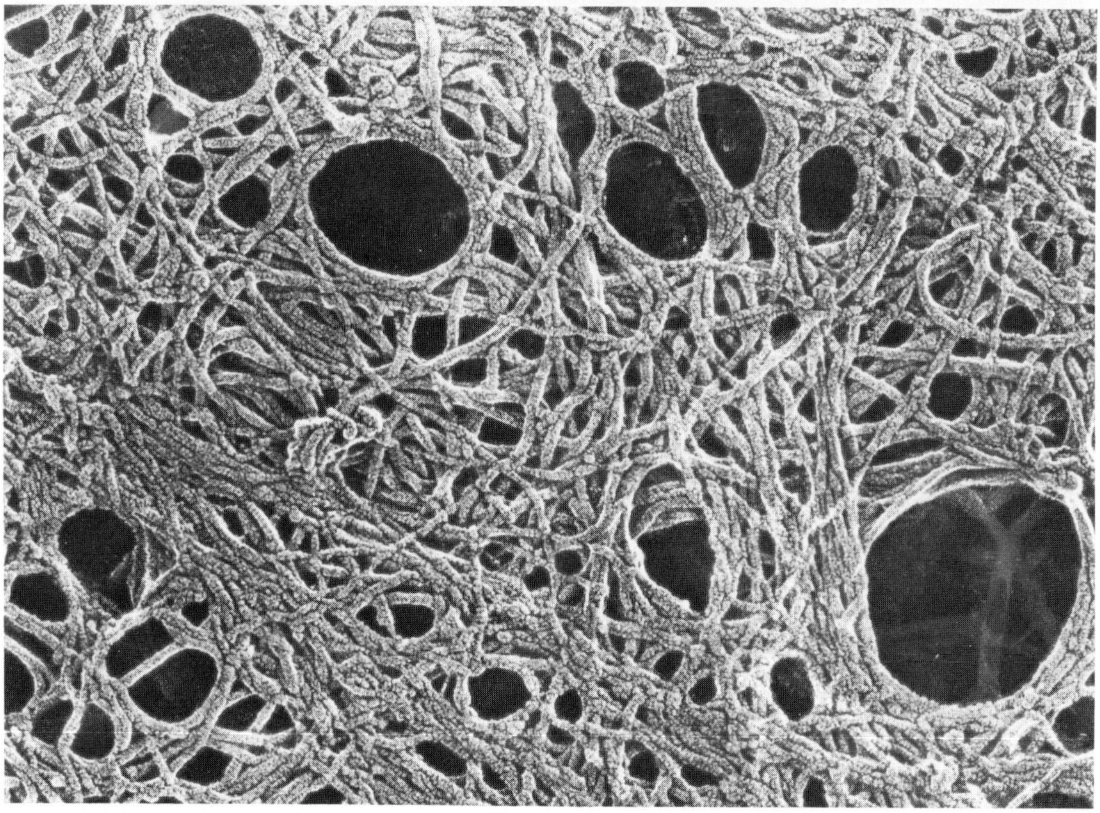

Figure 440, *upper* **Figure 441**, *lower*

Purified actin consists of a feltwork of filaments, which shows a periodicity similar to that exhibited by the trabecular meshwork of the cytoplasmic matrix.

In the lower figure, a preparation of actin has been decorated with heavy meromyosin, resulting in a rope-like helical disposition of the attached myosin fragments.

Figures 442 and 443. Purified preparation of actin undecorated and decorated with heavy meromyosin. (Micrographs courtesy of John Heuser.)

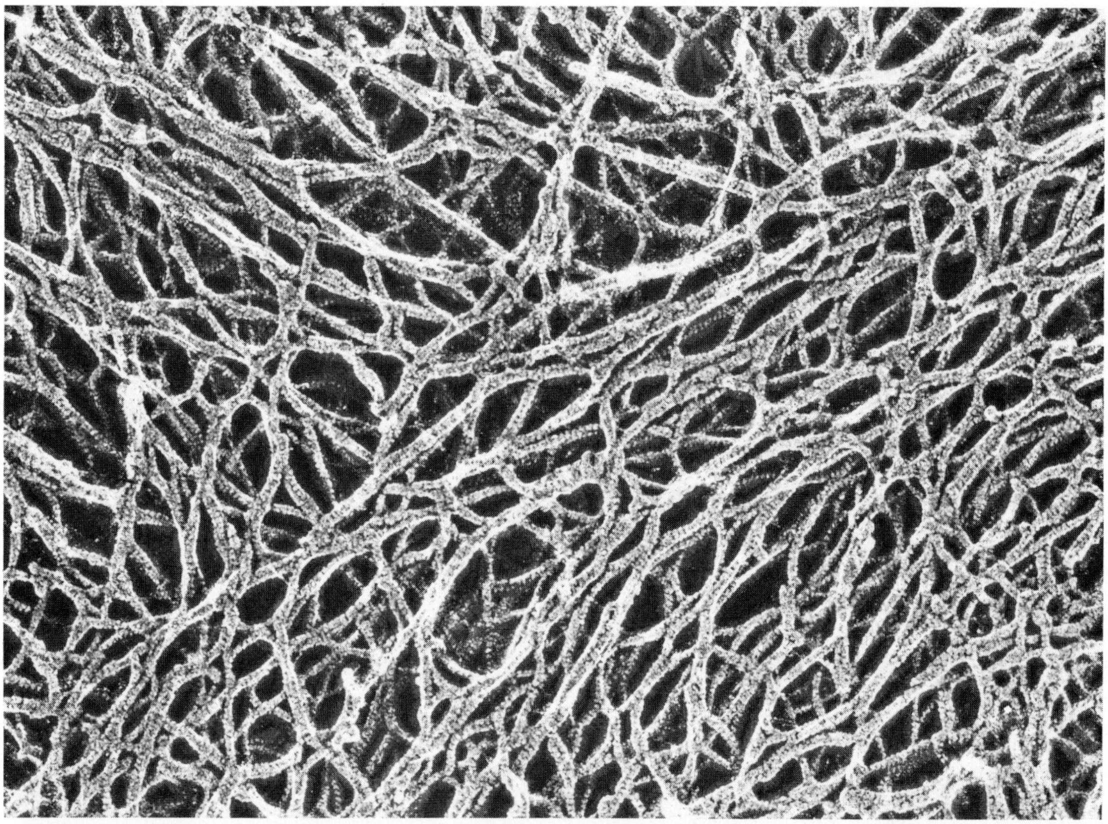

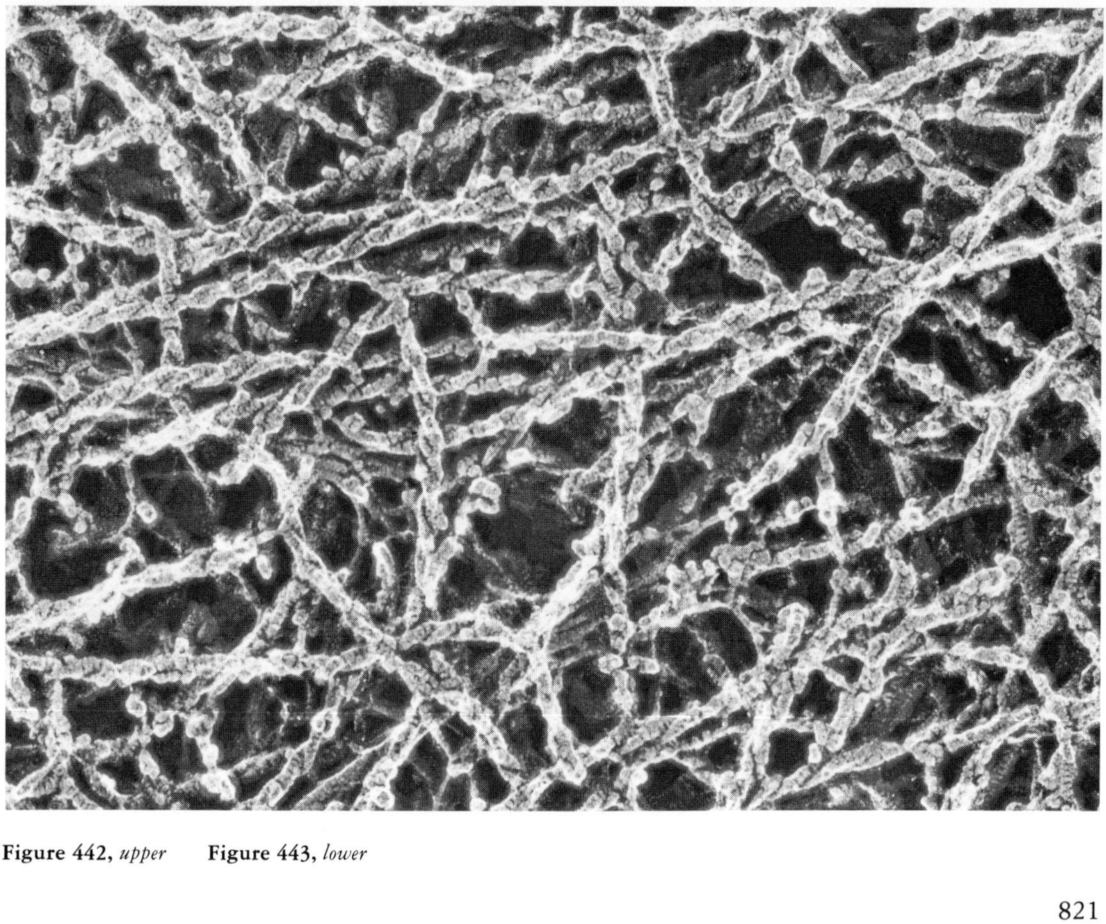

Figure 442, *upper* **Figure 443**, *lower*

It now seems likely that tropomyosin is also associated with actin filaments in nonmuscle cells. Immunocytochemical staining of various cell lines in culture using antitropomyosin antibody reveals a linear pattern of fluorescing fiber bundles indistinguishable from that seen with fluorescent antibody against actin.

The accompanying photomicrograph of a fibroblast exposed to antitropomyosin antibody closely resembles the comparable photomicrograph (Fig. 435) showing the distribution of actin in the same cell type. Therefore both antibodies are apparently binding to the same filamentous component of the cytoplasm.

Figure 444. Human skin fibroblast in culture stained by indirect immunofluorescence using tropomyosin antibody. (From Elias Lazarides, J. Cell Biol. 65:549–561, 1975.)

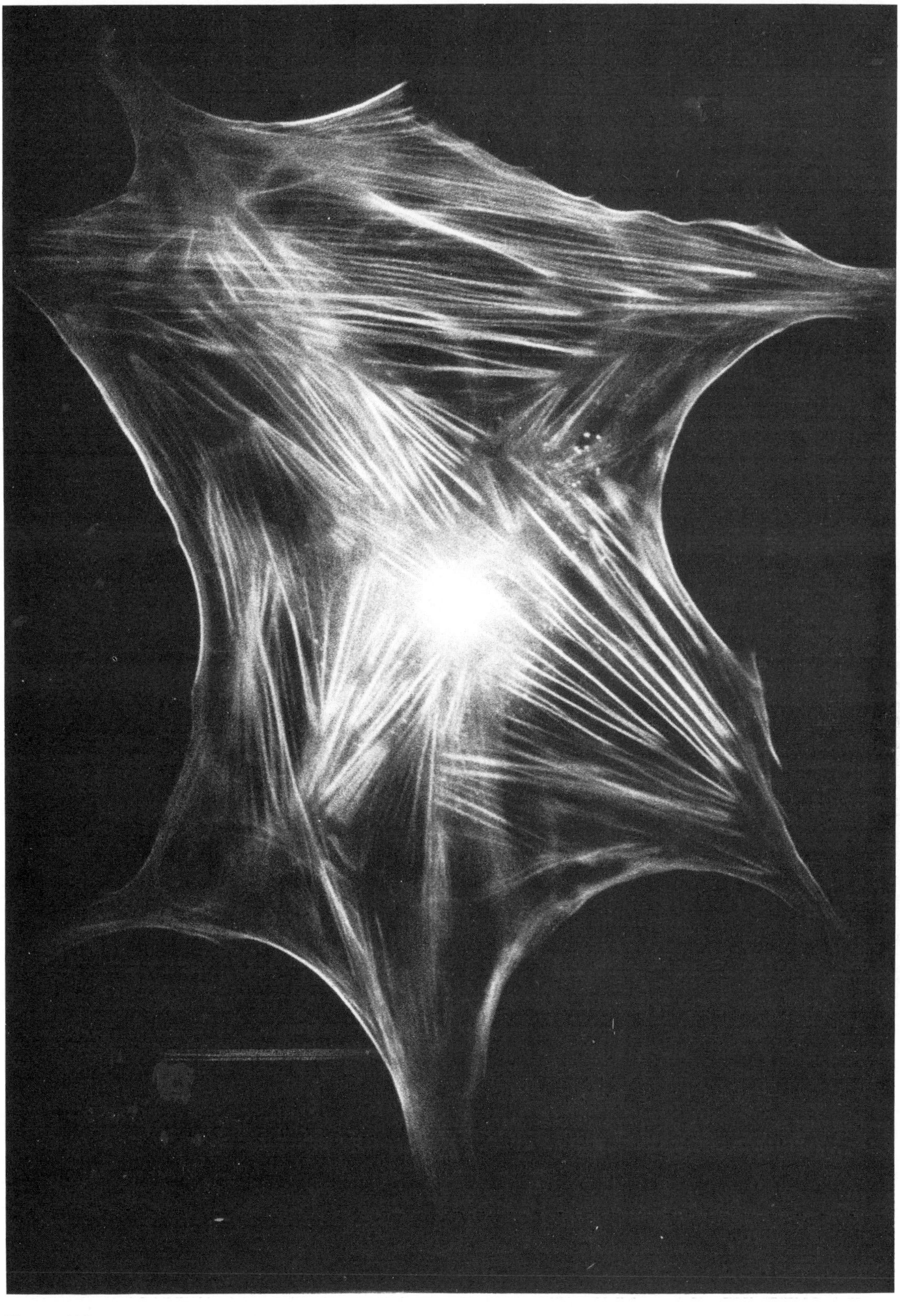

Figure 444

Association of Myosin with Actin Filaments

In electron micrographs of thin sections, myosin filaments usually cannot be identified in nonmuscle cells. However, their presence can be demonstrated with fluorescent antimyosin antibodies.

In the two cells illustrated here, antibody against platelet myosin delineates fibers 0.5 to 1 μm in width and up to several hundred micrometers in length. These correspond in distribution to straight fiber bundles visible in the same cells with the phase contrast microscope and interpreted as stress fibers. The staining is discontinuous, appearing as a series of bright spots at more or less regular intervals along the length of the fibers.

Stress fibers staining in this manner with antimyosin are most prominent in cells that are widely spread on the substrate and judged not to be actively motile. In cells that are actively moving at the time of fixation, myosin staining is diffuse throughout the cytoplasm. In dividing cells, it is highly concentrated in the mitotic spindle at metaphase and in the vicinity of the contractile ring during cytokinesis (Fujiwara and Pollard, 1976). Thus, the distribution of myosin varies with the activity of the cell, sometimes being clearly associated with the actin in stress fibers, and at other times diffuse or associated with configurations of actin that are not visible with the light microscope.

Figure 445. HeLa cells in culture stained with rhodamine-conjugated immune globulin. (Photomicrograph courtesy of Keigi Fujiwara.)

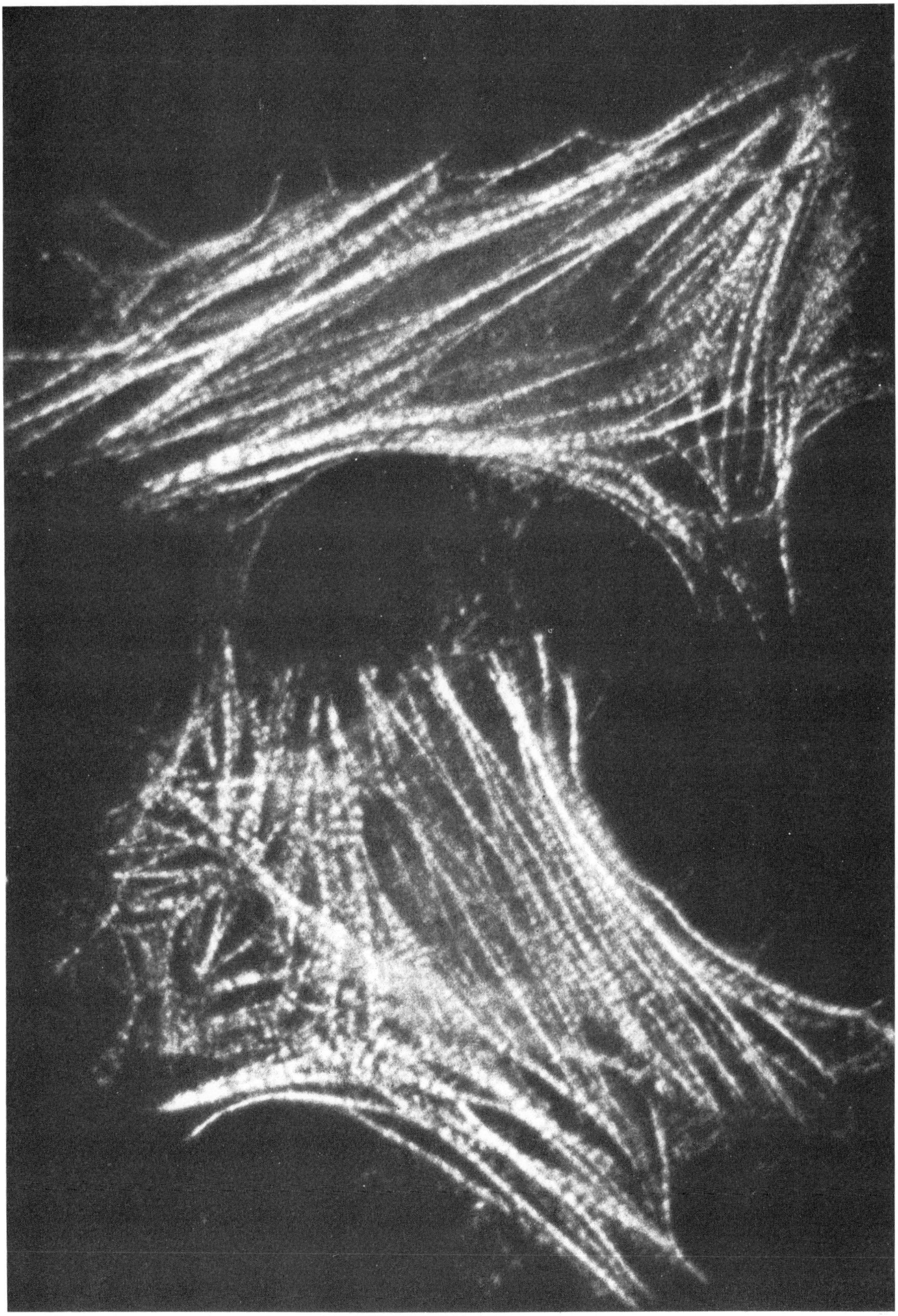

Figure 445

Relationship of Actin Filaments to the Cell Membrane

For the interaction of actin and myosin to bring about motility or change in cell shape, one or the other of these components must be attached either directly or indirectly to the surface membrane. There are several examples in electron micrographs where microfilaments appear to end at the plasmalemma. In cardiac muscle, the actin filaments of the myofibrils terminate at the intercalated discs in a dense layer beneath the sarcolemma. In smooth muscle, bundles of actin filaments end in similar densities in the peripheral cytoplasm. The microfilaments of the terminal web in various epithelia end in the dense component of the zonula adherens. In all of these examples, the exact relationship of the filaments to the membrane is obscured by a layer of dense amorphous material in the subsurface cytoplasm. However, there is suggestive evidence from freeze-fracture studies that the filaments reach the membrane and may attach to some of its integral protein particles (McNutt, 1978). Even in cell types where filaments are not seen ending at the surface in electron micrographs, actin is often found to be associated with the centrifugally isolated cell membrane fraction (Pollard and Korn, 1973).

Actin filaments associated with the plasma membrane were first clearly identified in the brush border of intestinal epithelial cells (Ishikawa et al., 1969) and isolated brush borders have since been studied in considerable detail. A bundle of longitudinal filaments that can be decorated with heavy meromyosin is found in the core of each microvillus. At the tip, the filaments end in a layer of dense material associated with the inner surface of the membrane. At high magnifications, slender cross bridges can be seen at 33 nm intervals extending laterally from the filament core to the membrane. At the base, the actin filaments extend downward into the apical cytoplasm, where some fan out and mingle with the transversely oriented filaments of the terminal web.

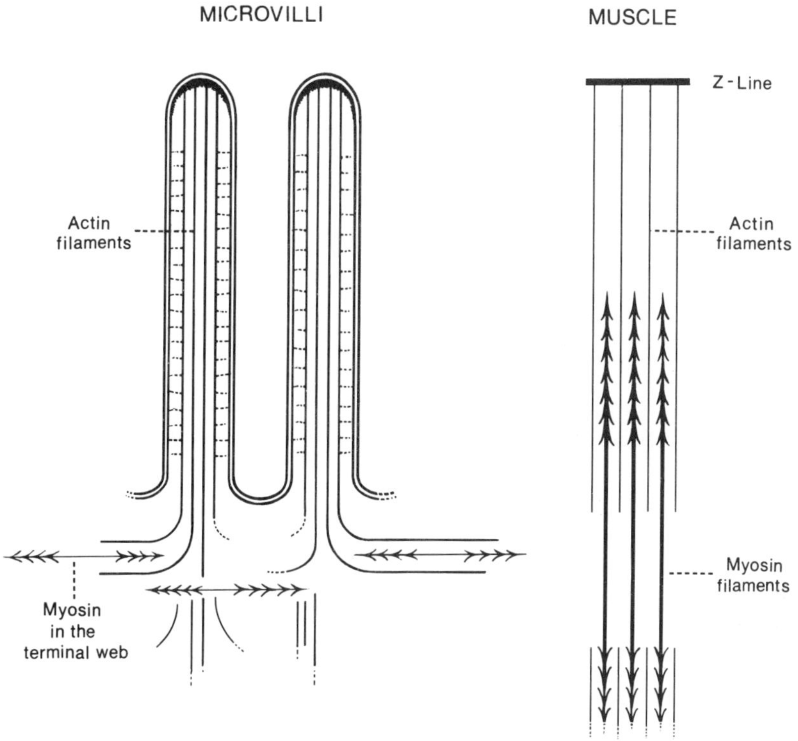

A model proposed for the functional organization of actin and myosin in the intestinal brush border (left) compared with that of striated muscle (right). Shortening is attributed to sliding of actin filaments with respect to myosin in the terminal web. The lateral projections from filaments shown here represent cross bridges projecting from intact myosin molecules and not the arrowheads formed on actin filaments exposed in vitro to the S-1 fragment of myosin. (Modified after Mooseker and Tilney, J. Cell Biol. 67:725–743, 1975.)

When decorated with the S-1 fragment of myosin (heavy meromyosin), the arrowhead pattern is in the same direction on all actin filaments and indicates a polarity away from the membrane at the villus tip. These findings are consistent with the hypothesis that movement of microvilli results from a sliding mechanism in which actin filaments anchored to the villus tip and to the lateral membrane interact with myosin and accessory proteins in the terminal web to produce shortening in a manner comparable to that responsible for contraction of skeletal muscle (Mooseker and Tilney, 1975).

It was originally speculated that the dense matrix at the villus tip and the cross bridges attaching the filament core to the membrane might consist of α-actinin, a protein found at the Z-line of skeletal muscle where it is believed to crosslink the actin filaments (Mooseker and Tilney, 1975). However, immunofluorescence microscopy with antibody against α-actinin has failed to localize this protein in the microvilli. Instead, another polypeptide of 95,000 dalton molecular weight, called *villin*, has been purified from isolated microvillus cores (Bretschner and Weber, 1979). Indirect evidence suggests that it is this protein, peculiar to the microvilli of the brush border, which links the actin cores to the inner surface of the membrane.

The highly ordered actin filament cores of the microvilli appear to lack associated proteins other than villin, but myosin, tropomyosin, filamin, and α-actinin can all be localized in the terminal web region. Thus the terminal web not only provides structural support for the microvillus cores but its filamentous components are also believed to be capable of interacting with them to produce shortening of the microvilli.

The accompanying micrographs of longitudinal and transverse sections of intestinal brush border show the actin filaments that form the core of each microvillus. At the villus tip, the relation of the filaments to the membrane is obscured by a layer of dense amorphous material associated with the inner aspect of the membrane.

Figures 446 and 447. Microvilli of the brush border of cat intestinal epithelium. (Micrographs courtesy of Susumu Ito.)

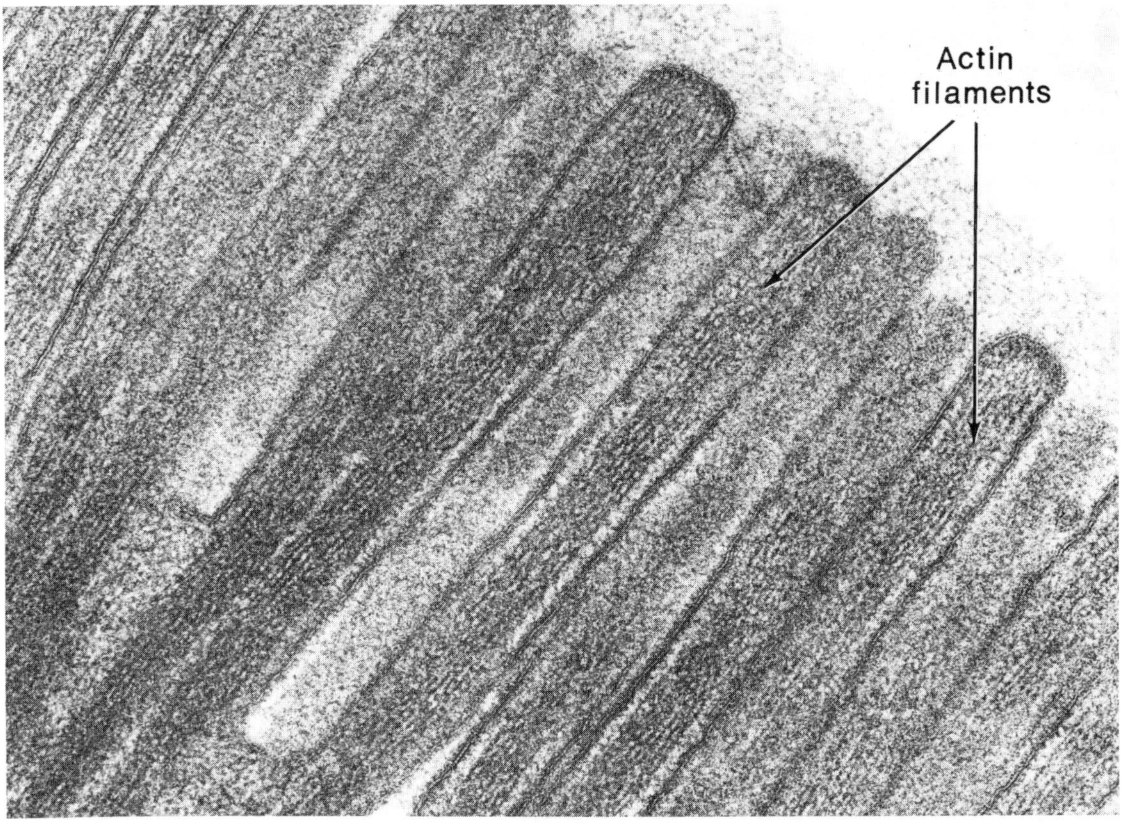

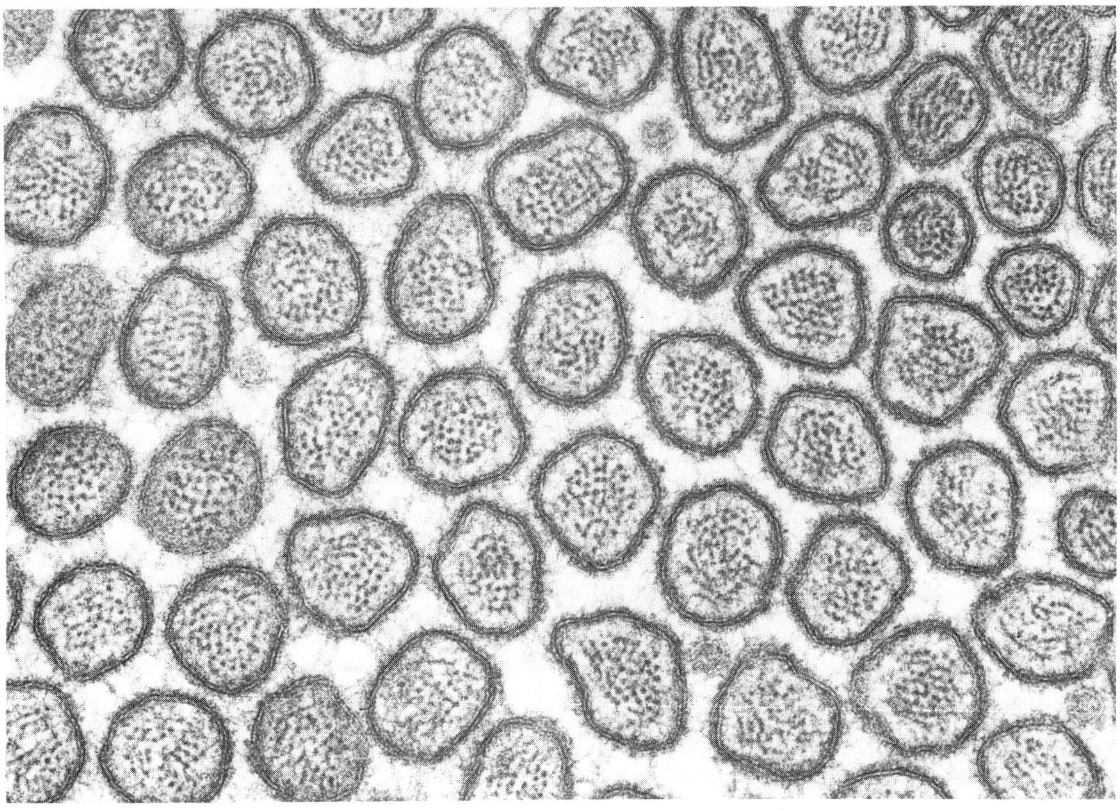

Figure 446, *upper* **Figure 447,** *lower*

The termination of the filaments in a density at the villus tip is well shown in a heavily stained thin section (upper left). At higher magnification (upper right) slender cross bridges can be seen at 33 nm intervals along the length of the filament bundle, attaching it to the limiting membrane of the villus. The actin filaments extend into the apical cytoplasm, where a few of the transverse microfilaments of the terminal web are visible.

The polarity of the actin filaments is clearly demonstrated in the lower figures showing brush borders exposed to heavy meromyosin and fixed in the presence of tannic acid. The arrowhead pattern of filament decoration is constant in direction on all filaments and indicates a polarity away from the membrane.

Figure 448 *A* and *B*. Electron micrograph of isolated brush border from chicken intestine. (From Mooseker and Tilney, J. Cell Biol. 67:725, 1975.)

Figure 449 *A* and *B*. Isolated brush border from chicken, decorated with heavy meromyosin and fixed in the presence of tannic acid. (From Begg, Rodewald and Rehbun, J. Cell Biol. 79:846, 1978.)

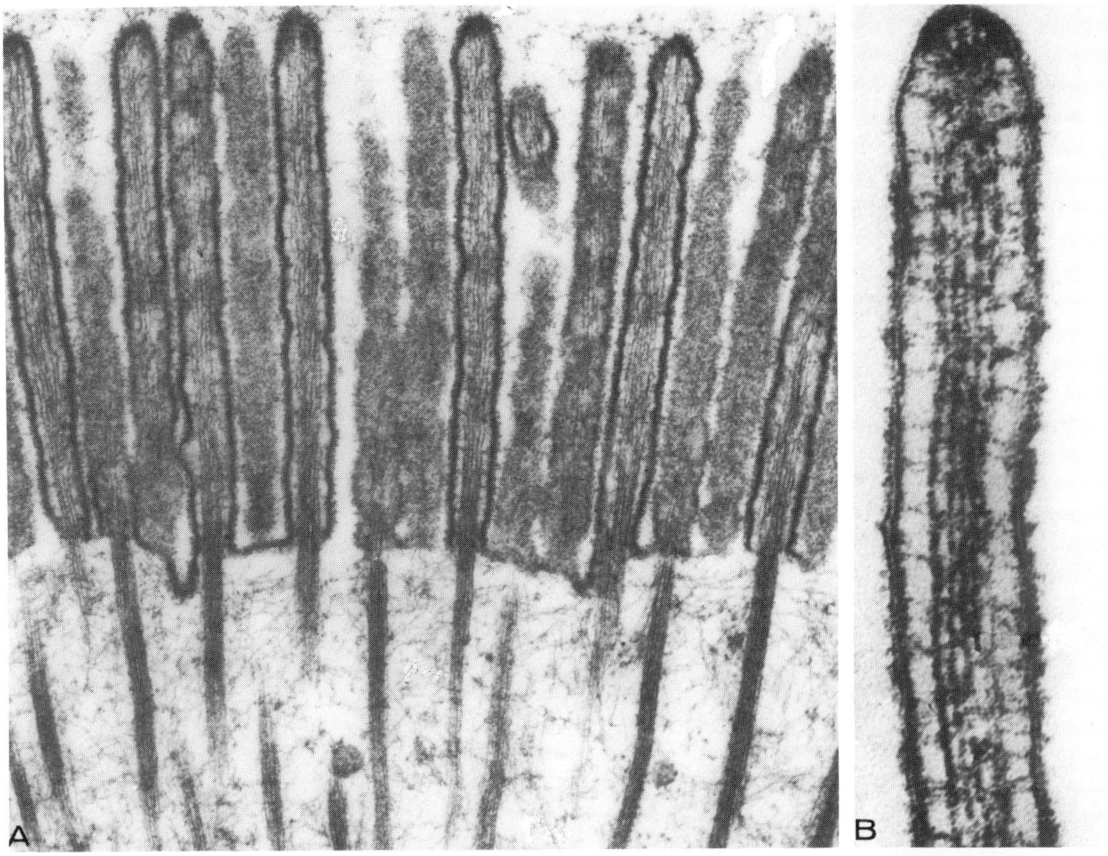

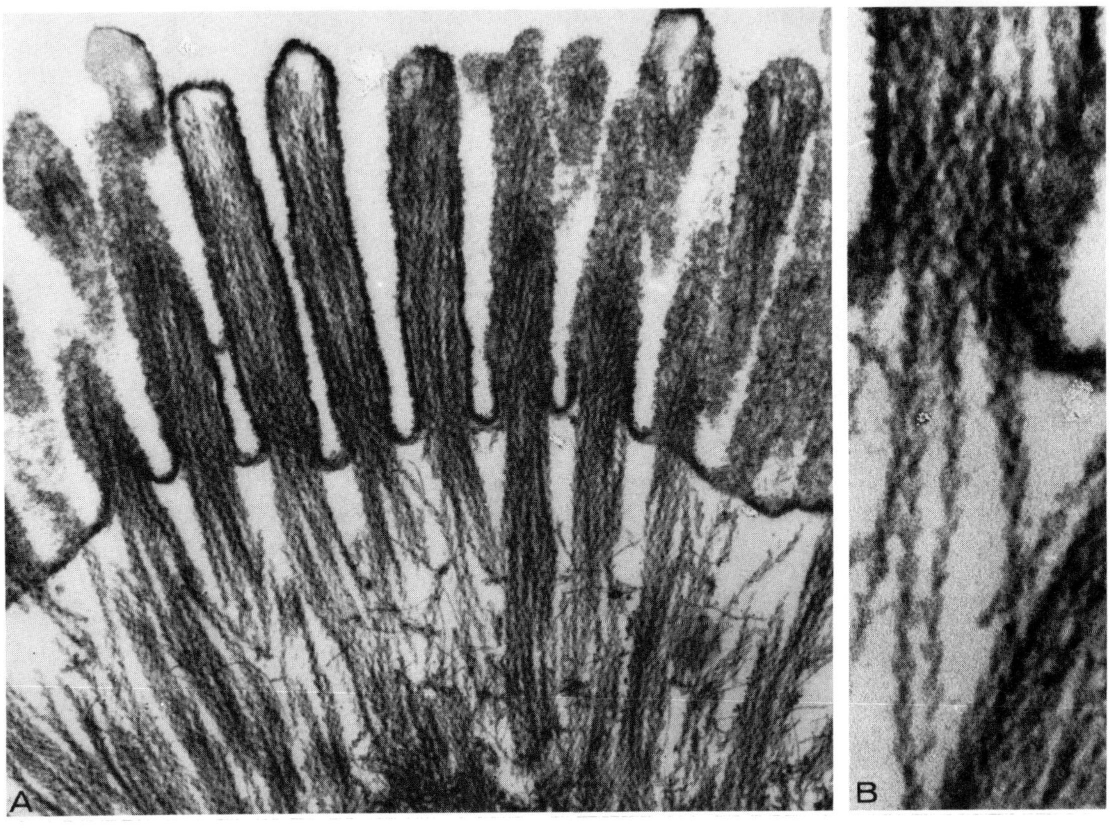

Figure 448, *upper* **Figure 449**, *lower*

Ten Nanometer Cytoplasmic Filaments

In addition to the 6 nm actin microfilaments and the 25 nm microtubules, most cell types also contain long unbranched cytoplasmic filaments about 10 nm in diameter. The distribution and composition of these intermediate-sized filaments have received much investigative attention. Unfortunately they have been given different names in several of the tissues in which they have been studied. Elucidation of their chemical nature has progressed slowly and major discrepancies persist concerning their constituent polypeptides.

In neurones, where they are called *neurofilaments,* they appear tubular in cross section with a 2.5 to 3.5 nm light core (Wuerker, 1970). This ultrastructural feature is less apparent in the so-called *glial filaments* of astrocytes. A subunit of glial filaments comigrates on gels with one of the major neurofilament proteins, has a similar amino acid composition, and exhibits some immunological cross reactivity (Dahl and Bignami, 1976; Yen et al., 1976; Schlaepfer, 1977). Nevertheless significant differences are believed to exist between the two classes of intermediate filaments in the central nervous system.

The 10 nm filaments of smooth muscle form an integrated cytoskeletal network held together at nodal points by the amorphous material of the dense bodies that are characteristic of these cells. The major component of the filaments is a 50,000 dalton protein named *skeletin* (Small and Sobierzek, 1977). A subunit of similar molecular weight from intermediate filaments of avian skeletal and cardiac muscle has been called *desmin* (Lazarides and Balzer, 1978). Filaments that comprise the tonofibrils of stratified squamous epithelia and many other epithelial cell types consist of *keratin,* a class of α-helical polypeptides of molecular weights ranging from 46,000 to 68,000 daltons (Culbertson and Freedberg, 1977; Sun, Shik and Green, 1979). The major protein of the 10 nm filaments in several fibroblast cell lines has a molecular weight of about 58,000 daltons and has been named *vimentin* from the Latin for wicker or lattice (Franke et al., 1978; Hynes and Destree, 1978).

There are many inconsistencies and unsolved problems in this rapidly advancing area of cell biology and the terminology is confusing. The number of chemically distinct types of intermediate-sized filaments remains unclear. The majority of those studied have subunits with molecular weights in the range 50,000 to 58,000 daltons. It appears likely that there are at least three and possibly four immunologically distinct types: one comprising the tonofibrils of epithelia (keratin), one typical of mesenchymal cells (vimentin), and another found in muscle (desmin, skeletin). Those in nervous tissue may constitute a fourth category. Although some cross reactivity has been reported, the bulk of the evidence indicates that neurofilament proteins are not identical with those of other intermediate-sized filaments. It seems reasonable to assume that all cytoplasmic filaments in this size range have some degree of similarity in their primary structure and subserve a common cytoskeletal function.

The 10 nm filaments follow a straight or gently curving course through the cytoplasm. They may occur individually or associated in loosely organized bundles. Shown here is their typical appearance in a cultured 3T3 cell. It is not possible to gain from electron micrographs an impression of the overall distribution of these filaments within the cell. For this purpose use of fluorescent antibodies and light microscopy is more useful. Several examples of this approach are given in the figures that follow.

Figure 450. Cytoplasm of 3T3 cell showing typical 10 nm filaments coursing through the meshes of the endoplasmic reticulum. (Micrograph courtesy of Scott McNutt, J. Cell Biol. 50:691–708, 1971.)

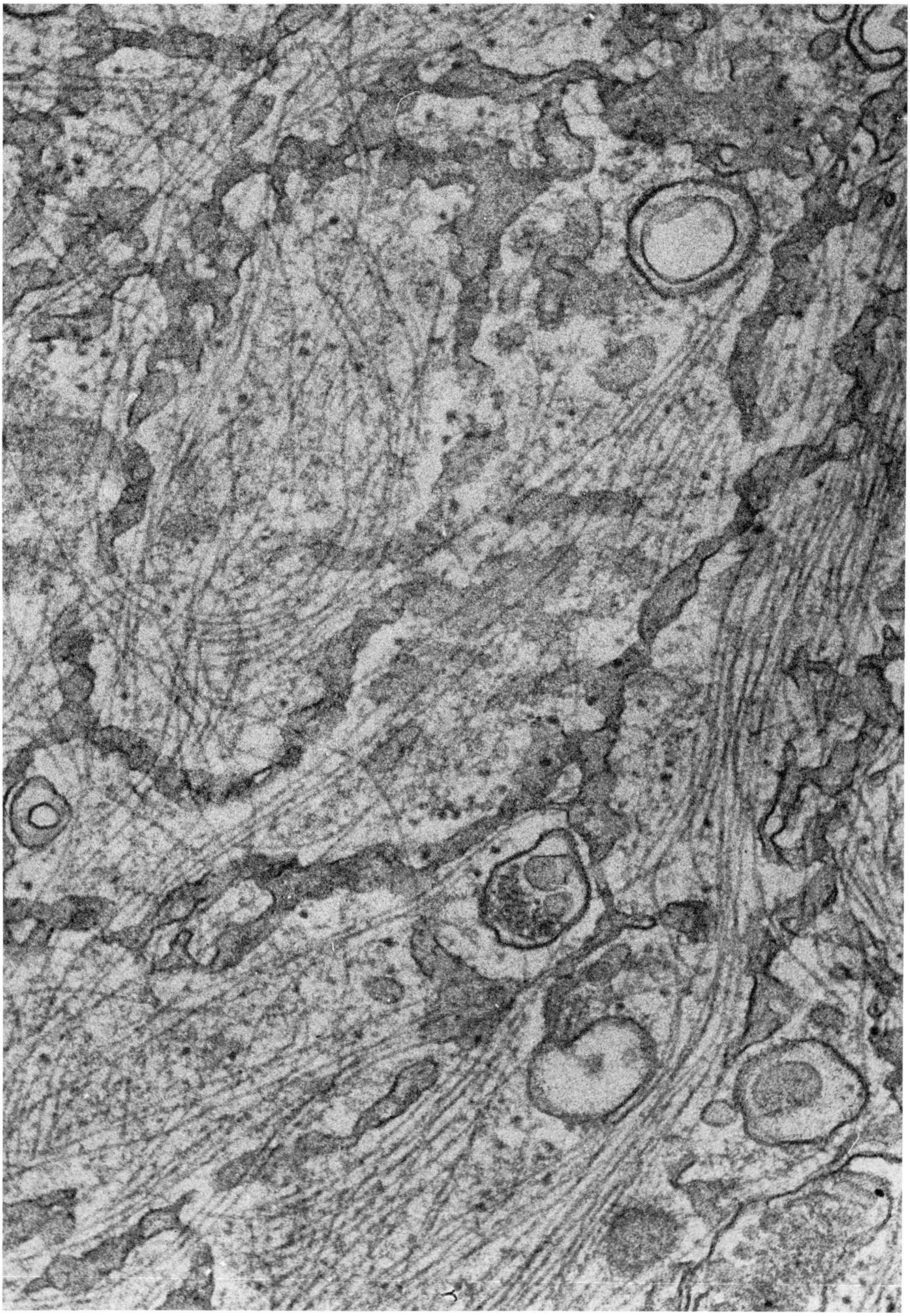

Figure 450

Antibodies raised in guinea pigs against a 57,000 dalton protein extracted from the cytoskeleton of 3T3 fibroblasts selectively bind to the 10 nm filaments of several mesenchymal cell lines. As previously mentioned, this cytoskeletal protein has been named *vimentin* (Franke et al., 1978). The upper micrograph of the facing page illustrates indirect immunofluorescence with antivimentin in cultured cells of mesenchymal origin derived from the wall of a vein. These filament bundles are not stained by antibody raised against keratin.

The lower figure is a cell from kangaroo rat renal epithelium (Strain Pt K_2) exposed to antibody against keratin. The conspicuous cytoskeleton in these epithelial cells is not stained by antibody against vimentin. The cells of muscle and brain show no significant fluorescence with either of these antisera. It is concluded that although the functionally related 10 nm filaments in various cell types have much the same ultrastructural appearance and similar solubility properties, they are immunologically distinct and therefore probably contain different polypeptides.

Figure 451. Immunofluorescence photomicrograph of cultured rat myofibroblasts from vein wall reacted with antibody against vimentin.

Figure 452. Immunofluorescence photomicrograph of a Pt K_2 cell reacted with antibody against prekeratin. (Micrographs from Franke et al., Proc. Nat. Acad. Sci. 75:5034–5038, 1978.)

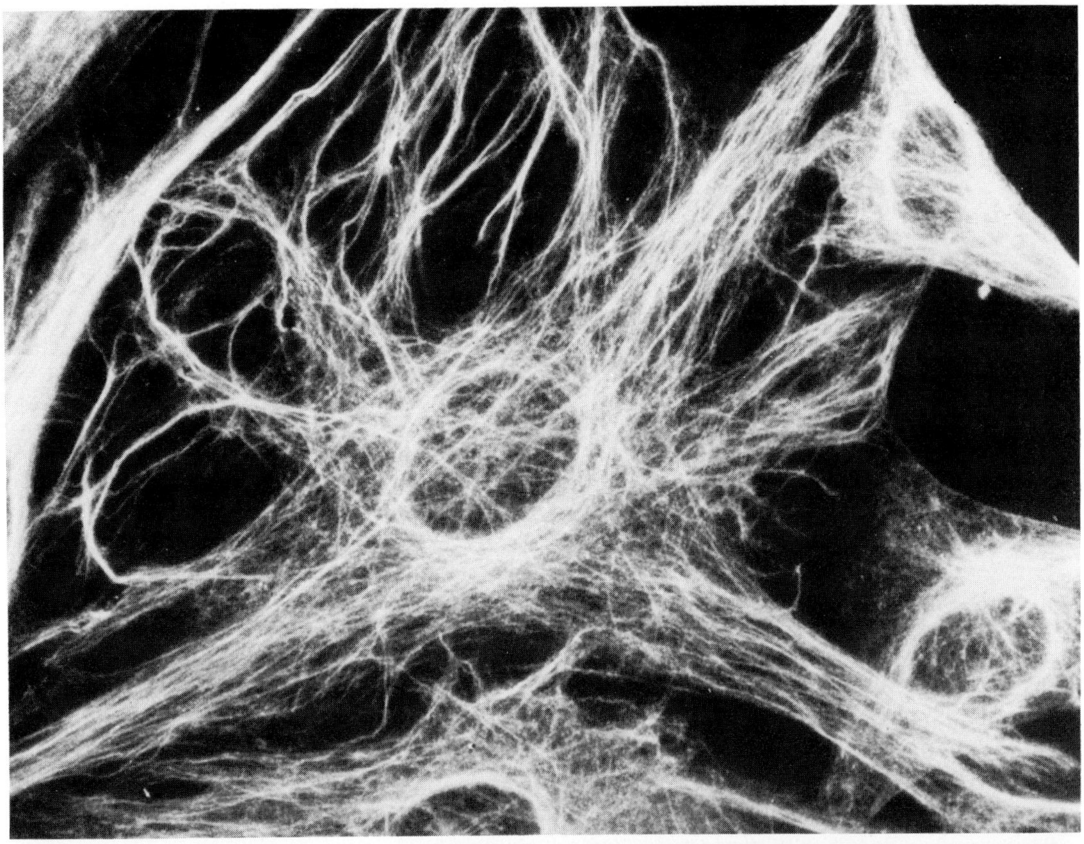

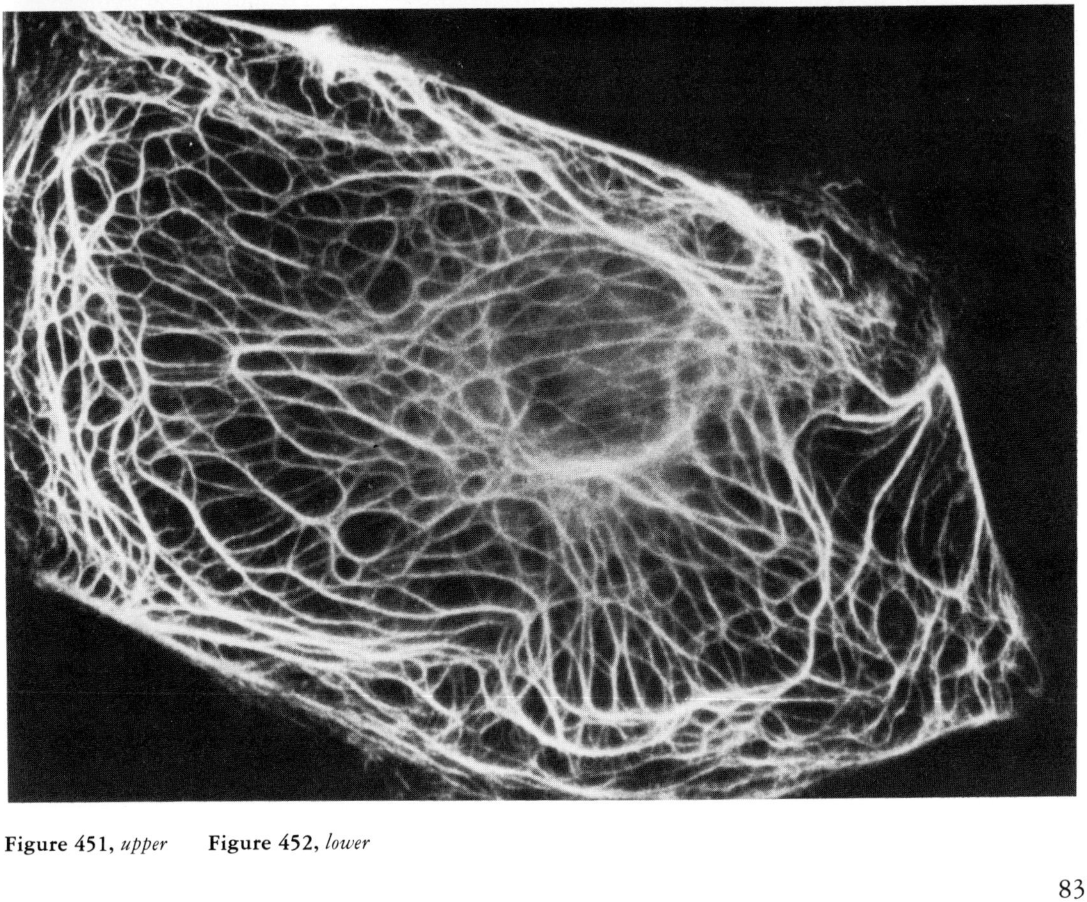

Figure 451, *upper* **Figure 452,** *lower*

Electron microscopy of thin sections provides little information about the three-dimensional arrangement of cytoskeletal elements. By immunofluorescence, it is possible to compare the distribution of microtubules and 10 nm filaments in cells spread upon a tissue culture substrate. In the accompanying photomicrographs, cultures of the same strain of hamster cells were stained with antitubulin (upper figure) and with antibody against a 58,000 dalton protein extracted from 10 nm filaments (lower figure).

The microtubules radiate outward from the cell center, extending to the very edge of the flattened cell periphery. The 10 nm filaments exhibit a rather similar fibrillar pattern, but they are mainly in the thicker central portion of the cell and seldom extend to its thin periphery. The patterns may vary somewhat in different cell strains.

More distinctive than the pattern of distribution of these two cytoskeletal elements is their different response to treatment with colchicine. Dissolution of microtubules after exposure to the drug eliminates most of the fibrillar fluorescence with antitubulin. The 10 nm filaments persist and aggregate into conspicuous thick coils around or adjacent to the nucleus.

Figure 453. Fluorescence photomicrograph of hamster N1L8 cells exposed to antitubulin.

Figure 454. Photomicrograph of hamster N1L8 cells stained with antiserum to 58 K protein of 10 nm filaments. (Figures from Hynes and Destree, Cell 13:151–163, 1978.)

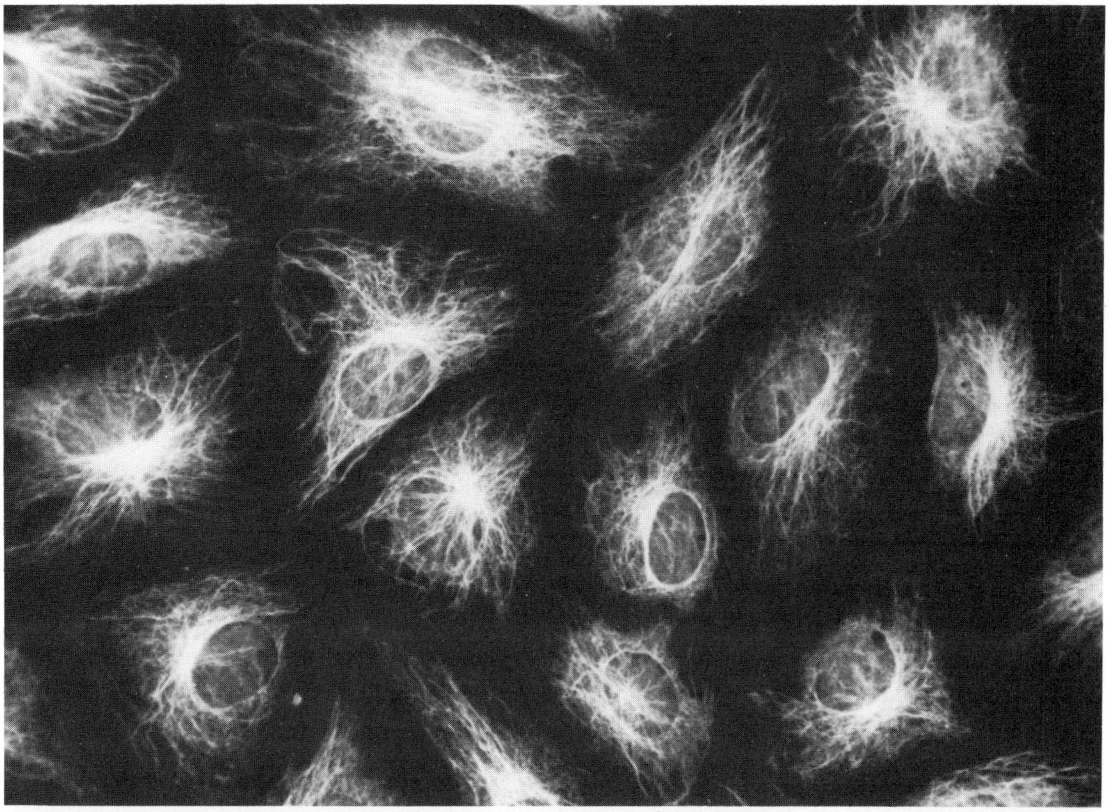

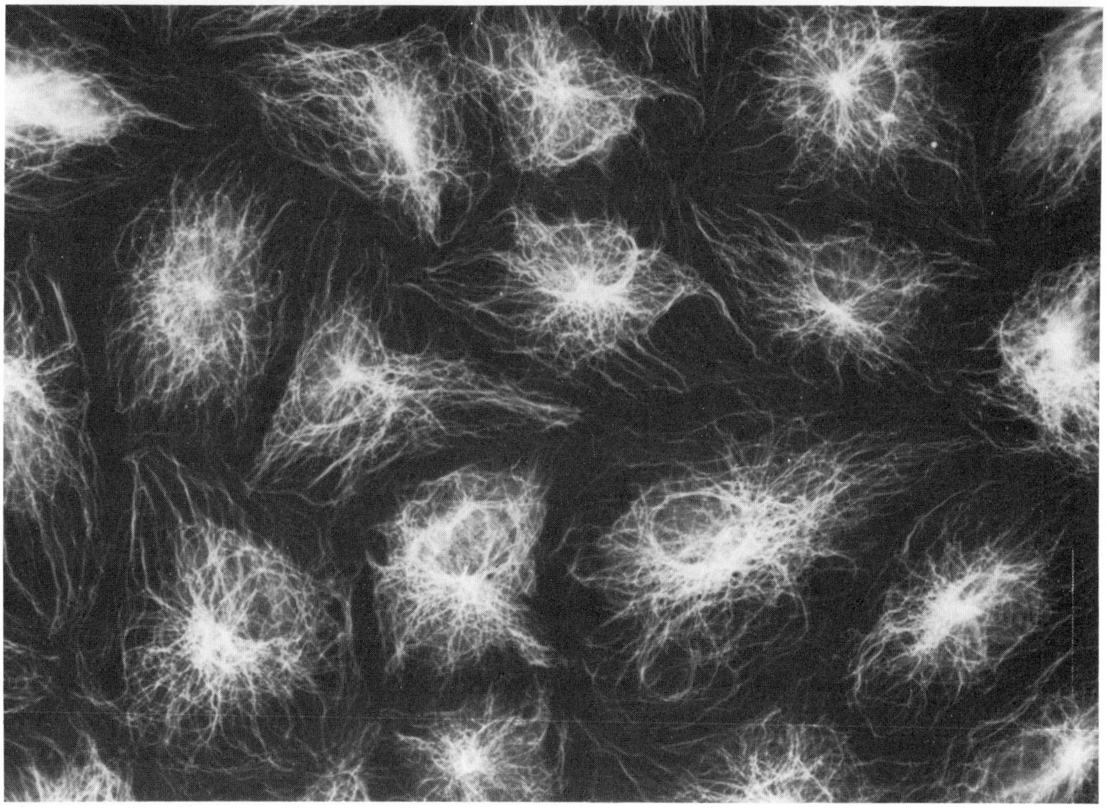

Figure 453, *upper* **Figure 454**, *lower*

Keratin Filaments

Cells of the epidermal epithelium in all species are rich in cytoplasmic filaments. The accompanying micrograph illustrates their extraordinary abundance in a basal cell of lamprey skin. Keratin filaments are about 9 nm in diameter and appear to consist of several different polypeptide subunits with molecular weights in the range of 40,000 to 63,000 daltons (Matoltsy, 1976; Brysk, Gray and Bernstein, 1977). Keratin filaments can be assembled *in vitro* from this group of polypeptides (Steinert and Guillino, 1976; Steinert et al., 1977). Fluorescent antibodies against a purified keratin fraction from human skin stain intensely the epidermal cells of several other species, indicating that they all possess homologous amino acid sequences. The cells contain a fluorescent network of fibers corresponding to aggregations of tonofilaments seen in electron micrographs of the same cells. Application of antikeratin to a variety of different cell strains of mesenchymal origin yields negative results, indicating that the 10 nm filaments of fibroblasts and other mesenchymal derivatives are not keratin (Sun and Green, 1978).

Figure 455. Basal cell in the epidermis of a larval lamprey, *Petromyzon fluviatilis.*

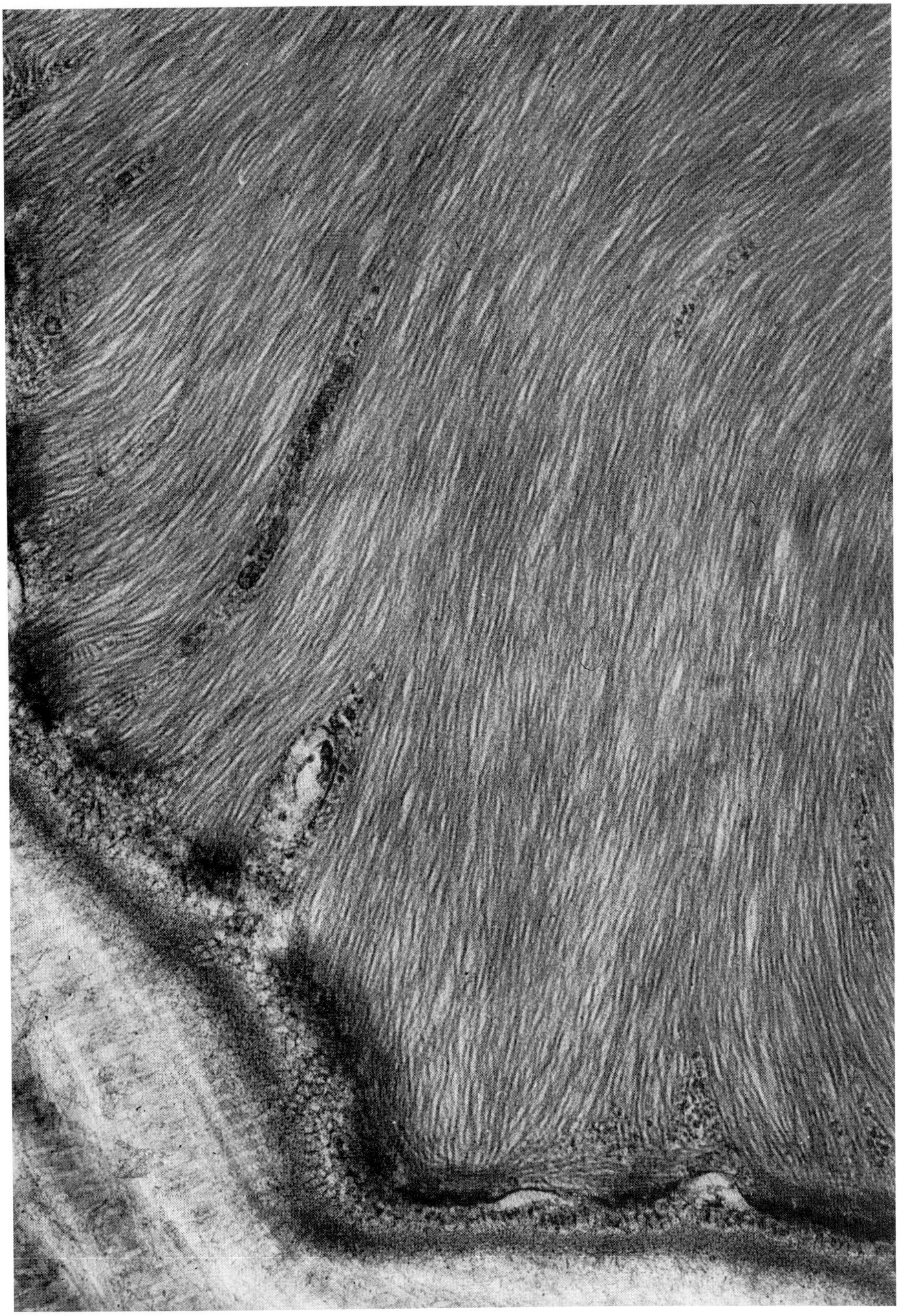

Figure 455

At higher magnification, the keratin filaments appear as punctate or elongate profiles depending upon their orientation in the section. They are very uniform in size, approximately 9 nm thick, and may have a beaded appearance along their length as though composed of repeating subunits. Among the filaments microtubules (at arrows) are occasionally seen.

Figures 456 and 457. Basal cell from the epidermis of the ammocoete of *Petromyzon fluviatilis.*

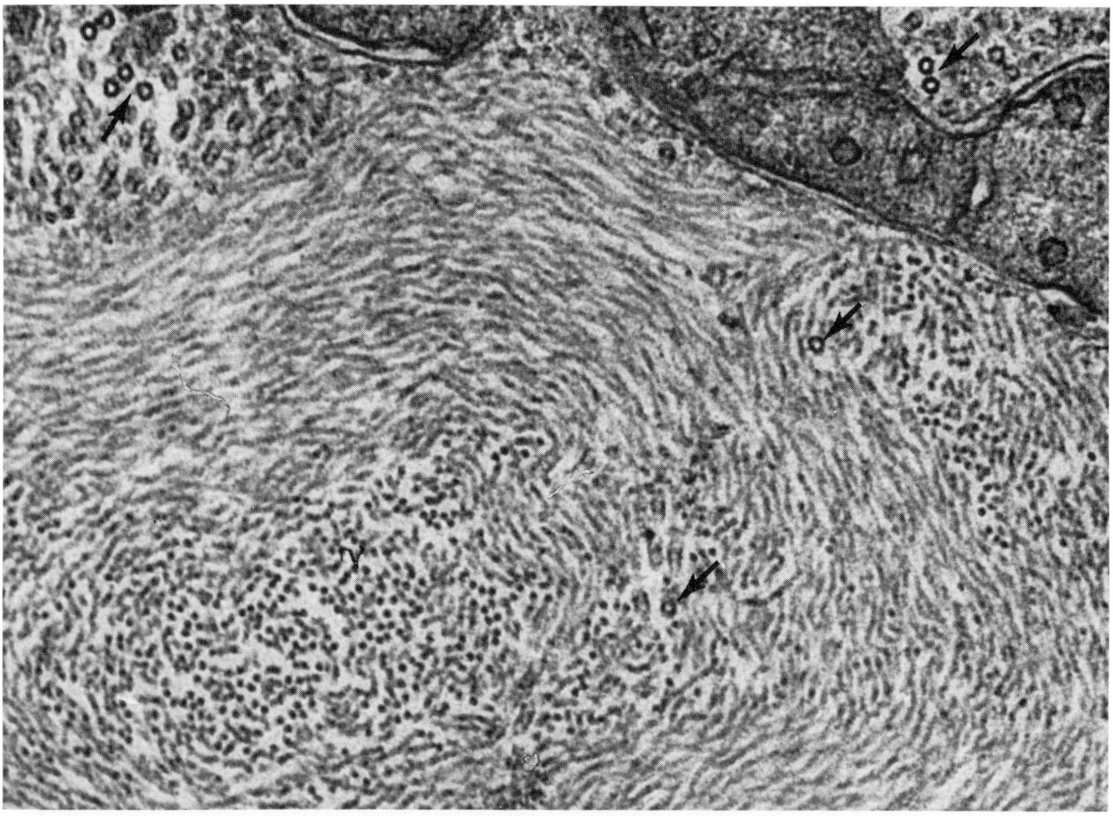

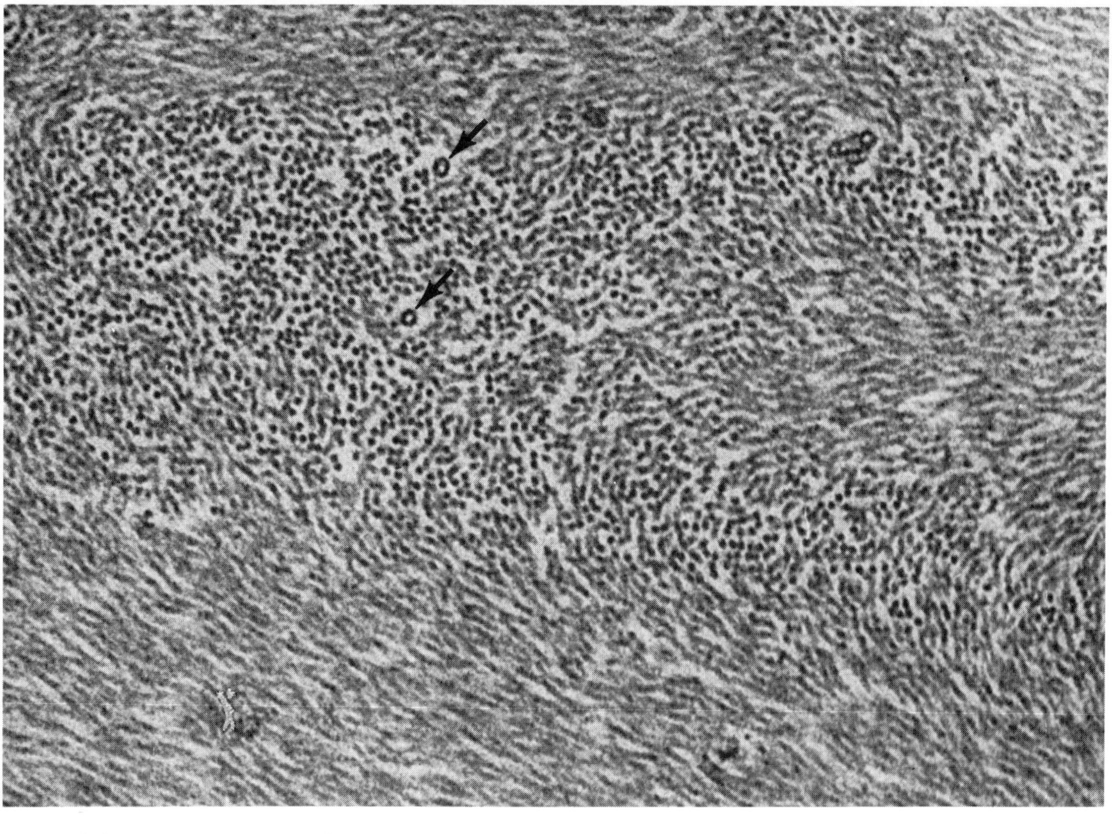

Figure 456, *upper* **Figure 457,** *lower*

Under certain circumstances, 10 nm filaments may accumulate in high concentration. The supporting cells of the seminiferous epithelium have a normal complement of filaments and microtubules throughout the spermatogenically active portion of the seminiferous tubules, but near their junction with the tubuli recti and rete testis, where there are very few germ cells, the Sertoli cytoplasm in some species is crowded with filaments. Numerous microtubules are also found among the 10 nm filaments (at arrows). The physiological significance of this striking local increase in the concentration of cytoskeletal elements is not understood.

Figure 458. A portion of the nucleus and adjacent cytoplasm in a monkey Sertoli cell from the terminal segment of the seminiferous tubule. (Micrograph courtesy of Martin Dym.)

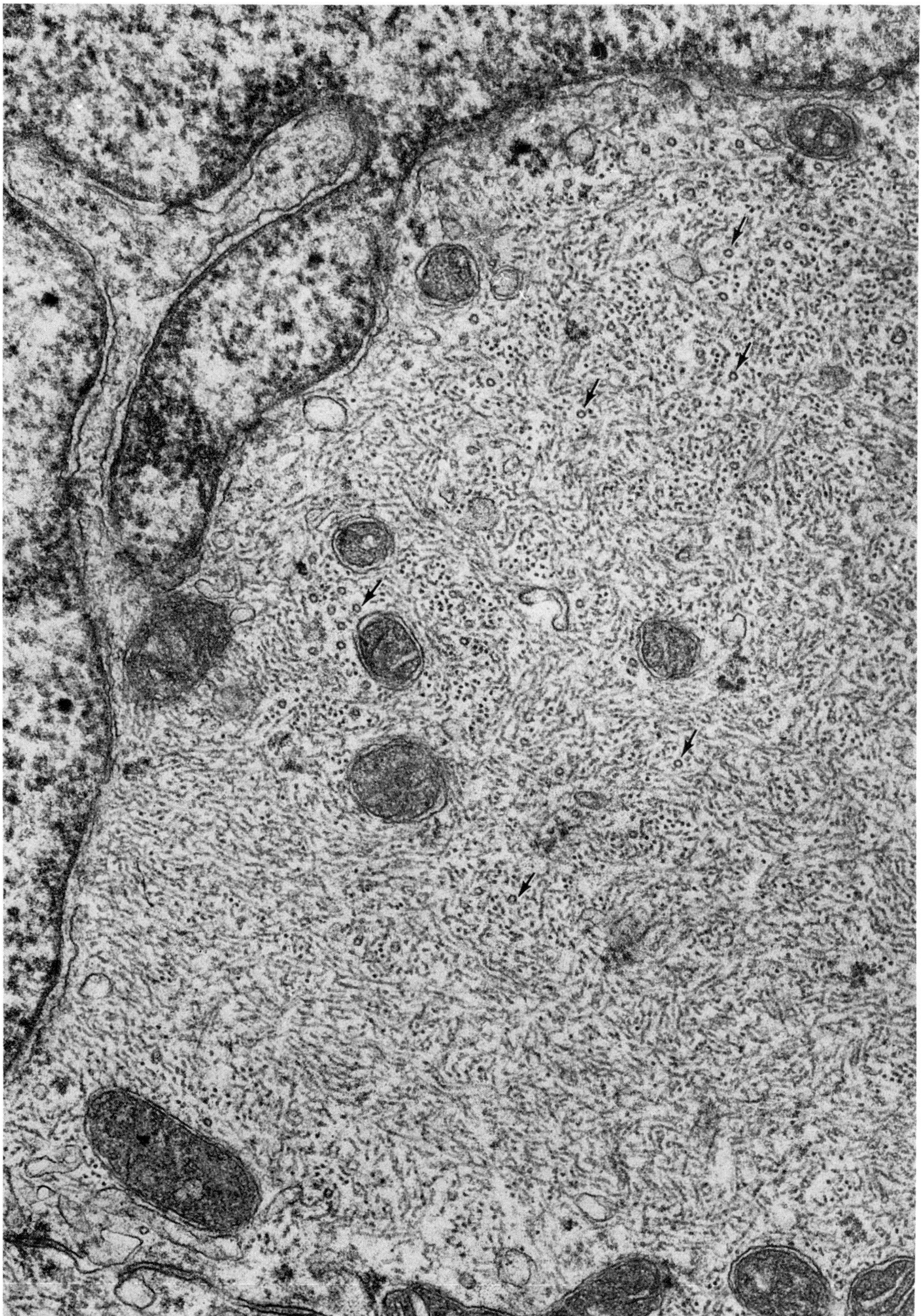

Figure 458

Neurofilaments are about 10 nm in diameter and of indefinite length. Unlike other cytoplasmic filaments of comparable dimensions, they appear in transverse section to have a clear center surrounded by a 3 nm thick dense wall (Peters and Vaughn, 1967; Weurker and Palay, 1969). They have therefore been interpreted as tubular. A single protein subunit called *filarin* has been reported by various authors to have a molecular weight between 50,000 and 60,000 daltons (Shelanski, 1973; Davison and Winslow, 1974). Neurotubules of invertebrates are reported to differ from those of vertebrates in appearance and in the molecular weight of their subunits.

The accompanying micrograph shows axoplasm in a cell from the ciliary ganglion containing abundant neurofilaments oriented, for the most part, parallel to the long axis of the axon.

Figure 459. Axon from the ciliary ganglion of a chicken. (Micrograph courtesy of Enrico Mugnaini.)

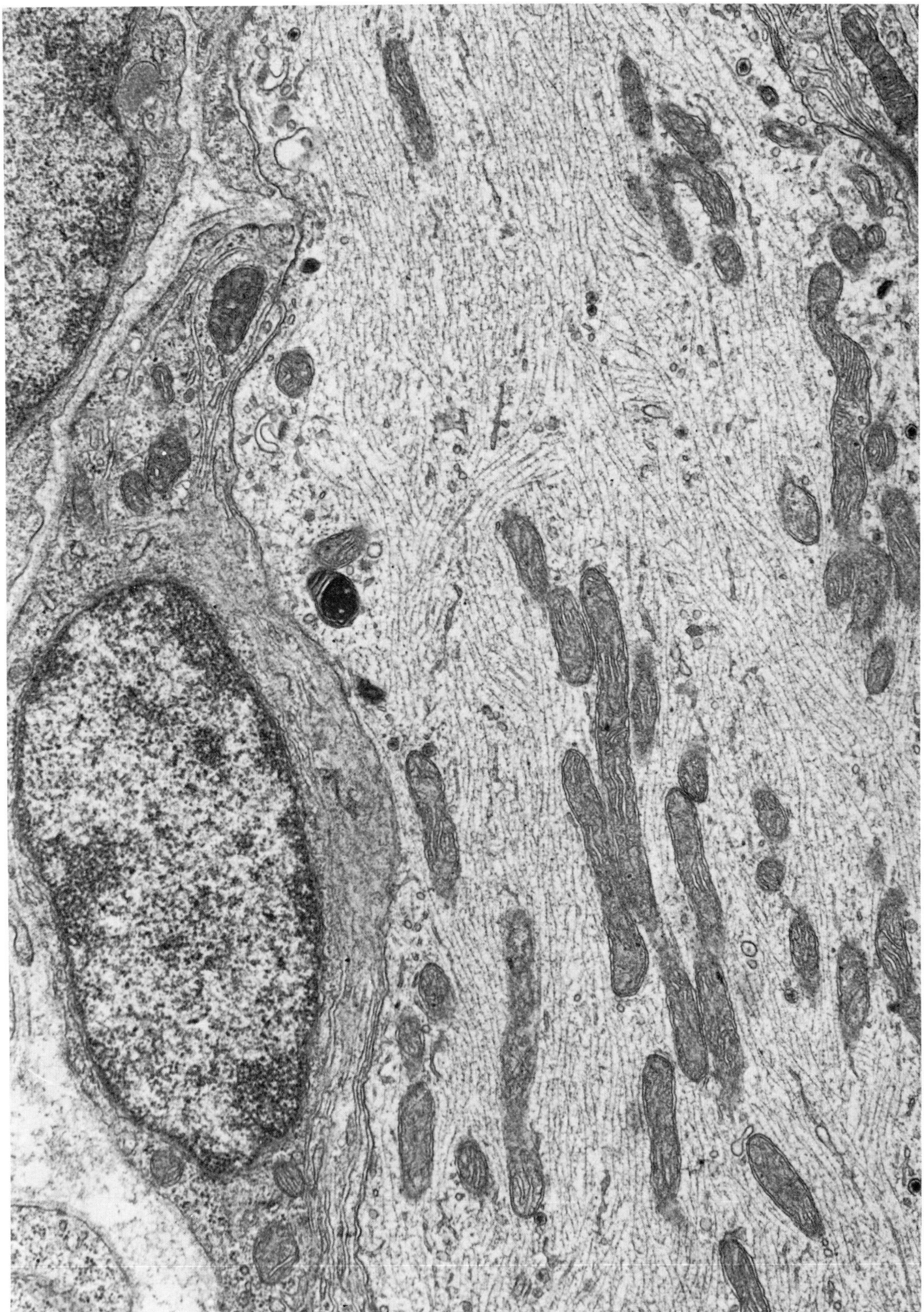

Figure 459

The processes of neuroglial cells in the nervous system of both invertebrates and vertebrates are crowded with filaments that closely resemble the tonofilaments of epithelial cells. They are generally oriented parallel to the long axis of the cell process, and some of them, like tonofilaments of epithelia, terminate in desmosomes.

Figure 460. Axial nerve cord of the annelid *Aphrodite aculeata.*

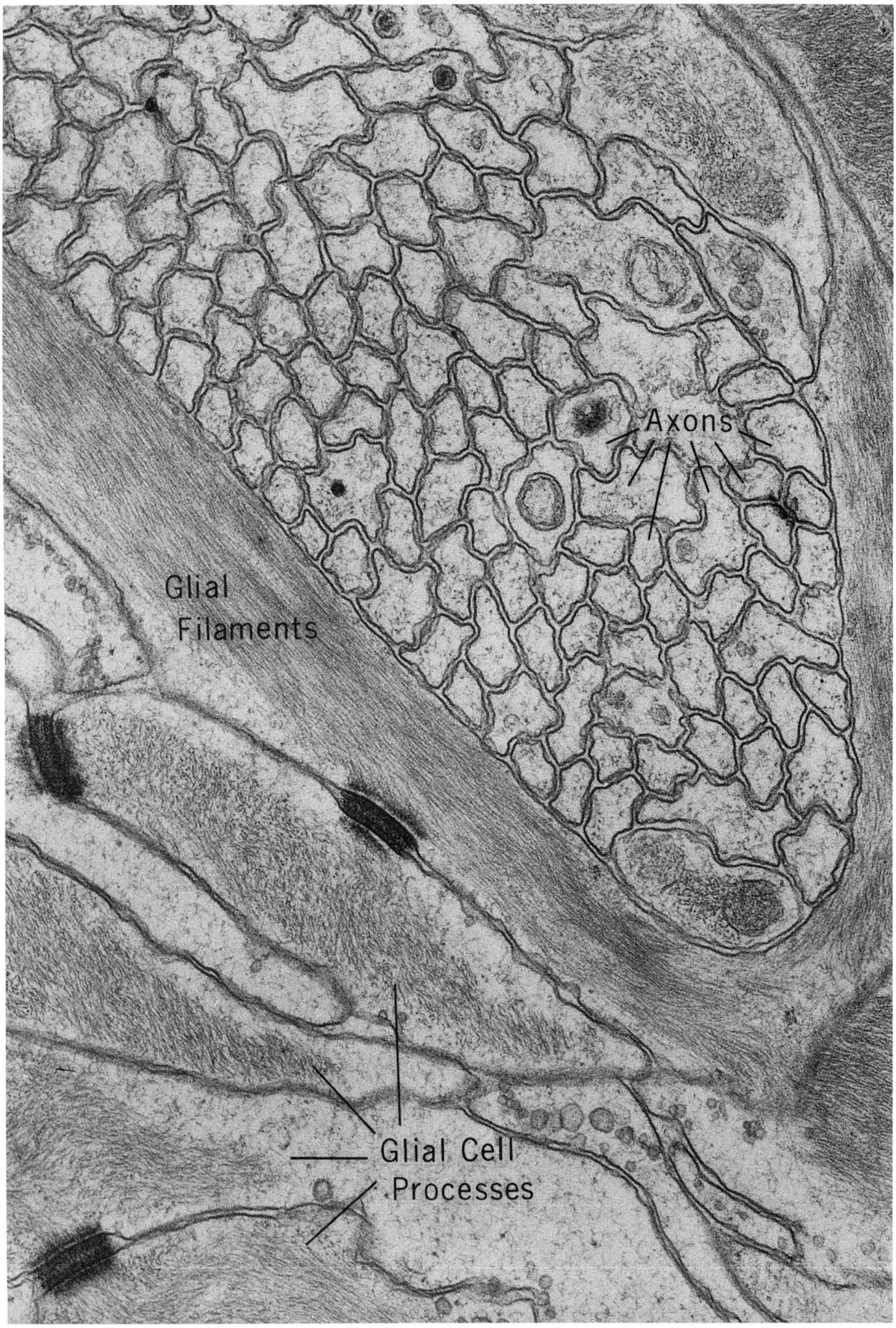

Figure 460

A prominent feature of one class of astrocytes is the presence of large numbers of 9 to 10 nm cytoplasmic filaments occurring throughout the perikaryon and extending into the cell processes. They are said to be slightly smaller than neurofilaments and instead of being distributed individually, they tend to be aggregated in conspicuous bundles. In the classical neurocytological methods for demonstration of astroglial filaments, gold was deposited on filament bundles and not upon the individual filaments.

A large bundle of closely packed filaments coursing along a mitochondrion in a glial cell process is shown in the accompanying micrograph.

Figure 461. Filaments in an astrocyte from the dorsal cochlear nucleus of rat brain. (Micrograph courtesy of Enrico Mugnaini.)

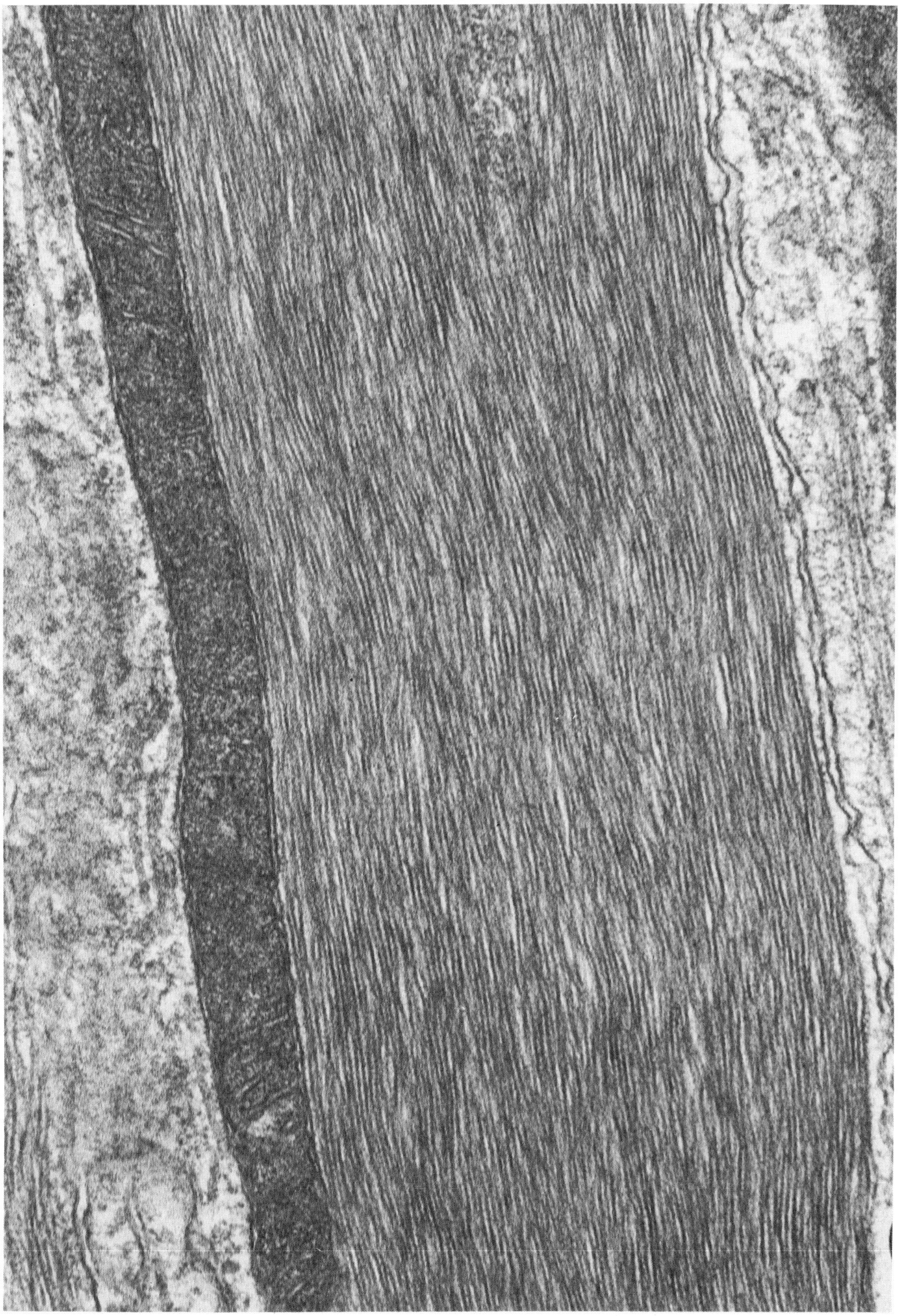

Figure 461

REFERENCES

Microtubules

Barclay, N. E. Marginal bands in duck and camel erythrocytes. Anat. Rec. 154:313, 1966.

Behnke, O. Incomplete microtubules observed in mammalian blood platelets during polymerization. J. Cell Biol. 34:697–701, 1967.

Behnke, O. Microtubules in disc-shaped blood cells. Int. Rev. Exp. Pathol. 9:1–92, 1970. (Review)

Bikle, D., L. G. Tilney and K. R. Porter. Microtubules and pigment migration in the melanophores of *Fundulus heteroclitus*. Protoplasma 61:322–345, 1966.

Binder, L. I. and J. L. Rosenbaum. The in vitro assembly of flagellar outer doublet tubulin. J. Cell Biol. 79:500–515, 1978.

Borisy, G. and J. Olmsted. Nucleated assembly of microtubules in porcine brain extracts. Science 177:1196–1197, 1972.

Borisy, G. G., J. M. Marcus, J. B. Olmsted, D. b. Murphy and K. A. Johnson. Purification of tubulin and associated high molecular weight proteins from porcine brain and characterization of microtubule assembly in vitro. Ann. N.Y. Acad. Sci. 253:107–119, 1975.

Burnside, B. The form and arrangement of microtubules: An historical, primarily morphological review. *In* Ann. N.Y. Acad. Sci. 253:14–26, 1975. (Review)

Byers, B. and K. R. Porter. Oriented microtubules in elongating cells of the developing lens rudiment after induction. Proc. Nat. Acad. Sci. 52:1091–1099, 1964.

Cande, W. Z., E. Lazarides and M. J. McIntosh. A comparison of the distribution of actin and tubulin in the mammalian mitotic spindle as seen by indirect immunofluorescence. J. Cell Biol. 72:552–567, 1977.

Connolly, J. A., V. I. Kalnins, D. W. Cleveland and M. W. Kirschner. Intracellular localization of high molecular weight microtubule accessory protein by indirect immunofluorescence. J. Cell Biol. 76:R781–R786, 1978.

Dehler, A. Beiträge zur Kenntnis des feinerem Baues der roten Blutkörperchen beim Hühnerembryo. Arch Mikr. Anat. 46:414–430, 1895.

Fawcett, D. W. and F. Witebsky. Observations on the ultrastructure of nucleated erythrocytes and thrombocytes, with particular reference to the structural basis of their discoidal shape. Z. Zellforsch. 62:785–806, 1964.

Freed, J. J. and M. M. Lebowitz. The association of a class of saltatory movements with microtubules in cultured cells. J. Cell Biol. 45:334–354, 1970.

Harris, P. Some structural and functional aspects of the mitotic apparatus in sea urchin embryos. J. Cell Biol. 14:475, 1962.

Heggeness, M. H., M. Simon and S. J. Singer. Association of mitochondria with microtubules in cultured cells. Proc. Nat. Acad. Sci. 75:3863–3866, 1978.

Inoué, S. The effect of colchicine on the microscopic and submicroscopic structure of the mitotic spindle. Exp. Cell Res. Suppl. 2:305–318, 1952.

Inoué, S. and H. Sato. Cell motility by labile association of molecules. The nature of mitotic spindle fibers and their role in chromosome movements. J. Gen. Physiol. Suppl. 50:259–288, 1967.

Junquiera, L. C., E. Raker and K. R. Porter. Studies on pigment migration in the melanophores of the teleost *Fundulus heteroclitus*. Arch. Histol. Jpn. 36:339–366, 1974.

Kim, H., L. I. Binder and J. L. Rosenbaum. The periodic association of MAP_2 with brain microtubules in vitro. J. Cell Biol. 80:266–276, 1979.

Ledbetter, M. and K. R. Porter. A microtubule in plant cell fine structure. J. Cell Biol. 19:239–250, 1963.

Lockwood, A. H. Tubulin assembly protein: Immunochemical and immunofluorescent studies on its function and distribution in microtubules and cultured cells. Cell 13:613–627, 1978.

McIntosh, J. R. et al. Model for mitosis. Nature 224:659–663, 1969.

Meves, Fr. Die Hünefeld-Hensenchen Bilder der roten Blutkörperchen der Amphibien. Anat. Anz. 24:465–476, 1904.

Meves, Fr. Gesammelte Studien an den roten Blutkörperchen der Amphibien. Arch. Mikr. Anat. 77:465–540, 1911.

Murphy, D. B. and G. G. Borisy. Association of high molecular weight proteins with microtubules and their role in microtubule assembly in vitro. Proc. Nat. Acad. Sci. 72:2696–2700, 1975.

Murphy, D. B. and L. G. Tilney. The role of microtubules in the movement of pigment granules in teleost melanophores. J. Cell Biol. 61:757–779, 1974.

Murphy, D. B., R. B. Vallee and G. G. Borisy. Identity and polymerization-stimulatory activity of the nontubulin protein associated with microtubules. Biochemistry. 161:2598–2605, 1977.

Olmsted, J. B. and G. G. Borisy. Microtubules. Ann. Rev. Biochem. 42:507–540, 1973. (Review)

Pease, D. C. The ultrastructure of flagellar fibrils. J. Cell Biol. 18:313, 1963.

Porter, K. R. Motility in cells. *In* Cell Motility, Book A (R. Goldman, T. Pollard and J. Rosenbaum, eds.), pp. 1–28, Cold Spring Harbor Conferences on Cell Proliferation, Vol. 3, 1976.

Ranvier, L. Récherches sur les élements du sang. Arch. Phys. D.S. 2:1–15, 1876.

Roth, L. E. and E. W. Daniels. Electron microscopic studies of mitosis in Amebae. II. The giant ameba *Pelomyxa carolinensis*. J. Cell Biol. 12:57, 1962.

Schliwa, M. and J. Bereiter-Hahn. Pigment movements in fish melanophores: Morphological and physiological studies. II. Cell shape and microtubules. Z. Zellforsch. 147:107–125, 1973.

Schliwa, M., M. Osborn and K. Weber. Microtubule system of isolated fish melanophores as revealed by immunofluorescence microscopy. J. Cell. Biol. 76:229–236, 1978.

Shelanski, M. L. and E. W. Taylor. Isolation of a protein subunit from microtubules. J. Cell Biol. 34:549–554, 1967.

Slautterback, D. L. Cytoplasmic microtubules. I. Hydra. J. Cell. Biol. 18:367–388, 1963.

Sloboda, R. D., W. C. Dentler and J. L. Rosenbaum. Microtubule-associated proteins and the stimulation of tubulin assembly in vitro. Biochemistry 15:4497–4505, 1976.

Smith, D. S., U. Järlfors and R. Beránek. The organization of synaptic axoplasm in the lamprey central nervous system. J. Cell Biol. 46:199–219, 1970.

Smith, D. S. On the significance of cross-bridges between microtubules and synaptic vesicles. Phil. Trans. Roy. Soc. B. 261:395–405, 1971.

Smith, D. S., U. Järlfors and M. L. Cayer. Structural cross-bridges between microtubules and mitochondria in central axons of an insect (*Periplaneta americana*). J. Cell Sci. 27:255–272, 1977.

Stephens, R. E. Microtubules. *In* Biological Macromolecules (S.N. Timasheff and G. D. Fasman, eds.), pp. 355–391, vol. 4, Dekker, New York, 1971. (Review)

Stephens, R. E. and K. T. Edds. Microtubules: Structure, chemistry and function. Physiol. Rev. 56:709–777, 1976.

Taylor, E. The mechanism of colchicine inhibition of mitosis. I. Kinetics of inhibition and the binding of H^3-colchicine. J. Cell Biol. 25:145–160, 1965.

Tilney, L. G. Origin and continuity of microtubules. *In* Origin and Continuity of Cell Organelles (J. Reinert and H. Ursprung, eds.), pp. 222–259, Springer Verlag, New York, 1971. (Review)

Tilney, L. G. and K. R. Porter. Studies on microtubules in Heliozoa. I. Protoplasma 60:317, 1965.

Tilney, L. G. and K. R. Porter. Studies on the microtubules in Heliozoa. II. The effect of low temperature on the formation and maintenance of axopodia. J. Cell. Biol. 34:327–343, 1967.

Weidenreich, Fr. Studien über das Blut und die blutbildenden und — zerstörenden Organe. III Uber den Bau der Amphibienerythrocyten. Arch. Mikr. Anat. 66:270–288, 1905.

Weingarten, M. D. and A. H. Lockwood, S-Y. Hwo and M. W. Kirschner. A protein factor essential for microtubule assembly. Proc. Nat. Acad. Sci. 72:1858–1863, 1975.

Weisenberg, R. Microtubule formation in vitro in solutions containing low calcium ions. Science 177:1104–1105, 1972.

Weisenberg, R. C., G. G. Borisy and E. W. Taylor. The colchicine binding protein of mammalian brain and its relation to microtubules. Biochemistry. 7:4466–4479, 1969.

Wilson, L. and J. Bryan. Biochemical and pharmacological properties of microtubules. Adv. Cell Molec. Biol. 3:22–71, 1973. (Review)

Cytoplasmic Filaments

Filaments of Muscle Cells

Astbury, W. T. and S. Dickinson. X-ray studies of the molecular structure of myosin. Proc. Roy. Soc. Lond. sB 129:307, 1940.

Bruche, E. Untersuchungen über den Bau der Muskelfasern mit Hülfe des polarisitren Lichtes. Denkschr. Kaiserl. Acad. Weis. Math.–Naturev. 15:69, 1858.

Cohen, C. The protein switch of muscle contraction. Sci. Am. Dec., pp. 36–45, 1975.

Hall, C. E., J. A. Jakus, and F. O. Schmitt. An investigation of cross striations and myosin filaments in muscle. Biol. Bull. 90:32, 1946.

Huxley, H. E. Muscle cells. *In* The Cell: Biochemistry, Physiology, Morphology (J. Brachet and A. E. Mirsky, eds.), p. 365, vol. 4, Academic Press, New York, 1960. (Review)

Huxley, H. and J. Hanson. Changes in the cross-striations of muscle during contraction and stretch and their structural interpretations. Nature 173:973–976, 1954.

Huxley, H. E. and J. Hanson. The molecular basis of contraction in cross-striated muscles. *In* Structure and Function of Muscle (G. H. Bourne, ed.) p. 183, vol. 1, Academic Press, New York, 1960.

Kelly, R. E. and R. V. Rice. Localization of myosin filaments in smooth muscle. J. Cell Biol. 37:105–116, 1968.

Lowy, J., F. R. Poulsen and P. J. Vibert. Myosin filaments in vertebrate smooth muscle. Nature 225:1053–1054, 1970.

McDougall, W. On the structure of cross-striated muscle and a suggestion as to the nature of its contraction. J. Anat. Physiol. 31:410, 539, 1897.

McDougall, W. A theory of muscle contraction. J. Anat. Physiol. 32:187, 1898.

Meigs, E. B. Contributions to the general physiology of smooth and striated muscle. J. Exp. Zool. 13:497, 1912.

Rollet, A. Untersuchungen über den Bau der quergestreiften Muskelfasern. Wien. Math.–Nat. Classe 49:81, 1885.

Small, J. V. Studies on isolated smooth muscle cells: The contractile apparatus. J. Cell Sci. 24:327–349, 1977.
Small, J. V. and J. M. Squire. Structural basis of contraction in vertebrate smooth muscle. J. Molec. Biol. 67:117–149, 1972.
Small, J. V. and A. Sobieszek. Studies on the function and composition of the 10 nm filaments of vertebrate smooth muscle. J. Cell Sci. 23:243–268, 1977.
Somlyo, A. P. Structural characteristics and mechanism of contraction, innervation and proliferation of smooth muscle cells. Ultrastructure and function of vascular smooth muscle. Adv. Exp. Med. Biol. 57:1–80, 1975.
Somlyo, A. P., C. F. Devine, A. V. Somlyo and R. V. Rice. Filament organization in vertebrate smooth muscle. Phil. Trans. Roy. Soc. B 265:223–229, 1973.
Squire, J. M. General model of myosin filament structure. III. Molecular packing arrangements in myosin filaments. J. Mol. Biol. 77:291–323, 1973.
Weber, A. and J. M. Murray. Molecular control mechanisms in muscle contraction. Phys. Rev. 53:612, 1973. (Review)

Filaments of Cells Other than Muscle

Begg, D. A., R. Rodewald and L. I. Rehbun. The visualization of actin filament polarity in thin sections. J. Cell Biol. 79:846–852, 1978.
Bignami, A. and D. Dahl. Astrocyte-specific protein and neuroglial differentiation. An immunofluorescence study with antibodies to the glial fibrillary acidic protein. J. Comp. Neurol. 153:27–38, 1974.
Bretschner, A. and K. Weber. Villin: The major microfilament associated protein of the intestinal microvillus. Proc. Nat. Acad. Sci. 76:2321–2325, 1979.
Cooke, P. H. A filamentous cytoskeleton in vertebrate smooth muscle fibers. J. Cell Biol. 68:539–556, 1976.
Dahl, D. and A. Bignami. Immunogenic properties of the glial fibrillary acid protein. Brain Res. 116:150–157, 1976.
Davison, P. F. and B. Winslow. The protein subunit of calf brain neurofilament. J. Neurobiol. 5:119–137, 1974.
Franke, W. W., E. Schmidt, M. Osborn and K. Weber. Different intermediate sized filaments distinguished by immunofluorescence microscopy. Proc. Nat. Acad. Sci. 75:5034–5038, 1978.
Franke, W. W., B. Appelhaus, E. Schmid and C. Freudenstein. The organization of cytokeratin filaments in the intestinal epithelium. European J. Cell Biol. 19:255–268, 1979.
Franke, W. W., C. Grund, E. Schmid and E. Mandelkow. Paracrystalline arrays of membrane-to-membrane cross bridges associated with the inner surface of plasma membrane. J. Cell Biol. 77:323, 1978.
Fujiwara, K. and T. Pollard. Fluorescent antibody localization of myosin in the cytoplasm, cleavage furrow, and mitotic spindle. J. Cell Biol. 71:848–875, 1976.
Hynes, R. O. and A. T. Destree. 10 nm filaments in normal and transformed cells. Cell 13:151–163, 1978.
Ishikawa, H., R. Bischoff and H. Holtzer. Formation of arrowhead complexes with heavy meromyosin in a variety of cell types. J. Cell Biol. 43:312–328, 1969.
Lazarides, E. Two general classes of cytoplasmic actin filaments in tissue culture cells: The role of tropomyosin. J. Supramol. Structure 5:531–563, 1976.
Lazarides, E. The distribution of desmin (100 Å) filaments in primary cultures of embryonic chick cardiac cells. Exp. Cell Res. 112:265–273, 1978.
Lazarides, E. and D. R. Balzer. Specificity of desmin to avian and mammalian muscle cells. Cell 14:429–438, 1978.
Lazarides, F., and K. Weber. Actin antibody. The specific visualization of actin filaments in non-muscle cells. Proc. Nat. Acad. Sci. 71:2268, 1974.
Matoltsy, A. G. Desmosomes, filaments, and keratohyaline granules: Their role in the stabilization and keratinization of the epidermis. J. Invest. Dermatol. 65:127–142, 1975.
Maupin-Szamier, P. and T. D. Pollard. Actin filament destruction by osmium tetroxide. J. Cell Biol. 77:837–852, 1978.
McNutt, N. S. A thin section and freeze-fracture study of microfilament-membrane attachments in choroid plexus and intestinal microvilli. J. Cell Biol. 79:774–787, 1978.
Mooseker, M. S. Brush border motility: Microvillar contraction in isolated brush-border models. J. Cell Biol. 63:231a, 1974.
Mooseker, M. S. Brush border motility. Microvillar contraction in Triton-treated brush borders isolated from intestinal epithelium. J. Cell Biol. 71:417–433, 1976.
Mooseker, M. S. and L. G. Tilney. Organization of an actin-filament membrane complex. Filament polarity and membrane attachment in the microvilli of intestinal epithelial cells. J. Cell Biol. 67:725–743, 1975.
Mukherjee, T. M. and L. A. Staehelin. The fine structural organization of the brush border of intestinal epithelial cells. J. Cell Sci. 8:573, 1971.
Pollard, T. D. Functional implications of the biochemical and structural properties of cytoplasmic contractile proteins. *In* Molecules and Cell Movement (S. Inoué and R. E. Stephens, eds.), pp. 259–286, Raven Press, New York, 1975.

Pollard, T. D. and E. D. Korn. Electron microscopic identification of actin associated with isolated amoeba plasma membranes. J. Biol. Chem. 248:448–450, 1973.

Pollard, T. D. and R. R. Weihing. Actin and myosin in cell movements. Crit. Rev. Biochem. 2:1, 1974.

Porter, K. R. and J. J. Wolosewick. Stereo high voltage electron microscopy of whole cells of the human diploid line Wl–38. Am. J. Anat. 147:303–323, 1976.

Schlaepfer, W. W. Immunological and ultrastructural studies of neurofilaments isolated from rat peripheral nerve. J. Cell Biol. 74:226–240, 1977.

Schlaepfer, W. W. and R. G. Lynch. Immunofluorescence studies of neurofilaments in the rat and human peripheral and central nervous system. J. Cell Biol. 74:241–250, 1977.

Schollmeyer, J. V., D. E. Goll, L. G. Tilney, M. Mooseker, R. Robson and M. Stromer. Localization of α-actinin in nonmuscle material. J. Cell Biol. 63:304A, 1974.

Shelanski, M. L. Chemistry of the filaments and tubules of brain. J. Histochem. Cytochem. 21:529–539, 1973.

Tilney, L. G. and M. S. Mooseker. Actin filament membrane attachment: Are membrane particles involved? J. Cell Biol. 71:402–416, 1976.

Weber, K. and U. Gröeschel-Stewart. Antibody to myosin: The specific visualization of myosin-containing filaments in non-muscle cells. Proc. Nat. Acad. Sci. 71:4561–4564, 1974.

Yen, S.-H., D. Dahl, M. Schachtner and M. L. Shelanski. Biochemistry of the filaments of the brain. Proc. Nat. Acad. Sci. 73:529. 1976.

Keratin Filaments

Brysk, M. M., R. H. Gray and I. A. Bernstein. Tonofilament protein from newborn rat epidermis: Isolation, localization and biosynthesis of marker of epidermal differentiation. J. Biol. Chem. 252:2127–2133, 1977.

Matoltsy, A. G. Keratinization. J. Invest. Dermatol. 67:20–25, 1976.

Steinert, P. M. and M. Guillino. Bovine epidermal keratin filament assembly in vitro. Biochem. Biophys. Res. Commun. 70:221–227, 1976.

Steinert, P. M., W. W. Idler and S. B. Zimmerman. Self-assembly of bovine epidermal keratin filaments in vitro. J. Molec. Biol. 108:547–567, 1977.

Sun, T. and H. Green. Immunofluorescent staining of keratin fibers in cultured cells. Cell 14:469–476, 1978.

INDEX

NOTE: *Italicized* numbers indicate illustrations.